11th International Seminar on Industrial Engineering and Management 2018

Technology and Innovation Challenges Towards Industry 4.0 Era

IOP Conference Series: Materials Science and Engineering
Volume 528

Makasar, Indonesia
27 – 29 November 2018

ISBN: 978-1-5108-8916-3
ISSN: 1757-8981

Printed from e-media with permission by:

Curran Associates, Inc.
57 Morehouse Lane
Red Hook, NY 12571

Some format issues inherent in the e-media version may also appear in this print version.

Printed by Curran Associates, Inc. (2019)

For permission requests, please contact the Institute of Physics at the address below.

Institute of Physics
Dirac House, Temple Back
Bristol BS1 6BE UK

Phone: 44 1 17 929 7481
Fax: 44 1 17 920 0979

techtracking@iop.org

Additional copies of this publication are available from:

Curran Associates, Inc.
57 Morehouse Lane
Red Hook, NY 12571 USA
Phone: 845-758-0400
Fax: 845-758-2633
Email: curran@proceedings.com
Web: www.proceedings.com

Table of contents

Volume 528

11th ISIEM (International Seminar on Industrial Engineering & Management, "Technology and Innovation Challenges Towards Industry 4.0 Era"

27–29 November 2018, Makasar, South Sulawesi, Indonesia

Accepted papers received: 4 April 2019

Preface

Papers

Decision Analysist and Information System

Industrial Management

Operation Research

Quality Engineering and Management

Suply Chain Management

PREFACE

Bismillahirrahmanirrahim,
Assalamu'alaikum Warrahmatullah Wabarrakatuh,

This issue is published in line with the Eleventh International Seminar on Industrial Engineering and Management (11th ISIEM). The articles cover a broad spectrum of topics in Industrial Engineering and Management, which are Quality Engineering Management, Decision Support System and Artificial Intelligent, Ergonomics, Supply Chain Management, Production System, Operation Research, and Industrial Management. These articles provide an overview of critical research issues reflecting on past achievements and future challenges. Those papers were selected from 120 abstracts, and we send these papers to IOP to be published there as an Open Access Proceeding. This statistic shows the high competition to get published on this proceeding.
This issue and seminar become special as more delegates come and join from various country as well as universities. We host 90 delegates both from abroad and local.

The 11th ISIEM is hosted by seven universities, which are Universitas Pasundan, Universitas Esa Unggul, Universitas Trisakti, Universitas Tarumanagara, Universitas Al-Azhar Indonesia, Atma Jaya Catholic University of Indonesia, and Universitas Pancasila.

This is the thirteenth years of the collaboration of those universities, and the second time we had MOU with IOP in United Kingdom to publishing the papers that is indexed by Scopus.

In this occasion, let us give special thanks to Prof. Dr. Md. Mamun Habib, from BRAC Business School (BBS), BRAC University, Bangladesh, Editor-in-Chief, International Journal of Supply Chain Management (IJSCM) UK, Prof. Dr. Dzuraidah Abd. Wahab from Center for Integrated Design of Advanced Mechanical Systems Faculty of Engineering and Built Environment, University Kebangsaan Malaysia, for their contribution as keynote speakers. We are also grateful to all reviewers, for their commitment, effort and dedication in undertaking the task of reviewing all of the abstracts and full papers. Reviewing a large number of submissions in a relatively short time frame is always challenging. Without their help and dedication, it would not be possible to produce this proceeding in such a short time frame. I highly appreciate all members of committees (advisory, steering, and organizing committees) for mutual efforts and invaluable contribution for the success of seminar.

Wassalamu'alaikum Warrahmatullah Wabarrakatuh.

Ir. Wahyukaton, MT.
Chairman

The 11th International Seminar on Industrial Engineering and Management (11th ISIEM)
Swiss-Belhotel Makassar, Sulawesi Selatan-Indonesia, November 27-29, 2018

IOP Conf. Series: Materials Science and Engineering **528** (2019) 011002 doi:10.1088/1757-899X/528/1/011002

Photographs from the 11th International Seminar on Industrial Engineering & Management (11th ISIEM)

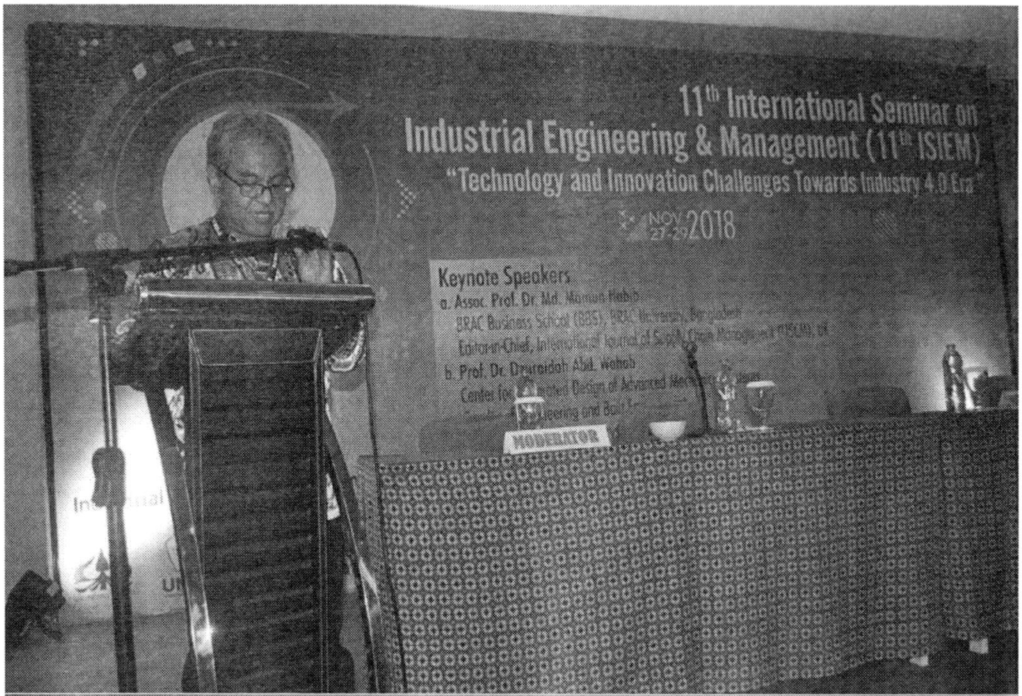

IOP Conf. Series: Materials Science and Engineering **528** (2019) 011002 doi:10.1088/1757-899X/528/1/011002

IOP Conf. Series: Materials Science and Engineering **528** (2019) 011002 doi:10.1088/1757-899X/528/1/011002

IOP Conf. Series: Materials Science and Engineering **528** (2019) 011002 doi:10.1088/1757-899X/528/1/011002

IOP Conf. Series: Materials Science and Engineering **528** (2019) 011002 doi:10.1088/1757-899X/528/1/011002

IOP Conf. Series: Materials Science and Engineering **528** (2019) 011002 doi:10.1088/1757-899X/528/1/011002

IOP Conf. Series: Materials Science and Engineering **528** (2019) 011003 doi:10.1088/1757-899X/528/1/011003

Peer review statement

All papers published in this volume of *IOP Conference Series: Materials Science and Engineering* have been peer reviewed through processes administered by the proceedings Editors. Reviews were conducted by expert referees to the professional and scientific standards expected of a proceedings journal published by IOP Publishing.

A Fuzzy Analytic Hierarchy Process Approach for Determining the Criteria Success Factors of MRT Parts' e-Procurement: the Case of Jakarta MRT Project

T Hartanto[1], **M Ginting**[1] **and O Sunardi**[1,a]

[1] Department of Industrial Engineering, Krida Wacana Christian University, Jakarta 11470, Indonesia
[a]oki.sunardi@ukrida.ac.id

Abstract. The Mass Rapid Transit (MRT) project is conducted by the central government in cooperation with the provincial government of DKI Jakarta and supported by the Japanese government through Japan International Cooperation Agency (JICA). The government goal is to create good e-Governance for the MRT project, and together initiate the use of e-Supply Chain Management (e-SCM), especially the e-Procurement, for the entire project. Technological innovation which is collaborated within the e-SCM becomes a very important aspect to support the e-Governance. This paper aims to determine the criteria and sub-criteria of success of the implementation of MRT parts' e-Procurement, and to determine the priorities of the MRT parts' e-Procurement process for the provincial government of DKI Jakarta. Fuzzy Analytic Hierarchy Process was employed to explore and simulate the level of importance of criteria and sub-criteria, in relations with the successful implementation of e-Procurement in MRT project. Top management commitment was found to be the most influential criteria by 0.334, followed by government policy and regulation (0.305), technology infrastructure (0.176), value acquisition (0.103), and suppliers' relations (0.082).
Keywords: MRT, e-SCM, e-Procurement, Fuzzy AHP, Criteria Success Factors.

1. Introduction

The Mass Rapid Transit (MRT) project development process requires several related agencies to fulfil the need for components. One important component is the drilling machine known as the Tunnel Boring Machine (TBM) which consists of the drill component (cutter head), body block, and motor house. TBM is produced by JTSC (Japan Tunnel Systems Corporation). In addition, required related components include mechanical and electrical, and infrastructure systems (e.g. accommodation facilities such as escalators for underground access, gate systems, railway tracks, carriages along with moving trains, human resources as control aspects, and maintenance-related spare parts).

The MRT project is conducted by the central government in a cooperation with the provincial government of DKI Jakarta and supported by the Japanese government through Japan International Cooperation Agency (JICA). The government goal is to create good e-Governance for the MRT project, and together initiate the use of e-Procurement for the entire project. Based on previous experience in other countries, Government to Business (G2B) collaboration sometimes generate challenges on how to relate and integrate each process during the projects [1]. E-Procurement is important because it is proven to improve transparency, efficiency, effectiveness, accountability, fairness, interoperability, and data security assurance. However, the implementation of e-Procurement in DKI Jakarta is far from optimal, especially in the Jakarta provincial government agency in MRT project. E-Procurement becomes the basis for initiating transparency for many government projects. In addition, the main challenge in implementing e-Procurement's success is in terms of determining which criteria to be systematically taken during the project execution [http://www.jakartamrt.com, accessed October 9, 2016].

The aims of this study is to determine the criteria and sub-criteria of success of the implementation of MRT parts' e-Procurement, and to determine the priorities of the MRT parts' e-Procurement process for the provincial government of DKI Jakarta.

2. Literature review

2.1. e-Supply Chain Management and e-Governance

Technological innovation which is collaborated within the supply chain management (e-SCM) becomes a very important aspect to support e-Governance. In well developed countries, e-Governance has proven to increasing the efficiency and technical capacity of work process, shortening the purchasing and transaction process, and increasing the information disclosure during and after the procurement process [2]. Further, e-SCM facilities can reduce costs, increase demand and create a new business model. It is very potential in providing benefits to all stakeholders in reducing costs and improve product quality and information [3, 4]. Further, e-SCM have proven to provide real time communication amongst supply chain members, better real time forecasting decisions, and improve partnerships [5]. Stakeholders involved in policy making can help build a trust relationship between government and society, and leverage the quality of the policy. In this case, good policy could influence the sustainability of e-SCM [6]. In the long term, e-SCM is hoped to be the main driver of good e-Governance. Further, e-Governance should be the basis for leveraging internal efficiency and supporting the interfaces with citizens [7]. This is the main idea of creating good government governance.

2.2. e-Procurement

E-Procurement refers to the use of internet-based information (integrated) and communication technologies in reaching an individual or a process in the procurement activities included in search, resources, negotiations, reservations, receipts and payments [8]. E-Procurement can be considered as one of the technological solutions where an organization procures the goods via the internet [2]. In other words, e-Procurement can be explained as a technology utilization, especially internet and communication facilities, in every procurement process from the beginning to the end of the process. In term of MRT existence in DKI Jakarta, E-Procurement is considered very important because with an electronic procurement system, it will automatically improve transparency, efficiency, effectiveness, accountability, fairness and non-discrimination, open and healthy competition, interoperability, and data security assurance. Supervision is also an advantage that exists in the application of e-Procurement, because any incoming information can be monitored and filtered. In the implementation of e-Procurement in the government sector, risk analysis and future strategies for implementation on e-Government projects are required. In addition, a clear guideline may function as to reducing corruption in public procurement activities [9].

2.3. Criteria and sub-criteria of e-Procurement

E-Procurement is one of the main topics of e-Government, where many organizations need advice and guidance on using the new platform. Without these critical success criteria and sub-criteria, it is almost impossible to get the process and interpret the success of e-Procurement, especially in the public sector [7, 10]. Several criteria have been explored from various contexts to explore how e-Procurement can contribute to the success and sustainability of organizations, such as government policy and regulation [11-14, 17], top management commitment [11, 13, 15, 16], suppliers relationships [13, 18, 19], value acquisition [13, 20], and technology infrastructure [1, 21, 22]. Based on the literature study, the researchers raised the hypothesis of the criteria and sub-criteria of success in e-Procurement which is represented in the form of a supportive model to answer the questions in this study, whereas this model is limited in accordance with the reference to considerations related to the procurement of MRT parts.

The above criteria are established through several sub-criteria such as the role of central government, the participation of local government, government commitment, business owner commitment, accountability, mutual trust, price competitiveness, conformity to specification, webpage system and internet connection quality. Figure 1 represents the relationship between sub-criteria, criteria and successful e-Procurement implementation.

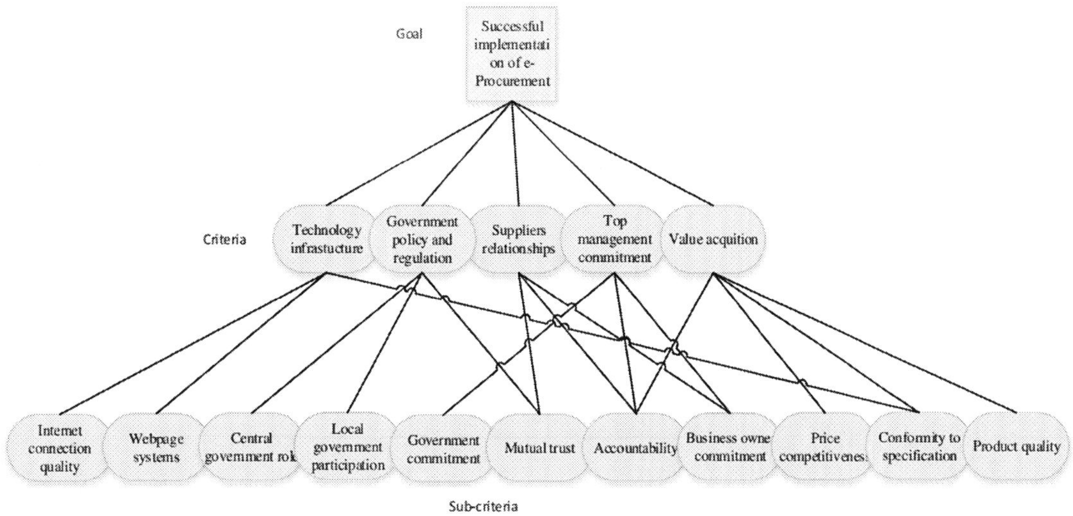

Figure 1. Sub-criteria, criteria and goal

3. Methods

By context, this research focuses on the criteria and sub-criteria related to MRT e-procurement spare part in DKI Jakarta. In addition, after determining the criteria and sub-criteria, the researchers continues the research process by weighting each criteria and sub-criteria with the Fuzzy Analytic Hierarchy Process (FAHP) [23], where data were obtained based on three experts from academic, business and government. The experts should possess the following criteria: minimum five years of experience in public transportation procurement process; and, involves in the current MRT project, as consultant, researcher and or team member. In this study, the three experts consist of Mr. Timbul F. Sitompul, Head of Administration LPSE Management Unit of DKI Jakarta Province, Mr. Tulus M. Sihombing, Researcher at STIMLOG, and Ms. Calvina Anastasia, Procurement Specialist at PT. MRT Jakarta. Each expert represents the government, academic, and business.

In the use of AHP, the results of the decisions given by the experts are not deterministic, but rather resemble linguistic perceptions. So, the use of AHP in multi-criteria decision making (MCDM) is less able to overcome the uncertainty experienced by decision makers when they are asked about the definite value in pairwise comparison [24]. In this case, the use of FAHP is more appropriate. Information in FAHP, as well as conventional AHP, employ 1 to 9 scales in the form of pairwise comparison. In addition, FAHP uses Triangular Fuzzy Number (TFN) for fuzzification of crisp comparison matrix. The fuzzification on AHP scale creates a new scale called the AHP fuzzy scale [25].

The next step is designing the questionnaires to assess each criteria and sub-criteria of e-Procurement. Data are then processed using FAHP to determine the weight of each criterion. The questionnaire uses five valuation scales: Equal Importance (EI), Moderate Importance (MI), Strong Importance (SI), Very Strong Importance (VSI), and Extreme More Importance (EMI). Each of the fuzzy rating scales has different membership functions, such as EMI (8,9,10), VSI (6,7,8), SI (4,5,6)), MI (2,3,4), and EI (1,1,2). These membership functions were used to calculate the weight of each criterion.

The initial process of fuzzy data processing was by creating matrix pairs of data. Further, three fuzzy numbers were converted into two numbers. Finally, the data were converted into one number using alpha cut method so as to get the crisp value (defuzzification process). The data were then combined from the three experts using the mean geometric method, so that the combined crisp value can be derived. In processing the data using the FAHP method, it is necessary to calculate the consistency test, where the consistency test is useful to determine whether (or not) the resulting pairwise comparisons are consistent. The consistency measurement of pairwise comparison matrices was based on the greatest Eigen value [26]. The consistency index of "n" order matrix was obtained by the following formula:

$$CI = (\lambda maks - n) / (n-1) \quad (1)$$

Where, CI = Consistency deviation ratio (consistency index); λmax = the largest eigenvalue of the "n" order matrix; and n = order matrix. After getting the value of CI, the next step was to calculate the CR value to discover the consistency of the data being processed. The formula of calculating CR was as followed:

$$CR = CI / RI \quad (2)$$

Where, RI value was based on "n". Table 1 described several RI value based on n (1 to 10).

Table 1. RI value

n	1	2	3	4	5	6	7	8	9	10
RI	0	0	0.58	0.90	1.12	1.24	1.32	1.41	1.45	1.49

After the fuzzy phase has completed, the next step was to allocate weights to determine the priorities of the criteria and sub-criteria [27].

4. Findings and discussions

4.1. Criteria assessment

Assessment of the importance weight of each criterion were done by three experts. There are several criterion that utilized as indicators: SIT (technology infrastructure)), RJS (suppliers relationships), DKMA (top management commitment), RK (government policy and regulation), and EN (value acquisition). The results obtained through the questionnaires were in the form of comparisons between criterions. All the results of the questionnaire were arranged in the form of a pairwise comparison matrix containing the fuzzy of numbers.

In the relative importance assessment of two elements, a reciprocal axiom means that if the element in column one is assigned 3 times more important than the element in column two, then the elements in column two should be equal to 1/3 times more important than the elements in column one. In data retrieval, researchers have made revisions after knowing pairwise comparison data were inconsistent. In this case, there was a mistake from the interpretation or distribution of opinion of each respondent or experts in interpreting the comparison on the criteria and sub-criteria so that we need to revise the weight on the criteria and sub-criteria. Pairwise Comparison Matrix between criterions obtained from three experts can be seen in Table 2-4.

Table 2. Assessment from Expert 1

	SIT			RJS			DKMA			RK			EN		
SIT	1	1	2	2	3	4	1/4	1/3	1/2	1/4	1/3	1/2	2	3	4
RJS	1/4	1/3	1/2	1	1	2	1/4	1/3	1/2	1/6	1/5	1/4	2	3	4
DKMA	2	3	4	2	3	4	1	1	2	1	1	2	2	3	4
RK	2	3	4	4	5	6	1/2	1	1	1	1	2	4	5	6
EN	1/4	1/3	1/2	1/4	1/3	1/2	1/4	1/3	1/2	1/6	1/5	1/4	1	1	2

Table 3. Assessment from Expert 2

	SIT			RJS			DKMA			RK			EN		
SIT	1	1	2	2	3	4	1/6	1/5	1/4	4	5	6	4	5	6
RJS	1/4	1/3	1/2	1	1	2	1/4	1/3	1/2	1/8	1/7	1/6	1/6	1/5	1/4
DKMA	4	5	6	2	3	4	1	1	2	1/8	1/7	1/6	4	5	6
RK	1/6	1/5	1/4	6	7	8	6	7	8	1	1	2	6	7	8
EN	1/6	1/5	1/4	4	5	6	1/6	1/5	1/4	1/8	1/7	1/6	1	1	2

Table 4. Assessment from Expert 3

	SIT			RJS			DKMA			RK			EN		
SIT	1	1	2	2	3	4	1/4	1/3	1/2	1/6	1/5	1/4	1/4	1/3	1/2
RJS	1/4	1/3	1/2	1	1	2	1/4	1/3	1/2	1	1	2	1/4	1/3	1/2
DKMA	2	3	4	2	3	4	1	1	2	2	3	4	2	3	4
RK	4	5	6	1/2	1	1	1/4	1/3	1/2	1	1	2	1	1	2
EN	2	3	4	2	3	4	1/4	1/3	1/2	1/2	1	1	1	1	2

4.2. Sub-criteria assessment

Assessment of the weight of interest of each sub-criteria was done by three experts. There were several sub-criteria that were employed as indicators: JI (internet connection quality), SWP (webpage systems), KS (conformity to specification), AK (accountability), HK (mutual trust), KPB (business owner commitment), KP (government commitment), PPP (central government role), PPD (local government participation), HT (price competitiveness), and KL (Quality). Table 5 shows the overall results of the expert assessment for sub-criteria under criterion "technology infrastructure" (sub-criteria 1) as example.

Table 5. Assessment from Expert 1 for sub-criteria 1

	JI			SWP			KS		
JI	1	1	2	1	1	2	2	3	4
SWP	1/2	1	1	1	1	2	2	3	4
KS	1/4	1/3	1/2	1/4	1/3	1/2	1	1	2

Similar steps were conducted to assess the other sub-criteria and criteria. The next step was to convert the numbers to a TFN (Triangular Fuzzy Number) scale. Fuzzy numbers are usually shown in the form of three numbers namely (l, m, u). The parameter represents the smallest possible value, the promising value, and the largest value representing the fuzzy problem. Tables 6 summarizes the converted fuzzy numbers.

IOP Conf. Series: Materials Science and Engineering **528** (2019) 012001 doi:10.1088/1757-899X/528/1/012001

Table 6. Converted assessment from Expert 1

	SIT		RJS		DKMA		RK		EN	
SIT	1.00	1.50	2.50	3.50	0.29	0.42	0.29	0.42	2.50	3.50
RJS	0.29	0.42	1.00	1.50	0.29	0.42	0.18	0.23	2.50	3.50
DKMA	2.50	3.50	2.50	3.50	1.00	1.50	1.00	1.50	2.50	3.50
RK	2.50	3.50	4.50	5.50	0.75	1.00	1.00	1.50	4.50	5.50
EN	0.29	0.42	0.29	0.42	0.29	0.42	0.18	0.23	1.00	1.50
Total	6.58	9.33	10.79	14.42	2.63	3.75	2.66	3.87	13.00	17.50

Similar steps were conducted for the assessment from Expert 2 and 3. The next step was the assessment on each sub-criteria that has been completed by the three experts and converted into two fuzzy numbers.

Before defuzzification, the assessment matrix were combined into one assessment matrix, by the following formula:

$$A_g(l,u) = \sqrt[3]{(l_1 * l_2 * l_3),(u_1 * u_2 * u_3)} \quad (3)$$

Where, A_1= (l_1, u1); A_2= (l_2, u2); A_3= (l_3, u3). From formula (3), a joint assessment matrix was formed. Table 7 summarizes the combined assessment matrix for criteria (as example). This step was also conducted for the sub-criteria.

Table 7. Combined matrix for the criteria

	SIT		RJS		DKMA		RK		EN	
SIT	1	1.5	2.50	3.50	0.25	0.34	0.62	0.80	1.49	2.00
RJS	0.29	0.42	1	1.5	0.29	0.42	0.29	0.37	0.51	0.69
DKMA	3.04	4.07	2.50	3.50	1	1.5	0.69	0.93	3.04	4.07
RK	1.27	1.63	2.80	3.46	1.12	1.46	1	1.5	3.08	3.96
EN	0.51	0.69	1.49	2.00	0.25	0.34	0.26	0.33	1	1.5
Total	6.12	8.31	10.29	13.96	2.92	4.06	2.87	3.94	9.12	12.22

The defuzzification was then utilized to convert the fuzzy values into "crisp" value (CFCS, Converting Fuzzy into Crisp Scores). The calculation of crisp value at $\alpha = 0.5$ was based on formula 4.

$$\text{Crisp Value} = 0.5*l + (1-0.5)*u \quad (4)$$

Table 8 summarizes the calculation of combined crisp value (as example).

Table 8. Combined crisp value for criteria

	SIT	RJS	DKMA	RK	EN
SIT	1	3.00	0.29	0.71	1.74
RJS	0.35	1	0.35	0.33	0.60
DKMA	3.56	3.00	1	0.81	3.56
RK	1.45	3.13	1.29	1	3.52
EN	0.60	1.74	0.29	0.30	1
Total	6.96	11.87	3.24	3.15	10.42

IOP Conf. Series: Materials Science and Engineering **528** (2019) 012001 doi:10.1088/1757-899X/528/1/012001

It is necessary to conduct consistency test to find out whether the matrix of pairs is consistent or not. The consistency test was performed by using formula (1) and (2) as the median values in the fuzzy matrix (Table 9). The limit value of CR is 0.1 for matrix bigger than 4x4[26].

Table 9. Eigen value computation for all criterion

	SIT	RJS	DKMA	RK	EN	Eigen Vector	Rank
SIT	0.14	0.25	0.09	0.23	0.17	0.176	3
RJS	0.05	0.08	0.11	0.11	0.06	0.082	5
DKMA	0.51	0.25	0.31	0.26	0.34	0.334	1
RK	0.21	0.26	0.40	0.32	0.34	0.305	2
EN	0.09	0.15	0.09	0.09	0.10	0.103	4

In addition, consistency test results show that all matched matrix combined in this study were consistent with CR ≤ 0.1. Consistency test is very important because the reliability of a study is also considered from the reliability of data used as a source in subsequent processing. In this case, consistency test has an important role in showing the reliability of data from the research results obtained. Table 10 summarizes the consistency test for criteria and sub-criteria.

Table 10. Consistency test for all criterion

	λmax	C.I.	R.I.	C.R.	Note
Criteria	5.310	0.077	1.120	0.069	Consistent
Criterion 1	3.075	0.037	0.580	0.064	Consistent
Criterion 2	3.057	0.028	0.580	0.049	Consistent
Criterion 3	3.095	0.047	0.580	0.082	Consistent
Criterion 4	3.087	0.044	0.580	0.075	Consistent
Criterion 5	4.213	0.071	0.900	0.071	Consistent

Crisp values for the criteria and sub-criteria become the starting point in calculating the weight of each criterion or sub-criteria. In performing weight calculations, researchers use Microsoft Excel in the calculation process. Weight calculation was conducted by normalizing the crisp value on criteria and sub-criteria. Table 11 shows the AHP weight calculation (as example).

Table 11. Weight index for criteria

	SIT	RJS	DKMA	RK	EN	Weight	Rank
SIT	0.14	0.25	0.09	0.23	0.17	0.176	3
RJS	0.05	0.08	0.11	0.11	0.06	0.082	5
DKMA	0.51	0.25	0.31	0.26	0.34	0.334	1
RK	0.21	0.26	0.40	0.32	0.34	0.305	2
EN	0.09	0.15	0.09	0.09	0.12	0.103	4

Table 12. Recapitulation of AHP weight index

Criteria	Weight	Sub-criteria	Weight	Final weight
SIT	0.176	JI	0.554	0.097
		SWP	0.278	0.049
		KS	0.168	0.029
RJS	0.082	AK	0.555	0.045
		HK	0.171	0.014
		KPB	0.274	0.022
DKMA	0.334	AK	0.435	0.145
		KPB	0.186	0.062
		KP	0.379	0.127
RK	0.305	HK	0.137	0.042
		PPP	0.625	0.191
		PPD	0.239	0.073
EN	0.103	AK	0.323	0.033
		HT	0.197	0.020
		KS	0.265	0.027
		KL	0.215	0.022

Table 12 summarizes the recapitulation of the weighting criteria and sub-criteria along with the final weighting which is the multiplication of the weight of the criterion by the weight of each sub-criterion. The table summarizes all the rank for criteria in MRT spare parts e-procurement: top management commitment, government policy and regulation, technology infrastructure, value acquisition, and suppliers' relationships.

5. Conclusion

The FAHP computation results show that the criterion of top management commitment has the greatest importance weight by 0.334, followed by government policy and regulation (0.305), technology infrastructure (0.176), value acquisition (0.103), and suppliers' relations (0.082). Each criterion should focus on certain sub-criteria. In the MRT e-Procurement case, technology infrastructure should focus on internet connection quality, followed by webpage systems, and the effort to meet the conformity to required specification. For the second criterion, the sequence of focus should be building accountability, creating mutual trust relationships, and developing business owners' commitment. The third criterion should pay attention to accountability, commitment from business owners, and commitment from the government. As for the fourth criterion should focus on building mutual trust, exploiting central government role, and engaging the participation of local government. The value acquisition criterion should focus on building accountability of each stakeholder, maintaining price competitiveness, conform to specification of materials and product, and make sure of the product and service quality.

Acknowledgement

Thank you to Mr. Timbul F. Sitompul, Head of Administration LPSE Management Unit of DKI Jakarta Province, Mr. Tulus M. Sihombing, Researcher at STIMLOG, and Ms. Calvina Anastasia, Procurement Specialist at PT. MRT Jakarta, for their role as experts' panel to this study.

6. References

[1] Aman A and Kasimin H 2011 *Transf. Gov.: People, Process and Policy* **5** 330
[2] Piera C, Roberto C, Giuseppe C and Teresa M 2014 *IERI Procedia* **6** 8

[3] Singh M and Byrne J 2005 *Elec. J. Inform. Syst. Eva.* **8** 71
[4] Giménez C and Lourenço H R 2008 *Int. J. Log. Manage.* **19** 309
[5] Ivanovska L P and Kaleshovska N 2013 *TEM Journal* **2** 314
[6] Deakins E, Dillon S, Al Namani H and Zhang C 2010 *Int. J. Publ. Sec. Manage.* **23** 520
[7] Suri P K and Sushil 2017 Strategic Planning and Implementation of E-Governance (Singapore: Springer) p 25
[8] Croom S R and Jones A B 2005 *J. Publ. Sec.* **5** 367
[9] Neupane A, Soar J, Vaidya K and Yong J 2014 *Transf. Gov.: People, Process and Policy* **8** 500
[10] Vaidya K, Sajeev A S and Callender G 2006 *J. Publ. Proc.* **6** 70
[11] Saatçioglu Ö Y, Deveci D A and Cerit A G 2009 *Transf. Gov.: People, Process and Policy* **3** 144
[12] Walker H and Brammer S 2012 *Int. J. Prod. Econ.* **140** 256
[13] Hwang B N and Lu T P 2013 *Int. J. Phys. Dist. Log. Manage.* **43** 657
[14] Khorana S, Boucher K F and Kerr W A 2014 *J. Common Market Studies* **53** 1
[15] Gunasekaran A and Ngai E 2008 *Int. J. Prod. Econ.* **113** 159
[16] Tzempelikos N 2015 *J. Bus. Int. Mark.* **30** 32
[17] Nurmandi A and Kim S 2015 *Int. J. Publ. Sec. Manage.* **28** 198
[18] Rockson S B, Anane E O and Sey K A 2017 *J. Log. Manage.* **6** 26
[19] Kim M, Suresh M C, Hillmer C K 2015 *J. Bus. Ind. Mark.* **30** 1
[20] Lundberg S, Marklund P O, Strömbäck E and Sundström D 2015 *Env. Econ. Policy Studies* **17** 487
[21] Popa S, Acosta P S and Loukis E 2016 *Program* **50** 118
[22] Choi H, Park M J, Rho J J and Zo H 2016 *Telecommunications Policy* **40** 644
[23] Saaty T L 2008 *Int. J. Serv. Sci.* **1** 83
[24] Ordoobadi S M 2009 *Supply Chain Manage.: an Int. J.* **14** 314
[25] Jindal A and Sangwan K S 2015 *Benchmarking: an Int. J.* **22** 393
[26] Saaty T L 1980 *The Analytic Hierarchy Process* (New York: McGraw-Hill) p 224

Supply Chain Risk Management Analysis using the Development of Fuzzy Reasoning Methods and Analytical Network Process (ANP) at Wooden Toys Industries

W N Tanjung[1], S S Asti[1], S Hidayat[1], E Ripmiatin[2], S A Atikah[1] and R S Khodijah[1]

[1]Industrial engineering department, Faculty of science and technology, Al Azhar University of Indonesia, Sisingamangaraja Road, Jakarta 12110, Indonesia
[2]Informatics engineering department, Faculty of science and technology, Al Azhar University of Indonesia, Sisingamangaraja Road, Jakarta 12110, Indonesia
selvy.sekar@gmail.com

Abstract. The supply chain is the relationship between material flows or services, money flow and information ranging from suppliers, producers, distributors, warehouses, retailers to customers. The smooth flow of material in the supply chain affected in the fulfillment of customer needs. If the problem occurs, it caused losses and risks. To minimize risks, it can be handled with Supply Chain Risk Management. This research was conducted in the supply chain in Wooden Toys Industries by identifying risks using the Fuzzy Reasoning method to calculate the amount of opportunity and impact that would arise if the risk occurred. To reduce the opportunity and impact of risk, a mitigation strategy is needed to consider and can be done using the Analytical Network Process method. Followed the results, the competitor's moves have a large chance and risk. Thus, risk mitigation can be proposed is necessary to conduct market observations and identify competitors at (31.3%).
Keywords: analytical network process, fuzzy reasoning, supply chain, supply chain risk management, risk.

1. Introduction

In the increasingly broad industrial world, where in an industry there are activities to convert raw materials into finished goods. In making changes from raw materials to finished goods, the supply chain must be able to be controlled so that the relationship between material flows or services, cash flow and information ranging from suppliers, producers, distributors, warehouses, retailers to end customers is able to meet the needs or demands with smoothly. The role of the supply chain in principle is to add value to the product, by moving it from one location to another, or by making changes to it [1]

Where every activity carried out by a company will not be separated from unplanned events or uncertainties that can affect the flow of materials or components in the supply chain. Unplanned events or uncertainties can be said to be risks that will occur [2]. Risk is an operational inhibiting factor in the supply chain, where risks can occur from upstream, factory, distribution and downstream of the distributor. The potential risk if it really happens will certainly have an impact on the performance of the company's supply chain management [3], from these risks can lead to losses; risks cannot be avoided but can be minimized or eliminated by handling appropriate risks. Handling for risks that have the potential to occur in the supply chain can be handled with Supply Chain Risk Management (SCRM). SCRM is a systematic identification, assessment and mitigation of potential disturbances in logistics networks with the aim of reducing the negative impact on the performance of the supply chain network [4]. By using SCRM, the risks that have potential can be classified.

The industry that has significant impact in national economic development is small and medium industries like a wooden toy companies where focused on children's toys, made from wood. They are also has risks and need some approximation to mitigate their risks. SCRM can be used to answer that problem by developing Fuzzy reasoning (FR) and Analytical Network Process (ANP). FR is a systematic tool to deal with qualitative and quantitative data and information that arise in an activity [5]. This method is used to find out the magnitude of the opportunity and the impact of risk. Furthermore, Analytical Network Process (ANP) is an extended method from Analytical hierarchy Process (AHP) which can be used for relative measurement and useful to reduce the ratio of priority from individual ratio [6]. Thus, ANP is used to determine a mitigation strategy that has a large percentage in order to minimize the risks. Wooden Toy Industry is industry small and medium where focused on children's toys

2. Methods

In the first stage, it is done by identifying risks that have an emerging opportunity and a large impact using the fuzzy reasoning (FR) method, in the first FR method, namely identifying risks in the supply chain in a wooden toy company by means experts who have been determined, identification of risks in terms of factors and variables, then after understanding what risks occur in a wooden toy company, the factors and risk variables are described in the form of a hierarchy. The stages in this study can be seen in figure 1.

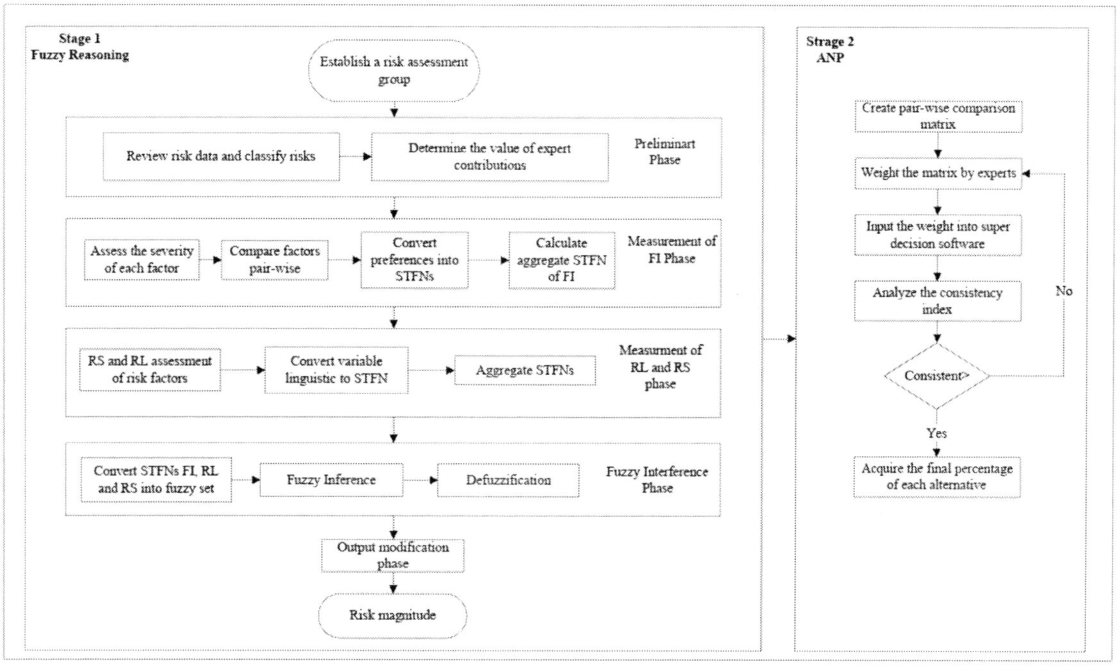

Figure 1. Fuzzy reasoning and ANP method flowchart

To assess the factors and risk variables, expert judgment is needed, where there are 3 experts who have different backgrounds. Expert weighting is carried out based on their respective positions, background and expertise. To measure the FI phase, an assessment of the likelihood of an existing variable is needed, the assessment is done by linguistic variables which are then translated into fuzzy grouping, then, the comparison between sub-factors or risk variables in each index factor influences them. The calculation is done by comparison, which is then carried out with aggregated STFNs on each risk variable. And after aggregating STFN on each risk variable then calculate the overall FI weight. To get the final value of FI can be done using equation (1) to equation (9).

$$\mu_{A*}(x) = \begin{cases} \frac{x-a^1}{a^m-a^1} \; for \; a^1 \le x \le a^m \\ 1 \\ \frac{a^u-x}{a^u-a^n} for \; a^m \le x \le a^n \\ 0 \; for \; otherwise \end{cases} \tag{1}$$

$$S_i^* = S_{t1}^* \otimes c_1 \oplus S_{t2}^* \otimes c_2 \oplus \ldots \oplus S_{tm}^* \otimes c_m \tag{2}$$

$$a_{ij}^* = a_{ij1}^* \otimes c_1 \oplus a_{ij2}^* \otimes c_2 \oplus \ldots \oplus a_{ijm}^* \otimes c_m \tag{3}$$

$$a_{ij}^* = \frac{a_{ij1}^* \otimes c_1 \oplus a_{ij2}^* \otimes c_2 \oplus \ldots \oplus a_{ijm}^* \otimes c_m}{1-\sum c_r} \tag{4}$$

$$a_{ij} = \frac{a_{ij}^1 + 2\left(a_{ij}^m + a_{ij}^n\right) + a_{ij}^m}{6} \tag{5}$$

$$\mathcal{A} = a_{ij} = \begin{matrix} \\ F_1 \\ F_2 \\ F_3 \\ F_4 \end{matrix} \begin{bmatrix} F_1 & F_2 & \cdots & F_n \\ 1 & a_{12} & \cdots & a_{1n} \\ 1/a_{12} & 1 & \cdots & a_{2n} \\ \vdots & \cdots & \ddots & \vdots \\ 1/a_{Ln} & 1/a_{2x} & \cdots & 1 \end{bmatrix}, i,j = 1,2,\ldots,n \tag{6}$$

$$w_i = \frac{1}{n}\sum_{j=1}^{n} \frac{a_{ij}}{\sum_{k=1}^{n} a_{kj}} i,j = 1,2,\ldots\ldots,n \tag{7}$$

$$w_i' = w_i x \prod_{i=1}^{t} w_{section}^{(i)} \tag{8}$$

$$FI^* = \sum_{i=1}^{n} S_i^* w_i' \; \text{i} = 1, 2, \ldots, \text{n} \tag{9}$$

Information:

S_i^*	= *Fuzzy* aggregated		a_{ijn}^*	= Corresponding STFN scale of Fi comparing Fj
S_{tn}^*	= STFN score			
c_n	= CFs allocated to expert		w_i	= Weight FI
a_{ij}^*	= Aggregated *Fuzzy* scale of fuzzy comparing to Fi		w_i'	= Final weight of FI
			FI^*	= *Fuzzy* of FI
F	= Risk factor in one section			

The first measurement of RL and RS is done by conducting an expert study of the police from the experience. The assessment is carried out with linguistic variables, after being carried out further to correct the STFN and carry out the STFN aggregate. To get the RL and RS values can be calculated using formulas (10) and (11).

$$RL^* = RL_1^* \otimes c_1 \oplus RL_2^* \otimes c_2 \ldots\ldots\ldots \oplus RL_m^* \otimes c_m \tag{10}$$

$$RS^* = RS_1^* \otimes c_1 \oplus RS_2^* \otimes c_2 \ldots\ldots\ldots \oplus RS_m^* \otimes c_m \tag{11}$$

And for fuzzy interface phase, it begins by converting aggregated STFN from each risk factor into a fuzzy set. To convert aggregated STFN that is by taking the intersection value between STFN and MF whose parameters are interconnected, then defuzzification is done to convert the fuzzy result into a suitable numerical value that can represent RM. To get a combination of RM* it can be done with equation (12) and from the combination results obtained the Risk Magnitude (RM*) with equation (13) to (15). Then do the conversion into the fuzzy set to find out the percentage of the value of RM * with equation (16).

$$R^k: if \; FI^* \text{are} \; \mu_{FI*}^{k}, \text{dan} \; RL^* \text{are} \; \mu_{RL*}^{k}, \text{and} \; RS^* \; \text{are} \; \mu_{RS*}^{k}$$
$$\text{Next} \; RM^* \text{is} \; \mu_{RM^*}^{k} \tag{12}$$

$$\mu_R^k(x,y) = \mu_{FI*}^k(x_1) \wedge \mu_{RL*}^k(x_2) \wedge \mu_{RS*}^k(x_3) \wedge \mu_{RM*}^k(y) \quad (13)$$

$$\mu R\ (\chi, y) = \vee_{K=1}^{K} R^k(\chi, y) \quad (14)$$

$$\mathrm{RM}^* = \mathrm{RP}^* \mathrm{o}\ \mathrm{R}(\chi, \mathrm{y}) \quad (15)$$

$$RM = \frac{\sum_{i=1}^{q} Y_i \mu_{RM^*}(y_i)}{\sum_{i=1}^{q} \mu_{RM^*}(y_i)} \quad (16)$$

Information:
μ_{FI*}^k = Denote the universe of FI*, RL*, RS* dan RM*
$\mu R\ (\chi, y)$ = Output fuzzy MF after aggregation

In the second stage, to mitigate the risks that has big opportunities and impacts by using the Analytical Network Process method. In the submission of a strategy to minimize risk, a literature study is conducted and the interview is returned to the expert and evaluates the comparative matrix of the proposed strategies. After getting an assessment from an expert, the 3 expert assessment results are converted into1 matrix using the geometric mean equation that is in equation (17).

$$Geometric\ Mean = \sqrt[n]{x_1 \times x_2 \times \ldots \times x_n} \quad (17)$$

From these inputs can find out how consistent the expert answers to the comparative matrix questionnaire that has been given, if consistent (<0.1) then the calculation can proceed to find out the final percentage of each alternative, if it is not consistent (> 0.1) then an expert reassessment against alternative strategies proposed in the form of a comparative matrix. The consistency calculation can be done using equation (18) and equation (19).

$$CI = \frac{\lambda_{max} - n}{n-1} \quad (18)$$

$$CR = \frac{CI}{RI} \quad (19)$$

3. Result and Discussion

3.1 Risk Analysis with Fuzzy Reasoning Method

3.1.1 Prelaminar Phase

For the initial step in this phase is risk identification, risk identification is carried out by means of knowing or determining the risk of what often happens or has the opportunity to occur in the supply chain of a wooden toy company. Risk identification was carried out with a literature study [7] and interviews with experts, after identifying 25 risk variables were grouped into 6 factors.

Table 1. Factor and variable risks in the wooden toys industries

Risk Factor	Risk Variable	Risk Factor	Risk Variable
Demand	Market saturation	Information	Information delays
	Competitor moves		Wrong choice of communication
	Delay in delivery to customer	Operation	Capacity inflexible
	Forecast error		Design change
Environment	Macroeconomic uncertainty		Disruption in production
	Nature disasters		Inventory risk
	Social uncertainty		Variability in production process
	Policy uncertainty	Supply	Dependency on single supplier
Finance	Cost/price risk		Inflexibility of supplier
	Exchange rate risk		Poor delivery performance
Information	Breakdown of IT infrastructure		Supplier poor quality
	Distorted information		Supplier bankruptcy
	Inadequate information security		

This stage begins with determining the value of the Contribution Factor to the expert who is the source of the research. Values are determined based on their position, background and expertise [7]. Table 1 is the allocation of CF values of each expert.

Table 2. CF of each expert

Expert	Background Expert	CFs
E1	CEO of Wooden Toys Industries	C1 = 0.5
E2	Finance that deals directly with suppliers and overseas production	C2 = 0.25
E3	Production is related to raw materials and runs in process	C3 = 0.25

For membership function index factor, risk likelihood, and risk severity are illustrated in Figure 2, while for risk magnitude is illustrated in Figure 3.

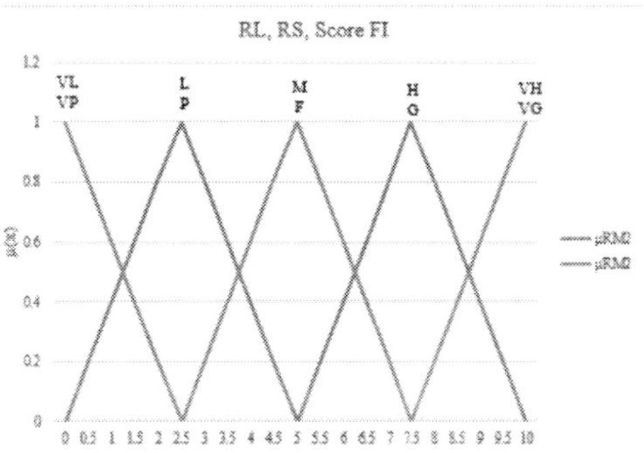

Figure 2. Graphic fuzzy membership functions FI, RS, RL

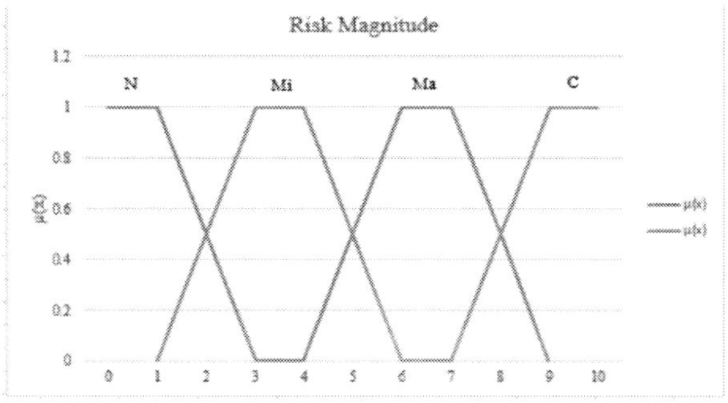

Figure 3. Graphic fuzzy membership functions RM

3.1.2 FI, RL and RS Measurement Phase

In the initial stage, which is the measurement of FI weight, for that purpose, an assessment of each risk is carried out and translates into fuzzy grouping from the assessment using linguistic variables which then make comparisons between risk variables in each of the index factors that influence it and then aggregated STFNs on each risk variable. And after aggregating STFN on each risk variable then calculate the overall FI weight. To determine the value of the FI, a calculation is made using equation (1) to equation (9).

Table 3. Measurement of FI, RL and RS

Risk Factor	FI*	RL*	RS*
Demand	(1.93091, 3.425524, 3.425524, 4.60216)	(4.375, 6.875, 6.875, 9.375)	3.75, 6.25, 6.25, 8.75)
Environment	(0.774526, 1.600278, 1.600278, 3.335869)	(3.75, 6.25, 6.25, 8.75)	(6.25, 8.75, 8.75, 10.0)
Finance	(1.07181, 2.296736, 2.296736, 3.368546)	(3.75, 6.25, 6.25, 8.75)	(4.375, 6.875, 6.875, 9.375)
Information	(1.819194, 4.004576, 4.004576, 5.782933)	(2.5, 5.0, 5.0, 7.5)	(3.75, 6.25, 6.25, 8.125)
Operation	(1.533914, 3.417977, 3.417977, 5.221739)	(4.375, 6.875, 6.875, 9.375)	(5.0, 7.5, 7.5, 10.0)
Supply	(1.934435, 3.638779, 3.638779, 5.020997)	(1.875, 4.375, 4.375, 6.875)	(3.75, 6.25, 6.25, 9.375)

For the measurement of RL and RS the first is done by conducting an assessment by an expert on the possibility of occurrence and the severity of the risk. Assessment is carried out with linguistic variables, after the assessment is carried out then converts the assessment to the STFN value to do the STFN aggregate calculation from the assessment results. To get the RL and RS values can be calculated using equations (10) and (11).

3.1.3 Fuzzy Inference Phase

At this stage, the first step is to convert aggregated STFNs FI, RL and RS into fuzzy sets, where to convert aggregated STFN by taking the intersection value between STFN and Membership Function (MF) whose parameters are interconnected, then defuzzification is performed to convert fuzzy results be a suitable numeric value that can represent RM. Fuzzy inference system produces mapping between input parameters FI *, RL *, RS * and RM * output. To do the RM * output value it can be done with equation (12).

The following are the results of STFN conversion into the fuzzy set for each risk factor.

Table. 4. STFN conversion into the fuzzy set

Risk Factor	FI*	RL*	RS*
Demand	{(Poor,0.7724), (Poor,3.6298), (Poor, 3.6298), (Fair, 0.8409)}	{(Medium, 0.75), (Medium, 0.75), (Medium, 0.75), (Medium, 0.75)}	{(Medium, 0.5), (High, 0.5), (High, 0.5), (High, 0.5)}
Environment	{(Very Poor, 0.6902), (Poor, 0.6401), (Poor, 0.6401), (Poor, 0.6657)}	{(Medium, 0.5), (Medium, 0.5), (Medium, 0.5), (High, 0.5)}	{(Medium, 0.5), (High, 0.5), (High, 0.5), (Very High, 0.5)}
Finance	{(Poor, 0.9187), (Poor, 0.9187)}	{(Low, 0.5), (High, 0.54), (High, 0.54), (High, 0.5)}	{(Medium, 0.75), (High, 0.75), (High, 0.75), (Very High, 0.75)}
Information	{(Poor, 0.7277), (Poor, 0.3982), (Fair, 0.6018), (Fair, 0.6018), (Fair, 0.6868)}	{(Low, 1.0), (Medium, 1.0), (Medium, 1.0), (Medium, 1.0)}	{(Low, 0.50, (Medium, 0.5), (Medium, 0.5), (High, 0.75)}
Operation	{(Very Poor, 0.3864), (Poor, 0.6328), (Fair, 0.3672), (Poor, 0.3864), (Fair, 0.9113)}	{(Medium, 1.0), (Medium, 1.0), (Medium, 1.0), (Very High, 1.0)}	{(Medium, 0.75), (High, 0.75), (High, 0.75), (High, 0.75)}
Supply	{(Poor, 0.7738), (Poor, 0.5445), (Poor, 0.5445), (Fair, 4555), (Fair, 0.9916)}	{(Low, 0.75), (Medium, 0.75), (Medium, 0.75), (High, 0.75)}	{(Low, 0.5), (Medium, 0.5), (Medium, 0.5), (Very High, 1.0)}

After converting, the next step is to form a combination of the conversion results with the "if-then-rules" function, the number of rules can be known by multiplying the number of FI, RL, and RS. From the combination results obtained the results of Risk Magnitude (RM*) with equations (13) to (15). Then do the conversion into the fuzzy set to find out the percentage of the value of RM * with equation (16).

After calculating to get the RM value, the RM value is converted into fuzzy sets by taking the intersection between lines in the figure 3 graph. Because the RM value on each risk factor has the same results, then to find out the risks that have great opportunities and big impacts can be seen in the FI* value on the risk variable, then the competitor moves is obtained with the magnitude of the impact that arises is a major of 75% and a minor of 25%.

3.2 Mitigation Strategy with Analytical Network Process

3.2.1 The Proposed Alternative

To make a good strategy to minimize the risk of competitor moves , there are 6 strategies such as The strategies are market observations and recognizing opportunities (S1), Creating a different product, highlighting the superiority of products (S2), Highlighting the superiority of products (S3), Learning competitors' strengths and weaknesses (S4), Competitive price offerings (S5), and Creating events for product promotion (S6). For input calculations geometric mean of experts are done by using equation (17) can be seen in Table 5.

Table 5. Aggregated of experts judgment using geometric means

Strategy for Competitor Moves						
	S1	S2	S3	S4	S5	S6
S1	1.00	1.91	1.33	2.47	4.22	5.13
S2	0.52	1.00	1.91	3.27	3.27	3.27
S3	0.75	0.52	1.00	1.91	3.27	3.56
S4	0.41	0.31	0.52	1.00	0.75	2.76
S5	0.24	0.31	0.31	1.33	1.00	1.33
S6	0.19	0.31	0.28	0.36	0.75	1.00

3.2.2 Data Consistency Analysis

In this stage calculated whether the experts' judgment is consistent or not. For this reason, the step is to find out the Consistency Ratio (CR) value by calculating the Consistency Index (CI) previously. The CI used equation (18) and the CR used equation (19) as be seen in examples below.

$$CI = \frac{\lambda_{max} - n}{n-1} = \frac{6.1951}{6-1} = 0.0390$$

$$CR = \frac{CI}{RI} = \frac{0.0390}{1.24} = 0.0315$$

The results obtained CR value <0.1, it can be said that the expert judgment is consistent or value according to the Saat Saaty statement, that CR is recommended to have a value equal to 0.1 [10].

3.2.3 Priority of Strategy

To find out which strategies have priority to recommend can be seen from the percentage results that have the highest number. The strategies are market observations and recognizing opportunities (31.30%), Creating a different product, highlighting the superiority of products (25.20%), Highlighting the superiority of products (19.70%), Learning competitors' strengths and weaknesses (10.10%), Competitive price offerings (8.10%), Creating events for product promotion (5.60%). So we can know the strategies the strategies are market observations and recognizing opportunities that have the largest percentage of 31.3%.

4. Conclusion

Based on the results of research and analysis, obtained the results of the risks in the Wooden Toys Industries consists of 6 risk factors, where each risk factor has its own risk variables. For demand risk factors there are 4 risk variables, namely late delivery to customers, forecast errors, market saturation, and competitor moves. In environmental risk factors there are 4 risk variables, namely macroeconomic uncertainty, natural disasters, policy uncertainties, and social uncertainties. On financial risk factors there are 2 risk variables, namely cost/price risk, and exchange rate risk. In information risk factors there are 5 risk variables, namely breakdown of IT infrastructure, information distortion, inadequate information security, information delays, and wrong communication options. In operational risk factors there are 5 risk variables, namely capacity inflexible, design changes, disruption in production, inventory risk, and variability in production process. And on operational risk factors there are also 5 risk variables, namely dependence on single suppliers, inflexible of supplier, poor delivery performance, supplier poor quality, and supplier bankruptcy.

And based on calculations using the fuzzy reasoning method, the results show that the risk variables of the competitor moves in the risk factor of demand are more likely to occur and have a large impact. From these results, mitigation strategies are carried out using analytical network process methods with the help of super decision software and the results of market observation strategies and recognizing good competitors are carried out to minimize the risk of competitor moves. This research is an enrichment of the method or has not been implemented. Therefore, there are no significant implications of the results.

Acknowledgments
This work was supported in part by Simlitabmas Ristekdikti followed SK No.3/E/KPT/2017 and contract with LP2M Al Azhar Indonesia University under agreement No. 064/SPK/A-01/UAI/IV/2018

5. References

[1] Janivier J A M 2012 A new Introduction to Supply Chains and Supply Chain Management: Definitions and Theories Perspective. *International Bussiness Research Journal* **5**(1) pp 194-207
[2] Handayani D I 2013 Identifikasi Risiko Rantai Pasok Berbasis Sistem Traceability pada Minuman Sari Apel. *Spektrum Industri* **11**(2) pp 117-242
[3] Zaroni 2015 *Manajemen Risiko Rantai Pasok Dalam Model SCOR.* https://www.supplychainindonesia.com/new/manajemen-risiko-rantai-pasok-dalam-model-scor/ [Accesed: 23-May-2018]
[4] Dittmann J P 2014 *Managing Risk in the Global Supply chain.* A report by the Supply chain Management Faculty at the University of Tennessee. (University of Tennessee)
[5] Himawan S I P S 2017 *Pengelolaan Risiko Rantai Pasok Menggunakan Metode Fuzzy House of Risk (FHOR) (Studi Kasus Industri Mainan Kayu).* (Jakarta: Universitas Al Azhar Indonesia)
[6] Simanjuntak S J 2013 *Analisis Manajemen Risiko Rantai Pasok Buah Manggis dengan Metode Analytic Network Process di PT Agung Mustika Selaras, Jawa Barat.* (Bogor: Institut Pertanian Bogor)
[7] Jafarnejad A, Ebrahimi M, Abbaszadeh M A and Abtahi S M 2014 *Risk Management in Supply Chain using Consistent Fuzzy Preference Relations. International Journal of Academic Research in Business and Social Sciences* **4**(1) pp 77-89
[8] Zeng J, An M and Smith N J 2007 Application of A Fuzzy Based Decision Making Methodology to Contraction Project Risk Assessment. *International Journal of Project Management* **25** pp 589–600
[9] Palupi A H, Tama, I P and Sari R A 2014 Evaluasi Dampak Lingkungan Produk Kertas dengan Menggunakan *Life Cycle Assessment* (LCA) dan Analytic Network Process (ANP) (Studi Kasus: PT X Probolinggo). *Jurnal Rekayasa dan Manajemen Sistem Industri* **2**(5) (Malang: Teknik Industri Universitas Brawijaya)
[10] Saaty T and Vargas L 2006 *Decision Making with the Analytic Network Process. Economic, Political, Social and Technological Applications with Benefits, Opportunities, Costs and Risks.* (Pittsburgh: RWS Publication)

Information system design using labor productivity measurement for construction

Abdullah 'Azzam, Suci Miranda, Sri Indrawati

Industrial Engineering Department, Faculty of Industrial Technology, Universitas Islam Indonesia, Yogyakarta, Indonesia
abdullah.azzam@uii.ac.id, suci.miranda@uii.ac.id, sriindrawati@uii.ac.id

Abstract. Labor utilization in construction project is the most critical because it costs 30-50% of the total cost. In many projects, the labors e.g craftman are hired outside the company. Contractors and owners need sufficient information related to those labors whom they will work with. It is necessary for clients understanding each labor productivity performance in order to achieve the project quality. This study aims to provide the infomation needed by designing an information system that can be easily accessed. This is also to support clients for decision making in terms of teamwork project involved. The level of productivity is measured by Labor Utilization Rate (LUR) using three variables: effective working time, contributory working time, and ineffective working time. Software Development Life Cycle (SDLC) is applied in designing the information system. The finding showed that information system is properly implemented to assist clients in determining workers using productivity calculation. The system can inform the recommended workers by calculating and sorting their level of productivity that support client's decision making towards construction workers.
Keywords: information system, labor productivity, construction, SDLC

1. Introduction

Good project management in construction must vigorously pursue the efficient utilization of labor, material, and equipment. Improvement of labor productivity should be a major and continual concern of those who are responsible for cost control of constructed facilities. The high cost associated with budget and schedule overruns in the heavy construction can often be traced to low worker productivity. Labor cost is less predictable and runs a greater chance of overruns than either equipment or material cost. A significant factor influencing construction costs is the wages of construction labor, including additional expenses such as travel costs, insurance, pensions, and other benefit of worker [1]. Workers in New York and Zurich have the extreme wage close to USD100/hour, compared to an average labor rate of USD28/hour accross all region. Labor costs can reach 50 % or more of construction project cost, management of the work process presents the greates opportunity to reduce project cost [2]. Another reported that the labor cost component of a building project ranges from 30 to 50% or even up to 60% of the overall project cost [3]. Since construction labors affect each part of the project, it becomes clear that construction labor is a critical section of a construction project.

Whereas, another problem of skill shortage has arisen within the construction industry. There are 24 of 43 markets analyzed in the skill shortage of labor including Singapore, Hong Kong, and Tokyo [1]. Skill shortage is linked to productivity (labor's production rate). It is defined as the number of units of work produced by a person in specified time. Production rates may also specify the time in man-hours or man-days required to produce a specified number of units of work. The time that a labor will consume in performing a unit of work varies between labors and between projects and with climatic conditions, job supervision, complexities of the operation and other factors. For instance, it requires more time for erect shutters for stairs than for foundations. [4] have reevaluated the factors affecting the productivity of workers in the construction field such as cooperation factors, economic conditions, physical conditions and socio-physiological factors. The existing of high labor productivity at every stage of a project's development plays a significant role in a project success [5]. A standardization of guidelines for measuring labor productivity for building construction work with exploratory methods through the identification of needs and problems in the field [6]. However, productivity measurements using daily, weekly and monthly reports that are always made by building supervisors or consultants are easier to obtain information compared to methods commonly used in manufacturing such as time study, time and motion study and work sampling as it requires measurement actual productivity in the field [7]. In addition, [8] emphasized that manufacturing and construction have differences which companies engaged in manufacturing are serial and repetitive, and have short production cycles as well. This productivity measurement is very useful for companies as a reference to improve the performance of companies or groups of workers in the construction field. Likewise, in order to meet the consumers, it also required a good performance provided by companies or groups of workers in construction.

Regarding the productivity information of the construction workers, a client needs an information system providing a comprehensive information related to performance of workers in the construction field. Thus, a company can evaluate their workers as well as get the availability of workers based on their productivity as a key indicator for a company to hire the worker for a project. Software Engineering is the area which is constantly growing [9]. It is very interesting subject to learn as all the software development industry based on this specified area. There exist various models to develop software. But most of the existing Software Development Models pay less or very little attention towards client satisfaction. Client satisfaction is crucial, not only to the client, but even more to the developer because it costs far less to retain a happy client than it does to find a new client. Satisfying client is an essential element for staying in this modern world of global competition. Client satisfaction is very necessary for the acceptance and delivery of the software product. In terms of client satisfaction of worker productivity information, this study is to design an information system of worker productivity of a project using SDLC (Software Development Life Cycle) approach. The SDLC is initiated by the client's needs. In the beginning, these needs are in the mind of the client. The software developer by using a software development model has to identify, discover, understand and fulfill the requirement of the client in order to satisfy the client.

2. Method

Productivity in construction is often defined as output per labor hour. Since labor constitutes a large part of construction cost and the quantity of labor hours in performing a task in construction is more susceptible to the influence of management than are materials or capital, this productivity measure is often referred to as labor productivity. However, it is important to note that labor productivity is a measure to the overall effectiveness of an operating system in utilizing labor, equipment, and capital to convert labor efforts into useful output, and is not a measure of the capabilities fo labor alone. The research conceptual model explains every detail of this reseach that can be seen in Figure 1.

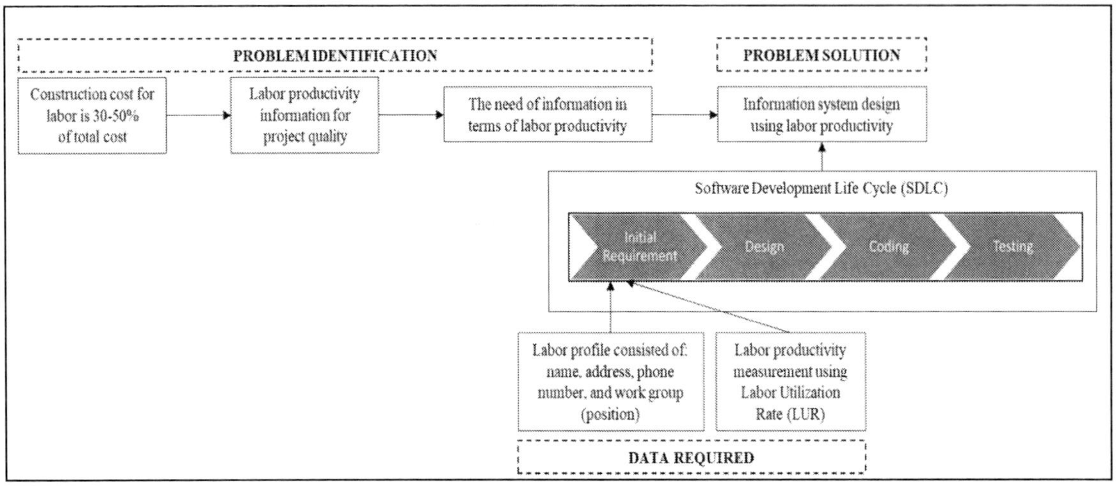

Figure 1. Research conceptual model

Four major steps are employed in designing the information system using SDLC (Software Development Life Cycle) approach.

- The first is initial requirement. It is to determine the factors required within the system developed. Regarding the labor profile, several factors that must put into the information system: name, address, phone number, work group (position), and the labor productivity. The productivity measurement employs Labor Utilization Rate (LUR). There are 3 data needed for calculating LUR: effective work, essential contributory work, and ineffective works. As the example of the calculation, previous research data [10] measuring the productivity of 5 workers is used. The observation was conducted for 3 days to identify the duration of each category of work. The work group or position refers to the work group explained in previous Figure 1: foreman and craftman (carpenter, bricklayer, etc)
- The second is design. It aims to change the results of the needs analysis stage into a structure [11]. For this reason, a number of tools are needed including data flow diagrams (DFD), entity relationship diagrams (ERD), and user interface designs. There are 3 external entities involved within the system designed namely the administrator, foreman, and user. The administrator is in charge of managing and giving access to every foreman who registers to the system. The foreman served as a construction project supervisor who then assessed and recorded the level of productivity of each craftman. The user is the end user or client of the system created such as the project managers. The information system framework is proposed as follows.

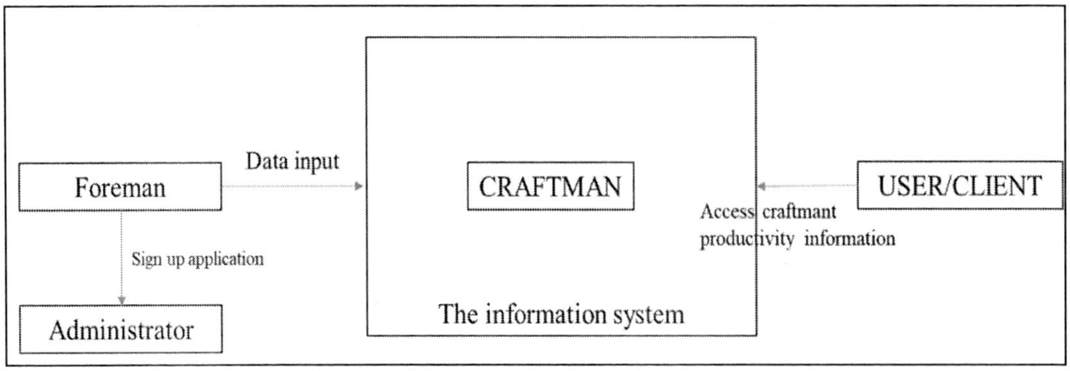

Figure 2. Information System Framework

- The third step is coding. This section focuses on the algorithm used to calculate LUR that is implemented into the Java programming language for android applications on mobile phones. The following algorithms are used for the implementation of LUR /worker/day calculations:
    1. Enter the total effective working time, total contibutory working time, and total ineffective working time;
    2. Calculate total time (total effective working time + total contibutory working time + total ineffective working time);
    3. Calculate the numerator (total productive working time + [¼ x total contributory working time worked]);
    4. Calculate LUR = Total time / numerator

  Calculating the average LUR is by adding the LUR / day value divided by the number of working days. Ordering the value of productivity can be carried out after getting the average value of LUR for each worker or craftman.
- The fourth is testing. It is required to test the results of the calculation of productivity of each worker or craftman. LUR calculation using both the system generated and manually calculated must have similar result. Likewise, the system can order the value of worker productivity from the biggest to the lowest value.

Based on the provisions set forth in the SKKNI published by the Construction and Human Resources Development Agency, the Construction and Competency Training and Development Center (BPKSDM-KPK), the classification of workers involved in a construction project is distinguished based on one's ability based on knowledge, skills and work attitudes include experts and skilled presence personnel. This study employs 2 categories of labor: foreman and craftman. Foreman is the supervisor who will examine the craftman's performance by assessing his productivity and record the calculation into the information system.

2.1. Labor Utilization Rate

LUR is the percentage obtained from the sum of effective work added with ¼ essential contributory work, then divides the sum by the total observations [12].

$$\text{LUR} = \frac{Effective\ Work + (\frac{1}{4}\ Essential\ Contributory\ Work)}{Total} \quad (1)$$

LUR is used for productivity measurement where the input used as the output comparator is the time variable. The calculations used are relatively easy, so it can simplify the process of calculating and developing the system. Yet, the factors affeting the low productivity of workers cannot be known through this formula. In construction projects, there are 3 category of activities to be considered [12]:

1. Productive/effective activities are contributing directly to the output expected from an activity, for example: pouring fresh concrete on the place to be cast, installing red bricks for wall pairs, and others.
2. Contributory activities do not directly affect the expected output of a construction project, but are often crucial or must be conducted in order to support productive activities carried out, for instance: reading work drawings to be carried out, receiving instructions or orders from foreman, etc.
3. Productive/Ineffective are idle activities, waiting, or other activities that do not contribute positively to the progress of the project being worked on, such as: smoking during working hours, or talking useless topics which are not related to the tasks given.

3. Results and Discussion

This study aims to build an information system related to the availability of construction workers or craftman as well as to provide construction workers productivity in order to prevent hiring less producticity workers. Several information needed to display the availability of construction workers in this system including profiles of companies or groups of workers providing construction services (name,

IOP Conf. Series: Materials Science and Engineering **528** (2019) 012003 doi:10.1088/1757-899X/528/1/012003

phone number, address, expertise/position), level of productivity (average level of productivity for each worker and groups of workers, the order of productivity levels for each worker or group of workers). The process of calculating the level of productivity can be seen in the picture below:

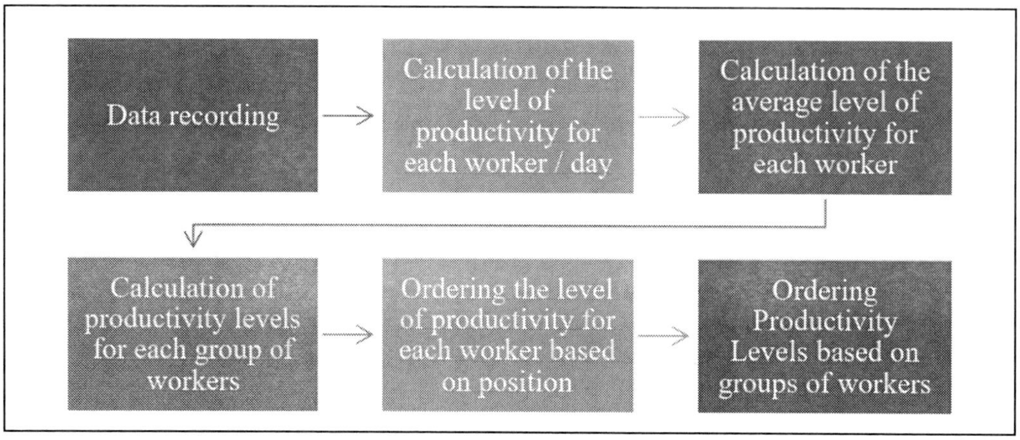

Figure 3. Productivity Level Calculation Process

Foreman is responsible to input 3 variables of each worker or craftman assessed into the system: total effective working time, total contributory working time, and total ineffective working time. In order to test the information system, the secondary data taken from research conducted by [10] is applied. LUR calculation for each worker on the first day and 3 days observation can be seen in following Table 1 and Table 2. For example, Tohir, his LUR is:

$$\text{Tohir LUR}=\frac{339+(\frac{1}{4}\times 58)}{(339+58+23)}=\frac{353.5}{420}=\mathbf{0.841667}\ \times \mathbf{100\%}=\mathbf{84.17\%}$$

Table 1. Calculation of LUR for day 1

Name	Total Effective Working Time (minute)	Total Contributory Working Time (minute)	Total Ineffective Working Time (minute)	LUR (%)
Tohir	339	58	23	84.17
Pudin	283	127	10	74.94
Wahyu	309	91	20	78.99
Sigit	261	126	33	69.64
Heri	299	95	26	76.85

Table 2. Calculation of LUR for 3 working days

Name	Day 1 (%)	Day 2 (%)	Day 3 (%)	Average (%)
Tohir	84.17	78.15	70.89	77.74
Pudin	74.94	59.58	66.9	67.14
Wahyu	78.99	82.56	75.71	79.09
Sigit	69.64	65.48	75.48	70.2
Heri	76.85	74.7	81.19	77.58

Using the results obtained from LUR calculation above, then the system created will sort the LUR to show the productivity level of each worker from the highest to the lowest productivity. This will facilitate the client in choosing construction workers or craftman according to his productivity. The productivity sequence is showed in Table 3 below.

Table 3. LUR rank of each worker

Rank	Name	Average (%)
1	Wahyu	79.09
2	Tohir	77.74
3	Heri	77.58
4	Sigit	70.2
5	Pudin	67.14

The output of LUR rank will be provided in information system. Contex diagram describes the system in general which is then downgraded to level 1 DFD. Then each sub system at level 1 (account management and productivity information management) is broken down again into level 2 DFD. Designing DFD, ERD, and User Interface are the next phase describing below.

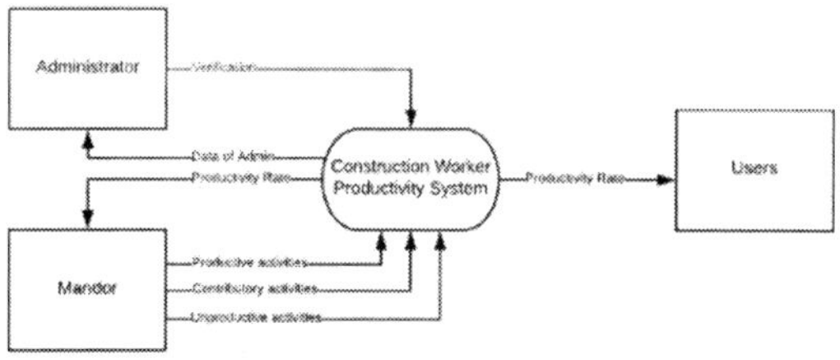

Figure 4. Context diagram

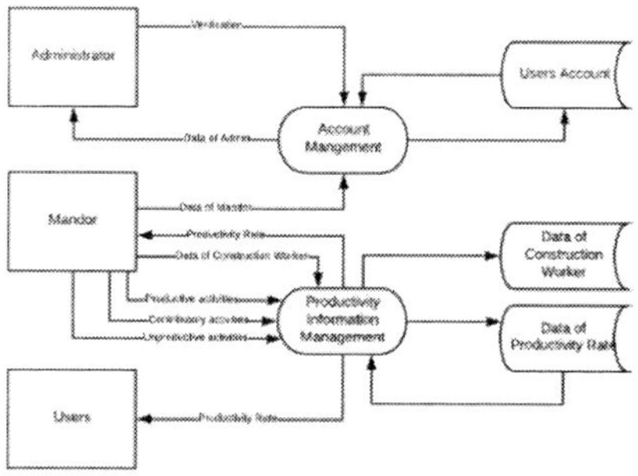

Figure 5. DFD level 1

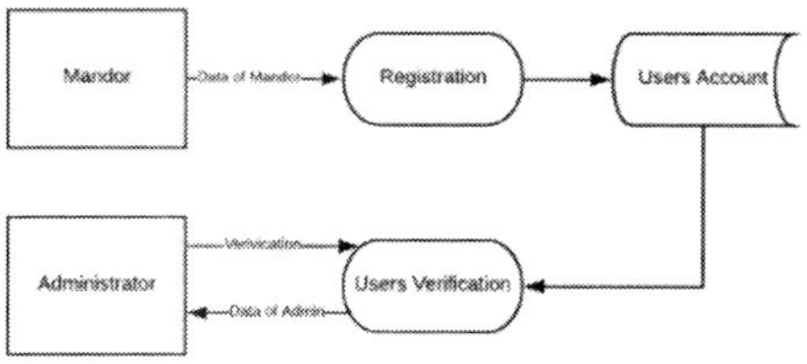

Figure 6. DFD level 2 of account management

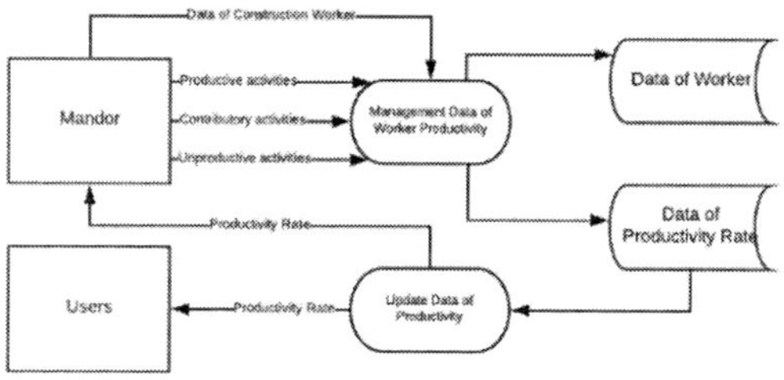

Figure 7. DFD level 2 of productivity information management

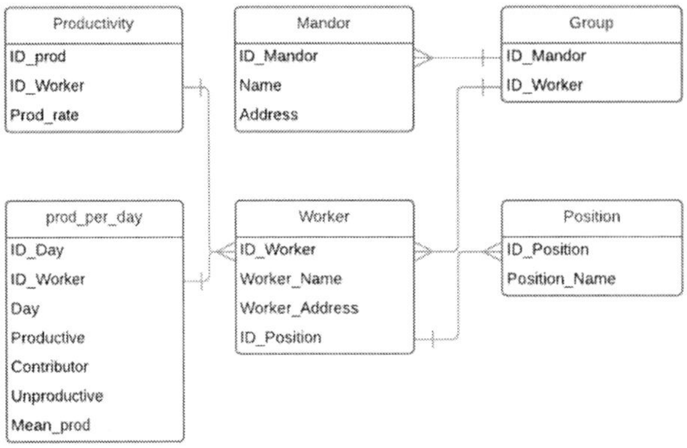

Figure 8. ERD Information System for Construction Worker Availability

User Interface (UI) design is important in a system design and has a lot of attention in recent years [13]. It is useful as an interaction between humans and computers that can increase the usefulness of a product [14]. If the UI is made worse, the ability to absorb the capabilities of the system will be reduced (Susanty et al, 2016). In this paper, we discuss some of the things in the UI that are made, namely input of labor productivity, results from productivity and sorting and search for workers.

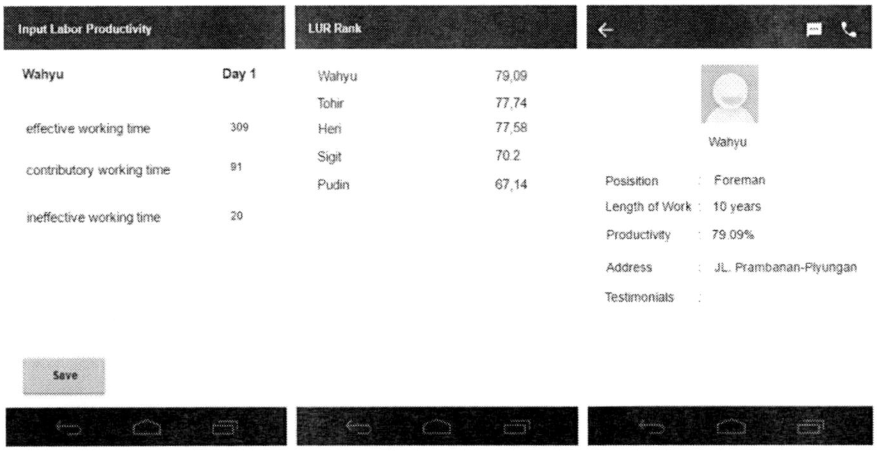

Figure 9. User Interface

4. Conclusion

This paper has presented the results of designing an information system of construction workers availability in terms of productivity level using mobile phone media. This system employs LUR formula in order to obtain the productivity level of each worker. Hereinafter, the information system given to the client can also inform the worker performance so foreman, as a supervisor, will evaluate and motivate those whose productivity level is below the minimum standard (LUR <50%). The LUR method has been easily operated by foreman to record worker productivity since the variables used are few (effective working time, contributory working time, and ineffective working time) and the unit used is time. In addition, it also facilitates the design of the system. However, there is still a shortage of LUR method, that can not capture and record the detail factors led to uneffective activities and workers as well. Further research can apply more advanced methods for productivity-based system design linked to internet era things. So that, it can be real time monitored by workers who can be integrated with the system designed.

Acknowledgement
Researchers would like to express our gratitude to the support of Industrial Engineering Department, Faculty of Industrial Technology, Universitas Islam Indonesia.

5. References

[1] Cost of labour in the construction industry, International Construction Market Survey 2017 retrieved from: http://www.turnerandtownsend.com/en/insights/international-construction-market-survey-2017/cost-of-labour-in-the-construction-industry/
[2] *Measurement and Analysis on Heavy Construction Projects* 2013 PicardConsultLLC
[3] Elbegati Emad *Cost estimating: Cost of Construction Labor and Equipment* (Egypt: Mansoura University/Faculty of Engineering) chapter 4 pp 64-96
[4] Kazaz A Ululbeyl S Acikara T Er B 2016 *Procedia Engineering* **164** pp 28 – 34
[5] Nasirzadeh F Nojedehi P 2013 *International Journal of Project Management* **31** pp 903–911
[6] Wuryanti Wahyu 2010 *Prosiding PPI Standardisasi*– Banjarmasin, 4 August 2010
[7] Wulfram I. Ervianto *Jurnal teknik sipil* **9** pp 31 – 42
[8] Loera I Espinosa G Enríquez C Rodriguez J 2013 *Procedia Engineering* **63** pp 947 – 955
[9] Kumar N, Zadgaonkar A Z and Shukla A 2013 *International Journal of Soft Computing and Engineering (IJSCE)* **3** pp 2231-2307
[10] Aprilian T 2010 Analisis Produktivitas Tenaga Kerja Pada Pekerjaan Struktur Rangka Atap Baja. (Surakarta: Universitas Sebelas Maret)
[11] Aggarwal K K and Singh Y *Software Engineering* 3rd Ed.
[12] Olomolaiye P O, Jayawardane A K W and Harris F C 1998 *Construction Productivity Management.* (Singapore: McGraw-Hill, Inc.)
[13] Hsiao S W, Lee C H, Yang M H and Chen R Q 2017 User interface based on natural interaction design for seniors. *Computer in Human Behavior* 75.147-159.
[14] Finstad K 2010 The usability metric for user experience. *Interacting with Computers* **22**(5) 323-327.

Development of Information Systems as a Means to Improve Sharia Cooperative Services

Riri Safitri[1], Dody Haryadi[2], Endah Sulisthyani[3], and Vareza Noorliko[4]

Informatics Engineering of University of Al Azhar Indonesia
[1]riri@uai.ac.id, [2]dodyharyadi@uai.ac.id

Abstract. Increasing and empowering the economy of the community is an important thing that is a shared responsibility. This is the basis for the LAZ Al Azhar Empowerment "Sejuta Berdaya" program that makes Jampang Village as one of the target groups in the form of Kelompok Swadaya Masyarakat (KSM) which are named KSM Pelita Jampang Gemilang. The community in Jampang village has a variety of jobs with a low economic level. The foundation provides funding to the community through a sharia microfinance institution that collects and distributes public funds in accordance with Islamic law. The Jampang Sharia Cooperative is still operating manually. Registration process, member information, amount of funds and reports are still done manually and stored in a notebook. This causes the operational process not to run properly, because the data is not organized and stored neatly. Based on these conditions, a financial information system is needed, which is a web-based sharia cooperative, so that the operational processes and reports to members and administrators can run better and faster, so that they can help effectively and on target. Information systems are created using SDLC through the stages of user requirements, analysis and implementation.
Keywords: information system, web-based system, sharia, microfinance, cooperation

1. Introduction

The information system is one of the programs that is widely used by many agencies in the current era of globalization, both government and private agencies in carrying out daily activities related to the presentation of data quickly and accurately. Information systems (IS) are combinations of hardware, software, and telecommunications networks that people build and use to collect, create, and distribute useful data, typically in organizational settings [1]. Knowing the potential impact of information systems and having the ability to put this knowledge to work can result in a successful personal career and organizations that reach their goals [2]. Various studies on cooperative information systems have been carried out, but most of these studies still apply conventional financial systems, namely the percentage of interest. In Indonesia with a Muslim majority, it is in desperate need of a sharia-based financial system in accordance with Islamic Sharia. Islamic financial principles are premised on the general principle of providing for the welfare of the population by prohibiting practices considered unfair or exploitative [3]. The expected results with the implementation of information systems is that the data that has been stored and managed manually can begin to be computerized so as to increase the effectiveness of cooperative performance.

Sharia Cooperative in Jampang Village is an organization of Kelompok Swadaya Masyarakat (KSM) Pelita Jampang Gemilang with a profit-sharing system oriented to small and medium enterprises which is a combination of the concept of cooperatives with the concept of sharia based on Islamic sharia economic system. This institution is one of the sharia-based microfinance institutions using sharia patterns to meet the needs of financial transactions. One of the activities carried out by the KSM Pelita Jampang Gemilang cooperative is to provide savings and loan facilities to its members. The savings and loan procedures that are carried out are still manual, such as the recording process, the document storage process, and the calculation process, so that it becomes a significant obstacle to the speed and accuracy of the data flowing.

Figure 1. Member savings book

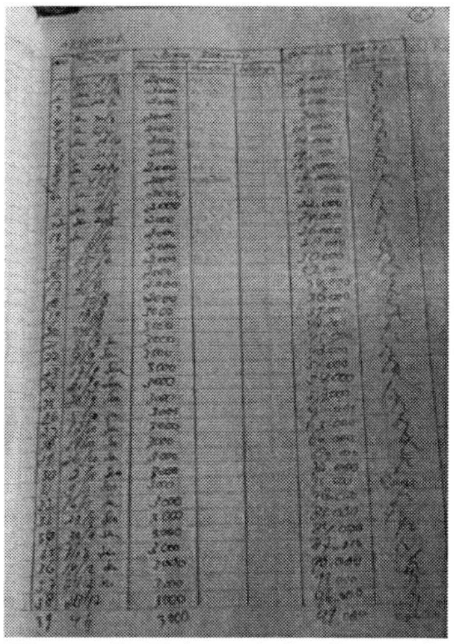

Figure 2. Tabarru' fund records

In addition, the report presented is very slow because managed data often accumulates the consequences of the length of time needed in data management and data recording process. With a large number of members and a very limited number of cooperative administrators, it will lead to a length of performance which results in dissatisfaction among members of the KSM Community. Management techniques that are applied to KSM Pelita Jampang Gemilang are sharia management techniques that are able to adapt to the growth and development of the environment. Figure 1 and Figure 2 show the process of recording tabarru' deposits and funds at KSM Jampang which is still done manually. Tabarru 'is one of the contracts in Islamic finance that does not take advantage in the transaction process [4].

KSM Jampang as a sharia microfinance institution under the auspices of LAZ Al Azhar must provide a report to LAZ Al Azhar regarding the economic development of the surrounding communities who are members. This is because LAZ Al Azhar has a basic scheme of economic empowerment programs called "Sejuta Berdaya", by grouping into 4 stages, namely rescue, strengthening, development and resilience [5]. This stage is based on the level of basic needs and income of the community that is fostered. This program was a community economic empowerment program in groups and revolving with a strong mentoring process and rolled out was not large and not comparable to the prospective beneficiaries, the achievement targets that had been running were not maximized.

2. Methods

System development methods are carried out with SDLC with the stages of requirements, analysis, design, implementation and testing.

2.1. System Analyst

Information systems for financial institutions are very important and have been developed. In a study, information technology in the field of financial loans is used to store and process data, and make it easier for the administration to correct existing and new customers' data [6]. The cooperative information system created is based on the needs of Jampang village in managing finances to develop the potential and economy of the community. The system is made of web-based, so that the survey process, member

IOP Conf. Series: Materials Science and Engineering **528** (2019) 012004 doi:10.1088/1757-899X/528/1/012004

registration can be done correctly and real-time. System requirements are based on system business processes that will be built as follows:

- Prospective Members come to the cooperative to register as members of the cooperative.
- Prospective Members get information from admin requirements to become members and requirements for making loans.
- If the registration form has been filled in completely, the admin copies the member data.
- Then prospective members pay a principal deposit of Rp.50,000 to the treasurer.
- Treasurer submits a deposit book to members.
- Members who make savings transactions or withdraw deposits directly to the treasurer, then the treasurer records transactions in the passbook and the cooperative cashbook.
- Members who apply for loans must complete the requirements that have been informed.
- Admin submits member loan submission data to the survey team.
- The survey team surveyed members' homes and assets.
- Submit data from the survey conducted by the survey team
- The chairman of the cooperative considers the loan proposed.
- Admin confirmation of loan approval from the chairman of the cooperative to the treasurer
- Treasurer submits loan disbursement to members.
- Members make loans to repay loans to the treasurer.

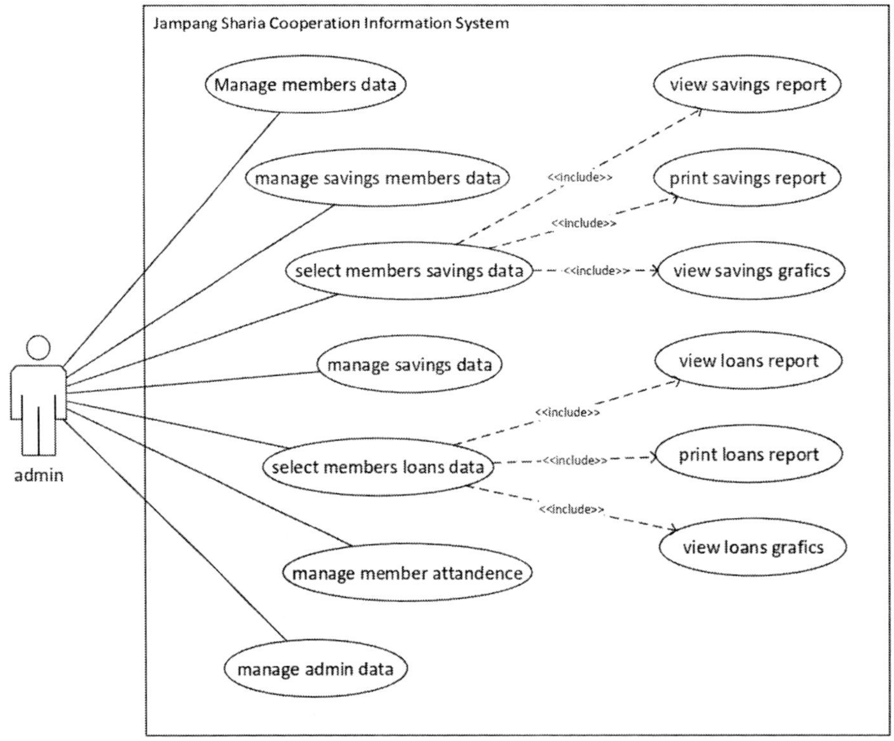

Figure 3. Use case diagram

2.2. System Design

KSM Jampang cooperative information system design is made using UML Language. The Unified Modelling Language (UML) is a general-purpose visual modelling language for systems. Although UML is most often associated with modelling Object Oriented software systems, it has a much wider application than this due to its in-built extensibility mechanisms [7]. UML notation is also used as a visual tool for graphical representation component model [8]. UML notation can simplify the process and adjust the system model to be built, so that complex problems can be well presented. The Use Case Diagram of this information system can be seen in Figure 3.

The design of the system database is made in the form of Entity Relationship Diagram (ERD). ERD is a modelling of data in a diagram based on data (entities) in the real world and the relationships between the data [8]. ERD of the system can be seen in Figure 4.

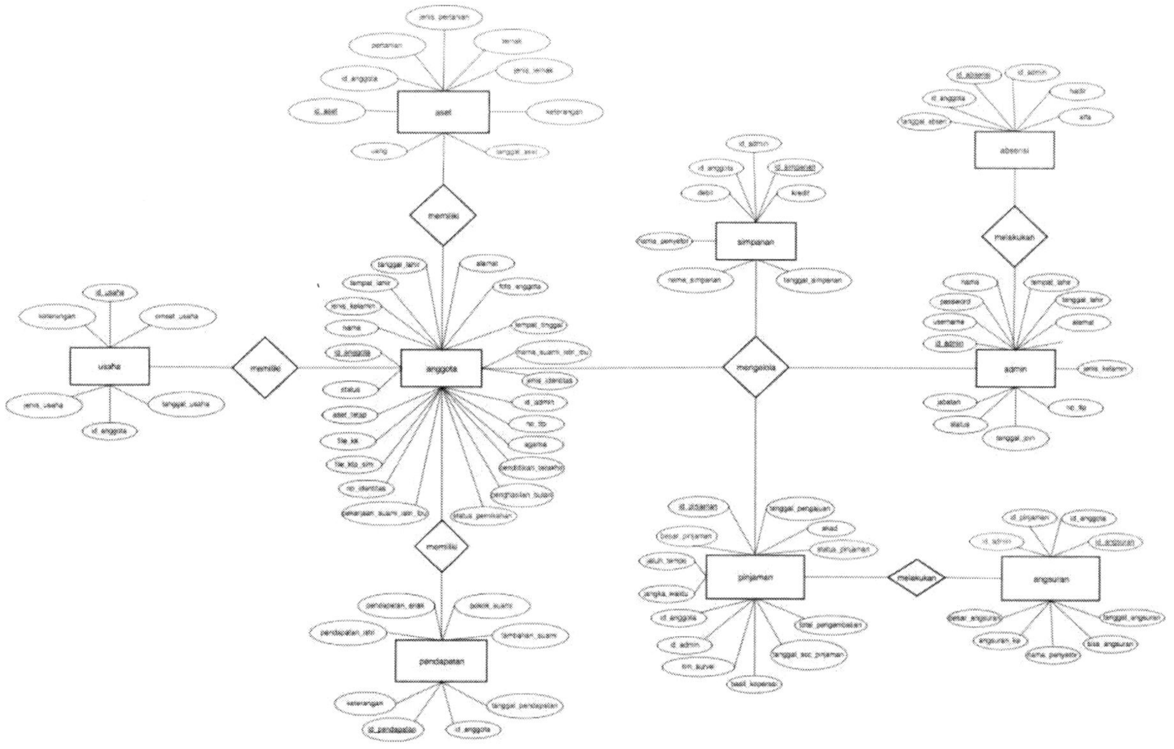

Figure 4. Entity Relationship Diagram

The development of this management information system certainly requires a variety of tools as support. There are also several configurations proposed to support the completion of the development of the system.

Hardware

- Processor: 2.5GHz Intel Core i5-6200U
- Memory: 4 GB
- Hard drive: 1 TB

Software

- Operating System: Windows 10
- Framework: CodeIgniter
- Programming Language: PHP 5
- DBMS: MySQL phpMyAdmin
- PHP Engine: XAMPP v3.2.2

IOP Conf. Series: Materials Science and Engineering **528** (2019) 012004 doi:10.1088/1757-899X/528/1/012004

3. Result and Discussion

Islamic cooperative of KSM Jampang information system is built in several main menus, namely admin menu, members, savings, loans and reports. The home menu display can be seen in Figure 5.

Figure 5. Home menu display

To improve KSM services to members, there are two special menus, namely members and reports. on the members menu, for member registration beforehand a survey was conducted to see the economic level of members so that they can be grouped based on level of basic needs and income.

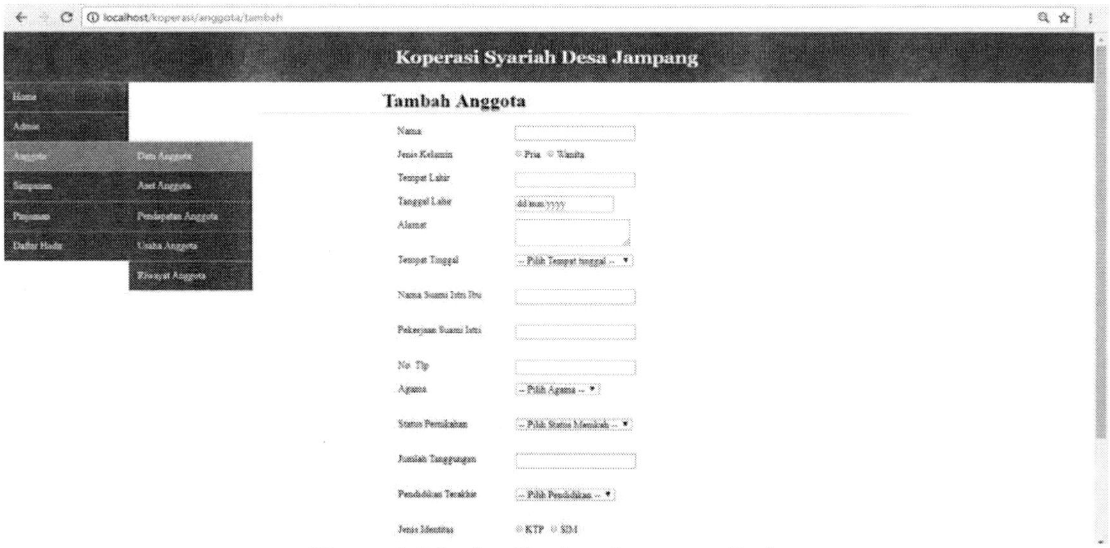

Figure 6. Member Registration menu display

When members have joined and get loans to increase business capital and make deposits, it is expected that there will be an increase in members' assets and income. This data can be seen in the member assets sub menu as shown in Figure 7.

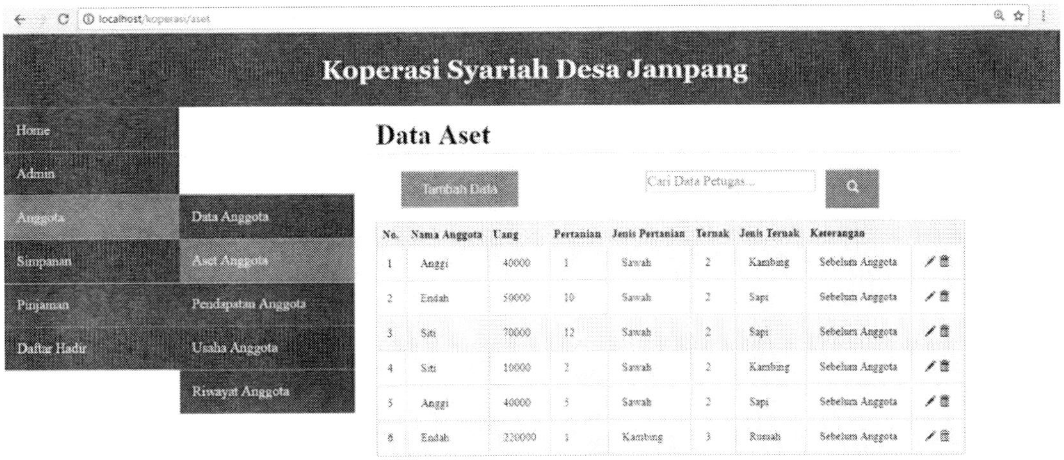

Figure 7. Member asset menu display

Weekly meetings are mandatory activities for cooperative members. In this meeting education will be given to members regarding the importance of joining a sharia cooperative and avoiding usury culture and moneylenders. In this meeting also carried out collection of compulsory members whose amount has been determined. Attendance at weekly meetings will be a consideration in helping and savings to each member.

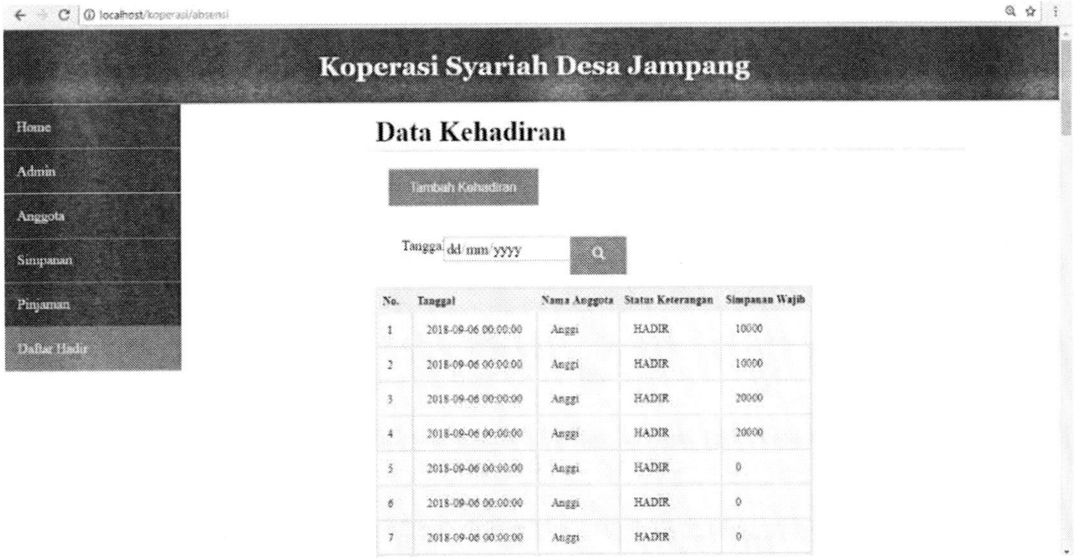

Figure 8. Member attendance menu at weekly meetings

As a cooperative that is based on Islamic values, providing loans to members is divided into two types of loan contracts. The first loan was Qordhul Hasan, on this loan there was no profit sharing and additional costs. Qordhul Hasan's loans were solely intended for social assistance. This loan is given to new members who join and consider the members 'economy through data on members' assets and income. The second loan agreement is profits sharing. In this loan, the value of profit sharing must be allowed by the member, whose value is determined based on the agreement of the members and cooperatives.

Figure 9. Loans menu display

On the report menu, the admin can view the savings and loan reports of each member, so that KSM can view and supervise member history. KSM Jampang as a care institution from LAZ Al Azhar was given a report to KSM Jampang every 6 months.

No.	Akad	Tanggal Pinjaman	Nama Anggota	Jumlah Pinjaman	Sisa Pinjaman	Jatuh Tempo	Keterangan	Nama Admin
1	Qa'ardul Hasan	2018-09-02 00:00:00	Anggi	Rp 20.000	0	2018-10-04	Lunas	Anggi
2	Qa'ardul Hasan	2018-09-03 00:00:00	Anggi	Rp 20.000	0	2018-10-03	Lunas	Anggi
3	Qa'ardul Hasan	2018-09-02 00:00:00	Endah	Rp 10.000	10000	2018-10-02	Lunas	Anggi
4	Bagi Hasil	2018-09-04 00:00:00	Siti	Rp 40.000	30000	2018-10-04	Lunas	Anggi

Figure 10. Loans report display

The report will later group the KSM members based on their economy, namely on assets owned before and after joining. The information system built provides convenience to Jampang cooperatives in managing data and financial transactions in the cooperative. Besides this system also helps cooperatives to provide good and fast service to members. This is in accordance with the benefits of the Information System in a company, namely be more productive, competitive gain advantage, reach more customers and improve service to their customers [9].

4. Conclusion

Based on the development and implementation that has been carried out, it can be concluded as follows:

- The information system built is an information system for the Shariah-compliant Islamic microfinance institution KSM Jampang with SDLC development
- Sharia cooperative information systems apply the concept of Islamic finance with high flexibility for loan methods and instalment payments

- With the KSM Jampang sharia information system, the management of cooperatives is better organized and computerized, the data of members and the history of members are recorded in a systematic and
- With this Sharia cooperative information system, KSM can manage members by grouping based on the level of income and assets owned by members

Acknowledgments
Acknowledgments we convey to **RISTEKDIKTI** for the assistance of research funds and publications that have been given based on Decree Number **029/KM/PNT/2018** and Agreement/ Contract Number **067/SPK/A-01/UAI/IV/2018** and **Institution of Research and Public Services of University of Al Azhar Indonesi**a for Publication Grant Funding.

5. References

[1] Valacich J, Schneider C and Jessup L 2010 *Information Systems Today - Managing in the Digital World.* 4th Ed. (United States: Pearson Education, Inc.)
[2] Stair R and Reynolds G 2015 *Fundamentals of Information Systems* pp 128-161 Accessed from https://www.radford.edu/mhtay/ITEC110/Fundamental_Info_Sys/Lecture/ch01_5e.pdf at Sept 2018.
[3] El-Zoghbi M and Tarazi M 2013 *Trends in Sharia-Compliant Financial Inclusion. Journal of Brief.* Washington, DC: CGAP. pp 1-11 Accessed from https://www.cgap.org/sites/default/files/Focus-Note-Trends-in-Sharia-Compliant-Finanicial-Inclusion-Mar-2013.PDF at Sept 2018.
[4] Fauzan A 2016 *Prinsip Tabarru' Teori dan Implementasi di Perbankan Syariah.* Al-Amwal **8**(2) pp 400-411. Accessed from http://syekhnurjati.ac.id/jurnal/index.php/amwal/article/view/1115 at Sept 2018.
[5] Forum Komunikasi Al-Azhar 2016 *Bakti Sosial Forum Komunikasi Jam'iyyah Al Azhar dalam Upaya Pengentasan Kemiskinan di KSM Pelita Jampang Gemilang 'Sejuta Berdaya'.* (Jakarta: Al-Azhar Peduli Umat)
[6] Masruchin S 2014 *Perancangan Aplikasi Sistem Informasi Pinjaman Keuangan Anggota Koperasi Syariah Mandiri.* (Bandung: Fasilkom Universitas Dian Nuswantoro) Accessed from http://eprints.dinus.ac.id/13246/1/jurnal_14196.pdf at May 2018.
[7] Arlow J and Neustadt I 2004 *UML and The Unified Process: Practical Object-Oriented Analysis and Design.* (Great Britain: Addison Wesley, An imprint of Pearson Education)
[8] Noor M and Masykurinnisa A 2017 Aplikasi Ayatul Ahkam Berbasis Web. *Jurnal Sains dan Informatika Politeknik Negeri Tanah Laut* **3**(2) Accessed from https://jsi.politala.ac.id/index.php/JSI/article/view/107/67 at Sept 2018
[9] Quizlet 2018 *Chapter 1: Managing in The Digital World MIS 304.* Quizlet Inc. *Accessed from https://quizlet.com/82784670/chapter-1-managing-in-the-digital-world-mis-304-flash-cards/* at November 2018.

Best Concept Selection for Dry-Soybean Cracking Machine Process Optimization using TOPSIS method

R A Anugraha[1], N M Darmawan and M Iqbal

[1]School of Industrial and System Engineering, Telkom University, 40257 Bandung, West Java, Indonesia
[2] rinoandias@telkomuniversity.ac.id

Abstract. Tempe is a traditional Indonesian food with processed soybean as the main ingredients. Indonesia is the largest tempe producer in the world with hundreds of thousands of traditional tempe maker. Currently the traditional production process is wasteful that requires the amount of water as much as 1321 liters/production of 37,5 kg tempe. Innovative dry production process is still less desirable because the dry soybean cracker machine is not optimal with a defect of 14,4% per production. In addition, the machine is unable to separate the husk and soybeans from the outlet. There are 12 alternative production concepts to produce highest quality tempe. TOPSIS method was used for selecting best production concept which reduces defect to 6,5% while manual separation process is eliminated. Water consumption is reduced to 800 liters/production of 35,125 kg tempe.
Keywords: TOPSIS, concept selection, dry-soybean

1. Introduction

SME (Small Medium Enterprise) sector which contribute big enough for Indonesia Gross Domestic Product in food field is tempe craftsmen. The average tempe consumption per person per year in Indonesia has reached 6,95 kilograms in 2014 [1]. The level of tempe consumption in Indonesia can be seen from the high volume of soybean imports by Indonesia as shown in Figure 1 [2].

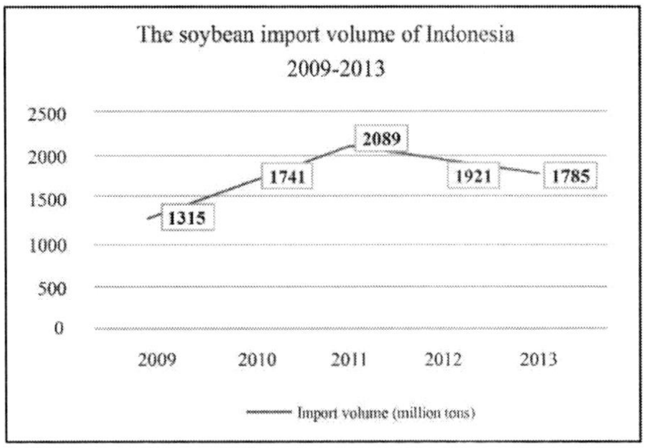

Figure 1. The soybean import volume of Indonesia, 2009-2013

Production process commonly done by tempe craftsmen in Indonesia is wet production process. This process requires a total water supply of 1321 liters per production, with the description as in Table 1 below.

Table 1. Total water supply for wet production process.

Wet Process	Amount of water (liter)
Soybeans boiling	300
Soybeans soaking	300
Soybeans cracking	471
Soybeans washing to separate husks	250
Soybeans drying	0
Giving yeast	0
Fermentation	0
Total	1321

To minimize the amount of water that is used along the production, there is another production process that is dry process. This process requires only 800 liters of water per production. Dry production process is shown in Figure 2.

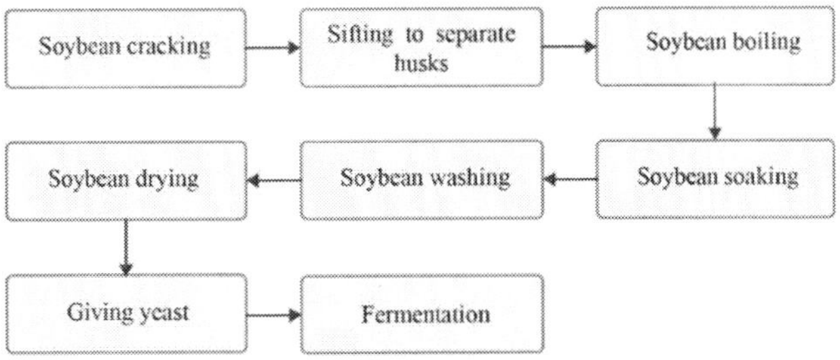

Figure 2. Flow process of tempe dry production

Based on the dry production process flow diagram of tempe in Figure 2, it is known that the initial step is cracking the soybean into two parts, which is an important process in making tempe. This step is important because when the soybean is cracked, there will be several conditions that will happen and affect the next processes. First condition as targeted, soybeans are divided into two and the husk are detached. Secondly, soybeans may be divided into two but the husks still attached, this condition will inhibit the fermentation process. And thirdly, soybeans are crushed into powder that can't be used for tempe production. By using the existing machine, there are two main problems for this process that cause the low implementation of this process in tempe production in Indonesia. The first problem is that there is a defect of 14.5% and the husks are still mixed 100% with soybean cracking results out of the outlet, so it needs extra process to separate the husks manually by sifting. The separation of the husks of soybean is important because it aims to make the yeast infiltrate the entire soybean in order to produce good tempe [3].

In order to optimize the performance of dry crusher machine, then these studies do the selection of the concept of crusher machine that can be done optimally. In product development, there are various stages to be implemented. However, the conceptualization stage of the product is a key stage in the product development which is divided into two activities; concept generation and concept selection [4]. In this research, the method used for concept selection is TOPSIS (Technique for Order of Preference

IOP Conf. Series: Materials Science and Engineering **528** (2019) 012005 doi:10.1088/1757-899X/528/1/012005

by Similarity to Ideal Solution) method. TOPSIS method is chosen because it considers both ideal and anti-ideal concepts [5].

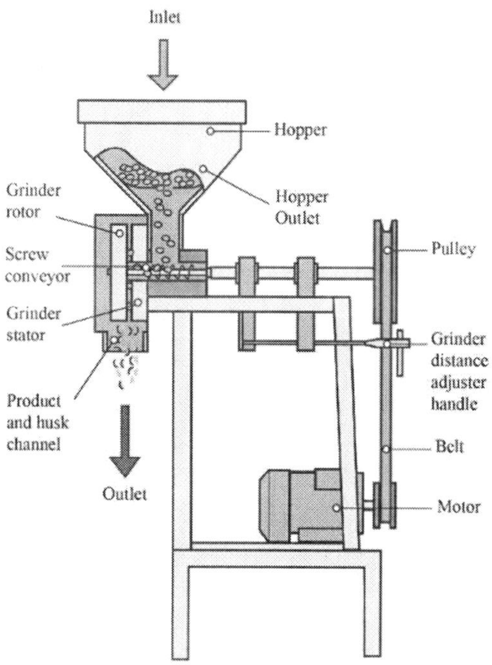

Figure 3. Existing dry soybean cracking machine

2. Methods

2.1. TOPSIS (Technique For Order Of Preference By Similarity To Ideal Solution)
Multiple Criteria Decision Making or Multiple Criteria Decision Analysis (MCDA) was developed by Hwang dan Yoon. MCDM used for decision making with various criteria, which are usually contradictory [6]. This method was developed to choose the best alternative based on various criteria which is the integration of finance factor and external factor related to performance. There are many approaches to MCDA, each involving different scientific procedures for generating inputs, structures to represent, combining algorithms, and processes for interpreting and using formal results in the context of providing advice or decision making [7]. One method used as an approach for multiple criteria decision making is TOPSIS (Technique for Order of Preference by Similarity to Ideal Solution) method. Alternative selection using TOPSIS method is based on the assumption that the best alternative has the shortest Euclidean distance from the positive ideal solution (consisting of the best value for each attribute regardless of the alternative) and has the furthest distance from the negative ideal solution (comprising the worst value) [8]. TOPSIS method has been widely used in various fields ranging from those related to machine tool selection [9], sustainable energy alternatives [10], to car beam bumper [11].

2.2. TOPSIS method's stages
In general, TOPSIS method has three main stages involved in the evaluation phase [12], which are determining the criteria of selection and alternative design concepts, determining the importance of each criterion and the effect of alternatives on the criteria, and calculating the ranking of all alternatives using the TOPSIS algorithm.

(1) At this stage, alternative combinations of concepts from several sub-solutions to the design objectives are generated. After generate the sub solutions, not all of the concepts are screened and scored but only the feasible concepts [13]. The combination of possible concepts is 12 concepts

from concept A to L. After that, the selection criteria are determined based on the design objectives, and the weights based on the level of importance given by the customer / user via questionnaire 14].

(2) The initial stage of TOPSIS method is to create a table that contains the value of each concept against each criterion. Because there are three assessment criteria that have a type of qualitative assessment. Thus, it needs to be converted to quantitative by determining a range of relative values from one to five [12]. The values given in Table 2 for each criterion can be written using each unit or relative performance's rating [12].

(3) Normalization of decision matrix can be done by determining normalized rating like the following formula.

$$n_{ij} = \frac{x_{ij}}{\sqrt{\sum_{j=1}^{m} x_{ij}^2}} \quad i = 1, \dots, m, \; j = 1, \dots, n \tag{1}$$

(4) The weighting for normalized decision matrix can be done with the following formula.

$$V = N_D \cdot W_{nxn} = \begin{vmatrix} V_{1i,\dots} & V_{1j,\dots} & V_{1n} \\ \vdots & \vdots & \vdots \\ V_{m1} & V_{mj} & V_{mn} \end{vmatrix} \tag{2}$$

Where wj is the weight of the i-th criterion, and $\sum_{j=1}^{n} w_j = 1$.

(5) To calculate the value of positive ideal solution (PIS) and negative ideal solution (NIS), using the following formula.

$$A^+ = \left\{ \left(\overset{max}{j} \, v_{ij} | i \in I \right) \left(\overset{min}{j} \, v_{ij} | i \in J \right) | i = 1,2, \dots, n \right\} \tag{3}$$

$$A^- = \left\{ \left(\overset{min}{j} \, v_{ij} | i \in I \right) \left(\overset{max}{j} \, v_{ij} | i \in J \right) | i = 1,2, \dots, m \right\} \tag{4}$$

It is known that I is for criteria related to profit or benefit, and J is for criterion related to cost.

(6) The distance of each alternative from the positive ideal solution is calculated as follows.

$$d_{i^+} = \left\{ \sum_{j=1}^{n} \left(v_{ij} - v_j^+ \right)^2 \right\}^{1/2} \tag{5}$$

Then, the distance of each alternative from the negative ideal solution is calculated as follows.

$$d_{i^-} = \left\{ \sum_{j=1}^{n} \left(v_{ij} - v_j^- \right)^2 \right\}^{\frac{1}{2}} \tag{6}$$

(7) The relative closeness value for alternative Ai to A + is defined by the following equation.

$$cl_{i^+} = \frac{d_{i^+}}{(d_{i^+} + d_{i^-})}, 0 \le cl_{i^+} \le 1, \tag{7}$$

After getting the relative closeness value, then all alternatives are sorted according to the highest to lowest value as in Table 7.

Table 2. Decision matrix

Weight		0,18	0,24	0,29	0,29
No.	Concept	The machine could crack the soybean optimally	The hopper could discharge the soybean optimally	The machine could separate the cracked-soybean and husk	The machine could work optimally with efficient electricity power
1	A	4	4	4	0,33
2	B	4	4	5	0,73
3	C	5	4	4	0,33
4	D	5	4	5	0,73
5	E	4	4	4	0,33
6	F	4	4	5	0,73
7	G	5	4	4	0,33
8	H	5	4	5	0,73
9	I	4	5	4	0,33
10	J	4	5	5	0,73

11	K	5	5	4	0,33
12	L	5	5	5	0,73

Table 3. Normalized decision matrix

No.	Concept	The machine could crack the soybean optimally	The hopper could discharge the soybean optimally	The machine could separate the cracked-soybean and husk	The machine could work optimally with efficient electricity power
1	A	1,37,E-03	1,48,E-03	1,37,E-03	8,16,E-03
2	B	1,37,E-03	1,48,E-03	1,71,E-03	1,80,E-02
3	C	1,71,E-03	1,48,E-03	1,37,E-03	8,16,E-03
4	D	1,71,E-03	1,48,E-03	1,71,E-03	1,80,E-02
5	E	1,37,E-03	1,48,E-03	1,37,E-03	8,16,E-03
6	F	1,37,E-03	1,48,E-03	1,71,E-03	1,80,E-02
7	G	1,71,E-03	1,48,E-03	1,37,E-03	8,16,E-03
8	H	1,71,E-03	1,48,E-03	1,71,E-03	1,80,E-02
9	I	1,37,E-03	1,85,E-03	1,37,E-03	8,16,E-03
10	J	1,37,E-03	1,85,E-03	1,71,E-03	1,80,E-02
11	K	1,71,E-03	1,85,E-03	1,37,E-03	8,16,E-03
12	L	1,71,E-03	1,85,E-03	1,71,E-03	1,80,E-02

Table 4. Weighted normalized decision matrix

Concept	The machine could crack the soybean optimally	The hopper could discharge the soybean optimally	The machine could separate the cracked-soybean and husk	The machine could work optimally with efficient electricity power
A	2,47,E-04	3,55,E-04	3,98,E-04	2,37,E-03
B	2,47,E-04	3,55,E-04	4,97,E-04	5,23,E-03
C	3,09,E-04	3,55,E-04	3,98,E-04	2,37,E-03
D	3,09,E-04	3,55,E-04	4,97,E-04	5,23,E-03
E	2,47,E-04	3,55,E-04	3,98,E-04	2,37,E-03
F	2,47,E-04	3,55,E-04	4,97,E-04	5,23,E-03
G	3,09,E-04	3,55,E-04	3,98,E-04	2,37,E-03
H	3,09,E-04	3,55,E-04	4,97,E-04	5,23,E-03
I	2,47,E-04	4,44,E-04	3,98,E-04	2,37,E-03
J	2,47,E-04	4,44,E-04	4,97,E-04	5,23,E-03
K	3,09,E-04	4,44,E-04	3,98,E-04	2,37,E-03
L	3,09,E-04	4,44,E-04	4,97,E-04	5,23,E-03

Table 5. Positive Ideal Solution (PIS) and Negative Ideal Solution (NIS)

Concept	The machine could crack the soybean optimally	The hopper could discharge the soybean optimally	The machine could separate the cracked-soybean and husk	The machine could work optimally with efficient electricity power
PIS	4,46,E-04	3,88,E-04	4,46,E-04	2,12,E-03
NIS	3,57,E-04	3,11,E-04	3,57,E-04	4,69,E-03

Table 6. Distance every alternatives with ideal solution

Concepts	**A**	**B**	**C**	**D**	**E**	**F**
d_{i+}	2,16,E-08	8,24,E-06	1,78,E-08	8,23,E-06	2,16,E-08	8,24,E-06
d_{i-}	8,22,E-06	9,89,E-09	8,23,E-06	1,37,E-08	8,22,E-06	9,89,E-09
Concepts	G	H	I	J	K	L
d_{i+}	1,78,E-08	8,23,E-06	1,37,E-08	8,23,E-06	9,89,E-09	8,22,E-06

d_i-	8,23,E-06	1,37,E-08	8,23,E-06	1,78,E-08	8,24,E-06	2,16,E-08

Table 7. Relative closeness value

Concept	A	B	C	D	E	F
cl_i+	9,97,E-01	1,20,E-03	9,98,E-01	1,66,E-03	9,97,E-01	1,20,E-03
Concept	G	H	I	J	K	L
cl_i+	9,98,E-01	1,66,E-03	9,98,E-01	2,15,E-03	9,99,E-01	2,62,E-03

3. Result and Discussion

Based on data processing and analysis, result is obtained, refers to the purpose of research, which is to support the process of dry soybean breaking to be more optimal, then the concept of dry soy crusher machine selected is the concept of K with the value of relative closeness 9.99, E-01. Which means concept K is relatively considered has the most ideal closeness value to ideal and non-ideal solution, compared to other alternative concepts. The K concept has a conical hopper shape for the mass flow product outflow connected with a single pitch, single flight screw conveyor. The diameter of the grinding wheel is 200 mm with 1 mm distance that adjust by T-handle type of adjuster fitted with a distance label that makes the distance adjustment process more accurate. The grinding wheel of the rotor is driven by a ¼ HP electrical motor and power is transmitted using a belt drives transmission system. In order for the product and the residue to be separated simultaneously upon the outlet, the machine is provided with a blower separator of the 2 "hole diameter output hole. The product and the residue will come out on a separate path because the channel is branched off.

4. Conclusion

The existing dry soybean cracker machine used by most traditional tempe craftsmen still can not work optimally. So, in this study, the selection of design concept of dry soybean breaker machine to get the concept with a combination of components that are more optimal than the existing machine is conducted. The K concept has been chosen after being compared to the other eleven concepts. This concept has an adjustable distance for achieving optimum parameters with ease, a hopper that has an optimum diameter so that the soybean distribution becomes better. The selected machine is also able to separate soybean products and husks simultaneously so that manual separation process can be eliminated. Although the addition of the separator blower to separate the product and the skin means adding to the required electrical power requirements, but compared to existing machines on the market, the total power required is still lower. With the proposed machine, the defect percentage of tempe production using dry-soybean cracking machine could be decreased to 6,5% from 14,4% and the amount of water needed compared to wet process is decreased also to 800 liters/production of 35,125 kg tempe from 1321 liters/production of 37,5 kg tempe. In addition, the authors suggest that subsequent studies should be tested for selected concepts in order to verify the consequences of a combination of selected concept components that may occur.

5. References

[1] Kementrian Pertanian Republik Indonesia 2015 *Outlook Komoditas Pertanian Sub sektor Tanaman Pangan* (Jakarta: Pusat Data dan Sistem Informasi Pertanian)

[2] Kementrian Pertanian Republik Indonesia 2016 *Outlook Komoditas Pertanian Sub Sektor Tanaman Pangan* (Jakarta: Pusat Data dan Sistem Informasi Pertanian)

[3] Suhendra and B. Setiawan 2012 *Model Efisiensi Mesin Pengupas dan Pembelah Biji Kedelai Tipe Piringan Menggunakan Program Powersim* (Pontianak: POSITRON) pp 25-32

[4] Prasad K D, Subbaiah K V, Gireesh C H and Koushik U 2017 *Evaluation of Conceptual Product Design Solutions using House of Quality-TOPSIS Integrated Methodology* (India: SSRG International Journal of Mechanical Engineering) pp 206-212

IOP Conf. Series: Materials Science and Engineering **528** (2019) 012005 doi:10.1088/1757-899X/528/1/012005

[5] Vinodh S, Mulanjur G and Thiagarajan A 2013 *Sustainable Concept Selection using Modified Fuzzy TOPSIS: A Case Study* (London: Taylor and Francis) **5** pp 109-116

[6] Yavuz M 2016 *Equipment Selection by Using Fuzzy TOPSIS Method* (Semarang: IOP Conference Series: Earth and Environmental Science)

[7] Huang I B, Keisler J and Linkov I 2011 *Multi-criteria Decision Analysis in Environmental Sciences: Ten Years of Applications and Trend* (Amsterdam: Elsevier) pp 3578-3594

[8] Locatelli G and Mancini M 2012 *A Framework for The Selection of The Right Nuclear Power Plant* (London: Taylor and Francis) pp 4753-4766

[9] Tho N H, Dawal S Z M, Yusoff N, Tahriri F and Aoyama H 2013 *Selecting a CNC Machine Tool Using The Intuitionistic Fuzzy TOPSIS Approach for FMC* (Zürich: Trans Tech Publication)

[10] Afsordegan A, Sanchez M, Agell N, Zahedi S and Cremades L V 2016 *Decision Making Under Uncertainty Using a Qualitative TOPSIS Method for Selecting Sustainable Energy Alternatives* (London: Taylor and Francis)

[11] Davodi M M, Sapuan S, Ahmad D, Aidy A, Khalina A and Jonoobi M 2011 *Concept Selection of Car Bumper Beam with Developed Hybrid Bio-composite Material* (Amsterdam: Elsvier) pp 4857-4865

[12] Yang S, Ong S and A Nee 2016 *A Decision Support Tool for Product Desin for Remanufacturing* (Amsterdam: Elsvier) pp 144-149

[13] Meizianno A, Anugraha R A and Atmaja D S E 2016 *Design of Supporting Equipment in Orthodoks Black Tea Drying Process Using Rational Product Design and SCADA at PTPN VIII Rancabali* (Bandung: e-Proceeding of Engineering) p 2322

[14] Lestari I R, Anugraha R A and Iqbal M *Perancangan Material Handling Equipment pada Proses Penggilingan ke Oksidasi Enzimatis Bubuk Teh Menggunakan Metode Perancangan Produk Rasional Pada PT Perkebunan Nusantara VIII Rancabali* (Bandung: Jurnal Rekayasa Sistem & Industri)

Decision Making Strategy For Decreasing The Potential Hazards of Work Accidents at Division R&D Using SWOT And AHP Methods

D Rimantho[1], A Elistiani[2], S Sundana[3], and AS Sundari[4]

Industrial Engineering Department, Pancasila University, Jagakarsa – Jakarta 12640 Indonesia
[1]dino.rimantho@univpancasila.ac.id

Abstract. PT. XYZ has various divisions, one of them is the R&D division. This division is responsible for research and innovation. The highest impact on accidents caused by a worker injured and bleeding around 14.29%. In addition, the highest potential hazards are raw material spill around 14.29%. Thus, the purpose of this paper is to identify the factors potentially that lead to the high potential hazards in the R&D division. In addition, it also formulates the alternative to decision-making strategies. The research method used the SWOT method and AHP to determine the priority to selecting strategies for reducing the potential for occupational hazards. This study used the questionnaires and distributed among five experts in the company. The result shows that the strategy of Weakness-Opportunity (WO) is the highest weight around 3.38. Then, the calculation by using the AHP method obtained that monitoring and evaluation of occupational health and safety through improvement of Work Instruction became the main priority is equal to 0.32. The AHP method also showed the consistency value of approximately 0.09. Therefore, further research is needed to evaluate the alternatives chosen.
Keywords: Analytic Hierarchy Process (AHP), hazard, OHS, Strategy, SWOT

1. Introduction

Occupational Health and Safety issues in Indonesia are still often overlooked. This is indicated by the high rate of workplace accidents. The number of occupational accidents that occur to Indonesia is still increasing. For example, in 2009 there were around 83,714 accidents occurring in the Indonesian work industry. The total cost to compensate for losses reach 219.7 billion Rupiah. A few years later, in 2013 the number of accidents hit 99,491 cases. That number increased by 18% from the value of 2009 [1].
PT. XYZ is one of the cosmetic manufacturers. One division of the company is Research and Development (R&D). The R&D division has the potential to cause work accidents when doing activities in work and causing losses to workers and to the company. Data onto the R&D division shows that the percentage of the impact/consequence of work accidents at PT. XYZ Division R&D in the category of worker shows that the occurrence of injuries and bruises around 14.30%, hearing loss 7.14%, eye irritation 14.30%, skin burn 7.14 %, respiratory disorders 7.14%, skin irritation 7.14%, fell 7.14%, electric shock 3.57%, bleeding 21.42%, burn 3.57%, damaging tools 3.57%, and WRMSD 35.7%.

Work accidents are catastrophes that occur in the work environment that may arise due to unsafe working environment conditions or because of human error and must be controlled [2]. Therefore, the anticipatory action is needed to prevent work accidents and occupational diseases. The Government of

Indonesia through the Ministry of Manpower always calls for the application of the Occupational Health and Safety Management System (OHS) and OHSAS: 18001[3].

Propos strategies to control and reduce the consequences and potential hazards of work accidents have been carried out by many previous researchers to implement OHS. For example, identification of potential hazards, assessment, and control of risk management of companies in preparing OHS plans must involve general OSH experts, namely people that have been certified by the Ministry of Labor in the field of occupational safety and health in the cigarette industry [4]. In addition, the application of the use of personal protective equipment as an effort to protect workers in companies [5]. In addition, proposals and corrective actions for Job Safety Analysis for companies and facilitate work safety facilities such as safety devices or personal protective equipment [6].

In order to implement the OHS, different strategies and decisions are needed in each company. One strategy in making a decision can be done by applying the Weakness Opportunity Threat (SWOT) Strength method. Some profit and non-profit organizations have used SWOT as one of their analyzes, such as steel production organization [7]. In addition, to obtain a good strategy, a SWOT analysis is needed to identify the strengths, weaknesses, opportunities, and threats that exist. According to Kotler & Armstrong describes that SWOT analysis is a comprehensive assessment of the strengths, weaknesses, opportunities, and threats of a company [8]. This analysis is needed to determine several strategies that exist on the company so that decisions can be made so that are appropriate to the conditions of the company.

In order to be able to take the right decision in applying the strategy produced by the SWOT analysis, a method is needed to make/analyze decision priorities. AHP is a multi criteria decision-making technique that can help express general decision operations by describing complex problems with multi-purpose, as well as alternative hierarchical structures [9]. In addition, there are many methods that can be applied to analyze the priority to an OHS problem decision, for example Evaluating and Prioritizing of Performance Indicators of Health, Safety and Environment Using Fuzzy Technique for Others Reference by Similarity to Ideal Solution (TOPSIS) [10]. Furthermore, Application of Fuzzy Decision Making Trial. And Evaluation Laboratory (DEMATEL) Method of Analyzing Occupational Risks on Construction Sites [11]. AHP is an effective method of making decisions, especially when there is subjectivity and this is very suitable for solving problems where decision criteria can be arranged into a sub-criteria hierarchy [12]. Thus, this article aims to analyze the right strategies in making decisions to minimize and reduce the potential dangers associated with the security of workers in the R & D division.

2. Methodology

Data collection was carried out in various ways such as field observations, and interviews with five experts such as academics, practitioners, and companies. Questionnaire method is done by giving a list of questions both opened and closed. The questionnaire distribution was divided into two stages. The first stage is shown to determine the internal and external factors whose results are sub-criteria. The first questionnaire will be analyzed using the SWOT method. Furthermore, the second questionnaire are to obtain priority strategies through the proposed alternative strategy in reducing the potential hazard of work accidents. In addition, secondary data will be obtained from various library sources and documents related to research. Based on the data that has been obtained, an analysis is performed using the SWOT method.

Several alternatives obtained through SWOT analysis have not been fully implemented in a company, and not necessarily all can be done simultaneously due to various limitations. Thus, it requires a sequence of policy priorities that will be applied for advanced analysis by applying decision-making techniques using Analytical Hierarchy Process (AHP). The following stages are analyzed using the AHP method.

IOP Conf. Series: Materials Science and Engineering **528** (2019) 012006 doi:10.1088/1757-899X/528/1/012006

3. Result and discussion

Identification of the factors related to the strategy for making decisions on reducing potential hazards in the R & D division were done by studying various literature, documents, as well as direct interviews and filling out questionnaires with informants that were believed to know the problem being studied. Based on the identification of the indicators of internal and external factors, then the SWOT questionnaire was prepared to obtain an assessment of weight and rating. Based on the assessments conducted by experts, the Internal Strategy Factor Analysis System (IFAS) was obtained and the External Strategy Factor Analysis System (EFAS) shown in tables 1, 2, 3, and 4 below.

Table 1. Internal Strategy Factor Analysis System (IFAS)

Strength	Weight	Rating	Weight x Rating
Work safety knowledge of workers	0.05	3	0.15
The availability of competent trainers	0.05	4	0.20
Having good cooperation and relationships	0.04	2	0.08
The number of workers is adequate	0.04	2	0.08
Good arrangement and placement of raw materials / samples	0.04	3	0.12
A good laboratory waste management process	0.04	3	0.12
The raw material is always closed properly	0.05	3	0.15
Replacement of raw materials label routinely	0.04	2	0.08
Total Strength			0.98

Table 2. Internal Strategy Factor Analysis System (IFAS)

Weakness	Weight	Rating	Weight x Rating
Lack of Personal protective equipment	0.05	4	0.20
OHS management systems have not been managed properly	0.06	4	0.24
Lack of IK (Work Instructions)	0.05	3	0.15
Lack of the process of taking raw materials	0.04	2	0.08
Scale calibration for weighing equipment in the laboratory has not been carried out routinely	0.03	2	0.06
The absence of routine and scale checks on machines in the laboratory	0.04	3	0.12
Laboratory temperature is unstable	0.05	3	0.15
Air circulation is not good	0.05	4	0.20
Noise in the laboratory area	0.04	2	0.08
Excessive use of materials	0.03	2	0.06
The raw materials that exceed their useful life	0.04	2	0.08
The machine / past the age limit usage	0.05	3	0.15
Machines often have problems	0.05	4	0.20
Incorrect installation and placement of machine locations	0.05	3	0.15
Total Weakness			1.92

Table 3. External Strategy Factor Analysis System (EFAS)

Opportunity	Weight	Rating	Weight x Rating
Support from management related to Human Resources (HR) training	0.07	3	0.21
There is support from management related to OSH training on a scale basis	0.08	4	0.32
Support from the government	0.06	3	0.18
The good physical condition of workers	0.06	2	0.12
Age of worker	0.06	2	0.12
The level of education of qualified workers	0.05	2	0.10
Government policy on OHS management systems	0.06	3	0.18
There is support regarding training in the Work Instruction method (IK)	0.07	3	0.21
Total Opportunity			1.46

Table 4. External Strategy Factor Analysis System (EFAS)

Threats	Weight	Rating	Weight x Rating
Decreasing opportunities for companies to innovate	0.09	3	0.27
The environmental conditions of workers outside the company are less conducive	0.06	2	0.12
Uncontrolled the behavior of workers outside the company	0.05	2	0.10
Other companies are more aware of the provision of Personal Protective Equipment (PPE)	0.06	4	0.24
The higher cost of weighing calibration	0.08	3	0.24
Higher engine maintenance costs	0.06	3	0.18
There is a risk of fire or explosion	0.06	3	0.18
Decreasing opportunities for companies to innovate	0.07	3	0.21
Total Threats			1.38

Translation of the formulation of strategies that have been made in the form of a SWOT matrix in table 5 below.

Table 5. SWOT Matrix

	Strength (S)	**Weakness (W)**
	Work safety knowledge of worker	Lack of Personal protective equipment
	The availability of competent trainers	OHS management systems have not been managed properly
	Having good cooperation and relationships	Lack of IK (Work Instructions)
	The number of workers is adequate	Lack of the process of taking raw materials
	Good arrangement and placement of raw materials/samples	Scale calibration for weighing equipment in the laboratory has not been carried out routinely
	A good laboratory waste management process	The absence of routine and scale checks on machines in the laboratory

IOP Conf. Series: Materials Science and Engineering **528** (2019) 012006 doi:10.1088/1757-899X/528/1/012006

Table 5. SWOT Matrix (cont.)

	Strength (S)	**Weakness (W)**
	The raw material is always closed properly Replacement of raw materials label routinely **Total Weight = 0.98**	Laboratory temperature is unstable Air circulation is not good Noise in the laboratory area Excessive use of materials Raw materials that exceed their useful life The machine / past the age limit usage Machines often have problems Incorrect installation and placement of machine locations **Total Weight = 1.92**
Opportunity (O) Support from management related to Human Resources (HR) training There is support from management related to OHS training on a scale basis Supporting from the government The good physical condition of workers Age of worker The level of education of qualified workers Government policy on OHS management systems There is support regarding training in the Work Instruction method (IK) **Total Weight = 1.46**	**Strategy SO** **Total Weight S + Total Weight O = 2.24** Monitor the implementation of work safety procedures regularly (S5, S6, S7, S8, O8) Optimizing training and improving employee competencies (S1, S2, O2, O7) Develop and increase opportunities for employees to innovate (S3, O1, O4, O5)	**Strategy WO** **Total Weight W + Total O = 3.38** Work safety monitoring and evaluation through improving work instructions (W2, W3, W4, W5, W6, W10, W11, O2, O3, O7, O8). Commitment to the implementation of work safety (W2, W3, W4, W5, W6, W10, W11, O2, O8). Focus on providing personal protective equipment (W1, O8) Maintenance of facilities and infrastructure to support the implementation of work safety (W7, W8, W9, W12, W13, W14, O3, O7).
Threats (T) Decreasing opportunities for companies to innovate The environmental conditions of workers outside the company are less conducive Uncontrolled the behavior of workers outside the company Other companies are more aware of the provision of Personal Protective Equipment (PPE) The higher cost of weighing calibration Higher engine maintenance costs There is a risk of fire or explosion Decreasing opportunities for companies to innovate **Total Weight = 1.38**	Strategy ST **Total Weight S + Total Weight T = 2.36** Improving workers' welfare (S1, T4) Increased awareness of workers on hazards and risks at work (S1, T2, T3, T7)	Strategy WT **Total Weight W + Total Weight T = 3.30** Improve and prioritize work safety in the R & D Division (W1, T1, T4, T7) Periodically checking medical workers (W1, W7, W8, W9, T1, T2, T3) Working closely with the agencies engine maintenance service provider (W12, W13,W14,T5,T6)

Table 5 shows the alternative strategy with the highest weight is the Weakness-Opportunity (WO) strategy. Furthermore, to determine the position of the company's quadrant based on the value of the factors plotted on the X and Y axes.

Axis X = Weight of Strength - Weight of Weakness
= 0.98 - 1.92 = -0.94

Axis Y = Weight of Opportunity - Weight of Threat
= 1.46 - 1.38 = 0.08

Figure 1 provides information that the results of the R & D division SWOT analysis are in Quadrant III. This division faces opportunities in managing large work safety. However, on the internal side, there are several obstacles. This condition shows an effort to minimize weaknesses to get opportunities. Or in other words, how to strive for development with conditions that favor the weakest conditions but can be used to capture opportunities

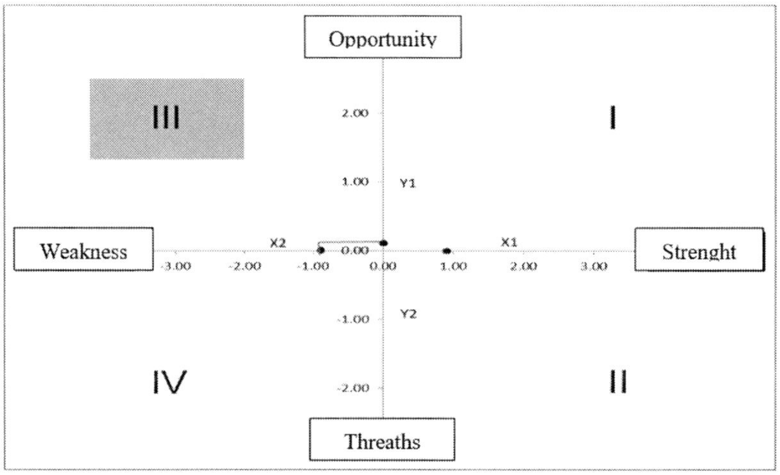

Figure 1. SWOT Analysis Diagram

Based on the SWOT analysis that has been done previously, the sub-criteria results are a WO strategy that is minimizing weaknesses to get opportunities to reduce the potential for work accidents in the R & D division. Thus, there are several alternative strategy proposals obtained from literature studies that are adjusted to the conditions in the R & D division. Several alternative strategies are shown in table 6 below.

Table 6. Alternative strategy

Code	Alternative Strategy
AS1	Monitoring and evaluating OHS through improvement of Work Instructions (IK)
AS2	Committing to OHS implementation not only for management but for the entire personnel
AS3	Improved provision of Personal Protective Equipment (PPE)
AS4	Maintenance and improvement of facilities and infrastructure to support the implementation of OHS

Based on table 10 above, the next step is to analyze the strategy selection using the AHP method. Thus the hierarchical structure of the selection of work accident reduction strategies is shown in Figure 2 below.

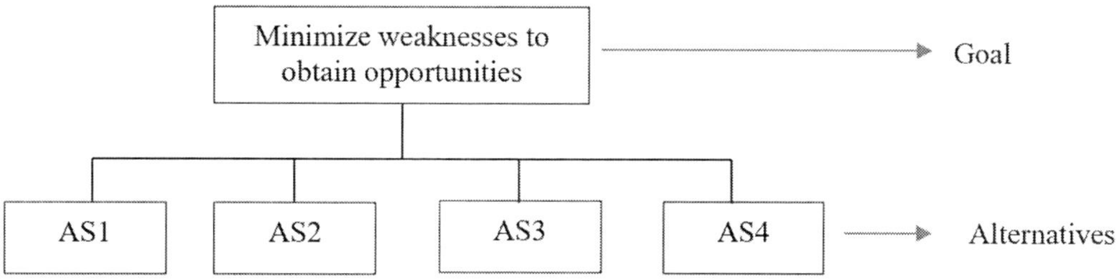

Figure 2. Hierarchical Structure of Alternative Decline in Work Accident Potential

Priority determination of alternatives is obtained from the second stage questionnaire, using pairwise comparisons between alternatives. The results of the assessment of the five experts with pairwise comparisons calculated using the geometric mean. Overall the mean geometric mean results are shown in table 7 below

Table 7. Results of Pairwise Comparison Matrix Using Geometric Mean

Alternatives	AS1	AS2	AS3	AS4
AS1	1.00	2.14	1.38	0.89
AS2	0.47	1.00	1.72	1.55
AS3	0.72	0.58	1.00	1.55
AS4	1.22	0.64	0.64	1.00
Total	3.32	4.37	4.74	4.99

Furthermore, the matrix normalized by dividing the number in each cell with the number of digits in the column in question. From the normalized matrix, the priority weights can be calculated which the mean values ($\dot{X}$) of each matrix row. The results of the matrix normalization are shown in table 8 below.

Table 8 Normalized Weight Pairwise Comparison Ratings

Alternatives	AS1	AS2	AS3	AS4	$\sum$	X
AS1	0.30	0.49	0.29	0.18	1.26	0.32
AS2	0.14	0.23	0.36	0.31	1.04	0.26
AS3	0.22	0.13	0.21	0.31	0.87	0.22
AS4	0.34	0.15	0.14	0.20	0.82	0.21
Total	1.00	1.00	1.00	1.00	1.00	1.00

Based on Table 8, the Consistency Ratio (CR) is calculated. If the Consistency Ratio (CR) is smaller or equal to 10%, the matrix meets the requirements of consistency [12]. Based on the calculation results, the CR value is around 0.09. The consistency value is 0.09 or equal to 9%, so because 9% ≤ 10%, the consistency value is acceptable.

Based on the results of data processing that the hierarchical structure in Figure 2 has been consistent. This is proven by the consistency ratio that is smaller than 10%. It can be seen from the four assessment weights of each alternative in reducing the potential hazard of work accidents, the biggest weight on the sub-criteria is monitoring the evaluation of OHS through the improvement of AS1 Work Instructions (IK) with a value of 0.32, then in the second order is committed to the implementation of OHS not only for management but for all AS2 personnel with a value of 0.26, then continued with an increase in the availability of AS3 Personal Protective Equipment (PPE) with a value of 0.22, and the last order is maintenance and repair of facilities and infrastructure to support the application of OHSwith value equal to 0.21.

In Figure 1 it is known that the R & D division is in Quadrant III or the Weakness-Opportunity (WO) strategy. This division faces opportunities in very large management. However, on the internal side, there are significant obstacles. This condition aims to minimize weaknesses to get opportunities. With another definition, how to seek development with conditions that favor the weakest conditions but are used to capture opportunities. Therefore, the company is required to constantly improve and improve internal problems, so that it can provide support for business development in the long term. Despite facing various threats, this division still has internal strength. This division faces enormous opportunities, but on the other hand, faces several internal constraints/weaknesses. The focus of this strategy is to minimize the company's internal problems so that it can rely on better opportunities (turn around). One effort that can be done is to carry out a monitoring and evaluation strategy through the improvement of IK. The study conducted by Yogie et al. shows the similar results where one of the Weakness-Opportunity (WO) strategies in his research is related to Occupational Health & Safety Evaluation [13].

In order to realize the monitoring and evaluation of the implementation of OHS through the improvement of the IK, there are several parties that are obliged to support this strategy, such as managers and supervisors in the R & D division as supervisors, OHS implementation team in the R & D division as monitors and evaluators of OHS implementation, and all workers in the R & D division which functions as a participant who contributes to support the implementation of monitoring and evaluation of the implementation of OHS through the improvement of IK in the reduction of potential hazards of work accidents in the R & D division. Therefore, the first step that will be taken is to form a special OHS implementation team that is tasked with monitoring and evaluating all activities of workers in the R & D division. With the monitoring of all activities of workers in the R & D division regularly. Thus, all activities are always within the criteria of work instructions that have been established in the long term. For example, monitor workers when they want to use or when they use tools and machines. Regular routine checks on machines in the laboratory are an effort to improve IK. For example, by checking and routinely cleaning the weighing equipment before and after being used for trial and recording the checking and cleaning of the weighing instrument into the log book.

If the monitoring has been carried out regularly and the IK has been carried out properly, then it is evaluated by the party who is obliged to evaluate the monitoring of the IK. like managers and supervisors and OHS special teams in R & D. If in the monitoring there is still a discrepancy in the procedure for implementing the IK, it will be subject to initial sanctions in the form of reprimand to workers. In addition, if similar things still occur, then there will be sanctions in the form of a Warning Letter (SP) for workers. Thus, the existence of sanctions against the non-conformity of the IK in the R & D division can be minimized to prevent potential hazards of work accidents.

4. Conclusion

From the results of the strategic decision-making process to reduce the potential hazards of work accidents using the SWOT analysis and AHP methods in the R & D division, it can be concluded that the factors that have the potential to cause work accidents in the R & D division are internal factors, such as weaknesses in the R & D division. Furthermore, the determination of the strategy to reduce the potential danger of work accidents in the R & D division was obtained by the Weakness-Opportunity (WO) strategy. By using the AHP method for the selection of strategies in reducing the potential danger of work accidents in the R & D division, the highest AS 1 value is around 0.32 with a consistency value of approximately 0.09. Several suggestions that can be given in this article such as, reviewing and monitoring work instructions especially in the R & D division, routine and periodic checks on machines and tools in the laboratory. In addition, further studies related to the problems faced need to be done in order to provide better references.

5. References

[1] Ministry of Labor and Transmigration of Indonesia 2013 *Tekan Angka Kecelakaan Kerja Depnaker Gandeng Tiga Universitas dari Jatim.* Retrieved on July 15, 2018, from http://www.depnakertrans.go.id

[2] Restuputri D P and Sari R P D 2015 Analisis kecelakaan kerja dengan menggunakan metode Hazard and Operability Study (HAZOP). *Jurnal Ilmiah Teknik Industri* **14**(1)

[3] Latief Y, Machfudiyanto R A, Arifuddin R, Setiawan R M F and Yogiswara Y 2017 Study of Evaluation OSH Management System Policy Based On Safety Culture Dimensions in Construction Project. *J. Phys.: Conf. Ser.* **877** 012028

[4] Ridwan A S S and Allessio F 2017 Implementasi SMK3 Pada PT. Nojorono Tobacco International. *Journal of Industrial Hygiene and Occupational Health* **1**(2)

[5] Rimantho D and Hanantya M W 2017 Enhancing the management of the noise level using six sigma method: a case study on the machining industry, *IOP Conf. Ser.: Mater. Sci. Eng.* **277** 012055

[6] Dewi D and Zulfan 2015 Usulan Perbaikan Keselamatan Kerja Untuk Meminimumkan Kecelakaan Kerja Dengan Pendekatan Job Safety Analysis (JSA) Pada Area Lantai Produksi Di PT. Alam Permata Riau, *Jurnal Sains, Teknologi dan Industri* **13**(1)

[7] Tervonen P, Haapasalo H and Niemelä M 2009 Evolution of Safety Management and Systems in a Steel Production Organization. *The Open Management Journal* **2** pp 17-27

[8] Kotler P T and Armstrong G 2008 *Principles of Marketing.* 12th Ed. (Pearson, Prentice Hall)

[9] Pattinaja A M and Rimantho D 2017 Strategy of the policy of sustainable housing development using AHP method at the village Ihamahu-Maluku, Indonesia. *International Journal of Applied Engineering Research* **12**(14)

[10] Yarahmadi R and Shima Sadoughi S 2012 Evaluating and prioritizing of performance indices of environment using fuzzy TOPSIS. *Indian Journal of Science and Technology* **5**(5)

[11] Seker S and Zavadskas E K 2017 Application of Fuzzy DEMATEL Method for Analyzing Occupational Risks on Construction Sites. *Sustainability* **9** 2083

[12] Saaty T L and Vargas L G 2012 *Models, Methods, Concepts & Applications of the Analytic Hierarchy Process* (New York: Springer)

[13] Yogie M, Wisnubroto P and Oesman TI 2017 Evaluasi Kesehatan & Keselamatan Kerja (K3) Untuk Penilaian Produktivitas Dengan Metode SWOT dan Balanced Scorecard Pada PT. Mitra Rekatama Mandiri, *Jurnal Rekavasi* **5**(1)

Risk Management Analysis Using FMECA and ANP Methods in the Supply Chain of Wooden Toy Industry

W N Tanjung[1], S A Atikah[1], S Hidayat[1], E Ripmiatin[2], S S Asti[1] and R S Khodijah[1]

[1]Industrial engineering department, Faculty of science and technology, Universitas Al Azhar Indonesia, Sisingamangaraja Road, Jakarta 12110, Indonesia.
[2]Informatics engineering department, Faculty of science and technology, Universitas Al Azhar Indonesia, Sisingamangaraja Road, Jakarta 12110, Indonesia.

Abstract. Risk is a factor that has a potential to occur in every part of an industry, including its supply chain. Every industry must consider and manage its supply chain risks in order to minimize losses while improving performance. In this study, a supply chain risk management (SCRM) approach was conducted in the supply chain of wooden toy industry as a way to minimize the impact of risk. In this case, SCRM method is done by identifying risks, assessing risks using the *failure mode effect criticality analysis* (FMECA) method to determine the risk priorities that must be mitigated, and establish strategies to mitigate the risk with *analytical network process* (ANP) method. Thus, the risk priority in the wooden toy industry's supply chain is the cost/ price risk with the highest WRPN value of 33,379. In order to mitigate the cost/price risk, conducting strategic accounting practices and financial planning are recommended to be considered as its strategies.
Keywords: ANP, FMECA, SCRM

1. Introduction

Industry is a complex system consisting of various elements that interact directly or non directly with one another in the system. Every industry has a system that describes the flow of products, money and information in a supply chain. The supply chain is a series of relationships between companies or activities that carry out the supply of goods or services from the place of origin to the place of the buyer or customer [1] which activities include material procurement, product manufacturing, and forward / reversed logistics [2]. However, in the supply chain, every element and activity that exists both internally and externally is vulnerable to risk. Risk is the probability of an event that results in a loss when the event occurs during a certain period [3]. Some examples of supply chain risks are raw material shortages, supplier failures, rising in material prices, machine breakdown, uncertain demand, inaccurate forecasting, order changes, and transportation failures [4].

Competition between industries focuses not only on the percentage of profit generated but also on more critical matters: how well the company manages the risks of its internal and external supply chains. Although the profit generated is still relatively small, if the foundation of the supply chain system is good, then the company will grow more stable. Supply chain risk management (SCRM) is risk management in the supply chain by approaching and making decisions that optimally align organizational processes and decisions to take advantage of opportunities while minimizing risk simultaneously [5].

One industry that has a significant contribution in national economic development is wooden toy industry (hereinafter WTI), which produces children's toys with a wide variety of designs and also distributes their products to major retailers. It is without exception that the supply chain of WTI is still vulnerable to risks that have the probability to emerge and disrupt the performance of the supply chain, this can result in delays in production activities, delays in distribution and ultimately delays in goods reaching customers. Responding to this, the SCRM approach was used by using the failure mode effects critical analysis (FMECA) method, which is a risk assessment method on a product or activity using three assessment factors, namely severity, occurrence, and detection. As well as using criticality analysis to sort the level of criticality of the existing risks. The result of the assessment then became the input for establishing strategies, then with Analytical Network Process (ANP) the strategy priorities would be determined. The research was carried out by conducting reference studies for the risk and strategies, then interviews on risks that could arise in the company's supply chain and interviews for taking expert judgements which were then processed into quantitative forms so that the level of risk and the recommended strategies could be identified.

2. Methods

The research was conducted in the wooden toy industry to indentify risk in its supply chain. For the risk identification is used the supply chain risks categorization by Jafarnejad [6], an interview was also done to check and compare the risks is obtained from the literature to the real conditions in the wooden toy industry. A risk assessment was done using three methods: AHP, FMECA, and ANP methods. The AHP method was used to obtain the risk weights which then processed further with the FMECA method to obtain the rank of the risks. In this research, three experts are chosen to judge the risks using pairwise comparison matrix, then their judgements became the input in AHP method that were calculated using Expert Choice software. Then FMECA method was done to determine the occurrence, severity, detection, and the criticality level of each risk. Criticality level helped us to filter which risks that necessary for WRPN calculation in order to rank the risks. The last step was to generate some recommended strategies for the most critical risk, strategies were obtained from literature study, and then were judged by experts before they were calculated with ANP method. The use of ANP method was necessary to help us knew which strategies to be considered for mitigation. Figure 1 shows the step of each method used for this research.

2.1. Analytical Hierarchy Process (AHP)

Analytic Hierarchy Process (AHP) is a theory of relative measurement with absolute scales of both tangible and intangible criteria based on the judgment of knowledgeable and expert people [7]. In this research the use of AHP method was to obtain risk weight, not to determine priority. The weight then were used to calculate the global weight and determine the WRPN value in the FMECA method.

2.2. Failure Mode Effect Criticality Analysis (FMECA)

FMECA method has a risk assessment scale [8] based on the value of the RPN that ranges from 0 to more than 324 as can be seen in table 1. The RPN is calculated by considered three factors, there are occurrence, severity, and detection. Each factor has a rating scale of 1 to 10 as in table 2.

IOP Conf. Series: Materials Science and Engineering **528** (2019) 012007 doi:10.1088/1757-899X/528/1/012007

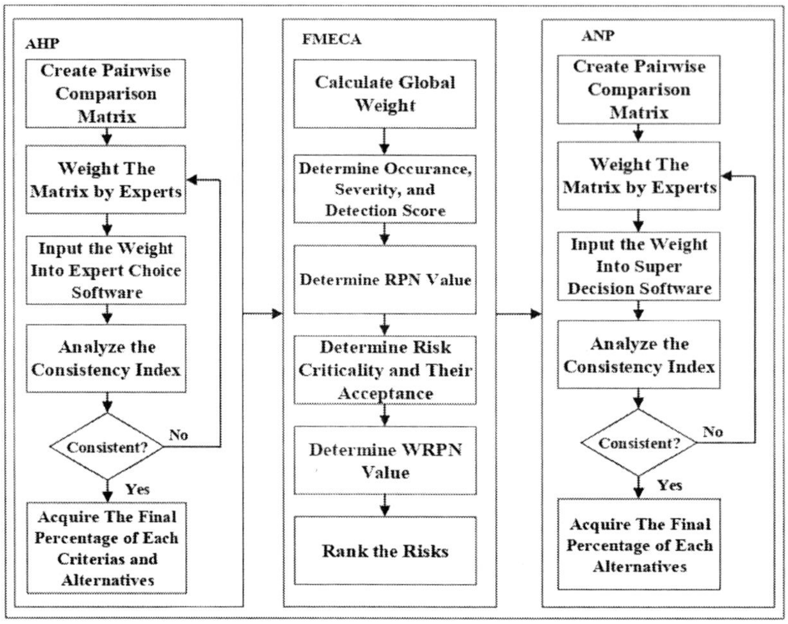

Figure 1. AHP, FMECA, and ANP method flowchart

Table 1. Criticality analysis

Criticality		Risk
Criticality Level	**Score**	**Acceptance**
Low	0 - 30	Acceptable
Moderate	31 - 60	Tolerable
High	61 - 180	
Very High	181 - 252	Unacceptable
Critical	253 – 324	
Very Critical	> 324	

Table 2. Rating scale of occurance, severity, and detection

Skor	*Occurrence*(O)	*Severity* (S)	*Detection* (D)
1	Remote probability	No effect	Almost certain
2	Very rare	Very small effect	Very easy
3	Moderately rare	Minor effect	Easy
4	Slightly rare	Very low	Fairly easy
5	Rare	Low	Moderate chance of detection
6	Slightly often	Moderate	Slightly difficult
7	Moderately often	High	Moderately difficult
8	Often	Very high	Difficult
9	Very often	Serious	Very difficult
10	Certain probability	Very severe	No chance of detection

After collecting RPN value from the experts, the next step of FMECA is to calculate global weight of each risk using risk weight we obtained in the AHP method. The equation of global weight is as shown below:

$$Global\ weight = Risk\ factor\ weight \times Local\ weight \tag{1}$$

After that, we also had to calculate Risk Priority Number (RPN) value which needed for determining risk criticality and its acceptance using the equation below:

$$\mathrm{RPN} = \mathrm{O} \times \mathrm{S} \times \mathrm{D} \quad (2)$$

The RPN value also used for calculating the Weighted Risk Priority Number (WRPN) that has an equation as shown below:

$$WRPN = RPN \times f(Wi) \quad (3)$$

2.3. Analytical Network Process (ANP)
ANP is a general theory of relative measurement used to derive composite priority ratio scales from individual ratio scales that represent relative measurement of the influence of elements that interact with respect to control criteria [9]. ANP method was used to assess strategies and determine which strategies that were recommended for mitigation. The ANP method began by calculating the geometric mean to get the overall value from the three experts' judgements. The equation is as shown below:

$$\text{Geometic Mean} = (a_1 \times a_2 \times \ldots \times a_n)^{1/n} \quad (4)$$

After the experts' judgement were calculated, then it was needed to also check and analyze the consistency ratio. Consistency ratio shows that the experts' judgements are consistent and can be use for further calculation. The expert's judgements are considered consistent if the CR value fall at or below 10%. CI and CR equation are as shown below:

$$\mathrm{CI} = (\lambda\mathrm{max} - \mathrm{n})/(\mathrm{n} - 1) \quad (5)$$

Information:
λmax : Maximum eigen vector
n : Number of alternative or criteria

$$\mathrm{CR} = \mathrm{CI}/\mathrm{RI} \quad (6)$$

Information:
CI : Consistency Index
RI : Ratio Index

3. Result and Discussion
This study aims to identify risks that have a high probability of occurrence and assess risks to determine their criticality rank. This study also aims to generate recommendation strategies that can be taken into consideration in mitigating the risk. Meanwhile, implementation and monitoring have not been carried out in real conditions.

3.1. Risk Identification
Risk identification was carried out with the aim of determining the risks that have emerged or have the possibility to appear in the supply chain. Identification of these risks was carried out with literature study by using supply chain risks categorization by Jafarnejad [6] and interview with experts, after which 25 risk variables were grouped based on the 6 risk factors that sheltered them:

3.2. Data consistency and risk weight
After the expert assessments have been done, weights for 6 risk factors and 25 local weights can be calculated. The local weight addressed a risk variable weight. The weight result can be seen in Table 4. Moreover, weight of each risk must be validated using consistency ratio (CR) value and followed the rule the CR value is 0.1 or more. The purpose of the consistency test was to find out whether the assessments of the three experts were consistent and also determine whether the data could be used for

the next calculation step or not. The results of CR obtained sequentially are: 0.04; 0.01; 0.04; 0.00; 0.07; 0.08; and 0.03. Thus, all the CR values are below 0.1 then the data was consistent.

Table 3. Wooden Toy Industry's supply chain risk factors and variables

Risk Factor	Risk Variable	Risk Factor	Risk Variable
Demand	Competitor moves	Information	Information delay
	Delays in delivery to customers		Wrong choice of communication
	Forcast errors	Operational	Capacity inflexibility
	Market saturation		Design changes
Environment	Macroeconomic uncertainty		Disruption in production
	Natural disasters		Inventory risk
	Policy uncertainty		Variability in production process
	Social uncertainty	Supply	Dependency on single supplier
Financial	Cost/price risk		Inflexibility of supplier
	Exchange rate risk		Poor delivery performance
Information	Breakdown of IT infrastructure		Supplier poor quality
	Distorted information		Supplier bankruptcy
	Inadequate information security		

3.3. Calculationg Global Weight

Here the FMECA method begun by calculating the global weight, which is the weight obtained by multiplying the local weight by the weight of the risk factor using equation (1). For example, look at the demand risk factor which has a weight of 0.22 and the competitor move risk variable which has a weight of 0.136. Using equation (1) the results are as follows, the rest of the global weight shown in Table 4:

$$\text{Global weight} = 0{,}22 \times 0{,}136 = 0{,}030$$

3.4. Calculationg RPN Value

The RPN value or Risk Priority Number was calculated to determine the value of the risk by considering the occurrence, severity, and detectability of risk. Furthermore, after completed the assessment of the three experts, the O (occurance), S (severity), and D (detection) scores were calculated to get the average value, so there would be only one value. The RPN was calculated with equation (2). Take an example of forecast error risk variable, the risk has a value of 8 for occurrence (often occurs), a value of 4 for severity (very low), and a value of 4 for detection (quite easily detected). With equation (2) it is produced:

$$\text{RPN} = 8 \times 4 \times 4 = 112$$

3.5. Determining Risk Criticality and Its Acceptance

After the RPN value had been determined, each risk variable critical level was categorized based on the RPN value, then the risk acceptance was classified based on the criticality level. Using this criticality analysis helped us to map the risk, take a look at Table 4, risk variable cost/price risk is categorized as critical because the RPN value is in the range of 253 to 324 according to table 2. This level of criticality has unacceptable risk acceptance that need corrective actions immediately.

3.6. Calculating the WRPN value and ranking the risks

The last stage in FMECA was risk ranking, based on criticality analysis we took only the risk that had a high criticality level to be sorted using WRPN (weighted risk priority numbers) calculation which could be obtained using equation (3), for example for policy uncertainty risk, the calculation is as follows:

$$\text{WRPN} = 150 \times 0{,}023 = 3{,}456$$

IOP Conf. Series: Materials Science and Engineering **528** (2019) 012007 doi:10.1088/1757-899X/528/1/012007

Table 4. Risk criticality and its acceptance

Risk Factor	Weight	Risk Variable	Local Weight	Global Weight	O	S	D	RPN	Criticality	Risk acceptance
Information	0,094	Breakdown of IT infrastructure	0,38	0,036	3	5	2	28	Low	Acceptable
Demand	0,22	Competitor moves	0,136	0,030	5	4	2	33	Moderate	Tolerable
Financial	0,234	Exchange rate risk	0,458	0,107	3	5	2	33	Moderate	Tolerable
Supply	0,191	Poor delivery performance	0,197	0,038	3	5	2	37	Moderate	Tolerable
Supply	0,191	Supplier bankruptcy	0,332	0,063	4	6	2	40	Moderate	Tolerable
Supply	0,191	Dependency on single supplier	0,156	0,030	3	5	2	41	Moderate	Tolerable
Operational	0,179	Variability in production process	0,157	0,028	4	3	3	43	Moderate	Tolerable
Information	0,094	Distorted information	0,221	0,021	4	5	3	49	Moderate	Tolerable
Demand	0,22	Delays in delivery to customers	0,344	0,076	5	5	2	50	Moderate	Tolerable
Information	0,094	Inadequate information security	0,183	0,017	3	4	4	58	Moderate	Tolerable
Operational	0,179	Design changes	0,151	0,027	6	6	2	67	High	Tolerable
Information	0,094	*Wrong choice of communication*	0,1	0,009	4	6	3	72	High	Tolerable
Supply	0,191	Inflexibility of supplier	0,084	0,016	5	5	3	76	High	Tolerable
Information	0,094	Information delay	0,116	0,011	4	5	3	77	High	Tolerable
Operational	0,179	Inventory risk	0,105	0,019	6	5	3	79	High	Tolerable
Demand	0,22	Market saturation	0,378	0,083	3	6	4	82	High	Tolerable
Supply	0,191	Supplier poor quality	0,231	0,044	5	7	3	104	High	Tolerable
Environment	0,082	Social uncertainty	0,169	0,014	5	6	4	112	High	Tolerable
Demand	0,22	Forcast errors	0,142	0,031	8	4	4	112	High	Tolerable
Operational	0,179	Disruption in production	0,419	0,075	5	7	3	122	High	Tolerable
Environment	0,082	Policy uncertainty	0,281	0,023	5	6	5	150	High	Tolerable
Operational	0,179	Capacity inflexibility	0,168	0,030	6	7	4	154	High	Tolerable
Environment	0,082	Natural disasters	0,268	0,022	4	6	8	186	Very High	Unacceptable
Financial	0,234	Cost/price risk	0,542	0,127	7	6	6	263	Critical	Unacceptable
Environment	0,082	Macroeconomic uncertainty	0,282	0,023	8	10	7	519	Very Critical	Unacceptable

Table 5. Ranked Risk

Risk Factor	Risk Variable	Global Weight	RPN	WRPN	Criticality	Rank
Financial	Cost/price risk	0,127	263	33,379	Critical	1
Environment	Macroeconomic uncertainty	0,023	519	11,996	Critical	2
Operational	Disruption in production	0,075	122	9,167	Critical	3
Demand	Market saturation	0,083	82	6,807	Critical	4
Operational	Capacity inflexibility	0,030	154	4,631	High	5
Supply	Supplier poor quality	0,044	104	4,576	High	6
Environment	Natural disasters	0,022	186	4,083	High	7
Demand	Forcast errors	0,031	112	3,513	High	8
Environment	Policy uncertainty	0,023	150	3,456	High	9
Operational	Design changes	0,027	67	1,807	High	10
Environment	Social uncertainty	0,014	112	1,552	High	11
Operational	Inventory risk	0,019	79	1,491	High	12
Supply	Inflexibility of supplier	0,016	76	1,217	High	13
Information	Information delay	0,011	77	0,840	High	14
Information	*Wrong choice of communication*	0,009	72	0,677	High	15

Then, the critical level for each risk variable was determined based on the comparison with the average WRPN, the average WRPN of the 15 risk variables at Table 5 is 5,946. So if the WRPN value in Table 5 exceeded the WRPN average, then it is categorized as a critical risk. According to Table 5, there are have four risks that are labeled as critical. But, in this research focused to develop strategies for the most critical risk which was cost/price risk.

3.7. Determine strategies priority

The main risk generated is price/cost risk. In the case in the wooden toy industry it is considered as the commodity cost volatility. Commodity cost volatility risk is a financial risk which in its handling involves effective and efficient management. Therefore, there are 8 strategies proposed to be assessed and prioritized. The proposed strategy consists of strategies for the commodity cost volatility risk and also strategies for effective management. Both are different things, so that the strategy is not combined in one matrix. Determination of strategy priorities was done using the ANP method. To input the valuation, the geometric mean of the experts' judgment using equation (4) must be done. The results are as shown in Table 6.

Table 6. Aggregated experts assessments using geometric mean equation

Strategies for cost/price risk									
Commodity Cost volatility						Effective Management			
	Track commodity price movements (T1)	Establish clear terms with suppliers (T2)	Understand the true quantity used (T3)	Differentiate savings goals (T4)	Combining wood with various commodities (T5)		Strategic accounting practices (M1)	Hiring external accountant (M2)	Financial planning (M3)
T1		0,843433	0,305711	0,36246	0,721125	M1		4,217163	1,709976
T2	1,185631		1,169607	0,36246	0,753947	M2	0,237126		0,36246
T3	3,271066	0,854988		2,758924	0,721125	M3	0,584804	2,758924	
T4	2,758924	2,758924	0,36246		0,753947				
T5	1,386723	1,326352	1,386723	1,326352					

According to Barosi and Busse [10], strategies like T1 to T4 could help on understanding exposure to raw material and how to manage the cost volatility. On the other hand, in order to achieve effective and efficient management could be done by practicing effective financial magement as Shallow [11] elaborated that strategies like M1 to M3 could be applied for financial management in a small business.

3.7.1. Data Consistency. This step has similarity with AHP method in part to consider the consistency expert's judgements. E*quation* (5) can be used to be found the Consistency Ratio (CR) value as an example bellow. There *is the calculation of CI for effective management:*

$$CI = (3{,}00139 - 3)/(3 - 1) = 0{,}000699$$

After that, calculating the consistency ratio using equation (6), and got the result as shown:

$$CR = (0{,}000699)/(0{,}52) = 0{,}00134$$

According to Saaty It is recommended that CR should be less than or equal to 0.1. Seeing that the result, the effective management factor was less than 0.1 then the experts judgements is consistent. But for commodity cost volatility factor, the experts' judgement exceeded the recommended CR and has the value of 11%, because of that the strategy priority assessment continued only with management factor.

*3.7.2. Strategies priority. I*n order to get the best strategy that were recommended, the highest percentage can be chosen as a decision. Thus, in this research, the recommended strategies for cost/price risk are strategic accounting practices (54%) and financial planning (33%).

4. Conclusion

Based on the risk analysis in the wooden toy industry's supply chain, it can be concluded that there are 6 risk factors and 25 risk variables in the wooden toy industry's supply chain. Those 6 risk factors are demand, environment, financial, information, operational, and supply risk. The most critical risk from the 15 risk variables is cost/price risk with the highest WRPN value of 33,379. For that critical risk, developed and assesed the recommended strategies for minimizing the cost/price risk based on ANP method which are obtained: conducting strategic accounting practices, and financial planning.

Acknowledgement

This work was supported in part by Simlitabmas Ristekdikti followed SK No.3/E/KPT/2017 and contract with LP2M Al Azhar Indonesia University under agreement No. 064/SPK/A-01/UAI/IV/2018

5. References

[1] Assauri S 2011 *Manajemen Produksi dan Operasi* (Jakarta: Lembaga Penerbit FEUI).
[2] Pourhejazy P and Kwon O K 2016 The New Generation of Operations Research Methods in Supply Chain Optimization: A Review. *Sustainability* **8** 1033 p 1
[3] Badariah N, Surjasa D and Trinugraha Y 2012 Analisa supply chain risk management berdasarkan metode failure mode effects analysis (FMEA). (Jakarta: Universitas Trisakti) *Jurnal Teknik Industri* p 110–118.
[4] Zaroni 2015 *Manajemen Risiko Rantai Pasok dalam Model SCOR*. Artikel Supply Chain, Retrieved from: http://supplychainindonesia.com/new/manajemen-risiko-rantai-pasok-dalam-model-scor/
[5] Zeng F 2011 *Models for Evaluation of Supply Chain Risk with Application to Healthcare Management*. (Arlington: University of Texas at Arlington) pp 1-2
[6] Jafarnejad A, Ebrahimi M, Abbaszadeh M A and Abtahi S M 2014 Risk Management in Supply Chain using Consistent Fuzzy Preference Relations. *International Journal of Academic Research in Business and Social Sciences* **4**(1) pp 77–89
[7] Saaty T and Vargas L 2006 *Decision Making with the Analitic Network Process. Economic, Political, Social and Technological Applications with Benefits, Opportunities, Costs and Risks*. (Pittsburgh: RWS Publication) p 347
[8] Yssaad B, Khiat M and Chaker A 2012 Maintenance Optimization for Equipment of Power Distribution System Based on FMECA Method. *ACTA ECTROTEHNICA*. (Mediamira Science Publisher) **53**(3) pp 218-223
[9] Saaty T 1999 *Fundamentals of The Analytic Network Process* (United States of America: University of Pittsburgh) p 1
[10] Barosi F and Busse P 2011 *Six Steps to Managing Raw Material Price Volatility* (New York: Alixpartners Publication) Retrieved from: https://legacy.alixpartners.com/
[11] Shallow K N 2017 *Strategies for Effective Financial Management in Vincentian Small Businesses* (America: Walden University) pp 70-92

IOP Conf. Series: Materials Science and Engineering **528** (2019) 012008 doi:10.1088/1757-899X/528/1/012008

Measuring Influence from Safety Climate to Safety Behavior in Bus Rapid Transit Drivers

D M Safitri[1], A Mediana[2] and W Septiani[3]

[1,2,3]Industrial Engineering Department, Laboratory of Work System Design and Ergonomics Faculty of Industrial Technology, Universitas Trisakti, Jakarta, Indonesia
[1]dianm@trisakti.ac.id

Abstract. Safety climate and safety behavior has an essential contribution to safety in all forms of the organizations. This research aimed to measure the impact of safety climate on safety behavior on bus rapid transit drivers in Jakarta. Bus rapid transit accident occurred inside and outside the bus lane. The incidence of the accident indicates that the risk is real. The measurement of the safety climate was using Nordic Safety Climate Questionnaire (NOSACQ-50). The NOSACQ-50 was adapted to fit the characteristics of transportation object. The questionnaire was distributed to 100 bus rapid transit drivers in Jakarta. Structural Equation Modelling – Partial Least Square (SEM-PLS) method was used to determine the influence safety climate to safety behavior among bus rapid transit drivers. The testing on the outer and inner model was conducted by bootstrapping and blindfolding process using Smart PLS software on the significance level of 0.05. The results of the study are that the workers' safety commitment and safety communication, learning, and trust in co-workers' safety competence had a positive and significant influence on safety behavior.
Keywords: safety climate, safety behavior, bus rapid transit

1. Introduction

The research model built through a literature study. Some studies suggested that the incidence of accidents was usually related to the safety behavior of workers [1], [2]. The safety climate also influenced safety behavior [3], [4]. Based on these findings, the literature study was conducted using a specific keyword that was the safety climate and safety behavior. This systematic literature study yields the outcomes of indicators for the safety climate and safety behavior research variables. These safety variables and indicators would form the research model in the form of the hypothetical model.

Safety was one indicator of the quality of transportation services so that public transport operators are very concerned about safety issues. The government as a regulator of public transport services was also very concerned about safety. Research on transportation safety has a high urgency because its services are responsible for more lives than private vehicles. Safety climate was the perception of workers against policy and all efforts of management or organization in seeking safety [2].

Safety climate was the worker's perception of concern for the safety of the organization in which he works. Previous research states that safety climate has a relationship with safety behavior [5]–[9]. This research aimed to assess the relationship between safety climate and safety behavior on bus rapid transit operator in Indonesia. Safety climate research has been done in many organizations and companies. Among them were in hospitals and health services [2], [10]–[12], construction [5], [13]–[16], manufacturing industry [17]–[23], and chemical industry [24]. More specifically, safety climate research was conducted on the public transport by involving the driver as respondents as done by [3], [6], [25]–[29].

Several major cities in Indonesia have bus rapid transit (BRT) services to support community mobility. Jakarta was one of the cities in Indonesia that has a BRT system called Transjakarta Busway. Transjakarta Busway was operated by private and state-owned companies. The Jakarta government was very concerned with the safety of public transport. In the year 2017, there were many accidents involving Transjakarta buses. The accident occurred inside and outside the bus lane. The incidence of the accident indicates that the risk of an accident was existing. So it was necessary to assess the factors that cause the accident.

Human factors were mentioned to be the primary cause of traffic accidents. Human error was usually referred to as an issue that was hard to be solved in an accident prevention effort. Blaming people was unfair because, in the system point of view, transport safety was a hierarchical system. Elements in the transport safety system include the organization and the human. Organizations had a role in transport safety systems. In the system, the organization occupies the highest hierarchy that has a significant influence on the safety of public transport services. The human factor held the authority beneath it. Organizational factors and human factors have interrelated relationships [30]. In the study, organizational factors were represented by safety climate variables, while safety behavior variables represented human factors. This study aimed to explain the relationship between safety climate and safety behavior in Bus Rapid Transit operators in Jakarta.

2. Methods

Literature studies show that several tools can be used to measure safety climate including NOSACQ-50 [31]. Other researchers designed the questionnaire for safety climate measurements on each object. The safety climate measurements in the study adopted the NOSACQ-50 survey. Adaptation of item tests on NOSACQ-50 needed to be done to fit the context of public transport objects. The terms for production and manufacturing were adjusted for the activities of the bus driver. At this stage also added the safety climate variables in the NOSACQ-50 variable set based on a literature study. According to [31], Safety Climate had 7 dimensions of; (1) management safety priority, commitment, and competence (SC1); (2) management safety empowerment (SC2); (3) management safety justice; as well as shared perceptions of (SC3) (4) workers' safety commitment (SC4); (5) workers' safety priority and risk non-acceptance (SC5); (6) safety communication, learning, and trust in co-workers' safety competence (SC6); and (7) workers' trust in the efficacy of safety systems (SC7).

The process of adaptation of the questionnaire was needed because the variables obtained from previous research were not in the field of public transport. Pilot studies were conducted to improve the constructs of the test items. Expert opinions were used to consider the process of adaptation of item test questionnaires from industrial objects to transportation objects. Each question item in the questionnaire was assessed using a Likert scale of 1 through 5. The value one stated strongly disagree, while the value 5 stated strongly agree with the statement in the questionnaire. The measurement of safety behavior in the study used a set of questions in the survey used by [9].

PLS-SEM is a variance- based multivariate analysis method which overcomes the shortcomings in the research that couldn't obtain relatively large samples [32]. Compared to Covariance-Based SEM, PLS-SEM is suitable for predictive oriented research with quite a high model complexity. PLS-SEM uses nonparametric testing, where the latent value variable is explicitly estimated [32], [33].

3. Result and Discussion

Testing the hypothetical model with SEM PLS consists of two stages. The first was the outer model test phase. At this stage, correlation of indicators with latent variables was analyzed. A valid indicator was determined using the loading factor value. The second stage was hypothesis testing on the inner model. Hypotheses Test using T-test. Data processing was done by using Smart PLS 3 software.

According to [34], in determining outer loading relevance testing, it was determined that if the outer loading value were below 0.4, then the inductor would be eliminated. If the outer loading value were above 0.7, the indicator would be maintained. If the value were between 0.4 and 0.7, then the value of the Average Variance Extracted (AVE) would be considered. If AVE were below 0.5, then the indicator would be discarded, and if above 0.5 will be maintained. Construct indicators SC1.1, SC1.3, SC1.5, SC3.1, SC3.2, SC5.1, SC5.2, SC6.4, SC7.1, SC7.2 had factor loading values below 0.6 and should be eliminated from the hypothetical model. In addition to the value of the loading factor, convergent validity could also be seen from the Average Variance Extracted (AVE). In the study, the AVE value of each construct was above 0.5. Variables SC1, SC3, SC5, and SC7 had the AVE value below 0.50; it indicates that the indicator has a higher average error rate [34]. However, in this study, these variables will still be used in the reliability test because the value of outer loading meets the validity requirements. Reliability Test was done by considering Cronbach's Alpha Value. The variable was declared reliable if it has Cronbach's alpha value above 0.6. Based on the Cronbach's alpha value, then the SC3, SC5, and SC7 latent variables will be excluded from the model. The final hypothetical model is as shown in Figure 1.

Table 1 Cronbach's Alpha and Average Variance Extracted Value

Variables	Number of items based on literature	Adapted number of items based on a pilot study	**Cronbach's Alpha**	**Average Variance Extracted (AVE)**
Safety Behavior	10	10	0.922	0.589
Management safety priority, commitment, and competence (SC1)	9	7	0.612	0.404
Management safety empowerment (SC2)	4	6	0.811	0.507
Management safety justice; as well as shared perceptions of (SC3)	4	4	0.195	0.407
Workers' safety commitment (SC4)	6	4	0.958	0.888
Workers' safety priority and risk non-acceptance (SC5)	7	4	0.422	0.418
Safety communication, learning, and trust in co-workers' safety competence (SC6)	8	4	0.782	0.617
Workers' trust in the efficacy of safety systems (SC7)	7	3	-1.910	0.456

The relationship of influence between variables was exposed to T-value written above the arrow line. T Value above 1.96 indicates that the impact of one variable on another variable was significant. Table 1 shows that the effect of all safety climate variables has a positive influence on safety behavior. A positive original sample value indicated the positive controls. The value of T-statistics presented the level of significance of the importance of independent variables on the dependent variable. Based on the table, 5% significance has a T value = 1.96. The T-Statistics score above 1.96 indicates that the effect was significant. Workers' safety commitment (SC4) and safety communication, learning, and trust in co-workers' safety competence (SC6) have a positive and significant influence on safety behavior. While management safety priority, commitment, and competence (SC1) and management safety empowerment although influence safety behavior (SC2), but the impact was not significant. The value of P-value can also indicate the level of significance of the influence. P-value values smaller than 0.05 indicate a meaningful relationship. From the output of the calculation, it was known that the conclusion of the P-value was in line with the result shown by the value of T-statistics. The Path Coefficient, T-Statistics, and P-Value are as shown in Table 3. The item test for assessing the relationship between safety climate and safety behavior for bus rapid transit drivers is as shown in Table 2.

PLS-SEM method was established for prediction determinations. The R2 values (i.e., coefficients of determination) represent the amount of explained variance of the endogenous constructs in the structural model. A well-built path model to explain specific fundamental target constructs should deliver satisfactorily high R2 values. In general, R2 values of 0.25, 0.50, and 0.75 for target constructs were considered as weak, medium, and substantial, respectively [32]. The calculation of the Q2 value for measuring predictive significance used the prediction error. The path model would have a predictive weight for a selected reflective endogenous construct if the Q2 value were above zero. From Table 4, the R2 and Q2 value indicate that the target construct was medium, and the path delivered an excellent predictive implication.

Table 2 Item Tests List

Latent Variable	Item Test
Management safety priority, commitment, and competence (SC1)	Management guarantees every driver receives the required information regarding his or her safety and passengers. Management prioritizes the safety of drivers and passengers on public transport services I am confident in management's ability to handle safety-related issues Management takes immediate action on any accident, and occupational health risks found during inspection/audit.
Management safety empowerment (SC2)	Management seeks to design routine activities to improve occupational safety and health Management involves drivers in decision-making related to safety The company recruits new drivers and is adequately trained to understand safety regulations and procedures. Safety issues are given high priority in training programs for drivers. Management encourages workers to attend safety training programs. The safety training provided to the driver was sufficient and allowed me to recognize the risk of danger.
Workers' safety commitment (SC4)	We work hard together to achieve a high level of safety We care about the safety of ourselves and others We help each other to work safely We are responsible for the safety of ourselves and others
Safety communication, learning, and trust in co-workers' safety competence (SC6)	We always discuss to find a solution if someone finds a potential risk of accident We seriously consider the advice and opinions of others regarding the prevention of accidents We have high confidence in the ability of colleagues to ensure safety
Safety Behavior (SB)	I adhere to safety procedures before, during and after driving. I am used to the use of safety equipment (seat belts, indicators on the dashboard, fire extinguishers, and so forth) in the vehicle I qualify with knowledge and job skills I am always cautious and alert when driving. I always check the vehicle before driving to make sure the bus condition I operate is well worth the way. I always do document checking and Standard Operation Procedure before driving. I adhere to traffic signs for my self and passenger safety. I will report according to the procedure if it finds unsafe condition. I set the temperature of the bus cabin where I work comfortably enough to stay concentrated. I make sure all the safety equipment in the vehicle is available.

Table 3 Path Coefficient, T-Statistics, and P-Value

	Original Sample (O)	T-Statistics	P-value
SC1 →SB	0.086	0.743	0.448
SC2 →SB	0.250	1.539	0.104
SC4 →SB	0.247	2.360	0.020
SC6 →SB	0.246	1.806	0.048

Table 4. R^2 and Q^2 Value of the Endogenous Latent Variable

Endogenous latent variable	R^2 value	Q^2 value
Safety behavior	0.527	0.245

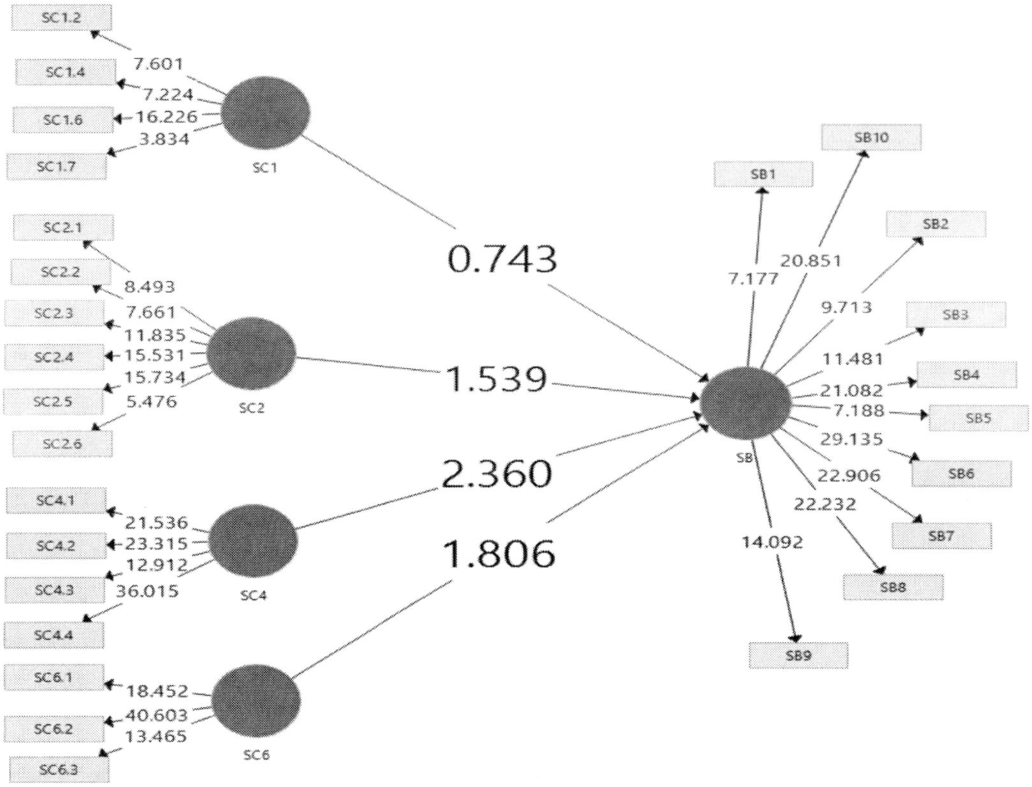

Figure 1 Graphics Output from A Hypothetical Test Of SEM-PLS Method using Smart PLS Software

4. Conclusion

Model of the relationship between safety climate to safety behavior in this research can represent a phenomenon in a real system. Two safety climate variables have a positive and significant influence on safety behavior. For prevention of accident, safety behavior improvement can be focused on improvement of safety climate, especially on the dimension of workers 'safety and safety communication, learning, and trust in co-workers' safety competence indicators. This research opens opportunities for development, especially in the addition of other latent variables related to transport safety. Safety behavior is not the only variable of the human factor that can cause an accident. Further study should include the influence of other variables such as safety awareness, safety participation, safety knowledge, and safety motivation.

Acknowledgment

Universitas Trisakti supported this work through the Internal Research Grant program for Industrial Technology Faculty of Academic Year 2017/2018. We would also like to express our gratefulness to PT Transportasi Jakarta for sharing their pearls of wisdom with us during the research

5. References

[1] Dommes A, Granié M A, Cloutier M S, Coquelet C and Huguenin-Richard F 2015 Red light violations by adult pedestrians and other safety-related behaviors at signalized crosswalks. *Accid. Anal. Prev.* **80** pp 67–75

[2] Neal A, Griffin M and Hart P 2000 The impact of organizational climate on safety climate and individual behavior. *Saf. Sci.* **34**(1–3) pp 99–109

[3] Fugas C S, Silva S A and Meliá J L 2012 Another look at safety climate and safety behavior: Deepening the cognitive and social mediator mechanisms. *Accid. Anal. Prev.* **45** pp 468–77

[4] Wills A R, Watson B and Biggs H C 2006 Comparing safety climate factors as predictors of work-related driving behavior. *J. Safety Res.* **37** pp 375–83
[5] Glendon I and Litherland D 2001 Safety climate factors, group differences and safety behavior in road construction. *Saf. Sci.* **39** (February) pp 157–88
[6] Amponsah-tawiah K and Mensah J 2016 The impact of safety climate on safety related driving behaviors. *Transp. Res. Part F Psychol. Behav.* **40** pp 48–55
[7] Cooper M D and Phillips R A 2004 Exploratory analysis of the safety climate and safety behavior relationship. *J. Safety Res.* **35**(5) pp 497–512
[8] Wills A R, Watson B and Biggs H C 2006 Comparing safety climate factors as predictors of work-related driving behavior. *J. Safety Res.* **37**(4) pp 375–83
[9] Jiang L, Yu G, Li Y and Li F 2010 Perceived colleagues' safety knowledge/behavior and safety performance: Safety climate as a moderator in a multilevel study. *Accid. Anal. Prev.* 42(5) pp 1468–76
[10] Neal A and Griffin M A 2006 A study of the lagged relationships among safety climate, safety motivation, safety behavior, and accidents at the individual and group levels. *J. Appl. Psychol.* **91**(4) pp 946–53
[11] Singer S, Lin S, Falwell A, Gaba D and Baker L 2009 Relationship of safety climate and safety performance in hospitals. *Health Serv. Res.* **44**(2P1) pp 399–421
[12] Zarei E, Khakzad N, Reniers G and Akbari R 2016 On the relationship between safety climate and occupational burnout in healthcare organizations. *Saf. Sci.* **89** pp 1–10
[13] Mohamed S 2002 Safety Climate in Construction Site Environments. *J. Constr. Eng. Manag.* **128**(5) pp 375–84
[14] Zhou Q, Fang D and Wang X 2008 A method to identify strategies for the improvement of human safety behavior by considering safety climate and personal experience. *Saf. Sci.* **46**(10) pp 1406–19
[15] Li Q, Ji C, Yuan J and Han R 2017 Developing dimensions and key indicators for the safety climate within China's construction teams: A questionnaire survey on construction sites in Nanjing. *Saf. Sci.* **93** pp 266–76
[16] Chan A P C, Wong F K W, Hon C K H, Lyu S and Javed A A 2017 Investigating ethnic minorities' perceptions of safety climate in the construction industry. *J. Safety Res.* **63** pp 9–19
[17] Zohar D and Luria G 2005 A Multilevel Model of Safety Climate: Cross-Level Relationships Between Organization and Group-Level Climates. *J. Appl. Psychol.* **90**(4) pp 616–28
[18] Milijic N, Mihajlovic I, Strbac N and Zivkovic Z 2013 Developing a questionnaire for measuring safety climate in the workplace in Serbia. *Int. J. Occup. Saf. Ergon.* **19**(4) pp 631–45
[19] Brondino M, Pasini M and da Silva S C A 2013 Development and validation of an Integrated Organizational Safety Climate Questionnaire with multilevel confirmatory factor analysis. *Qual. Quant.* **47**(4) pp 2191–223
[20] Yousefi Y *et al.* 2016 Validity Assessment of the Persian Version of the Nordic Safety Climate Questionnaire (NOSACQ-50): A Case Study in a Steel Company. *Saf. Health Work* **7**(4) pp 326–30
[21] Kim K W, Park S J, Lim H S and Cho H H 2017 Safety Climate and Occupational Stress According to Occupational Accidents Experience and Employment Type in Shipbuilding Industry of Korea. *Saf. Health Work* **8**(3) pp 290–95
[22] Murphy L A, Robertson M M and Carayon P 2014 The next generation of macroergonomics : Integrating safety climate. *Accid. Anal. Prev.* **68** pp 16–24
[23] Oah S, Na R and Moon K 2008 The Influence of Safety Climate, Safety Leadership, Workload, and Accident Experiences on Risk Perception: A Study of Korean Manufacturing Workers. *Saf. Health Work*
[24] Vinodkumar M N and Bhasi M 2009 Safety climate factors and its relationship with accidents and personal attributes in the chemical industry. *Saf. Sci.* **47**(5) pp 659–67
[25] Huang Y H, Lee J, Chen Z, Perry M, Cheung J H and Wang M 2017 An item-response theory approach to safety climate measurement: The Liberty Mutual Safety Climate Short Scales.

Accid. Anal. Prev. **103**(March) pp 96–104

[26] Wishart D, Somoray K and Evenhuis A 2017 Thrill and adventure seeking in risky driving at work : The moderating role of safety climate. *J. Safety Res.* **63** pp 83–9

[27] Öz B, Özkan T and Lajunen T 2010 Professional and non-professional drivers' stress reactions and risky driving. *Transp. Res. Part F Traffic Psychol. Behav.* **13**(1) pp 32–40

[28] Zohar D, Huang Y H, Lee J and Robertson M M 2015 Testing extrinsic and intrinsic motivation as explanatory variables for the safety climate-safety performance relationship among long-haul truck drivers. *Transp. Res. Part F Traffic Psychol. Behav.* **30** pp 84–96

[29] Wills A R, Biggs H C and Watson B 2005 Analysis of a Safety Climate Measure for Occupational Vehicle Drivers and Implications for Safer Workplaces. *Aust. J. Rehabil. Couns.* **11** pp 8–21

[30] Proctor T and Van Zandt R 1994 *Human Factors in Simple and Complex Systems. 2nd Ed.* (London: CRC Press)

[31] Kines P *et al.* 2011 Nordic Safety Climate Questionnaire (NOSACQ-50): A new tool for diagnosing occupational safety climate. *Int. J. Ind. Ergon.* **41**(6) pp 634–46

[32] Sarstedt M, Ringle C M and Hair J F 2014 PLS-SEM: Looking Back and Moving Forward. *Long Range Plann.* **47**(3) pp 132–37

[33] Nitzl C 2016 The use of partial least squares structural equation modelling (PLS-SEM) in management accounting research: Directions for future theory development. *J. Account. Lit.* **37** pp 19–35

[34] Hair J F, Hult G T M, Ringle C M and Sarstedt M 2014 *A Primer on Partial Least Squares Structural Equation Modeling (PLS-SEM).* (California: SAGE Publications, Inc.)

Design of Dust Collector on Sorting Machine Vibro Mesh Type Using Design for Assembly (DFA) Approach with Boothroyd and Dewhurst Method in PT. Perkebunan Nusantara VIII Ciater

C A Gulo[1], M Rahayu[2], S Martini[3] and M I Kurniawan[4]

Telkom University, Bandung, Indonesia
[1]christyardyanto@gmail.com, [2]mira.rahayu82@gmail.com

Abstract. PT. Perkebunan Nusantara VIII is a company that produce orthodox black tea in Indonesia with a land area of less than 400 hectares. In the production section, especially the sorting room in sorting machine, which produces dust contaminants are very disturbing operators who are working. For the maintenance process is scheduled for 2 hours on working hours every 1 month, if maintenance takes more than 2 hours it will cause the production process is delayed. To support the maintenance process on dust collector, Design of dust collector using design for assembly (DFA) aproachp using Boothroyd and Dewhurst method. DFA approach is chosen to simplify the process of maintenance dust collector, which required unloading dust collector every maintenance. There are 2 dust collector designs proposed to get the best assembly design. The design of dust collector with optimal assembly time, with the application of Boothroyd and Dewhurst is design 2 with total component of 82 pieces with assembly time of 552.62 sec with assembling efficiency of 24.43%. It is hoped that by using design 2, the dust collector maintenance process does not interfere with the production process and the shorter time of assembly.
Keywords: material transporters, musculoskeletal disorders, work accident, product design, Ergonomic Function Deployment

1. Introduction

PT. Perkebunan Nusantara VIII Ciater Lembang is a company engaged in the tea plantation, has a land area of approximately 400 hectares. PT Perkebunan Nusantara VIII has several work stations such as timber stations, milling stations, drying stations, sorting stations, and packing stations. Some of these stations produce dust particles when the production process is in progress. But the station that produces the most dust particles is in the sorting station. The sorting station is a station located after the drying station where the tea leaves which have been carried out are dried by burning process. Tea leaves that have been dried will go to the sorting station which will then be sorted based on size and weight. Sorting machines work by turning to push tea powder to filter quickly. This rotation causes the dust from the tea powder to rise into the air. The dust generated by the sorting machine spreads around the station resulting in dirty room and may compromise the respiratory system of the operator. Not only disrupt the respiration of the operator but also disturb the operator's visibility. PT Perkebunan Nusantara VIII will use a tool called a vacuum cleaner as a dust collector to be installed in the sorting machine. The purpose

IOP Conf. Series: Materials Science and Engineering **528** (2019) 012009 doi:10.1088/1757-899X/528/1/012009

of installation of this tool is to suck the dust generated by sorting machine but this tool requires regular maintenance to keep the system running properly. The main design of this tool has a working system such as a dust collector.

The design of the dust collector will have a plastic that serves to accommodate the dust that has been inhaled during the sorting process in the machine. The dust accommodated on the bucket can be reused as a dye batik. Cleaning the bucket is done by first turning off the sorting machine and vacuum cleaner then the bucket is removed and the dust that has been accommodated can be accommodated to the dust container. The bucket cleaning process will cause the production process to be suspended temporarily which will have an impact on the loss of production time. Therefore, the design of this vacuum cleaner using DFA to produce design with a fast assembly time so that the production process is not too long inhibited.

2. Methods

2.1. Dust Collector

Dust collector is a system used to improve the air quality generated from industrial and commercial processes by collecting dust and other impurities from air or other gases. The working principle of this tool is by lowering the pressure on the suction side under atmospheric pressure (free air). The air around the suction hole will enter into the suction hole which causes the dust contained in the air around the suction hole will go into the suction hole. The incoming air is then filtered using a filter to filter out the dust so that the air out of the dust collector system is completely clean [1].

2.2. Design for Assembly

The design for assembly approach was chosen with the aim of supporting the maintenance process by minimizing the number of assembly parts, optimizing assembly time when it will be dismantled and reinstalled, as well as the level of assembly efficiency in the design of local exhaust ventilation devices exhaust ventilation is recommended more often in checking LEV systems such as fan motors, air cleaners, hoods, pipes that can experience damage or corrosion [2]. This method depends on the design that will be evaluated and improve continuously [3]. There are 4 primaries expected in design for assembly results:

1. Reduced material cost.
2. Reduce labor and automatic assembly.
3. Reduced assembly cycle time.
4. Higher product quality and reliability.

2.3. Boothroyd and Dewhurst Method

Quantitative method to calculate efficiency assembly level, there are 2 steps to analyze efficiency assembly level [4].

$$E = NM.\ ta/TM \quad (1)$$

Where,

E = Theorical minimum item count
NM = Minimum part count
Ta = Actual assembly process time for the product
TM = Total assembly process time for all item.

3. Result and Discussion

3.1. Result

Researchers observe the area on the sorting machine, to decide the proposed design dust collector that will be applied. Researchers proposed 2 design in this research then make different assembly methods, and parts, then analyze 2 proposed design then evaluate the design after that choose the best proposed design.

A. Design 1

Determination of the various components obtained from the results of benchmarking carried out on the form of similar components through several media and market.

Table 1. Benchmarking Design 1

Part	Benchmarking	Description	Result
Support pillar	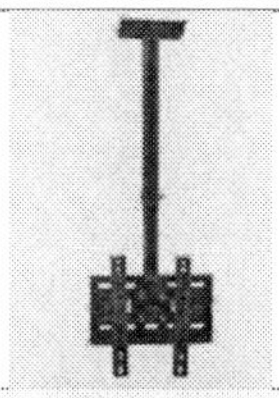	The support pillar uses iron plate material and is attached to the ceiling / foundation of the room.	The use of 3 support poles to hold from the main pipe, 2 pieces to hold the hood pipe
Ring support pillar		Ring support pillar use rubber material so that it can be stretched when inserting a pipe into the ring	Part of the support pillar ring that is assembled with a support pole using 2 bolts for tightening
Dynamo motor enclosure		The enclosure of an electric motor is shaped like a cage with a hole in the front and bottom for the pipe	This cage will be placed on the wall as a barrier
Flange	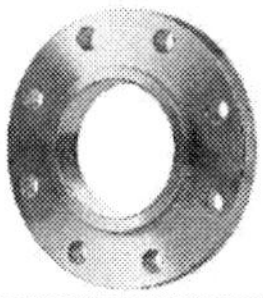	Circle shape using 8 holes.	Flange is used as a connection with between pipes and then tightened with screw

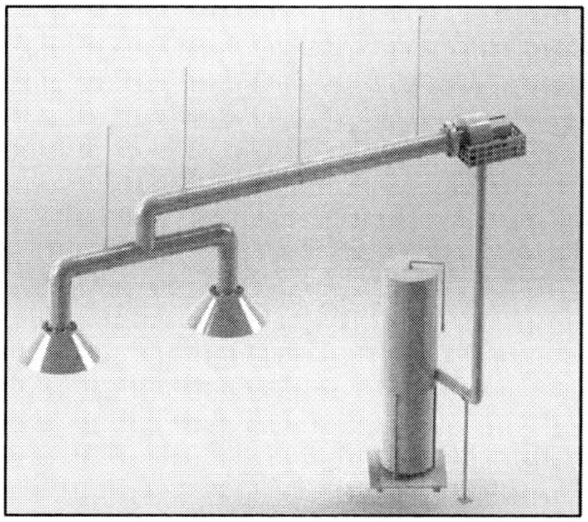

Figure 1. Proposed Design 1 Dust Collector

Proposed design 1, determine parts, dimension, and fastener based on benchmarking from literature, and others media.

Table 2. Component of Design 1

Part	Quantity	Description
Air Cleaner Circle Support	1	Base Part
Fan Cage	1	Base Part
Pole	5	Base Part
Hood	2	Base Part
Flange	2	Main Part
Hood Pipe	2	Main Part
T Pipe	1	Main Part
Main Pipe	1	Main Part
Outlet Pipe	1	Main Part
Fan	1	Main Part
Case Fan-front	1	Main Part
Case Fan-back	1	Main Part
Dinamo	1	Main Part
Bag Filter	1	Main Part
Air Cleaner Circle	1	Main Part
Bag Filter Cage	1	Main Part
Bucket	1	Main Part
Bag Filter Scraper	1	Main Part
Bolt 8	28	*Fastener*
Bolt 4	16	*Fastener*
Mur 8	24	*Fastener*

Based on the proposed design 1 there are 21 kind of parts consist of 103 components that needed for design 1. Next step is determine total operation for proposed design 1 that shown in table below.

Table 3. Total Assembly Operation of Design 1

Sub Assembly	Total Assembly Operation
Hood	11
Fan	4
Air Cleaner	7
Hood & Fan	1
Total	23

Boothroyd and Dewhurst Method for design 1. Total operation time for assembly 660.48 secs with efficiency assembly level 20.89 %

Table 4. Boothroyd and Dewhurst Method Design 1

A		B	G	H
ID	**Part Name**	**Number of times the operation carried out**	**Operation time (TM), B*(D+F)**	**Theorical minimum part**
			(sec)	**(NM)**
1		**Sub Assembly Hood**		
1.1	Hood	2	-	-
1.2	Bolt 8	16	56.96	8
1.3	Mur 8	16	128.96	8
1.4	Hood pipe	2	15	2
1.5	Bending pipe	2	14	0

Table 4. Boothroyd and Dewhurst Method Design 1 (cont.)

A		B	G	H
ID	**Part Name**	**Number of times the operation carried out**	**Operation time (TM), B*(D+F)**	**Theorical minimum part**
			(sec)	**(NM)**
2			**Sub Assembly Hood & Duct**	
2.1	Sub Assembly Hood	2	14	2
2.2	T Pipe	1	14	0
2.3	Bending pipe	1	7	0
2.4	Main Pipe	1	18	1
2.5	Support Pole	4	-	-
2.6	Bolt 4	16	128.96	8
3			**Sub Assembly Fan**	
3.1	Fan Cage	1	-	-
3.2	Case Fan - front	1	5	1
3.3	Fan	1	5	1
3.4	Case Fan - back	1	5	1
3.5	Motor	1	6	1
3.6	Bolt 8	4	32.24	4
4			**Sub Assembly Air Cleaner**	
4.1	Air Cleaner Circle Support	1	-	-
4.2	Air Cleaner Circle	1	5	1
4.3	Bucket	1	10.75	1
4.4	Bag Filter	1	7.6	1
4.5	Bag Filter Cage	1	7.6	1
4.6	Bag Filter Scraper	1	11.35	1
4.7	Bending Pipe	1	9.6	0
4.8	Outlet Pipe	1	18	1
4.9	Outlet pole	1	5	1
5			**Sub Assembly Hood, Duct & Fan**	
5.1	Sub Assembly Duct	1	-	-
5.2	Bolt 8	8	28.48	0
5.3	Mur 8	8	64.48	0
5.4	Flange	2	15	0
5.5	Sub Assembly Fan	1	14.5	1
6			**Final Assembly**	
6.1	Sub Assembly Hood, Duct dan Fan	1	-	-
6.2	Sub Assembly Air Cleaner	1	13	1
	Design 1	103	660.48	46
	Efficiency Formulation		(3*NM)/TM	
	Efficiency		20.89%	

B. Design 2

Determination of the various components obtained from the results of benchmarking carried out on the form of similar components through several media and market.

Table 5. Benchmarking Design 2

Part	Benchmarking	Description	Result
Simplifying Design		In the T pipe and hood pipe, pipe turns are no longer used because these two components will be put together	The integration of this design will minimize the time in the pipe hood and pipe T assembly
Type of Pipe Connection		The pipe is formed with a flange-like tip shape so that it can be directly connected	This connection comes from the vendor so that the assembly process only enters the pipe

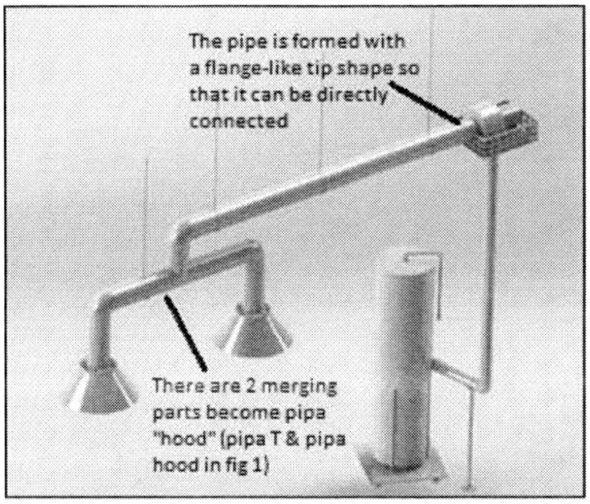

Figure 2. Proposed Design 2 Dust Collector

Proposed design 2, determine parts, dimension, and fastener based on benchmarking from literature, and others media.

Table 6. Component of Design 2

Part	Quantity	Description
Air Cleaner Circle Support	1	Base Part
Fan Cage	1	Base Part
Pole	5	Base Part
Hood	2	Main Part
Hood Pipe	2	Main Part
Main Pipe	1	Main Part
Outlet Pipe	1	Main Part
Fan	1	Main Part
Case Fan - Front	1	Main Part
Case Fan - back	1	Main Part
Dinamo	1	Main Part
Bag Filter	1	Main Part
Air Cleaner Circle	1	Main Part

Table 6. Component of Design 2 (cont.)

Part	Quantity	Description
Bag Filter Cage	1	Main Part
Bucket	1	Main Part
Bag Filter Scraper	1	Main Part
Bolt 8	20	*Fastener*
Bolt 4	16	*Fastener*
Mur 8	20	*Fastener*

Based on the proposed design 2 there are 19 kind of parts consist of 80 components that needed for design 2. Next step is determine total operation for proposed design 2 that shown in table below.

Table 7. Total Assembly Operation of Design 2

No	*Sub Assembly*	Total Assembly Operation
1	*Hood*	9
2	*Fan*	4
3	*Air Cleaner*	7
4	*Hood & Fan*	1
	Total	21

Boothroyd and Dewhurst Method for design 2. Total operation time for assembly 515.52 secs with efficiency assembly level 24.44 %

Table 8. Boothroyd and Dewhurst Method Design 2

A		B	G	H
ID	Part Name	Number of times the operation carried out	Operation time (TM), B*(D+F) (sec)	Theorical minimum part (NM)
1		**Sub Assembly Hood & Duct**		
1.1	Hood	2	-	-
1.2	Bolt 8	16	56.96	8
1.3	Nuts 8	16	128.96	8
1.4	T Pipe	1	14	0
1.5	Bending Pipe	1	7	0
1.6	Main Pipe	1	18	1
1.7	Support Pole	4	-	-
1.8	Bolt 4	16	128.96	8
2		**Sub Assembly Fan**		
2.1	Fan Cage	1	-	-
2.2	Case Fan - front	1	5	1
2.3	Fan	1	5	1
2.4	Case Fan - back	1	5	1
2.5	Motor	1	6	1
2.6	Bolt 8	4	32.24	4
3		**Sub Assembly Air Cleaner**		
3.1	Air Cleaner Circle Support	1	-	-
3.2	Air Cleaner Circle	1	5	1
3.3	Bucket	1	10.75	1
3.4	Bag Filter	1	7.6	1
3.5	Bag Filter Cage	1	7.6	1
3.6	Bag Filter Scraper	1	9.85	1
3.7	Bending Pipe	1	9.6	0
3.8	Outlet pipe	1	18	1
3.9	Outlet pole	1	5	1

Table 8. Boothroyd and Dewhurst Method Design 2 (cont.)

A		B	G	H
ID	**Part Name**	**Number of times the operation carried out**	**Operation time (TM), B*(D+F) (sec)**	**Theorical minimum part (NM)**
4	**Sub Assembly Hood, Duct & Fan**			
4.1	Sub Assembly Duct	1	-	-
4.2	Joint Pipe	1	7.5	0
4.3	Sub Assembly Fan	1	14.5	1
5	**Final Assembly**			
5.1	Sub Assembly Hood, Duct and Fan	1	-	-
5.2	Sub Assembly Air Cleaner	1	13	1
	Design 2	80	515.52	42
	Efficiency Formulation		(3*NM)/TM	
	Efficiency		24.44%	

3.2. Discussion

Table 9. Comparison

No	DFA Goals	Design 1	Design 2
1	Total Components (piece)	103	80
2	Total Assembly Time (sec)	660.48	515.52
3	Efficiency (%)	20.89%	24.44%

The comparison proposed design between both design changes with design 1 with 103 components can be reduced until 80 components in design 2 and assembly time 515.52 sec with efficiency level 24.44%. It's believed that design 2 is the best design to implemented in PT Perkebunan Nusantara with minimum assembly time for maintenance process. With minimize assembly time for dust collector, its hoped that maintenance process for all machine in PT Perkebunan Nusantara VIII will not caused extra time for maintenance process, and production will continue without delay.

4. Conclusion

Based on research that has been done in PT. Perkebunan Nusantara VIII Ciater, the design of dust collector with optimal assembly time, with application method Boothroyd and Dewhurst is design 2 with total of 82 pieces with assembly time 515.52 sec with assembling efficiency 24.44%. It is hoped that by using design 2, the dust collector maintenance process does not interfere with the production process and the shorter time of assembly.

5. References

[1] Nagyszalanczy S 2002 *Woodshop Dust Control, A Complete Guide to Setting Up Your Own System Completely Revised and Updated.* (Newtown, USA: The Taunton Press, Inc.)
[2] Health and Safety Authority 2014 *Local and Exhaust Ventilation Guidance.* (Irlandia: Health and Safety Authority)
[3] Yusri 2012 *Penerapan Design for Assembly (DFA) Untuk Mereduksi Biaya Produksi Suatu Produk.* (Padang: Politeknik Negeri Padang) pp 28-32
[4] Boothroyd G 1994 Product design for manufacture and assembly. *Computer-Aided Design* **26**(7) pp 505–520
[5] Latar M A 2014. *Tahapan Perancangan Sistem Ventilasi Lokal.* (Jakarta: Universitas Esa Unggul)
[6] Latar M A 2012 *Lokal Exhaust Ventilation/Ventilasi Pengeluaran Setempat.* (Jakarta: Universitas Esa Unggul)
[7] PT. Perkebunan Nusantara 2008 *SOP Pembuatan Teh Hitam Ortodoks.* (Bandung: PT. Perkebunan Nusantara)

IOP Conf. Series: Materials Science and Engineering **528** (2019) 012010 doi:10.1088/1757-899X/528/1/012010

Design of Wood Pellet Trolley using Finite Element Method (FEM) and Design for Assembly (DFA) Approach at PT. Perkebunan Nusantara VIII Ciater

R R Putra[1], M Rahayu[2], S Martini[3] and M I Kurniawan[4]

Telkom University, Bandung, Indonesia
[1]rikorizqullahputra@gmail.com,[2]mira.rahayu82@gmail.com,
[3]srimartini59m@gmail.com, [4]iccank.4596@gmail.com

Abstract. PT. Perkebunan Nusantara VIII is a company produces orthodox black tea. The wood pellets distribution in drying station still uses human power without any tools. The process is done as far as five meters with the transport time of each sack for three minutes and the weight is 30 kg. In the previous research has produced a design that used as reference in this research. The purpose of the research is to provide trolley design by applying Finite Element Method and Design for Assembly approach. The method is chosen because the trolley force should be considered so the trolley has good strength. The initially complex and inflexible designs are expected to be simpler and more efficient. The results of this research are wooden pellet trolleys on handle, platform, frame, and stairs chunk are considered safe from the results of the test. While the wheels chunk need to be re-election of material into titanium alloy because it is considered unsafe based on the test results. Then the design of wooden pellet trolleys results in a higher assembly efficiency rating of 27.77% with 127 total components and total processing time of 928.98 seconds so it can be concluded better than reference design.
Keywords: wood pellets, finite element method, assembly efficiency

1. Introduction

PT. Perkebunan Nusantara VIII Ciater, West Java is a company that produces 8 kinds of orthodox black tea [1]. The production of orthodox black tea consists of several processes as shown in figure 1. Each production process has its own room, in each room there is a Material Handling Equipment (MHE) which is used as a transportation of various types of materials or other objects. In the drying process, the decrease in water content of the tea powder uses a drying machine with heat flow to sterilize the possibility of bacteria. The drying machine requires wood pellets to fuel the machine. The need of wood pellets reached 3.6 tons/day. The wood pellet is distributed continuously every 10 minutes, which is done to maintain the stability of hot air temperature in the tea dryer which will affect the quality of tea.

IOP Conf. Series: Materials Science and Engineering **528** (2019) 012010 doi:10.1088/1757-899X/528/1/012010

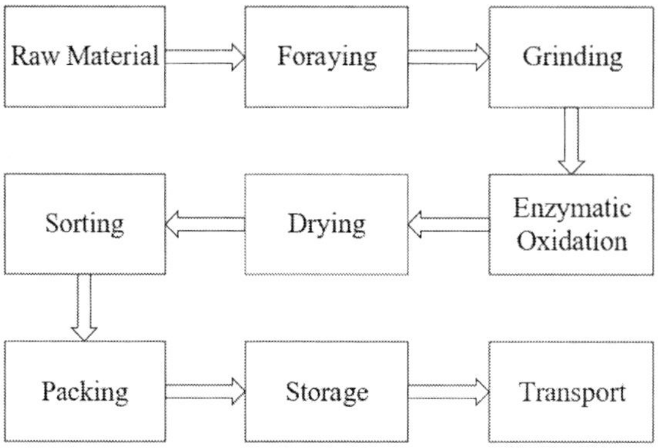

Figure 1. The Production Process

In the existing process, wood pellets are distributed from the transport truck to the temporary storage inside the drying station using human power as shown in figure 2. The process of transporting wood pellets by the operator is carried out by five meters with the transport time of each sack for three minutes and the weight of each wood pellet is approximately 30 kg.

Figure 2. Existing Condition of Wood Pellet Transportation

In a previous study, [2] has proposed a wood pellet trolley design as illustrated in figure 3 using the Ergonomic Function Deployment (EFD) approach to improve the productivity of the company. The result of the research is a concept of MHE design in the form of trolley which is expected to be a solution for the company to be able to increase work productivity.

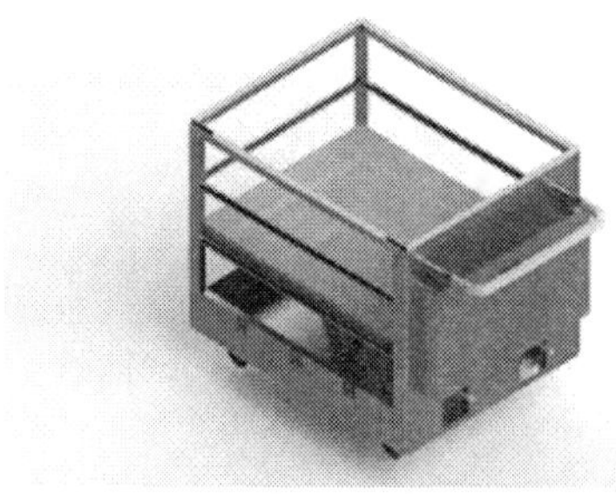

Figure 3. Reference Design of Wood Pellet Transport Trolley

In the design, the trolley frame strength must be really taken into account, so the design of the trolley is not only a good design, but also has good strength. Because if it is ignored it will be dangerous for the operator. Therefore, with Finite Element Method (FEM), calculations that cannot be known through the calculation manually can be known as the maximum voltage, the voltage at each node, and also the magnitude of the shift or deformation can be known. Improper product design leads to inefficient design, thereby reducing product reliability and increasing the cost of the assembly process [3]. An industry needs to design the product well in order to improve its competitiveness. Using the Design for Assembly (DFA) method, product production costs can be simplified without reducing product function. Initially complex and inflexible designs, they are simpler and more capable of being used elsewhere. DFA analysis refers to minimizing the components of the product. The use of the DFA method in redesigning the wood pellet trolley in order to get the trolley design to be more efficient to operate on the PT. Perkebunan Nusantara VIII.

2. Methods

2.1. Finite Element Method (FEM)
Finite Element Method (FEM) is a method to dividing complex objects into blocks or simple elements or dividing complex objects into smaller fragments that are easier to explore, called elements [4].

2.2. Design for Assembly (DFA)
Design for Assembly (DFA) is a process to improve product design to be easily assembled and at low assembly cost, focused on functional aspects and assembly of a product [5] [6]. Quantitative methods of assessing efficiency levels by Boothroyd and Dewhurst will calculate the estimated time of assembly between product components. There are two ways to analyse the efficiency level of assembling a product:

1. Possibility to minimize components
2. Possibility to redesign every component is easy to assemble

Reducing the number of components in the assembly is a step to improve assembly efficiency of a product. This can be done by removing unnecessary components and merging several components into one component.

2.3. Assembly Efficiency
Assembly efficiency is a ratio of the ideal assembly time to the actual assembly or the initial design of a product.

$$E = \frac{N_m \times t_a}{T_m} \tag{1}$$

Where,

E = Assembly efficiency
N_m = Theoretical minimum item count
t_a = Actual assembly process time for the product
T_m = Total assembly process time for all item

Calculating the efficiency of assembly can be done by finding the code and time of both handling and insertion, which is then included in the DFA analysis table. The assembly efficiency formulation is basically the ratio between ideal assembly time and real assembly time. Ideal time above is determined by the number of minimum components that become factors in minimizing costs.

2.4. Manual Assembly
Manual Assembly is an assembly activity of product components by manual, manual assembly is divided into two, namely:

a. Handling (acquiring and grasp, moving, and orienting the part)
   - Whether components can be taken with: one hand, one hand with the aid of a tool, two hands, or two hands with the help of another.

- Orienting (Part Symmetry)
  What degree of component can be rotated perpendicular to the axis (α) or in line axis (β) for its reorientation.
- Ease the components to be picked up and manipulated.
- The thickness of the product.
- Component size.

b. Insertion and Fastening
Insertion is the activity of merging of one component with other components. Fastening is to integrate a component with other components permanently or locked. Factors that affect at the time of this merger there are three, namely:
- Components are not locked at the merge.
- Components are immediately locked at the merge
- Components require additional operations for assembly to be locked.

3. Result and Discussion

3.1. Finite Element Method (FEM)
The trolley design is tested using Finite Element Method (FEM). The FEM test is a test aimed to detecting damage caused by stress and displacement. FEM tests are performed on handle, platform, frame, wheels, and ladder chunk. The chunk is chosen to be tested because they are receiving the most force and load when operated and are the main component of trolley. The test consisted of:

a. von Mises
von Mises was done to get the biggest load on the simulation result.

b. URES
URES is done to see how far the displacement is.

c. Factor of Safety
FOS is conducted with a series of preliminary studies aimed to conduct security analysis of materials and previous research concepts.

After obtaining the value of the test for each component, then the value will be analyzed to know the strength of the product material, the summary of the test result is described in table 1. If the test result shows good result, no change of material required. However, if the test results are not in accordance with the expected, it is necessary to change the material with the constraints to be determined.

Table 1. Test Result of Finite Element Method (FEM)

No.	Part	Material	Test	Maximum Value	Minimum Value	Yield Strength Value
			von Mises	2.087e+000N/m^2	8.477e+004N/m^2	
1	Handle	Stainless Steel	URES	0.000e+000mm	7.049e-005mm	1.723e+008
			FOS	2.03E+03	8.26E+07	
			von Mises	3.766e+002N/m^2	3.091e+006N/m^2	
2	Platform	Aluminum Alloy	URES	0.000e+000mm	1.968e-002mm	2.757e+007
			FOS	8.92E+00	7.32E+04	
			von Mises	1.056e+001N/m^2	5.927e+006N/m^2	
3	Frame	Aluminum Alloy	URES	0.000e+000mm	2.664e-001mm	2.757e+007
			FOS	4.65E+00	2.61E+06	
			von Mises	0.000e+000N/m^2	3.151e+008N/m^2	
4	Wheels	Aluminum Alloy	URES	0.000e+000mm	1.170e+000mm	2.757e+007
			FOS	8.75E-02	1.00E+16	
			von Mises	2.764e+000N/m^2	1.442e+007N/m^2	
5	Ladder	Stainless Steel	URES	0.000e+000mm	1.681e-002mm	1.723e+008
			FOS	1.20E+01	6.24E+07	

Based on the result of FEM analysis, wheel is categorized as unsafe because in von Mises test obtained maximum value exceeds yield strength value and in FOS test obtained FOS value ≤ 1. Therefore, it is necessary to re-election material on wheel. In this case selected titanium alloy because it has the best strength and is still within the limits of material selection. The material will be tested by the load. Based on the test results, test values obtained for chunk wheels with titanium alloy material as in table 2 below.

Table 2. Test Results for Wheels with Titanium Alloy

Test	Minimum Value	Maximum Value
von Mises	3.766e+002N/m^2	3.091e+006N/m^2
URES	0.000e+000mm	1.968e-002mm
Factor of Safety	8.921e+000	7.321e+004

3.2. Design for Assembly (DFA)

3.2.1. Reference Design. The reference design of wood pellet trolley has three components as base part, 14 components as main part, and 10 components as fastener. As well as having seven subassemblies, namely platform sub assembly, wheels sub assembly, frame sub assembly, ladder sub assembly, bottom frame sub assembly, top frame sub assembly, and final sub assembly. The number of assembly operations for reference design is described in table 3 below.

Table 3. Assembly Operations Total for Reference Design

No.	Sub Assembly	Number of Assembly Operations
1	Platform	14
2	Wheels	4
3	Frame	12
4	Ladder	26
5	Bottom Frame	4
6	Top Frame	1
7	Final	2
	Total	63

3.2.2. Proposed Design. In planning the proposed design (Figure 4), evaluation and identification of components may be combined, modified or eliminated. Proposed design changes are obtained through a benchmarking process to existing trolleys that already on the market to ensure design changes can be applied to the proposed design. The number of assembly operations for proposed design is described in table 4 below.

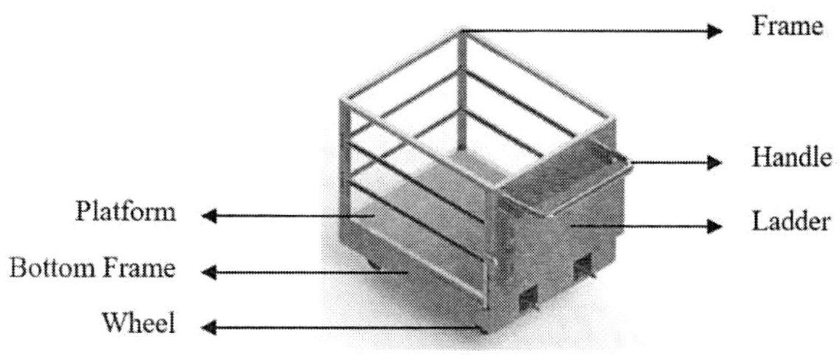

Figure 4. Proposed Design of Wood Pellet Transport Trolley

IOP Conf. Series: Materials Science and Engineering **528** (2019) 012010 doi:10.1088/1757-899X/528/1/012010

Table 4. Assembly Operations Total for Proposed Design

No.	Sub Assembly	Number of Assembly Operations
1	Platform	14
2	Wheels	4
3	Frame	3
4	Ladder	26
5	Bottom Frame	4
6	Top Frame	1
7	Final	2
	Total	54

After the proposed design assembly process is known, the next step is to calculate the efficiency value of assembly on the proposed design using DFA analysis table. Filling the table is done by identifying manual handling code and manual insertion code for each component. Calculations are performed to determine the overall assembly process time resulting in assembly efficiency values. Calculation of total assembly process time and its efficiency value can be seen in table 5.

Table 5. Manual Assembly Proposed Design

Item ID	Item name	Number of items	Total operation time, s	Figures for min. parts
1		**Platform Sub Assembly**		
1.1	Hydraulic Base Sheet	4	29.20	2
1.2	Hydraulic Arm	8	58.40	4
1.3	Hydraulic Spiner	2	7.90	1
1.4	Hydraulic Pin 1	8	60.00	4
1.5	Hydraulic Pin 2	8	57.04	4
1.6	Hydraulic Nut	2	23.00	1
1.7	Hydraulic Rod	2	23.00	1
1.8	Socket Set Screw	10	125.70	5
1.9	Platform Base (Base Part)	1	-	-
1.10	Platform	1	13.00	1
1.11	M10 Bolt	8	82.40	4
1.12	M10 Nut	8	79.44	2
2		**Wheels Sub Assembly**		
2.1	Wheels Bracket(Base Part)	4	-	-
2.2	Wheels	4	26.52	4
2.3	Wheels Connector	4	12.52	4
3		**Frame Sub Assembly**		
3.1	Main Frame (Base Part)	1	-	-
3.2	Frame Door	1	10.50	1
3.3	Storage Container Door	1	12.10	1
3.4	Frame Door Lock	1	3.50	1
4		**Ladder Sub Assembly**		
4.1	Ladder Pillar	12	37.56	12
4.2	Connecting Rubber	14	53.20	14
4.3	Step Footing	6	21.00	6
5		**Bottom Frame Sub Assembly**		
5.1	Platform Sub Assembly (Base Part)	1	-	-
5.2	Wheels Sub Assembly	4	10.92	4
5.3	M10 Nut	16	158.88	8
6		**Top Frame Sub Assembly**		
6.1	Frame Sub Assembly (Base Part)	1	-	-
6.2	Ladder Sub Assembly	1	6.10	1
6.3	Handle	1	6.10	

Table 6. Manual Assembly Proposed Design (cont.)

Item ID	Item name	Number of items	Total operation time, s	Figures for min. parts
7	**Final Sub Assembly**			
7.1	Bottom Frame Sub Assembly (Base Part)	1	-	-
7.2	Top Frame Sub Assembly	1	11.00	1
	Total	**127**	**928.98**	**86**
	Efficiency Formula		**3x(NM/TM)**	
	Efficiency Value		**27.77%**	

3.3. Comparison

The proposed design has fewer number of components i.e. 127 components, compared to the reference design that has 185 components. The proposed design requires less total assembly operation time of 928.98 seconds, compared to the reference design that requires total assembly operation time of 1475.93 seconds. The proposed design has a higher assembly efficiency value of 27.77%, compared to the reference design of the trolley having an assembly efficiency value of 22.97%. It can be concluded that by using the proposed design can minimize the number of components and assembly operation time with the principles of Design for Assembly (DFA) Boothroyd and Dewhurst method. Comparison of number of components, total assembly operation time, and efficiency value in each design can be seen in the table 6.

Table 7. Comparison

No.	DFA Goals	Reference Design	Proposed Design
1	Number of Components (pcs)	185	127
2	Total Assembly Operation Time (sec)	1475.93	928.98
3	Efficiency Value (%)	22.97	27.77

4. Conclusion

Based on the results of research, can be obtained some conclusion that is expected to be problem solving in company. The conclusions that can be taken include:

1. Test results of the main components of the wood pellet trolley using Finite Element Method (FEM), handles, platforms, frames and ladders chunk are categorized safe and can operate well based on stress, displacement and factor of safety tests. As for the wheel chunk is categorized as unsafe so it needs to be re-election of material into titanium alloy.
2. After comparison of reference design with proposed design using Design for Assembly (DFA) approach, proposed design has better assembly system with assembly efficiency value of 27.77% which increased from the previous design by 22.97%. Then the proposed design has a shorter total processing time that is 928.98 seconds down from the previous design that is for 1475.93 seconds.

5. References

[1] PT. Perkebunan Nusantara 2008 *SOP Pembuatan Teh Hitam Ortodoks*. (Bandung: PT. Perkebunan Nusantara).
[2] Pradani W 2018 *Perancangan Troli Pengangkut Pelet Kayu dengan Menggunakan Pendekatan Ergonomic Function Deployment (EFD) untuk Meningkatkan Produktivitas Kerja*. (Bandung: Universitas Telkom)
[3] Wahjudi D 1999 *Penilaian Desain Produk dengan Assembly Analysis and Line Balancing Spreadsheet dan Ullman 13 Guidelines untuk Meningkatkan Kinerja Perakitan*. (Surabaya: Universitas Kristen Petra)
[4] Akin J 2009 *Finite Element Analysis Concepts via Solid Works*. (Texas: World Scientific).
[5] Boothroyd G, Dewhurst P and Knight W 2011 *Product Design for Manufacture and Assembly*. (Florida: Taylor & Francis Group)

[6] Boothroyd G, Dewhurst P and Knight W 2002 *Product Design for Manufacture and Assembly*. (New York: Marcel Dekker)

Design of Wood Pellets Carrier using Ergonomic Function Deployment (EFD) Approach to Increase Productivity of Work: A Research at PTPN VIII Ciater

W R Pradani[1], M Rahayu[2], S Martini[3] and M I Kurniawan[4]

Telkom University, Bandung, Indonesia
[1]windipradani@gmail.com, [2]mira.rahayu82@gmail.com, [3]srimartini59m@gmail.com, [1]Iccank.4596@gmail.com

Abstract. This study aims to design a wood pellets carrier using Ergonomic Function Deployment (EFD) approach to improve productivity of work. This research was conducted at PT. Perkebunan Nusantara VIII, Ciater. There is a manual transport activity of wood pellets where to carry 30 tons of wood pellets is done by 5 workers each of which must carry a maximum of 60 kg of wood pellets. In the process of designing the trolley, knowing the worker complaints using NORDIC Body Map questionnaire and then analysed the value of the existing work posture. In addition, the making of needs statement for the EFD approach is based on the ergonomics concept is ESHCE (Effective, Safe, Healthy, Comfortable and Efficient) and another product design stage to generate an ergonomic trolley design. After generated the ergonomic trolley design then carried out analysis of push pull and productivity. By comparing the existing productivity and proposal productivity, it can be said that this wooden pellet transport trolley can increase work productivity in the drying work station.
Keywords: wood pellets carrier, Ergonomic Function Deployment, Ergonomic trolley design

1. Introduction

Indonesia is a tropical country and also an agrarian country that has a fertile soil structure that makes it possible to be utilized as agricultural land or plantation. One of the important plantation commodities for Indonesia today is the tea commodity. Tea plantations become a leading business sector that can absorb a lot of workforce. PT. Perkebunan Nusantara VIII which is one of the State-Owned Enterprises in the agricultural sector where the company produces various types of plantation products in Indonesia and among them are plantation and tea leaf production. Tea produced one of them is Orthodox tea. The stages of making black tea are as follows:

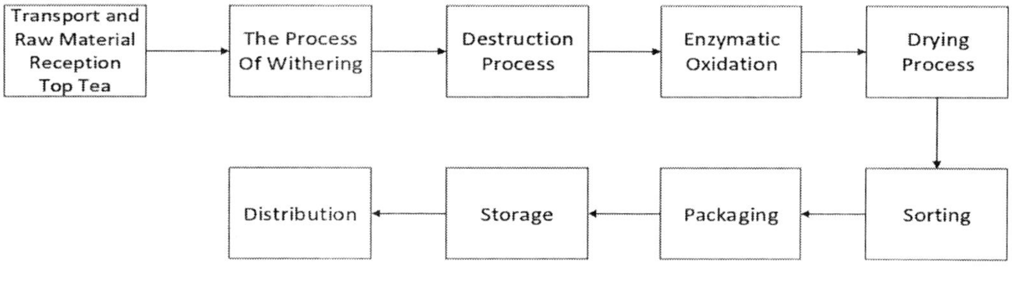

IOP Conf. Series: Materials Science and Engineering **528** (2019) 012011 doi:10.1088/1757-899X/528/1/012011

Figure 1. The Stages of Making Black Tea at PTPN VIII Ciater

In the drying process, a drying machine with a temperature of 100°C - 120°C which is fueled by wood pellets is required. The wood pellets are imported from the supplier and then brought into the room where the drying process takes place. In the existing process, the wood pellets are removed by manual transport by the five workers. In the process of transportation, workers must carry two bags of wood pellets weighing 30 kg / sack with a distance of five meters. Frequency of transport into the drying work station is approximately 100 times back and forth within a maximum of three hours.

Figure 2. Documentation of workers during the wood pellets transporting activities

To find out the complaints of the workers while transporting is to make observations, interview directly with workers and by distributing questionnaires NORDIC Body Map to the workers concerned. After knowing the complaints from workers, then the next step is to assess the work attitude. Appropriate approaches to performing a work attitude assessment are the Posture Evaluation Index (PEI) derived from integration of Lower Back Analysis (LBA), Ovako Working Posture Analysis System (OWAS) and Rapid Upper Limb Assessment (RULA).

Table 1. Recapitulation of PEI Score of Workers

LBA Safety Limit (N)	LBA Score (N)	OWAS Safety Limit	OWAS Score	RULA Safety Limit	RULA Score	PEI Safety Limit	PEI Score
≤ 3.400	5971	≤ 2	2	≤ 4	7	≤ 2	3.68

According to the above exposure, the problem is not very supportive to the aspect of ergonomics. It strongly opposes the concept of ergonomics that has the concept to create an EASNE work system (effective, safe, convenient, healthy and efficient). Therefore, it is necessary to design the Material Handling Equipment (MHE) which resembles the trolley so that later can be used to reduce the workload of the operator and also can increase work productivity by using Ergonomic Function Deployment (EFD) approach. So that the needs of operators can be met and can increase work productivity in drying stations especially for the process of transporting wood pellets.

2. Methods

2.1. Productivity
Productivity in general is the ratio between the inputs used with the output produced. One of the steps taken by a company in order to improve its productivity is by making improvements of tools (technology) or by increasing human resources. The formula of productivity is as follows:

$$\text{Productivity} = \frac{\text{Effectiveness}}{\text{Efficiency}} \quad (1)$$

2.2. Ergonomic
Ergonomics is a science that studies and examines the limitations and advantages of humans, then the information obtained will be used to design products, machinery, facilities, environment and work system. The main objective of an ergonomic application is the application of good aspects of health, safety, and working comfort to achieve good quality of work. At a higher level of ergonomics aims to create optimal working conditions. Ergonomics and K3 (Occupational Safety and Health) are two things that can not be separated because they have the same goal that is in improving the quality of work [1].

2.3. Ergonomic Function Deployment (EFD)
Ergonomic Function Deployment (EFD) is a development of Quality Function Deployment (QFD) by adding new aspect that is by adding new relationship between consumer desire and ergonomic aspect of product [2][3]. The relationship is made with the House of Quality (HOQ) format but since this method is in the ergonomic domain, it is called House of Ergonomic (HoE). The description of the HoE is as follows [4].

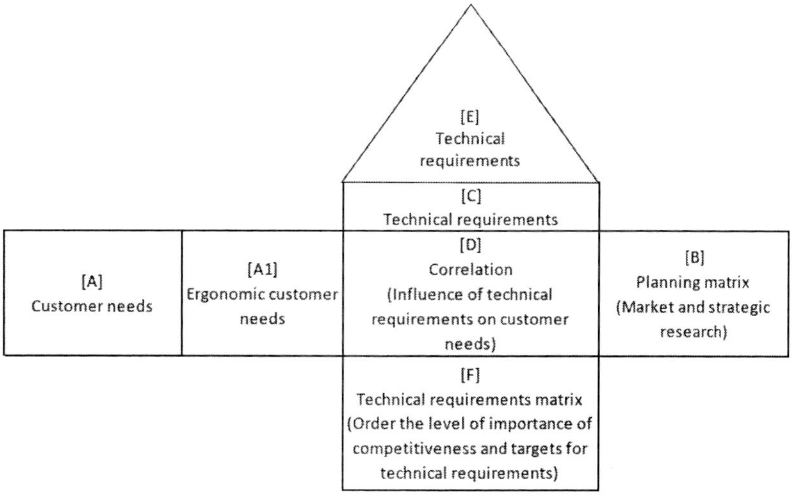

Figure 3. House of Ergonomic (HOE)

2.4. Conceptual Model
This research designs a trolley that will be used to transport wood pellets using EFD approach. To design the trolley, the first thing to do is to identify the activity of transporting the existing wood pellets and to identify what the complaints are often experienced by the workers by looking at scores produced by LBA, OWAS and RULA [6]. In this EFD approach, the formation of needs statement is based on the concept of ergonomics is ESHCE (Effective, Safe, Healthy, Comfortable and Efficient). After that will be analyzed in terms of ergonomic and after emerging an MHE design or trolley design, will be analyzed whether the trolley can increase productivity and how many forces are required to

push the trolley by using push pull analysis. The results of the trolley design will then be made technical drawings to know the dimensions in each part.

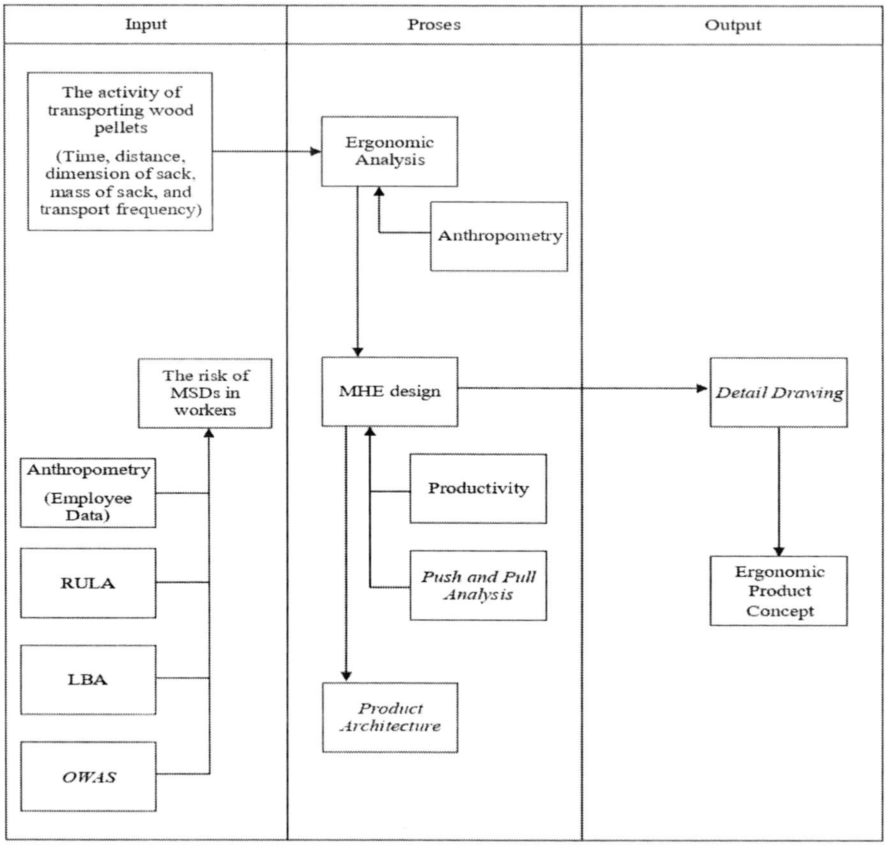

Figure 4. Conceptual Model of Research

3. Result and Discussion

3.1 Identifying Needs Statement

In the stage of identifying needs statement will be done by using statement needs statement which is determined from EFD variable that is EASNE. So get the results recapitulation needs statement as follows.

Table 2. Recapitulation Needs Statement

No.	Needs Statement
1	The product has a maximum carrying capacity.
2	The product is safe when in use.
3	Products can reduce the workload of workers and reduce the risk of work accident.
4	The product has a comfortable size when in use.
5	The product has an ergonomic design.
6	Products can be raise and lower.
7	Products can minimize the power released by the user.
8	The product can minimize the number of required operators.
9	The product has tools.

3.2 Create the HoE

HoE is a method used to find information that contains product identification as a consideration in developing a product. HoE consists of the relationship between the needs statement and the needs matrix and the relationship between the needs matrix. After establishing the HoE and determining other aspects required in the EFD is determining the final specification that will be used to make the wooden pellet transport trolleys design. The final specification in this study is as follows.

Table 3. Final Specification of Trolley

No.	Needs Matrix	Part Specification	Target Value	Unit
1	Maximum weight of trolley	Maximum weight of trolley	360	kg
2	Product material standard	Product material standard	-	List
3	Trolley locking system	Wheels locking system	Yes	Binary
		Trolley frame locking system	Yes	Binary
4	Transport time	Transport time	180	Sec
5	High of trolley	High of paltform	15	cm
		High of frame	93	cm
		High of handle	5	cm
		Diameter of wheels	15	cm
6	Width of trolley	Width of platform	110	cm
		Width of frame	110	cm
		Length of handle	80	cm
7	Length of trolley	Length of platform	138	cm
		Length of frame	138	cm
		Width of handle	15	cm
8	Diameter of handle	Diameter of handle	5	cm
9	Product desgin adjusts the dimensions of the worker	Product desgin adjusts the dimensions of the worker	-	List
10	Hydraulic system	Hydraulic system	Yes	Binary
11	Number of operators	Number of operators	1	Person
12	High of tool	High of aids	220	cm
13	Width of tool	Width of tool	11.5	cm
14	Length of tool	Length of tool	84	cm

After obtaining the desired final specification, then made a wooden pellet transport trolleys design. The following is the result of the wooden pellet transport cart design.

Figure 5. The Design of Wooden Pellet Transport Carrier

As mentioned earlier that the wooden pellet sacks coming from the supplier in once arrived at PT. Perkebunan Nusantara VIII, Ciater brings 20 - 30 tons or a maximum of 1000 bags of wood pellets. The transportation activity of this sack of wood pellets is carried out by 5 (five) operators with the load to carry each operator is a maximum of 6000 kg with the frequency of transporting as much as 100 times back and forth. The total time required for manual handling activities from truck to sack in a drying work station is 3 hours. With the design of this wooden pellet transport trolley is expected to increase productivity at the drying station. This trolley can accommodate a maximum of 360 kg or 12 bags of wood pellets so that the frequency of transport approximately 83 times back and forth. So the total time it takes the workers to transport the wooden pellet sacks is 2 hours 49 minutes.

To find out how many forces are required by the worker to push the carrier's carrier cart. In doing the calculation of push pull analysis can be done by using 2 (two) ways that is with Table Snook and Ciriello and also with analysis of MHL in software Jack 8.2. Below is the result of push pull analysis using Snook and Ciriello Table.

Height	Percent	2.1 m push, One push every							7.6 m push, One push every							15.2 m push, One push every							30.5 m push, One push every					45.7 m push, One push every					61.0 m push, One push every			
		6 s	12 s	1 min	2 min	5 min	30 min	8 hr	15 s	22 s	1 min	2 min	5 min	30 min	8 hr	25 s	35 s	1 min	2 min	5 min	30 min	8 hr	1 min	2 min	5 min	30 min	8 hr	1 min	2 min	5 min	30 min	8 hr	2 min	5 min	30 min	8 hr
Initial forces																																				
144	90	20	22	25	25	26	26	31	14	16	21	21	22	22	26	16	18	19	19	20	21	25	15	16	19	19	24	13	14	16	16	20	12	14	14	18
	75	26	29	32	32	34	34	41	18	20	27	27	28	28	34	21	23	25	25	26	27	32	19	21	25	25	31	16	18	21	21	26	16	18	18	23
	50	32	36	40	40	42	42	51	23	25	33	33	35	35	42	26	29	31	31	33	33	40	24	27	31	31	38	20	23	26	26	33	20	22	22	28
	25	38	43	47	47	50	51	61	27	31	40	40	42	42	51	31	35	37	37	40	40	48	28	32	37	37	46	24	27	32	32	39	23	27	27	34
	10	44	49	55	55	58	58	70	31	35	46	46	48	49	58	36	40	43	43	45	46	55	32	37	42	42	53	28	31	36	36	45	27	31	31	39
95	90	21	24	26	26	28	28	34	16	18	23	23	25	25	30	18	21	22	22	23	24	28	17	19	22	22	27	14	16	19	19	23	14	16	16	20
	75	28	31	34	34	36	36	44	21	23	29	29	[illegible]	32	39	24	27	28	28	30	30	36	21	24	28	28	35	18	21	24	24	30	18	21	20	26
	50	34	38	43	43	45	45	54	26	29	36	36	[illegible]	40	48	29	33	35	35	37	38	45	27	30	35	35	44	23	26	30	30	37	22	26	26	32
	25	41	46	51	51	54	55	65	31	35	45	45	[illegible]	48	58	35	40	42	42	45	45	54	32	36	42	42	52	27	31	36	36	45	27	31	31	38
	10	47	53	59	59	62	63	75	35	40	52	52	55	56	66	40	46	49	49	52	52	62	37	41	48	48	60	32	36	41	41	52	31	35	35	44
64	90	19	22	24	24	25	26	31	13	14	20	20	21	21	26	15	17	19	19	20	20	24	14	16	19	19	23	12	14	16	16	20	12	14	14	17
	75	25	28	31	31	33	33	40	16	19	26	26	27	28	33	19	21	24	24	26	26	31	18	21	24	24	30	16	18	21	21	26	15	18	18	22
	50	31	35	39	39	41	41	50	20	23	32	32	34	35	41	23	27	30	30	32	33	39	23	26	30	30	37	20	22	26	26	32	19	22	22	28
	25	38	42	46	46	49	50	59	25	28	39	39	41	41	50	28	32	36	36	39	39	47	28	31	36	36	45	24	27	31	31	39	23	26	26	33
	10	43	48	53	53	57	57	68	28	32	45	45	47	48	57	32	37	42	42	44	45	54	32	36	41	41	52	27	31	36	36	44	26	30	30	38
Sustained forces																																				
144	90	10	13	15	16	18	18	22	8	9	13	13	15	16	18	8	9	11	12	13	14	16	8	10	12	13	16	7	8	10	11	13	7	8	9	11
	75	13	17	21	22	24	25	30	10	13	17	18	20	21	25	11	13	15	16	18	18	22	11	13	16	18	21	10	11	13	15	18	9	11	13	15
	50	17	22	27	28	31	32	38	13	16	22	23	25	27	32	14	17	20	20	23	24	28	15	17	20	23	28	12	14	17	19	23	12	14	16	19
	25	21	27	33	34	38	40	47	16	20	28	29	32	33	39	17	20	24	25	28	29	34	18	21	25	29	34	15	18	21	24	28	15	17	20	24
	10	25	31	38	40	45	46	54	19	23	32	33	36	39	46	20	24	28	29	33	34	40	21	25	29	33	39	18	21	24	28	33	17	20	23	28
95	90	10	13	16	17	19	19	23	8	10	13	13	15	15	18	8	10	11	12	13	13	16	8	10	12	13	16	7	8	9	11	13	7	8	9	11
	75	14	18	22	22	25	26	31	11	13	17	18	[illegible]	21	25	11	13	15	16	18	18	21	11	13	16	18	21	9	11	13	15	18	9	11	12	15
	50	18	23	28	29	33	34	40	14	17	22	23	[illegible]	27	32	14	17	19	20	23	23	28	15	17	20	23	27	12	14	17	19	23	12	14	16	19
	25	22	28	34	35	40	41	49	17	21	27	29	[illegible]	33	39	18	21	24	25	28	29	34	18	21	25	28	33	15	18	21	24	28	15	17	20	23
	10	26	33	40	41	46	48	57	20	24	32	33	37	38	45	20	25	28	29	32	33	40	21	25	29	33	39	17	20	24	27	32	17	20	23	27
64	90	10	13	16	16	18	19	23	8	10	12	13	14	15	18	8	10	11	11	12	13	15	8	9	11	13	15	7	8	9	11	13	7	8	9	10
	75	14	18	21	22	25	26	31	11	13	17	17	19	20	24	11	13	14	15	17	17	21	11	13	15	17	20	9	11	12	14	17	9	10	12	14
	50	18	23	28	29	32	33	39	14	17	21	22	25	26	31	14	17	19	19	22	22	27	14	16	19	22	26	12	14	16	18	22	12	14	15	18
	25	22	28	34	35	39	41	48	17	21	26	27	31	32	37	18	21	23	24	27	28	33	17	20	24	27	32	14	17	20	23	27	14	17	19	22
	10	26	32	39	41	46	48	56	20	25	30	32	36	37	44	21	25	27	28	31	32	38	20	24	28	32	37	17	20	23	26	31	16	19	22	26

Figure 6. The Result of Push Analysis [1]

Height	Percent	2.1 m pull, One pull every							7.6 m pull, One pull every							15.2 m pull, One pull every							30.5 m pull, One pull every					45.7 m pull, One pull every					61.0 m pull, One pull every			
		6 s	12 s	1 min	2 min	5 min	30 min	8 hr	15 s	22 s	1 min	2 min	5 min	30 min	8 hr	25 s	35 s	1 min	2 min	5 min	30 min	8 hr	1 min	2 min	5 min	30 min	8 hr	1 min	2 min	5 min	30 min	8 hr	2 min	5 min	30 min	8 hr
Initial forces																																				
144	90	14	16	18	18	19	19	23	11	13	16	16	17	18	21	13	15	15	15	16	17	20	12	13	15	15	19	10	11	13	13	16	10	11	11	14
	75	17	19	22	22	23	24	28	14	15	20	20	21	21	26	16	18	19	19	20	20	24	14	16	19	19	23	12	14	16	16	20	12	14	14	17
	50	20	23	26	26	28	28	33	16	18	24	24	25	26	[illegible]	19	21	22	22	24	24	29	17	19	22	22	27	15	16	19	19	24	14	16	16	20
	25	24	27	31	31	32	33	39	19	21	28	28	29	30	[illegible]	22	25	26	26	28	28	33	20	22	26	26	32	17	19	22	22	28	16	19	19	24
	10	26	30	34	34	36	37	44	21	24	31	31	33	33	40	24	28	29	29	31	31	38	22	25	29	29	37	20	22	25	25	31	18	21	21	27
95	90	19	22	25	25	27	27	32	15	18	23	23	24	24	29	18	20	21	21	23	23	28	16	18	21	21	26	14	16	18	18	23	13	16	16	19
	75	23	27	31	31	32	33	39	19	21	28	28	29	30	36	22	25	26	26	28	28	33	20	22	26	26	32	17	19	22	22	28	16	19	19	24
	50	28	32	36	36	39	39	47	23	26	33	33	35	35	42	26	29	31	31	33	33	40	24	27	31	31	38	20	23	27	27	33	20	23	23	28
	25	33	37	42	42	45	45	54	26	3	39	39	41	41	49	30	34	36	36	38	39	46	27	31	36	36	45	24	27	31	31	38	23	26	26	33
	10	37	42	48	48	51	51	61	30	33	43	43	46	47	58	33	38	41	41	43	44	52	31	35	40	40	50	27	30	35	35	43	26	30	30	37
64	90	22	25	28	28	30	30	36	18	20	26	26	27	28	33	20	23	24	24	26	26	31	18	21	24	24	30	16	18	21	21	26	15	18	18	22
	75	27	30	34	34	37	37	44	21	24	31	31	33	34	40	24	28	29	29	31	32	38	22	25	29	29	36	19	22	25	25	31	19	21	21	27
	50	32	36	41	41	44	44	53	25	29	37	37	40	40	[illegible]	29	33	35	35	37	38	45	27	20	35	35	43	23	26	30	30	37	22	26	26	32
	25	37	42	48	48	51	51	61	30	34	44	44	46	47	[illegible]	34	39	41	41	43	44	52	31	35	41	41	50	27	30	35	35	43	26	30	30	37
	10	42	48	54	54	57	58	69	33	38	49	49	52	53	63	38	43	46	46	49	49	59	35	39	46	46	57	30	34	39	39	49	29	34	34	42
Sustained forces																																				
144	90	8	10	12	13	15	15	18	6	8	10	11	12	12	15	7	8	9	9	10	11	13	7	8	9	11	13	6	7	8	9	10	6	7	7	9
	75	10	13	16	17	19	20	23	8	10	13	14	16	16	19	9	10	12	12	14	14	17	9	10	12	14	16	7	9	10	11	14	7	8	10	11
	50	13	16	20	21	23	24	28	10	13	16	17	19	20	23	11	13	14	15	17	17	20	11	13	15	17	20	9	11	12	14	17	9	10	12	14
	25	15	20	24	25	28	29	34	12	15	20	20	23	24	28	13	15	17	18	20	21	24	13	15	18	20	24	11	13	15	17	20	11	12	14	17
	10	17	22	27	28	32	33	39	14	17	22	23	26	27	32	14	17	19	20	23	24	28	15	17	20	23	27	12	14	17	19	23	12	14	16	19
95	90	10	13	16	17	19	20	24	8	10	13	14	16	16	19	9	10	12	12	14	14	17	9	10	12	14	17	7	9	10	12	14	7	9	10	12
	75	13	17	21	22	25	26	30	11	13	17	18	20	21	25	11	14	15	15	18	18	22	12	13	16	18	21	10	11	13	15	18	9	11	13	15
	50	16	21	26	27	31	21	37	13	17	21	22	25	26	31	14	17	19	19	22	23	27	14	17	19	22	26	12	14	16	19	22	12	14	16	18
	25	19	26	31	33	37	38	45	16	20	26	27	30	31	37	17	20	22	23	26	27	32	17	20	23	27	32	14	17	19	22	26	14	16	19	22
	10	22	29	36	37	42	43	51	18	23	29	31	34	36	42	19	23	26	27	30	31	37	19	23	27	31	36	16	19	22	25	30	16	19	21	25
64	90	11	14	17	18	20	21	25	9	11	14	15	17	17	20	9	11	12	13	15	15	18	9	11	13	15	18	8	9	11	12	15	8	9	10	12
	75	14	19	23	23	26	27	32	11	14	19	19	22	22	26	12	14	16	17	19	19	23	12	14	17	19	23	10	12	14	16	19	10	12	13	16
	50	17	23	28	29	32	34	40	14	18	23	24	27	28	33	15	18	20	21	23	24	28	15	18	21	24	27	13	15	17	20	23	12	14	16	20
	25	20	27	33	35	39	40	48	17	21	27	28	32	33	39	18	21	24	25	28	29	34	18	21	25	28	33	15	18	21	24	28	15	17	20	25
	10	23	31	38	40	45	46	54	19	24	31	32	37	38	45	20	24	27	28	32	33	39	21	24	28	32	38	17	20	24	27	32	17	20	23	27

Figure 7. The Result of Pull Analysis [4]

So, obtained the value of initial push force limit of 40 kg and sustained push force limit of 26 kg. Then for the initial pull force limit of 35 kg and sustained pull force limit of 25 kg.

4. Conclusion

Based on the results of data processing and analysis, then obtained some conclusions that can solve the problems in this study. Here is the conclusion that has been obtained.

1. The product of wooden pellet carrier trolleys is able to meet the objectives of the research that can design a product that can improve work productivity by applying the concept of ergonomics (ESHCE) in the process of designing.
2. In designing the wooden pellet transport trolley, the material which will be applied to the product is Alluminium Alloy for platform, frame, wheel and hydraulic while Stainless Steels for handle and ladder.
3. Based on the calculation of push pull analysis found that to operate the trolley requires power to push for 40 N and power to attract 35 N.

5. References

[1] Yassierli *et al.* 2007 The influence of age on isometric endurance and fatigue is muscle dependent: A study of shoulder abduction and torso extension. *Ergonomics.* **50**(1) pp 26–45 doi: 10.1080/00140130600967323.

[2] Ulrich K T and Eppinger S D 2012 *Product Design and Development.* 5th Ed. (New York: McGraw-Hill)

[3] Surya R Z and Badruddin R 2015 Aplikasi Ergonomic Function Deployment (Efd) Pada Redesign Alat Parut Kelapa Untuk Ibu Rumah Tangga. *Jurnal Ilmiah dan Penerapan Keteknikan Pertanian* **8**(1) https://doi.org/10.17969/rtp.v8i1.2687.

[4] Adrianto R, Desrianty A and M F H 2014 *Usulan Rancangan Tas Sepeda Trial Menggunakan Metode Ergonomic Function Deployment (EFD).* **02**(02) pp 353–63

[5] Snook S and Ciriello V M 2002 Liberty Mutual Tables for Lifting, Carrying, Pushing and Pulling

also known as the Snook Tables. *Ergonomics* **34** pp 1197–213
[6] Bintang A N and Dewi S K 2017 Analisa Postur Kerja Menggunakan Metode OWAS dan RULA. *Jurnal Teknik Industri* **18**(01) pp 43-54. https://doi.org/10.22219/JTIUMM.Vol18.No1.43-54

The Design of Material Transporter for Paper Sack in Packaging to Decrease The Risk of Muscoloskeletal Disorders using Ergonomic Function Deployment (EFD) Approach: A Research at PT. Perkebunan Nusantara VIII Ciater, West Java

M I Kurniawan[1], M Rahayu[2] and S Martini[3]

Telkom University, Bandung, Indonesia
[1]Iccank.4596@gmail.com, [2]mira.rahayu82@gmail.com, [3]srimartini59m@gmail.com

Abstract. The Tea packaging paper sack process at PT. Perkebunan Nusantara VIII is done manually using human labor to lift paper sack weighing 40-60 kg. This condition makes it harder for the labor and increases the risk of Musculoskeletal Disorders (MSDs). In the existing condition, the RULA score is seven which have high risk of work accidents. To minimize the risk of MSDs, a proposed work tool is designed using the Ergonomic Function Deployment approach to produce ergonomic tools. The proposed work tool was developed with the ESHCE principle (Effective, Safe, Healthy, Comfortable, and Efficient) which became a variable in meeting the Ergonomic Function Deployment (EFD) approach. The research results showed that the proposed work tools could help workers in carrying out packaging activities with the decrease in the RULA value on the labor's posture and could reduce processing time because it united all activities in the packaging area. So it can be concluded that the proposed work tools can reduce the risk of MSDs and can help workers to process the packaging of tea with minimum time.
Keywords: Material Transporters, Musculoskeletal Disorders, Work Accident, Product Design, Ergonomic Function Deployment.

1. Introduction

PT. Perkebunan Nusantara VIII located in Ciater, West Java is a company that produces orthodox black tea. There are several steps that must be passed to produce orthodox black tea, from the process of transporting and receiving raw materials, the process of foraging, grinding, enzymatic oxidation process, drying, sorting process, packing, and storage process to the warehouse. The stages of producing orthodox black tea have a goal to produce the best tea products, but to distinguish between competitor tea products and orthodox tea products owned by PT. Perkebunan Nusantara VIII can be seen from the packaging.

In general, the packaging process is done using the machine so that the processing time is shorter and can minimize errors commonly done by humans. However, some activities at PT. Perkebunan Nusantara VIII at the packaging work station still uses human power, this can increase the risk of work accident. The following is the flow on the packaging work station, which is illustrated in Figure 1 below.

IOP Conf. Series: Materials Science and Engineering **528** (2019) 012012 doi:10.1088/1757-899X/528/1/012012

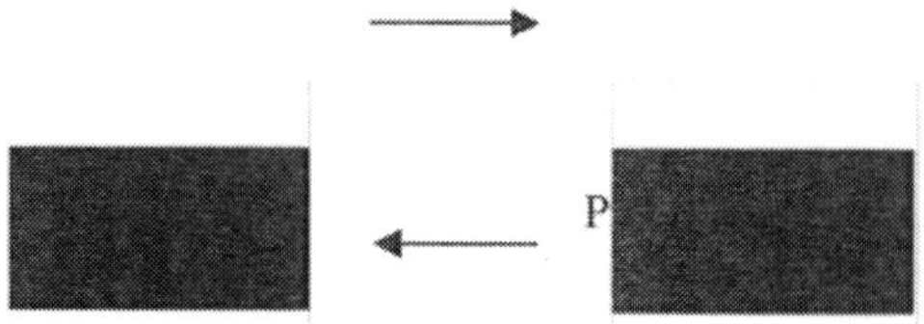

vity On Packag ion

In the packaging process carried out some paper sack transport activity by lifting a paper sack weighing 40-60 kg using human power, the activity gives a very high impact MSDs, so it takes work tools that reduce the risk of MSDs in workers. The previous work tool produced a RULA score of seven that was simulated on the CATIA software as shown in Figure 2.

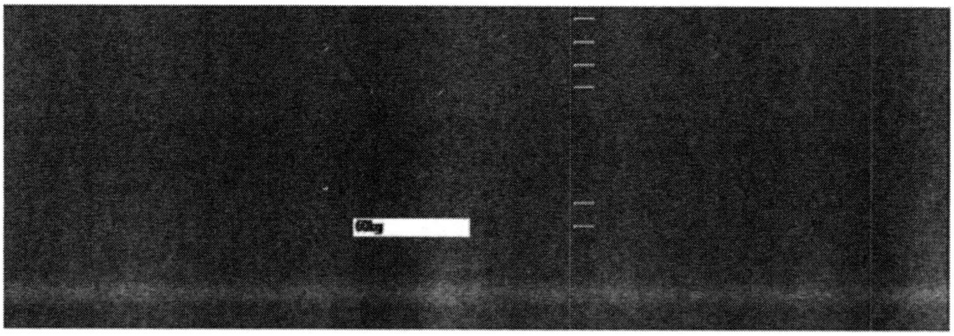

Figure 2. Simulation of Paper Sack Lifting Condition at Packaging Workstation

Based on the results that have been obtained will be conducted ergonomic evaluation of the existing work tools in the company so that when the use of tools of the proposal work can facilitate the operator in completing the work and can reduce the risk of MSDs.

2. Methods

2.1 Ergonomics

According to [1] states that ergonomics is a systematic branch of science to utilize information about the nature, abilities and limitations of humans in designing a work system so that people can live and work on the system well, that is to achieve the desired goal through the work, effectively, safely, healthy and comfortable.

2.2 Anthropometry

According to [2] Anthropometry comes from the word *Anthropos*, which means man, and *metricos*, which means measurement. In short, anthropometry is a science that deals with aspects of human physical size.

2.3 Formatting author affiliations

Ergonomic Functional Deployment (EFD) is the development of Quality Function Deployment (QFD) by adding new relationship between consumer desire and ergonomic aspects of the product [3][4]. This relationship will complement the shape of the house of quality matrix which also translates into desirable aspects of ergonomics.

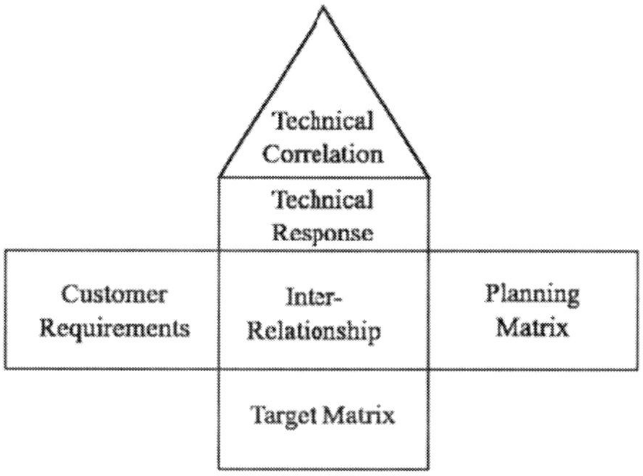

Figure 3. House of Ergonomics

The EFD approach is used to achieve EASNE principles of Effective, Safe, Healthy, Convenient and Efficient. With the achievement of the principle of EASNE, ergonomic product improvements can be achieved and can reduce the accident that occurs in the operator doing packaging activities.

2.4 Conceptual Model

In solving the problem, a framework is needed to solve the problem in the study which is poured into the following conceptual model:

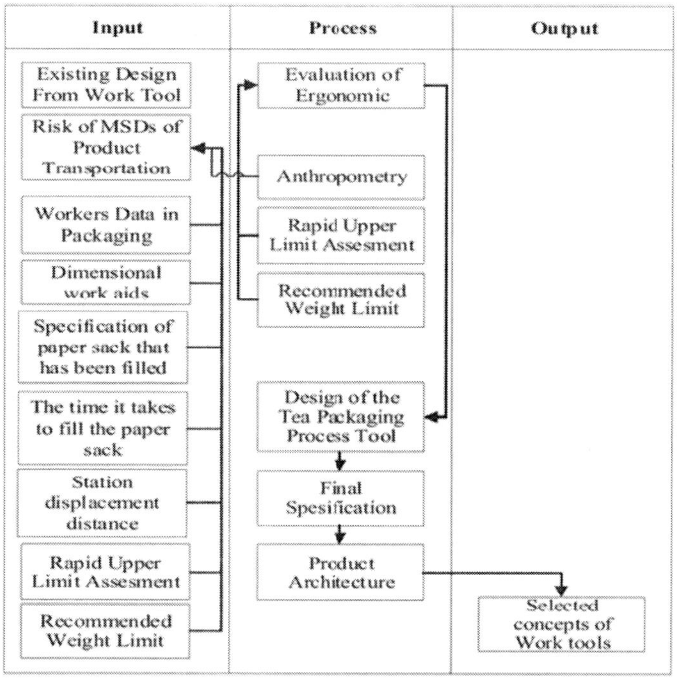

Figure 4. Conceptual Model

IOP Conf. Series: Materials Science and Engineering **528** (2019) 012012 doi:10.1088/1757-899X/528/1/012012

3. Result And Discussion

3.1 Ergonomic Evaluation of Existing Work Tool

The ergonomic evaluation is performed by determining the RULA score when the operator does the job of moving the paper sack for the next activity in Table 1 and knowing the recommended loads that the operator can carry out in doing the work that can be seen in Table 2.

Table 1 The Result Of RULA Workers For Existing Condition

Body Posture of Distribution Activity Paper Sack Without Work Tool										
Posture A						Posture B			Final Score	
Upper Arm		Forearm		Wrist and Arm		Neck	Trunk	Leg		
R	L	R	L	R	L				R	L
2	2	1	1	7	7	7	7	7	7	7

Based on Table 1, the results of RULA from the condition of the worker's exciting with the value 7 which means that the working conditions using the tool can cause high MSDs and must be made directly changes to the work performed by the operator.

Table 2. Recommended Weight Limit for Existing Condition

Object Weight		Hand Location (cm)				Vertical Distance (cm)	Asymetrix Angle		Frequency Rate	Duration	Object Coupling
		Origin		Destination			Origin	Destination	Lifts/min	(Hours)	
L (avg)	L (Max)	H	V	H	V	D	A	A	F		C
60	60	37.5	89	-	-	175	0	-	0,2	<1	Fair

Here is the recommended weight limit calculation (RWL) for lifting objects in the existing working tool condition used:

a. $RWL = LC \times HM \times VM \times DM \times AM \times FM \times CM$ (1)

$$= 23 \times \frac{25}{37.5} \times ((1\text{-}0{,}0032(89-75)) \times (0{,}82 + \frac{4.5}{175}) \times (1-(0.0032 \times 0)) \times 1 \times 1$$

$$= 23 \times 0.67 \times 0.95 \times 0.84 \times 1 \times 1 \times 1$$

$$= 12.31$$

b. $Lifting\ Index = \frac{Object\ Weight}{RWL}$ (2)

$$= \frac{60}{12.31} = 4.84$$

Based on the result of Recommended Weight Limit obtained value of 12.31 which means that the recommendation of the load that can be done by the operator with the existing condition should be 12.31 kg, because the burden of the existing condition of the value generated from the Lifting Index of 4.84 which means that the work is risky to do.

3.2 House Of Ergonomic

In making House of Ergonomic is one way to determination of ergonomic design by knowing the related relation between design specification and consumer requirement to work tool which will be proposed by researcher. HoE can be seen in Figure 5 [5].

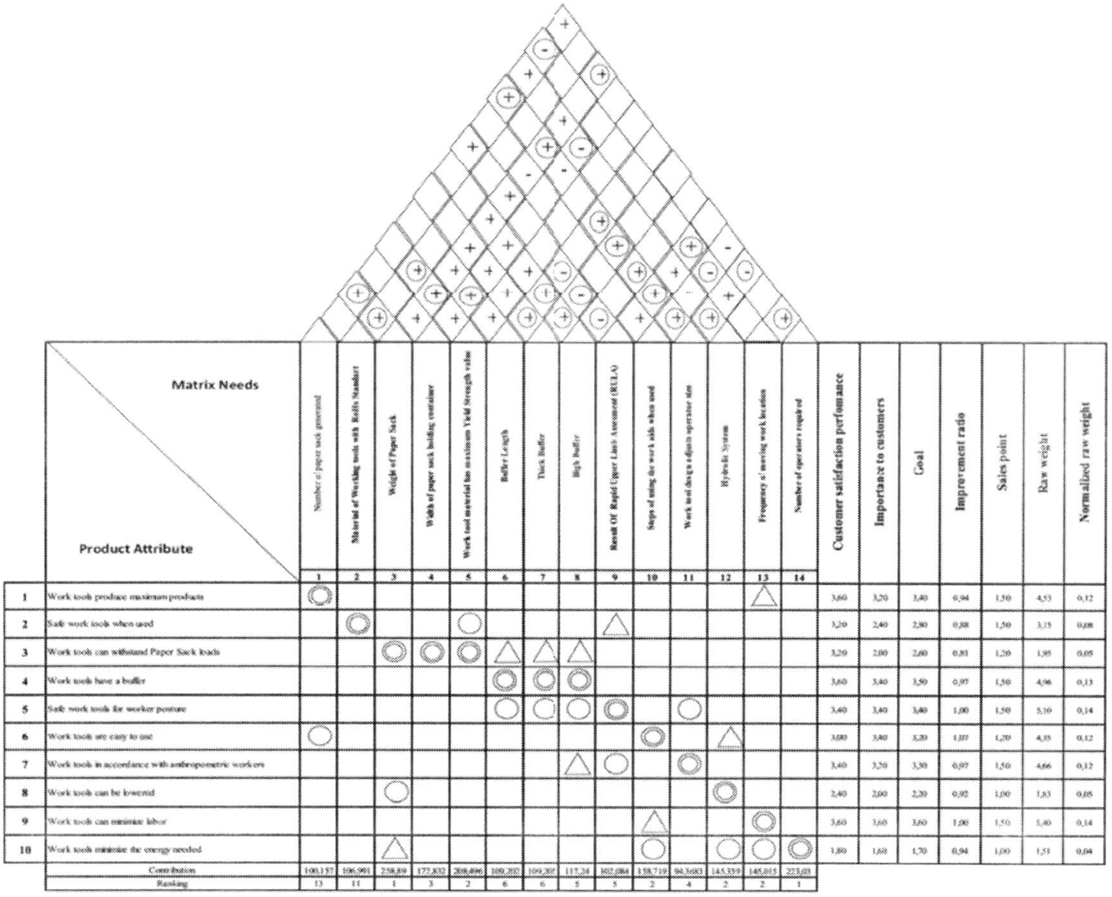

Figure 5. House of Ergonomic

Based on House of Ergonomic Figure can be specified final specifications for ergonomic devices. The final specifications are used to determine the size dimensions of the proposed packaging work aids, and the determination of the material to be used in the proposed work tool. The final specification table can be seen in Table 3.

Table 3. Final Specification of Proposed Work Tool

No	Matrix Needs	Part Spesificatiom	Target Value	Unit
1.	Number of paper sack generated	Number of paper sack generated	>10	*Sack*/hr
2.	Material of Working tools with RoHs Standart	Material of Working tools with RoHs Standart	-	List
3.	Weight of Paper Sack	OP	40	kg
		BOP	50	kg
		DUST	60	kg
4.	Width of paper sack holding container	Length Of Container	60	cm
		Width Of Container	20.5	cm
5.	Work tool material has maximum Yield Strength value	Work tool material has maximum Yield Strength value	-	List
6.	Buffer Length	Buffer Length	60	cm
7.	Thick Buffer	Thick Buffer	15	cm
8.	High Buffer	High Buffer	115	cm

IOP Conf. Series: Materials Science and Engineering **528** (2019) 012012 doi:10.1088/1757-899X/528/1/012012

Table 3. Final Specification of Proposed Work Tool (cont.)

No	Matrix Needs	Part Spesificatiom	Target Value	Unit
9.	Result of Rapid Upper Limb Assesment (RULA)	Result of Rapid Upper Limb Assesment (RULA)	4	Score
10.	Steps of using the work aids when used	Steps of using the work aids when used	4	Step
11.	Work tool design adjusts operator size	Work tool design adjusts operator size	-	List
12.	Hydrolic System	Hydrolic System	Yes	Binary
13.	Frequency of moving work location	Frequency of moving work location	1	Times
14.	Number of operators required	Number of operators required	1	People

Based on the final specification table we get the dimensions to design the ergonomic tools, and the material used is Nickel Alloy this is because the ergonomic tool requires a strong and durable material.

3.3 Analysis

In the analysis of the proposed tool aids were conducted to determine the condition of the operator doing his job using the better proposal work tool and did not give the effect of risk of Musculoskeletal Disorders (MSDs). The concept of the Ergonomic Proposed Working Tool can be seen in Figure 6.

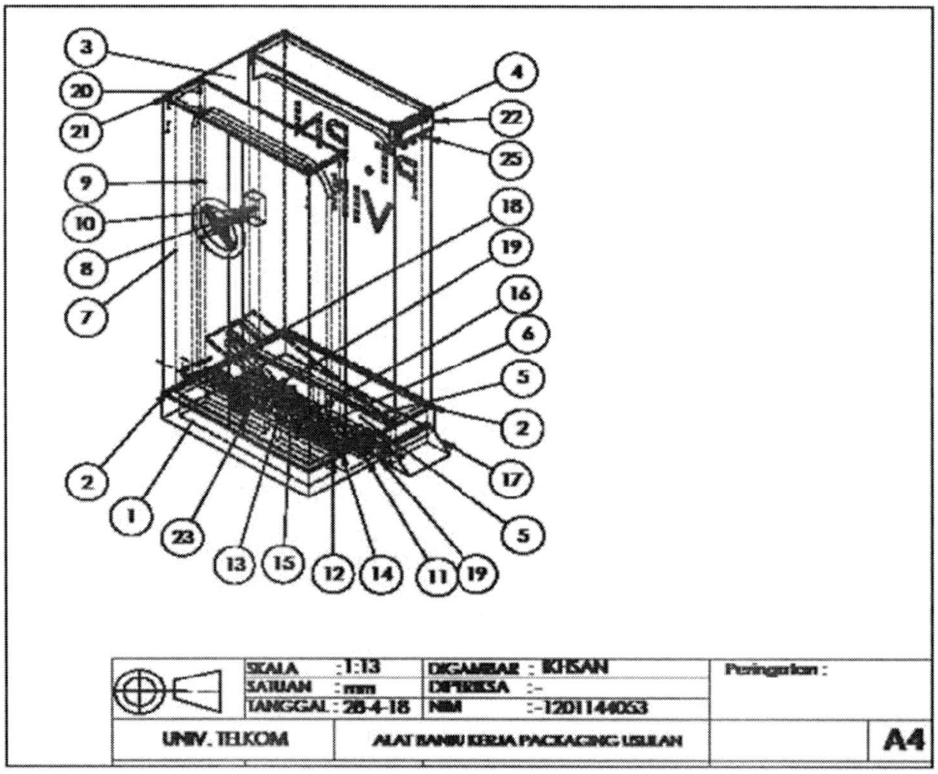

Figure 6. Concept of Proposed Work Tools

From the concept that has been designed to do work posture analysis using RULA analysis to know the condition of Working Tools Proposed when used by the operator so that it can know the relationship with the existing condition. Analysis of differences in posture collapse can be seen in Table 4.

Table 4. Comparative Analysis of RULA Existing Work Tools and Proposed Work Tools

	Existing Work Tool	Ergonomic Work Tool
Body Posture of Worker		
Score	7	4
Analysis	Risk activity to do, so it takes investigation and make changes directly	Activity may be risky to do, so further investigation is needed and changes may be possible

Based on Table 4 the comparison between the existing work tools and the proposed work tools that have been designed to have significant differences can be seen clearly when the existing condition to move the paper sack is done by lifting, while the working aid is not charged for lifting but just stepping on the pedal to lower the paper sack without lifting, so it can affect the decline in the value of RULA which can prevent the occurrence of Musculoskeletal Disorders. As for the paper sack transport activities are reduced, this is because the tools of the proposed work can complete all activities at one work location and not fully use human labor to operate or move paper sack to the next work site, because the worker only stepped on the pedal that was useful to lower the paper sack from the next work tool the worker just pulled the paper sack with a small load.

4. Conclusion

Based on the results of data processing and analysis obtained some conclusions to solve problems at the work station packaging to reduce the risk of Musculoskeletal Disorders, among others:

1. Analysis of the RULA score is seven which means that the value can cause high work fatigue so that it must be made changes directly on the work posture. With the existence of work tools packaging proposal that has been designed to make the work posture on the operator to be ergonomic, it can be marked by the decline in the value of RULA score into four so as to reduce the risk of MSDs
2. Work tools proposed for tea packaging can reduce processing time from packaging work because all activities will be carried out in one location without the transfer of work activities that can increase work time and cause MSDs risk when operators lift loads that should not be done repeatedly.
3. In the selection of materials for the design of the working tools of the packaging of tea selected Nickel Alloy material because the material can be said ergonomically when the usage time long enough so that can be defined that the material required is a strong material and durable.

5. References

[1] Sutalaksana I, Anggawisastra R and Tjakraatmaja J 2006 *Teknik Perancangan Sistem Kerja.* 2nd Ed. (Bandung: ITB)

[2] Iridiastadi H and Yassierli 2014 *Ergonomi Suatu Pengantar.* (Bandung: PT Remaja Rosdakarya)

[3] Ulrich K T and Eppinger S D 1995 *Perancangan dan Pengembangan Produk (Translate).* (Jakarta: Salemba Teknika)

[4] Surya R Z and Badruddin R 2015 Aplikasi Ergonomic Function Deployment (Efd) Pada Redesign Alat Parut Kelapa Untuk Ibu Rumah Tangga. *Jurnal Ilmiah dan Penerapan Keteknikan Pertanian.* **8**(1) https://doi.org/10.17969/rtp.v8i1.2687.

[5] Adrianto R, Desrianty A and M F H 2014 *Usulan Rancangan Tas Sepeda Trial Menggunakan Metode Ergonomic Function Deployment (EFD).* **02**(02) pp 353–363

IOP Conf. Series: Materials Science and Engineering **528** (2019) 012012 doi:10.1088/1757-899X/528/1/012012

[6] Boothroyd G, Dewhurst P and Knight W A 2011 *Product Design for Manufacture and Assembly.* (Florida: Taylor & Francis Group)
[7] Akin J 2009 *Finite Element Analysis Concepts via Solid Works.* (Texas: World Scientific).

Workplace Ergonomic Risk Assessment Toward Small-Scale Household Business

Dessi Mufti[1], Aidil Ikhsan and Tri Marta Putri

Industrial Engineering, Faculty of Industrial Technology, Bung Hatta University, Gajah Mada Street No 19, Gunung Pangilun, West Sumatera, Indonesia
[1]dessimufti@bunghatta.ac.id

Abstract. The purpose of this research is to assess work risk caused by non-ergonomic of both facility layout and workplace at some Small Scale Medium Enterprises (SMEs) at nagari Batu Taba, sub-district IV Angkek, West Sumatera. The research samples are traditional food industrial and convection. The research participants are those who have been worked for more than ten years. Amongst the grievances experienced by the worker are tiredness and musculoskeletal disorders. This evaluation adopts the ergonomic facility layout and workplace through using Workplace Ergonomic Risk Assessment (WERA). The assessment result by implementing WERA shows that the worker does experience risk with final score 30 at medium level, as for the worker who cut the clothes pattern is at high level, 45. Task is not accepted and immediately changes. It means that some recommendations to improve the workplace are of vital importance. The ergonomic design introduced in this research has potential to reduce the risk at some small scale industrial at nagari Batu Taba, sub-district IV Angkek.
Keywords: MSDs, WERA, Facility Design

1. Introduction

Recently, the health and safety issues in the workplace have been a must [14]. These are now become a priority and are not to be abandoned by small scale, household and big scale industries. In fact, nowadays, some household industries in Indonesia are often neglecting the safety issue in the workplace. Small scale enterprises played a crucial role in the industrial economics. Low modal intensity and high worker production capacity are among the benefits of this small scale enterprise. According to the ergonomic principle, some of the small scale enterprises do not consider the facility layout and workplace when designing the industrial workstation. The activity of production process takes place in the non-specific place. Most of these small scale enterprises do not separate the production room, raw material room and finished goods room. The call to consider ergonomic principle when designing industrial workstation may reduce the musculoskeletal disorders. It is also to prevent injury toward the industrial operator [13].

The basic philosophy in every business organization is to produce particular goods and to sell it in order to obtain profit as to satisfy the consumer requirement. The useful system of performance evaluation depends on the working environment [10]. The goal of every company is to obtain maximum result by employing all facilities they owned [6]. Hence, it is necessarily important for the company to employ the ergonomic facility to avoid the worker for experiencing risk as to achieve the maximum result.

A very ergonomic workplace is of vital importance for all kind of industries, including small scale, big scale and household business. Some of the workers do not aware about the workplace condition.

Sometimes, the workplace is not properly design so that it may cause accident or injury at work in which it may harm the worker's physical. The appearance of the health issue as mentioned above give rise to the workplace risk assessment toward all kind of industries. The aim is to create safe and comfort workplace. It is all about arranging anything related with workstation, including the equipment and the environment to be properly design in order to avoid injury or any other negative effect toward health. The layout and workstation design have a significant impact toward health for both long and short terms. The factor to be considered in forming the cell or group work is decision in designing human (worker) along with the working system//working facility [26]

The existence of uncomfortable feeling resulted from inappropriate layout of the workstation may lead to the work injury and others negative impact toward worker's health. For example, inappropriate layout can cause eye fatigue, neck stiff, back pain, arm pain and the like. Improved good facility layout can reduce the comprehensive risk level on the upper body and increase productivity by 194% compared to the existing layout [1]. Inappropriate chair level adjustment may contribute stagnancy of blood flow to the leg part. Inappropriate sitting posture can cause injury at the lower back. Identification implemented toward traditional food making by using manual working equipment can cause injury risk toward neck, shoulder, back, wrist and leg [16]. The repetitive work particularly can cause tendon shoulder disorders [5]. Workplace ergonomic assessment assures that work system (equipment, tool, workstation and policy) is properly design in order to improve worker safety and health and optimized work business including productivity, efficiency, quality and profitability [24]. The main idea related with workstation planning is to reduce material maintenance cost and this plays a crucial role in the operational safety and productivity [12]. Ergonomic is one of the main components of safety program. The company has started to implement ergonomic as an effective program in the workplace [7]. Nowadays, it is much better to implement ergonomic workplace rather than cost savings like infestation toward the worker, machine and method [13]. The implementation of the ergonomic is of vital importance toward the industry, hence ergonomic assessment have to be the initial step in the safety and health assessment process [7,8].

Review shows that at small scale industry in particular, almost there is no ergonomic principle implemented. Most of the task is manually conducted in this kind of industry [13]. The practice of safety management is not appropriately conducted in most of the small scale industry, particularly at the facility implementation used by small scale industry (SMEs) due to financial problem, lack of awareness, rejection toward change, and lack of worker training, those are the main problems ([25]. Both the internal and external factors influence the fluency of small and middle class business (UKM). The daily operation may risk the UKM in terms of production, safety, and the business itself [11]. One of the form of the safety is the use of aid tool or supporting tool employed during the production process

Small scale and Medium scale industries (SMEs) play a very dominant role in subdistrict Ampek Angkek. Industry and craft are the second superior sector after agriculture sector in sub-district Ampek Angkek. In these two sectors there are 39.6% people who work at convection, embroidery, overlay and *suji,* small scale industry (food) and others [4]. This research conducted in Nagari Batu Taba which is one of the country located in subdistrict IV Angkek. Small scale industry in Nagari Batu Taba is traditional food production: Sakura cake and convection.

One of the steps in the making of sakura cake is printing process of the batter made of flour, egg, butter or margarine and salt. This printing process consumes the biggest portion of making time comparing with other processes such as stirring, frying and caramelizing. The cake printing process is one of the process which requires worker to bend as the result of traditional equipment being used and sitting position which may cause injury risk toward the worker. The purpose of this research is to identify the worker posture/position during work hour at Kue Cangking small scale industry in order to provide solution to improve the worker posture/position, so that it may reduce the musculoskeletal disorder which is often experienced by operator at work. This research use WERA method. The result of the initial Nordic Body Map survey indicates pain at neck, shoulder, wrist, back and leg [16].

The making process of the clothes pattern still conducted manually and using inappropriate working facility. The worker position is sitting on the floor, bent back and folded legs. The disorders appeared due to inappropriate workplace and work facility. In many industry and service sectors, inappropriate working posture can cause problems to many parts of the body, particularly toward musculoskeletal disorders [2]. In order to avoid health and safety issue at workplace, appropriate method and well-designed equipment for the worker are among the solutions. For example by implementing risk estimation tools which designed especially for small scale and medium scale enterprises [15].

Musculoskeletal symptom is the main cause of the inefficient working time and cost expansion. Inappropriate body posture is the most important risk factor which indicates the musculoskeletal symptom existence related with work and workplace [3]. WERA provides a good indication toward musculoskeletal disorders related with the job which probably reported as pain, illness or discomfort at relevant parts of the body [17]. The operator who works for more than 8 hours on repetitive work type with leg position sustaining the operator's body tend to experience fatigue at hip, thigh and leg [12].. The experts and management team agree that WERA tool prototype is easy and fast to be used. This prototype can be used as workplace to variety of job/task and is precious in the workplace [19].

Observation is the most common one to be used in assessing musculoskeletal disorders related with work (WMSDs). The reason is it is cheap and practice to use in different kinds of workplace. However, there is no equipment which includes many physical risk factors in the workplace. Hence, a new observation technique is develop. This new observation technique is well known as WERA method [18]. As a tool, WERA is a pen and paper technique without necessarily using any special equipment. This method can be implemented in every room in the workplace without disturbing the worker. WERA includes many physical risk factor such as posture, repetition, strength, vibration, contact stress and task duration and involving five main body parts that has to be assessed (shoulder, wrist, back, neck and leg) [18].

Musculoskeletal prevalent symptoms are high at neck, hip and activity at hand. The best strategies to reduce musculoskeletal symptom are depend on the activity, appropriate working method, appropriate seat and ergonomic tool [3].

2. Methodology

This research focus on small scale industry about traditional food making and convection at Nagari Batu Taba. The observed workers have been working for more than 10 years with repetitive working pattern. This is a direct observation. The researcher records the working process and interview the worker. The main focus at this primary data collection is to know the production process of sakura cake and the arrangement of fabric that will be turn into pattern. The making of sakura cake takes time until 10 hours per day. Meanwhile, it took 5 hours to move the pattern onto the fabric. These activities cause fatigue and give rise to work risk among the worker. Ergonomic static posture analysis conducted by capture the production process picture using camera. The worker body posture and the movement of the other body parts captured and analyzed by using WERA method. WERA assessment signifies good working indication of musculoskeletal disorders toward pain and discomfort relevant parts of the body [18] No 6, particularly the musculoskeletal disorders related with workplace, especially in the printing making industry [21].

WERA covers the risk of 6 physical factors including posture, repetition, strength, vibration, contact stress and task duration. WERA also involve 5 main parts of the body (shoulder, wrist, back and leg). As a tool, WERA is a pen and paper technique without necessarily using any special equipment. This method can be implemented in every room in the workplace without disturbing the worker. It has assessment system and the act level which provide guidance about risk level and requires act to do a detail assessment [19].

2.1. Existing Condition

In the initial conditions of workers in the household business do not pay attention to the work position that can cause the risk of injury as shown in Figure 1.

Figure 1. Leg position which causes fatigue and numb

3. Result and Discussion

The worker conducts the food making process and the cut process of the pattern right on the floor. These cause the worker's leg to experience pain after working. During working, the worker tends to choose not to lean against the wall but bent to the front. Table 1 shows the possible risk of injury to each part of the body caused by work position and does not pay attention to ergonomic devices.

Table 1. Worker Risk Factor (Traditional food making)

No	Risk Factor	Worker position picture	Explanation	Score
				R
1	Shoulder/ neck		The shoulder is moderate bent up with hand position closer to the chest. Hence the score is medium. Next, the repetition does exist. However, pause position during working does exist as well, hence the score is light.	3
2	Wrist		Wrists are moderate bent up or bent down at about $>15^0$ at the cake making process. The position at printing process is by pressing it. This position is at the moderate working risk position. The repetition does exist in this kind of position at about 11-20 times/minutes. Therefore this position is categorized as moderate.	4
3	Back		The back is moderate bent to the front between $0\text{-}20^0$ and 20^0 -60^0. This position last for 4-8 minutes. Hence, this position is categorized as moderate.	4
4	Neck		The neck is low bent between $0\text{-}20^0$. In one printing, worker may rest for 60 seconds. Hence this position is categorized as low.	4

IOP Conf. Series: Materials Science and Engineering **528** (2019) 012013 doi:10.1088/1757-899X/528/1/012013

Table 1. Worker Risk Factor (Traditional food making) (cont.)

No	Risk Factor	Worker position picture	Explanation	Score R
5	Leg		The next position is leg. The worker's legs are extremely bent forward or sitting with feet do not touch floor. This position last for 4 minutes. This position is categorized as high working risk.	5
6	Forceful	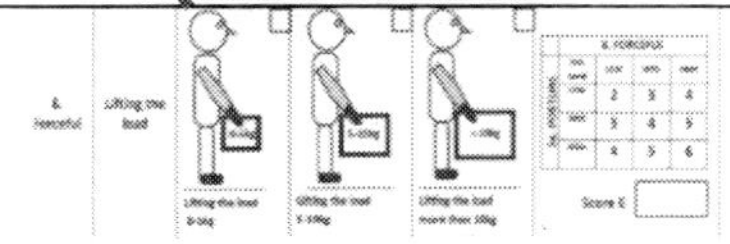	The forceful experienced by the worker during one time production process is at about 0-5 kg. At this position the worker is categorized as low working risk.	2
7	Vibration	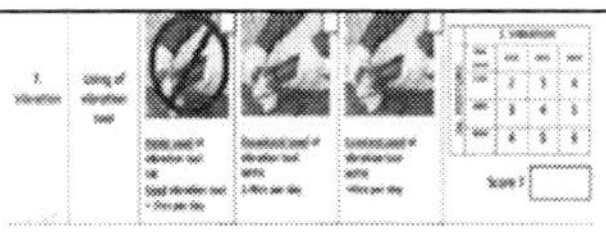	There is no vibration caused by the tool since the worker never uses the vibration tool.	2
8	Stress Contact	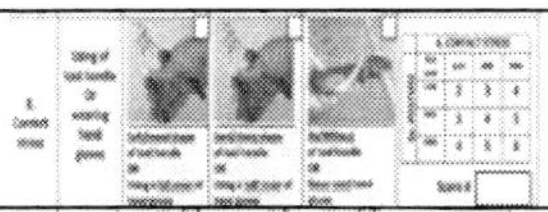	Stress contact caused by the conditions of workplace environment such as: small, dirty and messy.	2
9	Task Duration	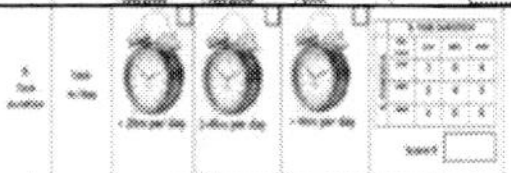	In the production process of dry cake making, worker only work 3-6 hours per day.	4

Table 2. Worker Risk Factor (Fabric pattern cut)

No	Risk Factor	Worker position picture	Explanation	Score R
1	Shoulder/ neck		The shoulder is moderate bent up with hand position closer to the chest. Hence the score is medium. Next, the repetition does exist. However, pause position during working does exist as well, hence the score is moderate.	4
2	Wrist		Wrists are moderate bent up or bent down at about $>15^0$ at the cake making process. The position at printing process is by pressing it. This position is at the moderate working risk position. The repetition does exist in this kind of position at about 20 times per minute.	6

Table 2. Worker Risk Factor (Fabric pattern cut) (cont.)

No	Risk Factor	Worker position picture	Explanation	Score R
3	Back		The back is moderate bent to the front between 0-20^0 and 20^0 -60^0. This position last for 9-12 minutes. Hence, this position is categorized as high.	6
4	Neck		The neck is extremely bent forward at about >20^0, and movement with some pauses. Hence, this position is categorized as moderate.	5
5	Leg		The next position is leg. The worker's legs are extremely bent forward at about >60^0. This position last for one hour. This position is categorized as high working risk.	6
6	Forceful	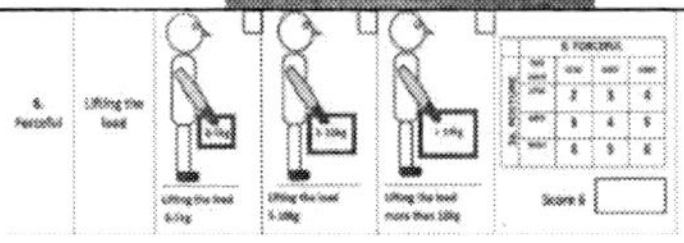	The forceful experienced by the worker during one time production process is at about 0-5 kg. At this position the worker is categorized as low working risk.	2
7	Vibration	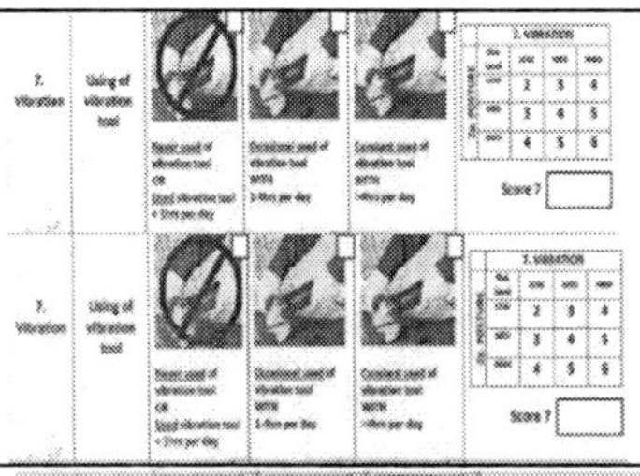	Occasional used of vibration tools with 1-4 hours per day and posture is categorized as high.	5
8	Stress Contact		Soft/sharp shape of tool handle with posture categorized as high.	5
9	Task Duration	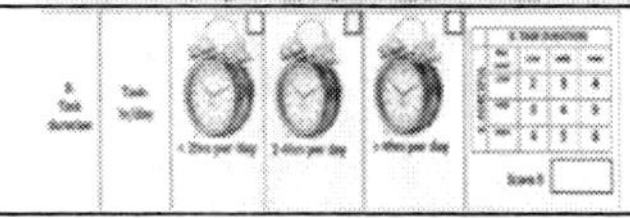	The task last for more than 4 hours per day with forceful categorized as high.	6

The scores above show that the risk experienced by the worker with final score 30 are at medium level. Hence, working posture improvement shall be conducted immediately, so the worker will experience no more grievances toward body posture in conducting continuous work in a long term which can cause damage toward the worker's body. The common working position performed by the printing worker is sitting. Most the workers experience grievance toward shoulder, neck, back, wrist due to inappropriate working facility. The leg position where one leg is bent/fold and the other leg is straight in the printing process should be evaluated. The risk experienced including paralyzed and painful up to the waist particularly when it is about to stand up.

Meanwhile the score obtained by the worker who works to cut and make the cloth pattern is 45. This score is at high level and action task is not accepted, immediately change. It is due to the sitting position where two legs are fold and the position of the back is extremely bent when cutting the fabric. The worker legs position are extremely bent forward as >60^0. This position last for 1 hour, hence it is a high risk work. The worker back position is bent forward between 0-20^0 and 20^0-60^0. Next, the worker neck position is >20^0, neck is extremely bent forward and movement with some pauses/moderate.

The result of WERA assessment would be used to assess developed work ergonomic risk to provide the method of fast filtering task on physical risk factor exposure in relation with musculoskeletal disorders. WERA sees object directly and is able to measure the physical risk factor related with the WMSDs in the workplace including body posture, repetition, weight lift, vibration, contact voltage and task duration [20]. In the cake printing process, WERA provides score toward the development of pain discomfort feeling felt at shoulder, wrist, back, neck and leg. This demonstrates that WERA assessment provides a good indication about musculoskeletal disorders related with the work that reported as grievance, illness, or discomfort in the relevant part of body. Following are the final scores of WERA body parts that have been analyzed in the cake making and cutting the fabric pattern:

Table 3. WERA Body Parts Final Score

No	Item WERA	Score		No	Item WERA	Score	
		Cake making	Cutting the fabric			Cake making	Cutting the fabric
1	Shoulder	3	4	6	Forceful	2	2
2	Wrist	4	6	7	Vibration	2	5
3	Back	4	6	8	Contact Stress	2	5
4	Neck	3	5	9	Task Duration	4	6
5	Leg	6	6		Total	30	45

The Head of Industry, Agam's UMKM Industry and Trade Cooperative Office, Ediwar in Lubuk Basung, said on Monday that 5,850 SMIs consisted of 2,297 food or food industries, 2,635 clothing or clothing industries. "This industry is spread in 16 districts with a workforce of 24,975 people. In order for the early industry to remain active, it will provide guidance and provide training on how to produce good products and have selling points.

In principle, the attention is not only to the quality of the product, but to how the equipment used by workers does not pose a risk of injury to their limbs. Research conducted on SMEs shows that workers do not pay attention to the risk of injury caused by the use of non-ergonomic work tools.

4. Conclusion

The purpose of working posture assessment is to provide act level toward worker body posture during activity. Act level is obtained by filling the NBM questionnaire which later processed by using WERA measurement. Score determination is a stage to know the risk level of analyzed factor, so that the appropriate act can be discovered in order to reduce the risk factor.

In the cake making process and the cut of fabric pattern, the worker conducts the activity on the floor. The worker has to bend and the two legs are extremely bent. These may cause injury. The worker would suffer pain after working. The scores obtained shows that the risk experienced by the worker with final score 30 are at medium level. Hence, immediate working posture improvement is of vital importance. The purpose is to avoid worker grievance on the body part in conducting a continuous work in a long term which can cause disorders to the worker's body part.

The common work position performed by the printing worker is sitting. Most of the workers experience grievance toward shoulder, neck, back, wrist due to the inappropriate working facility. Like the traditional print tools made of coconut shells and workplace facilities there are on the floor or on the table without thinking of the risks that will occur if this continues to continue. The legs position in this printing process is indeed to be considered in which one leg is folded and the other one is straight. The possible risks in relation to these legs position are paralyzed and painful up to the waist when it is about to stand up.

Meanwhile the score obtained by the worker who works to cut and make the cloth pattern is 45. This score is at high level and action task is not accepted, immediately change. The high score was found at leg, back, and neck positions. Both the workplace and these working position are really risky and dangerous toward the workers.

Hence, improvement may take place toward the facility such as the workbench/desk. This desk has to be adjusted with the body posture of the worker. The height of this desk has to be adjusted with the elbows. Meanwhile, the width and the length of the desk has to be in line with the hand position. This design is to facilitate the worker to conduct the operation of the working activities. Moreover, this design is also to reduce the grievance particularly toward neck, leg, shoulder, wrist and back which is caused by the non-ergonomic position during working.

The printing worker sits for all the time in conducting the job. Most of the worker experience grievance or pain at shoulder, neck, back, wrist and leg since the worker still has to deal with inappropriate facility. For example, the use of traditional printing tool which is made of coconut shell, working on the floor or on the desk without aware about the long term effect.

The 2 SMEs taken in this research show that the workplace, the work facility and the worker work risk caused by tend to be neglected. These used to be considered as common and generally accepted. Hopefully with this paper, researchers, SME, and government can pay more attention to the aspect of ergonomics and health to the perpetrators in this industry. It is expected that in the future research, the practitioners of the SMEs would be able to determine the working productivity of their companies. However, the performance indicator of the SMEs should be first determined since the work indicator of the SMEs is slightly different from the big companies. Furthermore, this research can also be developed to examine the use of work facilities and their impact on the overall worker's limbs.

Acknowledgement

This research is supported by the Ministry of Research and Technology of Higher Education (Kemenristekdikti) in the scheme of research on applied products academic year II 2018.

5. References

[1] Abad J D 2018 Ergonomics and simulation-based approach in improving facility layout. *Journal of Industrial Engineering International* (Springer Berlin Heidelberg)

[2] Abdulrahman B 2015 Ivestigation of work-relataed Musculoskeletal Disordess (MSDs) in warehouse workers in Saudi Arabia. *Proc. 6th International Conference on Apllied Human Factors (AHFE 2015) and the Affiliated Conferences* (Saudi Arabia)

[3] Aghilinejad M, Ehsani A A, Talebi A, Koohpayehzadeh J and Dehghan N 2016 Ergonomic Risk Factors and Musculoskeletal Symptoms In Surgeons With Three Types Of Surgery: Open, Laparoscopic and Microsurger. *Medical Journal of the Islamic Republic of Iran (MJIRI)* (Iran: University of Medical Sciences)

[4] Biro Pusat Statistik 2017 *Ampek Angkek Subdistrict in Figures 2017* (Agam: Biro Pusat Statistik Kab Agam)

[5] Bernard B P 1997 *Musculoskeletal disorders and workplace factors. A critical review of epidemiologic evidence for workrelated musculoskeletal disorders of the neck, upper extremity, and low back.* (Cincinnati, OH: National Institute for Occupational Safety and Health-NIOSH)

[6] Bagherinejad J and Niknam A 2017 Solving the competitive facility location problem considering the reactions of competitor with a hybrid algorithm including Tabu Search and exact method. *Journal of Industrial Engineering International* (Springer Berlin Heidelberg)
[7] Brodie D M 2008 Ergonomics Risk Assessment: Determining When, Why, Who and How You Should Perform One. *Occupational Hazard - Web E Blast.*
[8] Burdorf A 2010 The role of assessment of biomechanical exposure at the workplace in the prevention of musculoskeletal disorders. *Scand J Work Environ Health* **36**(1) pp 1-2
[9] Cherkos T, Zegeye M, Tilahun S and Avvari M 2017 Examining significant factors in micro and small enterprises performance: case study in Amhara region, Ethiopia. *Journal of Industrial Engineering International* (Springer Berlin Heidelberg)
[10] Farooquie P, Gani A, Zuberi A K and Hashmi I 2012 An empirical study of innovation-performance linkage in the paper industry. *Journal of Industrial Engineering International* (Springer Berlin Heidelberg)
[11] Islam A and Tedford D 2012 Risk determinants of small and medium-sized manufacturing enterprises (SMEs) - an exploratory study in New Zealand. *Journal of Industrial Engineering International* (Springer Berlin Heidelberg)
[12] Kazerouni A M, Agard B and Chinniah Y 2012 The Application of Risk Assessment to Facilities Planning: A Synthesis of Risk Assessment Methods and Layout Design Models. (Canada: Department of Mathematical and **SIAS** Industrial Engineering, École Polytechnique de Montréal)
[13] Mali S C and VyavahareMaharastra R T 2015 An Ergonomic Evaluation of an Industrial Workstation: A Review. *International Journal of Current Engineering and Technology Research* **5** pp 1820-2826
[14] Sout M, Norhidayah M A M and Ismail A R 2015 Workplace Ergonomics Risk Assessment In The Mining Industry. *jurnal teniat* jun 2015_29122015
[15] Nyirenda V, Chinniah Y and Agard B 2015 Identifying Key Factors for an Occupational Health and Safety Risk estimation Tool in Small and Medium-size Enterprises. *IFAC Papers online* **48**(3) pp 541-546
[16] Putri T M, Mufti D and Suryani E 2107 Analisis Risiko *Musculoskeletal Disorders* Pada Pekerja Pembuatan Kue Cangking Dengan Metoda Wera. *Workplace Ergonomic Risk Assessment* (Padang: Universitas Bung Hatta)
[17] Rahman M N A, Abdul M N, Rani M R A and Rohani J M 2012 WERA Tool for Assessing Exposure Risk Factors of Work Related Musculoskeletal Disorders - A Reliability and Validity Study. *Proceedings of the 2012 International Conference on Industrial Engineering and Operations Management Istanbul* July 3 – 6 (Turkey)
[18] Rahman M N A, Rani M R A and Rohani J M 2011 Wera: An Observational Tool Develop To Investigate The Physical Risk Factor Associated With Wmsds. *J. Hum. Ergol.* **40** pp 19-36
[19] Rahman M N A, Rani M R A and Rohani J M 2011 Investigation of the Physical Risk Factor in Wall Plastering Job using WERA Method. *Proceeding of international multiconference of engineers and computers 2011* (Hongkong: IMECS) **II**(March 16 – 18)
[20] Rahman M N A 2014 *Development Of An Ergonomic Risk Assessment Tool For Work Postures.* A thesis submitted in fulfilment of the requirements for the award of the degree of Doctor of Philosophy (Malaysia: Faculty of Mechanical Engineering Universiti Teknologi Malaysia)
[21] Rahman M N A, Hui U, Haql R A H, Hassan M, ArifiuZ A M T, Yunos M Y and Adzil Z S2015 Musculoskeletal Discomfort Among Workers In Mould Making Manufacturing Industry. *ARPN Journal of Engineering and Applied Sciences* **10**(15)
[22] Budiprasetya R, Sugiono A, Yanuar R and Analiof E 2016 Musculoskeletal Disorders on Wiring Harness Work Station with WERA (workplace Ergonomic Risk Assesment. *Jurnal Rekayasa dan Manajemen Sistem Industri* **4**(10) (Malang: Teknik Industri Universitas Brawijaya)

IOP Conf. Series: Materials Science and Engineering **528** (2019) 012013 doi:10.1088/1757-899X/528/1/012013

[23] Takala E P, Pehkonen I, Forsman M, Hansson G A and Mathiassen S E 2010 Systematic evaluation of observational methods assessing biomechanical exposures at work. *Scand J Work Environ Health* **36**(1) pp 3-24

[24] Tompa E, Dolinschi R, Oliveira C D, Amick B C and Irvin E 2009 A Systematic Review Of Workplace Ergonomic Interventions With Economic Analyses. Accesed: November 5, 2009. *J Occup Rehabil* **20** pp 220–234

[25] Unnikrishnan S, Iqbal R, Singh A and Nimkar I M 2015 *Safety Management Practices in Small and Medium Enterprises in India, Industrial Safety and Environmental Management Group.* (Mumbai, India: National Institute of Industrial Engineering-NITIE) pp 46–55

[26] Vafaeinezhad M, Kia R and Shahnazari-Shahrezaei P 2015 Robust optimization of a mathematical model to design a dynamic cell formation problem considering labor utilization, *Journal of Industrial Engineering International* (Berlin: Springer Berlin Heidelberg)

Design of The Kansei Board Game to Motivate the Elementary School Student in Learning English

Selna Shalawati[1] and Hartomo Soewardi[2]

Industrial Engineering Department, Faculty of Industrial Technology
Islamic University of Indonesia Yogyakarta Indonesia
selna.shalawati@gmail.com[1] , hartomo@uii.ac.id[2]

Abstract. English is one of the widely used international languages for communication among the nations. Therefore people needs to master the language both in speaking as well as writing and reading. In Indonesia, the mastery of this foreign language is still poor. Thus it is imperative to teach students this language since the elementary school. But, at this time the existing methods of teaching and learning process in the school has not been able to produce the expected quality. Even many students are not interested to learn it. 64% of the students score are below standard, and 67% of the students are not motivated to study hard. The objective of this study is to develop the Kansei Board Game to improve student's motivation in learning English effectively. Kansei Engineering Type 1 method is used to determine design specification based on user's feelings and emotions. A survey was conducted to identify the words of Kansei, while a statistical analysis is used to validate the hypotheses of the proposed design. The study showed that the method for teaching and learning process by using the Kansei board games is valid to meet user criteria and can increase motivation for students in learning English.
Keywords: kansei engineering, board game, learning process, motivation.

1. Introduction

English is an international language used by various nations for ease of communication worldwide weather formally or informally in all walks of life [1]. In English there are four aspects of skills to possess by the students such as listening, speaking, reading, and writing. Listening and reading concerns with the process of receiving, whereas, writing and speaking deals with a production process, or another word, output [2]. This reality encourages people to excel at speaking, writing listening and reading in English.

Based on the regulation of the Minister of National Education no 22/2006, the objectives of teaching are: (1) Developing communication competence in limited oral form to accompany action in the school. (2) Having awareness of the nature and language to improve the nation's competitiveness in the global community [3]. Thus, it is very important to master English for Indonesian people through learning process since early stage.

However, currently in Indonesia, English is quite unpopular because many people have no English proficiency Based on the preliminary studies, it was found that 64% of students 'grade are still below standard of national grade score and 67% of students are not motivated to study hard. This fact shows that motivation of students in learning English is low.

[4] Revealed that children in elementary school prefer picture, fairy tales, and game to other media

for learning English. But a previous study also discloses that the most interesting learning method is a game [5]. Meanwhile [6] explained that game is very helpful to motivate students in expressing their opinion with English and to be more active, autonomous, and energetic. Furthermore [6] advised that games should have some characteristics such as fun, easy to use, attractive, encouraging, and easy to practice. Thus board games are the most popular tools for students to learn English effectively and efficiently as what [7] developed on the board game to increase speaking ability.

The objective of this study is to develop the Kansei board game to motivate the elementary student to learn English effectively.

2. Research Method

2.1 Survey

The survey was conducted by distributing questionnaires to 30 elementary school students of between 10- up to 12-year-old. They consist of 19 males and 11 females. This survey aims to identify the Kansei words about the attributes of board game that they are looking for.

2.2 Apparatus

Some tools are used to support the study. They are:

1. Questionnaire was developed into three parts. First part is to identification of Kansei words. Second is to determine the physical design specification. The last part is to validate the proposed design.
2. Anthropometer. This tool is to measure the body dimensions.
3. IBM SPSS software version 22 is used to process statistical data.
4. Solid work Software is used for designing a virtual prototype of the board game.

2.3 Kansei Engineering Method

Kansei Engineering (KE) is used as a Method to determine some specification based on user's feelings and emotions and to incorporate them into product design [8].

In this study, we used Kansei Engineering Type 1. This method solves product problems into more detailed concepts to several levels, and is defined into the physical design of product (see figure 1) [8].

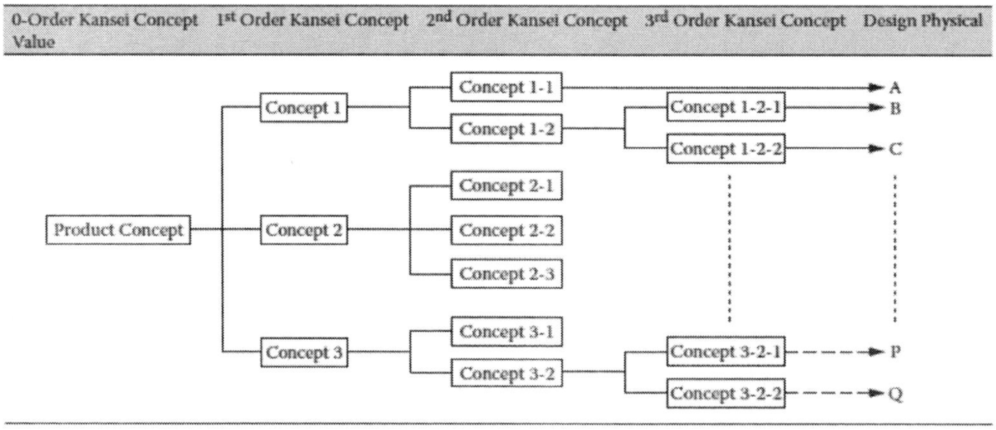

Figure 1. Conceptual map of Kansei Engineering Method Type I.

Steps KE type 1 as follow:

1. Identification of target.
2. Determination of Product Concept.
3. Breaking Down the Product Concept.

4. Deployment to Physical Design Characteristics.
5. Translation to Technical Specifications

2.4 Anthropometry

Anthropometry is the study of the dimensions and certain other physical characteristics of the human body. Anthropometry measurements are a critical element use in equipment and workplace/ work-space in their design [9].

The determination of anthropometric dimension in this study was done when Kansei process number 2 (collecting the word Kansei) was completed. Thus, designing the board game on the grip section for the board game in this study required body dimensions such as FW (finger width) and PW (palm width).

To achieve an appropriate fit between the operator and the design it requires steps in anthropometric processing as follow [9]:

1. Define the equipment's potential user population.
2. Choose the proportion of the population to be accommodated by the design.
3. Determine the body dimensions important in the design.
4. Determine the percentile values of the dimensions.
5. Determine the type of clothing and personal equipment worn by the users.

2.5 Statistic analysis

Non-parametric statistical analysis was implemented in this study. Cronbach Alpha was used to test reliability [10] while Spearman's correlation was employed to test the validity of Kansei words [11].

Furthermore, to calculate the normality in the body dimensions resulted from anthropometry Kolmogorov - Smirnov was used to know the data required and the dimensions of normal data so that it can be used.

Test of Marginal homogeneity analysis was used to validate Kansei word with board game suggestion design. Descriptive non parametric statistical evaluates usability and motivation of students [11].

3. Results and Discussion

3.1. Result of Survey

Based on the survey that was conducted in the preliminary study, it lead to 7 Kansei words that presented consumers voices. It was tested by using fault tolerance of 5% as mentioned in table 1.

Table 1. Kansei words need to break down.

Selected Kansei Words	Description	Validity Score	Reliability Score
Pull	Interesting shape.	0.640	0.805
Unique	Fiture offered that differentiates from previous product.	0.851	0.798
No Noise	Produce noisy sound when heard (dB).	0.728	0.797
Comfortable	Provides comfort when used, in model form.	0.851	0.797
Easy to carry	Models have a mild impression.	0.869	0.865
Easy to clean up	Easy to clean up after playing	0.654	0.806
Strong	Using quality materials	0.851	0.792

Appointment on r- table. Data are valid if r calculation > 0.361 and data are reliable if Cronbach's Aplha > 0.6.

3.2. Result of Mapping Process for Design

Table 2. Pull Mapping Concept.

Kansei Word	Sub Concept Level 1	Sub Concept Level 2	Sub Concept Level 3	Sub Concept Level 4	Design Specification
Pull	Have design Motif	Classic Motif	Wood	Button Part, Body case, drawer part	Classic Motif for Button Part, Body case, drawer part  
	Have design color	Classic color	Wood	Button Part, Body case, drawer part	Classic color for Button Part, Body case, drawer part.
	Have design shape	Body Case	Box	Size 30 cm x 17 cm	Bodycase shaped box size 30 cm x 17 cm. 

Table 3. Unique Mapping Concept.

Kansei Word	Sub Concept Level 1	Sub Concept Level 2	Sub Concept Level 3	Sub Concept Level 4	Design Specification
Unique	There is a place to make a sound	Speaker	Type of sound	Mezzo-soprano (Tone voice medium)	Mezzo-soprano (Tone voice medium). [12].

Table 4. No Noise Mapping Concept.

Kansei Word	Sub Concept Level 1	Sub Concept Level 2	Sub Concept Level 3	Sub Concept Level 4	Design Specification
No Noise	Decibels (medium)				1000 Hz (30 dB). [12].

IOP Conf. Series: Materials Science and Engineering **528** (2019) 012014 doi:10.1088/1757-899X/528/1/012014

Table 5. Comfortable Mapping Concept.

Kansei Word	Sub Concept Level 1	Sub Concept Level 2	Sub Concept Level 3	Sub Concept Level 4	Design Specification
	The latest Boardgame buttons are easy to use.	Easy button pressed.	Shape	Round	Button round shape diameter size 2.2 cm and thickness 3 mm. 
Comfortable	The latest Board game posts easy to read.	Font type	Times New Roman		Type Times New Roman. 
		Font Size	10		Font Size 10
	Handle design.	Shape	Box	Size Length 9 cm, Width 2 cm	Design the handle of the box shape Size Length 9 cm, Width 2 cm. 

Table 6. Easy to carry Mapping Concept.

Kansei Word	Sub Concept Level 1	Sub Concept Level 2	Sub Concept Level 3	Sub Concept Level 4	Design Specification
Easy to carry	Weight Body Case	Medium weight	500 gram - 1 Kg		500 gram - 1 Kg.
	Shape Board game	Box			Board game shaped box. 

Table 7. Easy to clean up Mapping Concept,

Kansei Word	Sub Concept Level 1	Sub Concept Level 2	Sub Concept Level 3	Sub Concept Level 4	Design Specification
Easy to clean up	How to Spruce Up	The latest board game can be folded.	Lots of folds	Fold 2	Board game folded. 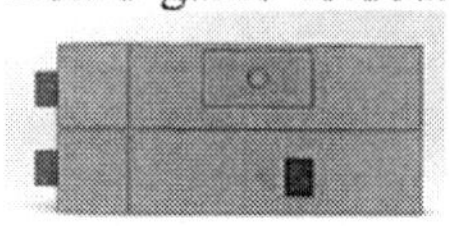
		The latest board game is a place to store dice (drawer)	Drawer	Easy to open and close	Size 5 x 5 cm (drawer) box shape. 

Table 8. Strong Mapping Concept.

Kansei Word	Sub Concept Level 1	Sub Concept Level 2	Sub Concept Level 3	Sub Concept Level 4	Design Specification
Strong	Body Case	Body case made of Wood material.	Plywood teak	Thick 3 mm	Body case made of plywood material with a thickness of 3 mm. 
	Button	Button made of Wood material	Plywood teak	Thick 3 mm	Buttons made of plywood material with a thickness of 3 mm. 
	Handle	The handle is made of foam material	Foam PC	Thick 2 mm	The handle is made of PC foam material with a thickness of 2 mm.  

3.3. Result of the Specification Design

1. Design specification for Pull design of board game are: having a Classic Motif and Classic color for Button Part, Body case, drawer part, Body case shaped box size 30 cm x 17 cm.
2. Design specifications for unique design are; having sound type Mezzo-soprano (Tone voice medium).
3. Design specifications for No noise design are 1000 Hz (30 dB).
4. Design specification for comfortable design are Button round shape size 2.2 cm and thickness 3 mm, Type Times New Roman Font size 10, Design the handle of the box shape Size Length 9 cm, Width 2 cm.

5. Design specification for Easy to carry design are Weight Body Case 500 gram - 1 Kg, board game shaped box. Design specifications for Easy to clean up design are Board game folded the two, Size drawer 5 x 5 cm box shapes.
6. Design specification for Strong design are Body case and Buttons made of plywood material with a thickness of 3 mm, The handle is made of PC foam material with a thickness of 2 mm.

3.4. Virtual Concept Design
The last step is constructing board game virtually, for overall design is presented on figure 2.

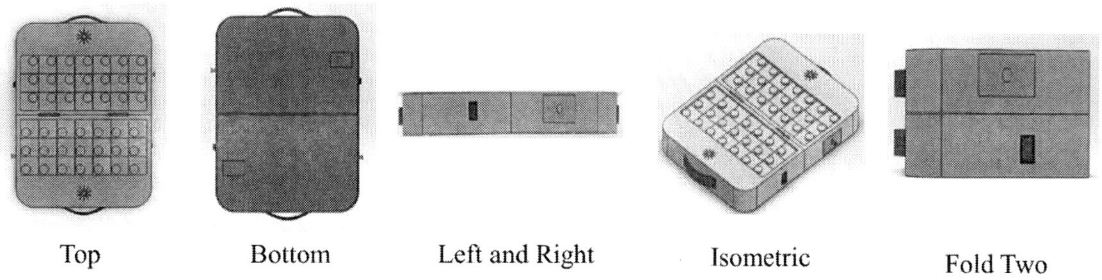

Figure 2. Board game.

3.5. Result of Statistical Analysis
Test of Marginal homogeneity was implemented to validate board game. The test was on hypotheses whether there was a significant difference between user's requirements and board game design. The hypotheses were developed as follow:
H0: There is no a significant difference between the user's requirements and board game design
H1: There is a significant difference between the user's requirements and board game design.
As shown at table 9. The z values ranged from 0.201 to 0.670. Since $z > 0.05$ the null hypothesis can be accepted. This means that the board game design meets user's requirements.

Table 9. Stuart Maxwell test of Marginal Homogeneity results

Kansei Word	***Marginal homogenity***
Pull	0,456
Unique	0,201
No Noise	0,670
Comfortable	0,257
Easy to carry	0,577
Easy to clean up	0,593
Strong	0,435

As shown at table 9 the z values ranged from 0.201 to 0.670. Since $z > 0.05$ the null hypothesis can be accepted. This means that the board game design meets user's requirements.

4. Conclusion
Based on the research, it is conclusive that:
1. Kansei words' characteristics that user require from board game design are Pull, Unique, No Noise, Comfortable, Easy to Carry, Easy to clean up, Strong.
2. The physical specification board game is body case have box shape with the size 30cm x 17cm. Button round shape size 2.2 cm and thickness 3 mm, Type Times New Roman Font size 10, Design the handle of the box shape Size Length 9 cm, Width 2 cm. Board game folded the two,

Size drawer 5 x 5 cm box shapes. Body case and Buttons made of plywood material with a thickness of 3 mm, The handle is made of PC foam material with a thickness of 2 mm.

3. The board game design meets user's requirements. Because the Kansei word > 0.05 of significant level as shown at table 9.

5. References

[1] Smith G G et al. 2013 Play games or study? Computer games in eBooks to learn English vocabulary. *Computers & Education* **69** pp 274–286.

[2] Zhang and Bei 2013 An Analysis of Spoken Language and Written Language and How They Affect English Language Learning and Teaching. *Journal of Language Teaching and Research* **4**(4) pp 834-838

[3] Rita and Situmorang J 2014 Pengembangan Pembelajaran Multimedia Interaktif Berbasis Internet Pelajaran Bahasa Inggris. *Jurnal Teknologi Informasi & Komunikasi dalam Pendidikan* **1**(2), p-ISSN: 2355-4983; e-ISSN: 2407-7437.

[4] Pratiwi V U and Setiyono M S 2015 Aplikasi 'Guessing Games' Untuk Meningkatkan Prestasi Siswa Dalam Speaking Bagi Siswa Sd Kelas 2. ISSN.

[5] Chen and Yen H 2012 Collaborative Cross Number Puzzle Game To Enhance Elementary Students Arithmetic Skills. *The Turkish Online Journal of Educational Technology* **11**(2)

[6] Khonmohammad H, Gorjian B and Eskandari Z 2014 Using Games To Affect Learners' Motivation In Learning English Grammar. *International Journal of Language Learning and Applied Linguistics World* **5**(3)

[7] Fung Y M 2016 Effects of Board Game on Speaking Ability of Low-proficiency ESL Learners. *International Journal of Applied Linguistics & English Literature* **5**(3)

[8] Nagamachi M and Lokman A M 2011 *Innovations of Kansei Engineering*. (New York: Taylor & Francis Group)

[9] Tayyari P C F 2003 *Occupational Ergonomics Princip and Applications*. (Boston/Dordrecht/London: Kluwer Academic)

[10] Vehkalahti K and Puntanen S 2006 *Estimation of realibility: a better alternative for Cronbach's alpha*. (Finland: Departement Mathemathics and Statistic University of Helshinki)

[11] Sheskin D J 2004 *Handbook of Parametric and Non Parametric Statistical Procedures*. 3rd Ed. (Washington: Chapman & Hall/CRC)

[12] Berg R E and Stork D G1995 *The Physics of Sound* (New Jersey: Prentice-Hall Inc.)

Internet Addiction among Indonesia University Students: Musculoskeletal Symptoms, Physical and Psychosocial Behavior Problems

G B Pratama, A Widyanti*

Department of Industrial Engineering, Bandung Institute of Technology (ITB), Indonesia
*widyanti@mail.ti.itb.ac.id

Abstract. The explosive growth of the internet has led to pathological use, also known as internet addiction. A lot of countries have reported increased number of internet addiction among teenagers and the negative effects of it. Purpose of this study is to observed prevalence of internet addiction and the musculoskeletal symptoms (MSS), physical, and psychosocial behavior problems among sample of Indonesian university students. Seven hundred and fifteen Indonesian university students (*mean* age = 20.9 year, *SD* = 2.52 year, 290 female) participated voluntarily in this study by filing out a set of questionnaires consist of internet addiction test (IAT) questionnaire, Nordic body map to observe MSS, and a questionnaire about physical and psychosocial problems. The prevalence of internet addiction among Indonesian university students were 3.2 %, whereas 61% of the samples were categorized as having problem with internet. As predicted, higher internet addiction was accompanying with higher MSS. Most participants (> 50%) were reporting MSS in upper neck, lower neck, right shoulder, back, and waist. higher internet addiction also shows higher physical and psychosocial behavior problems. This study confirms number of general prevalence of internet addiction in teenagers. Result of this study also strengthen the fact of negative effects of intense use of internet not only for physical and musculoskeletal, but also for psychosocial wellbeing.
Keywords: internet addiction, musculoskeletal, disorder, Nordic body map, psychosocial

1. Introduction

The internet, which originally designed for communication and research activity, shows dramatic increase in term of user for decades. Recently, internet is not only used for communication and searching information, but also used for social entertainment (games, songs, and video) and shopping [1]. Moreover, internet is also used as a psychological way to relieve depression through getting rid of bad mood and exchange of funny messages or jokes [2].

The explosive growth of the internet has led to pathological use, also known as internet addiction. Internet addiction can be defined as inability of individual to control the use of the internet, which eventually causes difficulties in a person's life [3]; [4]. Another study defined internet addiction as too much dependency on internet use [5]. In general, internet addiction is characterized by an excessive and uncontrollable behavior to be involved in internet activities that lead to several consequences [6].

Internet addiction have been associated with various problems, such as academic and professional problems and lack of sleep [7], psychiatric symptoms [8], psychological problem [9], behavioral problems [10], psychological well-being [11], educational harms like wasting of time [12], decrease in academic performance [13], and communication problems with peers [14]; [15]. Moreover, in the worst condition, the internet addiction requires professional intervention to cure [16].

Problem of internet addiction are commonly found in young people (i.e., teenagers and young adult) or millennial generation (defined as adults ages 18 to 29, [17]). It is because the internet user are mostly teenagers and young adult, in contrast with elder people who reluctant or low in the use of internet. The commonly found of internet addiction among teenagers is coupled with the fact that they are more vulnerable than mature adults to the harmful effect of different kinds of addiction such as internet addiction [14].

The prevalence of internet addiction among teenagers has been reported in both developed and developing countries with various magnitude. In developed countries, Kormas [18] state a number of 1.5% of internet addiction among teenagers in Greek, Kaltiala-Heino [19] found 1.6% in Finnish, Liu [20] state a number of 4% among US adult; 3.7% in UK[21]; and 3.9% in Italian [22]. The number is even higher in developing countries, for example 8% in China [23], 10.7% in South Korea [24], and 8.5% in Jordan [25].

Indonesia, one of the emerging developing countries has been facing explosive use of internet as well. In 2016, The Association of Internet Service Providers in Indonesia (APJII) announced that internet penetration in Indonesia has now reached 40 percent of the population, or 100 million of internet users (http://www.indonesia-investments.com/id). Most of the internet user are teenagers including university students. However, research about internet addiction in Indonesia is very limited, even zero. However, effort to observe similar kind of addiction in relation with computer matter, not specifically with internet user, has been started to be investigated [26].

Purpose of this study is to observe prevalence of internet addiction among Indonesian university students. In addition, this study also aimed to assess effect of the internet addiction to Musculoskeletal symptoms (MSS), physical, and psychosocial behavior problem of internet addiction.

2. Method

Seven hundred and fifteen Indonesian university students participate in this study voluntarily (*mean* age = 20.9 year, *SD* = 2.52 year, 290 female) by filling out a set of questionnaire. Participants were selected based on convenience sampling. A paper-and-pencil survey was used that include sections of demographic data, internet use, internet addiction, psychological, psychosocial, musculoskeletal disorder. Overall, 15 minutes were needed to complete the whole questionnaires. The demographic data include age and gender of the respondents were captured. Respondents were also asked to state the frequently and place where they used the internet and whether they are supervised when using it. In addition, pictures of the number of hours spent on the internet a week and most predominate online activity were also figured out. The form of the online activities that were assessed divided into: information search on the internet, online communications (including email, chat rooms), and online games (see Chak and Leung 2004 for reason of dividing the online activities).

Internet Adiction Test (IAT, [27]) was used in this study. The IAT consist of 20 items using 5-point Likert scale ranging from 1 (not at all) to 5 (always). The internet addiction test questionnaire was adapted into Indonesian, following back translated procedure proposed by International Commission Test (ICT, see [28]; [29]; [30] for examples). First, two Indonesian bi-lingual with TOEFL score minimum of 550 was translating the original version of the IAT questionnaire (in English) into Indonesian. The best Indonesian version was decided based on the two Indonesian versions through consensus of authors. Later, third bilingual Indonesian person was back translating the Indonesian version into English in which this person has never seen the original version of the IAT questionnaire. Finally, the original and back-translated version of the IAT questionnaire was compared, necessary revision to the Indonesian version was conducted by authors if needed. The IAT items covers psychological dependence, tolerance, compulsive use and withdraw, as well as problems related to schnool , sleep, familiy, and health/time management. The final score of IAT fall between 20 to 100.

The IAT score were categorized as average user for score of 20-39, frequent problems due to internet usage for 40-49, and addict use of the internet for score of 70-100 [27].

To observe musculoskeletal symptom, Indonesian version of Nordic body map questionnaire [31] was used. The questionnaire allowing comparison of low back, neck, shoulder and general complaints for use in epidemiological studies. The original questionnaire consists of 40 items identifying areas of the body causing musculoskeletal problems. A body map is available to assist respondents to indicate nine symptoms of musculoskeletal disorder in the neck, shoulders, upper back, elbows, low back, wrist/hands, hips/thighs, knees and ankles/feet. Four likert scale ranging from 0 (not hurt) to 4 (very hurt) is used to capture the level of MSS suffer by the participants.

Physical and psychosocial behavior problem is observed by [32] questionnaire that is intended to assess the influence of internet addiction on daily life. The physical behavior problems are assessed with 10 questions including nutrition sleeping, and activity behavior, whereas psychosocial behavior problem is assessed using 12 questions in relation with friend and family relations. The responses to the questions are 1 (never), 2 (sometimes), and 3 (always).

3. Results

Socio-demographic of participants and the use of internet as well as prevalence of internet addiction among Indonesian university students can be seen in Table 1.

Table 1. Socio-Demographic and Internet Addiction Data.

Demographic data	All (N=715)	Male (N = 425)	Female (N = 290)
Age (mean, SD)	20.90 (2.52)	20.88 (2.90)	20.87 (2.53)
Internet addiction category:			
- Average user	256 (35.8%)	142 (34.9%)	113 (38.6%)
- Frequent problem	436 (60.9%)	246 (60.4%)	177 (60.4%)
- Internet addict	23 (3.2%)	19 (4.7%)	3 (1.0%)
Duration of internet use (in hours) based on internet addiction category (mean, SD):			
- Average user			
- Frequent problem	7.74 (4.52)	7.58 (4.87)	7.94 (4.05)
- Internet addict	8.42 (4.41)	7.99 (4.17)	9.05 (4.68)
	11.48 (4.73)	11.15 (4.85)	13.67 (3.79)
Prevalence of MSS (reported by participants for all 28 body parts) based on internet addiction category (mean, SD):			
- Average user			
- Frequent problem	11.55 (9.76)	10.97 (10.09)	12.28 (9.33)
- Internet addict	14.46 (11.72)	13.81 (11.78)	15.41 (11.59)
	23.35 (19.79)	21.45 (20.09)	36.00 (14.00)
Internet use for:			
- Academicals matters	619 (86.57%)*	355 (87.22%)	252 (86.01%)
- Social media (chat, blog, ig., fb.)	472 (66.01%)	228 (56.02%)	236 (80.55%)
- Entertainment (music, video, game)	593 (82.94%)	347 (85.26%)	233 (79.52%)
- Online shopping	14 (1.96%)	9 (2.21%)	5 (1.71%)

Note. Percentage from total answer in the specific category, one participants can give more than 1 answer

Further statistical analysis was conducted using SPSS for windows version 20. ANOVA shows that there are no significant differences between male and female participants on age, duration of internet use, and prevalence of MSS (all $p > 0.01$). Overall, female participants reported longer duration of internet use and prevalence of MSS. In-depth analysis based on internet addiction categories reveals significant differences among the three categories of user (i.e. average user, frequent problem, and internet addict) on internet duration of use (F(2, 714) = 7.99, p = 0.01, MSE = 19.91) as well as prevalence of MSS ($F(2,714) = 13.73, p = 0.01$, MSE = 130.14).

Prevalence of MSS in each part of the body in relation with level of internet addiction can be seen in the Figure 1.

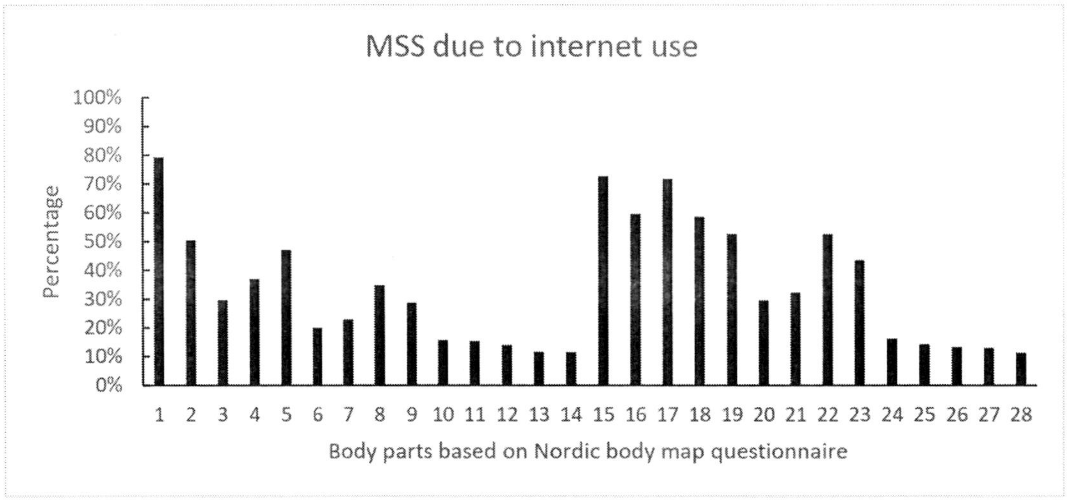

Figure 1. Prevalence of MSS among Sample of Indonesian University Students.

Note.

1	Upper neck	11	Left knee	21	Right lower arm
2	Left shoulder	12	Left calf	22	Right wrist
3	Left upper shoulder	13	Left ankle	23	Right hand
4	Right upper shoulder	14	Left sole	24	Right thigh
5	Hip	15	Nape	25	Right knee
6	Left elbow	16	Right shoulder	26	Right calf
7	Left lower shoulder	17	Back	27	Right ankle
8	Left wrist	18	Wrist	28	Right sole
9	Left hand	19	Buttock		
10	Left thigh	20	Right elbow		

The reported physical and psychosocial behavior problem can be seen in Table 2.

Table 2. The Physical and Psychosocial Behavior Problems in Relation with Internet Use

No	Physical Behavior Problems	Never	Sometimes	Always
1	Recently, I'm having sleeping problems	33%	48%	19%
2	There are times when I go to bed late of the Internet	22%	66%	12%
3	There are times when I skip or delay meals when I am online	39%	52%	9%
4	There are times when I eat my meals in front of the computer	11%	65%	23%
5	I eat snacks (e.g. junk food) when I play games on the computer	35%	53%	12%
6	My body hurts due to the long period of internet use	14%	59%	28%
7	I have eye problem due to the long period of internet use	23%	53%	24%
8	I play sports in real life instead of sports games on the Internet	16%	43%	40%
9	I play active games with my family and friends	16%	58%	26%
10	When I spend time on the Internet or on the computer for a long time, I go on although I am aware that my body receives radiation	10%	50%	41%

Table 2. The Physical and Psychosocial Behavior Problems in Relation with Internet Use (cont.)

No	Psychosocial Behavior Problems	Never	Sometimes	Always
1	The Internet decreases my relationships with family/friends	39%	51%	9%
2	Apart from the Internet and computer games, I spend time with my family with various activities	7%	45%	47%
3	I can easily strike up communication and make friends with others on the Internet	20%	48%	32%
4	I argue with my parents about the time I spend on the Internet	70%	27%	3%
5	I prefer to spend time on the Internet instead of spending time outside with others	49%	45%	6%
6	In everyday life (e.g. at school, in the street, hanging around, etc.) I can easily strike up communication and make friends.	7%	52%	40%
7	When there is no internet in my life, my life becomes boring	35%	50%	15%
8	When somebody disrupts me when I am on the Internet, I feel anger and resentment for that person	79%	20%	1%
9	I suffer from conditions such as restlessness, heart palpitations, or tremors when I cannot connect to the Internet	85%	12%	3%
10	Nowadays, I feel unhappy, alone, and uneasy	55%	36%	9%
11	When I play Internet/computer games for a long time, I have annoying or frightening thoughts in my mind before I sleep	59%	33%	8%
12	There are times when I go to school without doing my homework because I play games on the computer for a long time	61%	36%	3%

Additional analysis using ANOVA is applied to observe the influence of internet addiction category to physical and psychosocial behavior problems (based on average value of all statements of physical and psychosocial behavior problems). Significant differences are found in both physical and psychosocial problems based on different internet addiction level ($F(2,707) = 53.87, p < 0.001, MSE = 458.33$ and ($F(2,707) = 87.23, p < 0.001, MSE = 773.22$ for physical and psychosocial behavior problem respectively). Analysis of each statement of physical and psychosocial behavior problems based on internet addiction category reveals similar pictures of significant differences of each statement (all $p < 0.01$). As expected, higher internet addiction category shows significant higher physical and psychosocial behavior problems.

4. Discussion

Aim of this study is to observe prevalence of internet addiction among sample of Indonesian university students and its effect in musculoskeletal symptoms, physical and psychosocial behavior problems. Result shows that prevalence of internet addiction among sample of Indonesian university students was about 3.2%. This result is congruent with results of previous studies in relation with the prevalence of internet addiction in both developed and developing countries. Furthermore, result of this study is also in line with the fact that Indonesians ranked as the world's most addicted smartphone using population, spending on average more than three hours a day on the devices, on average of 181 minutes a day, according to a recent study by Millward Brown AdReaction [33].

Significant gender differences in internet addiction was found in this study. The proportion of male participants addict to internet is higher than those of female participants (4.7% vs. 1.0 % for male and female participants respectively). Furthermore, female participants show slightly higher internet duration in all category of internet use (normal, problem and addict) than male participants. Previous studies found mixed result in relation with gender differences in internet duration of use. Lam [33] and Gur [32] found that male tent to spend longer time in internet than female. However, another study reported no gender differences in internet addiction and duration of internet use [35]. The mixed result can be explained based on the different use of internet between male and female. In this present study, beside academic matter, female spent time on internet in using social media whereas male spent time on internet on entertainment (i.e., games). As stated by Gur [32] in general, the differences between male and female in internet addition was caused by the fact that male students were playing online games and technology more than female student, and the fact that males students percept computers as toys.

5. Conclusion

This present study shows that there is a relation between internet use and MSS. As predicted, the higher the level of the internet addict, the higher MSS that are reported. The most body parts that suffer from MSS due to internet use are upper neck, lower neck, right shoulder, back, and waist. As far as acknowledged by authors, only limited study is observed the relation between them. An exception is a study conducted by Dol [36] who observed fatigue and pain related to internet usage. The study found that regions with highest fatigue scores are eyes, neck, and shoulder, whereas regions with highest pain scores are neck, shoulder, and waist. Thus, this present study is in line with result of Dol' study.

With regards to the internet addiction and physical and psychosocial behavior problem, result of this present study support result of previous study such as studies conducted by Yeonsoo [37] and Cao and Su [38] who stated that internet addiction decreased physical activities. In this present study, most participants reported excessive time in front of computer using internet, thus decreasing their physical activities. They often skip meal time or having meal in front of computer as well as reported pain in several parts of the body. In relation with psychosocial behavior problem, the present study is supporting the previous studies in which internet addiction is related with psychological problem [38], behavioral problems [10], as well as psychological well-being [11].

This study has several limitations. First, there is a possibility of response bias due to self-report questionnaire of IAT. Although multiple assessment by means of clinical procedure may minimize the bias, the self-reported questionnaire give advantage in its easy of use and practicability in assessing a large number of respondent. Second, assessment of internet addiction was only conducted in urban area of Indonesia. There is a mixed result in the differences between urban and rural area in internet addiction, in which some studies found the differences (e.g., [39]; [40]) whereas other studies do not ([41]; [6]). There is a possibility that similar findings will be found. Therefore, assessment of internet addiction in rural area of Indonesia is worth to conduct.

In sum, this study shows prevalence of internet addiction among Indonesian university students. Despite the limitations mentioned above, this present study act as a first step in observing prevalence of internet addiction among Indonesian university students. Consistent with literature, this study confirms number of general prevalence of internet addiction in teenagers. Result of this study also strengthen the fact of negative effects of intense use of internet not only for physical and musculoskeletal, but also for psychosocial wellbeing. Result of this study is fruitful as a step in advance understanding about the effect of internet addiction as well as prevention of internet addiction and its negative effects.

6. References

[1] Chou C and Hsiao M C 2000 Internet addiction, usage, gratification, and pleasure experience: the Taiwan college students' case. *Computers & Education* 35(1) pp.65-80

[2] Lin S S and Tsai C C 2002 Sensation seeking and internet dependence of Taiwanese high school adolescents. *Computers in human behavior* **18**(4) pp 411-426

[3] Davis R A 2001 A cognitive-behavioral model of pathological Internet use. *Computers in human behavior* **17**(2) pp 187-195

[4] Young K S and Rogers R C 1998 The relationship between depression and Internet addiction. *Cyberpsychology & behavior* **1**(1) pp 25-28

[5] Ko C H, Yen J Y, Yen C F, Chen C S and Chen C C 2012 The association between Internet addiction and psychiatric disorder: a review of the literature. *European Psychiatry* **27**(1) pp 1-8

[6] Bruno A, Scimeca G, Cava L, Pandolfo G, Zoccali R A and Muscatello M R 2014 Prevalence of internet addiction in a sample of southern Italian high school students. *International Journal of Mental Health and Addiction* **12**(6) pp 708-715

[7] Chou C, Condron L and Belland J C 2005 A review of the research on Internet addiction. *Educational Psychology Review* **17**(4) pp 363-388

[8] Shapira N A, Goldsmith T D, Keck Jr P E, Khosla U M and McElroy S L 2000 Psychiatric features of individuals with problematic internet use. *Journal of affective disorders* **57**(1-3) pp 267-272
[9] Cao F and Su L 2007 Internet addiction among Chinese adolescents: prevalence and psychological features. *Child: care, health and development 33*(3) pp 275-281
[10] Sung J, Lee J, Noh H M, Park Y S and Ahn E J 2013 Associations between the risk of internet addiction and problem behaviors among Korean adolescents. *Korean journal of family medicine* **34**(2) pp 115-122
[11] Cardak M 2013 Psychological well-being and Internet addiction among university students. *Turkish Online Journal of Educational Technology-TOJET* **12**(3) pp 134-141
[12] Griffiths M 2000 Does Internet and computer" addiction" exist? Some case study evidence. *CyberPsychology and Behavior* **3**(2) pp 211-218
[13] Kubey R W, Lavin M J and Barrows J R 2001 Internet use and collegiate academic performance decrements: Early findings. *Journal of communication* **51**(2) pp 366-382
[14] Gross E F, Juvonen J and Gable S L 2002 Internet use and well-being in adolescence. *Journal of social issues* **58**(1) pp 75-90
[15] Morahan-Martin J and Schumacher P 2000 Incidence and correlates of pathological Internet use among college students☆. *Computers in human behavior* **16**(1) pp 13-29
[16] King D L, Delfabbro P H, Griffiths M D and Gradisar M 2012 Cognitive-behavioral approaches to outpatient treatment of Internet addiction in children and adolescents. *Journal of clinical psychology* **68**(11) pp 1185-1195
[17] Lenhart A, Purcell K, Smith A and Zickuhr K 2010 Social Media & Mobile Internet Use among Teens and Young Adults. Millennials. *Pew internet & American life project*
[18] Kormas G, Critselis E, Janikian M, Kafetzis D and Tsitsika A 2011 Risk factors and psychosocial characteristics of potential problematic and problematic internet use among adolescents: a cross-sectional study. *BMC public health* **11**(1) p 595
[19] Kaltiala-Heino R, Lintonen T and Rimpelä A 2004 Internet addiction? Potentially problematic use of the Internet in a population of 12–18 year-old adolescents. *Addiction Research & Theory* **12**(1) pp 89-96
[20] Liu T C, Desai R A, Krishnan-Sarin S, Cavallo D A and Potenza M N 2011 Problematic Internet use and health in adolescents: data from a high school survey in Connecticut. *The Journal of clinical psychiatry* **72**(6) p 836
[21] Kuss D J, Van Rooij A J, Shorter G W, Griffiths M D and van de Mheen D 2013 Internet addiction in adolescents: Prevalence and risk factors. *Computers in Human Behavior* **29**(5) pp 1987-1996
[22] Bruno A, Scimeca G, Cava L, Pandolfo G, Zoccali R A and Muscatello M R 2014 Prevalence of internet addiction in a sample of southern Italian high school students. *International Journal of Mental Health and Addiction* **12**(6) pp 708-715
[23] Cao H, Sun Y, Wan Y, Hao J and Tao F 2011 Problematic Internet use in Chinese adolescents and its relation to psychosomatic symptoms and life satisfaction. *BMC public health* **11**(1) p 802
[24] Park S K, Kim J Y and Cho C B 2008 Prevalence of Internet addiction and correlations with family factors among South Korean adolescents. *Adolescence* **43**(172)
[25] Atoum A and Al-Hattab L 2015 Internet addiction and its relation to psychosocial adaptation among Jordanian high basic stage students. *Journal of Psychology and Behavioral Science* **3**(1) pp 96-104
[26] Jap T, Tiatri S, Jaya E S and Suteja M S 2013 The development of Indonesian online game addiction questionnaire. *PloS one* **8**(4) p e61098
[27] Young K S 1998 *Caught in the net: How to recognize the signs of internet addiction--and a winning strategy for recovery.* (John Wiley & Sons)
[28] Brislin R W 1970 Back-translation for cross-cultural research. *Journal of cross-cultural psychology* **1**(3) pp 185-216
[29] Johnson A and Widyanti A 2011 Cultural influences on the measurement of subjective mental workload. *Ergonomics* **54**(6) pp 509-518

IOP Conf. Series: Materials Science and Engineering **528** (2019) 012015 doi:10.1088/1757-899X/528/1/012015

[30] Widyanti A, Johnson A and de Waard D 2013 Adaptation of the rating scale mental effort (RSME) for use in Indonesia. *International Journal of Industrial Ergonomics* **43**(1) pp 70-76
[31] Kuorinka I, Jonsson B, Kilbom A, Vinterberg H, Biering-Sørensen F, Andersson G and Jørgensen K 1987 Standardised Nordic questionnaires for the analysis of musculoskeletal symptoms. *Applied ergonomics* **18**(3) pp 233-237
[32] Gür K, Yurt S, Bulduk S and Atagöz S 2015 Internet addiction and physical and psychosocial behavior problems among rural secondary school students. *Nursing & health sciences* **17**(3) pp 331-338
[33] Amin K 2014 *Indonesians spend most time on smartphones in the world.* Jakarta, Indonesia: The Jakarta Post. Retrieved from: http://www.thejakartapost.com/news/2014/06/05/indonesians-spend-most-time-smartphones-world.html.
[34] Lam L T, Peng Z W, Mai J C and Jing J 2009 Factors associated with Internet addiction among adolescents. *Cyberpsychology & behavior* **12**(5) pp 551-555
[35] Petrie H and Gunn D 1998 December. Internet" addiction": The effects of sex, age, depression, and introversion. In *British psychological society London conference* **15** p 1998
[36] Dol K S 2016 Fatigue and pain related to internet usage among university students. *Journal of physical therapy science* **28**(4) pp 1233-1237
[37] Kim Y, Park J Y, Kim S B, Jung I K, Lim Y S and Kim J H 2010 The effects of Internet addiction on the lifestyle and dietary behavior of Korean adolescents. *Nutrition research and practice* **4**(1) pp 51-57
[38] Cao F, Su L, Liu T and Gao X 2007 The relationship between impulsivity and Internet addiction in a sample of Chinese adolescents. *European Psychiatry* **22**(7) pp 466-471
[39] Johansson A and Götestam K G 2004 Internet addiction: characteristics of a questionnaire and prevalence in Norwegian youth (12–18 years). *Scandinavian journal of psychology* **45**(3) pp 223-229
[40] Yen C F, Ko C H, Yen J Y, Chang Y P and Cheng C P 2009 Multi-dimensional discriminative factors for Internet addiction among adolescents regarding gender and age. *Psychiatry and clinical neurosciences* **63**(3) pp 357-364
[41] Poli R and Agrimi E 2012 Internet addiction disorder: prevalence in an Italian student population. *Nordic journal of psychiatry* **66**(1) pp 55-59

An Initial Study into Indonesian Consumer Awareness of the Ergonomic Product

G B Pratama, N Nurfitrisari and A Widyanti*

Department of Industrial Engineering, Bandung Institute of Technology (ITB), Indonesia
*widyanti@mail.ti.itb.ac.id

Abstract. Ergonomic product refers to a product which is designed based on the principle of human centered design in order to accomodate user's characteristics and capacities. Despite the benefits of an ergonomic design, there are not many industrial products labeled as an ergonomic one. This indicates the consumers are still unfamiliar with the existence of the ergonomic products and its utilization. The aim of this study was to investigate the consumers' awareness toward an ergonomic product. A survey to 16 Indonesian university students and 24 employees was done utilizing a set of questionnaire representing six parameters comprising consumer awareness of ergonomic product, perceived of design, perceived of comfortability, perceived of social image, perceived ease of use, and perceived usefulness. The results revealed that only 55% of respondents were aware of ergonomic product. There were indications of positive appeal from ergonomic product for its consumer in term of its design, comfortability, social image, and usefulness.
Keywords: ergonomic-labeled, design, product, perception

1. Introduction

Ergonomic product is an entity designed with respect to the principle of human centered design so that it focuses on the use of natural movements and characteristics of the user in order to minimize the risk of product using [1]. The other study defined ergonomic product as a form of product differentiation which is developed through an affective design oriented to the new user which does not have any prior experience of using the product [2]. This concept used in order to increase comfort and reduce the risk of workplace accidents or diseases that can reduce work productivity. Not only from consumers' point of view, ergonomics aspects are also beneficial if applied to the work system design [3]. Some studies were also attempted to develop objective criteria to assess the ergonomic aspect of a product. For instance, to be labelled as ergonomics product, a design must fit the user, be usable in a neutral posture, be easy to use and appropriate for the task, and improve health and safety must be fulfilled. However, there are still many others who say that ergonomics is a subjective entity.

In fact, today there are not many household and industrial products produced by including the label "Ergonomic Brand". The company is reluctant to include the label "Ergonomic Brand" or more information which explains the product from the ergonomic side because they does not know the further impact caused by the use of "Ergonomic Brand" on their products. This indicates that the Indonesian consumers are still unfamiliar with the existence of "Ergonomic Brand" and its benefit. A further impact felt by producers of "Ergonomic Brand" is the difficulty of marketing their products in the market segment of consumers in Indonesia. Furthermore, Sutarjo [4] mentioned that stakeholders' awareness is one of critical problems which Indonesia have regarding the application of ergonomics. Nevertheless, it should be noted that ergonomic claim of a product does not always mean that it actually is.

The good news is that some companies have an interest in improving the public's understanding of ergonomics, and therefore constantly on the lookout for ergonomic claims in the marketplace, doing the best to help customers separate fact from fiction. While finding an ergonomic product may not be as simple as just reading a label, consumers can follow a few simple guidelines to determine if the product they are interested in has ergonomic features that work for them. In short, ergonomic label is often included in a product brand and plays an important role in success of product.

The objective of the present study was to investigate the awareness of consumer toward an ergonomic product. This research also aimed at examining the consumer perception of the ergonomic-labeled product. Eventually, the results of this study is expected to be a basis for a larger study of consumers' acceptance and for companies' consideration to decide "what to do" for their user-centered based product.

2. Methods

Forty Indonesian people (*mean* age = 28.24 year, SD = 8.99) consists of twenty males and twenty females, participated in this study voluntarily by filling out a set of questionnaire. The survey included four questions of demographic data and 26 questions regarding consumers' perception and willingness to buy of ergonomic-labeled product. Overall, 15 minutes were needed to complete the whole questionnaire.

A set of questionnaire comprised of five constructs from varied literature was used in this study. The set of questions was adapted into Indonesian, following a standard back translate procedure by International Commision Test (ICT, see [5]; [6]; [7] for examples). The variables' constructs and the asscociated source utilized in this research were specified in Table 1. Before, answering the questionnaire items, the participants were shown three examples of ergonomic products (mouse, keyboard, and chair).

Table 1. Constructs and Sources

No.	Construct	Definition	Literature Source
1	Perceived design	Utilization of design elements in order to create a pleasing appearance of a product	Hwang [8]
2	Perceived of comfortability	A pleasing state of physiological, psychological and physical harmony between a user and its environment	Pijls et al. [9]
3	Perceived of social image	The extent to which users may derive respect and admiration from peers in their social network as a result of their product usage	Moore and Benbasat [10]

Table 1. Constructs and Sources (cont.)

No.	Construct	Definition	Literature Source
4	Perceived ease of use	The difficulty of understanding the feature of product, and in such way lead to less confidence in usability, it is expected that products which have high complexity are likely to be low on perceived functionality and this will translate into lower levels of utilitarian attitude	Gima [11]
5	Perceived usefulness	The degree that a technology increases consumer's job performance	Davis [12]; Shin [13]

3. Results

Socio-demographic of participants and people's awareness of ergonomic product among Indonesian can be seen in Table 2. In this initial study we conduct the survey to 16 students and 24 employee and most of them (70%) had received knowledge of ergonomics beforehand. However, it is interesting that the percentage is lower (55%) on the consumer awareness of the ergonomic brands.

Table 2. Socio-Demographic and Ergonomic Product Awareness Data.

Demographic data	All (N=40)
Age (mean, SD)	27.70 (8.40)
Employment status	
- Student	16 (40%)
- Employee	24 (60%)
Having knowledge of ergonomics (Yes)	28 (70%)
Consumer awareness of ergonomic brands (Percentage of people know about ergonomic product)	22 (55%)

This study used five constructs adapted from several literature mentioned in previous section which include perceived design, perceived of comfortability, perceived of social image, pereived ease of use, and perceived usefulness. These constructs were expected as the objective representation of The reported consumer perception toward the ergonomic product **can be seen in Table 3**.

Table 3. Consumer perception toward the ergonomic product

No	Questionnaire Item	1	2	3	4	5	6	7
1	Perceived Design	0%	1%	0%	4%	24%	**61%**	12%
2	Perceived of Comfortability	0%	0%	2%	8%	13%	**61%**	18%
3	Perceived of Social Image	1%	0%	10%	29%	15%	**40%**	5%
4	Perceived Ease of Use	2%	17%	17%	**30%**	17%	21%	0%
5	Perceived Usefulness	0%	0%	2%	15%	31%	**42%**	12%

Note.

1: Strongly disagree
2: Disagree
3: Slightly disagree
4: Neutral
5: Slightly agree
6: Agree
7: Strongly agree

4. Discussion

Aim of this study is to investigate the consumer awareness of an ergonomic product. There are several points which are interesting to discuss. Firstly, the result showed that 70% of respondents had received knowledge about ergonomics either from a formal academic program or a non-fomal source. This fact supports a presumption that most of respondents already had a basic definition of ergonomics in their mind. It is logical, because all of the respondents in this initial case were university students and employees which assumed had received a high education level, at least diploma or bachelor degree. However, we did not verify the eligibility of their understanding of ergonomics.

Second, there were indication that the ergonomic products are still unfamiliar among a group of consumers. There were only 55% of participants responded that they knew about the ergonomic product. The percentage could mean that there is only a small number of the ergonomic-labeled product play in the current market, with the result that the ergonomic-labeled products are quite difficult to find by the consumer. The other logical explanation is that the existing consumers are not really care or interested in using an ergonomic product. It is possible that the ergonomic label of a product does not have a significant effect to be appealing enough for the user candidate, because not every ergonomic-labeled product felt positevely different to the general product or actually ergonomic afterall.

Furthermore, a set of questionnaire items were utilized to investigate the consumer perception of an ergonomic product. Among five constructs, there are indication of positive appeal of ergonomic-labeled products for its user in term of design, comfort, social image, and usefulness. It was indicated that majority of respondent agree (6) with these constructs. However, the result showed that most of participant responded a neutral answer on perceived ease of use. This may indicates that a number of consumers are still not sure about how easy the product functions or feature to be understood, in spite of its advantages.

5. Conclusion

In summary, this study shows that there are a number of consumer which are not familiar with the ergonomic-labeled product. In parallel, most of survey participants agreed that ergonomic product has some particular appeals in term of its design, comfort, effect to users' social image, and usefulness. Result of this study is fruitful as a step in advanced understanding about the effect of consumer awareness to the willingness to buy an ergonomic product. Further study is expected to be a basis for a larger study of consumers' acceptance and for companies' consideration to decide "what to do" for their user-centered based products.

This study has two limitations. First, there is a possibility of response bias of the paper-based questionnaire. However, the paper-based questionnaire gave advantage in its easy of use and practicability in assessing a large number of respondent. Second, assessment of consumer awareness was only conducted in a small number of participants, this is the reason that this study was only named as an initial study. Moreover, this study only investigated how the consumers perceive the ergonomic aspect of a certain product using the tendency on several parameters. Further study is expected to examine the relationship among those parameters so that it is possible to define a conceptual model which describe how the user candidate thinks about an ergonomic product.

Ultimately, this research was expected to play role as a part of bigger study in the future, both in term of the number of respondents, as well as how deep each of construct is investigated. As future study, it is possible to explore the relation among the investigated constructs. In such wise, there would be a conceptual model which can describe the relevancy among used constructs. It would be also interesting if the relation among variable can also be compared by conducting a simple experiment to investigate the ergonomic products and non-ergonomic products.

6. References

[1] Karwowski W 2005 Ergonomics and human factors: the paradigms for science, engineering, design, technology and management of human-compatible systems. *Ergonomics* **48**(5) pp 436-463

[2] Horn D and Salvendy G 2006 Consumer-based assessment of product creativity: A review and reappraisal. *Human factors and ergonomics in manufacturing & service industries* **16**(2) pp 155-175

[3] Neumann W P, Dixon S M and Nordvall A C 2014 Consumer demand as a driver of improved working conditions: the 'Ergo-Brand'proposition. *Ergonomics* **57**(8) pp 1113-1126

[4] Sutarjo U S 2007 Ergonomics policy in Indonesia. *Journal of human ergology* **36**(2) pp 57-61

[5] Brislin R W 1970 Back-translation for cross-cultural research. *Journal of cross-cultural psychology* **1**(3) pp 185-216

[6] Johnson A and Widyanti A 2011 Cultural influences on the measurement of subjective mental workload. *Ergonomics* **54**(6) pp 509-518

[7] Widyanti A, Johnson A and de Waard D 2013 Adaptation of the rating scale mental effort (RSME) for use in Indonesia. *International Journal of Industrial Ergonomics* **43**(1) pp 70-76

[8] Hwang G J, Kuo F R, Chen N S and Ho H J 2014 Effects of an integrated concept mapping and web-based problem-solving approach on students' learning achievements, perceptions and cognitive loads. *Computers & Education* **71** pp 77-86

[9] Pijls R, Groen B H, Galetzka M and Pruyn A T 2017 Measuring the experience of hospitality: Scale development and validation. *International journal of hospitality management* **67** pp 125-133

[10] Moore G C and Benbasat I 1991 Development of an instrument to measure the perceptions of adopting an information technology innovation. *Information systems research* **2**(3) pp 192-222

[11] Atuahene-Gima K 1995 An exploratory analysis of the impact of market orientation on new product performance: a contingency approach. *Journal of Product Innovation Management: an international publication of the product development & management association* **12**(4) pp

275-293
[12] Davis F D 1989 Perceived usefulness, perceived ease of use, and user acceptance of information technology. *MIS quarterly* pp 319-340
[13] Shin D H 2007 User acceptance of mobile Internet: Implication for convergence technologies. *Interacting with computers* **19**(4) pp 472-483

The Survey on the Perception of Safety and Security Culture for Nuclear Workers

Atyanti Dyah Prabaswari[1], Rini Dharmastiti[2]

[1] Universitas Islam Indonesia, Jl. Kaliurang, DI Yogyakarta, Indonesia
[2] Universitas Gadjah Mada, Jl. Grafika No. 2, DI Yogyakarta, Indonesia
[1]Atyanti.dyah@uii.ac.id

Abstract. A survey was developed to evaluate employee perceptions of the safety culture and culture of nuclear security. The survey was made up of 52 questions divided into 13 factors: leadership, involvement of employee, work environment, employee competence, management commitment, procedures, leader attitude, training and qualifications, employee attitude, implementation and operation, integrated safety, self-assessment, belief. The survey was conducted on 125 employees working in nuclear area. A survey were conducted between February and May 2016. A numerical value was given to each question, so that the results of the survey were represented as a score. The data reveal that age group 60-69 years dominate by having the best perception on 6 factors. The overall value of factors, the company is included in rank B. The relationship between the long term period of working and to the value of rank does not show the relationship trend. MANOVA analysis showed that if multivariate test was done together, the thirteen factors did not show any differences on the group lenght of working. However, if analyzed individually, it turns out the Training and Qualification factors have different effects on the group of long term period of working.
Keywords: safety and security culture, nuclear security system, survey, questionnaires

1. Introduction

In a research, safety culture receiving more attention in industries including in nuclear power operations [1]. Even some examples of accidents where safety culture is identified as contributing to accidents, including BP's Texas City Refinery Explosion in 2005 (The Chemical Safety Board, 2007), The Washington Metropolitan Area Transit Authority Rail Collision in 2009 (National Transportation Safety Board , 2010), The Deepwater Horizon Oil Spill in 2010 (United States Coast Guard, 2011), The Upper Big Branch Mine Explosion in 2010 (The Mine Safety and Health Administration, 2011), and The Fukushima Nuclear Accident in 2011 (National Diet of Japan, 2012) [1].

The Fukushima nuclear accident happened because of not understanding the security culture although the system has the sophisticated protection based on technology. Nuclear security culture consisting of series characteristics, attitude, and behaviors of individuals, organizations, and institutions that serve to support and enhance nuclear security [2]. The result of the most incidents nuclear is lack of awareness nuclear security culture and in addition, nuclear security culture became an important subject in the nuclear security summit at Hague with 53 countries as participants including Indonesia [3].

The relationship between security culture and safety culture is when the both consider the risk of human error, but nuclear security has an additional on deliberate action intended to cause harm [2]. The security requires different attitudes and behaviour such as confidentiality of information and the prevention of malicious actions. The nuclear security culture refers to the dedication, accountability, and understanding of all individuals who involved in activities that have an influence on the security nuclear activities.

Measurements of safety culture were conducted using questionnaires. Similar with measurements of security culture also using questionnaires to surveys of personnel at nuclear power plant [4]. This is because a questionnaire survey is an effective way to test the level of awareness of nuclear security. The questionnaire that used to survey should cover all aspect of nuclear security. *WINS (World Institute of Nuclear Security)* publish of a simplied questionnaires but the questionnaires is not enough to analyse all of elements of nuclear security [4]. Several countries have developing questionnaire that reflect of all elements of their security culture.

ROK (nuclear power in South Korean) initiated a study to developing a questionnaire consisting of 20 questions of security nuclear culture in 2009 [4]. But besides South Korean, the number of countries that using nuclear security culture is not known yet. Indonesia has a questionnaire the safety culture that contained in the [5] for Implementation of Safety Culture.

The result of the literature study is indicating the existing questionnaire do not include all elements associated with the safety culture and security culture. It becomes necessary to develop a new survey tool be in the form questionnaire to evaluate the safety culture and the security culture. A survey examining awareness of safety and security culture in nuclear area with developing questionnaire.

2. Methods

2.1 Basic Theory

2.1.1 Safety Culture. The first appeared term of safety culture is in the OECD Nuclear agency Report of 1987 which published in INSAG of the year 1988 because of accident from Chernobyl in 1986 [6]. Later on, this term accepted as a culture where safety issues is top priority in the company.

Utal (1983) says that the definition of safety culture is a part of a trust and a value which integrated with control system and organizational structure that constitute behavioural norms [7]. Safety culture is a combination of characteristics and attitude whithin organization and individuals that establish as a top priority [8]. Besides that, safety culture is a part of the organizational culture which can influence attitude of members who can belief in safety and health performance [6].

2.1.2 Security Culture. Security culture is a set of characteristics, attitude, and behaviours of individuals, organizations and institutions can support and enhance nuclear security [2]. The nuclear security aims to ensure that implementation of security nuclear receives a certain amount.

2.1.3 The relationship Safety and Security Culture. The nuclear safety is the achievement of appropriate operating conditions, accident prevention, or mitigation of concequence of accidents, which may affect in the protection of workers, society, and the environment from undue radiation [9]. The security in nuclear is prevention and detection, responding theft, sabotage, unauthorized access, illegal transfer, or other malicious acts involving nuclear material, other radioactive substances or facilities related [9]. Between safety and security are built upon the legal and regulatory framework. the framework should define the responsibilities of the state, government regulations, and operators.

2.2 Research Method

2.2.1 Developing Questionnaire. Factor affecting security and safety nuclear was derived from 11 literature review on the safety measurement, security culture/climate, and guidelines about safety and security culture/climate in nuclear area. The literature is used for getting the most factor that commonly used. The literature about safety take from guideline (internatioanal safety advisor group) [8], assessing safety culture in nuclear engineering laboratory [10], identifying the common features (measuring safety climate such as in nuclear industries, nuclear power plant, 23 energy industries in England) [11], guidelines from OHSAS [12], SCART guidelines [13], the factors which take effect occupational health and safety in radiology area and in a company [7] [14]. The literature about security take from nuclear security series [2] and survey on personnel at nuclear power plants [4].

2.2.2 The Questionnaire. The result of research, the number of factors is 13 factors that influence safety and security culture in the nuclear area [15]. The answer to the questions fall into five categories: strongly disagree, disagree, neutral, agree, and strongly agree.

2.2.3 The 13's Factor

- Leadership
  This factor consists of 10 questions. These questions include organization, compliance, acknowledgment, command and direction, involvement, acceptance of suggestions and inputs, policies, responsibilities, and procedures related to the safety culture and culture of nuclear security.
- Involvement of Employee
  This factor consists of 4 questions. These four questions include discussion, and active involvement by workers.
- Work environment
  This factor consists of 6 questions. The six questions related to the safety culture and culture of nuclear security are policy transparency, organizational performance measurement, performance indicators, convenience and facilities provided.
- Employee competence
  Questions that fall into this factor are related to support, examples from leadership, training, competence, evaluation, emergency response systems, and reporting media. Questions on this factor amounted to 7 pieces.
- Commitment of management
  The number of questions on this factor consists of 2 pieces. The question is related to coordination and task assignment, roles and responsibilities to improve nuclear safety and security.
- Procedures
  This factor consists of 4 questions. The four questions relate to easy to find work procedures, existing procedures covering all systems at all levels of the organization, procedures for retention of outdated procedures, roles and responsibilities for procedures.
- Leaders attitude
  The number of questions on this factor there are 4 pieces. these four questions are related to the evaluation of the leader, the support of employees, the lack of awareness, and a good example of understanding and knowledge about safety and security risks.
- Training and qualifications
  The question on this factor consists of 2 questions. The question involves a medical examination, and the principle of nuclear security is as important as nuclear safety.
- Workers attitude
  This factor has 2 questions. These questions include workers' compliance with requirements and procedures and objectives and targets to be achieved according to the procedure.
- Implementation and operation
  Questions on this factor amounted to 3 pieces. these questions include the training schedule of the organization, the organization's evaluation of an event, the ease of understanding the guidelines or procedures.
- Integrated of safety

Factor include the presence of a committee for safety and nuclear security and activeness leader to finding the root of the problem.

- Self assessment
  The number of questions on this factor there are 3 pieces. Questions on these factors relate to self-learning of self-errors or co-workers, occupational safety and security concerns, and workers and co-workers remind each other about the procedures.
- Belief
  This factor consists of 3 questions. Existing questions relate to the belief that every worker has an important role in safeguarding or strengthening nuclear safety and security in the organization. Existing questions also include belief in threats that can come from within and outside the country. In addition, there are also questions related to regular emergency response exercises.

Table 1. The Question in Leadership Factor

No	Questions	S. D	D	N	A	S.A
1	Organizations always planning the good procedures including handling prevention and incident					
2	Organizations always evaluate compliance with regulations concerning nuclear safety and security both within and outside the country					
3	Leaders provide recognition and rewards to employees who trying to improve nuclear safety and security					
4	Leaders provide the orders and direction for employees to raising awareness and behavior toward safety and security					

3. Result and discussion

This survey was conducted in February 2016 - May 2016. Respondents who fill out the questionnaire are required to had attended Radiation Protection Training for New Employees as the basic of Occupational Health and Safety knowledge. A total of questionnaires is 125 but the questionnaire which met requirements for processing is 77 questionnaires.

To compare the survey result before Februari 2016 was conducted in 2014. The result declare that the employee has Score C and the questionnaire which that used not cover the security culture.

3.1. Scoring Age by Group

The results of each age associated with their point of views on each factor that have different values. The value is taken from the mean value in each age group. The answer strongly disagree has a value of 1, disagree has a value of 2, neutral has a value of 3, agree has a value of 4, strongly agree has a value of 5.

Figure 1 explains that for the age group 30-39 years old have the best perception of self-assessment factors, and training and qualifications. The 40-49 years old has the best perceptions of beliefs, employee competencies, and procedures. The 50-59 years old have the best perception of commitment of management, and implementation and operation. The 60-69 years old have the best perceptions of worker involvement factors, work environment, leader attitude, worker attitude, integrated safety and leadership. The age group 60-69 years dominates by having the best perception on 6 factors. In Figure 1 it is explained that the number of samples (n) for each age group has different amounts.

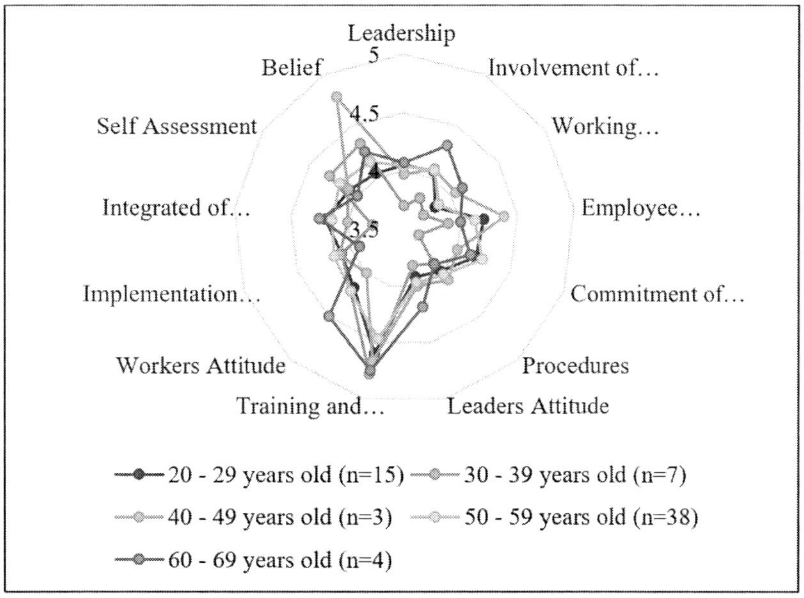

Figure 1. Scoring Age by Group

3.2 Overview by Grade

The index level is determined on the difference of the highest value of 5 and the lowest value is 1 of the likert scale, then divided into 5 assessment scales as in Table 2. This index level is useful to know the general description of the application of safety culture and security in an organization.

Table 2. The Grade

Grade	Indeks
A	$\geq 4{,}2$
B	$3{,}4 \leq X < 4{,}2$
C	$2{,}6 \leq X < 3{,}4$
D	$1{,}8 \leq X < 2{,}6$
E	$< 1{,}8$

The average score on the employee perception level based on the index level for each factor is ranked B except the Training and Qualification factors which score 4.54 (rank A). The overall score is at 4.08 index in rank B.

3.3 Analysis the relationship of Long Term Period of Working to perception value

The regression to analysis of the relationship is shown equation model could not be used to see the actual condition. The result of significance does not show the trend of relationship between the the long term period of working with value. When viewed from the analysis, the relationship between the long term period of working with the value score of employee perceptions could not be modeled into an equation because it has no relationship trend. This is similarly to [4] research.

3.4 Analysis the influence of Long Term Period of Working to Value

The conclusions from the MANOVA analysis (dependent variable is the long time period of working, and independent variable is the 13 factors) were that if the multivariate test were conducted together, the thirteen factors did not show differences in the period lenght of working group (Multivariate test has sig. >0,05). However, if analyzed individually, it turns out the Training and Qualification factors have different effects on the period lenght of working group. Because all of Factors has sig. > 0,05, except Traing and Qualification which has sig. 0,047. Based on the analysis, the perception of Training and Qualification factors at the company have different effect because longer the long term period of working make the training and the qualifications will also change. That will change the qualification from the employee.

4. Conclusion

The questionnaire consisted of 13 factors, leadership, employee involvement, work environment, employee competence, management commitment, procedures, leader attitude, training and qualifications, workers attitude, application and operation, integrated safety, self-assessment, confidence. A total of 125 people working in the nuclear agency/company/institution were surveyed. The level of perception of a culture of nuclear safety and security is expressed numerically.

Scoring age by group has the result is the ages of 20 - 29 years as the youngest age group have not had a prominent perception for nuclear safety and security culture. The age group 60 - 69 years old dominates by having the best perception on 6 factors.

The average score on the employee perception level based on the index level for each factor is ranked B except the Training and Qualification factors which score 4.54 (rank A). The overall score is at 4.08 index in rank B.

The conclusions from the MANOVA analysis were that if the multivariate test were conducted together, the thirteen factors did not show differences in the lenght working group. However, if analyzed individually, it turns out the Training and Qualification factors have different effects on the period lenght working group.

5. References

[1] Morrow S L, Koves G K and Barnes V E 2014 "Exploring The Relationship Between Safety Culture And Safety Performance In U.S. Nuclear Power Operations. *Safety Science* **69** pp 37-47

[2] IAEA 2008 *IAEA Nuclear Security Series No.7 Nuclear Security Culture*. (Vienna: International Atomic Energy Agency)

[3] European Union 2014 *The Hague Nuclear Security Summit Communiqué*. (Hague: Council of European Union)

[4] Yoo H and Lee J H 2015 Result of Nuclear Security Culture Survey on Personnel at Nuclear Power Plants. *Journal Annals of Nuclear Energy* **85** pp 398-402

[5] BATAN 2016 Peraturan Kepala Badan Tenaga Nuklir Nasional Nomor: 200/KA/X/2012 Tentang Pedoman Pelaksanaan Penerapan Budaya Keselamatan. (Yogyakarta: Batan)

[6] Cooper M D 2000 Towards a model of safety culture. *Safety Science*

[7] Khoiri M 2010 *Budaya Keselamatan, Perilaku K3, dan Indikator Kinerja Keselamatan Dalam Pelayanan Radiologi di Instalasi Radiologi RSUP. Dr. Sardjito*. (Yogyakarta: Universitas Gadjah Mada)

[8] IAEA 1991 *Safety Culture (Internatioanal Safety Advisor Group. Safety-Series 75-INSAG -4)*. (Vienna: International Atomic Energy Agency)

[9] IAEA 2010 *The Interface Baetween Safety and Security at Nuclear Power Plants INSAG-24*. (Vienna: International Atomic Energy Agency)

[10] Ostrom L, Wilhelmsen C and Kaplan B 1993 Assessing Safety Culture. *Nuclear Safety* **34**(2)

[11] Flin et al. 2000 Measuring Safety Climate: Identifying The Common Features. *Safety Science* **34** pp 177-192

[12] ISO 18001 2007 Sistem Manajemen Keselamatan dan Kesehatan Kerja. *Occupational Health and Safety Management Systems* (OHSAS).

[13] IAEA 2008 *SCART Guideline Reference Report for IAEA Safety Culture Assessment Review Team*. (Vienna: International Atomic Energy Agency)

[14] Puspitasari D 2012 *Pengaruh Faktor-faktor Budaya Keselamata Terhadap Perilaku K3 di Sebuah Perusahaan.* (Yogyakarta: Universitas Gadjah Mada)
[15] Prabaswari A D 2016 Pengembangan Instrumen Pengukuran Persepsi Karyawan Terhadap Budaya Keselamatan dan Budaya Keamanan Kerja di Bidang Nuklir. *Prosiding Seminar Nasional Teknik Industri.* (Yogyakarta: Universitas Gadjah Mada)
[16] Santoso S 2015 *Panduan Lengkap SPSS Versi 23.* (Jakarta: Elex Media Komputindo)

The Mental Workload Analysis of Staff in Study Program of Private Educational Organization

Atyanti Dyah Prabaswari[1], Chancard Basumerda[2] and Bagus Wahyu Utomo[3]

[1,2] Universitas Islam Indonesia, Jl. Kaliurang, DI Yogyakarta, Indonesia
[3]Sekolah Tinggi Teknologi Adisutjipto, Jl. Janti Blok R, DI Yogyakarta, Indonesia
[1]Atyanti.dyah@uii.ac.id

Abstract. The mental workload affects productivity workers to accomplish the task. Staff has many tasks that not only main task but additional task. The additional task is from superior in the same place or the other place. The SOP for duties and distribution of task is not clear so the fatigue can affect to productivity. The method to measure mental workload is NASA-TLX. NASA TLX has six indicators: mental demand, physical demand, temporal demand, performance, frustration, and effort. The analysis not only based on result of NASA TLX but also using statistic test. The result is performance indicator is high that described the staff not satisfied with their job has done. The staff not satisfy but the effort has been taken out is very high. The result of Mann Whitney is not different with two group of the period long term of working which divided in < 7 years and >7 years. The result of correlation test is if the staff make more effort, the satisfaction for their tasks has been done is bigger. The regression test produces an equation but cannot be used to calculation or prediction the staff mental workload. Because regression coefficient is not significant (not usable). The solution is staff must divide the task. Dissatisfaction can reduce motivation and the members cannot perform well.
Keywords: NASA TLX, Mental Workload, Staff, Productivity

1. Introduction

The working hours, mental, and frustration when facing certain tasks can be considered as risk factors for work stress [1]. The demands of task must be arranged so that someone is not too loose or even the task is to burden. The arrangement is to ensure long term safety, health, comfort, and efficiency of productivity [2].

Physical fatigue, mental fatigue, and emotional reaction such as headaches, indigestion, and irritability can be caused by excessive workloads [3]. Workload is a factor between education level and cognitive decline because of the people who has low education level more often have monotonous jobs with only little challenge and cognitive so that low mental workload demands are related to cognitive abilities [4]. The decrease of level of work productivity can caused by fatigue condition experienced by workers in long term period [5].

The components that affect workers productivity include individual factors that including age (25%), gender (15%), period long term of working (10%), the remaining is influenced by workload and fatigue (50%) [6]. The external factor that has affect to productivity based on environment work such as stress of work (23,5%), motivation (25,5%), nutrition (25,5%), and workload (25,5) [7]. The evaluation of workload is as key to research and develop human machine interface, as well as to find levels of comfort, satisfaction, efficiency, and security in the workplace [2].

University has an organization. The organization divided to some faculty based on majors. The faculty divided to some study of program based on concentration program. The organization has system and standard operation procedure (SOP) to arranging the mechanism administration process. Staff has task to arranging administration that has correlation to educational. Every study of program has staff to handle the administration.

In existing condition SOP to arranging administration that has related to education. The process to doing that tasks have been clear. However, the staff not only doing administrative work as their main tasks, but also do additional work. The type of additional work is short term and support educational activities such committee of event, etc. The additional work not only from superior in their place, but also from superior in other place where still same faculty or in higher level in university. The additional work not only one and has different duty in at once time. The SOP of staff task is not clear to doing additional work, so the fatigue can affect to productivity.

The result of the literature study and observation is indicating the existing condition of staff must be analysed. Because that can be affect to their administrative work or additional work. The method to measure mental workload is NASA-TLX. NASA TLX is the most widely used method and provided good results [8].

2. Methodology

Observation is important to find solution from the problem staff. Not only observation, but also ask the staff about their tasks in that time. The literature study to find about method that can be produce the result of workload.

Collecting data using NASA TLX that has six indicators [8]: (1) Mental demand (MD) is measuring activities mental and perceptual to see, remember, and find. Not only that but also to classify the tasks into difficult, simple, or complex. The rating is low to high (2) Physical demand (PD) is measuring the number of physical activities that needed (example: pull, push, rotate, etc.). The rating is low to high (3) Temporal demand (TD) is measuring time pressure that felt during work whether the work can be done slowly, or quickly so that is feels tiring. The rating is low to high (4) Performance (P) is the success of workers in carrying out their duties and how they are satisfied with the results of their work. The rating is perfect to failure (5) Frustration (FR) is how many workers feel insecure, desperate, offended, or disturb when doing their work. The rating is low to high (6) Effort (EF) is amount of hard work that workers need to achieve the required level of performance. The rating is low to high. NASA TLX has steps [8]: (1) Weighting is a section to respondents choose one indicator to cause a mental workload (2) Rating (3) Scoring is multiplying the rating and weighting for each indicator (4) Weighted workload (WWL) is sum of score all indicators (5) Average of WWL (6) Interpretation score in Table 1.

Table 1. The Interpretation Score of NASA TLX [8]

Workload	Value
Low	0-9
Medium	10-29
Somewhat high	30-49
High	50-79
Very high	80-100

Processing data using score of NASA TLX and test statistic. Test statistic that used in this research is Mann Whitney test, correlations and regression test. Analysis to this research is based on result of NASA TLX and test statistic. Flow chart of research in Figure 1.

IOP Conf. Series: Materials Science and Engineering **528** (2019) 012018 doi:10.1088/1757-899X/528/1/012018

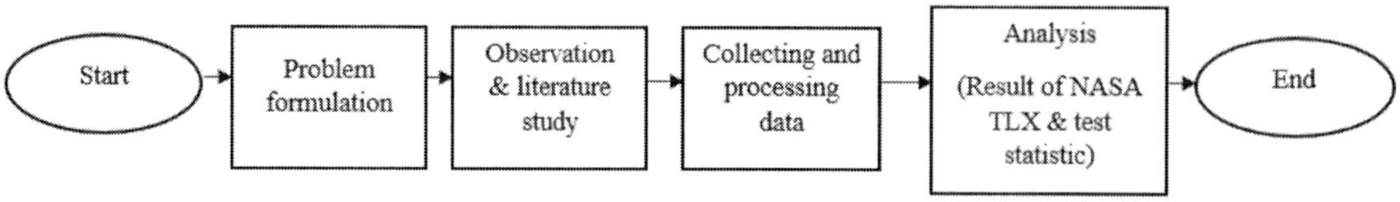

Figure 1. Flow Chart of Research

3. Result and discussion

Total of respondent is 13 that profile can be find in Table 2. Collecting and processing data using NASA TLX has the result in Table 3.

Table 2. Respondent Profile

Gender	Category			Total
	Age			
	21-30 year	**31-40 year**	**41-50 year**	
Female	4	6	1	11
Male	1	1	0	2
	Period the long term of working			
	≤ 7 year	> 7 year		
Female	6	5		11
Male	1	1		2

Table 3. The Result of Interpretation Score

Category	Total	Average
MD	2025	155,8
PD	600	46,2
TD	3800	292,3
P	3860	296,9
EF	3360	258,5
FR	695	53,5
Average of WWL	73,5	
Interpretation of score	High	

MD has average score 155,8 that described the staff need high mental and perceptual activities. Because the staff must do some of administration to educational in different type such as schedule, letter, etc. Besides that, must doing additional work which not only one and has different duty in at once time. PD has average score 46,2 that described the staff not doing activities that must moving. The staff just doing their tasks while sitting and not moving around. PD score has the lowest score. TD has average score 292,3 because time pressure to accomplish their task in fast time not only their main tasks but also their additional work. TD score is in second rank in six categories.

P has the highest score (296,9) that described the staff not satisfied about the result. Because the P value is interpreted as good if it is close to a value of 0 or the average total is small. FR has 53,5 that described the staff not has feel insecure, desperate, offended, or disturb when doing their work. EF has average score 258,5 that described the staff must accomplish all the tasks in at once time. That needs the big effort.

The interpretation all six indicator is the staff not satisfied about the result (P indicator) but the the EF indicator that is taken out is high. Not only the EF but also the TD indicator has high score that described high time pressure.

The Mann Whitney Test [9] using two group of the period long term of work: (1) ≤ 7 years (2) > 7 years. The result is sig. 0,518 which bigger than probability score (0,05). Interpretation result is not having differences score between two group. Each group has the same mental workload in NASA TLX method.

Correlation and regression test have a goal to knowing about how big the relation between EF, TD, PD, FR, MD and P indicators. The correlation test has result in Table 4. The MD and FR has significant result because score significant <0,05. The score is -0,621 that indicates the bigger frustration make mental demand is smaller. Because of frustration so the needs to finding, remember, etc. is smaller.

Table 4 Correlations of EF, MD, PD, TD, FR and P Indicators

		P	**EF**	**MD**	**PD**	**TD**	**FR**
Pearson Correlation	P	1.000	-.393	-.092	.055	-.334	-.180
	EF Score	-.393	1.000	.210	-.196	-.360	-.144
	MD	-.092	.210	1.000	.328	-.293	-.621
	PD	.055	-.196	.328	1.000	-.117	-.407
	TD	-.334	-.360	-.293	-.117	1.000	.411
	FR_Level	-.180	-.144	-.621	-.407	.411	1.000
Sig. (1-tailed)	P	.	.092	.382	.429	.133	.278
	EF Score	.092	.	.245	.261	.114	.319
	MD	.382	.245	.	.137	.166	.012
	PD	.429	.261	.137	.	.352	.084
	TD	.133	.114	.166	.352	.	.081
	FR_Level	.278	.319	.012	.084	.081	.

Regression test has resulted a formulation that calculations from Table 5. The formulation is in equations (1).

$$Y = 658{,}317 - 0{,}676\,X1 - 0{,}176X2 - 0{,}359X3 - 0{,}431X4 - 0{,}311X5 \quad (1)$$

The definition from equation: (1) Y = P (2) X1 = EF (3) X2 = MD (4) X3 = PD (5) X4 = TD (6) X5 = FR. Significant test to testing the equation . The significant test using t-test two sides that has hypothesis: (1) H_0 = Regression coefficients is not significant (2) H_1 = Regression coefficients is significant. The result of significant test is H_0 that means regression coefficient is not significant (not usable). This is verified the equation cannot be used to calculation or prediction the staff mental workload.

Table 5. Coefficients of Regression Significant Test

Model		Unstandardized Coefficients		Standardized Coefficients	t	Sig.
		B	Std. Error	Beta		
1	(Constant)	658.317	152.273		4.323	.003
	Effort Score	-.676	.350	-.603	-1.934	.094
	MD	-.176	.278	-.227	-.632	.547
	PD	-.359	.743	-.153	-.483	.644
	TD	-.431	.260	-.531	-1.662	.140
	FR Level	-.311	.475	-.252	-.656	.533

The result from NASA TLX is P indicator has the biggest score. The biggest score is not good because the good P indicator is having smallest score near to zero. Score E and TD indicator indicating employee need high time and effort to finish their task but they are not satisfied. Solutions of the problem and the result is dividing the task. The important elements in the functioning of social system of any kind and size is the distribution of tasks and division [10]. The distribution ensured orderly carried out and available resources are optimally used. The staff will be satisfied with the task distribution. has effective function [10] said the dissatisfaction can reduce motivation and the members cannot perform well.

4. Conclusion

Performance indicator has the highest score that means the staff not satisfy with their task has been done. They do the tasks with high effort and time pressure. Analysis with statistic test not producing significant formulas. The formulas can not usable to prediction the productivity staff. The other of analysis statistic test has result that two groups period the long term of working has not difference. Every staff has the same mental workload to doing their task. Beside that, the higher score of FR make score MD is smaller. The SOP for every staff must be clear. If there is more than one staff then should be clear division of task. Who does internal affairs and who does external affairs. Or who does main task and who does additional task.

5. References

[1] González-Muñoz E L and Gutiérrez-Martínez R E 2007 *Contribution of mental workload to job stress in industrial workers* **28**(4) Work (Reading, Mass) pp 355-361
[2] Rubio S, Díaz E, Martín J and Puente J M 2004 Evaluation of subjective mental workload: A comparison of SWAT, NASA-TLX, and workload profile methods. *Applied Psychology* **53**(1) pp 61-86.
[3] Manuaba A and Wignjosoebroto S 2000 *Ergonomi untuk keselamatan, kesehatan kerja dan produktivitas.* (Surakarta: Uniba Press)
[4] Bosma H, Van Boxtel M P J, Ponds R W H M, Houx P J H and Jolles J 2003 Education and age-related cognitive decline: the contribution of mental workload. *Educational gerontology* **29**(2) pp 165-173.
[5] Ulfah N, Nurcahyo J and Dwiandhono I 2013 Model Kuantitatif Manajemen Kelelahan dan Beban Kerja untuk Peningkatan Produktivitas Pekerja Penggilingan Padi. *Jurnal Kesehatan Masyarakat Nasional* 7
[6] Katsuro P C, Gadzirayi, Taruwona M and Suzanna M 2010 Impact of occupational health and safety on worker productivity: a case of Zimbabwe food industry. *African journal of Business*

Manajemen 4(13) pp 2644-51
[7] Reniati R A 2006 *Faktor-faktor yang mempengaruhi produktivitas kerja pada pekerja bagian pembuatan drum unit drum plant PT X* (Semarang: Universitas Diponegoro)
[8] Hancock P A and Meshkati N 1988 *Human Mental Workload* (Elsevier)
[9] Santoso S 2015 *Panduan Lengkap SPSS Versi 23* (Jakarta: Elex Media Komputindo)
[10] Mikula G and Freudenthaler H 2002 Division of task and duties and the perception of injustice: The case of household chores. *Psychologische Beitrage* **44** pp 567-584.

Selection process of sustainable production indicators using eco-quality function deployment

Pregiwati Pusporini[1] and Iwan Vanany[2]

[1]University of Muhammadiyah Gresik, East Java, Indonesia
[2]Institute of Sepuluh Nopember (ITS) Surabaya, East Java, Indonesia
[1]pregiwati_pusporini@umg.ac.id, [2]vanany@ie.its.ac.id

Abstract. Sustainable production has to become a necessity for all industries for last decades. While the industry has to improve the quality of the product, they also need to consider three essential indicators (economic, social and environment). To concern for sustainable production, there are many indicators that can be set. Every industry may have different indicators with others depends on their policy, goal, and circumstances. The objective of this research is to propose a practical approach to determine appropriate sustainable production indicators. The selection process of sustainable production indicators will be challenging. The proposed method used a framework of eco-quality function deployment. The construction of house of quality in this concept is derived from three main perspectives of sustainable development such as, economic, social and environment. The assessment of eco-quality function deployment involves expert judgments. The linguistic terms in fuzzy concept is applied. The result showed that the combination of fuzzy eco-QFD could be a useful tool for the selection process of sustainable production indicators that is flexible enough to be modified and can accommodate many varieties of indicators.
Keywords: sustainable production, sustainable indicators, eco-quality function deployment

1. Introduction

Sustainable development has to become a primary issue and major concern over the world in the last decades. The concept of sustainable production as a central part of sustainable development emerged at the United Nations Conference on Environment and Development in 1992 [1-2].This issue forces any organization to concern sustainability effort into their business and activities. Sustainability is derived from three principal requirements namely: social, economic and environment. A fundamental concept of sustainable production is defined by The Lowell Center for Sustainable Production (LCSP) as a process to create products and services using processes and systems that are non-polluting, preserving energy and natural resources, economically feasible, harmless and healthy for employees, societies and customers, and socially and creatively rewarding for all working people [1-3]. A number of considerations and approaches have been developed relating to sustainability production awareness. The primary barrier of sustainability consideration is the lack of an applicable framework that can be used to select the appropriate indicators and suitable with the organization's circumstances. As a result, a systematic approach for the selection process of sustainability production indicators is proposed.

Researchers categorized main indicators of sustainable production into six categories which involve energy and material use (resources), natural environment (sinks), economic viability, community development and social justice, workers, and products [1-3]. The identification of these indicators may vary and differ from one company to another as it depends on the type of their products, processes, organisation's management and policy. The identification of them is important and indicates how much the company is concerned on sustainability issues. Not all of the indicators are good; some of them are too hard to measure, others have no meaning to the local community, and some are difficult to understand. Some criteria of sustainable production indicators have been suggested by now, such as measurable, relevant, understandable, reliable/usable, data accessible, timely manner, and long-term-oriented [1]. To promote the using of sustainable indicators as a part of the continuous process of setting goals and measuring performance, the Lowell Center for Sustainable Production (LCSP) has been developed a framework of sustainable production indicators. This framework was classified into five levels namely: Level 1 (facility compliance/conformance indicators), level 2 (facility material use and performance indicators), level 3 (facility effect indicators), level 4 (supply chain and product life-cycle indicators), and level 5 (sustainable system indicators) [1-5]. Many researchers have identified different categories of sustainable production indicators but there is a lack of guidelines, or even recommendations, on how a specific organization will prioritize and apply the significant indicators to fulfil their stakeholder's requirements.

Although many initiatives and organizations are currently trying to develop environmental, social or sustainability indicators for companies, there are still drawbacks related to the selection process of sustainable production indicators, for example: most indicator frameworks are still under development, no framework is applicable as a whole to assess sustainable production, no existing indicator framework equally addresses all six aspects of sustainable production, most framework effort to address economic performance (traditional economic indicators), most indicator frameworks are developed for internal management purposes only (not involve stakeholders requirements in the process of indicator formulation and evaluation)[5]. Another shortcoming of the selection process of sustainable indicators is every organization may have different indicators from one to another. It is very challenging to determine which indicators are appropriate (the same indicator may be useful at one company and ineffective at another). Therefore, each indicator has to be considered on an individual basis to reflect the specific characteristics of different organizations.

The objective of this research is to propose a practical approach to determine appropriate sustainable production indicators. Once sustainable production indicators have been developed the organization will need to determine what measures and metrics of these indicators are appropriate for their particular business conditions and specific circumstances. A comprehensive, flexible and adaptable framework for the selection process of sustainable production indicators is needed to address the lack of consensus which accommodates various indicators and is flexible enough to be applied and support continuous improvement activities.

2. Methods

This paper proposes the eco-quality function deployment for the selection process of sustainable production indicators. The main reason why employ eco-quality function deployment (eco-QFD) to prioritize sustainable production indicators is that the concept of QFDE has been verified through several studies and other QFD based methods incorporating the environmental issues have been developed. This indicates that the effectiveness of QFD based methods for designing and creating environmentally friendly products is widely recognized [7-12]. The extension of eco-QFD is needed in the proposed methodology. The construction of a house of quality in the proposed methodology will be derived from stakeholder's perspectives which consists of three dimensions namely: economy, social, and environment. By integrating these dimensions into the construction of a house of quality, it can aid the management team in a company to identify stakeholder's requirements and indicators within the production process and prioritize them according to the company's circumstances. Figure 1 presents the construction of a house of quality for the selection process of sustainable production indicators.

Sustainable production indicators (*hows*) / Stakeholders requirements (*whats*)		Indicator 1	Indicator2	Indicator3	Indicator 4	Indicator 5	Indicator 6
Social	Social requirements 1	Relationship matrix between stakeholders requirements (*whats*) and indicators (hows)					
	Social requirements 2						
Economic	Economic requirements 1						
	Economic requirements 2						
Environment	Environment requirements 1						
	Environment requirements 2						

Figure 1. HOQ for selection process of sustainable production indicators

The first important step in this stage is to identify stakeholder's requirements referred to as *whats* (as called in QFD), i.e. what extent the products or processes meet stakeholder's requirements (*SR*). To create a sustainable production, stakeholder's requirements need to be embedded into QFD approach whose quantification can be structured via a variety of methods based on literature review, analysis of customer complaints, surveys, an operational report, historical data, discussion, and brainstorming with some experts in a company. The use of these approaches will lead to a clear understanding of how critical the consideration of stakeholder's requirements are in the sustainable production processes.

The second step of QFD approach is identification of *technical attributes* or *engineering characteristics (hows*, as called in QFD*)*, i.e. the identification of sustainable production indicators (*hows*) which are related to stakeholder's requirements. Each stakeholder's requirement (*whats*) are correlated with sustainable production indicators (*hows)*. Indicators, as a basic and important part of sustainable production process, will facilitate the collection of meaningful data and allow accurate expression of sustainable information. Identification of indicators should be based on significant sustainable aspects associated with the organization's business activities. Approaches like literature review, stakeholder demands, market pressure, customer complaints, government policy, rules and regulations can be used as a guide to determine sustainable indicators.

After identifying *whats* and *hows*, the next step is to construct house of quality (HOQ) including the relationship between *whats* and *hows*. The expert team in a company needs to determine the weight of each stakeholder's requirement and the relationship between *whats* and *hows*. The data analysis for the QFD method is commonly based on human assessment that uses a particular rating scale. Rating scales 1 to 9 are used to identify the weight of each stakeholder's requirement (*whats*). Rating scale 1 indicates that this stakeholder's requirement is very unimportant and rating scale 9 indicates that the stakeholder's requirement is very important. Similar rating scale is also used to identify the relationship between stakeholder's requirements (*whats*) and sustainable production indicators (*hows*), for example rating scales 1, 3, 5 and 9 where 1 indicates that the stakeholder's requirement and sustainable production indicator are weakly related, while 3 is the indicative of a medium relationship and 9 of strong relationship. It is well recognised that the use of qualitative information including human assessment is always subjective and vague and this is the main reason to integrate the fuzzy concept into the QFD method to account for this subjectivity and vagueness.

Once stakeholder's requirements and sustainable production indicators established, the next step is translating them into the house of quality (HOQ). This step involves expert judgments. Using fuzzy logic, expert judgments will be transformed into fuzzy triangular numbers (TFNs). TFNs are chosen due to their computational simplicity and usefulness in processing the fuzzy information [11-14]. A stepwise algorithm of fuzzy QFD approach is developed herein:

Step 1: identify the linguistic terms as the triangular fuzzy number to represent the weight of stakeholder's requirements and their relationship with sustainable production indicators.

Easy-to-understand, these are five linguistic terms to represent the weight of environmental requirements, namely *very unimportant (VU), unimportant (U), medium (M), important (I), and very important (VI)*. These linguistic terms are transformed into a triangular fuzzy number (TFNs) as seen in Table 1.Similarly, linguistic terms are employed to identify the relation matrix between environmental requirements (*whats*) and environmental indicators (*hows*) as shown in Table 2.

Table 1 Linguistic terms for stakeholder's requirements

Linguistic Term	Crisp value	Triangular Fuzzy Number
Very unimportant	1	(1,1,2)
Unimportant	3	(2,3,4)
Medium	5	(4,5,6)
Important	7	(6,7,8)
Very important	9	(8,9,9)

Table 2 Linguistic terms for relationship matrix between stakeholder's requirements (whats) and sustainable production indicators (hows)

Linguistic Term	Crisp value	Triangular Fuzzy Number
Very weak	1	(1,1,2)
Weak	3	(2,3,4)
Medium	5	(4,5,6)
Strong	7	(6,7,8)
Very strong	9	(8,9,9)

Step 2: Determine and calculate aggregate fuzzy importance weight of stakeholder's requirements based on expert knowledge

During the construction of HOQ in QFD, experts will be involved to quantify their assessments by scoring the linguistic terms in Table 1 from which the *aggregate fuzzy importance weight* of stakeholder's requirements.

Step 3: Determine and calculate the relationship between stakeholder's requirements (whats) and sustainable production indicator (hows)

Similarly, experts will be involved to quantify their assessment by scoring via the linguistic terms in Table 2 from which the *aggregate fuzzy relationship* between stakeholder's requirements (*whats*) and sustainable production indicators (*hows*).

Step 4: Prioritize the sustainable production indicators

The final result of the purposed methods is to prioritize the sustainable production indicators. Then, the management team at a company would focus on the highest scores that indicate these indicators should be considered to related to their product, process characteristics and circumstances.

3. Result and Discussion

A case study is needed to apply the proposed method. The expert team at a company responsible for sustainable production process involves experts from cross divisional areas at company who have skill, knowledge and experience related to the production processes, and sustainable production process. At the company data was collected using interviews, direct observation and literature research. Furthermore a semi structured questionnaire was used as a guidance to direct the conversation during interview with the expert team to collect information on current key production indicators, stakeholder's requirements, sustainable indicators, and metrics for sustainable indicators. The main reason for using a semi structured interview was to avoid the rigid interview structure and also to gather any possible information connected with the sustainable production indicators. Questions on a specific topic were posed to the expert team; however they were also given the chance to express their own ideas and opinions related to the topic. The questionnaire consisted of four groups of questions including company background, production processes, key production indicators used and sustainable production indicators.

For an illustrative example, there are 10 stakeholder's requirement and 9 sustainable indicators listed based on questionnaire's result. Table 3 shows the full result for the hypothetical example and ranking of sustainable indicators. The result of the house of quality (HOQ) in Table 3 demonstrated the weight of stakeholder requirements (*SR*) and the priority of sustainable indicators (*SI*).The result of house of quality (HOQ) in Table 3 demonstrates that the energy used has highest level of importance (17.94) followed by solid waste (14.19), reprocessed materials (12.89), emission (10.95), water pollution (10.48), materials used (10.27), water used (10.05), harmful materials (7.29) and noise/vibration (5.95) respectively. The expert team would focus on the highest scores such as energy consumption, solid waste, and reprocessed materials.

Based on the HOQ analysis, the integration of fuzzy concept and eco-quality function deployment can provide a set of appropriate sustainable production indicators related to design of sustainable production process. This provides more accurate evaluation which can be used to improve a specific improvement due to organizational sustainable goals and circumstances. The assessment from the experts on the expert team can determine measures which are relevant and appropriate with the company's circumstances and conditions. Based on their knowledge, skill, and experience in the company, they can determine these measures and by using common sense, human thinking and judgment as expert team members. Furthermore, the flexibility of the proposed model lies in that it can accommodate many categories of stakeholder's requirements such as economic, social, and environment to furnish sustainable development. The proposed model can also be further developed associated with the sustainable business goals.

4. Conclusion

This paper focused on the developing of a systematic approach for selecting sustainable production indicators. The integration of fuzzy and eco-quality function deployment into process selection of sustainable production indicator was presented. In the proposed methodology, the construction of a house of quality is derived from stakeholder's requirement which consists of three dimensions namely: social, economy and environment. Then a set of selected indicators can be determined based on the stakeholder's requirements. A concept of fuzzy logic was applied to determine the priority of selected indicators. Using fuzzy QFD in the selection process of indicators for sustainable production has the following two significant advantages:

(1) Adoption of eco-QFD can help the design team at the company to consider on three essential perspectives (economy, social, and environment) at the early stage of product and development process as a main part of sustainable product development.

(2) Adoption of linguistic terms and fuzzy concept can allow the organization to prioritize sustainable production indicators that involve expert's judgments.

The overall results showed that the combination of fuzzy QFD could be a useful tool for the selection process of sustainable production indicators which are suitable and appropriate for the management goal.

Acknowledgement, this work is supported by the Directorate of Research, Technology and Higher Education Indonesia and the University of Muhammadiyah Gresik, Indonesia.

Table 3: The example of fuzzy QFD result to prioritize sustainable production indicators

	Stakeholders req. (SR)	**Weight of SR**	***SI_1* Materials used**	***SI_2* Harmful Mat.**	***SI_3* Rep.mat**	***SI_4* Energy used**	***SI_5* Water used**	***SI_6* Emission**	***SI_7* Noise/vib.**	***SI_8* Water pollution**	**SI_9 Solid waste**
SR_1	Waste reduction	7.58	48.03	-	61.93	57.51	48.03	42.97	42.97	48.03	61.93
SR_2	Pollution impact reduction	7	-	57.17	-	57.17	39.67	39.67	39.67	49	44.33
SR_3	Less material usage	6.92	56.49	-	52.45	43.81	-	-	-	-	48.42
SR_4	Less energy usage	8.17	57.17	-	66.69	71.46	-	51.72	-	-	-
SR_5	Less water usage	5	-	-	-	-	40.83	-	-	31.67	-
SR_6	Easy to process	5.67	42.97	-	35.89	35.89	-	-	-	-	-
SR_7	Safer to landfill	6.33	-	-	-	-	-	-	-	-	48.03
SR_8	Safer emission	6.33	-	-	-	51.72	-	48.03	-	-	-
SR_9	Less harmful to the living environment	6.33	-	48.03	40.11	40.11	35.89	35.89	35.89	40.11	40.11
SR_{10}	Disposable	6.33	-	40.11	-	-	35.89	-	-	40.11	40.11
	Total		**204.65**	**145.31**	**257.08**	**357.66**	**200.31**	**218.28**	**118.53**	**208.92**	**282.93**
	Percentage		**10.27**	**7.29**	**12.89**	**17.94**	**10.05**	**10.95**	**5.95**	**10.48**	**14.19**
	Rank		**6**	**8**	**3**	**1**	**7**	**4**	**9**	**5**	**2**

5. References

[1] Krajnc D and Glavič P 2003 Indicators of sustainable production. *Clean Techn Environ Policy* **5** p 279
[2] Azapagic A and Perdan S 2000 Indicators of Sustainable Development for Industry: A General Framework. *Process Safety and Environmental Protection* **78**(4) pp 243-261
[3] Veleva V, et al. 2001 Indicators of sustainable production. *Journal of Cleaner Production* **9**(5) pp 447-452
[4] Samuel V B, Agamuthu P and Hashim M A 2013 Indicators for assessment of sustainable production: A case study of the petrochemical industry in Malaysia. *Ecological Indicators* **24** pp 392-402
[5] Veleva V and Ellenbecker M 2001 Indicators of sustainable production: framework and methodology. *Journal of Cleaner Production* **9**(6) pp 519-549
[6] Mascarenhas A, et al. 2010 The role of common local indicators in regional sustainability assessment. *Ecological Indicators* **10**(3) pp 646-656
[7] Pusporini P and Abhary K 2015 Environmental performance evaluation (EPE) for small medium industries. *2015 IEEE International Conference on Industrial Engineering and Engineering Management (IEEM)*
[8] Pusporini P, Abhary K and Luong L 2012 Environmental performance as key performance indicators in the lean six sigma methodology. *Advanced Materials Research* **488-489** pp 1082-1086
[9] Pusporini P, Abhary K and Luong L 2013 Integrating environmental requirements into quality function deployment for designing eco-friendly product. *International Journal of Materials, Mechanics and Manufacturing* **1**(1) pp 80-84
[10] Pusporini P, Abhary K and Luong L 2013 Development of environmental performance model using design for six sigma (DFSS). *International Journal of Materials, Mechanics and Manufacturing* **1**(1) pp 102-106
[11] Baidya R, Dey K P and Gosh S K 2018 Strategic maintenance technique selection using combined quality function deployment, the analytic hierarchy process and the benefit of doubt. *International Journal of Advanced Manufacturing Technology* **94**(1-4) pp 31-44
[12] Suryaningrat I B 2016 *Implementation of QFD in food supply chain management: A case of processed cassava product in Indonesia. International Journal on Advanced Science Engineering Information Technology* **6**(3) pp 302-305
[13] Pusporini P, Abhary K and Luong L 2013 Intelligent design of environmental performance evaluation using fuzzy expert system. *Applied Mechanics and Materials* **330** pp 719-723
[14] Pusporini P and Vanany I 2017 Challenges in implementing cleaner production: Barriers and strategies in the Indonesian seafood processing industry. *2017 IEEE International Conference on Industrial Engineering and Engineering Management (IEEM)*, Singapore pp 1520-1524
[15] Tseng M L 2013 Modeling sustainable production indicators with linguistic preferences. *Journal of Cleaner Production* **40** pp 46-56

Workload analysis of chemical analyst at Analytical Research Laboratory PT.UPA – Orang Tua Group

N Y Hidayah[1,3], I Purwaningsih[2,4] and K R Ririh[1,5]

[1]Department Industrial Engineering Pancasila University, Jakarta-Indonesia
[2]PT UPA – Orang Tua Grup, Jl. Lingkar Luar Barat Kav. 35-36, Cengkareng 11740, Jakarta-Indonesia.
[3]nurhidayah@univpancasila.ac.id, [4]irowatip@gmail.com,
[5]kirana.ririh@univpancasila.ac.id

Abstract. The accuracy of the job completion is very important to increase productivity. PT.UPA-Orang Tua Group has a chemical analysis laboratory (Analytical Research Laboratory) which is tasked analyzing samples from Formulator Divisions. In carrying out the sample chemical analysis, the company has 10 working days lead time. But it was seen that a number of samples were handled beyond the 10-day deadline. This is thought to be a chemical analyst's workload that exceeds its capacity. In this study, workload will be calculated using the Workload Analysis method and also set a standard time for chemical analysis in which the company does not have it. The object of this research is to determine the standard time taken from the preparation of chemical analysis of each parameter. There are 8 chemical parameters that will be calculated. Then observations of work time were carried out by a chemical analyst with good abilities and moderate work rhythms. Determination of the standard time for chemical analysis is 24.07 hours per sample. And the workload for each analyst is 7 samples per month, so it takes 11 chemical analysts to complete 639 samples that will come to meet the 10-day work target.
Keywords: AR Laboratory, chemical analyst, standard time, workload analysis

1. Introduction
Companies can win the competition if they are always looking for ways to move forward. One of the factors to find out whether a company is said to be advanced or not, can be seen from its productivity growth. Productivity is the number of units of goods or services made by a worker in a certain time using various types of machinery and equipment available at his place of work [5]. Productivity cannot be measured if a company does not have a standard. A standard is usually the first step in determining the needs of workers. A good standard must show the level of optimal achievement that must be performed by workers. Standards are often equated with standard time because the size to determine a standard can be seen from the standard time. Taylor defines that standard time is the main basis for showing productivity [3]. Standard time is the time needed by the average worker to do work under normal conditions [6]. So that if the time standard is correct, then the time standard can represent the time needed by a worker to carry out his work.

Orang Tua Group is one of the companies engaged in food and beverages in Indonesia. The Orang Tua Group subsidiary which is in control of producing food commodities such as wafers, biscuits,

IOP Conf. Series: Materials Science and Engineering **528** (2019) 012020 doi:10.1088/1757-899X/528/1/012020

chocolate, candy and nuts is PT. Ultra Prima Abadi. PT. Ultra Prima Abadi has a research laboratory centre (Analytical Research Laboratory / AR Lab) which is tasked with analyzing the chemistry of every product that has been released.

The AR laboratory has 7 analysts in charge of analyzing sample chemistry. AR laboratory customers are all divisions that are developing company products. These divisions are called Front End divisions. All products that have been sent by Front End will be analyzed by chemical analysis consisting of various analysis parameters.

AR laboratory has a deadline for working on the sample analysis for each application for analysis is a maximum of 10 working days, but the company does not have a standardized working time analysis for each analysis parameter. With this deadline every month, AR laboratories are still experiencing some delays.

During the period June 2016-March 2017, the average sample that came every month was 1916 samples. During this period there were several samples that were not completed on time. The average sample that was successfully analyzed on time (On Due Date Sample) during that period each month was only 75.33%. This amount has not met the company's expectation of 100% On Due Date Sample.

Delay in analysis can cause large number of samples queued to be tested in the AR laboratory. The sample accumulation that caused the Front End division also experienced delays in the process of developing new products so as to provide a significant impact on the company's productivity in the form of customer complaints. The average number of customer complaints to AR laboratories during the period June 2016-March 2017 is 60 complaints each month. When compared with the number of overdue samples (the sample completed is more than the 10-day deadline) then the average customer complaint that is included consists of 7-8 overdue samples per customer. This large number of complaints must be immediately addressed by efforts to increase work productivity in the AR laboratory.

In an effort to increase work productivity in the AR laboratory, it is necessary to measure workload. Productivity will not increase if the workload of each employee is uneven. Workloads that are too heavy or light will have an impact on work inefficiency [3]. Workload that is too light means that there is an excess of labor which causes the organization to pay more employees with the same productivity resulting in cost inefficiencies. Conversely, if there is a shortage of labor, it can cause physical and psychological fatigue for employees, so that employees become unproductive because they are too tired.

Until now, AR laboratories have never calculated workload analysis through standardization of work time. The standard time must be determined in the calculation process of the chemical analyst workload analysis. Therefore, it is expected that by researching and analyzing the analyst workload will produce optimal output so that AR laboratory productivity will increase and the distribution of workload between analysts is evenly distributed.

The purpose of this study is to determine the standard time of sample analysis in the AR laboratory, to find out the maximum workload that can be given to chemical analysts based on workload analysis, and knowing the number of analysts needed.

2. Methods

The data needed in this study include:

a. The measurement of standard time of the analysis was only carried out on 8 chemical parameters, that are pH analysis, moisture content, reducing sugar content, total sugar content, fat content, protein content, total acid levels and salt levels.
b. The observation of work time required by a chemical analyst in completing the analyze using stopwatch.
c. Number of samples received in the Analytical Research (AR) laboratory for analysis of 8 parameters during the period June 2016 - March 2017. This data will be used to estimate the number of samples arrivals in the AR laboratory during April-December 2017
d. Working hours available for April - December 2017

e. Job description of chemical analyst in the AR laboratory.
f. Number of AR laboratory chemistry analysts.

Data processing and analysis are carried out with the following stages:

a. Determine the cycle time, normal time and standard time of the chemical analysis process. At this stage, the calculation of the cycle time has passed the uniformity and adequacy test data. Normal time is obtained from the adjustment of cycle time to the skills, effort, and consistency of analysts and the work atmosphere.
b. Estimating the number of sample arrivals per month during April - December 2017. In this study the forecasting method used is a geometric increase with a formula [1]:

$$P = P_0\,(1 + \frac{i}{100})^n \quad (1)$$

P forecasting value of the n-th period, P_0 the value of the n-th period, i average increase in percent, n number of periods predicted

Determine analyst workload. In this stage the number of samples that can be done by one chemical analyst per day will be calculated. The purpose of this calculation is to determine a realistic target so that there is no excessive sample load received by chemical analysts Analysts workload calculation using the formula [7]:

$$\text{Standard workload} = \frac{\text{Time available}}{\text{average completion time}} \quad (2)$$

Determine the number of chemical analysts based on the following equation [7]:

$$\text{Number of analysts needed} = \frac{(\text{standard time per unit} \times \text{number of units})}{\text{time available per person}} \quad (3)$$

From these calculations it can be seen whether the company needs to increase the number of analysts so that the work target can be achieved without giving excessive workload in accordance with the arrival of the number of samples.

3. Result and Discussion

3.1. Job description of chemical analysts in the AR laboratory

The details of the responsibility of the chemical analyst in the AR laboratory are:

a. Conduct routine chemical analysis for product samples from the Front End division.
b. Calculate and input the results of chemical analysis into the analysis report.
c. Evaluate and ensure the actual analysis is in accordance with the work instructions that apply in the AR laboratory.
d. Record the use of chemical reagents and equipment used in the chemical analysis process in the log book of use.

3.2. Number of chemical analysts in the AR laboratory

The number of chemical analysts in the AR laboratory is 7 analysts. All analysts will be distributed samples every day with the same quota for each person according to the arrival of the sample number

3.3. Determination of standard time

The work carried out in the AR laboratory is the analysis of the content of chemical compounds contained in the sample (consisting of 8 test parameters), data calculation, and report making. In the completion of the chemical analysis, each sample requires the preparation of a simultaneous work system, so some analysis can be done at the same time. The working cycle time of the sample is obtained from the longest cycle time of the parameters tested where the longest cycle time occurs in the analysis of protein content for 23.07 hours. In the protein analysis there is a treatment of sample curing for 20 hours in the machine which causes the protein analysis has the longest cycle time.

IOP Conf. Series: Materials Science and Engineering **528** (2019) 012020 doi:10.1088/1757-899X/528/1/012020

In the process of chemical analysis, the first stage is to prepare equipment for water content analysis that requires a process on the machine for 1.5 hours. During the process on the machine, the chemical analyst will use the time to work on the preparation for the analysis of fat and protein then proceed with the analysis of total acid. When the process in the machine occurs in the analysis of protein and fat, then the time will be used to analyze the pH and salt content in the sample. Continued with the analysis of reducing sugars and analysis of total sugar. Then complete the analysis of fat and protein. The last stage is data calculation and report making. So that the standard time for the completion of chemical analysis is the sum of the standard time's protein analysis (the longest standard time), the calculation of the data, and the making of the report that is equal to 1480.28 minutes or 24.67 hours per sample. To explain the activities of the analysis carried out simultaneously, it can be seen at fig.1 Gantt Chart of the chemical analysis process in the AR laboratory. The grey color identify that the activities or operations are done by machineries and the yellow color identify that the activities or operations are done by operators.

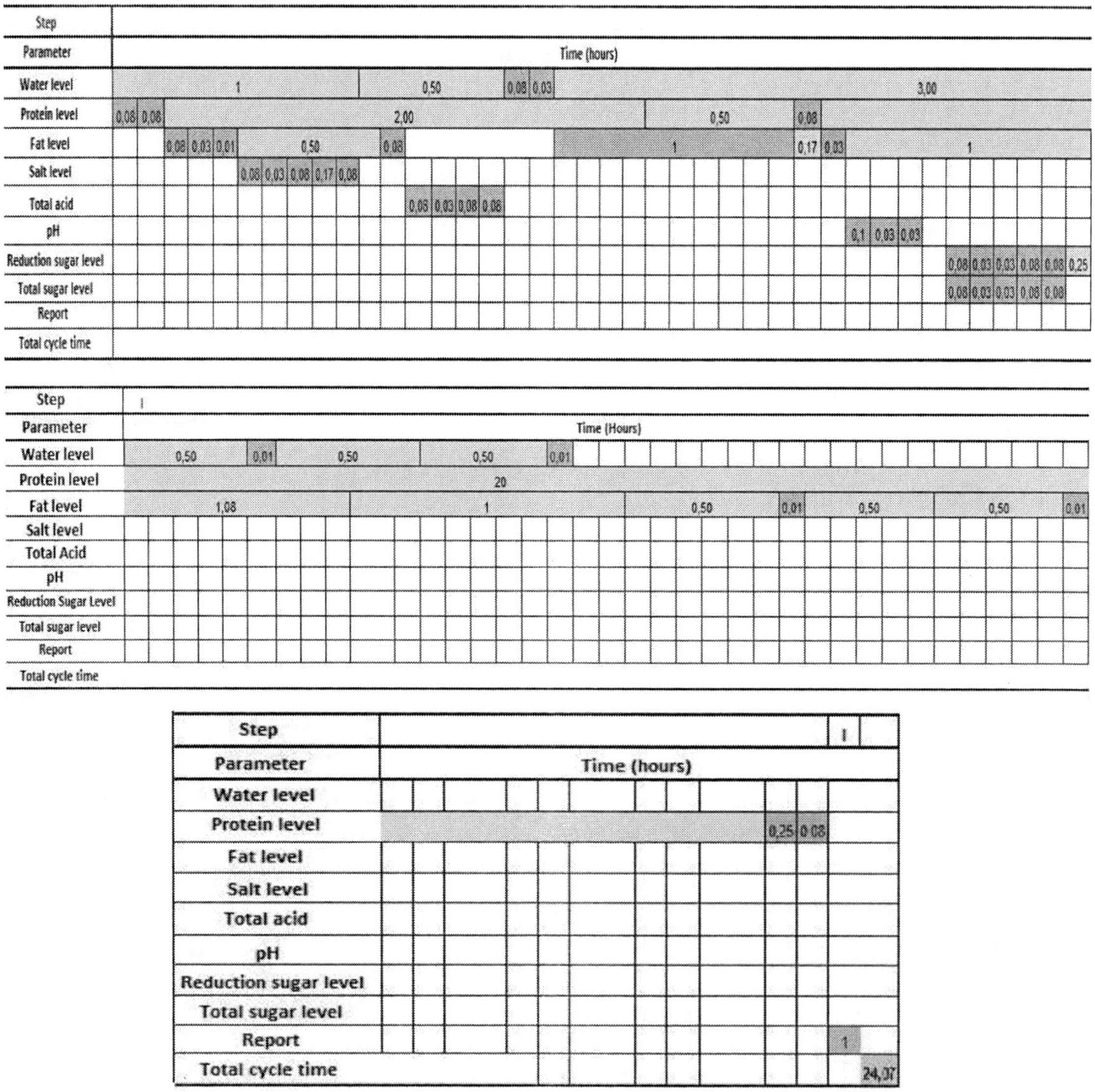

Figure 1. Gantt Chart of Chemical Analysis Process

IOP Conf. Series: Materials Science and Engineering **528** (2019) 012020 doi:10.1088/1757-899X/528/1/012020

3.4. Forecasting the number of samples

Forecasting the number of samples to be received in the AR laboratory during the April - December 2017 period is carried out using the geometric increase forecasting method. Forecasting results can be seen in table 1.

Table 1. Estimated sample arrivals for April - December 2017

Month	Number of sample
April	63
May	65
June	67
July	69
August	71
September	73
October	75
November	77
December	80
Total	639
Average	71

3.5. Workload analysis

Determination of workload based on the calculation of the number of samples that can be done by one chemical analyst in 8.5 working hours with the standard time required to complete one sample is 24.07 hours. Based on these data, the maximum number of samples that can be done by a chemical analyst can be calculated using equation (2) as follows:

$$\text{Workload} = \frac{\text{8,5 hours per day } x \text{ 20 day per month}}{\text{24,67 hours per sample}} = 7 \text{ sample/month/analyst}$$

Calculation of chemical analyst needs in the AR laboratory to be able to complete all samples received using equation (3) as follows:

$$Number\ of\ chemical\ analysts\ needed = \frac{24{,}67\ hours\ per\ sample\ x\ 639\ sample}{178\ days\ x\ 8{,}5\frac{hours}{day}\ per\ analyst}$$

$$= 10{,}42 \sim 11\ chemical\ analyst$$

The number of chemical analysts required by calculation is 11 chemical analysts, while the number of chemical analysts is currently only 7 people. This amount is inadequate when compared to the number of samples that will be received during the April-December 2017 period.

4. Conclusion

a. Based on research conducted in the laboratory of Analytical Research PT. UPA - Orang Tua Group can be concluded as follows:
b. The estimated number of sample arrivals during the April-December 2017 is 639 samples.
c. Based on the results of processing data from observations of chemical analysis work obtained a standard time of 24.07 hours or 2.9 days per sample with 8 parameters.
d. The maximum workload that an analyst can work on is 7 samples per month.
e. The number of analysts needed to complete the samples that will be come during the April-December 2017 period were 11 chemical analysts. With the availability of the current number of chemical analysts, there are still four additional chemical analysts needed to complete work targets.

5. References

[1] Gawatre 2016 Comparative Study of Population Forecasting Methods *IOSR Journal of Mechanical and Civil Engineering (IOSR-JMCE) e-ISSN: 2278-1684 p-ISSN: 2320-334X* **13**(4V) (July - Aug. 2016) pp 16-19

[2] Isman 2013 *Analisis Beban Kerja di Laboratorium Finger Print PT. PPLi* (Jakarta: Program Studi Teknik Industri Universitas Pancasila)

[3] Ko C S, Cha M S and Rho J J 2007 A Case Study for Determining Standard Time in a Multi-pattern and Short Life-cycle Production System. *Computers & Industrial Engineering* **53**(2) pp 321-325

[4] Mangkuprawira 2004 *Manajemen Sumber Daya Manusia Strategik* (Bogor: PT. Ghalia Indonesia)

[5] Pardede 2007 *Manajemen Operasi dan Produksi: Teori, Model dan Kebijakan* Revisi ed (Yogyakarta: Penerbit Andi) p 373

[6] Russel R S and Taylor B W 2011 *Operations Management* 7th Ed. (River Street NJ: John Wiley & Sons) p 348

[7] Wignjosoebroto S 2000 *Ergonomi Studi Gerak dan Waktu Teknik Analisis untuk Peningkatan Produktivitas Kerja.* 1st Ed. (Surabaya: Penerbit Guna Widya)

IOP Conf. Series: Materials Science and Engineering **528** (2019) 012021 doi:10.1088/1757-899X/528/1/012021

Design of Mechanical Roasting Machine using Ergonomic Approach to Increase Productivity of Small Medium Enterprise: A Research at Opak Craftsmen in Sumedang

W Tripiawan[1], Husni Amani[12], R M El Hadi[3], B H Sagita[4], and S Martini[5]

Industrial Engineering Program, Telkom University, Bandung, Indonesia
[1]wawan.tripiawan@gmail.com, [2]husni.amani @gmail.com, [3]rosadmeh2014@gmail.com, [4]bobyhs.telkomu@gmail.com, [5]srimartini59m@gmail.com.

Abstract. The object of this research is small and medium micro enterprises that make sticky "Opak" with a manual process. The process of ripening food by roasting, where food is placed on a ram wire that is placed on top of the fire stove. The quality of the opaque produced is very dependent on the skills of the employees and the accuracy of flipping through the opaque to obtain good results. In fact, the opaque quality of roasting results varies. Besides, limited production results, causing high demand from customers often cannot be fulfilled by the "Opak" craftsmen. To overcome this problem, the research team designed ripening of glutinous "Opak" with roller method. This sticky rice opaque machine with roller method uses electric power with a low-speed level, so the craftsmen can easily control the maturity level of food ingredients. With the making of the "Opak" sticking machine in the roller method, it is expected that the ripening process can be done with a machine and the maturity level can be uniform; so that the production process is not limited by employee skills; and customer requests can be met.
Keywords: Roasting, Applied Technology, Ergonomics, Optimization

1. Introduction

Small industries are often seen as a backward part of the economic structure, considered unable to be expected to make a meaningful contribution to economic growth and need to get help for a sense of social justice. This view is not always approved by many parties. But some people consider the small industry as an absolute part that must exist in a healthy economic structure. Until today, most of the industries are still implementing the conventional process to produce their product. One of these is still depends on the skill of its employee.

Opak is one of the traditional foods in West Java that is quite popular. This food is usually consumed as a snack or "snack" food type of crackers. The main raw material in making this opaque is glutinous rice while other commonly used ingredients are coconut milk. Opak is made by baking on coals. Glutinous rice that has been soaked overnight is cooked until it becomes rice after it is mixed with coconut milk and spices [1].

The need for a wide area as a place of drying. Spacious of drying land owned by a small-medium enterprise (SME) can also affect the amount of product that can be produced. If a small-medium enterprise (SME) has a wide area as a place of drying, more products can be dried and produced, and vice versa. So, when the market demand for the product has increased, small medium enterprise (SME) needs to consider a wide area of drying, the amount of product to be dried, and the length of time for drying the product until it is completely dry. It would be very unfortunate if a large market opportunity cannot be fulfilled by the small medium enterprise (SME) because of the limited area of drying. [2].

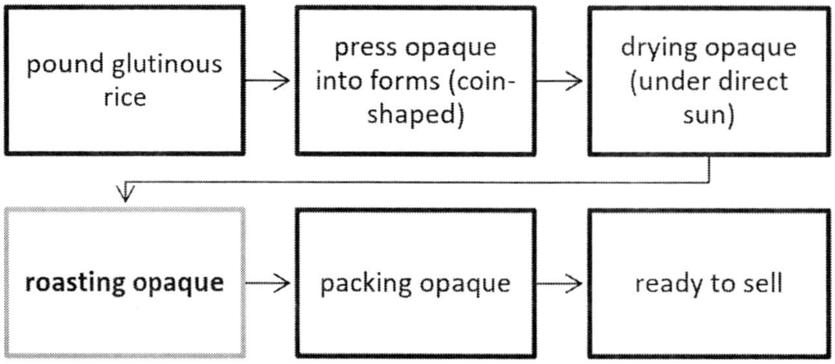

Figure 1. The Stages of creating "Opak Ketan" at Opak Craftsmen

This research focused on the roasting process. In the roasting process, a drying technique with a temperature of 100°C - 120°C which is fuelled by firewood or charcoal is required. In the existing process, the roasting process is done by the manual technique by three to five workers at a time. The stove was made of dry clay, that box-shaped or round-shaped filled by charcoal/firewood. The process for roasting may take 10 – 15 minutes of roasting to make the opaque at a good level of baked. The criteria of a well baked opaque can be seen in the color of the opaque result. The color of the good level of baked is yellow-brownish.

Figure 2. Traditional Process of Roasting

In each roasting workstation, one worker only gets about 100 pieces of baked opaque, that need 10 – 15 minutes processing time.

2. Methods

2.1. Productivity

Productivity, in general, is the ratio between the inputs used with the output produced. One of the steps taken by a company in order to improve its productivity is by making improvements of tools (technology) or by increasing human resources.

IOP Conf. Series: Materials Science and Engineering **528** (2019) 012021 doi:10.1088/1757-899X/528/1/012021

2.2. Ergonomic

Ergonomics is a science that studies and examines the limitations and advantages of humans, then the information obtained will be used to design products, machinery, facilities, environment, and work system. The main objective of an ergonomic application is the application of good aspects of health, safety, and working comfort to achieve a good quality of work. At a higher level of ergonomics aims to create optimal working conditions. Ergonomics and K3 (Occupational Safety and Health) are two things that cannot be separated because they have the same goal that is in improving the quality of work [3]. The definition of ergonomics can be described in a focusing, objectives, and the approach of ergonomics where the explanation is mentioned as follows [4]:

1. Ergonomics focuses on human interactions with the product, equipment, facilities, procedures, and environments where human daily life and work.
2. The goal of ergonomics is to increase the effectiveness and efficiency, improved safety, reduced fatigue, and so on.
3. Ergonomics approach is the application of information about human limitations, capabilities, behavior characteristics and motivation to design procedures and the environment in which human activity daily.

2.3. Product Development Stage

A product is a thing which can be sold by a company for a customer to fulfil their needs. Product development is a set of activity, begin from perception analysis, then end with production, selling, and distribution. The product development process generally comprises the steps of an early-stage or often also referred to as phases. According to [5] in his book, general product development process consists of 6 phases, these are:

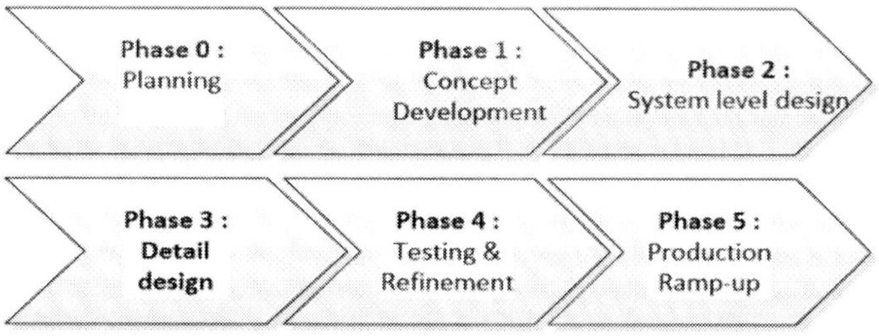

Figure 3. Phases of Product Development

- Planning Phase
  The planning phase is an activity called 'zero phase' because this activity precedes the approval of the project and the process of launching the actual product development.
- Product Development Process
  In the development phase, we have to identify the target market, market needs, and concept alternative and evaluate the chosen concept for future development. This concept is a description of shape, function, and design product, and it consists of product specification, product analysis, competitor analysis, and economic project analysis.
- System Level Design
  In this phase, we are defining the product architecture and detail product become a subsystem and component. The outputs for this phase are product shape, a functional specification from every subsystem, then flow diagram to define the ending of the process.
- Detail Design

Detail design phase includes full specifications of shape, material, and tolerance of all component units of the product and identification of all standard components purchased from suppliers. Otherwise, the process plan and equipment designed for each component are made, in the production system. The output of this phase is to control the recording of the products, images for each component of the product and production equipment, the specification of components that can be purchased, as well as plans for the manufacture and assembly of products

- Testing and Refinement
  In this phase of testing will be conducted to test the components of critical components such as material test of endurance.
- Production Ramp-up
  At this stage, the product was launched on a small scale in order to get feedback. A product is usually presented in the company's internal scale as input for further evaluation.

2.4. Conceptual Model

This research designs a roasting machine that will be used to roast/bake opaque using ergonomic approach. To design the machine, the first thing to do is to identify the activity of roasting opaque with the existing technique and to identify what the complaints are often experienced by the workers. After that will be analyzed regarding ergonomic and create design. The results of the roasting machine design will then be made technical drawings to know the dimensions in each part.

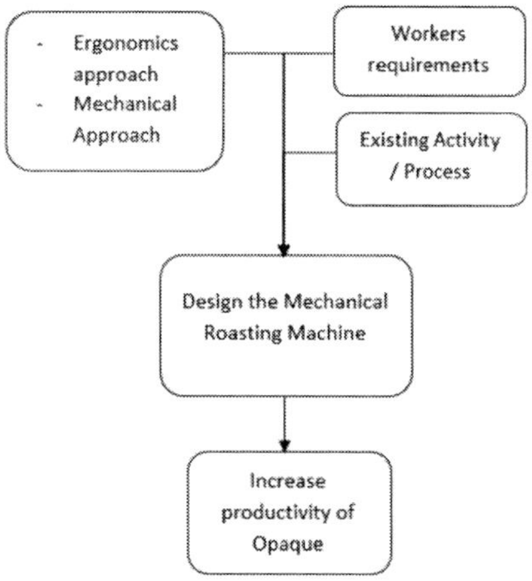

Figure 4. Conceptual Model of Research

3. Result and Discussion

3.1 Identifying Needs Statement

In the stage of identifying needs statement will be done by direct interview with the workers and the craftsmen owner. So, we get the results recapitulation needs statement as follows.

Table 1. Recapitulation Needs Statement

No.	Needs Statement
1	The product has a maximum carrying capacity.
2	The product is safe when in use.
3	Products can reduce the workload of workers and reduce the risk of work accident.
4	The product has a comfortable size when in use.
5	The product has an ergonomic design.
6	The product can minimize the number of required operators.
7	The product has tools.

3.2 Create the design

In the design process, we interpret from the needs statement into the design stage. We create the roasting machine design into several parts:

1. Main frame
   The main frame is the basic frame for the machine. In this frame, all another part will be attached, using bolt and nuts. The dimensions of this part are measured based on anthropometrics from average Indonesia people. The height of the main frame only a half from the average height of adult people in Asia which around 162 cm [6]. So, the dimensions of the main frame are 85 cm, 125 cm width, and 200 cm of length.
2. Hexagonal-shaped roller frame
   The hexagonal-shaped roller frame will replace the existing main process of roasting: roasting and flipping the opaque. The hexagonal-shaped are considered as the best fit shaped to make the opaque can flipped by itself while get baked. This will reduce the dependency on worker accuracy and speed during the flipping process.
3. Move-able charcoal stove
   To reduce the use of drying/roasting area, we design that the charcoal stove is attached below the main frame, and it is adjustable depends on the function. If it needs to remove or change the charcoal, the worker can easily slide the stove outside. When the workers need to roast the opaque, then they can slide in the stove back again.
4. Puller Stick
   The puller stick is stuck in the stove using bolts, so the workers not exposed to heat from the grill. The long of the puller stick can be customized based on the requirement. In this case, we design the length of the stick around 75 cm.

Table 2. Final Specification of Roasting Machine

No	Item	Height	Width	Long
1	Main frame	85 cm	125 cm	200 cm
2	Hexagonal-shaped frame	50 cm (diameter)	100 cm	-
3	Move-able charcoal stove	30 cm	100 cm	120 cm
4	Puller stick	-	-	75 cm

After obtaining the desired final specification, then we made the roasting machine design. The following is the result of the mechanical roasting machine design.

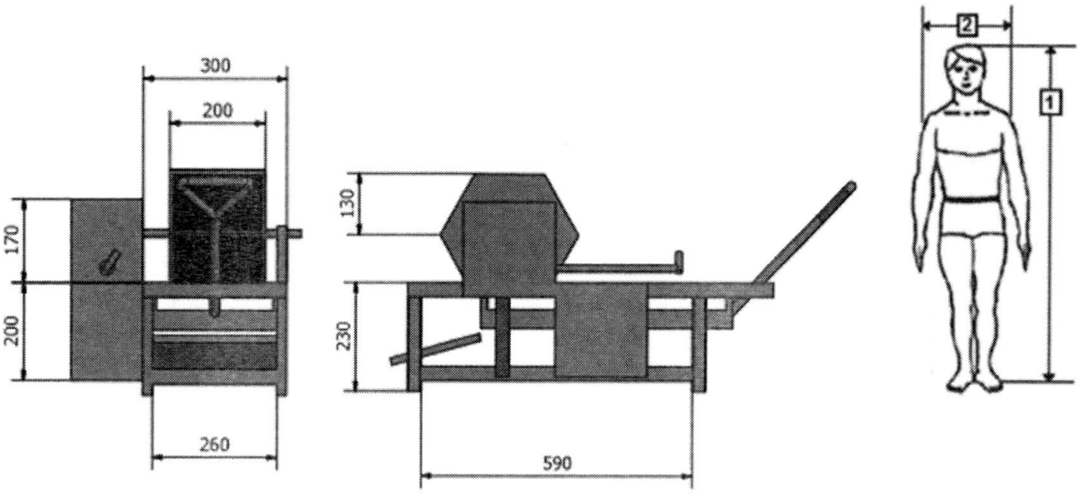

Figure 5. The Design of Mechanical Roasting Machine

IOP Conf. Series: Materials Science and Engineering **528** (2019) 012021 doi:10.1088/1757-899X/528/1/012021

3.3 Create the Prototype

After the design process is approved, we create the mechanical roasting machine prototypes on local iron craftsmen, so that the workers and the owner can evaluate that the use of the roasting machine improved their productivity. Before using the machine, each worker will be baked approximately 10-15 minutes for 100 pieces of opaque, and they have to continuously and manually flip the opaque. After using the machine, one worker can bake about 200-300 pieces of opaque, without continuously and manually flipping the opaque to get the good level of baked opaque.

Figure 6. The Final Assembly of Mechanical Roasting Machine

A comparison results of production before and after using the roasting -machine can be seen in Table below.

Table 3. The comparison results of production before and after using the roasting-machine

Variable	Before using machine	After using machine
Time	10 - 15 minutes	10 – 15 minutes
Capacity	100 pcs/stove	250 – 300 pcs/stove
Flipping method	Manual by worker	Auto flipping by the hexagonal-shaped frame
Charcoal stove	Static stove	Moveable stove

Based on the data above, the productivity level is increased, because the craftsmen can produce more quantity in the same production time. At the consideration, the other workers need to improve their skill to operate the machine.

4. Conclusion

Based on the results of data processing and analysis, then obtained some conclusions that can solve the problems in this study. Here is the conclusion that has been obtained.

1. The product of the mechanical roasting machine can meet the objectives of the research that can design a product that can improve work productivity by applying the concept of ergonomics in the process of designing.
2. In designing the mechanical roasting machine, the material which will be applied to the product is steel plate for the platform, frame, move-able stove, the wire mesh for the hexagonal-shaped frame, and iron pipe rod/iron stick for the puller stick.

IOP Conf. Series: Materials Science and Engineering **528** (2019) 012021 doi:10.1088/1757-899X/528/1/012021

5. References

[1] Listyani, Alinna dan Zubaidah, Elok. (2015), '*Formulasi Opak Bekatul Padi (Kajian Penambahan Bekatul Dan Proporsi Tepung Ketan Putih: Terigu)*', Jurnal Pangan dan Agroindustri. Vol. 3 No 3 p.950-956.

[2] N W Soraya et al , (2017), '*Feasibility study analysis for multi-function dual energy oven (case study: tapioca crackers small medium enterprise)*', IOP Conf. Series: Materials Science and Engineering 277.

[3] Yassierli *et al.* (2007) '*The influence of age on isometric endurance and fatigue is muscle dependent: A study of shoulder abduction and torso extension*', *Ergonomics*, 50(1), pp. 26–45. doi: 10.1080/00140130600967323.

[4] Widodo, Lamto *et al.* (2014) '*Designing of carton box stacking tool to increase the efficiency of pond machine at PT. Supreme Tirta Larisindo*', *Proceeding 7th International Seminar on Industrial Engineering and Management*, pp. ER35–42.

[5] Ulrich, K. T. and Eppinger, S. D. (2012) *Product Design and Development*, *Product Design and Development*. doi: 10.1016/B978-0-7506-8985-4.00002-4.

[6] Deros, B.M. *et al.* (2009). *Recommended Chair and Work Surfaces Dimensions of VDT Tasks for Malaysian Citizens*. European Journal of Scientific Research. 34. 156-167.

Situation Awareness Analysis on Motorcycle Riders using Quantitative Analysis of Situational Awareness

Hartomo Soewardi[1] and Amalia Diah Ayu Kiranti[2]

Industrial Engineering Department
Faculty of Industrial Technology
Islamic University of Indonesia Yogyakarta Indonesia
hartomo@uii.ac.id[1], amalliadiahak@gmail.com[2]

Abstract. A Traffic accident is caused by the negligence and the low situation awareness of the motorcycle riders. Based on the Indonesian Police traffic data, until May 2018, there were 32,127 motorcycle accidents with 6,009 people died. The highest accidents happened to the riders between 15-19 years old. Thus, feasibility of the motorcycle riders becomes crucial. The purpose of this study is to determine the level of awareness of the motorcycle riders as well as to determine the minimum age level to ride motorcycle. There were 75 respondents participating in this study who were about 17 to 19 years old. The Quantitative Analysis of Situational Awareness method was used to achieve the objective. The results of this study show that 17 years old riders have better situation awareness level than 18 years old ones. Meanwhile, the best situation awareness level is at age 19. It is because of the value of calibration (12.22%) which is smaller than 17 years old (17.52%) and 18 years old riders (26.48%). Therefore, this study recommends the feasibility of motorcycle riders should be 19 years old as the minimum requirement.
Keywords: Quantitative Analysis, Situational Awareness, Calibration, Level of Awareness

1. INTRODUCTION

The number of motorcycle accidents is caused by the rider's negligence and low situation awareness. Many accidents are caused by riders' low situation awareness [2]. Based on the Indonesian Police traffic data, until May 2018, there were 32,127 motorcycle accidents with 6,009 people died, and the rest experienced minor and severe injury [3]. The accidents were dominated by riders between 15-19 years old [3]. Most of them had not had a driving license even though they were able to ride. However, the facts show that there are still many accidents happened. It indicates that the riders' level of awareness has not satisfied the requirements. Thus, the feasibility of motorcycle riders is considered crucial.

Many studies have been done related to the situation awareness like those done by [2], [4], [7-8]. Febrianti, et al. (2013) found that a driver has lack of confidence even though the skill of driving is good for the public car driver. While [2] investigated the situation awareness of young car drivers. He found that most of the young drivers have low of situation awareness. It is because they tend to feel confident but the technique of driving is still unstable. In addition, numerous causes of accidents at the pilot including the perception, understanding of the situation, and anticipation of the future event had been investigated by [8]. Fang, et al. (2018) studied the assessment of operator's situation awareness for the smart operation of mobile cranes. It was identified that to increase the performance, the situation awareness should be improved. High self-confidence in driving that is incomparable to the appropriate driving method will endanger the drivers themselves and those around them, which will lead to high accident rates [2] [4].

This study focuses on assessing the actual accuracy value and perceived accuracy value to identify the characteristic of situation awareness and to determine the level of awareness of the motorcycle riders. This study also develops a recommendation and requirement of the minimum age for the feasibility of riding a motorcycle.

2. RESEARCH METHOD

2.1 Survey

In this study, determination of the sample can be determined using the following formula [1]:

$$n = \frac{N}{1+N(e)^2} \quad (1)$$

And: n = number of samples
N = population
e = critical value / desired accuracy limit

The critical value of 0.2 or 20% is used. Based on the data from the population of Sleman, Yogyakarta, in 2016, the number of residents age 15-19 amount up to 90,833 people (BPS, 2017). Therefore the calculation results in 75 respondents. Participants have an ability in riding motorcycle. They are divided into three groups with 25 people for each group. Group 1 is for 17 years old riders, group 2 is for 18 years old ones, and group 3 is for 19 years old ones. Each respondent is instructed to answer a situation awareness questionnaire consisting of 28 statements. The answer has two types of scores those are the actual accuracy score and perceived accuracy score.

2.2 Apparatus

The tools used to support the research are:

1. Questionnaire for testing the situation awareness
2. IBM SPSS software version 22 that offers advanced statistical analysis.
3. Microsoft Excel to calculate the value of research results.

2.3 Situation Awareness

There are three levels of situation awareness in the Endsley model i.e. perception of elements in the environment related to the task and purpose, understanding of the importance of the elements and the formation of mental models, and the predictions of the future [4] [5].

2.4 Quantitative Analysis of Situational Awareness (QUASA)

Quantitative Analysis of Situational Awareness (QUASA) is a technique that combines both objective questions (true/false probes) and subjective self-confidence ratings for each response [6]. Calibration of Situation Awareness is conducted to assess the level of suitability between actual accuracy and perceived accuracy as a proportion of error. Calibration can be determined by giving a trust rating. Trust ratings can be binary rank, ordinal multi-category rank, or sustainable scale [5]. The result is shown in a curve which consists of Y axis for actual accuracy (%) and X axis for perceived accuracy (%) [2] [7] [6]. The criteria of the calibration curve can be classified as follow [7]:

1. **Over-confident** means that the riders have self-confidence in riding, but the riding skill is still unstable.
2. **Under -confident** means that the riders have lack of self-confidence in riding, but the riding skill is stable.

As for the steps of data analysis by using this method is as follow:

1. Calculating the total score for actual accuracy and perceived accuracy is by equation [2]:

$$\text{Total score} = \text{number of questions} \times \text{number of respondents} \times 100\% \quad (2)$$

2. Determining the percentage for actual accuracy and perceived accuracy through equations [2]:
   To determine the actual accuracy value:
   $$\text{Actual accuracy} = \frac{\text{number of respondent answers correctly}}{\text{total score}} \times 100\% \quad (3)$$
   To determine the value of perceived accuracy:
   $$\text{Perceived accuracy} = \frac{\text{total respondent's confidence score}}{\text{total confidence level score}} \times 100\% \quad (4)$$
   Where the confidence score is calculated as below [6]:
   Confidence score = (k1 × 100%) + (k2 × 75%) + (k3 × 50%) + (k4 × 25%)
   And: k1 = number of respondents who answered very confidently
   k2 = number of respondents who answered confidently
   k3 = the number of respondents who answered was pretty sure
   k4 = number of respondents who answered less confidently
3. Developing a calibration curve is based on the results of step 2 (Y-axis that is actual accuracy and the X-axis is perceived accuracy).
4. Determine the level of awareness for each group of age is below [2]:
   $$\text{Level of Awareness} = \frac{(\sum (\text{confidence level score} \times \text{correct answer score}))}{\text{total score}} \times 100\% \quad (5)$$
5. Analyzing the results and the recommendations.

3. RESULTS AND DISCUSSION

3.1 Result of Statistical Analysis

Determination of the validation of each item statement through validity testing using Pearson Product Moment shows that 27 statements are valid, yet 1 statement is invalid, so that the reliability test performed using Cronbach's Alpha is 27 statements with high reliability.

3.2 Result of Calibration of Situation Awareness

The result of this study can be seen in Table 3.1 and Figure 3.1, Figure 3.2, and Figure 3.3. Table 3.1 shows the value of actual accuracy, perceived accuracy, and an error value for each age. While Figure 3.1, Figure 3.2, and Figure 3.3 are the result of calibration of situation awareness calculation for 17 years old, 18 years old, and 19 years old respectively.

Table 3. 1 Perceived Accuracy and Actual Accuracy Values

Age	Actual Accuracy	Perceived Accuracy	Error Value
17	62.96%	80.48%	17.52%
18	51.11%	77.59%	26.48%
19	66.52%	78.74%	12.22%

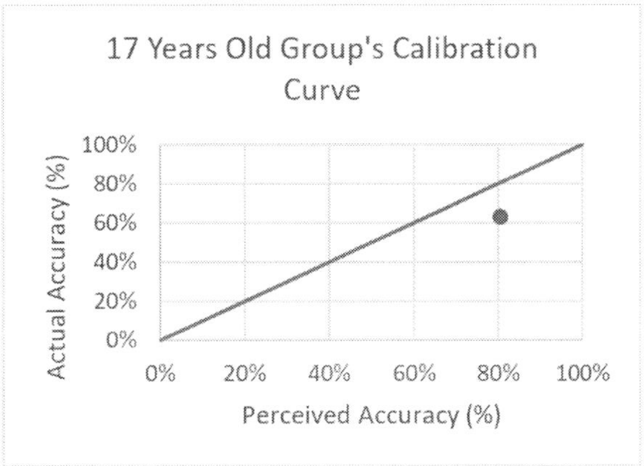

Figure 3. 1 Calibration curve for 17-year-old riders

Based on Figure 3.1 and Table 3.1, the actual accuracy value is 62.96% which is lower than the perceived accuracy value of 80.48%, it means the respondent's level of self-confidence is higher than what is actually experienced, then the error value is 17.52% which means the curve does not show well calibrated. So that, the characteristic of 17 years old riders is over-confident which means that the riders have self-confidence in riding, but the riding skill is still unstable.

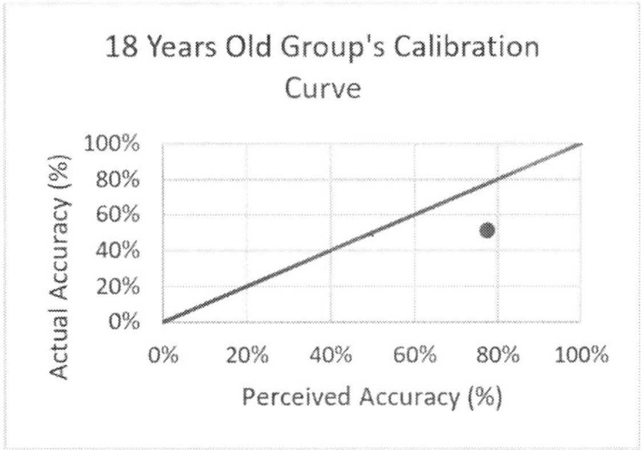

Figure 3. 2 Calibration curve for 18-year-old riders

Based on Figure 3.1 and Table 3.2, the actual accuracy value is 77.59% which is lower than the perceived accuracy value of 51.11%, it means that the respondent's level of self-confidence is higher than what is actually experienced, then the error value is 26.48% that means the curve does not show well calibrated. This figure is much lower compared to 17 years old riders. So that, the characteristic of 17 years old riders is over-confident which means that the drivers have a self-confidence in riding, but the riding skill is still unstable.

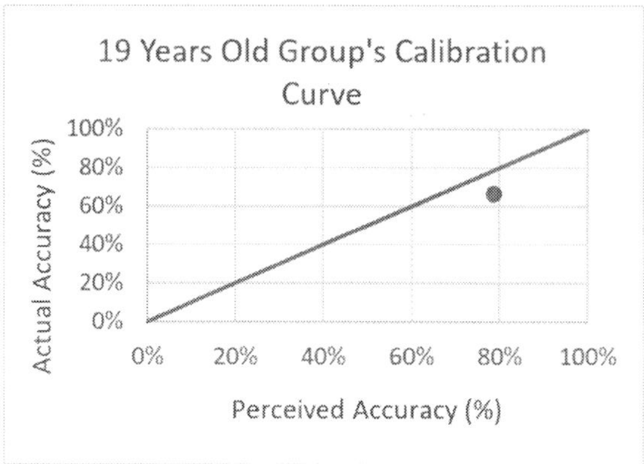

Figure 3. 3 Calibration curve for 19-year-old riders

Based on Figure 3.1 and Table 3.3, the actual accuracy value is 66.52% which is lower than the perceived accuracy value of 78.74%, it means the respondent's level of self-confidence is higher than what is actually experienced, then the error value is 12.22% and the calibration curve almost well calibrated. This figure is much higher compared to 17 years old riders and 18 years old ones. So that, the characteristic of 17 years old riders is over-confident which means that the riders have self-confidence in riding, and the riding skill is still unstable but better than 17 years old and 18 years old riders.

The best comparison between actual accuracy and perceived accuracy is if the two values are balanced to form a well-calibrated curve. Of the three error values, it is known that the smallest error value is owned by 19 years old group, which is 12.22%, so the group has better riding characteristics than the groups of 17 and 18 years old ones. The error value also shows that 17 years old riders have higher level of confidence compared to 19 years old riders.

3.3 Result of Level of Awareness

The results of the level of awareness in each age group are presented in table 3.2:

Table 3. 2 the level of awareness

Age (Years Old)	Level of Awareness
17	53.04%
18	41.15%
19	53.63%

Based on Table 3.2, 17 years old riders have 53.04% level of awareness, 18 years old ones have 41.15% level of awareness, and 19 years old riders have 53.63% level of awareness. The highest level of awareness is owned by 19 years old group. The level of awareness is still not at a safe level, the low value can be due to many traffic behaviors that are not good and tend to underestimate the risks that can harm the riders themselves and others.

Based on the results, 19 years old is better criteria to be the minimum age to have a driving license to reduce the number of accidents. According to [9], it is stated that age can influence safe driving behavior, but there are other factors including education, knowledge, and attitude. Thus, riders with high school education background still have an attitude of willingness to be cared for and mimic the behavior of other riders by riding the pace of the vehicle without considering the risk of accidents [10]. However, according to government regulations, 17 years old citizens can have a valid driving license even though they are still prone to traffic accidents. Thus, supervision from more experienced people to provide guidance and a good understanding of safe driving ethics as well as the provision of tools safety are necessary to minimize the risk of traffic accidents.

4. CONCLUSION

Based on the results, the comparison between the actual accuracy value and the perceived accuracy value for each age are as follows:

1. The actual accuracy value of age 17 is 62.96% and the perceived accuracy value is 80.48%, so the error value is 17.52%.
2. The actual accuracy value of age 18 is 51.11% and the perceived accuracy value is 77.59%, so the error value is 26.48%.
3. The actual accuracy value of age 19 is 66.52% and the perceived accuracy value is 78.74%, so the error value is 12.22%.

The characteristics of the riders of the three ages are over-confident, which means that the riders have self-confidence in riding, but the riding skill is still unstable. The level of awareness of 17-year-old motorcycle riders is 53.04%, 18-year-old riders is 41.15%, and 19-year-old riders is 53.64%. 19 years old is better and feasible to be minimum age of driving license to reduce the number of accidents.

5. REFERENCES

[1] S. Rahayu and I. S. Lingga 2009 "Pengaruh modernisasi sistem administrasi perpajakan terhadap kepatuhan wajib pajak," Jurnal Akuntansi, vol. 1, no. 2, pp. 119-138.

[2] A. S. Prawito, Yuniar and A. Desrianty 2014 "Pengukuran tingkat kewaspadaan pengemudi mobil usia muda di Kota Bandung menggunakan Quantitative Analysis of Situational Awareness (QUASA)," Reka Integra, vol. 1, no. 4, pp. 169-179.

[3] K. Polri 2018 "Kecelakaan di Indonesia selama triwulan terakhir," [Online]. Available: http://korlantas-irsms.info/graph/ageInjuryData. [Accessed 2 Mei 2018].

[4] Y. Fang, Y. K. Cho, F. Druso and J. Seo 2018 "Assessment of operator's situation awareness for smart operation of mobile cranes," Automation in Contruction, vol. 85, pp. 65-75.

[5] J. Rose, C. Bearman and J. Dorrian 2018 "The Low-Event Task Subjective Situation Awareness (LETSSA) technique: Development and evaluation of a new subjective measure of situation awareness," Applied Ergonomics, vol. 68, pp. 273-282.

[6] B. McGuinness 2004 "Quantitative analysis or situational awareness (QUASA): applying signal detection theory o true/false probes and self-ratings," Bae Systems.

[7] A. Febrianti, A. Desrianty and Yuniar 2013 "Tingkat kewaspadaan dan perilaku pengemudi angkutan kota berdasarkan karakteristik jarak tempuh trayek menggunakan metode QUASA dan Driver Behaviour Questionnaire," Seminar Nasional V Manajemen & Rekayasa Kualitas , vol. V, pp. 1-8.

[8] C. M. Muehlethaler and C. P. Knecht 2016 "Situation awareness training for general aviation pilots using eye tracking," IFAC, vol. 49, no. 19, pp. 066-071.

[9] R. Ariwibowo 2013 "Hubungan antara umur, tingkat pendidikan, pengetahuan, sikap terhadap praktik safety riding awareness pada pengendara ojek sepeda motor di Kecamatan Banyumanik," Jurnal Kesehatan Masyarakat, vol. 2, no. 1, pp. 1-8.

[10] L. Ambarwati, H. Sulistio, G. H. Negara and Z. Hariadi 2010 "Karakteristik dan peluang kecelakaan pada mobil pribadi di wilayah perkotaan," Jurnal Rekayasa Sipil, vol. 4, no. 2, pp. 124-135.

IOP Conf. Series: Materials Science and Engineering **528** (2019) 012023 doi:10.1088/1757-899X/528/1/012023

Analysis of Machine Maintenance Processes by using FMEA Method in the Sugar Industry

Hartomo Soewardi[1], Siska Ari Wulandari[2]

Industrial Engineering Department, Faculty of Industrial Technology
Islamic University of Indonesia Yogyakarta Indonesia
hartomo@uii.ac.id[1], siskaarii@gmail.com[2]

Abstract. Sugar is the daily basic need of human life. It is made from cane trees and manufactured using a milling machine in the early stages. This machine is operated 24 hours per day; consequently, it should be in a good condition and the company must be able to ensure the machine can run well by scheduling regular maintenance. This study aimed to investigate the potential failure of manufacturing process to provide some recommendations for improvement. Failure Mode and Effect Analysis (FMEA) method was applied to analyze the situation, and Logic Tree Analysis (LTA) was implemented to classify the types of improvement. This study found 13 failure modes, including a loose hammer tip, a broken end of the cutter, a non-balanced hammer rotation, a broken hammer handlebar, a broken wheel bearing, a loose switch, a loose split pen, a loose fastener bolt, an overload, a destroyed rubber seal, a broken turbine coupling, a broken screw-bolt, and a hollow roller. Meanwhile, the classification of improvement consisted of 3 categories namely safety problem (A), outage problem (B), and economic problem (C). Subsequently, the A category recommended setting up the machine in accordance with SOP while the B category recommended inspecting the machine routinely and preparing parts. In addition, the C category recommended performing accurate installation, lubrication, and cleaning.
Keywords: Milling Machine, FMEA, LTA, Preventive Maintenance, Failure Mode

1. INTRODUCTION

One of the operation elements that affect the quality of production is the machines used because a continuous machine use can cause a lot of trouble and wear. To prevent trouble, maintenance should be done [1] including repair, arrangement, and replacement of parts in accordance with the schedule [2].

In the sugar industry, the milling machine is the main machine in the initial sugarcane processing operated for 24 hours per day. This process uses the flow shop system operating from the beginning to the end in sequence. If there is any trouble in the milling station, it cannot proceed to the next production stage. This condition should not take place; therefore, preventive maintenance should be regularly performed.

Failure Mode and Effect Analysis (FMEA) method is a systematic process for identifying potential failures of designs and processes with the purpose of minimizing or eliminating these failures [3]. In this study, FMEA method was utilized as a tool to identify potential machine failures by considering the Risk Priority Number (RPN).

This paper focused on proposed preventive maintenance for the milling station to prevent potential machine failures. This study also provided some recommendations for improvement which focused not only on failure modes with the highest rating based on the Risk Priority Number (RPN) but also on all failure modes considered as trouble. This is because, if a failure mode even with the smallest Risk Priority Number (RPN) is ignored, problems such as high repair costs or threat to the operator and work environment will occur.
To identify the cause of the failure and failure modes that are the most influential on the machine, the FMEA method (Failure Mode and Effect Analysis) was employed while LTA (Logic Tree Analysis) was used to find out the failure mode category for the following repair steps that should be done to prevent failure [4].

2. METHODS

2.1 Survey

A survey by interviewing the supervisor was done to identify the ongoing production process of the milling machine as well as to find out the detailed conditions of the machine.
Further survey was conducted by distributing questionnaires to 15 workers for assessing the level of severity, occurrence, and detection of machine failure. The workers were machinist assistants, foremen, deputy foremen, and operators of the maintenance section. The rating scale should be used as a reference to determine the score of frequency of occurrence, degree of severity, and change of detection.

2.2 Failure Mode and Effect Analysis (FMEA)

Failure Mode and Effect Analysis (FMEA) is a method for evaluating the possibility of failure of a system, design, process, or service which can be quantified for determining the priority of a possible solution (Risk Priority Number) [5]. FMEA encourages higher reliability and quality and improves security. This can also be used to assess and optimize a maintenance plan [6]. In this study, such method was used to identify the failures that occurred, including:

1. Damage level (severity) or how serious the damage to the process was
2. Frequency (occurrence) or how much interference could cause a failure
3. Detection level (detection) or how failures could be identified before happening [5]

In FMEA, the Risk Priority Number (RPN) represents the relationship between three variables that include Severity (S), Occurrence (O), and Detection actions [7]. The RPN provides guidance to determine potential failures and then provides suggested actions for design/process changes to the severity so that appearances are lower [8].
The steps taken were:

1. Identifying the machines used in the production process
2. Identifying some potential failure modes of the machine, the effect of failure modes on the machine, and the cause of the failure modes
3. Assessing failure modes on the machine for severity (S), occurrence (O), and detection (D)
4. Calculating the Risk Priority Number (RPN) using the equation [5]:

$$RPN = S \times O \times D \quad (1)$$

2.3 Logic Tree Analysis (LTA)

Logic Tree Analysis (LTA) was used to determine the consequences and give priority to each failure mode [9]. It is a qualitative process to categorize each failure mode by answering questions as follows [10]:

1. Evident: Does the operator know that in normal conditions there has been disruption in the system?
2. Safety: Does the failure mode cause safety problems?
3. Outage: Does the failure mode cause the machine to stop?

The problem category was divided into four consisting of [10]:

IOP Conf. Series: Materials Science and Engineering **528** (2019) 012023 doi:10.1088/1757-899X/528/1/012023

1. A category or safety problem: if failure mode caused safety disturbance to the operator or the environment
2. B category or outage problem: if failure mode resulted in system failure that could cause significant economic losses
3. C category or economic problem: if failure mode did not affect the safety or failure of the system, but it caused relatively small economic losses for repair
4. D category or hidden failure: if the failure mode was difficult to detect because it was hidden from the operator's vision

3. RESULT AND DISCUSSION

3.1 Result of Failure Mode and Effect Analysis (FMEA)

The results of this study are shown in Table 3.1 and Table 3.2 with Table 3.1 describing the function of the machine and Table 3.2 explaining the results of failure modes, effects, and causes of failure modes for each machine as well as the Risk Priority Number (RPN).

For the unigrator machine, the failure modes identified were a loose hammer tip, a broken end of the cutter, a non-balanced hammer rotation, and a broken hammer handlebar. These failure modes resulted from the components that have passed the service life and the material that was not good, making the components slip and enter the machine and cause a stop. In the cane unloading crane machine, the failure modes identified were a broken wheel bearing and a loose switch caused by less lubrication and excess loads making the crane unable to move. Meanwhile, in the cane table and cane carrier machines, the failure mode found was a loose split pen due to wear, causing the chain to break. In the intermediate carrier, the failure modes included a loose fastener bolt and overload because it was not welded and due to low vapor pressure that made the machine motor drive stop working. In the hydraulic pump, the installation was too tight causing damage to the rubber seal, so it could not pump maximally. Meanwhile, in the milling machine, failure modes were also identified because the turbine coupling and screw-bolt were broken and the roller was hollow. Such failures occurred because the material used was not good and the part was hit by a heavy object, leading to less optimal milking process.

Table 3.1 Function of Machine

Name of Machine	Description
Unigrator	Chopping sugar cane
Cane Unloading Crane	Lifts sugar cane from Lorry then move it to Cane Table
Cane Table	Set sugar cane placed and put sugar cane on Cane Carrier
Cane Carrier	Transport sugar cane from Cane Table to Milling Machine
Intermediate Carrier	Transport sugar cane pulp from one Milling Machine to the next
Hydraulic Pump	Put pressure on mill roll to be stable
Milling Machine I-V	Milking sugar cane

IOP Conf. Series: Materials Science and Engineering **528** (2019) 012023 doi:10.1088/1757-899X/528/1/012023

Table 3.2 Failure Modes, Effect and Cause, Risk Priority Number (RPN)

Name of Machine	Failure Mode	Effect	Cause	S	O	D	RPN
Unigrator	a loose hammer tip	Component can enter the milling machine and cause a stop	The components exceed the service life	5.8	2.3	4.1	55
	a broken end of the cutter	Component can enter the milling machine and cause a stop	The material used is not good	5.5	4.6	3.7	92
	a non-balance hammer rotation	Bearing components are damaged	Installation is done with a different hammer weight	3.7	2.6	3.3	32
	a broken hammer handlebar	Component can enter the milling machine and cause a stop	The components exceed the service life	5.7	4.3	3.5	86
Cane Unloading Crane	a broken laker wheel	Crane cannot move	Less lubrication	4.0	5.3	3.9	82
	a loose switch	Motor drive stops working	Electric panels lift high loads	4.3	4.8	3.6	74
Cane Table	a loose split pen	Chain is broken	Wear	3.5	3.7	3.6	47
Cane Carrier	a loose split pen	Chain is broken	Wear	4.2	3.4	3.8	54
Intermediate Carrier	a loose fastener bolt	Component can enter the milling machine and cause a stop	Welding is not done	4.6	3.1	3.7	52
	overload	Motor drive stops working	Vapor pressure is low	5.9	2.6	3.8	67
Hydraulic Pump	sheal rubber destroyed	Pressure is not achieved so it cannot pump maximally	Rubber installation is too tight	3.4	4.3	4.3	64
Milling Machine I-V	the turbine coupling is broken	Gear box cannot be connected	The material used is not good	6.3	3.5	3.5	77
	the screw-bolt is broken	Component can enter the milling machine and cause a stop	The material used is not good	6.4	5.7	3.7	135
	the roller is hollow	Milking process is less than optimal	Heavy object hits	3.2	5.6	3.9	70

From the data in Table 3.2, an assessment of FMEA variables was made containing the severity, occurrence, and detection of failure modes in each machine followed by calculation of the value of Risk Priority Number (RPN).

The highest RPN value is obtained when the severity level of the failure modes has a significant effect in which the machine experiences downtime for the repair process. The high frequency of occurrence and high detection rate on the machine because of the failure mode that occurs can be detected by the machine operator with eyesight.

In contrast, a low RPN value is obtained because the severity of the failure modes has a mild effect which does not interfere with the production process and the machine continues to operate. The level of occurrence with a slight frequency and a high detection rate for damage due to failure mode can be detected by the operator through sound and vibration.

3.2 *Result of Logic Tree Analysis (LTA)*

Table 3.3 Logic Tree Analysis

Name of Machine	Failure Mode	Evident?	Safety?	Outage?	Category
Unigrator	a loose hammer tip	Yes	No	Yes	B
	a broken end of the cutter	Yes	Yes	Yes	B
	a non-balance hammer rotation	Yes	Yes	No	C
	a broken hammer handlebar	Yes	Yes	Yes	B
Cane Unloading Crane	a broken laker wheel	Yes	No	No	C
	a loose switch		Yes	No	A and C
Cane Table	a loose split pen	Yes	Yes	No	C
Cane Carrier	a loose split pen	Yes	Yes	No	C
Intermediate Carrier	a loose fastener bolt	Yes	Yes	Yes	B
	overload	Yes	No	Yes	B
Hydraulic Pump	sheal rubber destroyed	Yes	No	No	C
Milling Machine I-V	the turbine coupling is broken	Yes	No	Yes	B
	the screw-bolt is broken	Yes	Yes	Yes	B
	the roller is hollow	Yes	No	No	C

The classification of failure modes using LTA method was conducted by answering 3 questions in Table 3.3. The failure mode included in the A category was a loose switch, which is a safety problem. This damage can be identified by the operator when the machine is operating. If this damage is not repaired immediately, it can cause short-circuiting which can threaten the safety and work environment.

The failure modes of B category were a loose hammer tip, a broken end of the cutter, a broken hammer handlebar, a loose fastener bolt, a broken screw-bolt, and overload, categorized as an outage problem. Such damage causes the engine to stop operating for 30-60 minutes because at the time of repairs the operator looks for the broken components that enter the engine. The failure modes in C category consisted of non-balanced hammer rotation, broken wheel bearing, loose split pen, destroyed rubber seal, hollow roller, and loose switch. They do not cause the engine to stop since repairs can be done when the engine is running, so it does not interfere with the sugar production process.

3.3 Preventive Maintenance
Preventive maintenance is performed by carrying out inspections and actions that have been planned for the implementation time to restore the operating function of a machine to avoid severe damage to the engine [11].
Based on the analysis from Table 3.2, it is acknowledged that damage occurred because the component life exceeded the limit, and there was less lubrication, less cleanliness after use, and wear. From this, improvements should be made by checking conditions, scheduling component replacement, giving lubrication, and cleaning regularly. The condition of the machine should be checked every day, lubrication with oil is done once a week, component replacement is carried out after 1 milling season, and cleaning and maintenance of the machine should be performed weekly and monthly.

Table 3.4 Preventive Maintenance

Name of Machine	Failure Mode	LTA Category	Preventive Maintenance
Unigrator	a loose hammer tip	B	Checking the condition regularly, replacing components after one milling season (6 months)
	a broken end of the cutter	B	Checking defects during installation, checking the condition regularly, replacing components after two milling season (1 year)
	a non-balance hammer rotation	C	Installing with the same weight (17kg-19kg per hammer handlebar), checking the condition regularly
	a broken hammer handlebar	B	Checking the condition regularly, replacing bolts before they wear (4 to 5 months)
Cane Unloading Crane	a broken laker wheel	C	Giving regular lubrication with oil and Vaseline, replacing components after two milling season (1 year)
	a loose switch	A and C	Load lifting according to capacity (5, 10 and 16 tons), cleaning of dirt
Cane Table	a loose split pen	C	Checking the condition regularly, replacing components after one milling season (6 months)
Cane Carrier	a loose split pen	C	Checking the condition regularly, replacing components after one milling season (6 months)
Intermediate Carrier	a loose fastener bolt	B	Welding bolts during installation, checking the condition regularly
	overload	B	Checking vapor pressure regularly, maintaining vapor supply for a stable pressure
Hydraulic Pump	sheal rubber destroyed	C	Checking defects during installation, replacing sheal rubber after one milling seasons (6 months)
Milling Machine I-V	the turbine coupling is broken	B	Checking defects during installation, checking the condition regularly, replacing components after two milling season (1 year)
	the screw-bolt is broken	B	Checking the condition regularly, replacing bolts before they wear (4 to 5 months)
	the roller is hollow	C	Controlling heavy objects from grinding

4. CONCLUSION
Based on the results of the processing and discussion carried out, it can be concluded as follows:
The study using Failure Mode and Effect Analysis (FMEA) obtained such failure modes as broken screw-bolt, broken end of the cutter, broken hammer handlebar, broken wheel bearing, broken turbine

coupling, loose switch, hollow roller, overload, destroyed rubber seal, loose hammer tip, loose split pen, loose fastener bolt, loose split pen, and non-balanced hammer rotation.
The classification of failure modes using Logic Tree Analysis (LTA) in each machine found failure modes of A category (safety problem) that included loose switch, failure modes of B category (outage problem) including loose hammer tip, broken end of the cutter, broken hammer handlebar, loose fastener bolt, broken screw-bolt, and overload, and failure modes of C category (economic problem) consisting of non-balanced hammer rotation, broken wheel bearing, loose split pen, destroyed rubber seal, hollow roller, and loose switch.
The following preventive maintenance should be carried out to prevent machine failure based on the LTA category. For the A category (safety problem), it is recommended to operate the machine in accordance with the Standard Operating Procedure (SOP), for the B category (outage problem), recommendation includes regular machine inspection and preparation of parts for replacement of damaged components, and for the C category (economic problem), installation of appropriate components, lubrication, and cleaning of dirt is recommended.

5. REFERENCES

[1] R. R. Mansur A., "Analisis risiko mesin bagging scale dengan metode fuzzy failure mode and affact analysis (FUZZY-FMEA) di Area Pengantongan Pupuk Urea PT. Pupuk Sriwidjaja ," *Teknoin,* vol. 21, pp. 158-166, 2015.

[2] S. Assauri, Manajemen produksi dan operasi, Jakarta: Lembaga Penerbit Fakultas Ekonomi UI, 1993.

[3] K. M. K. Johnson K.G, "A study into the use of the process failure mode and effects analysis (PFMEA) in the automotive industry in the UK," *Journal of Materials Processing Technology,* vol. 139, pp. 348-356, 2003.

[4] D. Y. Munawir H., "Analisa penyebab kerusakan mesin sizing baba sangyo kikai dengan metode FMEA dan LTA (Studi Kasus di PT. PRIMATEXCO INDONESIA)," in *Seminar Nasional IENACO*, Surakarta, 2014.

[5] A. S. C. P. Andiyanto S., "Penerapan metode FMEA (Failure Mode and Effect Analysis) untuk kuantifikasi dan pencegahan resiko akibat terjadinya lean waste," *Jurnal Online Poros Teknik Mesin,* vol. 6, pp. 45-57, 2017.

[6] H. O. P. T. H. Arabian-Hoseynabadi, "Failure Modes and Effects Analysis (FMEA) for wind turbines," *Electrical Power and Energy Systems,* vol. 32, p. 817–824, 2010.

[7] Y. Z. D. M. Mauidzoh U., "Perawatan mesin kompresor udara dengan metode Reliability Centered Maintenance (Studi Kasus di PT Polidayaguna Perkasa," in *Prosiding SNTI dan SATELIT*, Malang, 2017.

[8] K. N.P, "Risk analysis by using Failure Mode and Effect Analysis for safe mining," *International Journal of Science and Research (IJSR),* vol. 3, pp. 2512-2515, 2014.

[9] B. P. Prihastono E., "Perawatan preventif untuk mempertahankan utilitas performence pada mesin cooling tower di CV. Arhu Tapselindo Bandung," *DINAMIKA TEKNIK,* vol. 10, pp. 17-27, 2017.

[10] N. A. Hidayah N.Y., "Analisis pemeliharaan mesin blowmould dengan metode RCM di PT.CCAI," *Jurnal Optimasi Sistem Industri,* vol. 16, pp. 167-176, 2017.

[11] F. A.M, "Perencanaan perawatan mesin injection molding dengan menggunakan metode Reability Centered Maintenance di PT. Victory Plastic," *JTM,* vol. 04, pp. 99-110, 2016.

Designing technical specifications for goyor woven fabric as material for civil servant official uniforms using quality function deployment and conjoint analysis

D Refinda[1], E Liquiddanu[1], I W Suletra[1],W Sutopo[1] and M Budijanto[1]

[1]Department of Industrial Engineering, Sebelas Maret University, Surakarta, Indonesia
deltarefinda@student.uns.ac.id, liquiddanu@gmail.com, suletra@staff.uns.ac.id, wahyudisutopo@staff.uns.ac.id, murman.budijanto@gmail.com

Abstract. Sragen regency has a traditional woven fabric called goyor woven, whose presence is disappearing. One way to revive its existence is to encourage its use as the material for official uniforms of Sragen civil servants. For this purpose, there is a need to research the most preferable design of the goyor woven fabrics by the civil servants. The aim of this research is to determine the technical specifications of the goyor woven fabric by combining the characteristics of the goyor fabric and the preferences of Sragen civil servants. The method used is integrated QFD with conjoint analysis. QFD is used to translate consumer voices into technical specifications while conjoint analysis is used to find the best combination of the proposed goyor woven fabric design with the technical specifications. The initial result of the research indicates that the technical specifications of the goyor fabric design that is formed are red and blue fabrics with floral and geometric patterns. Based on conjoint analysis the option that best suits the preferences of Sragen civil servants is the combination of blue goyor fabric with geometric pattern.
Keywords: Goyor woven fabric, Conjoint Analysis, Quality Function Deployment (QFD), Technical Spesifications

1. Introduction

One of the typical Central Java woven fabrics that penetrate the international market is goyor woven fabric. Goyor in Javanese means soft or falling, because if it is used the cloth falls and is not stiff. Goyor fabric is made using traditional looms and made from rayon yarn. Goyor woven fabric has a characteristic that is cool when used in hot weather but feels warm when used in cold weather.

In Central Java, there are two regions which are the center of goyor woven fabric production, it isTawangsari-Sukoharjo and in Kalijambe-Sragen. For the Tawangsari region, the production of goyor woven fabric is done to supply the export market, while in Kalijambe the production of goyor woven fabric has begun to disappear and leaves only a few workers who make goyor woven fabrics if there are orders from local businessmen, and even then the weaving process only.

Based on the field observations and data from central statistics agency, the decline in the production of goyor woven fabric in Kalijambe was inseparable from the declining level of the goyor woven fabric sales. The sale of goyor woven fabrics for the export market has been dominated by the Tawangsari area, while the goyor woven fabric itself has not been able to attract the attention of the local market. Based on this, one of the ways to revive the center of the goyor Sragen woven fabric is by targeting the local market [6]. To open the way of marketing goyor fabric in the local market is to make its use as the main material for civil servants uniforms in Sragen. With the local government enforcing its employees to wear a series of goyor woven fabrics, the demand for fabric production will increase and its existence is protected.

This research aims to determine the technical specifications of the design of goyor woven fabric as the official uniform material of the Sragen civil servants. Moreover, it is important to know the preferences of the needs and desires of civil servants in Sragen. After that , the translation of the consumer's voice into the technical voice is used as a reference in making the design, pattern, color and model of the goyor woven fabric that is used, so that the fabric design can be designed in such a way without reducing the high values contained in it and able to follow the current trends. The method used to identify customer needs is by Quality Function Deployment (QFD).

QFD would provide an overview of what the customer needs to be immediately produced by the company. QFD makes it possible to prioritize customer needs, find innovative responses to these needs and improve the production process [3][9].

Whereas to determine the proposed goyor fabric design, conjoint analysis is used. This analysis is used to determine consumer desires for the most preferred combination of fabric designs [2]. From this conjoint analysis , there would be stimuli or proposals for fabric design with several combinations. The combination formed is obtained from the QFD results. This analysis is needed because indicators such as colors and patterns cannot stand alone. These indicators are relative, depending on consumer preferences and how each indicator is combined [3].

2. Methods

The step of this research is strart by measure consumer preferences by using a questionnaire. The step aims to get VOC (Voice of Customer) that would be used in making QFD [5]. Respondents in this study were civil servants in Sragen Regency. The sampling technique used was sampling quota with a target of 150 respondents.

There are 3 stages to collect respondents' preferences: open questionnaire stage, closed discussion with experts, and closed questionnaire stage. The open questionnaire contains the characteristics of the goyor fabric and how consumers think about goyor fabric-based clothing. From this open questionnaire, VOCs are produced. In preparing open questionnaire questions based on apparel design framework [7]. The closed discussion stage was carried out with goyor woven fabric practitioners as well as exporters from Sukoharjo. The results of closed discussions are in the form of VOE (Voice of Engineer) where the activity aims to get technical specifications which are then made product proposals that are in accordance with VOC and VOE. While the closed questionnaire contains questions that have been directed to find out the response what if the fabric is goyor used as aofficial uniform for civil servants. In preparing closed questionnaire questions based on apparel design framework [7] and VOE results.

After knowing the preferences of the respondents, the next step is to make a combination of the proposed designs of goyor woven fabrics. Design combinations are made using conjoint analysis. Conjoint analysis help quantify the utility for consumers who would buy based on specific product attributes. Through it, the optimal utility of attributes can be identified and used to design products with attributes that are most preferred by consumers [4], [10].

The main advantages of conjoint analysis with other methods is the method is able to form stimuli of multi-level product attributes and set product attributes according to their level. In general, the basic model is conjoint analysis with the following linear regression approach [8]:

$$U(x) = \sum_{i=1}^{m} \sum_{j=1}^{ki} \alpha_{ij} x_{ij} \quad (1)$$

U(x) = total utility
αij = the value of the use attributes of the first level j
xij = attribute dummy variable to level j
m = number of attributes
ki = number of attribute levels

The value of the importance of an attribute, Ii is defined as follows:

$$Ii = \{ \text{Max } (\alpha ij) - \text{Min } (\alpha ij)\} \text{ for each } i \quad (2)$$

The relative importance of an attribute to other attributes:

$$Wi = \frac{Ii}{\sum_{i=1}^{m} Ii} \text{ where} \sum_{i=1}^{m} w_i = 1$$

The level of relative importance of each attribute uses the following formula [1] :

$$TKRi = \frac{UT_i - UR_i}{\sum_{j=1}^{k}(UT_i - UR_i)}$$

TKRi = the level of importance of the i-attribute
UTi = the highest usefulness of the i attribute level
URi = the lowest value of the i-attribute level
K = number of attributes

3. Result and discussion

Data collection and processing in this study was done by collecting consumer preferences using an open questionnaire. The results of the open questionnaire were validated by goyor woven practitioners through closed discussions, then the results were used as a reference in making a closed questionnaire. The result of an open and closed questionnaire is the VOC, while the results of closed discussions with practitioners are VOE. After knowing the VOC and VOE, HOQ (House of Quality) can be made which would produce technical specifications from the goyor woven fabric design. From the technical specifications that have been obtained, a proposal design is made using conjoint analysis.

3.1. Open questionnaire

The initial stage to find out the preferences of civil servants members is by making an open questionnaire. This questionnaire contains the characteristics of goyor fabric and how consumers think about goyor fabric-based clothing. The questionnaire is made with an open type where the respondent can write down the answers along with the reasonsas detailed as possible. The questionnaire is divided into 3 main sections: a brief definition, respondent data, and general questions. In the first section contains a brief definition of goyor woven fabric, the second part contains the identity of respondents describing geographical variables and demographic variables, while in the third part contains general questions containing 6 questions to translate functional variables, 2 questions to translate aesthetic variables, 4 question items to translate expressive variables. Thus the number of question items given to respondents is 12.

Table 1. Open questionnaire grid

Dimensions	Indicator	Description of questions	Instrument
Geographical	Consumer area	Where did the respondent come from	Questionnaire II (4)
Demographic	Age range of consumers	What is the age of the respondent	Questionnaire II (2)
Demographic	Consumer sex	What is the gender of the respondent	Questionnaire II (3)
Demographic	The ability of consumers to buy	What is the large range of respondents' income	Questionnaire II (8)
Functional	Fabric	Knowledge of goyor woven fabric	Questionnaire III (1)
Functional	Display	How is the impression togoyor woven fabric	Questionnaire III (4)
Functional	Display	What is liked and disliked from goyor woven fabrics	Questionnaire III (7-8)
Aesthetics	Color	What fabric color is preferred	Questionnaire III (5)
Aesthetics	Pattern	What fabric patterns are liked	Questionnaire III (6)
Expressive	Habit	How often wear clothes made from woven fabrics in official service	QuestionnaireIII (11)
Expressive	Job/Position	Do you agree if you pay more for products from goyor woven fabric	Questionnaire III (12)

After the questionnaire questions have been prepared, the open questionnaire is ready to be distributed to the respondent. Because this questionnaire is open, it is not necessary to test the variability or reliability tests in compiling the questions, because the function of the open questionnaire is to find out the respondents' opinions and free answers on what the researchers asked the questionnaire. Questionnaires were distributed to the regional offices of Sragen Regency with a target of 150 samples. After all the answers to the open questionnaires have been collected, the next step is to validate the answers with the fabric goyor practitioners in Tawangsari Village, Sukoharjo Regency.

3.2. Closed Questionnaire

Practical validation results will be used as a reference in making a closed questionnaire. Because this questionnaire is closed, respondents are not asked to express their opinions but only answer questions that have been directed to the proposed product in accordance with their preferences. The closed questionnaire has the same principle as the open questionnaire, the difference is that there is a product assessment section. The product assessment contains questions about respondents' preference for uniform clothes made from goyor woven fabric.

Table 2. Closed questionnaire grid

	Product Assesment
1	Respond of the respondent if the clothes are designed simply
2	Respond of the respondent with fabric that looks falling so that makes it flexible/ fit to body
3	Respond of the respondent if the clothes do not interfere with the user's movement
4	How agree the respondent if the clothes use comfortable materials
5	Respond of the respondent if the fabric color is dark
6	Respond of the respondent if the fabric pattern is geometric
7	Do respondents prefer to sew fabric into clothes instead of buying ready-made clothes
8	How agree the respondent if the clothes are comfortable to use even outdoors
9	How agree the respondent with clothes with easy maintenance
10	How agree the respondents to clothes that are durable even though they are often washed

3.3. HOQ

The next process is making a HOQ to find out the technical specifications of goyor woven fabric designs that are in accordance with consumer preferences. To compile the HOQ, the first step is to determine the VOC (Voice of Customer) obtained from the processing of open and closed questionnaire data as well as VOE (Voice of Engineer) obtained from a closed discussion process with practitioners.

Table 3.VOC

	Voice of Customer
1	Fabric color designs that tend to be dark like maroon and dark blue
2	Pattern fabric design are geometricand flora
3	Comfortable material to use
4	Comfortable clothes to use even at outdoors
5	Simple design
6	Does not interfere with user movements
7	Prefer to sew cloth into clothes instead of buying ready-made ones
8	Easy cloth maintenance
9	Clothes are durable even though they are often washed
10	Affordable price

Table 4.VOE

	Voice of Engineer
1	Rayon yarn used can be replaced with imported yarn from China to suppress COGS
2	Fabric design is made only 1 pattern (not using tumpal), so the weaving process is faster and more efficient because it doesn't change pattern
3	Pattern design follows the export market (geometric)
4	Fabrics don't need to be ironed
5	Patterns are made big so that the work is easier and faster
6	The yarn uses a standard size, which is 42/2
7	The size of one fabric was made 70x300 cm (size 1 long sleeve shirt), not following the size of the export market which tends to be larger

After the VOC and VOE are known, the next step is to combine them into a HOQ. The HOQ scheme consists of customer requirements and technical requirements. For the Customer Requirements section filled by the VOC that has been identified, while the Technical Requirement is filled in by the product material compose, the work process, and the proposed procedure of VOE. In this research, HOQ is limited by not including banchmarking with competing products because the focus of the research is to obtain technical specifications.

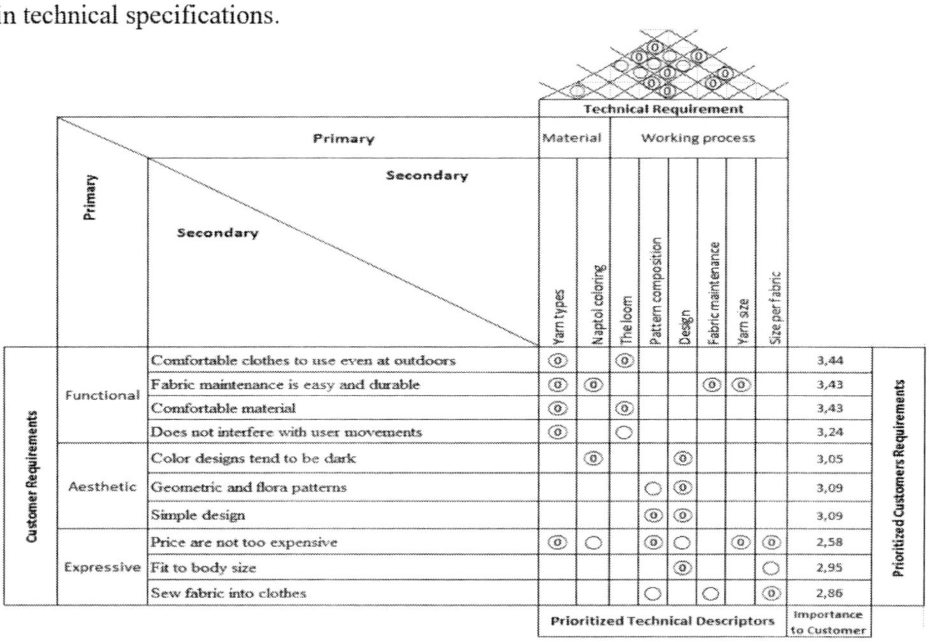

Figure 1. HOQ technical specifications for goyor fabric design

From the HOQ above, it can be concluded that the most important thing for consumers is uniform with materials that are comfortable to use even if they are outdoors, the prioritized customers requirement is prices that are not too expensive, while the prioritized technical descriptors is to make patterns and colors new ones that follow the local market.

Table 5.HOQ assessment results

	Customer requirements	Importance to customer	Prioritized customers requirements
1	Comfortable clothes to use even at outdoors	3,44	18
2	Fabric maintenance is easy and durable	3,43	36
3	Comfortable material	3,43	18
4	Does not interfere with user movements	3,24	12
5	Color designs tend to be dark	3,05	18
6	Geometric and flora patterns	3,09	12
7	Simple design	3,09	18
8	Prices are not too expensive	2,58	42
9	Fit to body size	2,95	12
10	Sew fabric into clothes	2,86	15

Table 6.HOQ assessment results

	Technical requirements	Prioritized technical requirements
1	Yarn types	24
2	Naptol coloring	12
3	The loom	30
4	Pattern composition	18
5	Design	42
6	Fabric maintenance	21
7	Yarn size	24
8	Size per fabric	18

The difference between importance to customer and prioritized customer requirements is where the score is come from. Scoring for Importance to customer is from the average of respondent's answer in close questionnaire. Scoring for Prioritized customer requirements and prioritized requirements is from corelation matrix on HOQ: ⓪ is +9 and◯ is +3.

3.4. Conjoint analysis

Conjoint analysis is used to determine consumer desires for a combination of the most preferred fabric designs and to identify the relative importance of each attribute. There are 2 attributes that would be analyzed using conjoint: Fabric color, whether red or blue and fabric pattern, whether flora or geometric.

The combination of attributes was obtained based on the results of the QFD where the majority of respondents liked the design of red and blue fabrics and fabric patterns in the form of flora and geometric. Of these two attributes, a combination that may be formed is 4 stimuli (design combinations) . Of the four existing fabric designs, a questionnaire was made to find out the proposed design of the goyor woven fabric that was most preferred by respondents. For this reason, each respondent is given a question to assess the combination of existing fabric designs. Because there are 4 stimuli, the respondent mustfilling out the questionnaire by giving a value from 1 (least preferred) to 4 (most preferred) in each fabric design. Respondents for this questionnaire amounted to 10 people and were civil servants of the Sragen Ministry of Industry and Trade.

Table 7. Utility estimate data processing stimuli

Attribute		Utility Estimate										
		Resp.1	Resp.2	Resp.3	Resp.4	Resp.5	Resp.6	Resp.7	Resp.8	Resp.9	Resp.10	Subfile summary
Patterns	Flora	-.500	1.000	.000	-1.000	-.500	.000	-.500	.000	.000	1.000	-.050
	Geometris	.500	-1.000	.000	1.000	.500	.000	.500	.000	.000	-1.000	.050
Colors	Merah	-1.000	.000	-.500	-.500	-1.000	-1.000	-1.000	-1.000	-.500	.000	-.650
	Biru	1.000	.000	.500	.500	1.000	1.000	1.000	1.000	.500	.000	.650
(Constant)							2.500					

From the data processing above, it is known that if the utility estimate is negative, the respondent does not like the design combination. But if the utility estimate is positive, it means that the respondent likes the stimuli. It can be seen that Subfile Summary is a general assessment, where the majority of respondents like the design of goyor fabric with geometric pattern and blue color (proposed design of fabric stimuli 4).

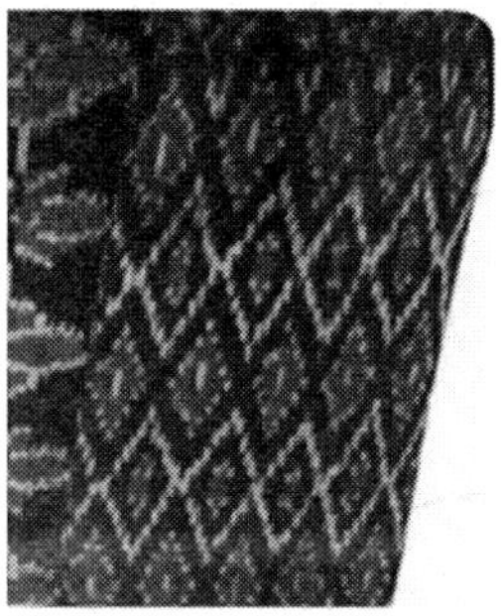

Figure 2. Propose Design of Fabric

Table 8. Final Technical Spesification

	Technical spesification
1	Local rayon yarn types
2	Local naptol coloring
3	Used traditional loom not machines
4	Just 1 pattern composition
5	Design follows local preferences
6	No need to be ironed
7	Using 42/2 yarn size
8	Size per fabric is 70x300 cm

4. Conclusion

The conclusion of this research is that respondents liked and agreed that the official uniform material made from goyor fabric. In addition to the fabric that is comfortable to wear, cloth is also the result of local production of Sragen. The respondent's impression of the majority of goyor woven fabric is good, beautiful, and attractive. Respondents like red clothes (dark red, maroon, brick red, red heart) and blue (dark blue, donker, navy). The pattern of the clothes that most respondents like is geometric motifs. The most preferred by respondents from goyor woven fabric is the very comfortable material. It feels cool when used in hot and warm weather when used in cold weather. The most disliked by respondents from clothes made from goyor woven fabric is the material that is too falling or "goyor" so that it is difficult to arrange. For the combination of goyor fabric design, the most preferred by respondents is the blue color and geometric pattern.

Acknowledgment

This research was funded by Universitas Sebelas Maret skim Program Pengembangan Produk Unggulan Daerah (PPPUD) (Contract No. 544/UN27.21/PM/2018)

5. References

[1] Aaker, D.A. and G.S. Day 1980. Marketing Research. New York: John Willey &. Son, Inc.
[2] Desrina Yusi Irawati and Octavia Ezrani 2018 Servqual dan Conjoint Analysis Dalam House of Quality untuk Layanan Ojek Online. *Jurnal Teknik Industri* **19**(1), 82-96.

[3] Gerald M. Katz 2003 PRACTIONER NOTE: A Response to Pullman et al.'s 2002 Comparison of Quality Function Deployment versus Conjoint Analysis, *Journal of Product Innovation Management* **21**(1), 61-63.
[4] Green, P.E. and Srinivasan 1978 Conjoint Analysis in Consumer Research: Issues and Outlook. *Journal of Consumer Research* **5**(2), 103-123.
[5] John R. Hauser, Songting Dong and Min Ding 2013 Self-Reflection and Articulated Consumer Preference, *Journal of Product Innovation Management* **31**(1), 17-32.
[6] Jeremy J. Michalek, Fred M. Feinberg and Panos Y. Papalambors 2004 Linking Marketing and Engineering Product Design Desicions via Analytical Target Cascading, *Journal of Product Innovation Management* **22**(1), 42-62.
[7] Lamb, J. M. and Kallal, M. J. 1992 A Conceptual Framework for Apparel Design. *Clothing and Textiles Research Journal* **10**(2), 42-47.
[8] Malhotra Naresh K 1993 *Marketing Research and Applied Orientation*. (USA: Prentice Hall International)
[9] Madeleine E. Pullman, William L. Moore, and Don G. Wardell 2002 A Comparison of Quality Function Deployment and Conjoint Analysis in New Product Design, *Journal of Product Innovation Management*, **19**(5), 354-364.
[10] Mennecke, A. M. Townsend 2007 A study of the factors that influence consumer attitudes toward beef products using the conjoint market analysis tool, *Journal of Animal Science* **85**(10), p. 2639-2659.

Empowering product development through creative culinary house design

Y Prasetyawan[1], M Suef[1] , H Supriyanto[1] and I O K Wardani[1]

Industrial Engineering Department, Institut Teknologi Sepuluh Nopember
yudhaprase@yahoo.com, m_suef@ie.its.ac.id, hariqive@yahoo.com, intan.oktasari@gmail.com

Abstract: The existence of creative culinary houses with thematic besides culinary has attracted a lot of public attention. In the business balance between supply and demand, sharpness of analysis is needed to see the potential on both sides. Although the supply of culinary business has been very numerous and varied in the city of , analysis of how to build a unique product and find a niche that can still be penetrated is very important. Culinary developments around ex-lokalisasi area are still limited to products that have been developed in the community such as crackers, chips, seaweed and the like. The superior products that are in demand by the community and have great potential are various heavy and light foods that can be consumed in daily life to get enough sales potential and business sustainability. The community gives a positive appreciation in the question of perception about the existence of culinary houses and gives positive interest to become consumers, but some still have negative perceptions about the ex-lokalisasi area and expect the stigma to be lost. In the perception as a producer, the public still doubts the potential for marketing and sales success and the limitations of the place. It is expected that this creative culinary house can have relationships with various other economic centers in the city .
Keywords: Creative Culinary Houses, Superior Product, Marketing and sales

1. Introduction
In this part, the research background, problem formulation, and research objective will be shown.

1.1 Research Background
The rapid development of the city of in the last decade has resulted in a significant increase in economic potential. The increase in the number of points of recreation facilities for individuals and families spur city residents to be more eager to enjoy life. The development of Ex-lokalisasi requires tremendous energy for the City Government and the entire city of , to change the perception of the region that has a negative connotation into a promising economic potential area. The movement of the citizens of the city of and its surroundings to want to visit this area also requires considerable economic and tourist attraction. Culinary developments in the city of show remarkable improvement, along with the development of improved urban infrastructure. The growth of the culinary center which is a structuring transformation of street vendors has been able to provide economic passion for micro entrepreneurs. The existence of creative houses is expected to be able to increase the number of micro business people to fill the niche of potential market needs even though the number of micro business food and beverage

(culinary) has been very much in the city of .Several creative houses have been established by several other agencies, among others, the Trade and Industry Office, the Agriculture Service and Mandiri Bank around Ex-lokalisasi area. The existence of creative houses with thematic besides culinary has attracted a lot of public attention. However, its existence requires visitor continuity, which among others will be able to increase market potential or the potential for purchasing products on display. Likewise, the culinary creative House that will be managed by the Cooperatives and Micro Enterprises Office is expected to be able to attract visitors who are willing to buy products or micro business people who will and are learning to develop a culinary business.

In the business balance between supply and demand, sharpness of analysis is needed to see the potential on both sides. Although the supply of culinary business has been very numerous and varied in the city of , analysis of how to build a unique product and find a niche that can still be penetrated is very important. It is expected that the culinary creative House will be able to sustainably develop this potential market with a variety of needs analysis of supporting facilities that can enhance market competitiveness.

1.2 Research Objective

Based on the problem formulation, the objective for the research are (1) Obtain the latest conditions about the products and markets of food and beverage commodities (culinary) in the city of , especially around the former Dolly localization area, (2) Formulating superior products based on market potential and public preferences in general, (3) Get a picture of people's perceptions as consumers who will buy creative culinary products, (4) Obtain a description of the public's perception as a producer (business actor) who will enliven the product market in the culinary house and participate in learning to develop the business, and (5) Formulate a general description, layout and operational design of the culinary creative House.

2. Methods

In this part, first the research procedure will be discused. Then the method of collecting data will be presented. After that an example of te result from data collection will be presented. This includes CIMOSA worksheet.

The procedure of conducting the research is pictured in the figure 1.

3. Result And Discussion

In this part, the result of the data collection will be presented.

3.1 Formulation of Creative Culinary Houses

The Creative House in question is expected to bring culinary business interest to the citizens of , especially for those with unemployment status. The number of unemployed in the city of can actually be completed by establishing 3 SMEs per RW with the capacity to absorb 5 people. Below is the design of Culinary Creative Houses .

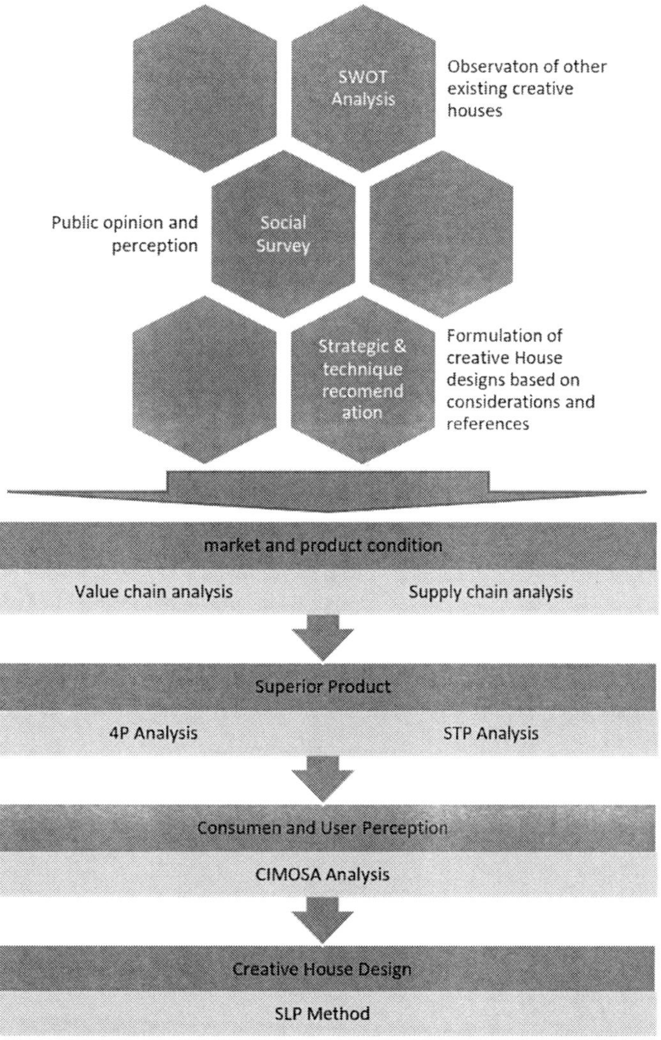

Figure 1. Research procedure to make creative house design start from observation until SLP method in the end. This perception based on public and user perception.

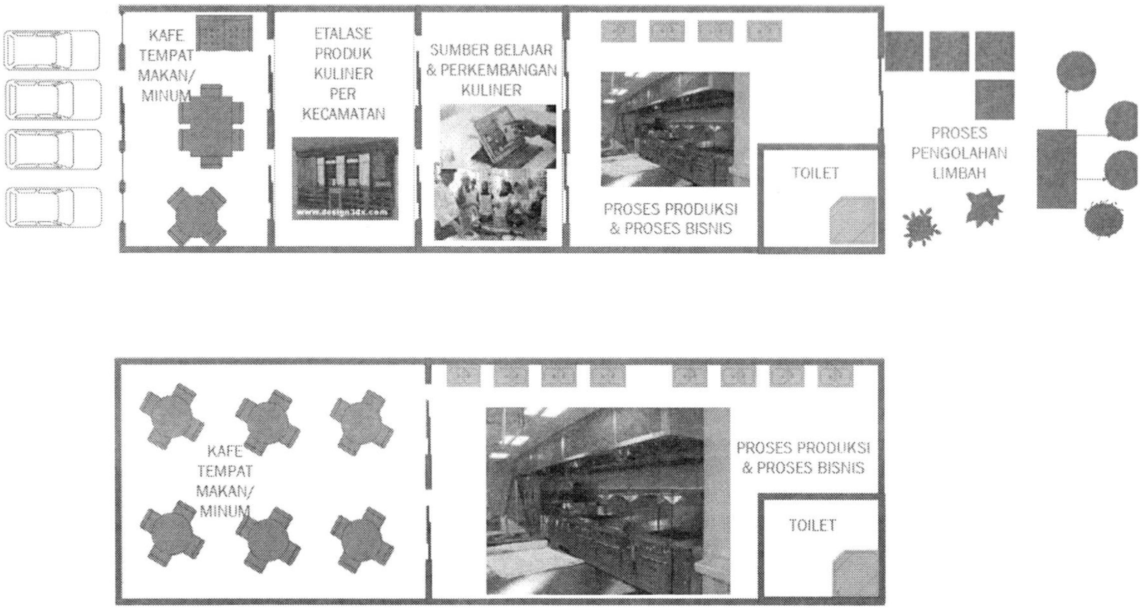

Figure 2. Culinary Creative Houses which collaborated with Coperation & SME Corporate 2017. This Design consist of four rooms, one toilet, and 1 waste recycle room.

3.2 Supply Chain Analysis

The supply chain is a series of flows of goods, information, and processes used to send products or services from the source location (supplier) to the destination (customer or buyer). In relation to this, the supply chain or supply chain contained in the products of food and beverage commodities that exist in Ex- lokalisasi mostly have the same plot. This is because most of the products produced are processed foods and drinks that have a longer expiration period than fresh foods and beverages. In some products the preservation or cooling process can also prolong the life of the product. The following is an illustration of the supply chain of food and beverage products in the ex-lokalisasi area. Below is supply chain ilustration.

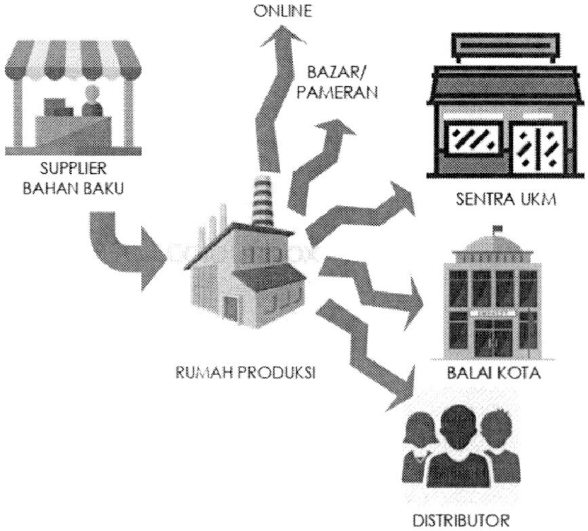

Figure 3. Supply Chain Illustration of Culinary Creative Houses that deliver to raw material supplier, SME, City hall, and distribution.

In general, the existing condition of supply chain for food and beverage products produced around the ex-lokalisasi area starts from the procurement of raw materials from suppliers, where the type and amount of raw material needed varies depending on the food or beverage produced.

3.3 4P Analysis

In the marketing mix strategy, the product is the most important element. The selection of the types of products that will be produced and marketed will determine the promotional activities needed, as well as the pricing and distribution methods. In the product strategy, things that need to be considered in the product are quality, service, features, and others, because when there are consumers who buy products, not just buy products but also the satisfaction offered on the product.

Some of the criteria considered to develop creative culinary House planning include culinary creative House design, culinary creative House facilities and infrastructure, and products sold in creative culinary Houses. Based on the results of the questionnaire, creative culinary house design consisting of cafes, culinary product storefronts, learning facilities, production processes and waste management are in demand by respondents. For the culinary creative facilities and infrastructure, the main one is having an adequate parking space while the products sold in the culinary creative house are traditional food and drinks typical of the district or city in East Java.

3.4 CIMOSA Analysis

Creative Culinary Houses must establish the goals and direction towards which this business will run. The purpose of this creative culinary house is as a means for Dolly locals to develop their economy. Then after the goal is obtained, then the best strategy is formulated that must be compiled by the creative House to achieve that goal. Strategy development can be done based on strength, weakness, opportunities, and threats.

In the core process part of developing a product, creative culinary houses must present products that are different from other creative houses. For example, creative culinary houses carry the theme of traditional snacks that are packaged in a modern form. The develop product section is made based on the preferences of consumers. In the get order section, the creative House can choose the right media to promote the community so that people know more about the creative culinary house. At present, with the development of technology, online media in the form of social media is more popular with the public. This can be used as a means for creative culinary houses to make promotions. Promotions can be made through social media such as Facebook or institutions that are currently popular with the public, as well as through the website. In the fulfill order section the culinary creative house fulfills the product to be sold in collaboration with the community at Dolly to make the food and drinks to be sold. Meanwhile, for food and beverage variations can be adjusted to current consumer preferences. By selling traditional food from other regions can be used as a means of preserving cultural heritage in the form of traditional food. In the support product section, it relates to what will be provided by the culinary creative House to support the business process.

In the support process section for Human Resources, creative culinary houses must have good human resources management to support the running of the business. Creative houses must at least have the resources as servants to serve customers and have cooks who are able to cook well so as to maintain the quality of the food sold. Human resources in the creative House must be right in order to meet the demands of customers who come. In the finance & accounting section, creative houses need to calculate the existing cash flow. Finance & Accounting can be done by taking into account financial flows in creative Houses so that costs become more efficient. In the information technolgy section, creative houses can utilize existing technologies such as utilizing online information media as a means of promotion. In addition, along with the development of creative houses the use of technology can be applied to regulate the inventory of products sold so that there is no demand from consumers that cannot be fulfilled.

3.5 BMC Analysis

Business Model Canvas will provide reference recommendations to formulate supplier and consumer balance, income and costs, as well as reference operational activities to process the resources owned effectively and efficiently. The Canvas Business Model (MBC) is formed with Nine Buildings Blocks which makes it easy for business people to frame and expand business. The following is a Business Model Canvas consisting of key activities, key partnerships, key resources, channels, Customer Relationships, Value Proposition, Customer. Segments, revenue streams and cost structures. The following is a culinary creative House business model:

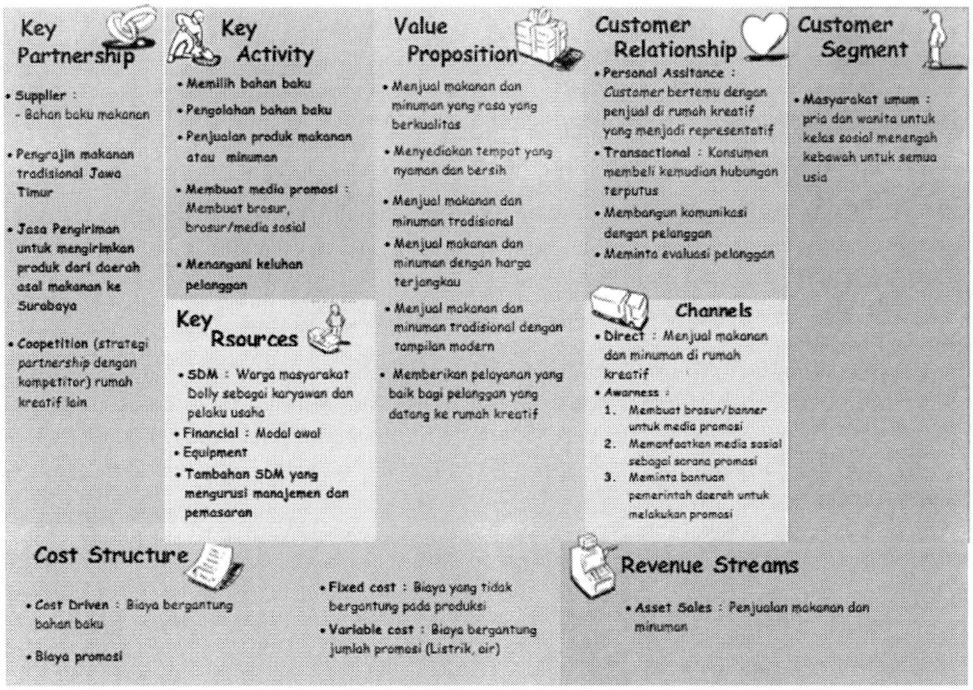

Figure 4. Culinary Creative House Business Model use BMC and CIMOSA Analysis.

4. Conclusion

Based on the fiding in the result and the discussion above, the conclusion for this research are:

1. Culinary developments around Ex-lokalisasi area are still limited to products that have been developed in the community such as crackers, chips, seaweed and the like.
2. The superior products that are in demand by the community and have great potential are various heavy and light foods that can be consumed in daily life to get enough sales potential and business sustainability.
3. The community gives a positive appreciation in the question of perception about the existence of culinary houses and gives positive interest to become consumers, but some still have negative perceptions about the ex-lokalisasi area and expect the stigma to be lost.
4. In the perception as a producer, the public still doubts the potential for marketing and sales success and the limitations of the place. It is expected that this creative culinary house can have relationships with various other economic centers in the city of .
5. The general description of culinary creative houses is technically consisting of sales marketing areas, culinary learning areas of the past, until the waste production and processing process has been described and can be continued with more detailed specifications.

5. References

[1] A'lima, R. N., 2016. *Perumusan Strategi Jangka Pendek dan Jangka Panjang bagi UMKM di Sentra Industri Logam Ngingas, Sidoarjo,* (Surabaya: Teknik Industri ITS)
[2] Anityasari, M., 2010. *Business Process, Handout mata kuliah: Pengantar Teknik dan Sistem Industri.* (Surabaya: Institut Teknologi Sepuluh Nopember)
[3] Armellia Utami, E. N. D. A., 2016. Analisis Potensi Objek Wisata Pantai Pagatan di Kecamatan Kusan Hilir Kabupaten Tanah Bumbu. *Jurnal Pendidikan Geografi,* III(5), pp. 1-14.
[4] David, F. R., 2011. *Strategic Management.* 11st ed. (Jakarta: Salemba empat)
[5] Gayatri, A. T. B. d. U., 2009. Penentu Sektor Unggulan dalam Pembangunan Daerah (Studi Kasus: Kabupaten Ogan Komering Ilir). *Jurnal Ekonomi dan Studi Pembangunan,* X(1), pp. 34-50.
[6] Harmoyo, D., 2011. *Analisis Manajemen Strategi Koperasi Jasa Keuangan Syariah Bantul Maat Wat Tamwil "Syariah Sejahtera",* (Surakarta: Sekolah Tinggi Agama Islam Negeri Surakarta)
[7] Hisyam, I., 2013. *Strategic Management: Developing Vission and Mission, Handout mata kuliah: Manajemen Organisasi dan Sumber Daya Manusia.* (Surabaya: Institut Teknologi Sepuluh Nopember)
[8] Imron, E. U. d. A., 2012. Perumusan Strategi Perusahaan Berdasarkan Competitive Advantage. *Jurnal Ilmiah Teknik Industri,* II(11), pp. 154-164.
[9] Jumayla, S., 2014. *Perancangan Instrumen Penilaian Tingkat Kemandirian UKM Tenant di Inkubator Industri ITS.* (Surabaya: Jurusan Teknik Industri ITS)
[10] Moubray, J., 1992. *Reliability Centered Maintenance (RCM).* 1st Ed. (Oxford: Butterwort Heinemann)
[11] Moubray, J., 1997. *Reliability Centered Maintenance (RCM) II.* 2nd Ed. (New York: Industrial Press Inc.)
[12] Priyanta, D., 2000. *Keandalan dan Perawatan.* (Jurusan Teknik Sistem Perkapalan)
[13] Yssaad, B., Khiat, M. and Chaker, A., 2013. Reliability Centered Maintenance Optimazation for Power Distribution Systems. *Electrical Power and Energy Systems 55,* pp. 108-115.

Usability Testing of Laboratory Website using a Participatory Design Approach

S Firdaus[1], A D Sari[2], M R Suryoputro[3] and A U Khasanah[4]

[1,2,3,4] Program Studi Teknik Industri, Fakultas Teknologi Industri, Universitas Islam Indonesia
salamanfirdaus@outlook.com[1], amarria@uii.ac.id[2], mragils@uii.ac.id[3], annisa.uswatun@uii.ac.id[4]

Abstract. Website is one of supporting facilities that used by universities as communication and marketing media. The Industrial Modelling and Simulation Laboratory website is one of the academic websites at the Islamic University of Indonesia. There are still problems found on the website such as not uniform display content, less attractive appearance, complex navigation and too many link layouts. It was indicated that the website has not been easy and effective to be used so that an evaluation and usability testing needs to be carried out to improve the function of the website. The research method used was participatory design, performance measurement and usability scale (SUS) system questionnaire. The purpose of this study was to determine the results of evaluation based on participatory design and the level of usability on the proposed website. The evaluation process identifies problems and suggestions for improvement through 2 iteration stages to produce a proposal design. In the process of usability testing the proposed website shows that the proposal for a new laboratory website has an effectiveness level of 78%, an efficiency level of 76% and a satisfaction level of 74% with excellent categories.
Keywords : Academic Website, Usability, Participatory Design, Performance Measurement, System Usability Scale (SUS).

1. Introduction

One of the most widely used website domains today is for academic purposes [1]. Information on a website itself depends on the profile of the visitor, the destination of the visitor, information expected by the visitor, the level of knowledge, expertise or interest of the visitor [2]. Improving the quality of website-based services is an effort that must be done continuously to convince users to continue to trust the services provided [3]. Industrial Modeling and Simulation Laboratories is one of the laboratories in the Industrial Engineering Universitas Islam Indonesia that used as information media, especially for students and academic interests so that it must provide good usability. This is because generally the usability level owned by most applications is still low so that it is difficult to be used and learned [4] whereas if the usability value of a website is good, it will increase convenience for users. On the contrary, if the value is bad it will make the users inconvenient and an evaluation of the website needs to be carried out [5]. The objective of evaluation on a human and computer interaction system is its design and implementation [6]. Research on the evaluation form of an online community using paper prototyping, as a fast participatory design technique that includes community members in collaborative online community design, so that members will use the system that has been created because it is designed based on user feedback [7] Research related to website usability is also conducted by [8] which aims to find out the quality of information systems from websites with usability testing using the Post-study System Usability Questionnaire (PSSUQ) package.

There are several problems on the laboratory website such as display of website content / information that is not uniform, display that is not attractive, complex navigation and too many link layouts. It was indicated that the website has not been easy and effective to be used so that an evaluation and usability testing needs to be carried out to improve the function of the website. The research method used is participatory design, performance measurement and usability scale (SUS) system questionnaire. The purpose of this study was to determine the results of evaluation based on participatory design and the level of usability on the proposed website.

2. Research Method

2.1 Research Object

The object of research in this study is the Industrial Modelling and Simulation Laboratory in Department of Industrial Engineering, Universitas Islam Indonesia.

2.2 Research Subject

In this study there are 3 methods used systematically. Participatory design method is used to evaluate and develop the existing website. Performance measurement method and System Usability Scale (SUS) questionnaire is used for usability testing on the proposed website. Each respondent who is a subject of different research is the same as stated by Nielsen [9].

2.3 Research Instrument

a. Participatory design was carried out with the help of prototyping papers that were tested to the parties involved or stakeholders from the laboratory.
b. Performance measurement was carried out in 2 stages, first, the pilot test of 2 respondents and playing test on 10 respondents with a total of 5 assignments.
c. The Usability Scale System (SUS) questionnaire was conducted online on 30 respondents who met the characteristics of the study.

3. Result and Discussion

3.1 Field Studies

Field studies are an effort to collect and then organize and analyze data about the problems that are the focus of the research **[9]**. Usability pro*blems found on the website* that as many as 18.18% of users complain about the layout of the links and information that has been inflicted. In addition, 24.25% of users stated that the display between pages and content of page content is not uniform, and 12.12% states that general information that should be displayed on the website is not found. Another problem is 15.15% complained about the complex navigation system and 30.30% of users responded that the display of the website was now not attract attention and curiosity.

3.2 Website Development

The participatory design process involves the developer, the person in charge and the user in discussing the problems in the initial website design (Figure 1).

Figure 1. Existing Design

Data collected from the first stage participatory design process was used as a reference as the first iteration process as shown in Table 3. After the proposal design was made, the second stage of participatory design was followed by the second iteration which can be seen in Table 4, from this stage to the proposed design laboratory website (Figure 2).

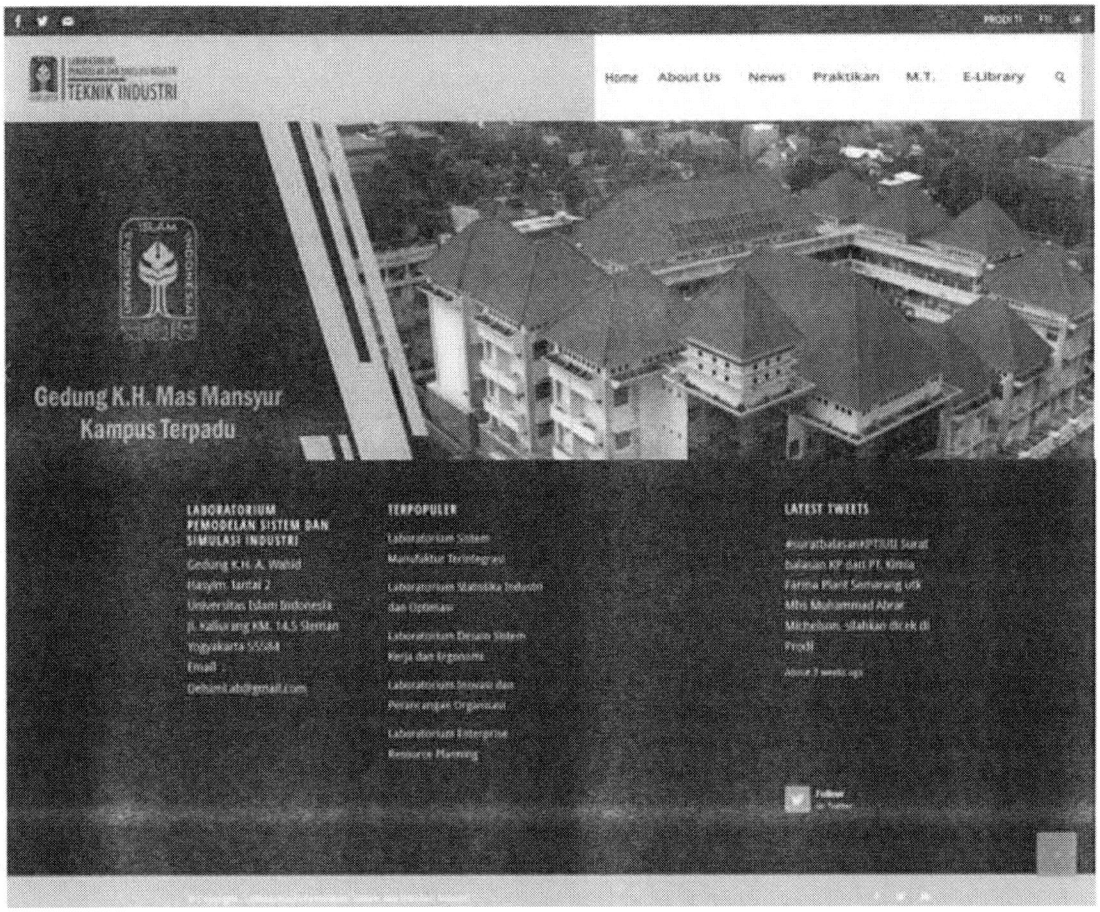

Figure 2. Proposed Design

3.3 Effectiveness

Effectiveness is measured based on how successful the respondent is in completing the task. Respondents were given 5 task scenarios to access the proposed website. From 10 respondents, only 3 respondents successfully completed the entire task scenario. The highest success rate achieved by one respondent was 100%, while the lowest was 60%. The average level of effectiveness of 10 respondents in completing the task is 78%, this shows that the effectiveness level of the proposed website was quite effective.

3.4 Efficiency

Efficiency is the time that users spend in completing task scenarios and efficiency indicators, ie the time that users need to complete one time (time completion rate) [10]. In the results of the tests that have been carried out showed that the average value of the respondents' efficiency in completing the task was 76% where the lowest percentage was 48% and the highest percentage of efficiency was 100%.

3.5 Satisfaction

The level of satisfaction is obtained from the results of the System Usability Scale (SUS) questionnaire by looking at the score generated [11]. Based on the results of the SUS questionnaire score, the proposed website was categorized as excellent because the average SUS score generated was 74 which was among the acceptable adjective rankings. It is indicated that the proposed website was classified as good quality and quite satisfying.

IOP Conf. Series: Materials Science and Engineering **528** (2019) 012026 doi:10.1088/1757-899X/528/1/012026

4. Conclusion

Based on the results of the research that has been done it can be concluded that:

1. The initial website design evaluation process is carried out with 2 stages of design iterations to produce new designs that are based on stakeholders' needs and satisfaction.
2. The development process identifies 6 problems and suggests improvements in the first iteration stage. Then in the second iteration process there are 3 problems along with suggestions for improvement and 2 suggestions for feature enhancements on the proposed website.
3. Usability testing results indicate the level of the proposed website's characteristic size is above 70%, so that it can be said that the user is able to achieve certain goals effectively, efficiently, and obtain satisfaction from the proposed website.

5. References

[1] Kothainayaki S, Sivakumaren K and Gopalakhrisnan S 2012 User Preferences on University Websites: A Study. *Library Philosophy and Practice* pp 788-795

[2] Sari A D, Suryoputro M R, Rochman Y A, Ulandari S and Puspawardhani E H 2015 Usability Analysis of Laboratory Website Design to Improve Learning Process. *Procedia Manufacturing* **3** pp 5504-511

[3] Napitupulu D B 2016 Evaluasi Kualitas Website Univeritas XYZ dengan Pendekatan Webqual. *Buletin Pos dan Telekomunikasi* **14** pp 51-64

[4] Capretz L, Ali A and Ouda A 2012 A Conceptual Framework for Measuring The Quality Aspects of Mobile Learning. *Bulletin of the IEEE Technical Committee on Learning Technology* **14** pp 30-34

[5] Arsyitahadi A H 2015 *Analisis Usability Website Akademik Perguruan Tinggi di Indonesia Menggunakan Metode Electree, Grey Relational Analysis, dan Weighted Product Model.* (Yogyakarta: Universitas Negeri Yogyakarta)

[6] Chandra T 2013 Evaluasi User Interface Desain Sistem Informasi Perpustakaan pada Perguruan Husni Thamrin Medan. *Jurnal TIME* **2** pp 1-6

[7] Osman A, Baharin H and Ismail M H 2009 Paper Prototyping as a Rapid Participatory Design. *Computer and Information Science*, 53-57.

[8] Lestari S 2014 Analisis Usability Web (Studi Kasus Website Umkm Binaan Bppku Kadin Kota Bandung). *Jurnal Ilmiah Teknologi Informasi Terapan* **1**(1) pp 46-51

[9] Usability Geek 2015 Usability Metrics – A Guide To Quantify The Usability Of Any System. (*online*): http://usabilitygeek.com/usability-metrics-a-guide-to-quantifysystem-usability/# (Accessed: 12 September 2017)

[10] Brooke J 2013 SUS: A Retrospective. *Journal of Usability Studies* **8**(2) pp 29-40

[11] Patton M Q 2009 *Metode Evaluasi Kualitatif.* (Jakarta: Pustaka Pelajar)

Failure Mode and Effect Analysis (Fuzzy FMEA) Implementation for Forklift Risk Management in Manufacturing Company PT.XYZ

Muhammad Ragil Suryoputro[1], Khairizzahra[2] and Amarria Dila Sari[3], Nawang Wahyu Widiatmaka[4]

[1,2,3,4]Industrial Engineering Department, Faculty of Industrial Technology, Universitas Islam Indonesia, Yogyakarta, Indonesia
[1]ragil.suryoputro@uii.ac.id, [2]rizzakhairizzahra@gmail.com, [3]amarria@uii.ac.id, [4]nawangwahyuwidiatmaka@gmail.com

Abstract. Company improvements and evaluations must always be carried out by companies to improve company performances. One of the factors that influence the performance was the smooth production on machine performance. Good risk management was one of the keys to maintaining a smooth production process. This research was conducted at a manufacturing company, PT. XYZ, to carry out risk management on machines that had low performance. From the results of the quarterly report to the four parts of maintenance, the forklift was a machine and facility with the lowest NPM (Quality Performance Score-QPS). The method used in this research was Fuzzy FMEA (Failure Mode and Effect Analysis), in order to obtain priority in repair and maintenance of forklifts. The result of this study was that the failure mode of the hydraulic hose rupture got the highest FRPN (Fuzzy RPN) value of 631 so that the hydraulic part with the failure mode of the broken hydraulic hose was recommended to be the main focus in the repair and maintenance of the forklift. The result of using Fuzzy FMEA was more proposed as a reference focus to get more objective risk management compared to conventional FMEA results in machine repair and maintenance.

Keywords: Risk Management, Forklift, Fuzzy FMEA, Maintenance Priority

1. Introduction

Manufacturing company would need a robust system to conduct improvement and evaluation in order to elevate the performance. The performance could be determined by the continuity and smoothness of the production process and machine performance [1]. In order to create a smooth process in the production process, a good risk management process was needed in solving the problem of machine damage or production equipment [2]. The amount of investment costs in purchasing machinery forced the company to try to maintain and repair the engine so that the machine can work optimally. Therefore, there was a need for risk management in this manufacturing company. Risk management is an approach that is carried out on risk by understanding, identifying and evaluating risks, then considering what will be done to the impacts that arise and reducing the risks that occur [3]. The purpose of risk management is to recognize the risks that can occur and develop strategies to reduce or even avoid them [4]. With good risk management, the company can minimize or even avoid risks that might occur so that the company's productivity will be elevated. PT. XYZ is a state-owned railway production industry that used a variety of machines and tools to produce train ordered and for the maintenance of the train itself.

According to the fourth quarterly report on maintenance, the forklift is a machine and production support facility with the lowest NPM (Value of Quality Performance/Quality Performance Score) value. Forklifts functioned as transportation and a device that could lift items [5]. Some factors that cause damage to forklifts are due to human error, the age of the old forklift, the capacity of the workload that exceeds the capacity of the forklift itself and the equipment used continuously. Based on these problems, a risk analysis of the damage that caused the broken forklift was made to be the focus of the proposed improvements in minimizing the risks to the train production process. The method chosen in this study was Fuzzy FMEA (Failure Mode and Effect analysis) rather than conventional FMEA. Conventional FMEA has several weaknesses which will give less accurate and vague results. Some weaknesses in conventional FMEA are explained in the next chapter. The difference in the results of risk management between conventional FMEA and Fuzzy FMEA will be presented and explained in the results and discussion chapter. It was as a confirmation in proving that Fuzzy FMEA was an accurate method to be used in risk management for forklifts at PT. XYZ.

2. Methods

2.1. FMEA (Failure Mode and Effect Analysis)

FMEA is a systematic method in identifying and preventing problems that occurred in products and processes [6]. The output of this method is RPN (Risk Priority Number). The RPN is a reference to provide recommendations to be used as a focus of maintenance and improvement from the biggest types of risk occurred based on its priorities, in this case the risk of the biggest failure that occurred was on the forklift. To obtain the score of RPN value, it was necessary to have input in the form of severity (S), occurrence (O) and detection (D) values. Severity is the value of the severity of the consequences that would lead to the continuation of the next process which also indirectly is detrimental; meanwhile, the occurrence is the value of how often the possible causes of failure occurred; furthermore, the detection is the value of how far the cause of failure can occur [7]. The assessment is carried out by the expert. Regarding the severity (S) there were 10 ranks with the description of the effects and criteria for each ranking. Ranking 1 means that it has no effect and is indicated by the criteria that does not result in anything. Whereas in rank 10 had dangerous effects without warning and has the criteria in the form of a machine that was not feasible to operate, because it could cause a sudden accidents and was contrary to occupational safety regulations [8]. On the other hand, at the value of occurrence (O) had 5 ranks with the presence of an explanation in the form of the probability of occurrence and the cause of failure and the probability of failure in each ranking. Ranking 1 showed if the chance of occurrence was very small with no failure associated with the process and the probability of failure was 1 in 1,500,000. Whereas rank 5 showed the probability of occurrence that was very high with failure almost unavoidable and the probability of failure was 1 in 2 [8]. Moreover, in detection (D) there were 10 ranks where each ranking could explain the detection group and its criteria. Ranking 1 was included in the detection group on *almost certain* with the criteria in the form of control almost certainly could detect failure. Whereas in the 10th rank included in the detection group *Absolutely impossible* with no control criteria to detect failure [8].

2.2. Fuzzy FMEA

Fuzzy logic was one method for analysing systems that involving uncertainties [9]. Research using Fuzzy logic would get more accurate results compared to using conventional FMEA methods [10]. According to [11] and [12], some Conventional FMEA weaknesses are:

1) Statements in FMEA were often subjective and qualitative explained in natural language.
2) Three levels of *severity* (S), *occurrence* (O) and *detection* (D) parameters that were assumed to have the same interests, it turned out that in practice the importance of the three parameters was not the same.
3) The same value of risk priority number (RPN) resulting from the multiplication of S, O and D levels may imply a different risk representation.

To overcome these weaknesses, methods based on Fuzzy logic were often used to manipulate linguistic terms that are used directly in making critical judgments. The process stages of the Fuzzy FMEA are described below:

1) Fuzzyfication, was a defining moment of membership functions and Fuzzy sets of severity, occurrence, detection and Fuzzy RPN (FRPN).
2) Evaluation of Fuzzy Rule, using IF-THEN rule where IF was a variable of Fuzzy input while THEN was a variable of Fuzzy output. All combinations must be grouped to produce Fuzzy rules. For example: IF Severity was Very High, And Occurrence was Very High, and Detection was Very High, THEN FRPN was Very High.
3) Defuzzyfication, with input defuzzyfication that was the Fuzzy set obtained from the composition of Fuzzy rules, while the output was the number in the Fuzzy set [9]. The technique used in this study was the Centre of Gravity (Centroid), which was the score value of the output obtained based on the centre of gravity of the decision making process. In this study, defuzzyfication was used to find the output value in the form of FRPN value based on the input that had been defined.

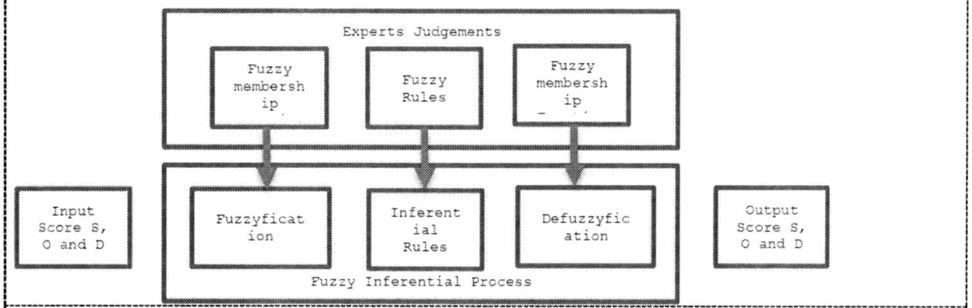

Figure 1. Fuzzy FMEA Stages [13]

For conventional FMEA, risk failure assessment was obtained by multiplying the value of severity, occurrence and detection. However, conventional FMEA did not pay attention to the importance of each input, so that in its calculation each input had the same level of importance. In addition, the assessment was subjective and qualitative, and also the RPN value was obtained only from the results of the third time input. Therefore, the severity, occurrence and detection values were assessed using Fuzzy Logic with the Mamdani (Min-Max) method to get the FRPN (Fuzzy Risk Priority Number) value [7]. The following was a table of number index categories for severity, occurrence, detection, and membership function parameters for input and output variables (Table 1., Table 2., Table 3. and Table 4.)

Table 1. Numeral Index of Category Severity, Occurrence and Detection [7]

Score			Category
S	O	D	
1	1	1	VL
2, 3	2, 3	2, 3	L
4, 5, 6	4, 5, 6	4, 5, 6	M
7, 8	7, 8	7, 8	H
9, 10	9, 10	9, 10	VH

Table 2. Parameters of Membership Functions of Input Variables [7]

Category	Curve Types	Parameter
VL	Trapezoid	[0 0 1 2.5]
L	Triangle	[1 2.5 4.5]
M	Trapezoid	[2.5 4.5 5.5 7.5]
H	Triangle	[5.5 7.5 9]
VH	Trapezoid	[7.5 9 10 10]

Table 3. Parameters of Membership Functions of Output Variables [7]

Category	Curve Types	Parameter
VL	Trapezoid	[0 0 25 75]
VL – L	Triangle	[25 75 125]
L	Triangle	[75 125 200]
L – M	Triangle	[125 200 300]
M	Triangle	[200 300 400]
M – H	Triangle	[300 400 500]
H	Triangle	[400 500 700]
H – VH	Triangle	[500 700 900]
VH	Trapezoid	[700 900 1000 1000]

Research data was obtained on 3-tons and 5-tons capacity forklifts to determine the RPN value of each type of damage. Assessment of severity (S), occurrence (O) and detection (D) was carried out by expert forklifts because there was not enough data on risk factors and uncertainty factors. In general, Microsoft Excel 2013 software was used as an instrument in processing research data. However, there was a specific instrument, Fuzzy Logic Toolbox, which is a MATLAB software program that was specifically used to calculate the RPN value on Fuzzy FMEA.

3. Results and Discussions

After observation, there are 12 failure modes categorized according to the appropriate equipment on the forklift carried out by the expert. FMEA assessment result is shown in the following table (Table 4.).

Table 4. FMEA Assessment Results

Information Worksheet		**System : Forklift (3 tons and 5 tons)**							
		Sub-system : Clutch tubes, Master clutches, Radiators, Tires, Solar Tanks, Hydraulic, Starters Motor							
		Sub-system function							
No	**Equipment**	**Function**	**Functional Failure**	**Failure Mode**	**Effect of Failure**	**S**	**O**	**D**	**RPN**
1	Clutch pad	To channel power from the engine to the transmission	Power is not achieved	Wear clutch pads	Forklifts are not road / road but lack power	8	10	6	480
2	*Master clutch*	Push hydraulic oil to release clutch pads from Clutch pads on flywheel	not release from flywheel	The piston master clutch wears out and the master clutch seal was broken	Gear did not want to be inserted	8	10	5	400
3	Radiator	Maintain the engine working temperature	The temperature was not maintained according to the	The radiator water is dirty, the radiator	Engine was overheat and does not start,	8	3	5	120

Information Worksheet	System : Forklift (3 tons and 5 tons)								
	Sub-system : Clutch tubes, Master clutches, Radiators, Tires, Solar Tanks, Hydraulic, Starters Motor								
	Sub-system function								
No	**Equipment**	**Function**	**Functional Failure**	**Failure Mode**	**Effect of Failure**	**S**	**O**	**D**	**RPN**
			working temperature	was rusted	component damage in the engine				
				Leaking radiator fins		8	3	5	120
				Hose radiator was broken		8	3	5	120
				Load was Overload		5	5	5	125
4	Tires	As a driving forklift	Tire leaking	The terrain was traversed by many plates, the operator does not pay attention to the road	Forklift was not moving	8	10	3	240
5	Solar tank	Solar Container	Forklift jam	The filters Clogged / dirty solar, dirty diesel tanks / lots of sediment, fuel pump clogged	The fuel pump damaged /Bosch pump	7	6	8	336
6	Hidraulic	Load Lifting	Unable to lift weights at capacity	*Leaking Seal* hydraulic	The fork drive power was	6	5	7	210

Information Worksheet		System : Forklift (3 tons and 5 tons)							
		Sub-system : Clutch tubes, Master clutches, Radiators, Tires, Solar Tanks, Hydraulic, Starters Motor							
		Sub-system function							
No	**Equipment**	**Function**	**Functional Failure**	**Failure Mode**	**Effect of Failure**	**S**	**O**	**D**	**RPN**
				The hydraulic hose ruptured	reduced, hydraulic does not work	8	8	7	448
7	Motor starter	Start the engine/ turn on the start engine	Engine failed to start	Run out Carbon brush starter motor	Could not be turned on	4	6	4	96
				Electric current is less than voltage working(12 V)		4	8	4	128

Based on the results of conventional FMEA calculation, the largest RPN value was in the failure mode of wore clutch with an RPN value of 480. In the S, O and D assessment using conventional FMEA, RPN was obtained from the results of the three multiplication of inputs without regard to the importance rating of each input. Whereas in Fuzzy FMEA the RPN results were obtained by paying attention to the degree of importance (degree of membership in Fuzzy Logic) from each input. The following were the priority results of Conventional FMEA and Fuzzy FMEA (Table 5).

Table 5. Priority Results of Conventional FMEA and Fuzzy FMEA

No	Failure Mode	RPN	Category	Priority	Fuzzy RPN	Category	Priority
1	Hydraulic hose breaks	448	High	2	631	High – Very High	1
2	Wear clutch	480	High	1	630	High – Very High	2
3	The piston and the master clutch seal were wore and damaged	400	Medium – High	3	613	High – Very High	3
4	The Solar filters were clogged/dirty, diesel tanks were dirty/many deposits, clogged fuel pumps	336	Medium	4	534	High	4
5	Overload	125	Low	8	533	High	5
6	The field were too many plates and operator was not having full attention	240	Low – Medium	5	489	High	6
7	The radiator water was dirty, the radiator was rusty	120	Low	9	439	Medium – High	7
8	The radiator grille was leaky	120	Low	10	439	Medium – High	8
9	Broken hose radiator	120	Low	11	439	Medium – High	9
10	Carbon brush motor starter was used up.	96	Very Low – Low	12	428	Medium – High	10
11	Leak hydraulics seal	210	Low – Medium	6	407	Medium – High	11
12	Electric current less than voltage working(12 V)	128	Low	7	261	Medium	12

In Conventional FMEA calculations, wear coupling pads had the highest RPN value of 480, while the broken hydraulic hose had the second highest RPN value of 448. In the calculation using Fuzzy FMEA, the highest RPN value was the broken hydraulic hose, as 631, while the wear clutch had the second highest RPN value as 630. The difference in results was due to the calculations using Conventional FMEA only multiplying the value of S (severity), O (occurrence) and D (detection) without regard to the importance of each input. This is due to its use only in linguistic terms in determining the S (severity scores), O (occurrence) and D (detection)in conventional FMEA [14]. Conventional FMEA only looked at the RPN value and treats risk factors in the same level of importance without calculating the relative importance of these risk factors [15]. In conventional FMEA the assessment of failure factors applied to natural language will obtain ambiguous and vague information [12]. Whereas with Fuzzy FMEA the value was generated by observing the importance of each input obtained from the results of processing Fuzzy. Based on these results, it can be concluded that the focus of attention in the maintenance and repair of the forklift was the hydraulic part with failure mode identified was the broken hydraulic hose. This was due to the failure mode has a high damage intensity, high severity, detection of the failure was tended to be difficult to detect with a high degree of importance based on the results of Fuzzy processing. Doing maintenance and repair certainly requires a lot of time and money. This priority was given as a way out to prevent damage that may occur on the forklift with limited costs as well as to speed up the process of handling forklift damage [13].
Hydraulic systems are important to be noted because they have an important contribution in lifting the load to be lifted by the forklift [16]. The hydraulic system determines the lifting weight of the forklift [17]. Lifting strength also affects the fork speed of the forklift in load and unload loads or goods. The lifting speed of the fork load on the forklift must be as minimal as possible in the lifting process or must be subtler in lifting the load which is sometimes ignored by the operator to achieve high performance. In this case, the hydraulic system also takes an important contribution to the forklift power control [18]. By knowing the biggest risk priority, it can be used as a focus in subsequent preventive maintenance on the forklift.

4. Conclusions and Recommendation

4.1. Conclusions
Based on the results of calculations using Fuzzy FMEA, the first priority scale that was used as the focus of maintenance and repair on the forklift was the hydraulic part with the failure mode identified was the broken hydraulic hose. This rupture of the hydraulic hose caused the power fork drive to decrease or even the hydraulic could not be functioning. Furthermore, this failure mode had a FRPN value of 631. The results using Fuzzy FMEA were more proposed as the reference focus for repair and maintenance due to the results pay attention to the importance of each input so that the results were more objective compared to the results of Conventional FMEA.

4.2. Recommendations
For further research, the recommendation is to conduct a deeper study of which should be the main focus between FMEA Conventional or Fuzzy FMEA in the application of risk analysis. Furthermore, it is expected to be able to use the integration of Fuzzy FMEA methods with other methods so that the results of the analysis of the risks that may occur can be more accurate.

5. References
[1] Rinawati D I and Dewi N C 2014 *Analisis Penerapan Total Productive Maintenance (TPM) Menggunakan Overall Equipment Effectiveness (OEE) dan Six Big Losses pada Mesin Cavitec di PT. Essentra Surabaya SNATIF* 21-26.
[2] Hapsari N Amar K and Perdana Y R 2012 *Pengukuran Efektivitas Mesin dengan Menggunakan Metode Overall Equipment Effectiveness (OEE) di PT. Setiaji Mandiri* Spektrum Industri Vol.10(2) pp 134-145
[3] Kerzner H 2001 *Project Management* (New York: John Wiley & Sons, Inc)

[4] Widerman M R 1992 *Project and Program Risk Management: A Guide to Managing Project Risk Opportunities* Amerika: Project Management Institute.
[5] Abdellatif, M., Shoeir, M., Talaat, O., Gabalah, M., Elbably, M., & Saleh, S. 2018 Design of an Autonomous Forklift Using Kinect. In *MATEC Web of Conferences* (Vol. 153, p. 04005). EDP Sciences.
[6] McDermott E R 2009 *The Basic of FMEA* (USA: CRC Press)
[7] Rusmiati E 2014. *Penerapan Fuzzy Failure Mode and Effect Analysis (Fuzzy FMEA) dalam Mengidentifikasi Kegagalan pada Proses Produksi di PT. Daesol Indonesia Jurnal Teknik dan Manajemen Industri.*
[8] Besterfield D 2006 *Total Quality Management* (New Jersey: Prentice Hall)
[9] Kusumadewi S 2002 *Analisis dan Desain Sistem Fuzzy Menggunakan Toolbox Matlab* (Yogyakarta: Graha Ilmu)
[10] Keskin G A and Ozkan C 2009 An Alternative Evaluation of FMEA: Fuzzy Art Algorithm *Journal of International Quality and Reliability Engineering* vol.1 p 647-661.
[11] Xu K Tang L C Xie M Ho S L Zhu M L 2002 Fuzzy Assessment of FMEA for Engine System. *Reliability Engineering and System Safety* Vol.1 p 17-29.
[12] Yeh R H and Hsieh M H 2007 Fuzzy Assessment of FMEA for a Sewage Plant. *Journal of the Chinese Institute of Industrial Engineers* Vol.1 p 505-512.
[13] Iqbal M Muflikhah L Setiawan N Y 2013 Penggunaan Fuzzy Failure Mode and Effect Analysis (Fuzzy FMEA) dalam Mengidentifikasi Resiko Kegagalan Proses Pemasangan dan Perbaikan AC *Jurnal Informatika/Ilmu Komputer* Vol.2 (7) p 1-6.
[14] Aisyah S 2017 Implementasi Failure Mode Effect Analysis (FMEA) dan Fuzzy Logic sebagai Program Pengendalian Kualitas *JIEMS (Journal of Industrial Engineering and Management Systems)* Vol.*4*(2) p 1-14.
[15] Mansur A and Ratnasari R 2016 Analisis Risiko Mesin Bagging Scale Dengan Metode Fuzzy Failure Mode and Effect Analysis (Fuzzy-FMEA) di Area Pengantongan Pupuk Urea PT. Pupuk Sriwijaja *Jurnal Teknoin* Vol.*21*(4) p 158-166
[16] Jimmy J Daywin FJ and Soeharsono G 2017 Perancangan Sistem Angkat Forklift Dengan Kapasitas Angkat 7 Ton *Jurnal Ilmiah Teknik Mesin* Vol.*12*(1) p 87-94
[17] Minav T Hanninen H Sinkkonen A Laurila L & Pyrhonen J 2014 Electric or Hyraulic Energy Recovery Systems in a Reach Truck *Journal of Mechanical Engineering p* 232-240
[18] Doci I & Imeri V 2013 Dynamic Analysis of Forklift during Load Lifting using Modeling and Simulation *International Journal of Current Engineering and Technology* p 342-347

Design of Tofu Cutting Tools to Improve Repetitive Tasks Using OCRA

Benedikta Anna Haulian Siboro[1], Vera Methalina Afma[2], Annisa Purbasari[3], Muhammad Qodri Kasim[4]

Department of Industrial Engineering, Universitas Riau Kepulauan
[1]anna@ft.unrika.ac.id, [2]vera.afma@gmail.com,[3]anice_nisa@yahoo.com

Abstract. Cutting process is one of processes in producing tofu before being packaged and marketed. Generally, cutting tofu is performed manually using a ruler with repetitive movements as 20-24 times, therefore it required longer time to cut and cause low back pain (LBP) in workers. Present study objective is to analyze the risk of work with repetitive task in the process of cutting tofu and designing tools to reduce these risks without decreasing the quality of the tofu. The methodology use OCRA and anthropometry which performed on two tofu factories (Mr. Joko and Mr. Udin) in Batam, as research subject. The results showed that repeated movements during tofu cutting were carried out for 33-35 seconds, with the percentage higher at hands work and waiting times in seconds. Furthermore, the OCRA risks of repetitive movements assessment is classified as medium-risk conditions (3.8-4.48). However, it is necessary to repair the cutting tools. Improvements are gained by designing cutting tools in accordance with the operator's hand anthropometry. The new tofu cutting tool produces only 2 times repetitive movements, which is lesser and using OCRA assessment the level of movement risk becomes optimal (≤ 1.5).
Keywords: repetitive tasks, OCRA, anthropometry, low back pain

1. Introduction

Tofu has the heftiest quality vegetable protein because of its most complete amino acid composition and is believed to have high digestibility (85% -98%), which make tofu as one of favourite foods for most people [1]. Tofu heaviest ingredient is soybean and Indonesia is the largest producer of tempeh and soybean markets in Asia [2]. In Indonesia, tofu is processed using various methods which generate varied product. The process also has no standardization as a reference for producing a superior tofu. Tofu is processed through several stages; soybean cleaning, milling, boiling, filtering, printing, tofu cutting and packaging. The Size of tofu is adjusted to market demand, thus the tofu factory usually produce several sizes. The tofu is cut using a cutting tool and measured by a ruler. The cutting of a tofu board generally took 20-24 times cutting movements. The process of cutting tofu is performed by several factories in Indonesia, including in Batam. To cut tofu, the worker's posture must bent and are carried out repeatedly during daily production. This repetition posture that carried out day to day will cause Low Back Pain (LBP). Several factors including anthropometric characteristics, the nature and severity of physical work, working postures, and methods of manual lifting/handling have been related to the severity of LBP [3]. In addition, other aspects such as lifestyle conditions and psychological factors may also be considered as independent risk factors for the risk of LBP [4].

There are numbers of risk assessment methods that can be used to assess work risks. One of them is OCRA (Occupational Repetitive Action). OCRA is an ergonomic tool that suitable to analyse work risks due to exposure of repetition activities. In other hand, OCRA method is widely known as a valid, scientifically founded tool, to detect the presence of risk factors for the upper limbs, likely to determine the level, in the medium –long term, of work related pathologies [5].

Countless researches are focus to improve cutting tool design. Budihamsyah and Putra proposed the design of tofu cutters at affordable prices used the Value Engineering method to be expected to increase time efficiency by 56.5% and reduce the disability rate by 88%. The proposed tool was designed using food grade stainless material in boxes according to the size of the tofu [6]. Siboro [7] had researched and designed the cutting tools based on cutting process only two times which quicker. However, her research was required to be developed as the base to lay the tofu was uneven and the thickness was varied thus the tofu had not completely and evenly cut. In addition, Izzhati [8] had also designed cutting tools used iron and one handle. Her design was set for large and thick tofu. Meanwhile, the purpose of this study is to analyse the work risk's caused by repetitive movements during the process of cutting tofu using OCRA method and designing tools to reduce the risk of repetitive movements.

2. Methods

2.1 Procedure

This research was carried out on the two tofu factories in Batam (Mr. Udin and Mr. Joko's factories). At the beginning of the study, interviews and observations were conducted to collect information and identify the work process of cutting tofu, investigated workers complaint on health issue and collected input to improve the work system in cutting tofu. Observations also include working time, rest time, duration of repetitive work, etc. to get accurate data for OCRA assessment. Observations were also used video recordings to identify technical actions and calculate the number of technical actions, work time cycles.

2.2 Work Risk Assessment with OCRA

The work-risk assessment will be carried out on two operators who frequently cut tofu at Pak Joko's Factory and one tofu operator at Mr. Udin Factory. Assessment aspects include 1) force of exertion, 2) frequency, 3) discomforted postures / movements, 4) lack of sufficient recovery, 5) duration, and 6) additional factors. For consistency between force/exertion intensity, estimation between methods, all ratters applied the Borg CR-10 scale [9] assessment which based on the items listed in the OCRA checklist. Furthermore, the work risk score will be calculated according to value of ATA (Actual Technical Actions), RTA (Recommended Technical Actions), and finally calculate OCRA index value.

$$\text{ATA} = \text{Frequency x Time total of repetitive task} \tag{1}$$

$$\text{RTA } \sum_{i=l}^{n} [CFX(Ffi \ x \ Fpi \ x \ Fci) \text{x Di}] x \ Fr \ x \ Fd \tag{2}$$

$$\text{OCRA}_{\text{index}} = \frac{nATA}{nRTA} \tag{3}$$

Where:

1,n = number of repetitive tasks performed
I = generic repetitive task
CF = constant of frequency of technical action per minute (=30)
Ff = frequent or high force exertions
Fp = uncomfortable posture factor or movement in each repetitive task
Fc = additional factor
D = net duration in minute of each repetitive task
Fr = multiplier for the risk factor lack of recovery period

Fd = overall duration factor of all repetitive tasks

Below is OCRA index to classify the activities, whether in optimal risk classification or not.

Table 1. OCRA Index

Area	OCRA Index Values	Risk Clasification	Consequent possible actions
Green	1.5 and lower	optimal	None
Green-Yellow	1.6 -2.2	acceptable	None
Yellow -Red	2.3-3.5	uncertain or very low	Check again, if possible improve work conditions
Red-Light	3.6 - 4.5	light	Improve work conditions, health surveillance, training
Red-medium	4.6-9.0	medium	Improve work conditions, health surveillance, training
Red-high	9.1 and higher	high	Improve work conditions, health surveillance, training

If the index value at OCRA is >2.2, it is necessary to re-established it for eliminate unwanted risk.

2.3 Tool Design
Tool design is carried out if the results of OCRA index are stated as risky and require improvements in order to reduce risk. In this study, tool design was carried out by measured the anthropometry of the worker's body. In stations, the worker should stand and most activities performed by hand. There are several dimensions that need to be considered as a guide in the measurement and retrieval of anthropometric data [10], namely: hand reach, elbow width, length palms, palms width, hand grip diameter. In determined those anthropometric data, normal distribution will generally be applied. In statistics, the normal distribution is formulated based on mean (x) and standard deviation (σ x) of the available data. Based on the calculation result, the "percentiles" can be appointed according to the normal distribution probability table. The percentile used in this study is 95th percentile which refer to 95% of the population, or below the mean. The percentile 5-th indicate that 5% of the population, or below, is valued and the 50th percentile indicate 50% of the population will be at the average value of that size.

3. Result and Discussion

3.1 Working Posture
Body postures during cutting tofu in both factories were dissimilar. Mr. Udin factory has a table to cut tofu which lower than the worker's body posture, while Mr. Joko's Factory already has a tofu cutting table that suitable with worker anthropometry. A bent posture in day to day basis will adversely affect the worker's health. Activities such repeated lifting, carrying, pulling and pushing heavy objects, frequent or prolonged bending, bowing, sitting or standing for a long time or other uncomfortable body postures, standing position prolonged will caused low back pain [11]. Research by Punnet [12] described 15% of adults, in general, suffering from frequent back pain or pain lasting more than two weeks, and 37% of LBP was attributable to occupational risk factors. In those both factories, it were identified that the repetition during the work cycle was the process of cutting tofu using a knife and ruler / mall as a size control (Table 1). High repetition of movements in manual tasks is related to musculoskeletal disorders, specifically on upper limb. Musculoskeletal disorders are among the most common working problems Buckle and Devereux [13] representing one-third of work injuries with significant economic and social consequences.

Table 2. Condition of Mr. Joko and Mr. Udin factories

Conditions	**Mr. Joko**	**Mr. Udin**
Body Posture		
	Body is not bent	Body is bent due to lower worktable
Tools	Knife, and *mal*	Knife and ruler
Times	25-30 seconds each table	30-35 seconds each table
Complaints	Pain on wrist, back	Pain on wrist, back, neck

3.2 OCRA Analysis before Design Improvement

The above conditions were supported by the results of OCRA calculations at each tofu factory. At Mr. Joko's factory, the number of actions taken in one work cycle were 29 activities by left hand, and 31 movements by right hand; generally activities include took and put the tofu mold, taking the size guidance, taking the knife and cutting tofu. At the Mr. Udin factory, one work cycle required 30 activities for the left hand and 33 activities by right hand, which involved pick and laid out the tofu mold, taking a ruler, picking up a knife and cut tofu. The Bhor score scale for each activity is determined by the duration of each activity. In both factories the longest time per activity was lifted and laid out the tofu mold (Bohr scale = 3), while other activities with a score = 1. The total score of the overall activities will influence the technical strength factor (FF) of the worker. Analysis of posture and movement in each single segments of the upper body (grip, wrist, elbow, shoulder) and associated with time (frequency and time period) static posture and dynamic movements were analyzed in Fp to determine the uncomfortable posture position. At the factory, Mr. Joko and Mr. Udin factories, the worker shoulder up 45 ° -80 ° when they lifted and placed the mold and when hold the knife to cut the hand was in pinch posture; both activities were uncomfortable posture. Another factor that influences the high OCRA index in these two factories is Fd, as the tofu cutting area is very close to the soybean milling station with a noisy machine level that affects tofu cutting work. Table below present both Mr. Joko and Mr Udin Factories have an OCRA index at the red light level which requires improvement of the work system, health and training for the workers.

Table 3. OCRA Index before Improvement

Calculation items	**Mr. Joko**		**Mr. Udin**	
	Right Hand	**Left Hand**	**Right Hand**	**Left Hand**
ATA	11160	10440	10080	9360
CF	30	30	30	30
FF	0,6786	0,6357	0,74464	0,54952
Fp	1	1	0,7	1
FC	0,8	0,8	0,8	0,8
D	360	360	360	360
Fr	1	1	1	1
Fd	0,5	0,5	0,5	0,5
RTA	2932	2746	2252	2374
OCRA	3,81	3,80	4,48	3,94

3.3 Design and OCRA after repair

From the OCRA index results, the next step is to improve the tofu cutting process by designing tofu cutting tools. Below is a design that equipped with size, disadvantages and advantages of each tool. The design of cutting tools for both factories will be dissimilar; according to actual conditions of the cutting process. In previous explanation, at Mr. Udin Factory, body posture of workers is bent during tofu's cutting thus it require table in accordance to the worker's posture. Therefore, in the design, tables and cutting tools are designed to fuse, so it more efficient. With the linear slider in the vertical and horizontal position, the operator will easier to cut the tofu. In the cutting tool for Mr. Joko, the cutting tool is designed not to be installed jointly with the table and cutting tools. It is also designed lightly to facilitate the cutting process of tofu.

Table 4. Improvement Design of Tofu Cutter

Design Items	**Mr. Joko**	**Mr. Udin**
Design	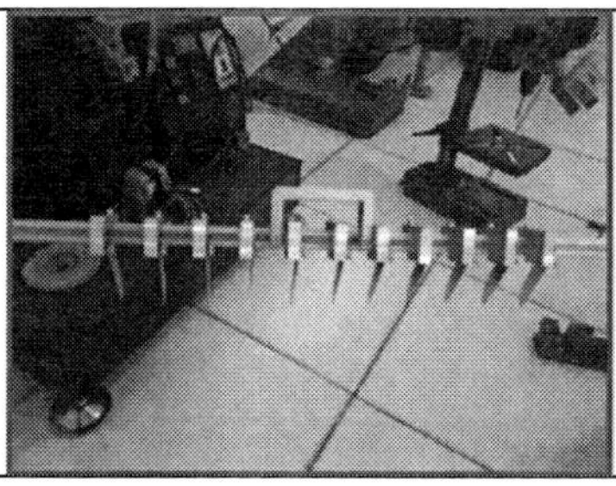	
Dimension	Hand reach: 77,6 cm (reference for length tofu cutting)	Hand reach: 77,6 cm (reference for length tofu cutting) Elbow width: 97,8 cm (reference for width of tofu cutting), Height of man: 95 cm (reference of table high of table)
Advantages	Cutting process only 2 times, no need extra places	Cutting process only 2 times
Disadvantage	Need more times to set straight line Need to lift 1 set cutter if cutting vertical or vice versa	Table take more places

Trial for the tools was carried out within one week to make sure that it could be used properly, and afterward the OCRA analysis was applied on the use of these tools. Table 5 below show the decrease in number of activities carried out (ATA) in one cycle of activities. The process of cutting tofu using the designed tool indicate positive impact on the efficiency of the cutting process time; the cutting process befell shorter 15-20 seconds after an average repair. For Fp, the uncomfortable posture in the tofu cutting process at Mr. Joko's and Mr. Udin's factories was remain endure during lifting and placing the mold where shoulder move to 45°-80°. Moreover, other additional factors such lighting and noise were remain present. The table 5 below display a significant decrease in the OCRA index to 0.48-0.9, where the score was based on Table 1 including the green level (optimal). Both of these tools can be used further without negative impact on the health of workers.

Table 5. OCRA Index after Improvement

Calculation Items	Mr. Joko		Mr. Udin	
	Right Hand	Left Hand	Right Hand	Left Hand
ATA	2520	1440	1440	1080
CF	30	30	30	30
FF	0,65	0,69	0,5	0,5
Fp	1	1	1	1
FC	0,8	0,8	0,8	0,8
D	360	360	360	360
Fr	1	1	1	1
Fd	0,5	0,5	0,5	0,5
RTA	2808	2981	2160	2160
OCRA	0,90	0,48	0,67	0,50

4. Conclusion

Based on above results we can conclude:

1. Repetitive activity and bending will impacting people health, particularly low back pain.
2. Before improvement, tofu cutting activity at Mr. Joko and Mr. Udin factories have risk potentiality. It is indicated by OCRA index level that showed 3.80-4.49 (red risk level). After improvement, OCRA index reach optimal level due to decrease of repetition activity from 20-24 times cutting per table to 2 times cutting process.

Acknowledge:

This research was supported by funding from Ministry of Research, Technology and Higher Education.

5. Reference

[1] Widyaningrum I 2015 Teknologi Pembuatan Tahu yang Ramah Lingkungan (Bebas Limbah). *Jurnal Dedikasi* **12**(1) pp 14-21

[2] Astawan M 2004 *Tetap Sehat dengan Produk Makanan Olahan* (Solo: Tiga Serangkai)

[3] Sowah D, Boyko R, Antle D, Miller L, Zakhary M and Straube S 2018 Occupational interventions for the prevention of back pain: Overview of systematic reviews. *Journal of Safety Research* **66** pp 39-59

[4] Duthey B 2013 Update on 2004 background paper 6.24 low back pain. (WHO)

[5] Boenzi F, Digiesi S, Facchini F G and Mummolo G 2016 Ergonomic improvement through job rotations in repetitive manual tasks in case of limited specialization and differentiated ergonomic requirements. *IFAC Conference Paper* **48**(12) pp 1667-1672

[6] Budihamsyah D and Putra B I 2017 Perbaikan Desain Alat Pemotong Tahu Dengan Pendekatan Rekayasa Nilai. *Prozima* **1**(2) pp 123-135

[7] Siboro B A H, Siregar R.A and Purbasari A 2017 Perancangan Alat Pemotong Tahu Untuk Mengurangi Gerak dengan Metode Motion Time Measurement (MTM)-Motion Time Study (Studi Kasus Pabrik Tahu Pak Joko). *Profisiensi* **5**(2) pp 115-122

[8] Izzhati D N 2010 Pengembangan Alat Pemotong Tahu yang Ergonomis Dengan Menggunakan Metode Rula. *Prosiding Seminar Nasional Sains dan Teknologi* **1**(1) pp 7-11

[9] Rosecrance J, Paulsen R and Murgia L 2017 Risk assessment of cheese processing tasks using the Strain Index and OCRA Checklist. *International Journal of Industrial Ergonomics* **61** pp 142-148

[10] Irdiastadi H and Yassierli 2014 *Ergonomi Suatu Pengantar* (Bandung: PT. Remaja Rosdakarya)

[11] Kusuma I F, Hasan M and Hartanti R I 2014 Pengaruh Posisi Kerja terhadap Kejadian Low Back Pain pada Pekerja di Kampung Sepatu,Kelurahan Miji, Kecamatan Prajurit Kulon, Kota Mojokerto. *Jurnal IKESMA* **10**(1) pp 59-66
[12] Punnett L 2005 Estimating the global burden of low back pain attributable to combined occupational exposures. *American Journal of Industrial Medicine* **48** pp 459-472
[13] Buckle P.W and Devereux, J.J 2002. The nature of work-related neck and upper limb musculoskeletal disorders. *Appl. Ergon.* **33**(3) pp 207–217

IOP Conf. Series: Materials Science and Engineering **528** (2019) 012029 doi:10.1088/1757-899X/528/1/012029

Fatigue-related differences in human facial dimensions based on static images

V Triyanti[1,2], Yassierli[1] and H Iridiastadi [1]

[1]Faculty of Industrial Technology, Institut Teknologi Bandung, Bandung 40132, Indonesia
[2]Department of Industrial Engineering, Atma Jaya Catholic University of Indonesia, Jakarta 12930, Indonesia
vivi.triyanti@student.ti.itb.ac.id[1]; vivi.triyanti@atmajaya.ac.id[2]

Abstract. Automatic fatigue recognition based on eye and mouth movements has been widely researched and used to detect human fatigue. However, there were only few studies that quantitatively examining fatigue status based on static images. This study was a pilot study that aimed to examine differences in human facial dimension between fresh and fatigue condition, based on photos. 4 photos from 8 subjects were taken, each photo depicted the subject in fresh condition with a neutral expression, fresh condition with a happy expression, fatigue condition with neutral expression, and fatigue condition with happy expression. Each photo was analyzed using Face Reader 7.1 software to detect the coordinates of the points around the eyes and mouth. 10 dimensions around the eyes were calculated for each situation. In neutral expressions, paired t-test with significance value of 0.05 proved that in 8 dimensions, value in fresh conditions were different from ones in fatigue conditions. But these results were not proven in the picture with happy expressions. Although further research is needed, this finding could be a first step for developing the knowledge to detect fatigue based on facial static images.
Keywords: Human fatigue, human facial dimension, static images

1. Introduction

Fatigue is a suboptimal condition of the body caused by exertion, where the context of fatigue depends on the condition and form of exertion, can be in the form of physical energy or mind [1]. Fatigue can be physical or mental. Conditions of fatigue can lead to changes in strategy in the use of resources so that the level of mental processing or physical activity can be maintained or can also be reduced [2]. The short-term effects of fatigue on individuals are performance disruptions including reduced ability to concentrate, make decisions, maintain alertness, and control emotions [3]. Fatigue can also increase error rates, slow down reactions, and thus increase the likelihood of accidents and injuries [4].

A subjective study states that the face when fatigue was different from the face in normal conditions and can be recognized from a photo. Hanging eye lid, glazed eyes, red eyes, dark circles under the eyes, pale skin, gloomy lips, and fine lines were signs of fatigue that appear in a collection of photos [5]. From this study, we may infer that there is possibilities that single or series of static images can be used to examine the status of a person's fatigue condition.

In addition to subjective research, so far there were not many researches related to facial recognition that uses quantitative measurements to detect fatigue based on a person's photos or static facial images. So far most studies used facial movement behaviors to detect fatigue, such as eye or mouth movements, or pupil size [6]. When humans were fatigue, especially when sleepy, the eyes tended to have specific movement behavior. For example, a faster rate of blinking, longer eye closure, slow eye gaze [7], and decreased levels of eye openness [8]. In drowsiness, sudden nodding movements occurred [9-11]. Some studies were also suggested that when fatigue, people had "declined" facial expressions [12-13]. However, quantitative explanations about these expressions have not been stated clearly.

In this paper, a pilot study was conducted to examine fatigue facial expressions based on static images. Detection of face points were done automatically using the facial expression analysis software FaceReader 7.1 from Noldus (www.noldus.com), in which one of the outputs of the software was XY coordinates for several points of the face. By processing the coordinates, we may analyze whether there were any differences in facial dimensions between fresh and fatigue conditions. This research can be an initial study for research related to fatigue detection using human photos, which can be used to detect worker's fatigue.

2. Methods

2.1. Data processing

All images were processed using Face Reader 7.1 software. Face Reader software generally aims to classify people's emotions into different basic emotional categories, namely happy, angry, sad, surprised, fear, disguised, and neutral [15]. In addition to the recognition of basic facial expressions, the Face Reader 7.1 outputs are X-Y coordinates for several points of the face (Figure 1) around the eyes, nose and mouth. There are a total of 49 coordinates (denoted by light blue dots). The coordinates have units of pixels and the values are measured from the upper left corner of the face image. From 49 coordinates, only 12 coordinates from points around eyes, eyebrows, and mouth that were used in this research, denoted by dark blue dots in Figure 1. Table 1 defined the position of each dot.

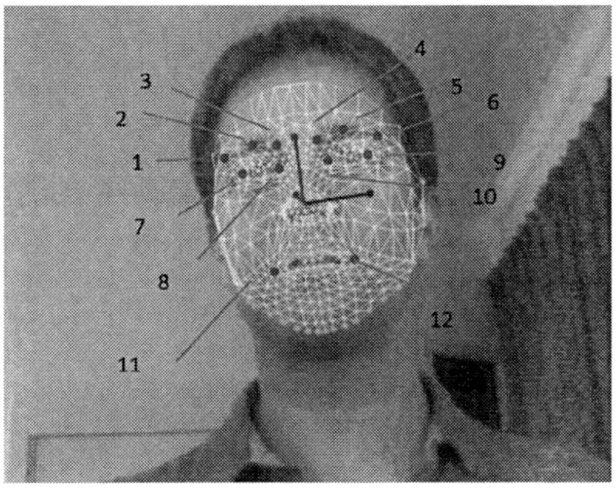

Figure 1. Coordinates of facial points, based on output from Software Face Reader 7.1

Table 1. Location of x,y coordinates on Face Reader 7.1 software based on Figure 1

Points	Facial parts
1-3	left eyebrow
4-6	right eyebrow
7-8	corners left eye
9-10	corners right eye
11-12	Mouth

IOP Conf. Series: Materials Science and Engineering **528** (2019) 012029 doi:10.1088/1757-899X/528/1/012029

Because in previous studies, one sign of fatigue was a "declined" face [12][13], there were several hypotheses were tried in this study. The calculation formula was determined based on the 12 points in Figure 1. The coordinates output from the 12 points were used to calculate 10 specific face dimensions, especially around eyebrows, eyes, and mouth (Figure 2), which were marked with capital letters.

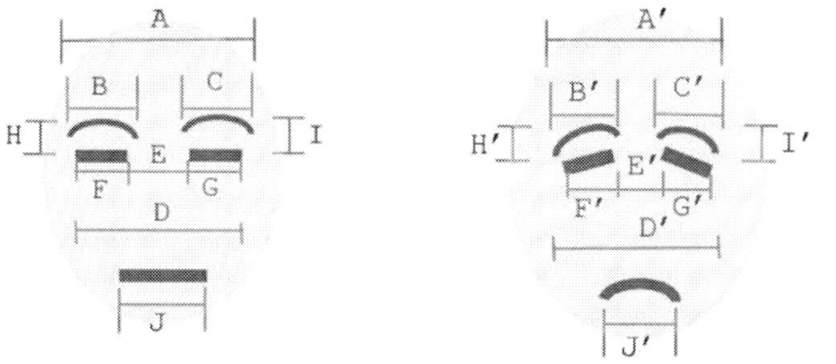

(a) Neutral expression (b) Fatigue expression

Figure 2. Face illustration and measured dimension
a) Neutral expression, illustrated with straight brow, eye, and mouth position
b) Fatigue expression, illustrated with declining brow, eye, and mouth position

The 10 face dimensions (Figure 2) could be categorized into 4 hypotheses, which were:

1. Horizontal distances of eyebrows in fresh conditions were higher than in fatigue conditions. The related dimensions were:
   A. Distance between outer points of eyebrows (P6 – P1)
   B. Left eyebrow length (P3-P1)
   C. Right eyebrow length (P6-P4)
2. Horizontal distances of eyes in fresh conditions were higher than in fatigue conditions. The related dimensions were:
   D. Horizontal distance between outer points of eyes (P10 – P7)
   E. Horizontal distance between inner points of eyes (P9 – P8)
   F. Left eye length (P7-P6)
   G. Right eye length (P10-P9)
3. In fatigue eyes, the eyelids tended to be more closed, therefore the vertical distance between the eye brow and upper eyelid was hypothesized further in fatigue than in fresh conditions. However, because Face Reader output did not display the coordinates of the eye lids, the eye points were approached with the coordinates of the eye tip. Based on this explanation, the related dimensions and the formula were
   H. Avg. of left eye-eyebrow vertical distance ((P7 – P2)+(P8-P2))/2
   I. Avg. of right eye-eyebrow vertical distance ((P9 – P5)+(P10-P5))/2
4. Horizontal length of mouth in fresh conditions was higher than in fatigue conditions.
   J. Length of mouth (P12 – P11)

For dimension A-G and J, with the hypotheses that the dimensions in fresh condition were greater than ones in fatigue condition, the differences were calculated as follows:

$$\text{Difference} = \text{distance of two points when fresh} - \text{distance of two points when fatigue} \quad (1)$$

For dimension H and I, the expected condition is the opposite, so the formula was:

Difference = distance of two points when fatigue - distance of two points when fresh (2)

2.2. Data collection
The subjects were 8 persons from various ages, where they were asked to stand or sit upright facing the camera. Then the subject was asked to show a neutral expression and continued with a happy expression with a smile [14]. At the time of measurement, a ruler was placed above the subject's head horizontally, to indicate the actual size relative to the image. All static facial images were taken using a smartphone. Images were taken in two times: in the morning (after the morning bath) and at night (just before going to bed at night). Both times were chosen to represent the state of fresh face (in the morning) and tired face (at night). In total there were 4 static images taken from each subject, which were:

a. Neutral expression – fresh condition (in the morning)
b. Neutral expression – fatigue condition (at night)
c. Happy expression – fresh condition (in the morning)
d. Happy expression – fatigue condition (at night)

In the initial data, the coordinates obtained were in pixel units, therefore they had to be converted into centimeters. Each distance data listed at point A-J of section 2.2 was calculated for each image. With the help of the distance obtained from the ruler, all images were interpolated into the same unit and measurement, so that all images could be compared. In this study, the images were taken from the fresh and fatigue condition in each expression. Based on the difference, the average and standard deviation values were calculated. A series of one way paired t-tests were performed to test that the average of difference between both values was more than 0 (it means that the distance of object in normal condition was higher than in fatigue condition).

3. Result and Discussion

3.1. Neutral expression
After all subjects performed neutral expression in fresh and fatigue conditions (Figure 3), each picture then was processed using Face Reader 7.1 software. Based on the output of the software, the eight subjects confirmed have been performed a neutral expression, although in all photos of subjects, neutral expression mixed with other expressions (Figure 4). The magnitudes of the 10 facial dimensions then were calculated and the difference between both conditions were further analyzed. The average and standard deviation of the differences, and the p-value of paired t-test are shown in table 2.

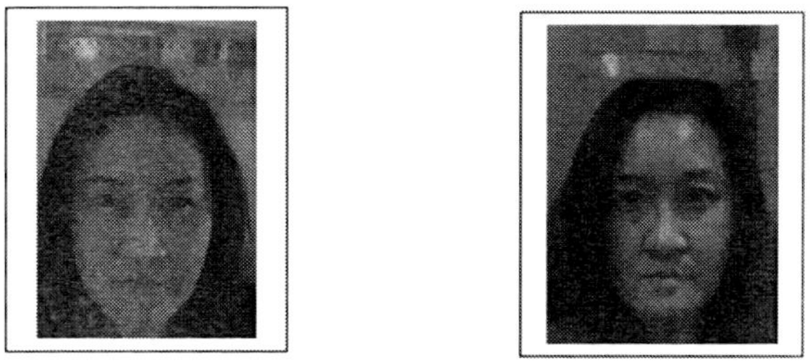

(a) Fresh (b) Fatigue

Figure 3. Example of subject's neutral expression in the fresh and fatigue condition

When the subject did not do any expression, or was called a neutral expression, the average differences in values of all measured dimensions showed positive values. There was even one participant who had a very large difference value between fresh and fatigue conditions, especially for the outer point of the eyebrow (P1) and the outer point of the eye (P7). These condition caused the difference value between fresh and fatigue conditions of the dimension "Distance between outer points of eyebrows (A)" and the dimension "Horizontal distance between outer points of eyes (D)" were also very large. As a result, for these two dimensions, the standard deviation of the eight samples became very high.

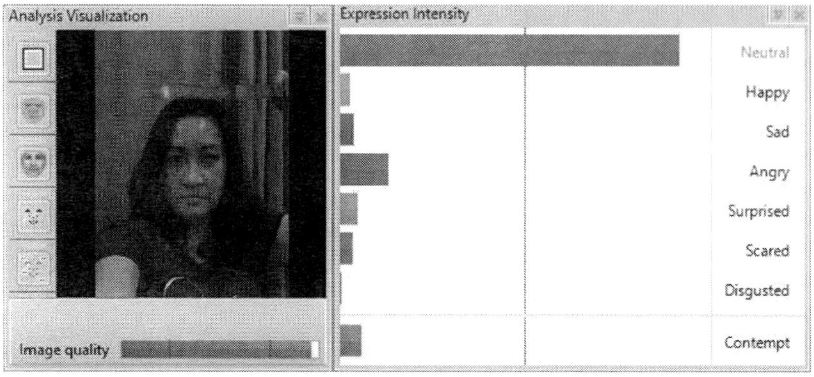

Figure 4. Example of software output for neutral expression and the expression intensity

Table 2. Average and Standard Deviation of dimension difference, and p-value of one way paired t-test for fresh and fatigue conditions in neutral expression

Measurement	Diff. Avr	SD	p value
Eye Brow			
A. Distance between outer points of eyebrows	2.000	2.117	0.016
B. Left eyebrow length	0.520	0.530	0.014
C. Right eyebrow length	0.678	0.696	0.014
Eye			
D. Hor. distance between outer points of eyes	1.491	1.562	0.015
E. Hor. distance between inner points of eyes	0.641	0.015	0.641
F. Left eye length	0.422	0.375	0.008
G. Right eye length	0.427	0.404	0.010
Eye brow-eye distance			
H. Avg. of left eye-eyebrow vertical distance	0.621	0.907	0.047
I. Avg. of right eye-eyebrow vertical distance	0.641	0.920	0.045
Mouth			
J. Length (horizontal) of mouth	0.599	0.905	0.052

The positive average differences value indicates that the value of the face dimension measured in fresh condition is higher than in fatigue condition, meaning that the eyes and eyebrows when in fatigue did look decline than when in fresh. When they were tested using a paired t-test, all dimensions showed significant differences, except the dimensions "Horizontal distance between inner points of eyes (E)" and the dimension "Horizontal length of mouth (J)". This results inferred that the difference between the closest distance of the eyes when in fresh and fatigue was not much different, likewise the length of the mouth. Originally it was predicted that when fatigue, the mouth tended to bend downward. Therefore it was hypothesized that the horizontal length of the mouth of fresh condition was greater than of fatigue condition. However, apparently this hypothesis was not proven. Although there was difference in mouth length, it could not be proven that the difference was significant.

3.2. Happy expression

In addition to neutral expressions, subjects were also asked to demonstrate happy expressions both at fresh condition (in early morning) and at fatigue (at night). Based on the Face Reader 7.1 software output, all objects were confirmed have made happy expressions in both conditions (Figure 5).

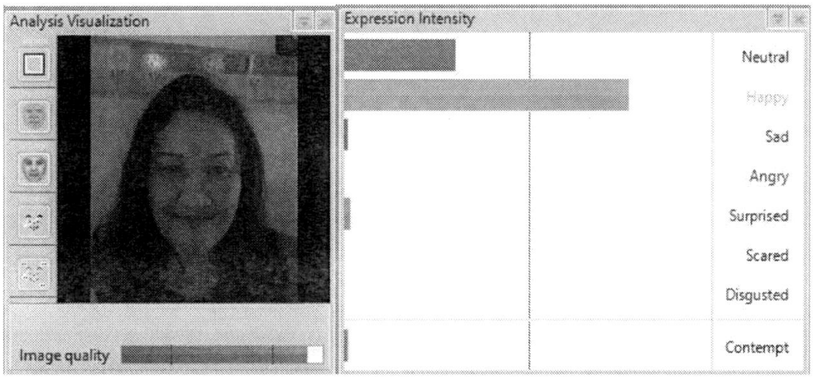

Figure 5. Example of software output for happy expression and the expression intensity

In contrast to neutral expressions, the results for happy expressions indicated dissimilarity. Of the ten dimensions calculated, all dimensions showed negative values. So it can be said that the dimensions value in fatigue were greater than in fresh conditions (and vice versa for dimensions H and I) (Table 3). However, based on the *p*-value it can be concluded that the differences is not significant. Values that tended to be greater when fatigue might be due to the subject's efforts to make happy expressions when in fatigue condition were greater than when in fresh conditions. This resulted in an unnatural expression from the subjects because of the perforce smile. The various happy expression might cause the location of the face points (and also the value of the measured dimensions) to be incompatible with the hypothesis.

Table 3. Average and Standard Deviation of dimension differences, and *p*-value of one way paired t-test for fresh and fatigue condition in happy expression

Measurement	Diff. Avr	SD	*p*-value
Eye Brow			
A. Distance between outer points of eyebrows	-0.877	3.939	0.274
B. Left eyebrow length	-0.488	1.288	0.160
C. Right eyebrow length	-0.316	1.323	0.260
Eye			
D. Hor. distance between outer points of eyes	-0.867	3.175	0.233
E. Hor. distance between inner points of eyes	-0.347	1.466	0.262
F. Left eye length	-0.283	0.872	0.194
G. Right eye length	-0.236	0.857	0.231
Eye brow-eye distance			
H. Avg. of left eye-eyebrow vertical distance	-0.084	0.797	0.387
I. Avg. of right eye-eyebrow vertical distance	-0.100	0.762	0.361
Mouth			
J. Length of mouth	-0.527	2.573	0.290

For happy expression, almost all subjects had a small difference average values, but had a large standard deviation. This meant the original data difference varied greatly, but was almost evenly distributed on the top and bottom of the average (resulting in a small average value). It can be said that there were subjects who gave large smile that affected the dimensions of the eyes and mouth, but there were also those who only smile slightly, so that the dimensions of the eyes tend to be the same as in neutral expression (Figure 6).

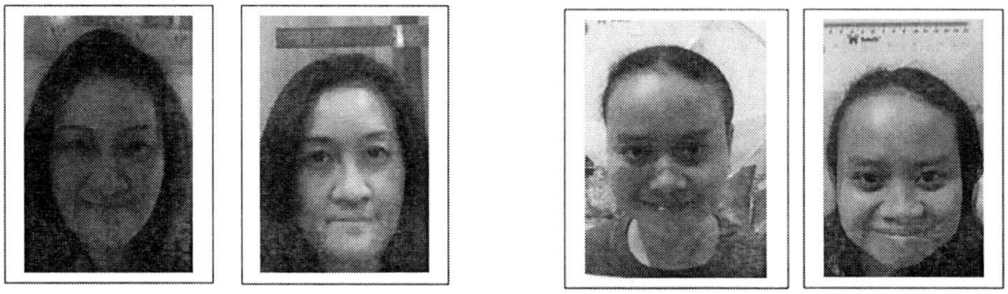

(a) Happy-fresh (b) Happy-fatigue (a) Happy-fresh (b) Happy-fatigue

Figure 6. Example of subject's happy expression in the fresh and fatigue condition

3.3. Discussion

This study indicated that fatigue face recognition based on static images (not video) was possible to be done automatically using software, especially if the subject only shows a neutral expression. This was a significant finding, considering that so far most studies had emphasized the facial movements (such as mouth and eye movements) to indicate fatigue and drowsiness. Furthermore, the results of this study could be used to gain more knowledge about the parameter and threshold of fresh and fatigue condition related to human facial. The knowledge would be valuable to the development of fatigue detection system based on static images that did not need continuous monitoring of object movements. For example by taking pictures of workers periodically (not continuously), we could assess the condition of their fatigue. Further study also could be done by analysing fatigue compared to varied expression (sad, anger, scared, etc.) and psychologies condition (stress, depressed, anxiety, etc.).

In spite of the positive finding, this pilot study still had many shortcomings. The main limitation was the small number of samples. More samples were needed to prove that for most population, the facial points were indeed changing according to fatigue conditions. In addition, there were also factor of cosmetics. In this study, subject was not restricted in the use of cosmetics. So in some subjects, pictures in fresh conditions were taken when they have used cosmetics. This might cause the software recognition of the facial point's location to be different. Another limitation that could not be ignored was the limitation of dimension measurement. In this study, most of measured dimensions were horizontal dimensions. Since the location of the facial points displayed by the software was very limited, there were limitations in measuring vertical dimensions. An example is the eye area. The software does not display the location of the eyelids or eyelids, so that the width of the eyes cannot be measured.

4. Conclusion

This research was a pilot study to examine the ability of software in recognizing the differences in facial points of subjects when in fatigue and fresh conditions, based on photos or static facial images. The results showed that of 10 dimensions, 8 dimensions around the eyes and mouth did show significant differences (using significance value of 0.05), especially if the subject only showed a neutral expression. Due to the faces appearance that tended to look "declined" when in fatigue condition, the value of the dimensions tended to be smaller when fatigue. However the same conclusion could not be generalized for other conditions. If the subject showed a happy expression, the differences in dimensions when fresh and fatigue were not significant. Because this study was still in early stages, further research is needed to validate these findings.

5. References

[1] Phillips R O 2014 What is fatigue and how does it affect the safety performance of human transport operators? *Fatigue in Transport Report* I (Oslo: Institute of Transport Economics Norwegian center of Transport Research)

[2] Galy E, Cariou M and Melan C 2012 What is the relationship between workload factors and cognitive load types? *Int. J. Psychophysiology* **83** pp 269-275

[3] Work Safe Victoria 2008 *Fatigue: Prevention in the workplace* 1. (Victoria: Victorian Work Cover Authority)

[4] Williamson A, Lombardi D A, Folkard S, Courtney T K and Connor J L 2011 The link between fatigue and safety. *Accident Analysis and Prevention* **43** pp 498–515

[5] Sundelin T, Lekander M., Kecklund G, Van Someren E J W, Olsson A and Axelsson J 2013 Cues of fatigue: effects of sleep deprivation on facial appearance. *Sleep* **36**(9) pp 1355–60

[6] Dawson D, Searle A K and Paterson J L 2014 Look before you sleep: Evaluating the use of fatigue detection technologies within a fatigue risk management system for the road transport industry. *Sleep Medicine Reviews* **18** pp 141-152

[7] Zhu W B, Yang H, Jin Y and Liu B 2017 A method for recognizing fatigue driving based on Dempster-Shafer Theory and Fuzzy Neural Network. *Mathematical Problems in Engineering* ID6191035

[8] Fernández A, Usamentiaga R, Carús J L and Casado R. 2016 Driver distraction using visual-based sensors and algorithms. *Sensors* **16**(11) 1805

[9] Ji Q, Zhu Z and Lan P 2004 Real time nonintrusive monitoring and prediction of driver fatigue. *IEEE Transactions on Vehicular Technology* **53**(4)

[10] Jabon M E, Bailenson J N, Pontikakis E, Takayama L and Nass C 2011 Facial-expression analysis for predicting unsafe driving behavior face and gesture recognition. *Pervasive computing* (October-December 2011)

[11] Alioua N, Amine A and Rziza, M 2014 Driver's fatigue detection based on yawning extraction. *Int. J. of Vehicular Technology* ID678786

[12] Hachisuka S, Ishida K, Enya T and Kamijo M 2011 Facial expression measurement for detecting driver drowsiness. *Eng. Psychol. and Cog. Ergonomics, HCII 2011, LNAI 6781* ed. Harris D (Verlag Berlin Heilderberg: Springer) pp 135–144

[13] Ying L T, Kanade T and Cohn J F 2011 Recognizing Action Units for facial expression analysis. *Handbook of Face Recognition* ed. Li S Z and Jain A K (London: Springer-Verlag London Limited)

[14] Slessor G, Bailey P E, Rendell P G, Ruffman T, Henry J D and Miles L K 2014 Examining the Time Course of Young and Older Adults' Mimicry of Enjoyment and Nonenjoyment Smiles. *Emotion* **14**(3) pp 532-44

[15] Ekman P and Rosenberg E L 2005 What the face reveals, basic and applied studies of spontaneous expression using the Facial Action Coding System (FACS). 2nd ed. (New York: Oxford University Press, Inc)

Ergonomic risk and work load analysis on material handling of PT. XYZ

L Widodo[1*], F J Daywin[1] and M Nadya[1]

[1]Industrial Engineering Department of Tarumanagara University, Jakarta, Indonesia
*lamtow@ft.untar.ac.id; mariscanadya30@gmail.com

Abstract. In Indonesia, some industries do manual work using human labor. Among other things is the process of raw materials handling. Likewise, this happened to PT XYZ, which has not used a tool in the material handling process. The material are steel plates and wood plates. This study aims to determine the physical complaints felt by workers, determine the workload classification, and analyze the work posture of workers. In accordance with the results of the Nordic Body Map questionnaire, there are complaints from workers, especially pain in the upper neck, pain in the left and right shoulder, pain in the left upper and right upper arm, backpain, low back pain, left and right forearm pain, left-right wrist pain, and right and left hand pain. Based on Rapid Entire Body Assessment (REBA) and Workplace Ergonomic Risk Assessment (WERA) analysis, all activities of carrying steel plates and wood plates are categorized as medium. Based on Ovako Working Posture Analysis System (OWAS), the elements of work that require immediate action are Taking Plate (Wood) and Carrying the Plate (Wood). The results of the calculation of energy consumption obtained by 5.94 kcal / minute and categorized as work heavy / heavy work (5.0 - 7.5 kcal / minute). Workers carr plates with a weight of 30 to 42 kg, while according to NIOSH Equation and Snook Table analysis the maximum load is 14.06 kg, so this is far beyond and dangerous.
Keyword: ergonomic, work load, material handling, WERA, OWAS

1. Introduction

Manual work, if not done ergonomically, can cause work accidents or what is called Over Exertion-Lifting and Carrying. Lifting and carrying loads by bending forward repeatedly causes chronic Low Back Pain [1]. That is the state of damage to body tissues due to excessive lifting load [1]. Lifting a load above 25 kg is potentially a high risk of physical injury to workers. It is advisable to form a lifting and carrying team if the load exceeds 25 kg [2]. Lifting activities that are not done properly, such as postures / positions that are not natural at work, can cause illness or injury to the spine in workers. This working condition also happens to workers in the activity of transporting iron plates and wooden plates at the XYZ Factory. Workers carry steel and wood plates using both hands from the raw material warehouse to the work station. The size of the steel plate varies, ranging from 16 kg to 42 kg. Meanwhile, the heaviest wood plate is 25 kg. These plates will be transported one by one by the number of workers ordered. A heavy burden to be lifted by the workers, so that they had complaints of pain in the waist, shoulders, neck, or hands.

Based on the above problems, an in-depth analysis of working conditions, work posture and workload is carried out. The analysis results are used as input in designing of ergonomic material handling tools. This tool is to facilitate the move the steel plates and wood plates so as to reduce fatigue in workers and avoid work-related diseases. This is in accordance with the purpose of ergonomics which is to maximize the design of products, tools and workplaces, in relation to integral anthropometry, so that getting a complete knowledge of the problems of human interaction with technology and its products, as well as the design of human systems can be optimal [3].

This study aims to determine the physical complaints of workers by using the Nordic Body Map questionnaire to find out the part of the worker's body that feels painful before and after doing work [4]. In addition to using the NBM questionnaire, worker complaints can be measured from the level of ergonomic risk with the Workplace Ergonomic Risk Assessment (WERA)[3] and Rapid Entire Body Assessment (REBA)[4][5][6]. The complaint in question is related to fatigue of musculoskeletal work (WMSDs) due to unnatural posture. Assessment of WERA [4], categorized into three categories: low (score 18–27: work acceptable), moderate (score 28–44: work needs further investigation and requires improvement), and high (score 45–54: job the observed is unacceptable, it is necessary to make immediate repairs). While the REBA assessment is categorized into five levels which can be seen in Table 1.

Table 1. REBA score and level of MSDs Risk [7]

REBA Score	Risk Level	*Action Level*	Action
1	Negligible risk	0	No action reqiured
2-3	Low risk	1	Change may be needed
4-7	Medium risk	2	Further investigation, change soon
8-10	High risk	3	Investigate and implement change
11-15	Very high risk	4	Implement change

In addition, there are also OWAS methods to evaluate and analyze uncomfortable work attitudes and result in musculoskeletal injuries[8][9][10] . Besides analyzing body posture, measurement of heart rate before and after work is also done. This measurement is to determine energy consumption during work [11][12][13]. Furthermore, the workload calculation is carried out by the Kamkalanan method to measure the amount of energy consumption[14]. To provide recommendations on workloads that can be safely lifted by workers for various job variables using NIOSH Lifting Equation [15][16] and snook table [17][18][19],with this method it will be known that the workload should be allowed to be manually lifted.

2. Methods

This research was conducted at PT. XYZ, a company that produces various types of furniture. The main raw material used are wood and steel. Research focused on material handling processes. The dimensions of wood and steel handled are quite large and heavy. Handling process is done manually. The study began with the provision of the Nordic Body Map questionnaire, a questionnaire to determine the subjective complaints of workers. Next is observation and analysis of work posture. The methods used are REBA, WERA and OWAS. Workload was analyzed by measuring heart rate. Calculation of workload using Kamalanan Formula. Results of all the analysis concludes that the condition of the workload and ergonomic risks of working system. This study resulted in recommendations to improve the work system for increased comfort and productivity

3. Result and Discussion

Data were collected by distributing questionnaires to workers at PT. XYZ. The results of the general questionnaire can be seen in Table 2 below.

Table 2. Summary of General Questionnaire

No	Question item	Results
1	Sex	male
2	Age	45 – 55 years
3	Weight	60 – 65 kg
4	Height	160-170 cm
5	Marriage status	married
6	Work duration	5-8 hours
7	Plate weight	5 – 45 kg
8	The amount of material carried in 1 car	100 – 150 pcs
9	Work experience	> 5 years
10	Frequent activities	stand
11	Complaints	Pain in arms, shoulders, waist, back
12	Suggestion	Provided handling tools

Furthermore, an analysis of the body parts that experienced the greatest physical complaints from the Nordic Body Map questionnaire was carried out. The results can be seen in Table 3 and marked in yellow.

Table 3. Analysis of physical complaints from the Nordic Body Map questionnair

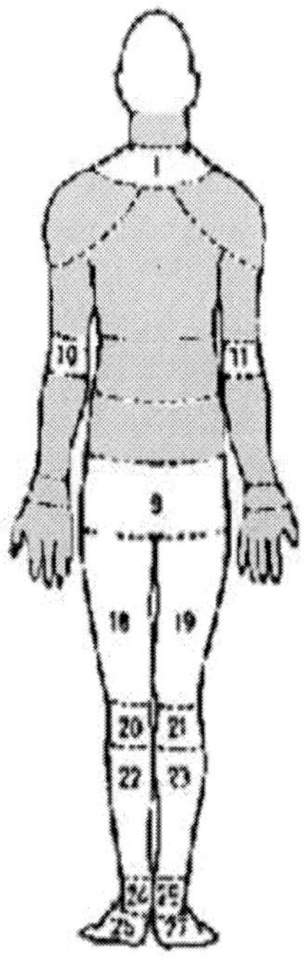

No	Rank of Physical Complaits	Causes
1	Pain in the upper neck	The position of the neck of the worker who must bow his head repeatedly when taking a wooden plate and lifting it.
2	Pain in the left and right shoulder	The process of lifting a steel plate and wood plate carried by placing it on the left or right shoulder repeatedly makes a complaint in this section.
3	Pain in the left upper and right upper arm	Position of the left upper arm and right upper arm that clamp the steel plate and wooden plate repeatedly during the process of transporting the material.
4	Back pain	Workers bend over and over when picking up wooden plates placed vertically on the warehouse floor causing the back of the worker to hurt.
5	Low back pain	Workers must take down and bring a wooden plate by transporting it from the raw material warehouse to the work station repeatedly causing complaints in this section.
6	Left and right forearm pain	The process of carrying a wooden plate and steel plate by holding it in the forearm left or right when taken.
7	Left-right wrist pain, right and left hand pain	The process of carrying an iron plate and wooden plate involves a lot of wrists and hands. The wrist and hand are used to hold the plate so as not to fall when taken from the raw material warehouse to the work station.

After obtaining the results of the analysis of the causes of physical complaints felt by workers, then the next is to do REBA and WERA analysis to determine the score when the worker moves the plates manually. The job position analyzed is when taking a plate, carrying the plate, and putting down the plate. The position of transporting plates can be seen in Figure 1 and Figure 2.

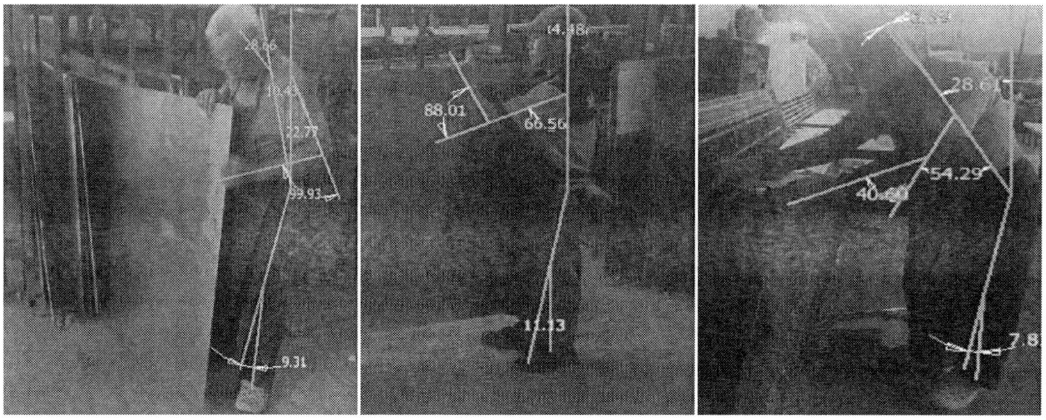

Figure 1. The process of carrying steel plates

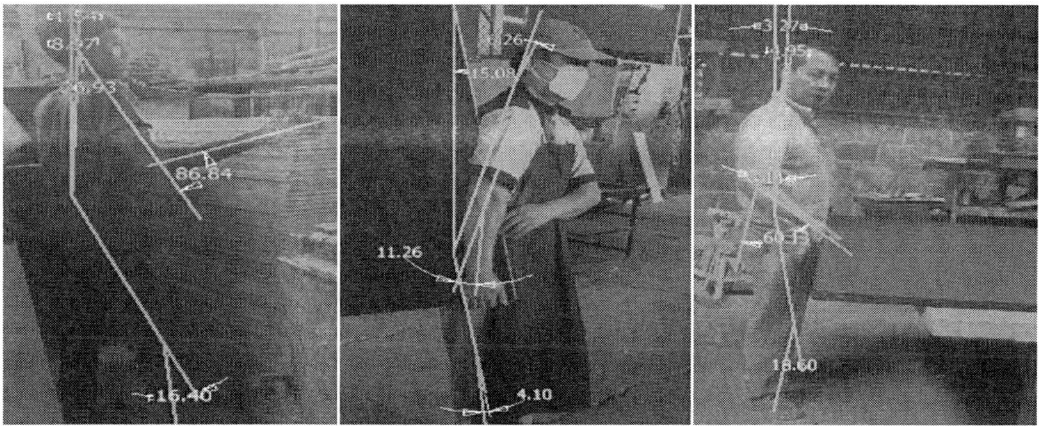

Figure 2. Process for carrying wood plates

From Figure 1 and Figure 2, the calculation of REBA and WERA scores is then performed. Calculation results are shown in Table 4 and Table 5.

Table 4. REBA score for each activity

No	Activity	REBA Score For Steel Plates	REBA score For Wood Plates
1	Taking Plate	4	5
2	Carrying the Plate	5	5
3	Putting down the Plate	6	5

Table 5. WERA score for each activity

No	Aktivitas	WERA Score For Steel Plates	WERA score For Wood Plates
1	Taking Plate	35	33
2	Carrying the Plate	31	32
3	Putting down the Plat	29	29

Based on REBA and WERA analysis, all activities of transporting steel plates and wood plates are categorized as medium risk. This means that the working conditions are further investigation, change soon, thus minimizing the risk of injury.

Furthermore, an OWAS method is analyzed to find out whether it is necessary to take corrective action or not. The results of the OWAS analysis are shown in Table 6.

Table 6. Risk Score Using the OWAS Method

Working Element	Score	Category	Action
Taking Plate (Steel)	1123	1	Not necessary
Carrying the Plate (Steel)	1273	1	Not necessary
Putting down the Plate (Steel)	2121	2	Need improvement in the future
Taking Plate (Wood)	2122	2	Need improvement as soon as possible
Carrying the Plate (Wood)	2172	3	Need improvement as soon as possible
Putting down the Plate (Wood)	1121	1	Not necessary

Next, measurement of heart rate, using Heart Rate Monitor (HRM). Heart rate measurement was done during work and at rest. Graphs of heart rate measurement can be seen in Figure 3 and Figure 4.

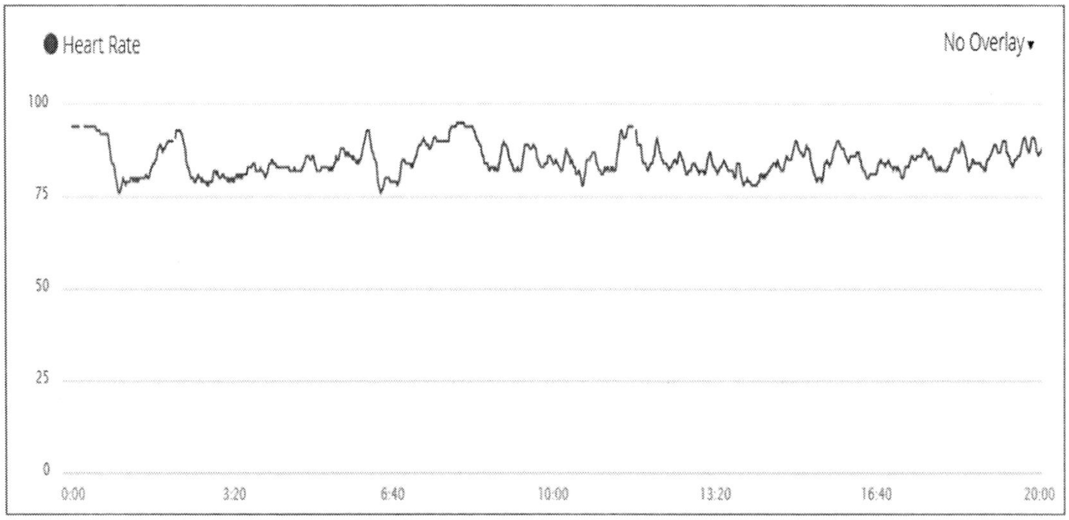

Figure 3. Heart Rate Graph (at rest)

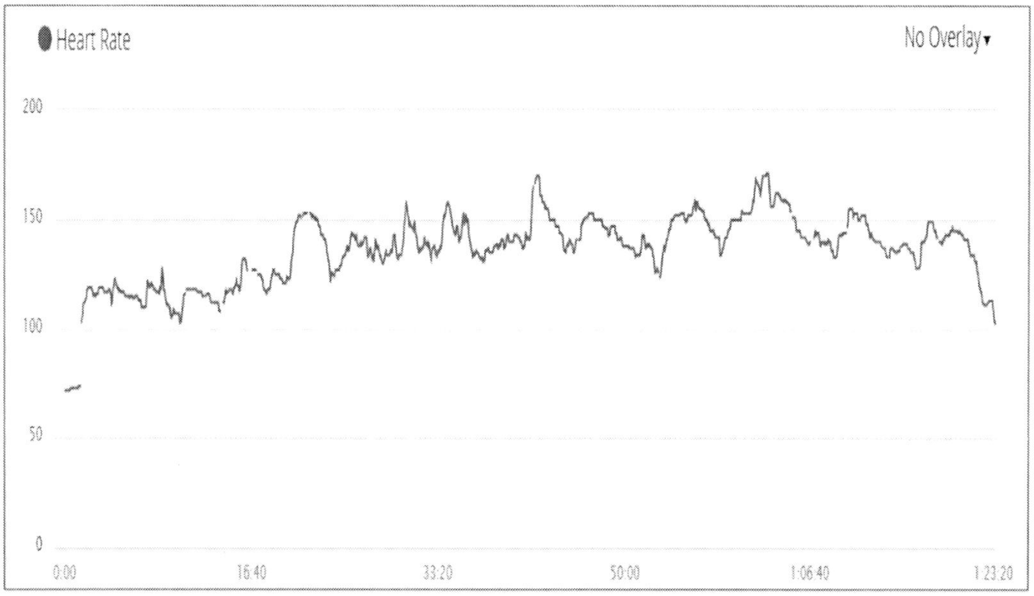

Figure 4. Heart Rate Graph (during work)

Calculation of workload using the equation stated by Kamkalanan as follows:

$$MWR = -1967 + 8{,}58HR + 25{,}1HT + 4{,}5A - 7{,}47RHR + 67{,}8G$$
$$= -1967 + 8{,}58 \times 136 + 25{,}1 \times 66{,}92 + 4{,}5 \times 48 - 7{,}47 \times 91 + 67{,}8 \times 0$$
$$= 415{,}802 \text{ watt} = 5{,}94 \text{ kcal/min}$$

The results of the calculation of energy consumption obtained by 5.94 kcal / minute and categorized as heavy work (5.0 - 7.5 kcal / minute).

Activities of lifting, lowering and carrying steel plates and wooden plates were analyzed with NIOSH Lifting Equation and Snook Table. The results are shown in table 7 below.

Table 7. NIOSH Lifting Equation, Snook Table Carrying and Snook Table Lifting/ Lowering

Analysis	Results
NIOSH *Lifting Equation*	34,01 lbs = 15,34 kg
Snook Table Carrying	31 lbs = 14,06 kg
Snook Table Lifting/lowering	22 lbs = 9,97 kg

Based on the NIOSH Lifting and Snook Table analysis, it can be seen that the maximum load limit allowed to carry a steel plate or wooden plate is 14.06 kg, while the lifting or lowering of a steel plate or wooden plate is 9.97 kg. However, workers at the XYZ Plant lift or lower steel plates up to 42 kg or wooden plates up to 30 kg. This exceeds the maximum load limit for lift / lower activity, so it can cause physical complaints to workers.

4. Conclusion

In accordance with the results of the Nordic Body Map questionnaire, there are complaints from workers, especially pain in the upper neck, pain in the left and right shoulder, Pain in the left upper and right upper arm, back pain, low back pain, Left and right forearm pain, and left-right wrist pain, right and left hand pain. Based on REBA and WERA analysis, all activities of transporting steel plates and wood plates are categorized as medium. This means that it has a high level of risk and changes are needed to reduce the risk of physical complaints to workers. Based on OWAS analysis, the elements of work that require immediate action are Taking Plate (Wood) and Carrying the Plate (Wood). The results of the calculation of energy consumption obtained by 5.94 kcal / minute and categorized as work heavy / heavy work (5.0 - 7.5 kcal / minute). Workers transport plates with a weight of 30 to 42 kg, while according to NIOSH Equation and Snook Table analysis the maximum load is 14.06 kg, so this is far beyond and dangerous.

5. References

[1] Holtermann A, Clausen T, Aust B, Mortensen O S and Andersen L L 2013 Risk for low back pain from different frequencies, load mass and trunk postures of lifting and carrying among female healthcare workers. *Int. Arch. Occup. Environ. Health*

[2] Visser S, *et al.* 2015 Lumbar compression forces while lifting and carrying with two and four workers. *Appl. Ergon.*

[3] Rahman M N A, Rani M R A and Rohani J M 2012 WERA Tool for Assessing Exposure Risk Factors of Work Related Musculoskeletal Disorders-A Reliability and Validity Study. *Proc. of the 2012 Int. Conf. on Industrial Eng. and Operations Management* (Istanbul, Turkey)

[4] Hignett S and McAtamney L 2000 Rapid Entire Body Assessment (REBA). *Applied Ergonomics* **31**(2) pp 201–205

[5] Ansari N A and Sheikh D M J 2014 Evaluation of work Posture by RULA and REBA: A Case Study. *IOSR J. Mech. Civ. Eng.* **11**(4) pp 18–23

[6] Kee D and Karwowski W 2007 A Comparison of Three Observational Techniques for Assessing Postural Loads in Industry. *Int. J. Occup. Saf. Ergon.*

[7] Hignett S and McAtamney L 2000 Rapid Entire Body Assessment (REBA). *Applied Ergonomics*

[8] Karhu O, Härkönen R, Sorvali P and Vepsäläinen P 1981 Observing working postures in industry: Examples of OWAS application. *Appl. Ergon.*

[9] Kivi P and Mattila M 1991 Analysis and improvement of work postures in the building industry: application of the computerised OWAS method. *Appl. Ergon.*

[10] Mattila M, Karwowski W and Vilkki M 1993 Analysis of working postures in hammering tasks on building construction sites using the computerized OWAS method. *Appl. Ergon.*

[11] Guidelines 1996 Guidelines Heart rate variability. *Eur. Heart J.* **17** pp 354-381

[12] Valentini M and Parati G 2009 Variables Influencing Heart Rate. *Prog. Cardiovasc. Dis.*

[13] Uusitalo A, Mets T, Martinmäki K, Mauno S, Kinnunen U and Rusko H 2011 Heart rate variability related to effort at work. *Appl. Ergon.*

[14] Kamalakannan B, Groves W and Freivalds A 2007 Predictive Models for Estimating Metabolic Workload based on Heart Rate and Physical Characteristics. *J. SH&E Res.*

[15] Waters T, Putz-Anderson V and Garg A 1994 Quick Guide for the NIOSH lifting equation. (Ohio: DHHS Publ.)

[16] Dempsey P G 2002 Usability of the revised NIOSH lifting equation. *Ergonomics*

[17] Snook S H and Ciriello V M 1991 The design of manual handling tasks: Revised tables of maximum acceptable weights and forces. *Ergonomics*

[18] Ciriello V M, Dempsey P G, Maikala R V and O'Brien N V 2007 Revisited: Comparison of two techniques to establish maximum acceptable forces of dynamic pushing for male industrial workers. *Int. J. Ind. Ergon.*

[19] Snook S and Ciriello V M 2002 Liberty Mutual Tables for Lifting, Carrying, Pushing and Pulling Also known as the Snook Tables. *Ergonomics*

IOP Conf. Series: Materials Science and Engineering **528** (2019) 012031 doi:10.1088/1757-899X/528/1/012031

Feasibility Analysis on the Development of Steel Sheet Zinc Plated and Galvalum Production Factory PT. S Steel

A Z Rahardja[1], E Chumaidiyah[2] and W Tripiawan[3]

Industrial Engineering Program, Telkom University, Bandung, Indonesia
[1]dzuraidaa@gmail.com, [2]endangchumaidiyah@yahoo.co.id,
[3]wawan.tripiawan@gmail.com

Abstract. PT. S Steel is a company producing zinc coated steel sheet (BJLS). The company runs from 1969 and is a joint venture company with Japan. The current production plant is located in an urban area, and the government of Makassar issued a government regulation in which the factory should not exist in the city area. So that PT. S Steel will build its new production plant in the industrial estate. In this feasibility study will be discussed about the construction of production plant PT. S Steel is reviewed in terms of market, technical aspects, legal and environmental aspects, as well as financial aspects. Market aspect analysis is done by using interview and historical data as reference. In the market aspect it is known that the amount of demand from zinc continues to decline while the galvalume rises as a new trend. Technical aspects analysis resulted in the output of facility layout and financial aspect analysis calculated based on calculation of NPV, BCR, and PBP with projection for 5 years period. financial analysis was assessed from several eligibility criteria, namely NPV with value Rp 487,404,605,133 and said to be feasible because NPV> 0, IRR 48.703% and said to be feasible because IRR> MARR 11%, and PBP worth 2.34. This feasibility study shows that the implementation of the construction of PT. S Steel is feasible to operate.
Key Word: Galvalume, Steel Zinc Plated Sheet, Feasibility Analysis, NPV, BCR, PBP

1. Introduction

The total population of Indonesia based on the 2010 Population Census reached 237,641,326 inhabitants, an increase of 15.21% compared to the population in 2000 which recorded 206,264,595 inhabitants. By 2015 the number of people based on the temporary census has reached 252,370,792 inhabitants.

The increasing number of residents, among others, led to an increase in the needs of homes caused by the number of young couples who married and need a new home. Besides, the housing demand is also driven by the large number of houses damaged due to natural disasters and natural conditions, the Ministry of Public Housing in 2010 estimated the needs of a simple and healthy national house on average reached 800,000 units per year. The demand for comfortable and cheap roofs is growing and the zinc coated steel sheet is one of the most promising options. Because in addition to the cheap price, durability of zinc coated steel sheet is also quite capable. In addition to the zinc coated steel roofing roof, the galvalume roof is also an inevitable choice. Galvalume roof made of mild steel, the public interest. Market demand for the roof of this galvalume continues to increase.

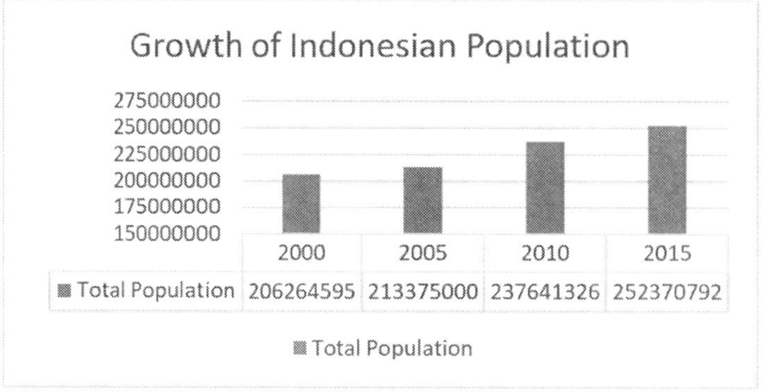

Figure 1. The Growth of Indonesia Population (2000-2015)
(source : BPS, 2017)

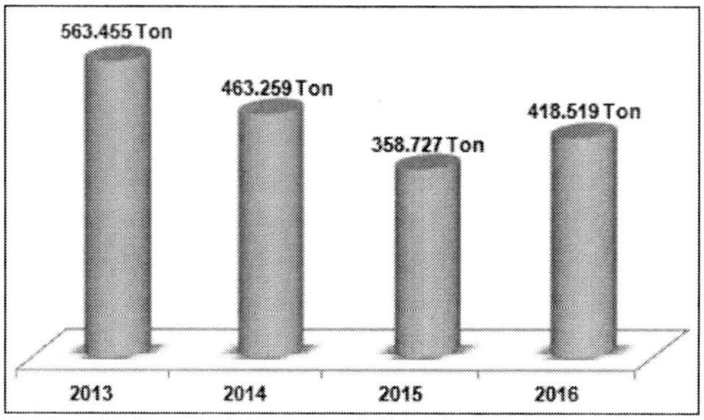

Figure 2. Indonesia Number of Galvalume Requests
(source : BPS, IZASI, KPPI, 2017)

Table 1. Sulawesi Number of Zinc Plated Request (ton)

No	City	Request	No	City	Request
1	LUWU	11978	13	BARRU	5877
2	MAKALE	1113	14	SELAYAR	4441
3	RANTEPEO	888	15	WATAMPONE	25486
4	MAMUJU	16260	16	PANGKEP	11053
5	MAJENE	5209	17	MAROS	11581
6	POLMAN	13674	18	SINJAI	8161
7	ENREKANG	6840	19	MAKASSAR	49321
8	SIDRAP	9890	20	GOWA	24481
9	PINRANG	12565	21	TAKALAR	9794
10	PARE2	4724	22	JENEPONTO	12192
11	WAJO	13528	23	BANTAENG	6291
12	SOPPENG	7789	24	BULUKUMBA	14073

(source : Interview with Distributor, 2017)

Based on Figure 2 and Table 1, There are high demand of sheet zinc plated in Indonesia especially in eastern part of Indonesia. So, it's required producers that capable of supporting these needs, PT. S Steel is a company engaged in the manufacture or production of zinc coated steel sheet located in Makassar. With the new prospect of galvalume roof, then PT. S Steel will also expand its production by adding

this galvalume roof as one of its production. The addition of this galvalume roof product to meet the needs of galvalume roof is increasing, and also to increase sales are declining because the demand for zinc roof has begun to be less desirable. Due to factory location and warehouse PT. S Steel located in Tello Baru, Makassar City violated Mayor Regulation No. 20 of 2010 concerning prohibition of warehouses and factories in the city as well as Local Regulation No.13 of 2009 regarding integrated warehouse and industrial area. Thus, the management section decided to build a new factory in the warehouse and integrated industrial area of KIMA. This needs to be re-examined because it is not at least the cost of removing the plant. Because if not moved by building new facilities will be subject to sanctions in the form of revocation of permits and the closure of facilities by force. However, if it is closed it will result in layoffs on all employees resulting in a lack of employment. In addition, if it is closed then the producers of zinc in the region of Sulawesi will experience shortages because zinc originating from Java cannot meet the amount of demand in Sulawesi and surrounding areas. The lost that can be borne can reach the number of billions rupiah, so the closure and revocationdic it's avoided by this company.
The purpose of this feasibility study is to assess the feasibility and business capability of PT. S Steel. And examine the aspects that affect the construction of this factory, such as market, technical, and financial aspects. If not feasible, then what needs to be done by the company in order to remain able to run its business. Sensitivity analysis is also conducted to determine the sensitivity of both raw material prices and selling prices later and anything that can affect in terms of price. This is done to assist corporate executives in making decisions as one of the material considerations.

2. Literature Review

2.1 Feasibility Study

The feasibility study is the study of business plans that not only analyse built business, but also when they are operated regularly to achieve maximum profit for an indefinite period [1]. In general, the aspects of the object in the feasibility analysis include [1]:

A. Market Aspects

The valuation of the market aspect is important because no successful business project without demand for goods or services generated by the project. Market aspect aims to determine the extent of market area, demand growth and market share [2]. Marketing includes all systems related to business activities aimed at planning, pricing, promoting and distributing goods or services that will satisfy the needs of the buyer or customer whether actual or potential [3].

B. Technical Aspects

Technical aspects are aspects related to the technical project development process and its operation after the project is completed. Based on this analysis can also be obtained preliminary assessment of investment cost plans including exploitation costs [4].

C. Financial Aspects

The financial aspect analysis aims to determine the investment plan through the calculation of expected costs and benefits, the ratio of expenditures and revenues, such as the availability of funds, the cost of capital, the ability of the project to repay the funds within a predetermined time and assess whether the project will be able to grow steadily [5]. This feasibility will be calculated by 4 factors, namely:

1. Net Present Value (NPV)
   NPV is a method for calculating the difference between the present value of the investment and the present value of the net cash income (cash flows of operations or cash flows of terminals) in the future.
2. Payback Period (PBP)
   The Payback Period method is the period required to cover the investment expense (core cash investment) using cash flow, in other words, the payback period is the ratio between

the initial cost of the investment and the cash inflows that result in units to time. The formula for calculating Payback Period is as follows:

$$Payback\ Period = \frac{Investment\ Value\ (outlay)}{Net\ Cash\ Inflow\ (Proceed)} x\ 1\ year$$

3. Internal Rate of Return (IRR)
   IRR is a method that calculates the interest rate equalizing the present value investment with the present value of future net cash receipts. The investment requirement can be said feasible if IRR> MARR. The formula for calculating IRR is as follows:

$$I_o = \sum_{t=1}^{n} \frac{CFt}{(1 + IRR)^t}$$

4. Benefit Cost Ratio (BCR)
   Benefit Cost Ratio (BCR) is defined as a ratio value equivalent to the benefit for an equivalent cost. The applicable equivalent size may be present value, annual value, or future value.

2.2 Sensitivity Analysis

Sensitivity analysis is an analysis carried out when calculating estimated future cash flows that are often faced with uncertainty. When we analyze estimates of future cash flows, we face uncertainty. As a consequence, the results of calculations can deviate from reality. Uncertainty can lead to reduced ability of a business project to operate to generate profits for the company [1].

3. Method

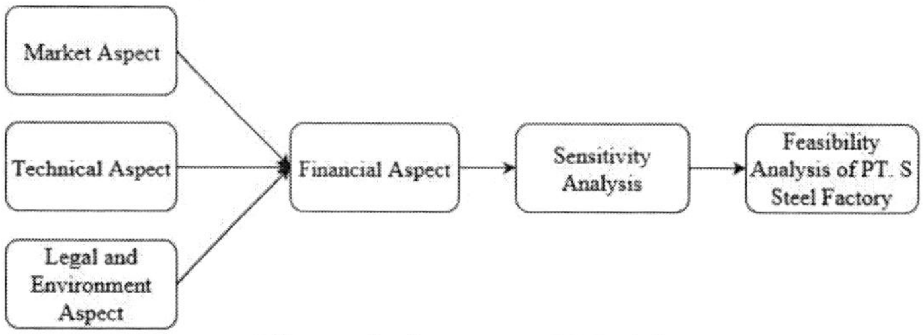

Figure 3. Conceptual Model

The feasibility analysis carried out in this study includes market aspects, technical aspects, financial aspects, legal and environmental aspects, as well as sensitivity and level of risk. These aspects have relevance to one another. So that results can be obtained that can produce a conclusion whether the establishment of a new production plant PT. S Steel is feasible or not.

Market aspect analysis is carried out at an early stage with the aim of knowing market opportunities related to consumer behavior regarding the use of prices and market competition for zinc and galvalum roofs in Indonesia. Technical aspect analysis is done to find out the location of the business, the layout of the building, labor and operational equipment at the new production plant of PT. S Steel. Legal aspect analysis aims to identify established business entities. Environmental aspects are useful for identifying environmental impacts that occur during the production process of the plant and how to process the waste. The results of the analysis of market aspects, technical aspects, legal and environmental aspects, will produce estimated costs of each aspect. These cost estimates will then be analyzed in financial aspects.

The results of the financial aspects will represent the projected feasibility of the implementation of the establishment of a new production plant of PT. S Steel using financial tools such as NPV (Net Present Value), Payback Period, IRR (Internal Rate of Return) and BCR (Benefit Cost Ratio). After conducting a number of market, technical and financial analyzes, decision making is an important part of the feasibility analysis process. The decision will be taken in accordance with the criteria for processing and analyzing data from this aspect. The end result is the feasibility of establishing a production plant of PT. S Steel.

4. Result and Discussion

Time series method is used to forecast the market. This is important to know the market aspect and know what technical aspect that suitable for the factory. The historical data pattern of sheet plated zinc is in the form of cyclical data pattern. Double moving average and Double exponential smoothing are used because this method is considered suitable for medium and long-term intersection as in this case. Double Moving Average Method 2 x 2 has the smallest MSE value, so the method is chosen to forecast. The query data of galvalum was obtained by conducting interviews to obtain available, potential, and targeted markets. In forecasting it is assumed there is an increase of 5% in line with the value of economic. In Table 2 can be see there is improvement total demand and that is market aspect for this company. Company only targeted eastern part of Indonesia because to reach western part of Indonesia still too hard as can be known big industry it's mostly in western part of Indonesia.

Table 2. Market Demands Forecast

Tahun	2018	2019	2020	2021	2022
Seng	7552,60	6994,02	6435,44	5876,85	5318,27
Galvalum	44.226	46.437	48.759	51.197	53.757

(Source: data processing 2018)

In technical aspect, the location of new production facility PT. S Steel is located in Takalar district, Patene industrial district and it's located 20 km from the previous production facilities of PT. S Steel. Owner of PT. S Steel already has land to build new facilities. The ground has a total length 300 m and width of 200 m. And to perform steel coating then it's required a Continuous Hot Dip Metal Coating Line machine. This machine is a composite of several basic engines for coating steel sheets. The Hot Dip Continuous Metal Coating Line machine that be use has a maximum capacity of 100,000 metric tons per year, with a maximum speed of 180 MPM. This is required because the demand and market aspect keep growth in the past 5 years, and this expected to fulfil all the demands. The weight of coil used it's 25 tons maximum, with a thickness of steel 0.6 mm and a maximum width of 1250 mm. Production capacity will only use at most 80% of engine capacity, so the engine capacity will be used real by PT. S Steel is 8,000 tons per month (accumulated from galvalum and zinc).

Table 3. Financial Aspect

Interest Rate	11,00%
NPV	Rp 487.404.605.133
PBP	2,34
IRR	48,703%
BCR	1,189

(Source: data processing 2018)

From the table 3 can be seen that the value of NPV is Rp 487.404.605.133. The IRR result is 48.703% which means the amount is greater than the MARR which is worth 11%, this indicates that the implementation of the construction of this new facility benefits the company. Based on the calculation of PBP, it can be seen that the time of return from the construction of facilities PT. S Steel is 2.34 ie the capital may return for 2.34 years (28 months). And the value of BCR is 1.189 (greater than 1) which indicates a greater profit than the required cost.

Sensitivity analysis is done on variable of raw material price increase, decrease of selling price, and increase of dollar price. The sensitivity of the increase of raw materials obtained by 17.56% indicates that raw material prices may rise and not cause losses during the percentage increase of not more than 17.56%. While the sensitivity to the selling price decrease is 49.82%, it indicates that the selling price can be lowered and will not cause loss during the percentage decrease not more than 49,82%. The increase in dollar price can affect the performance of the company when the dollar price reaches Rp 16,000 per US $ 1 it can be seen from the NPV value becomes negative when the dollar price is Rp 16,000.

5. Conclusion

Based on the result of feasibility analysis of construction of steel production factory of zinc coated sheet and galvalume of PT. S Steel, can be summarized as follows:

1. Target market used to know how big sales target of PT. S Steel in the next few years. To know the target market then done 2 ways that is using historical data and interview. The total market target of PT S Steel in 2018 amounted to 51,778 tons and continues to increase by 3% annually. This is because the decline in demand for zinc by 6% and galvalum rose by 5%.
2. Sales plan of PT. S Steel for a 5-year projection, the company does marketing by hiring booths on property-related events 6 times a year, and the company will strengthen relationships with existing distributors.
3. Technical aspects include technical requirements related to the manufacture and processing of sheet steel in PT. S Steel. The selection of machines and layout of PT S Steel can be said to be feasible because it can meet operational needs. The needs of raw materials and labor directly or indirectly have been able to meet the needs and can adjust to the production capacity. The total fund needed to build this production plant is Rp 339,862,680,000.
4. In this study, the financial aspects consist of investment costs, income, profit and loss, cash flow, and balance sheet. The components will be used to calculate the eligibility criteria such as NPV of Rp 487,404,605,133 and this is said to be feasible because of NPV> 0, IRR value of 48.703% and said to be feasible because IRR> MARR 11%, and PBP for 2.34 years.
5. Sensitivity analysis is done on variable of raw material price increase, decrease of selling price, and increase of dollar price. The sensitivity of the increase of raw materials obtained by 17.56%. While the sensitivity to the decrease selling price is 49.82%. The increase in dollar price can affect the performance of the company when the dollar price reaches Rp 16,000 per US $ 1 it can be seen from the NPV value becomes negative when the dollar price is Rp 16,000.
6. Suggestions for further research are Analyzing risk aspects that can occur both internally and externally.

6. References

[1] Umar, H. 2005. Studi Kelayakan Bisnis Edisi 3 Revisi. Jakarta: Gramedia Pustaka Utama.
[2] Muhammad, and Husnan. 2005. Studi Kelayakan Proyek. Yogyakarta: UPP AMPYKPN.
[3] Suliyanto. 2010. Studi Kelayakan Bisnis. Yogyakarta: Andi.
[4] Bustami, B. dan Nurlela. 2010. "Akuntansi Biaya" Edisi Kedua. Jakarta: Mitra Wacana Media.
[5] Darsono dan Ari Purwanti. 2010. Penganggaran Perusahaan Edisi Kedua. Jakarta: Mitra Wacana Media

Designing Self-Assessment Tool for Library Performance Measurement Adopting Malcolm Baldrige Framework (Case Study: Central Library of Andalas University)

N T Putri[1], D Jumeno[2], Henmaidi[3], E Wirdianto[4], P Fithri[5] and F Zulkhaira[6]

[1,2,3,4,5,6] Department of Industrial Engineering, Faculty of Engineering, Andalas University, Padang, West Sumatra Indonesia
nilda@ft.unand.ac.id[1]; destojmn@gmail.com[2]; henmaidi@ft.unand.ac.id[3];
eri_wirdianto@ft.unand.ac.id[4]; primafithri@eng.unand.ac.id[5];
fizu.khaira@gmail.com[6]

Abstract. Currently, competitions between universities at the national and international levels become tighter. Every university gives the best effort. They develop every aspect of the university's pillars quality to get more attention. These aspects include curriculum so that each student can learn more in an organized way. The same also applied for its facilities so that each student never lack of supporting tools and a delightful environment to learn. Even the lecturers are the best in their field so as to educate quality students. This happens not only in the national but also at the international level. In accordance with the proverbs often heard, 'library is the heart of the university'. Naturally the library becomes an important part of the university. Library of Andalas University has been awarded 'excellent' on a national scale by BAN-PT, according to SNI 7330.2009 and SNP 010: 2011. This achievement also requires library to maintain their quality. In line with Andalas University goal to obtain international recognition, library also need acquire ISO certification as in ISO 11620. The framework corresponding to the library's performance appraisal tool is MBNQA (Malcolm Baldrige National Quality Award), because MBNQA is the best management guide for organizations in order to achieve high quality and world-class certification. Thus, to obtain certification of international standards and maintain it, designing a new library performance appraisal tool is necessary. This new assessment tool will adopt MBNQA framework using national scale by BAN-PT and ISO 11620 indicator. The process to make new assessment tool will be assist by AHP (Analytical Hierarchy Process) method.
Keywords: ISO 11620, Library, MBNQA, Performance Assessment

1. Introduction

A library is an important existence for students. Especially for an academic library that is often considered as the 'heart' of an organization or institution. Usually, the role of academic library is to act as information service and the source of knowledge material for its user (student, lecturer and all of the parties in university society). As for the student, an academic library is an important existence in the areas of learning, teaching, research and service. The academic environment without a library also describe as a person in the absence of a brain. Indisputably library existence is determined by the user. In [1] state that conventional library lost its function because of modernism. Information is easy to get and its diverse of the source is outstanding so that more people use electrical source of information. Nowadays, people tend to browse the internet rather than screen through library shelves to find information they need. Thus, there are fewer visitors who want to visit library in the current time than in its glory days. To solve this problem, librarians must think harder to find a way so that library's existence stand firm [1].

According to letter by National Library Republic of Indonesia number 66/4.1.2/PPM.02/I.2016 on 13th January 2016 tell that library of Andalas University officially registered as grade A with 'very good' predicate. This became a new support for Andalas University that also already gain grade 'A' accreditation on January 2014. This result proves that library of Andalas University already get a very good accomplishment and success in maintaining their existence. It can be said that, library of Andalas University must maintain this accomplishment. It is not just to maintain this grade but if possible to achieve more by obtaining certified ISO as international standard. It is necessary to do so that current situation in compliance with future planning. In order to achieve those future planning librarians can do some activity such as performance assessment [2]. As [1] said that for an institution, performance assessment is a set of activities or evaluation task on its goal, design, policy, project, program and implantation of all activities that are systematic and objective for internal as well as external side. The purpose of performance assessment is to gain knowledge about library's current situation.

The library at Andalas University who already obtain 'very good' predicate in national scale is living up to SNI 7330.2009 and SNP 010:2011 evaluated by BAN-PT. Library need to maintain its quality because the next purpose is to gain ISO certification as in ISO 11620 for international scale. Thus, to achieve both standard and maintained, designing a new self-assessment tool for the purpose of quality assurance is needed. In [6] stated that (Malcolm Baldrige National Quality Award) MBNQA was envisioned as a standard of excellence that would help U.S organization achieve world-class quality. Malcolm Baldrige criteria have been accepted widely around the world as a standard for performance excellence. The Baldrige criteria for performance excellence are a comprehensive management framework that can be used to improve overall performance. This framework can give leaders all internal and external understanding of its organization to make the best decision, at the perfect time and in the right way [6]. Based on those reasons above, researcher thinks that it is need to design a self-assessment tool for quality assurance and performance assessment at the library of Andalas University using designed tool.

This new of design self-assessment tool is adaption of MBNQA framework using national scale by BAN-PT also ISO 11620 indicators. The process to make new assessment tool will be assist by AHP (Analytical Hierarchy Process) method. AHP method is a multi-parameter retrieval approach and was introduced by Saaty (1977 and 1994). This uses a multi-purpose, criteria, sub criteria, and alternative hierarchical structure. The data used is a set of pairwise comparisons. This comparison is to measure results, and performance of alternatives in terms of each individual decision. If the combination is not perfect, then it provides conditions for increasing consistency. In this study, which will combine many indicators into one framework, the AHP method is very helpful [8, 9].

2. Objectives

The Objective of this research is designing a self-assessment tool for library performance with combining indicators between ISO 11620 and Indonesian library accreditation standard and adapting MBNQA framework.

3. Methods

There are a number of tools for performance assessment. Some of them are for business or for-profit organization; there are also a number of development tools for non-profit organization in recent time [5]. The most common tool used for library assessment is LibQUAL+ that was developed from ServQUAL method by Association of Research Library (ARL). There are other methods such as Focus Group Discussion (FGD), interview, balanced scorecard and also MBNQA (Malcolm Baldrige National Quality Award). Each tool has its superiority and shortage. Many librarians try to find more

suitable tools to do a performance assessment so that the result can be more accurate [3]. The framework that is suitable to be adopt into new self-assessment tool is MBNQA (Malcolm Baldrige National Quality Award because MBNQA is the best management guidance for an organization in order to achieve high quality and world class certification.
As this new assessment tool will use to maintain library accreditation using Malcolm Baldrige is the correct choice, because the iterative process of review, planning, continuous improvement, and fundamental appraisal for institutional effectiveness and should be thoroughly integrated into the fabric of every establishment aspiring to excellence. Another reason that suggests Malcolm Baldrige framework provides a valuable measure of organizational effectiveness. A study found that the results obtained using this framework were consistent with increased job satisfaction, increased attendance, reduction of turnover, quality improvement, cost reduction, increased reliability, timely delivery, fewer errors, reduced waiting times (customers), increased satisfaction, fewer complaints, higher customer retention rate (profitability), better market share, and better financial indicators [7].
The method chosen in the process of designing a performance appraisal tool for library is AHP (Analytical Hierarchy Process). The use of this method is due to AHP has several advantages of other that fit with this research. In designing this new performance appraisal tool that necessary to define a complex method of decipher complicated issues and prioritizing the elements. The most important characteristic is the ability to decipher a rather large and complex task or problem into smaller, more manageable tasks or sub-problems. Most problems can be more easily understood and solved if they can be broken down into smaller parts. Some benefits of AHP over other are AHP doesn't require consensus but rather a combination of several different assessment results as well as consideration of the logical concessions need in the AHP method to determine priorities [4].

4. Designing Step for New Self-Assessment Tool

The process to make self-assessment tool divided into 3 major steps, they are:

a. Early step

The beginning step to designing self-assessment tool is to make a proposal of a new assessment tool for library performance before verified by expert. The framework which will be used as foundation for this self-assessment tool is adopting Malcolm Baldrige National Quality Award (MBNQA) framework. MBNQA framework consists of 7 criteria and 11 sub-criteria. The early steps of designing new self-assessment tool consist of:

i. Identified indicators from ISO 11620 and national library accreditation standard. From ISO 11620 there are 45 indicators meanwhile from national library accreditation standard there are 81 indicators.

ii. Group those indicators into Malcolm Baldrige criteria and sub-criteria that become proposed framework of self-assessment tool for library performance. Make questionnaire and give it to the experts. This questionnaire will help the experts into deciding how many indicators that will be used and where to place those indicators in new assessment tool based on Malcolm Baldrige framework.

iii. Determine experts that will be choose as respondent. The experts are must understand clearly about library performance, library accreditation as well as Malcolm Baldrige framework. Once expert selection is over, they will begin the filtering process of the indicators that are used in the new performance assessment tool. The experts are as follow:

- Ikhwan Arief, M.Sc as a lecturer at Industrial Engineering at Andalas University and also Chief of Library Committee at central library of Andalas University.
- Dr. Alexie Haryandie Bronto Adi as a lecturer at Industrial Engineering at Andalas University.
- Yose Rizal as head of user service area at central library of Andalas University.
- Monalisa Fitri Andes staff of processing and development of collection at central library of Andalas University.

- Armaida as coordinator of backup collection service at central library of Andalas University.
- Andi Saputra as staff of several administration and equipment sector at central library of Andalas University.
- Iswandi Syahrial Nupin head of processing and development of collection at central library of Andalas University.

This major step results in a new assessment framework that will use in next step. From overall early step 26 indicators chosen by experts is deleted. The decision to remove the indicator based on the discussions result of library staff as the experts. Deleted indicators are some points that not significantly affect performance such as TV, fan, sofa for guests, bulletin board and others. In addition to these reasons, several indicators have been represented by others because they have the same points. The indicators that deleted and moved can be seen at Table 1. and Table 2. After all indicators evaluated, each of them will put into a code system that consist 4 levels of this hierarchy; they are Level 0, Level 1, Level 2 and Level 3. The purpose of this hierarchy is place at level 0. In this case the purpose is self-assessment for library performance. Level 1 is the 7 criteria while Level 2 is 11 sub-criteria of Malcolm Baldrige framework. Meanwhile Level 3 consists of 97 indicator chosen from ISO 11620 and national accreditation standard for library. There are 44 indicators from ISO 11620 and 53 indicators from national accreditation standard remaining. Those indicators distribute into 7 criteria and 11 sub-criteria in accordance with experts opinion. The example of coded system for this new assessment tool can be seen at **Table 3.**

Table 1. Indicators that were deleted based on expert opinion

No	Source	Indicator	Reason Deleted
1	National Standard	Comparison of study program with Scientific Journal subscribed	This indicator will be combined with indicator 4.2.J
2	National Standard	Area of Collection (45%), consisting of: Referral area, Textbook area, periodic collection / printed area, Multi media collection area and Newspaper & clipping area	This indicator already represent and combined by indicator 2.2.C
3	National Standard	Area of Reader (25%) consisting of: Circulation Area, Circulation Area, Mixed Reading Area, Library Catalog Area, Cutting Area Display Area, Goods Area and Meeting Area	This indicator already represent and combined by indicator 2.2.C
4	National Standard	Area for Staff (20%) consisting of: Leadership Room, Administration Room, Procurement Room, Processing Room and Dining Room	This indicator already represent and combined by indicator 2.2.C
5	National Standard	Other areas (10%) consisting of: Lobby, Living Room and Toilet	This indicator already represent and combined by indicator 2.2.C
6	ISO 11620	Percentage of the total library lending to external users	Central library of Andalas Universuty does not apply this service yet
7	National Standard	The Power Status Library consists of	This indicator is include in indicator 1.1.B
8	National Standard	The Formal Languages of Formal Education consists of	This indicator is include in indicator 1.1.B
9	National Standard	Book processing / monograph	This indicator will be combined with indicator 4.2.E
10	National Standard	New magazine processing	This indicator will be combined with indicator 4.2.E

Table 2 Indicators that moved into another criteria and/or sub-criteria

No	Source	Indicator	From	Move To
1	ISO 11620	Ratio of acquisition expenditures to staff cost	1.1	1,2
2	ISO 11620	Percentage of expenditure on information provision spent on the electronic collection	1.1	1,2
3	ISO 11620	Percentage of library staff providing electronic services	1.1	1,2
4	ISO 11620	Percentage of Institutional Means Allocated to the Library	1.1	1,2
5	National Standard	Source of Budget	1.1	7,5
6	National Standard	Budget Amount / year	1.1	7,5
7	National Standard	Budget Allocation for libraries	1.1	7,5
8	National Standard	Survey of collegial needs	1.2	1,1
9	National Standard	Institutional Library	3.2	3,1
10	ISO 11620	Percentage of information requests submitted electronically	4.1	4,2
11	ISO 11620	Public seating occupancy rate	4.1	4,2
12	ISO 11620	Workstation use rate	4.1	4,2
13	ISO 11620	Median time of document acquisition	4.1	4,2
14	ISO 11620	Median time of document processing	4.1	4,2
15	ISO 11620	Shelving Accurancy	6.1	6,2
16	ISO 11620	Public access workstations per-capita	6.1	4,2
17	ISO 11620	Seats per-capita	6.1	7.4
18	National Standard	Total collection (print)	6.1	4,2
19	National Standard	Type of Reference Book owned (print)	6.1	4,2
20	National Standard	Electronic resources (monograph)	6.1	4,2
21	National Standard	Subscribe to electronic journals	6.1	4,2
22	ISO 11620	Cost per-loan	6.2	2,1
23	ISO 11620	Cost per-database session	6.2	2,1
24	ISO 11620	Cost per-content unit downloaded	6.2	2,1
25	ISO 11620	Cost-per-library visit	6.2	2,1
26	ISO 11620	Cost per-user	6.2	2,1
27	National Standard	Book loan system	6.1	.3.2
28	National Standard	Promotion type ever implemented	6.1	.3.2
29	National Standard	Special collection (research result, thesis, thesis and dissertation)	6.1	.3.2
30	National Standard	Access Electronic Journal	6.1	.3.2
31	National Standard	Information retrieval system	6.1	.3.2
32	National Standard	Book completeness	6.1	.3.2
33	National Standard	Computers for users with Internet access	6.2	.3.2
34	National Standard	Multi media device	6.2	.3.2
35	National Standard	AC	6.2	.3.2
36	National Standard	Inventory taking	6.1	7.3
37	National Standard	Weeding	6.1	7.3
38	National Standard	Homepage / library website	6.1	7.3
39	National Standard	Computers and printers for library management activities	6.2	7.3
40	ISO 11620	User area per-capita	6.1	7.4
41	National Standard	Percentage of core collections of the entire collection	6.1	7.4

Table 3. The Code of Each Element for Self-Assessment Tool

No.	Code	Criteria	Code	Sub-criteria	Code	Indicator	No
1	C.1	Leadership	1.1	Senior Leadership	1.1.A	Establishment of Library	1
					1.1.B	Institutional Library	2
			1.2	Governance and Societal Responsibilities	1.2.A	Information Literacy	3
					1.2.B	Added collection per year	4
					1.2.C	Library development activities carried out by library staff	5
					1.2.D	Utilization of funds	6
					1.2.E	Line of commands Head of Library	7
					1.2.F	Ratio of acquisition expenditures to staff cost	8
					1.2.G	Percentage of expenditure on information provision spent on the electronic collection	9
					1.2.H	Percentage of library staff providing electronic services	10
					1.2.I	Percentage of Institutional Means Allocated to the Library	11

b. AHP step

Designing self-assessment tool process is using Analytic Hierarchy Process (AHP) method to design hierarchy system for self- assessment tool, determining weight and determine experts' consistency in giving assessment value for this tool. The first step in the process of designing new tool is making a hierarchy for the self-assessment tool. To facilitate reading the hierarchy each element translates into some codes. This code system classified into three parts, they are criteria, sub-criteria and indicator. The steps are as follow:

i. Create a hierarchical structure of the problem.
The first step designing this new tool is arranging a hierarchy system for this self-assessment tool. Arranging hierarchy so it's easier to read each element translates into some codes. Construction of this hierarchical structure aims to elaborate the problem into more detailed parts. This code system classified into three parts, they are criteria, sub-criteria and indicator.

ii. Create a pair wise comparison matrix.
A pair wise comparison finished by providing an assessment of the importance between two indicators. The AHP method has 9 pairs of paired appraisal scores. Then distribute it to expert that already chosen. There are five experts.

iii. After that calculate each weight for criteria, sub-criteria and indicators based on AHP calculation until it lead to consistency value.
Weight assessment for each element is done by 5 experts with pairwise comparison. To unite all 5 experts' opinion, this calculation is using geometric average (geomean). After calculating geomean value, the next step is determining weight of each element. But in order to calculate weight of each element, matrix of pairwise comparison need to be made first. To fill in the matrix of pairwise comparison between same criteria, the value is 1. Meanwhile for result under the value 1 is 1/value upper.

iv. After each weight calculated the consistency of expert opinion need to be tested. The value of consistency ratio is related to the ability of the experts on performing pair wise comparisons to maintain consistency in their judgments. Even though it's hard to maintain perfect consistency, the Consistency Ratio (CR) value is expected $\leq 10\%$.

c. Designing matrix for self-assessment tool adopting national library accreditation system from BAN-PT into Malcolm Baldrige framework. Matrix performance appraisal for library is designed by adapting national library accreditation scoring system. In this matrix each indicator has a few standards that group into some value as can see in **Table 4.** For example for indicator 1.1.A is about status in establishment of library, appraiser can choose which scale the library belongs to. Then appraiser will get the scale value and multiplied it with total weight. The result from it will be final value from indicator 1.1.A and will be sum up from other indicator to get the value for library performance. The final value from library performance can be classified into some categories that can be seen at **Table 5.**

Table 4. Assessment Matrix for Sub-Criteria Senior Leadership

SUB-CRETERIA	CODE	INDICATOR	INFORMATION	SCALE				
				4	3	2	1	0
Senior Leadership	1.1.A	Establishment of Library	Status in establishment of library	The library was established based on the decree of the establishment of Diknas	The library was established based on the establishment decree of Diknas in the process	Library was established based on the decree of the founding of the rector	The library is in the process of its establishment	No SK of establishment of library
	1.1.B	Institutional Library	The organizational structure and the division of tasks from all parts of the library	The library has an organizational structure and job descriptions	The library has a complete organizational structure	The library has a simple organizational structure	The organizational structure of the library is in process	The library does not have an organizational structure

Table 5. Final Result Categories

Scale	
0 – 0,75	Early Development
0,76 – 1,25	Early Result
1,26 – 1,75	Early Improvement
1,76 – 2,25	Good Performance
2,26 – 2,75	Emerging Library Leader
2,76 – 3,25	Library Leader
3,26 – 3,75	Benchmark Leader
3,76 – 4,16	World Leader

5. Result and Discussion

From the process that already done in the design of self-assessment tool for library performance, it can be seen that each step determines the accuracy of measuring instruments made. The process is start from the selection of framework and indicators that will use, experts as respondents to the creation of library performance assessment matrix. After choosing the framework and indicators to be used, the compilation of the proposed tool is also very important. In the stages of grouping the indicators into the criteria and sub-criteria required serious consideration to determine the correct place. The initial plan determines the direction in the early stages of library performance measurement, because of the need for a very strong basis in grouping these indicators. Using the opinion of 7 experts, is expected to represent the requirement of library at Andalas University in terms of performance measurement.

Once expert's opinion collected, the result can be used as the basic form of performance measurement tools for libraries. The next step is to determine weight of each element for the criteria, sub-criteria and indicators prepared. The weights for each element in the measuring instrument are determined by the Analytical Hierarchy Process (AHP) method. The initial stage in AHP process is to hierarchies the self-assessment tool for measuring the weight. After that, make a questionnaire for the AHP process that is easily understand by respondents. In completing this questionnaire, the researcher must provide complete and easily understood information by the respondents because if the respondent doesn't understand the questionnaire then most likely the result of AHP calculation will be inconsistent. The result of inconsistent data is surely need repetitive retrieval that will require significant investment.

In the process of calculating the weight of each element required a very high accuracy to avoid errors. Especially in the manufacture of matrix pairwise comparison which is the results of AHP questionnaires, because this stage is the starting point in the process of calculating the weight of each element. Furthermore, no less important is calculation of consistency ratio for each element in the library performance measure. As said before, if the result shows an inconsistent data then retrieval process should be repeated. Tolerance limited for a consistency ratio is small equal than 10% or 0.1. From the calculation results can be seen that all the data doesn't exist beyond the limit of 0.1. However, some data have a consistency ratio value close to 0.1 such as comparison between 7 criterias with a value of 0.091 and some comparison indicators such as indicators for sub-criteria 2.2 with a value of 0.086 and an indicator for sub-criteria 6.1 with a value of 0.084. The ratio of the indicator with the lowest consistency ratio value is for sub-criterion 5.1 with 0.002. This is possible because there are only 3 indicators in the sub-criteria.

For comparisons between sub-criteria in one criterion, only one criterion meets the requirements in the calculation of consistency. This is because the criterion C.1 until C.6 only has 2 sub-criteria. So the consistency ratio doesn't need to be calculated because the results always consistent. Nor can it be calculate mathematically because the results are unlimited. This is because the index ratio for 2 elements is 0. Besides sub-criteria, it also occurs in the comparison of indicators for senior leadership sub-criteria (1.1) and workforce involvement (5.2) which only has 2 indicators in one sub-criterion.

For the total weight value of each indicator, the weight value of each indicator is multiplied by the weighted value of the criteria and the sub-criteria of the indicator are located. From the calculation results can be seen that the largest value of weight is for the indicator 5.2.B with a value of 0.092. Next is 7.5.D with 0.042 and 1.1.B with the same value. There are some indicators that have no value (N/A) in scoring matrix. This is because the University of Andalas library has not implemented the indicators so the value matrix for self-assessment cannot be made. For example, all indicators for cost such as indicator 2.1.A, 2.1.B, 2.1.C, 2.1.D and 2.1.E, require specific expenditure data but the cost budget for the Andalas University library is incorporated in the University's cost draft bill so that the costs incurred by the library not specified according to the needs for assessment of the indicator. Likewise, the indicators 7.1.C, 7.1.H, 7.2.A and 7.2.B are all indicators from ISO 11620. These indicators have not been applied yet by the Andalas University library, so the data required to measure each indicator is missing. Thus, there are 9 indicators in total that gas no available value in self-assessment matrix for library performance.

For overall process in designing assessment tool for library performance, some point should be discussed. The main point is, 'are the experts selected ideal for this research?' In this research choosing expert is an important point, because designing process of new assessment tool relied so much in experts' opinion credibility. Then, 'can this new assessment tool made based on expert opinion from the Andalas University Library be used for other libraries?' Even though this research use central library of Andalas University, researchers hopes this new assessment tool can be used in another library. Because of this new assessment tool designed using MBNQA framework and both national and international standard.

6. Conclusion

The conclusions of this research are as below:

a. Designing self-assessment tool for library performance finished with integrating national standard from National Library Accreditation Standard and international standard from ISO 11620:2008 using Malcolm Baldrige framework. Total initial elements for this assessment tool are 7 criteria, 11 sub-criteria and 145 indicators that come from 80 national standard indicators and 45 ISO indicators.
b. After selection process and discussion with 7 experts, there are 28 indicator deleted. Also there are 28 indicators that moved to another sub-criterion. The final framework for new library performance appraisal tool consists of 7 criteria, 11 sub-criteria and 97 indicators. For matrix performance appraisal is made with combining final framework of new library performance appraisal tool with assessment category that already approved by librarian of Andalas University. There are 9 indicators with (N/A) value because those indicators aren't applying yet in Andalas University library therefore there is no other way to assess those indicators.
c. This new assessment tool is expected to be able to maintain accreditation, maintain and improve the quality of the library both nationally and internationally. Nationally because it uses national accreditation standards while interatively because it adapts the ISO 11620 indicator. All indicators are embedded in a very good framework of malcolm baldrige framework.

7. References

[1] Ariomerebi M I and Nkiko C 2016 Annals of Library and Information Studies. *Performance assessment model for academic libraries: the Covenant University Library example* **63** (March) pp 7-15
[2] Azlan C, Warokka A and Hilman H 2012 Communications of the IBIMA. *The Library's Quality Management System and Quality Assurance in Higher Education: A Lesson from Southeast Emerging Educational Hub.* pp 1-11 871760 DOI: 10.5171/2012.871760.
[3] Baptista L and CesaltnaPires 2008 *Performance Evaluation of Academic Libraries Implementation Model.* Potugal

[4] Gonzales E, et all. 2013 *An AHP-based Approach to Prioritizing Resource for Highway Routine Maintanance*. University of Texas.
[5] McGregor F 2004 University of Wollongong. *Excellent Libraries: A Quality Assurance Perspective*. Advances in Librarianship **28** pp 17-53
[6] National Institute for Standards and Technology 2016 2015-2016 education criteria for performance excellence. (Gaithersburg, MD: Baldrige Performance Excellence Program)
[7] Ruben B D 2007 Wiley InterScience. *Higher Education Assessment: Linking Accreditation Standards and the Malcolm Baldrige Criteria* pp 59-83 DOI: 10.1002/he.246
[8] Triantaphyllou E and Mann S H 1995 Using The Analytic Hierarchy Process for Decision Making in Engineering Applications: Some Challenges. *Inter'l Journal of Industrial Engineering: Applications and Practice* **2**(1) pp 35-44
[9] Putri N T, Yusof S M and Irianto D The Delphi hierarchy process-based study of Quality Engineering in Malaysia and Indonesia Automotive Companies. *The TQM Journal* **26**(6) pp 566 – 576

Analysis daily newspaper distribution in Solo by Agent Based Simulation

Izatul Fitria Febriandini[1)], Yuniaristanto[2)], Wahyudi Sutopo[3)], Muhammad Hisjam[4)]

[1)] Laboratory Assistant Logistics Systems and Business, Industrial Engineering Department Sebelas Maret University, Indonesia
[2,3,4)] Research Group of Industrial Engineering and Techno Economic, Industrial Engineering Department, Sebelas Maret University, Indonesia

Abstract. Agent based simulation is a simulation model that can be used to describe the interaction between the involved agents. The interaction is generated from observations of human behavior. This study aims to provide an overview of business process PT Aksara Solopos and involved agents in it, especially for Koran Harian Solopos. In addition, to cope with the large number of paper returns is estimating the newspaper demand by simulation. Simulations done using NetLogo 6.0.3 applications. This research produced the number of clients who are subscribed by newspaper deliveryman and daily stock for the company, newsagents, newspaper deliveryman to complete the consumer demand.
Key Words: Agent Based Simulation, Daily Stock Newspaper, NetLogo

1. Introduction

Various industries have been affected by the development of technology especially in the area of information, and communication. One of those industries that affected by technology is newspaper industry. As the reader of newspaper is switching to digital based newspaper, the demand for printed newspaper become decreased. With the decreasing demand and many readers has switched to digital based newspaper, the evaluation for newspaper production need to be reassessed in order to adapt with the market condition.

Many simulation research have been done every year, such as :

Table 1. Previous research

No	Author	Content
1	Marzouk and Daour (2018)	presents a framework that aids contractors and safety managers in planning labor evacuation for construction sites using Building Information Modelling (BIM) and computer simulation using Agent-Based Simulation
2	Makinde, et.al. (2019)	Developed an agent-based model of a particular passenger rail system using an activity based simulation approach to predict the impact of public transport demand management pricing strategies.
3	Dong, et. al. (2019)	Simulating and optimizing Traffic Signal Control for real-time control system is investigated recently with development by the Internet of things (IoT)

4	Cloutier and Yang (2108)	Examines the robustness of three methods of GFO in the presence of variability in the contact locations and in the coefficients of friction between the hand and the object. A Monte Carlo simulation is used to determine the resulting probability of failure and sensitivity levels when variability is introduced
5	Vossos, et.al. (2018)	Monte Carlo simulation is used to compute the payback period and lifecycle cost savings of DC versus AC distribution systems. A future-market scenario is also examined, which evaluates how future efficiency improvements in power converters and changes in electricity tariffs may affect cost savings.

Agent-based simulation considers the interactions between the participating entities in detail, incurring substantial computational load. The agent state is represented by a set of variables associated with each agent (Yang, 2018). Agent-based simulation is an approach for simulation that also uses the notion of agents and the purpose of ABS compared to multi-agent systems is to gain insight into how global properties emerge from a system of local interacting processes (Larsen, 2018)

The decision to buy is a function of how close the expectations of the buyer on a product by the buyer on the perceived performance of the product (Oktina, 2002). This is one of the marketing strategy, by knowing the buyer and try to realize their need in products. Therefore companies need to assess whether the intended target market for this is correct or need to be evaluated to increase the amount of Koran Harian Solopos sales.

However, with the advent of digitalization era at this time, it could decrease Koran Harian Solopos sales and increase the newspaper return. To reduce the rate of return of newspapers, optimal stock of daily newspaper to complete the demand of customers every day have to be determined. So we need a simulation to forecast the demand of newspapers in Solo. Forecasting usually uses historical data (past data), but if forecasting doing with simulation, it can be done without using historical data (Zutshi, 2017)

Issues raised in the present study is an assessment of the extent to which the effectiveness of the distribution of Koran Harian Solopos in Solo.Based on the problems, explained into research objectives, such as:

- Identifying agents involved in the distribution process of Koran Harian Solopos.
- Identifying the amount of daily stock newspapers on the amount of a particular consumer.
- Identifying the factors that influence the behavior patterns of interaction performed by the agents involved in the business process of Solopos.

Limitation of problems used in this study are:

- Brands newspaper only used Koran Harian Solopos, which contains local news in Surakarta.
- Newspaper distribution area is Solo only.

2. Method

The assessment of the interaction between the involved agents starts from literature review that related to the issues of distribution of products in general. The next stages are collection and processing the data. The data collected comes from interviews some of the involved agents, and secondary data obtained from the literature review.

Processing data using Agent-Based Simulation, which uses a model to describe the events that actually occur. From the processing data, do conclusion and give advice to support future research.

Pseucode to describe the agent are :

IOP Conf. Series: Materials Science and Engineering **528** (2019) 012033 doi:10.1088/1757-899X/528/1/012033

```
breed [factories factory]
breed [distributors distributor]
breed [retailers retailer]
breed [customers customer]
```

Figure 1. The pseucode to defining agents

Pseucode to describe the demand stock are :

```
Factories-own [
  production_rate
  reorder_point
  orders
  available_stock
  clients
  daily_sales
  sales
  lost_sales
  forecast
  holding_cost
  order_cost
  total_cost
]
```

```
distributors-own [
  EOQ
  next_review
  reorder_point
  orders
  available_stock
  suppliers
  supplier_score
  supplier0
  clients
  daily_sales
  sales
  lost_sales
  forecast
  placed_orders
  holding_cost
  order_cost
  total_cost
]
```

```
retailers-own [
  EOQ
  next_review
  reorder_point
  suppliers
  supplier_score
  supplier0
  clients
  daily_sales
  sales
  lost_sales
  forecast
  placed_orders
  holding_cost
  order_cost
  total_cost
]
```

```
customers-own [
  next_review
  daily
  suppliers
  supplier_score
  supplier0
]
```

Figure 2. The pseucode to defining demand

3. Result and Discussion

Data used in this study is the primary data with the instruments are used interviews with the questionnaire. Resource is classified into two types; internal, Solopos sales division, and external companies that newsagents and newspapers deliveryman in the area of Solo.

Figure 3. Map of Solo

Interviewees Solopos internals are sales of its own, namely the Circulation Manager and Marketing Division (sales). Sales are elected as representatives of the company. Interviews were conducted to determine the background of the company, business processes thoroughly PT Aksara Solopos, sales flow of Koran Harian Solopos, and strategy that has been applied by the company to gain new consumers.

External resource are newsagents in Solo marked by yellow in Figure 2. Interviews with agents conducted to gather information related to the demand data of newspapers each day, the number of employee (a newspaper deliveryman), the consumers data who make a special subscription of Koran Hairan Solopos, average number of returned newspaper each month and the factors that influence the increase of sales of Koran Harian Solopos every day. While the interviews conducted by newspapers deliveryman are associated with a number of newspapers carried daily, the average number of newspapers sold each day, the number of consumers who subscribe, how to get a regular consumers, factors affecting sales of the newspaper every day and work hour each day.

Problem happens on a complex system. And complex systems not only have interconnected components, but also have process that is at a certain level which is derived from the interaction of the lower component (Vattam, 2011). Therefore, use a model to facilitate the research process.

Model design with Agent Based Simulation methods are simulated with NetLogo. Agent Based Simulation is a new modeling paradigm that is based on the utility of the agent. Which is entities or individuals interacting (Fuente, 2017).

NetLogo is a multi-agent programming language and environment modeling. A research tool and real-world simulations and complex social phenomena (Hitaswi, 2017). One example of an experiment of social phenomena is to develop research on smart city (Seidita, 2017). Based on this, four agents is used to model individual to resolve the issues raised, the company, newsagent, newspaper deliveryman and consumers.

'Breed' is a code defined for involved agents in the program. While the code 'own' is the code used to indicate the assignment of the involved agents (Pradopo, 2014)

The distribution process of existing newspaper which originated by the company to the newsagent that spread over an area of Solo. The newsagents has a few newspapers deliveryman to spread the newspapers to the final consumer in the form of institutions or individual consumers. Residential or consumers who pass by.

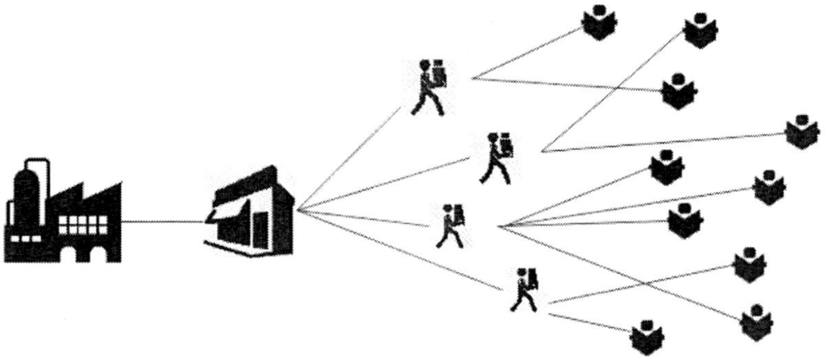

Figure 4. The description of the distribution process of newspapers in Solo

Based on interviews at several involved agents of the sale of Koran Harian Solopos, found that newspaper sales are influenced by several factors. These factors can be grouped into two, namely internal factors and external factors. Internal factors are affected by the activities sourced from within the company, such as delays in delivery. While external factors are sourced from all activities that have an impact on Solopos business processes, one of them is the news quality (headline). Other factors can be seen from a previous study conducted by Zakaria, et al (2005), regarding the market segment of Koran Harian Solopos that is based on the completeness of news, style, price, ease of obtaining, headline news, current news, and the news factuality.

Agent Based Simulation (AB) selected because AB can produce a complex system behavior, the behavior resulting from the interaction of the simple agents contained inside it (Macal et al, 2010). And according Muscallagiu (2014) agent based modeling is type of modeling that focuses representing agents and their interactions. The computer model used Net Logo 6.0.3.

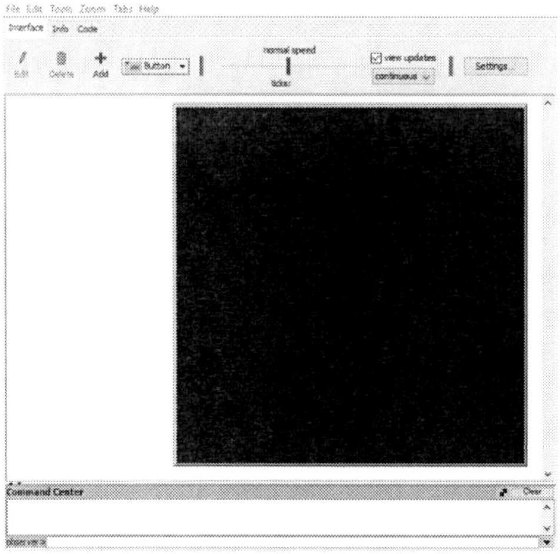

Figure 5. Initial View NetLogo

Users can make the adjustment to the number of companies, the lead time required, the number of newsagents (agen), number deliveryman (loper) and the desired number of consumers (pelanggan) through 4 pieces of the slider as in Figure 6.

Simulations run by making adjustments to all existing slider in accordance with the desired amount. Then press the setup button to erase and reset the program. Furthermore, for running the program, press 'Go' button (for running the program are continuous) or 'Step' button (for running the step by step).

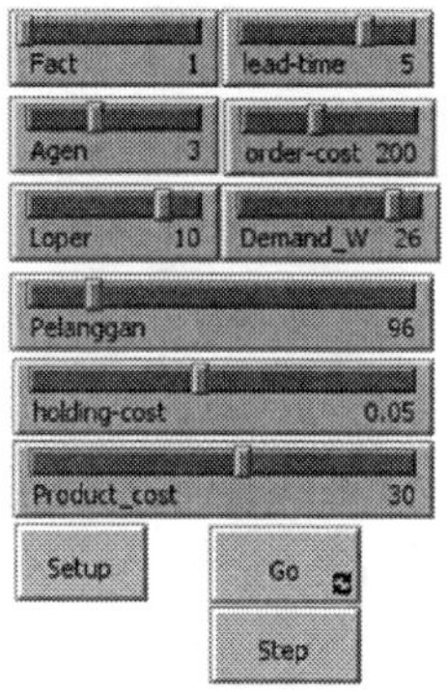

Figure 6. Slider to control the number of involved agents

Figure 7 is the display after NetLogo running the program. When the simulation is running, the location of all agents are assumed to equal (not move). The simulation was performed to determine the daily stock as needed.

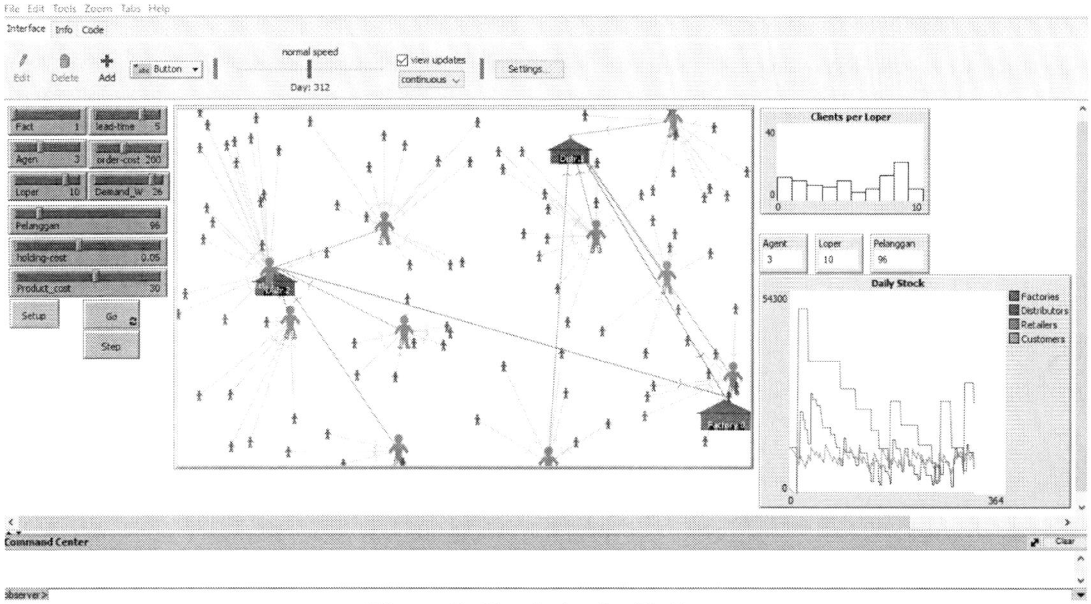

Figure 7. Simulation by NetLogo

Total newsagent in Solo is ± 128 agents. But in the program only simulated 3 newsagent (blue) and 10 newspaper deliveryman (green) and the number of customers are 96 people (grey). So we get the number of clients per deliveryman as follows:

Table 2. Simulation Results for Newspapers Deliveyman

Deliveryman	1	2	3	4	5	6	7	8	9	10
Client per deliveryman	8.1	2.2	3	3.96	4.93	6:08	6.92	7.89	8.94	9.91
Rounding	9	3	3	4	5	7	7	8	9	10

And the result is the number of the daily stock chart. And simulations generated from figure 8 is made during 146 days. From the graph in figure 8, it can be seen that the red color symbolizes the daily stock newspapers that are required by the company, the blue color is symbolizing the daily stock newspapers that are required by the newsagents, that in the simulation have 3 newsagents, and the green is symbolizing the daily stock newspapers that are required by newspapers deliveryman, in the simulation were 10 deliveryman. Newspaper stock levels line for companies should be above the line of the stock paper other agent. If the lines for companies under the line of another agent, it shows that the existing stock of the company could not complete the consumers demand, in this case, namely the newsagent and newspapers deliveryman.

The resulting model enables the visualization of real time metrics. During the processing time (run time) can be used control charts that could be updated in real time (Muscallagiu, 2014). It can be seen in figure 8.

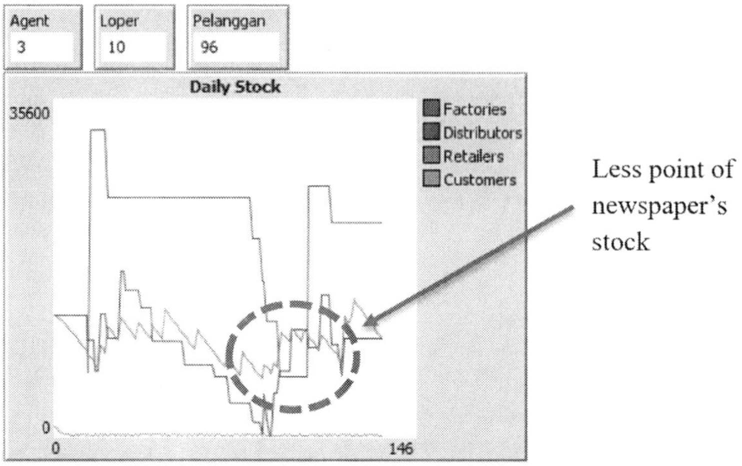

Figure 8. The simulation results of needed daily newspaper's stock

4. Conclusion

Agent Base Simulation is a concept that uses *agent* to determine the interactions made by involved agents. Based on the experimental results, we can conclude that the agent base simulation can be used to model the process of Koran Harian Solopos distribution to consumers in Solo. So we get the results of the output of a number of daily necessary stock of newspaper to complete customer demand with the number of newsagents, the number of deliveryman and the number of consumers can be adjusted. And NetLogo can be used as an application to modeling the simulations of the Koran Harian Solopos distribution. However, it should be underlined that the amount of demamd that occurs every day is very volatile, one of them is depends on the quality that news delivered.

4.1 Suggestion

- In this study does not present the optimal amount of the newspaper distribution on any given day. Need the optimization analysis to determine the optimal amount of demand of the newspaper.

Acknowledgments
The research is supported by Institute for Research and Community Service, Sebelas Maret University with Hibah Riset Fundamental UNS (RF-UNS) Research Program (Contract No. 543/UN27.21/PP/2018).

5. References

[1] Cloutier, A., & Yang, J. (2018). Examining the Robustness of Grasping Force Optimization Methods Using Uncertainty Analysis. *Journal of Risk and Uncertainty in Engineering Systems*, 1-8.
[2] Dong, H. V., Khanh, B. Q., & Lich, N. T. (2019). Integrating Multi-agent System, Geographic Information System, and Reinforcement Learning to Simulate and Optimize Traffic Signal Control. *Springer International Publishing AG*, 145-154.
[3] Fuente, D. d., Gomez, A., Ponte, B., & Costas, J. (2017). Agent-Base Prototyping for Business Management: An Example Based on the Newspaper Problem. *Journal of Economics, Business and Management*, 44-49.
[4] Hitaswi, N., & Chandrasekaran, K. (2017). Agent Based Social Simulation Model and Unique Identification Based Empirical Model for Public Distribution System. *International Conference on Recent Advances in Electronics and Communication Technology* (pp. 234-238). India: IEEE.

[5] Kerami, D. (n.d.). Konsep Umum Model dan Model Matematis. In *Permodelan Matematis* (pp. 1-41).
[6] Larsen, J. B. (2018). Agent Programming Languages and Logics in Agent-Based Simulation. *Springer International Publishing AG*, 517-526.
[7] Macal, C. M., Chan, W. V., & Young-Jun, S. (2010). Agent-Based Simulation Tutorial-Simulation of Emergent Behavior and Differences Between Agent-Based Simulation and Discrete-Event Simulation. *Proceedings of the 2010 Winter Simulation Conference*, (pp. 135-150).
[8] Makinde, O., Neagu, D., & Gheorghe, M. (2019). Agent Based Micro-simulation of a Passenger Rail System Using Customer Survey Data and an Activity Based Approach. *Springer Nature Switzerland AG*, 123-137.
[9] Marzouk, M., & Daour, I. A. (2018). Planning labor evacuation for construction sites using BIM and agent-based simulation. *Sadety Science*, 174-185.
[10] Muscallagiu, I., Emil, P. H., & Negru, V. (2014). Enhancing DisCSP-Netlogo from simulation to real-execution of agents in distributed constraints. *18th International Conference on Knowledge-Based and Intelligent Information & Engineering Systems - KES2014* (pp. 261-270). Romania: Elsevier.
[11] *NetLogo 6.0.4 User Manual*. (n.d.). Retrieved from http://ccl.northwestern.edu/netlogo
[12] Oktina, A. (n.d.). Analisis Hubungan Antara Karakteristik Konusmen dengan Atribut Produk dalam Pengambilan Keputusan Berlangganan Surat Kabar Solopos di Surakarta.
[13] Pradopo, S. T. (2014, April 18). *Panduan Pemrograman | Tutorial NetLogo*. Retrieved from Earth, Space & Beyond: http://tripradopo.web.ugm.ac.id
[14] Seidita, V., & Chella, A. (2017). Representing social intelligence: An agent-based modeling application . *Biologically Inspired Cognitive Architectures*, 35-43.
[15] Vattam, S. S., Goel, A. K., & Rugaber, S. (2011). Behavior Patterns: Bridging Conceptual Models and Agent-Based Simulations in Interactive Learning Environments. *11th IEEE International Conference on Advanced Learning Technologies* (pp. 139-141). USA: Computer Science.
[16] Vossos, V., Gerber, D., Bennani, Y., Brown, R., & Marnay, C. (2018). Techno-economic analysis of DC power distribution in commercial buildings. *Applied Energy*, 663-678.
[17] Yang, M., Andelfinger, P., Cai, W., & Knoll, A. (2018). Evaluation of Conflict Resolution Methods for Agent-Based Simulations on the GPU. *SIGSIM-PADS*, 129-132.
[18] Zakaria, R., Fahma, F., & Astuti, E. R. (2005). Perencanaan Strategi Pemasaran (Segmentation, Targeting dan Positioning) Koran Harian Solopos Berdasarkan Persepsi Konsumen. *Performa* (pp. 25-38). Surakarta: IDEC.
[19] Zutshi, A., Grilo, A., Nodehi, T., Mehrbod, A., & Goncalves, R. J. (2017). Simulation and forecasting of digital pricing models for an e-procurement platform using an agent-based simulation model. *Journal of Simulation*.

Multi Responses Optimization for the Sugar Content and Microbial Impurities of Carrot Syrup

O Isabella and Yurida Ekawati[1]

Industrial Engineering Department, Ma Chung University
[1]e-mail: yurida.ekawati@machung.ac.id

Abstract. Carrots are one type of agricultural products with large quantities in Indonesia. One of the carrot based products is carrot syrup. To determine the optimal process and composition of the carrot syrup, design of experiment using Taguchi method was carried out. This research have focused to optimize the carrot syrup with multiple quality using multi response Taguchi method based on the average quality losses. The quality dimension used are the sugar content and microbial impurities of the carrot syrup. The two quality dimensions have different responses, larger the better for sugar content and smaller the better for microbial content mold and yeast. The parameters determined to optimize the product quality dimensions are the duration of heating and the amount of sugar. The result of the optimal process and composition of carrot syrup is the duration of heat for 30 minutes and the amount of sugar of 210%.
Keywords: multi responses, Taguchi, carrot syrup

1. Introduction

Diversification is an attempt to make a product or service diverse. Product diversification can increase the value of a product. Diversification of products can be done on carrots, the agricultural products that have a relatively low durability and cannot be stored for a long time. Carrots are one type of agricultural products with large quantities in Indonesia. A previous research had been conducted to find carrot based products preferred by consumers and the attributes of consumer needs for the products [1]. One of the products found in the research was carrot syrup. The product design of carrot syrup had been developed to find the critical parts in producing carrot syrup [2]. The production process of carrot syrup which had already conducted can be used to produce carrot syrup [3]. However, to optimize the production of the syrup an experimental design need to be carried out. Robust design in quality engineering, as proposed by Taguchi, has significantly improved quality and yield in product and manufacturing process design [4]. Taguchi method can be used to optimize food products such as carrot syrup. As part of food products, carrot syrup has to follow the regulation or standard that have been set by the government. The standard allows an experimental design using more than one quality response. Several methods for addressing optimal factor levels for multi response quality characteristics have been proposed [5], [6], [7]. This study attempts to determine the optimization of the production process and composition parameters for optimal carrot syrup in terms of sugar content and microbial impurities. The method proposed by Wu [7] is used to optimize the multiple quality characteristics design for the carrot syrup.

2. Methods

The method for this research used the work of Wu [7] that offers a simple approach to optimize the multiple quality characteristics based on quality loss function.

2.1 Experimental design

The experimental design of Taguchi is conducted by stating the problems of concern, stating the objectives of the experiment, selecting the quality characteristics and measurement systems, selecting the factor which may influence the selected quality characteristics, identifying control factors and noise factors, selecting levels for the factors, and selecting the appropriate orthogonal array (OA) [8].

2.2 Perform the experiment and laboratory test

The experiment was done by making the combination of factor and level as described in OA. In this study, the sugar content and microbial impurities content is measured by laboratory test.

2.3 Data processing

Taguchi gives the following definitions for the average quality loss when *n* units of a product are measured [7].

For larger the better characteristic can be defined as:

$$\overline{L(y)} = k.\left(\frac{1}{n}\sum_{i=1}^{n} y_i^2\right)$$

For nominal the best characteristic can be defined as: (1)

$$\overline{L(y)} = k.\mu_0^2\left(\frac{s^2}{\bar{y}}\right) \quad (2)$$

And for smaller the better characteristic can be defined as:

$$\overline{L(y)} = k.\left(\frac{1}{n}\sum_{i=1}^{n}\frac{1}{y_i^2}\right) \quad (3)$$

Where k is quality loss coefficient, y_i is a measureable statistic of quality characteristic, $\mu_0$is the target value for nominal the-best characteristic, $\bar{y}$ is the sample mean of n units and $s^2 = \sum_{i=1}^{n}\frac{(y_i - \bar{y})}{(n-1)}$ is the unbiased estimator of process variance (σ^2). According to the concepts of Taguchi's quality loss function, the proposed optimization approach for multiple quality characteristic is given in the following steps [7]:

Step 1: Compute the total average quality loss, *Q*, for each experiment combinations based on the additive model.

$$Q = \left(\sum_{i=1}^{n} y_i\right) \quad (4)$$

Step 2: Transform the *Q* into the Signal-to-noise ratio (SNR), η, in decibels by the following equation.

$$S/N = -10\log_{10} Q \quad (5)$$

Step 3: Evaluate the effects of the control factors under consideration on η. The optimal factor settings are obtained by selecting the maximum η. Calculate he maximum η by the following equation.

$$\eta = \left(\frac{1}{n}\sum_{i=1}^{n} y_i\right) \quad (6)$$

3. Result and Discussion

The experiment used Standar Nasional Indonesia (SNI) as a main consideration in determining the factor and level in the experimental design because the product being tested was a food product. This research used SNI 3544:2013 standard for syrup [9]. SNI 3544:2013 establishes the terms, definitions, quality requirements, sampling, test methods, packaging and labeling syrup. Syrup itself is a beverage product made from a mixture of water and sugar with a minimum sugar content of 65% with or without other food and/or food additives permitted in accordance with applicable regulations.

3.1 Experimental design planning

The Taguchi method consists of three main stages: planning stage, conducting stage, and analysis stage [8]. Planning stage in this research began with the identification of quality dimensions to be improved. They were sugar content and microbial impurities content. SNI 3544:2013 standard for syrup contains coliform bacteria, Escherichia coli, Salmonella sp, Staphylococcus aureus, mold and yeast. Based on various kind of microbial impurities, there are two microbial that will be tested on this research, namely mold and yeast. Coliform bacteria and Escherichia coli are not included in the experiment since they are bacteria that usually occur in water. Salmonella sp is a group of bacteria commonly found in foods containing protein such as eggs and meat. So that type of microbial contamination has a small chance to grow in carrot syrup media.

Other types of microbial contamination are staphylococcus aureus microbes. Staphylococcus aureus microbes are microbes that can be contained in carrot syrup due to skin contact. The microbes grow in humans in the skin area, so hygiene process can minimize the carrot syrup contaminated by the microbes. Carrot syrup is the most suitable growth medium for microbial mold and yeast. High sugar content in the syrup will cause the growth of microbe mold and yeast, but that does not mean that the type of microbes come from the sugar content.

The planning stage was continued by selecting the parameters and their levels. Percentage of sugar content and microbial impurities in the carrot syrup were discussed with some experts in food technology. Based on the discussion, there were two control factors used in this experiment that can affect the quality attributes of carrot syrup. These factors were the duration of heating and the amount of sugar. The duration of heating and the amount of sugar affect the sugar content in carrot syrup. The sugar content in the syrup affects the growth of microbe mold and yeast as mention above.

The next step was determining the levels based on factors that had been determined, followed by the creation of an orthogonal array to determine the number of experiments to be performed. Both factors were determined using three levels. As mention above, the duration of heating and the amount of sugar can affect the sugar content in carrot syrup and the difference in sugar content can be easily distinguished by laboratory test, so there should be three levels that consider the lower, the middle and the upper level.

The level of the heating duration was determined based on previous research [3]. Based on the research the duration of heating was less than 30 minutes. Therefore, 30 minutes was chosen as the middle level and the difference with the lower and upper level was 10 minutes. The level of amount of sugar was determined based on SNI that stated that the minimum sugar content in the syrup will be at least 65% [9]. Based on the mathematical calculation to get sugar content of 65% the amount of sugar is at least 186%, but based on the preliminary laboratory test the sugar content can be greater or smaller than the calculation. Therefore the lowest level chosen was 200% increased by 5% for each level. The levels of each parameter is shown in Table 1.

The final stage of experimental planning was the determination of an orthogonal array. The orthogonal array was based on the number of factors and levels of each factor. This research used two control factors with three levels each, so the number of experiments to be carried out were nine experiments. The combination of the composition level of the experiment to be performed on carrot syrup can be seen on Table 2.

Table 1. The Level of Each Parameter

Factors	Levels		
	1	**2**	**3**
Duration of heating (A)	20 minutes	30 minutes	40 minutes
Amount of sugar (B)	200%	205%	210%

Table 2. Combination of Experiments

Experiment	Factors	
	A (Duration of Heating)	B (Amount of Sugar)
1	20 minutes	200%
2	20 minutes	205%
3	20 minutes	210%
4	30 minutes	200%
5	30 minutes	205%
6	30 minutes	210%
7	40 minutes	200%
8	40 minutes	205%
9	40 minutes	210%

3.2 Data processing

The data for the experiment were the sugar content and microbial impurities for mold and yeast of carrot syrup resulted from the combination of factors shown in Table 2. There were 9 combination and 2 replication so that the total samples were 18. The sugar content and microbial impurities were taken based on laboratory test. Table 3 shows the results for percentage of sugar and microbial impurities for mold and yeast on each combination experiment.

Table 3. Percentage of Sugar and Microbial Impurities for Mold and Yeast

Experiment	Sugar (%)		Microbial Mold (CFU/ml)		Microbial Yeast (CFU/ml)	
	1	2	1	2	1	2
1	60,57	77,89	1	0	10	3
2	65,49	76,75	0	0	37	1
3	74,87	74,76	0	0	1	2
4	60,23	74,6	0	0	17	16
5	67,99	77,68	0	0	2	2
6	73,91	75,69	0	0	2	1
7	74,87	77,07	0	0	70	2
8	75,57	83,1	0	0	6	2
9	80,75	75,4	0	0	0	1

The percentage of sugar content and microbial impurities data resulted from laboratory test then transformed into average quality loss for each response based on the quality characteristic. Sugar content used a larger the better characteristic and for microbial content mold and yeast used smaller the better characteristic. The average quality loss for sugar content was computed using equation (1) and the result can be seen in Table 4. The average quality loss for microbial mold and yeast were computed using equation (3) and can be seen in Table 4.

Table 4. Quality loss for each responses

Experiment	Sugar	Microbial Mold	Microbial Yeast
1	0,0002187	0,5	54,5
2	0,0002015	0	685
3	0,0001787	0	2,5
4	0,0002277	0	272,5
5	0,0001910	0	4
6	0,0001788	0	2,5
7	0,0001734	0	2452
8	0,0001600	0	20
9	0,0001646	0	0,5

From the data in Table 4, then the total average quality loss, Q, for each experiment was computed using equation (4) and transform the Q into the Signal-to-noise ratio (SNR), η, in decibels using equation (5). Table 5 shows the quality loss and SNR for the 9 experiments.

Table 5. Quality losses and SNRs of experiment for carrot syrup

Experiment	Factors		Q	SNR
	A	B		
1	1	1	55,0002187	-17,40364416
2	1	2	685,0002015	-28,35690699
3	1	3	2,500178658	-3,979710437
4	2	1	272,5002277	-24,35366869
5	2	2	4,000191025	-6,020807311
6	2	3	2,500178806	-3,979710693
7	3	1	2452,000173	-33,89520497
8	3	2	20,00015996	-13,01033469
9	3	3	0,500164629	3,008870245

From the data in Table 5. the maximum η was calculated using equation (6) and the result is presented in Table 6.

Table 6. Effects of the control factors under consideration η.

Level	Factors	
	Duration of heating	Amount of sugar
1	-16,58009	-25,21751
2	-11,45140	-15,79602
3	-14,63222	-1,65018

Based on all figures above, it can be seen levels with the highest score on each factor. This main effect is done to optimized carrots syrup based on sugar and microbial impurities, so it is found the optimal process and composition of carrot syrup is as follows:

Table 7. Optimal Level Result

Factors	Levels	Composition
Duration of heating	Level 2	30 minutes
Amount of sugar	Level 3	210%

Based on Table 7, the optimal composition for carrot syrup in term of sugar content and microbial impurities is the amount of sugar of 210% with duration of heating 30 minutes. The optimal composition is shown as experiment number 6.

4. Conclusion

A research to find the best composition and process of carrot syrup had been done using Taguchi method for multi responses. The quality dimensions for the carrot syrup are determined based on the SNI that states standard composition for syrup. The quality dimensions are sugar content and microbial impurities, with the quality responses for the two quality dimensions are larger the better for sugar content and smaller the better for microbial impurities. The parameters selected to optimize the product quality are the duration of heating and the amount of sugar with each parameter uses three levels. The result of the optimal process and composition of carrot syrup is the duration of heating for 30 minutes and the amount of sugar of 210%.

Acknowledgment

This work would not have been possible without the financial support from Directorate of Research and Community Service, Directorate General of Research and Development Strengthening, Ministry of Research, Technology and Higher Education of the Republic Of Indonesia.

5. References

[1] Halim V and Ekawati Y 2014 Perencanaan Produk Olahan Wortel Menggunakan Metode Quality Function Deployment. *Symbol* **1** pp 57-67

[2] Ekawati Y and Bazarado M 2016 Designing Food Products Based on Carrots Using the Product Design Phase of Quality Function Deployment. *ARPN Journal of Engineering and Applied Science* **11** pp 3109-16

[3] Ekawati Y, Noya S and Widjaja F 2017 Process Quality Planning of Quality Function Deployment for Carrot Syrup. *AIP Conferences Proceedings 1855 Green Process, Material and Energys*

[4] Derringer G and Suich R 1980 Simultaneous Optimization of Several Response Variables, *Journal of the Quality Technology* **12** pp 214-219

[5] Jeyapaul R, Shahabudeen P and Krishnaiah K 2006 Simultaneous Optimization of Multi-Response Problems in The Taguchi Method Using Genetic Algorithm. *International Journal of Advanced Manufacturing Technology* **30** pp 870-878

[6] Liao H C and Chen Y K 2003 Solving Multi-Response Problem in The Taguchi Method by DEA Based Ranking Method. *International Journal of Quality & Reliability Management* **19**(7) pp 825-837

[7] Wu F C 2008 Simultaneous Optimization of Robust Design with Quantitative and Ordinal Data. *International Journal of Industrial Engineering* **15**(2) pp 231-238.

[8] Ross P J 1996 *Taguchi Techniques for Quality Engineering* (New York: Mc Graw-Hill)

[9] Badan Standardisasi Nasional Indonesia 2013 *SNI 3544:2013*. (Jakarta: Badan Standardisasi Nasional)

IOP Conf. Series: Materials Science and Engineering **528** (2019) 012035 doi:10.1088/1757-899X/528/1/012035

Schedule Risk Analysis by Different Phases of Construction Project Using CPM-PERT and Monte-Carlo Simulation

Andrie Pasca Hendradewa

Industrial Engineering Department, Faculty of Industrial Technology, Universitas Islam Indonesia, Yogyakarta Indonesia
andrie.pasca@uii.ac.id

Abstract. Construction project is an important activity along with the increasing need of infrastructure and residential building. Construction projects with different investment value and complexity level affect the completion duration of construction work. The higher the value or the more complex the building, the longer time required to accomplish the construction project. During the construction process, it is possible to find risks both from internal and external management which potentially cause project overdue that will lead to cost overruns. This study focused on assessing and managing schedule risk in three main phases of construction project, such as: feasibility study, design, and construction. A certain sample of residential building project was choosen as study case. In this study, mathematical approaches are used by performing CPM-PERT and Monte-Carlo simulation method. The results of this study can be used to help project managers to minimize potential risks by projecting simulation results and to assess if a project can be completed in timely manner according to given schedule.
Keywords: schedule risk, construction project, CPM, PERT, monte carlo simulation

1. Introduction

Timeliness is one indicator to determine success of a project, in addition to cost and performance [1]. Project lateness will affect the total cost, since the workers need to be on duty in longer duration than the original schedule. Therefore, to minimize the possibility of project lateness, it is required a proper project scheduling. According to [2], here are many significant issues in project scheduling such as: the estimation of activity duration, resources allocation, and an unstructured scheduling. The main problem of project scheduling is the uncertainty, especially in resource availability. Uncertainty is a situation that can not be specified, which has many possibility that indicated by the lack of information about the result of an event [3]. The uncertain situation faced by the project manager is also known as risk, which if it happens, has positive or negative effect to duration, cost, scope or quality of project [4]. On a construction project, many factors like material procurement delay, human resource problem, weather, and bureaucracy have potential to cause risks [5]. Therefore, risk analysis and risk control technique are required to obtain effective and efficient project completion process. Risk control includes the identification, assessment, and prioritization of risks in a project. It allows the management to control over project schedule, estimation, and quality. By implementing risk analysis and management, the right decision can be made as early as possible to avoid lateness in the project. This research aims to build model from a complex project management system by considering potential risk factors, so the possibility of project lateness can be minimized.

1.1. Project Risk Management

Time is one measurement to determine project completion success. The inability to manage time, inavailability of technical skills to deal with sudden changes, and difficulty to visualize process makes it difficult for management to monitor project progress status [6]. Practically, both projects and the environment will continue to change in a sustainable manner, focusing on the importance of recognizing potential risks and opportunities, therefore quick decision making are needed on many alternative strategies [3]. According to Weick [7], project manager ability to understand situation on alternative action will determine how well the strategic decisions are made that impacting the company performance. Consequently, an advanced project management skill is needed to control the risk of uncertain situation.

1.2. Potential Risks of Construction Project

Project risks, as specified by Zou et al [8], are categorized into four different phase according to project life-cycle such as feasibility study, design, construction, and operation. Risks on feasibility phase, mostly related to clients and governmental agencies. Client need, strategic brief, and practical project feasibility study which also consider the price fluctuation of construction material are the main focus of this phase. The next stage is design phase where designers play the most important role to understand and satisfy clients' wants and need. Risks on this phase are related to government bureaucracy, design defect and change that come from both client and contractors as well. After the design is fixed, project get into the construction phase, the risks on this phase are likely to correspond with contractors and subcontractors, such as variation of project delay, unavailability of sufficient amount of skilled workers, lack of coordination, and occurence of safety accident. The possible risks on final phase after construction work is completed may only came from the funding for the project operation and difficult government regulation related to facility management, environment sustainability. Overall, the majority of risks occured in the pre-operation phase [8].

2. Methodology

There are several approaches that can be used in project management such as Critical Path Method (CPM) and Project Evaluation Technique (PERT). Those methods help management in generating deterministic decisions that is associated with project completion time [6]. By using CPM, project managers are enabled to understand the critical activities that determine overall duration during the completion of project. While PERT, emphasize on scheduling technique by estimating the time required for each task using three-point estimators. PERT assumed that the duration of each project activity is a random variable and follows Beta distribution [9]. Another approach for risk analysis is simulation technique. Monte Carlo is simulation method for risk analysis by using probabilistic approach from many uncertain activity combination to evaluate the uncertainty. According to Barraza [10], the advantage of Monte Carlo simulation is able to provide near-realistic estimation results by balancing the probability value of each activity as a critical value.

Monte Carlo simulation has been widely used for the purpose of risk analysis in many project management studies. Research from Tysiak and Sereseanu [11] implemented CPM and Monte Carlo simulation method for risk analysis in IT project. The estimation of timeliness possibility for construction schedule became the main focus of research from Ganame and Chaudhari [12]. While a research from Yuan et al [13] applied Monte Carlo simulation to develop an early warning system prototype for construction schedule.

This study is distinct from other approaches [11] [12] [13], which emphasize on the potential risk analysis by different phase of construction project schedule using CPM-PERT and Monte Carlo simulation. Since the majority risks of construction project occured in the pre-operation stage [8], it became the main focus of this study. By implementing this approach, project managers' assessment and decision can be more accurate to potential risks characteristics on pre-operation stage such as feasibility, design, and construction phase.

2.1. PERT and CPM

PERT assumed that the activity duration follows Beta distribution [14], so that three point estimator such as optimistic duration (*a*), pesimistic duration (*b*), and most likely duration (*m*) are used to calculate the mean (μ) and the variance (σ^2) for each activity duration from state *i* to *j* with the following formula:

$$\mu_{(i,j)} = \frac{a_{(i,j)} + 4m_{(i,j)} + b_{(i,j)}}{6}, \qquad \sigma^2_{(i,j)} = \frac{(b_{(i,j)} - a_{(i,j)})^2}{36} \tag{1}$$

Critical path which represent the longest duration from the first state (start) to the latest state (finish) of the project is the key to find the critical activities in PERT network by arranging project schedule that consist of earliest start time (ES), earliest finish time (EF), latest start time (LS), latest finish time (LF), and slack time (SL).

Earliest start time (ES) and earliest finish time (EF) are calculated forward from the first activity using the following formula:

$$EF_{(i,j)} = ES_{(i,j)} + \mu_{(i,j)}, \qquad ES_{(i,j)} = \max\left(EF_{(\forall,i)}\right) \tag{2}$$

Latest finish (LF) and latest start (LS) are calculated backward from the last activity with the formula below:

$$LF_{(last)} = EF_{(last)}, \qquad LS_{(i,j)} = LF_{(i,j)} - \mu_{(i,j)}, \qquad LF_{(i,j)} = \min\left(LS_{(j,\forall)}\right) \tag{3}$$

Slack time (SL) or the measurement for identifying critical activities is determined using the following formula:

$$SL_{(i,j)} = LS_{(i,j)} - ES_{(i,j)} \tag{4}$$

If there are project activities that have SL value equal to 0, then the activities are identified as critical that form critical path on the PERT network. Project completion time is calculated by adding the duration of all critical activities.

2.2. Monte Carlo Simulation

Monte Carlo simulation is a method that depends on the generation of random or pseudo-random number from given distribution [15]. As stated by Hong et al [16], Beta-PERT distribution is more suitable to used in Monte Carlo simulation for estimating construction activity duration. Beta-PERT is the improvement of Beta distribution which can better fit to uniform, normal, and lognormal distribution [17]. It relies on α and β values ($\alpha > 0$; $\beta > 0$) which determine the distribution shape that calculated using formula below [17]:

$$\alpha = \frac{(\mu - x_{min})(2x_{prob} - x_{min} - x_{max})}{(x_{prob} - \mu)(x_{max} - x_{min})}, \qquad \beta = \frac{x_{max} - \mu}{\mu - x_{min}}\alpha \tag{5}$$

The $x_{min}, x_{prob}, x_{max}$ and μ in formula (6) respectively represent the minimum, most probable, maximum and mean of activity duration estimation value.

In this study, we use 10000 replications for simulating various project completion time. Random numbers are generated from Beta inverse using the parameter α and β to create all replications. The generated random numbers or pseudonumbers are used as the simulated activity duration (x) which expressed as the following formula:

$$x = f^{-1}(p\,|\alpha,\beta) = \{x\!: f(x|\alpha,\beta) = p\} \tag{6}$$

The p value represents the probability of occurence that is generated randomly from uniform distribution with the interval 0 to 1 as shown as the formula below:

$$p = f(x|\alpha,\beta) = \frac{1}{B(\alpha,\beta)}x^{\alpha-1}(1-x)^{\beta-1}I_{[0,1]}(x) \tag{7}$$

Project completion time (T) is calculated by the critical path which consists of critical activities that already simulated for 10000 different project scenarios, therefore the T value also generated from T_1 to T_{10000}. Since the project completion time (T) is assumed to follow normal distribution [6], standard score of normal distribution $N(\mu, \sigma^2)$ is calculated as follow:

$$Z = \frac{T_i - \mu}{\sigma} \quad (8)$$

The mean (μ) and standard deviation (σ) are calculated using formula (1) for all simulated project completion time ($T_1, T_2, \ldots, T_{10000}$). The standard score of normal distribution is used to determine the timeliness possibility of a given project completion time as notated as $p(T)$ in cumulative distribution formula below:

$$p(T) = \Phi(Z) = \frac{1}{\sigma\sqrt{2\pi}} \int_{-\infty}^{x} e^{\frac{-(t-\mu)^2}{2\sigma^2}} dt \quad (9)$$

3. Results and Discussions

The activity list of a certain residential building project and the estimated duration of each activity such as optimistic duration (a), pesimistic duration (b), and most likely duration (m) are given in Table 1.

Table 1. List of Project Activity

Activity Code	State (*i* , *j*)	Activity Name	Immediate Predecessor	Estimated Duration (days) *a*	*m*	*b*	Phase
A	(1 , 2)	Strategic Brief & Survey	-	9	10	15	Feasibility
B	(2 , 3)	Budgeting & Feasibility Studies	A	10	11	13	
C	(3 , 4)	Concept Design	B	10	14	16	Design
D	(4 , 5)	Develop Design & Regulation Compliance	C	13	15	21	
E	(5 , 6)	Technical Design	D	9	12	14	
F	(6 , 7)	Enabling Works	E	8	10	11	Construction
G	(7 , 8)	Sub Structure Work	F	23	25	28	
H	(8 , 9)	Column & Curtain Wall	G	18	21	22	
I	(9 , 10)	Plumbing	H	5	8	9	
J	(9 , 12)	Ceiling	H	4	9	10	
K	(9 , 11)	Roof Installation	H	10	16	20	
L	(10, 12)	Electrical Work	I	6	9	10	
M	(11, 12)	Roof Isolation	K	4	5	7	
N	(12, 13)	Plastering & Mist Coat	L , M	12	14	18	
O	(13, 15)	Windows, Frames, and Doors Installation	N	7	11	19	
P	(15, 17)	Painting	P	15	18	22	
Q	(12, 14)	Tiling & Flooring	J , L	12	15	15	
R	(14, 16)	Decoration	O	8	12	13	
S	(16, 17)	Landscaping	R	6	7	10	
T	(17, 18)	Cleaning & Final Control	P , S	2	5	6	
U	(18, 19)	Project Closure & Post Review	T	6	7	10	

Based on the Table 1, the project activities then arranged to a network called PERT chart by considering its predecessor and successor, as shown in Figure 1.

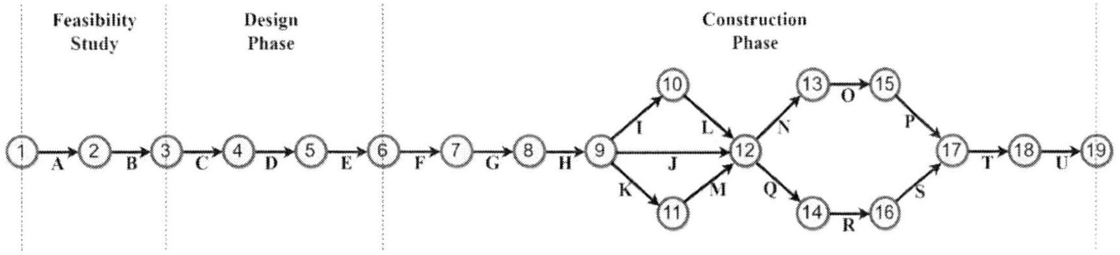

Figure 1. PERT Chart of Project

PERT chart only gives information about the sequence of project activities, but the critical path which has important role to determine the project duration need to be identified with CPM method. The critical path analysis is shown in Table 2.

Table 2. Critical Path Analysis

Activity Code	State (*i* , *j*)	Estimated Duration (days) *a*	*m*	*b*	Critical Path Analysis from PERT Chart (days) μ	*ES*	*EF*	*LS*	*LF*	*SL*	Critical Activity Identification
A	(1 , 2)	9	10	15	11	0	11	0	11	0	Critical
B	(2 , 3)	10	11	13	11	11	22	11	22	0	Critical
C	(3 , 4)	10	14	16	14	22	36	22	36	0	Critical
D	(4 , 5)	13	15	21	16	36	52	36	52	0	Critical
E	(5 , 6)	9	12	14	12	52	64	52	64	0	Critical
F	(6 , 7)	8	10	11	10	64	74	64	74	0	Critical
G	(7 , 8)	23	25	28	25	74	99	74	99	0	Critical
H	(8 , 9)	18	21	22	21	99	120	99	120	0	Critical
I	(9 , 10)	5	8	9	8	120	128	124	132	4	Non-Critical
J	(9 , 12)	4	9	10	8	120	128	133	141	13	Non-Critical
K	(9 , 11)	10	16	20	16	120	136	120	136	0	Critical
L	(10, 12)	6	9	10	9	128	137	132	141	4	Non-Critical
M	(11, 12)	4	5	7	5	136	141	136	141	0	Critical
N	(12, 13)	12	14	18	14	141	155	141	155	0	Critical
O	(13, 15)	7	11	19	12	155	167	155	167	0	Critical
P	(15, 17)	15	18	22	18	167	185	167	185	0	Critical
Q	(12, 14)	12	15	15	15	137	152	151	166	14	Non-Critical
R	(14, 16)	8	12	13	12	152	164	166	178	14	Non-Critical
S	(16, 17)	6	7	10	7	164	171	178	185	14	Non-Critical
T	(17, 18)	2	5	6	5	185	190	185	190	0	Critical
U	(18, 19)	6	7	10	7	190	197	190	197	0	Critical

Length of Critical Path (Project Duration) = $EF_{(last)} = LF_{(last)}$ = *197 days*

The critical activities can be determined based on the activity with zero slack time (SL). According to Table 2, the critical activities are: A-B-C-D-E-F-G-H-K-M-N-O-P-T-U. As a complement to default PERT chart, the critical path by project phase is shown in Figure 2.

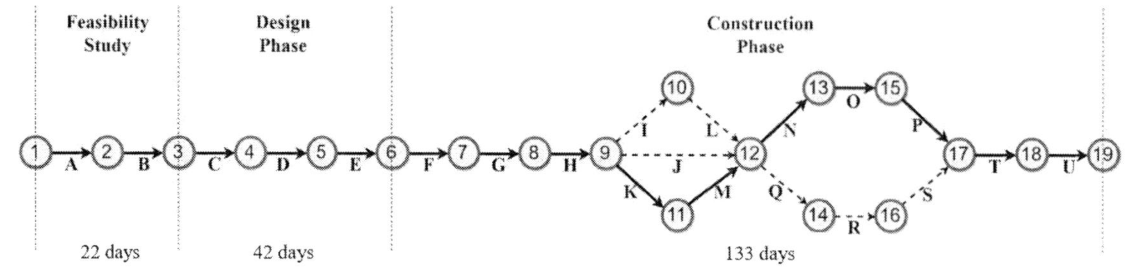

Figure 2. Critical Path of Project

According to critical path analysis using CPM-PERT method, the critical project duration is 197 days which consists of feasibility study phase (22 days), design phase (42 days), and construction phase (133 days). The duration represents the shortest time possible to complete the project, that means there are always uncertain conditions that make project completion duration become exceeded.

In this situation, simulation technique become useful to tackle the problem. By using simulation technique, the uncertain conditions are then simulated for a number of replications, so the management is able to understand any situation that possible to happen in the future. Since the critical path is already identified using CPM-PERT method, Monte Carlo is performed for 10000 replications to simulate the duration of critical activities which generated using Beta inverse function as expresed in formula (7). The simulation result is shown in Table 3 below.

Table 3. Monte Carlo Simulation Result

Critical Activity Code	Estimated Duration (days)			Simulated Duration of Critical Activities (days)			Average and Std Dev of Simulated Duration by Phase (days)	Histogram of Simulated Duration by Phase (days rounded)
	a	*m*	*b*	*Min*	*Avg*	*Max*		
A	9	10	15	9.0068	10.6674	14.4704	Avg=21.8296	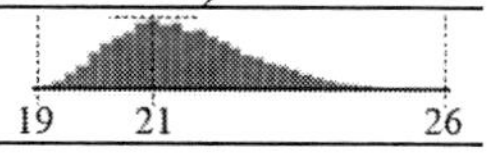
B	10	11	13	10.0216	11.1622	12.8689	StdDev=1.1580	
C	10	14	16	10.1936	13.6733	15.9775	Avg=41.1602	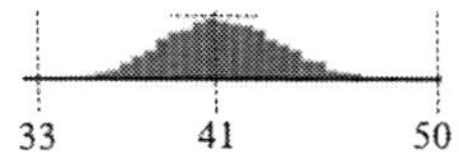
D	13	15	21	13.0307	15.6557	20.5089	StdDev=2.0271	
E	9	12	14	9.1523	11.8312	13.9382		
F	8	10	11	8.1396	9.8340	10.9698	Avg=132.5823	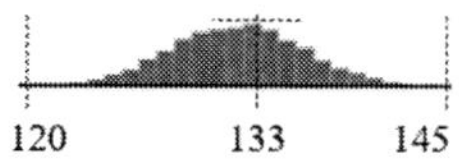
G	23	25	28	23.0273	25.1678	27.9006	StdDev=3.8088	
H	18	21	22	18.2968	20.6685	21.9922		
K	10	16	20	10.2076	15.6471	19.9261		
M	4	5	7	4.0241	5.1704	6.8536		
N	12	14	18	12.0123	14.3275	17.6958		
O	7	11	19	7.0460	11.6152	18.4261		
P	15	18	22	15.0912	18.1584	21.8300		
T	2	5	6	2.3330	4.6650	5.9904		
U	6	7	10	6.0115	7.3282	9.9174		
Simulated Project Duration (μ=195.5720 days ; σ=4.4982)								

From the simulation, 10000 various possible combinations of project duration are generated that plotted into histogram that shown in Figure 3 below.

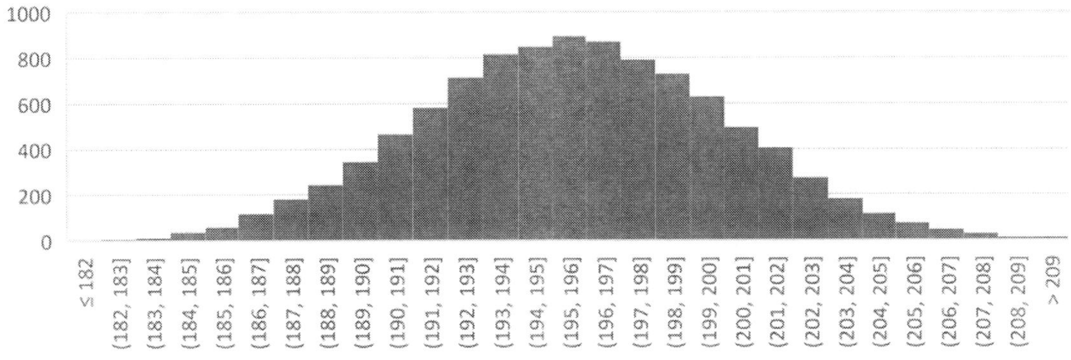

Figure 3. Histogram of Simulated Project Duration

Histogram that is shown on figure 3 looks rather similar to normal distribution pattern, which are symetrical and centered about its mean. It is supported by assumption [6] that project completion duration follows normal distribution. Through on the assumption, various schedule timeliness possibilities are generated by following cumulative normal distribution function with parameter μ dan σ which obtained from the simulation into formula (9) and (10). The graph of schedule timeliness possibilities both for each phase and overall project (with additional 5% and 95% possibility indicators) are shown in figure 4 below.

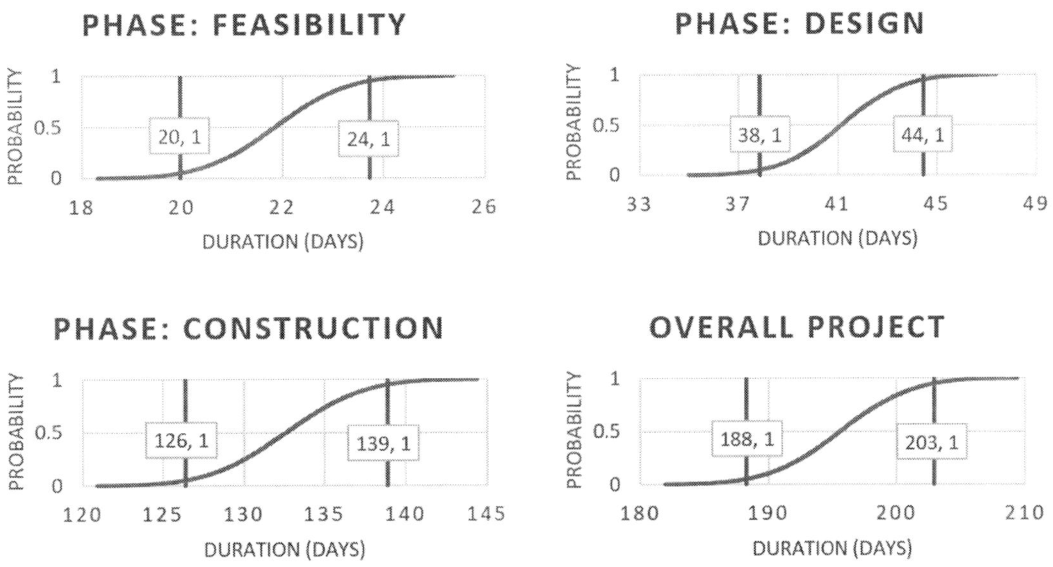

Figure 4. Schedule Timeliness Possibilities

The graphs shows that the shorter project deadline the less likely the project can be finished on time. The longer its deadline the more likely the project can be completed in timely manner. If project schedule given to management is based on the CPM-PERT result, which is 197 days, the timeliness possibility of the project is $p(197) = \Phi(0.3071) \approx 62.06\%$. In other words, the schedule or project deadline have potential risk of lateness 37.94%. According to the simulation result, timeliness possibilities for each phases duration can be identified as well, the critical duration for feasibility study is 22 days then the timeliness possibility for feasibility phase is $p(22) = \Phi(0.1472) \approx 55.85\%$, for 42 days duration of design phase has timeliness possibility $p(42) = \Phi(0.4143) \approx 66.07\%$, while the construction phase has $p(133) = \Phi(0.1097) \approx 54.37\%$ timeliness possibility.

4. Conclusion

The ability of a company to complete projects in timely manner become business competitive advantage, especially in the increasingly fierce business competition environment. CPM-PERT is a method that allows the management to estimate the duration of a construction project, so that the financial budget and resources can be planned properly. In this study, a simulation technique is used as a complementary to CPM-PERT method to give project duration estimation and the potential risk value of lateness from a given project schedule. Monte Carlo method is used to simulate each activity duration to analyze the completion timeliness possibility both for each phase and overall project. According to the simulation, the management has possibility to finish a construction project in 197 days (as it estimated by CPM-PERT) is 62.04%, in 203.1 ≈ 204 days is 95%. While 100% timeliness possibility can be obtained in minimum 208.6 ≈ 209 days. Those possibility range will help the management to determine schedule for their project, both for each project phase and or overall project. Potential risk analysis which explain the category and its severity will be the main focus for further research.

Acknowledgments
This research was supported by Departemen Penelitian dan Pengembangan Masyarakat (DPPM) Universitas Islam Indonesia. We are very grateful for the support.

5. References

[1] Larson E and Gray C 2011 *Project Management: The Managerial Process* (New York: McGraw-Hill Irwin)

[2] Zhou J, Love P, Wang X, Teo K, and Irani Z 2013 A Review of methods and algorithmes for optimizing construction scheduling *Journal of Operation Research Society* pp 1091-1105

[3] Perminova O, Gustafsson M, and Wikstrom K 2008 Defining uncertainty in projects - a new perspective *International Journal of Project Management* **26** pp 73-79

[4] Project Management Institute 2004 A Guide to The Project Management Book of Knowledge (PMBOK) 3rd, Newton Square

[5] Hsiau H and Lin C 2009 A Fuzzy PERT Approach to Evaluate Plant Construction Project Scheduling Risk Under Uncertain Resources Capacity *Journal of Industrial Engineering and Management* **2**(1) pp 31-47

[6] Karabulut M 2017 Application of Monte Carlo simulation and PERT/CPM techniques in planning of construction projects: A Case Study *Periodicals of Engineering and Natural Sciences* **5**(3) pp 408-420

[7] Weick K 1977 Entactment processes in organizations. In: Staw M, Salancik G R, editors., New Directions in Organizational Behavior (Chicago: St. Clair)

[8] Zou P, Zhang G, and Wang J 2007 Understanding the key risks in construction projects in China *International Journal of Project Management* **25**(6) pp 601 - 614

[9] Premachandra I 2001 An approximation of the activity duration distribution in PERT *Journal of Computer & Operations Research* **28** pp 443 - 452

[10] Barraza G 2011 Probabilistic estimation and allocation of project time contingency *Journal of Construction Engineering and Management* **137**(4) pp 259-265

[11] Tysiak W and Sereseanu A 2010 Project Risk Management Using Monte Carlo Simulation and Excel *International Journal of Computing* **9**(4) pp 362 - 367

[12] Ganame P and Chaudhari P 2015 Construction Building Schedule Risk Analysis Using Monte-Carlo Simulation *International Research Journal of Engineering and Technology* **2**(4) pp 1402-1406

[13] Yuan Z, Wang Y and Sun C 2017 Construction schedule early warning from the perspective of probability and visualization *Journal of Intelligent & Fuzzy Systems* **32** pp 877 - 888

[14] Huang J W and Wang X X 2009 Risk Analysis of Construction Schedule Based on PERT and MC Simulation in *International Conference on Information Management Innovation Management and Industrial Engineering* Xi'an China

[15] Rubinstein R Y and Kroese D P 2007 Simulation and the Monte Carlo method (2nd Edition) (New York: John Wiley & Sons)

[16] Hong K, Yu J, Liu Z, Zhao M, and Bi L 2015 Construction schedule risk analysis of diversion tunnel based on improved PERT *Journal of Tianjin University (Social Sciences)* **17** pp 122-128

[17] John Wiley & Sons 2000 The Beta-PERT Distribution [Online] Available: http://www.riskamp.com/beta-pert

The Optimization of Semi Finished Garment Distribution of Forboys SME in Pandeglang Area Using Genetic Algorithm

WN Tanjung [1], N Nurhasanah[1], CF Lutfia[1], B Aribowo[1], R Safitri[2], B Samiono[3], M Devana[1], P Kalifa[1], SW Fauzia[1], A Supriyanto[2]

[1]Industrial Engineering Department, Faculty of Science and Technology, Universitas Al Azhar Indonesia, Sisingamangaraja Street, Jakarta 12110, Indonesia
[2]Informatic Engineering Department, Faculty of Science and Technology, Universitas Al Azhar Indonesia, Sisingamangaraja Street, Jakarta 12110, Indonesia
[3]Management Department, Faculty of Economy, Universitas Al Azhar Indonesia, Sisingamangaraja Street, Jakarta 12110, Indonesia
widya@uai.ac.id, nunungnurhasanah74@gmail.com, cutfirdalutfia@gmail.com

Abstract. Distribution is one of crucial activities for a company because it will benefit the company from the sales of its distributed products. One way to facilitate distribution is determining an optimal route. Determining an optimal route is an important issue to solve since it can affect vehicle operating cost and time. Moreover, it is required to obtain an efficient route. This research deals with the optimization of distribution route of a garment SME located in Pandeglang, namely Forboys (FBS), who produces children's clothing. This research attempts to seek which route is best for distributing semi-finished goods from FBS Warehouse to each CMT in Pandeglang, so as to minimize the cost incurred. In order to distribute semi-finished goods quickly, an efficient distribution route from FBS Warehouse to each CMT is required. This research is conducted uses Genetic Algorithm method. Genetic algorithm which is reliable in producing optimal output that can be used to solve the most appropriate for this type of problem. Based on the results of the research, the optimum route is found to be the one starting the delivery from FBS Warehouse - CMT 3 - CMT 2 - CMT 1 and back to FBS Warehouse, with a total distance of 205.5 km and the cost incurred for each distribution of IDR 386,500.
Keywords: distribution, genetic algorithm, optimal route

1. Introduction

Each company must be good at cost management, as well as at its distribution, by spending outcome as little as possible but still getting income as optimal as possible. Distribution is an activity of moving goods or service from producer to the final customer through distribution channels on time [1]. Companies must distribute their products to the customers quickly with minimal cost. Point depot and the determination of the route is the problems that commonly encountered in distribution [2].

In this case, a matter that can be considered is the determination of the distance or route taken during the distribution.

The determination of the optimal route is one of the problems encountered by Forboys (FBS) SME that has a depot or distribution centre located in Pandeglang. Currently, FBS SME distributes semi-finished goods to Pandeglang (Petir and Baros area). Each collector of each CMT (Cut, Make and Trim or tailor) pick up the goods to be sewn and subsequent to sewing, each CMT will send back the goods to depot 1 or FBS Warehouse. The distribution is carried out every Friday, including sending finished goods and semi-finished goods to sew for a week ahead. Thus, FBS SME wants to have an optimal route and count its distribution cost if it is done from FBS Warehouse to each CMT located in Pandeglang. In this research, it is necessary to solve the problem of the optimization of the distribution route to each CMT in Pandeglang. Therefore, the determination of the optimal route for FBS SME is performed using Genetic Algorithm method.

2. Methods

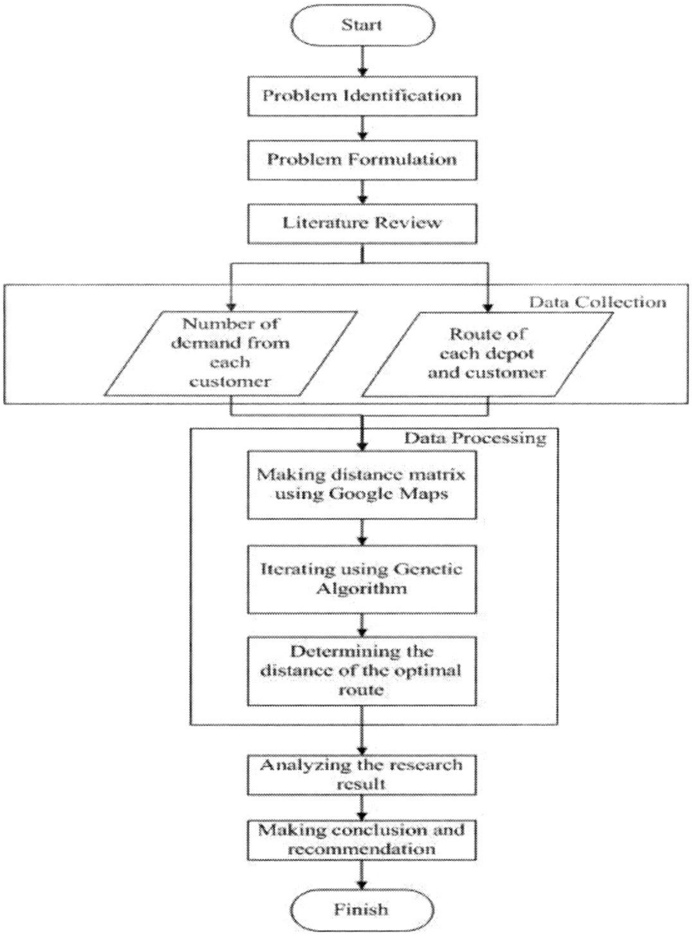

Figure 1. Research Metodology

The first stage performed in this research is identifying the problem by synthesizing ones occurring in the distribution of Forboys' semi-finished goods to every CMT in Pandeglang. Afterwards, we formulate the

IOP Conf. Series: Materials Science and Engineering **528** (2019) 012036 doi:10.1088/1757-899X/528/1/012036

problems and determine how the literature review shall be conducted. Furthermore, all data concerning the number of demands from each customer and the route of each depot are collected to determine customer demand. The next stage is forecasting routes for a month (March) as many as 20 PO and collecting data of route for each depot and customer by finding out the location of each depot and CMT. The data collection stage includes making distance-matrix using Google Maps by means of inputting the distance travelled from each depot to the customer and between customers. After collecting the entire data, calculation is made using Genetic Algorithm. This Genetic Algorithm stage includes making initial population, objective function evaluation, adjusting the criteria of the optimization, selection, crossover and mutation, determining new population, and determining the best individual to find the optimal route of the depot. The analysis of the results of the calculation of the optimum route distance for the distribution of semi-finished goods from the depot to each CMT of Forboys SME in Pandeglang is done in this research using Genetic Algorithm. At the end of the research, the best route for the distribution from the depot to each CMT can be concluded.

3. Results and Discussion

3.1 Forecasting

At this stage, forecasting is done using Exponential Smoothing method. It is done by collecting data six month earlier, from September to February 2018. Forecasting is done to determine how many orders will be made in March 2018. Here are the forecasting results of each CMT:

Table 1. Forecasting of CMT 1

Month (2018)	Time Index	Actual Demand	Forecasting
September	1	1767	5040
October	2	1131	2094
November	3	2592	1227
December	4	11907	2456
January	5	8388	10962
February	6	4452	8645
March	7		4871
Total		30237	

Table 2. Forecasting of CMT 2

Month (2018)	Time Index	Actual Demand	Forecasting
September	1	0	2477
October	2	0	248
November	3	0	25
December	4	1440	2
January	5	9504	1296
February	6	6396	8683
March	7		6625
Total		17340	

Table 3. Forecasting of CMT 3

Month (2018)	Time Index	Actual Demand	Forecasting
September	1	0	2793
October	2	0	279
November	3	1548	28
December	4	7542	1396
January	5	3600	6927
February	6	4068	3933

March	7	4054
Total	16758	

Table 4. Number of Customer Demands for each CMT

The name of CMT	Demand			Total maximum capacity
	PO	DZ	PCS	PO
CMT 1	6,00	405,92	4871	
CMT 2	8,00	552,08	6625	20
CMT 3	5,00	337,83	4054	
Total	19,00			

3.2 Genetic Algorithm

At this stage, the calculation using Genetic Algorithm requires the data of the distance-matrix from the depot to the customer and between customers. The distance matrix in Table 5 is obtained using Google Maps app. Here is the distance matrix for each route:

Table 5. Distance Matrix from FBS Warehouse to each Customer

Distances Matrix	Depot 1 (FBS Warehouse)	CMT 1	CMT 2	CMT 3
From/To	**1**	**2**	**3**	**4**
1	0	84,7	102	98,4
2	84,7	0	10,8	22,4
3	102	10,8	0	11,6
4	98,4	22,4	11,6	0

3.2.1 Initial Population

At this stage, the determination of initial population is conducted using encoding technique. Encoding technique used in this research includes permutation encoding and value encoding. In permutation encoding, there is a chromosome consisting of a set of numbers representing the sequence of the route's position. While in value encoding, each depot and CMT is represented by a number; the number is a gene contained in each chromosome. The following is the value encoding in this research:

Table 6. Value encoding of depot and CMT in Pandeglang

Encoding	
FBS Warehouses	1
CMT 1	2
CMT 2	3
CMT 3	4

After determining encoding as shown in Table 6, the chromosome of each number or location is determined for the route formation. FBS warehouse is not included in the genes, assuming that the initial town is also the final town of the distribution. The following is Table 7, showing the initial population of Genetic Algorithm completion:

Table 7. The Formation of Initial Population

Chromosome -	Chromosome form		
1	2	3	4

2	3	2	4
3	3	4	2
4	2	4	3
5	4	2	3
6	4	3	2

3.2.2 Selection

At selection stage, the first thing to do is to calculate fitness value. Fitness value in Table 8 is obtained from the total distance traveled according to the chromosomes in the initial population. Here are the Fitness values for the initial population:

Table 8. Fitness values of the Initial Population

Fitness Value	
Chromosome [1]	205,5
Chromosome [2]	230
Chromosome [3]	220,7
Chromosome [4]	220,7
Chromosome [5]	233,6
Chromosome [6]	205,5

The calculation to get the population inverse values from the values in Table 9 is as follows [3]:

$$Q(i) = \frac{1}{Fitness\ (i)} \tag{1}$$

Table 9. Inverse Values of the Initial Population

Q [1]	0,005
Q [2]	0,004
Q [3]	0,005
Q [4]	0,005
Q [5]	0,004
Q [6]	0,005
Total Q	0,027

Further more, the calculation to get Probability values of Fitness Values is as follows:

$$P(i) = \frac{Qi}{Total\ Q} \tag{2}$$

Table 10. Probability of Fitness Values

Probability	
P [1]	0,177
P [2]	0,159
P [3]	0,165
P [4]	0,165
P [5]	0,156
P [6]	0,177

3.2.3 Cross Over

At this crossover stage, the first thing to do is to accumulate the probability value of the fitness values from chromosome [1] to chromosome [6]. The following is Table 11 showing the results of the accumulation of probability values:

Table 11. Cumulative Values of Fitness Probability

Cumulative value	
C [1]	0,177
C [2]	0,336
C [3]	0,501
C [4]	0,666
C [5]	0,823
C [6]	1,000

After obtaining cumulative values, random numbers are generated for all existing chromosomes. The following are the results of the random numbers:

Table 12. Random Values

Random values	
R [1]	0,861
R [2]	0,170
R [3]	0,641
R [4]	0,484
R [5]	0,543
R [6]	0,141

The first stage to determine the chromosome required for the crossover is comparing the cumulative probability values and random values. The chromosome will be crossed if the random number is smaller than the cumulative probability number. Subsequent to determining the pair, the determination of the crossing site or the part to be crossed is performed by generating random numbers from 1 to m-1 in which m is the length of the sub chromosome, with a crossover probability value of 0.5. The value of the probability of crossover that good is ranged from 0.85% to 95% [4]. After determining these, the next thing to do is to cross. The following is the process after each chromosome is crossed:

Table 13. The Crossover Process

The crossover process																
Chromosome [1]	=	Chromosome [1]	><	Chromosome [2]	=	**4**	3	2	><	3	2	4	=	**4**	2	3
Chromosome [2]	=	Chromosome [2]	><	Chromosome [6]	=	**3**	2	4	><	2	3	4	=	**3**	4	2
Chromosome [6]	=	Chromosome [6]	><	Chromosome [1]	=	**2**	3	4	><	4	3	2	=	**2**	4	3

The results of the crossover performed in Table 13 shall produce new chromosomes called offspring chromosomes. The following is the result of the crossover that has been conducted:

Table 14. New Population After Crossover

Population after crossover			
Chromosome [1]	4	2	3
Chromosome [2]	3	4	2
Chromosome [3]	2	4	3

Chromosome [4]	3	4	2
Chromosome [5]	4	2	3
Chromosome [6]	2	4	3

3.2.4 Mutation

At this stage, mutation is conducted by generating random numbers for each existing chromosome. These random numbers are obtained by limiting random numbers to the number of genes of all chromosomes. The process of mutation is done by exchanging the mutated part with the location on the right. The following are the results after the mutation:

Table 15. Initial chromosome that has been mutated

The mutation process			
Chromosome [1]	3	2	4
Chromosome [2]	3	4	2
Chromosome [3]	3	4	2
Chromosome [4]	4	3	2
Chromosome [5]	4	2	3
Chromosome [6]	4	2	3

3.2.5 The Distance of the Optimal Route

At the Genetic Algorithm stage, the distance of the optimal route is the final stage for determining which chromosome is strongest in surviving in each iteration performed. In this research, the optimum route result is obtained after 8 iterations. The following is the iteration values obtained:

Table 16. Iteration 1

Fitness value					
Chromosome [1]	3	2	4	=	233,6
Chromosome [2]	3	4	2	=	234,4
Chromosome [3]	3	4	2	=	220,7
Chromosome [4]	4	3	2	=	205,5
Chromosome [5]	4	2	3	=	233,6
Chromosome [6]	4	2	3	=	233,6

Table 17. Iteration 2

Fitness value					
Chromosome [1]	4	2	3	=	233,6
Chromosome [2]	4	2	3	=	233,6
Chromosome [3]	3	2	4	=	233,6
Chromosome [4]	3	2	4	=	233,6
Chromosome [5]	3	4	2	=	220,7
Chromosome [6]	4	3	2	=	205,5

Table 18. Iteration 3

Fitness value					
Chromosome [1]	4	2	3	=	233,6
Chromosome [2]	4	3	2	=	205,5
Chromosome [3]	4	2	3	=	233,6
Chromosome [4]	3	2	4	=	233,6
Chromosome [5]	4	2	3	=	233,6
Chromosome [6]	4	2	3	=	233,6

Table 19. Iteration 4

Fitness value					
Chromosome [1]	2	4	3	=	220,7
Chromosome [2]	2	4	3	=	220,7
Chromosome [3]	3	4	2	=	220,7
Chromosome [4]	3	2	4	=	233,6
Chromosome [5]	4	3	2	=	205,5
Chromosome [6]	3	2	4	=	233,6

Table 20. Iteration 5

Fitness value					
Chromosome [1]	2	4	3	=	220,7
Chromosome [2]	4	2	3	=	233,6
Chromosome [3]	4	3	2	=	205,5
Chromosome [4]	3	4	2	=	220,7
Chromosome [5]	2	4	3	=	220,7

Table 21. Iteration 6

Fitness value					
Chromosome [1]	4	2	3	=	233,6
Chromosome [2]	3	4	2	=	220,7
Chromosome [3]	4	3	2	=	205,5
Chromosome [4]	3	4	2	=	220,7
Chromosome [5]	2	3	4	=	205,5

Chromosome [6]	4	3	2	=	205,5

Chromosome [6]	4	2	3	=	233,6

Table 22. Iteration 7

Fitness value					
Chromosome [1]	2	4	3	=	220,7
Chromosome [2]	2	3	4	=	205,5
Chromosome [3]	4	3	2	=	205,5
Chromosome [4]	2	3	4	=	205,5
Chromosome [5]	3	4	2	=	220,7
Chromosome [6]	4	2	3	=	233,6

Table 23. Iteration 8

Fitness value					
Chromosome [1]	3	2	4	=	230
Chromosome [2]	3	4	2	=	220,7
Chromosome [3]	4	3	2	=	205,5
Chromosome [4]	4	2	3	=	233,6
Chromosome [5]	3	4	2	=	220,7
Chromosome [6]	3	4	2	=	220,7

After performing iterations, it is found that chromosome [3], with 1-4-3-2-1 route, is the best chromosome with a distance of 205.5 km since the location of its optimum route, from iteration 5 to iteration 8, does not change.

After obtaining the optimal route for the distribution of semi-finished goods, the cost needed for each distribution can be calculated. The following is the cost incurred for the delivery from FBS Warehouse to Pandeglang:

Table 24. Distribution Operating Cost

The type of car	Box truck "Colt Diesel 110ps Mitshubisi"	
Gasoline	IDR	154.500
Toll	IDR	82.000
Driver	IDR	150.000
Total	IDR	386.500

4. Conclusion

According to the results of the calculation using Genetic Algorithm, the shortest and the most optimal route for distributing FBS's semi-finished goods to three CMTs, spread in Pandeglang, can be concluded as follows:

1. The shortest optimal route obtained from the results of the calculation is 1-4-3-2-1 with a total distance of 205.5 km, starting from Depot 1 (FBS Warehouse) - CMT 3 - CMT 2 - CMT 1 - back to Depot 1 (FBS Warehouse).
2. The distribution cost incurred by FBS SME for the distribution from FBS Warehouse to all CMTs and back to FBS Warehouse is IDR. 386,500.

5. Acknowledgement

Thanks to Directorate of Research and Community Service of the Ministry of Research and Techonolgy of the Republic of Indonesia which funded this research activity in 2018 with an Applied Product Research scheme.

6. References

[1] Assauri, Sofjan. 2004. *Manajemen Pemasaran.* (Jakarta: Rajawali Press)

[2] Harry S. & Syamsudin N. 2011. *Penerapan Supply Chain Management pada Proses Management Distribusi dan Transportasi untuk Meminimasi Waktu dan Biaya Pengiriman.* (Jurnal Poros Teknik.Vol. 3, No.1)

[3] Adi, Dewa. 2012. *Penentuan Jarak Terpendek pada Jalur Distribusi Barang di Pulau Jawa dengan Menggunakan Algoritma Genetika.* (Surabaya: Institut Teknologi Sepuluh November)

[4] Obitko, Marek. 2008. *Genetics Algorithms.* [online] (http://www.obitko.com/tutorials /genetics-algorithms/)

Utilizing project management software in project scheduling: a case study

Suci Miranda and M Sugarindra

Industrial Engineering Department, Faculty of Industrial Technology, Universitas Islam Indonesia, Yogyakarta, Indonesia
suci.miranda@uii.ac.id, sugarindra@uii.ac.id

Abstract. Project scheduling is the application of skills, techniques, and intuition acquired through knowledge and experience to develop effective schedule models. It is integrated with various project components, such as activities, resources, and logical relationships. This is a case study of project scheduling under resource-constrained in an aircraft industry. In order to successfully run a project should focus on the project accomplishing effort related to the deadline. When the schedule execution of activities does not meet the project scheduling plan, the entire project may take longer duration to achieve the deliverables. This paper presents a rearranged schedule compared to the current set. Utilizing project management software (PMS), the project can be finished in 91 days employing 12 operators. Microsoft Project is preferred for further evaluating and managing project scheduling solution. The data is treated automated and the leveling method is applied to compress the completion project time related to the scope of the study is the limited resources allocated. Leads and lags have been identified regarding the new schedule results. The project can be accomplished quickly enough using the rearranged schedule.
Keywords: project management software, project scheduling, limited resource

1. Introduction

A project is a temporary endeavor undertaken to create a unique product, service, or result [1]. More companies are clearly seeing the benefits of managing project in order to build organizational project management expertise: lower costs, greater efficiencies, improved customer and stakeholder satisfaction, and greater competitive advantage [2]. Project management performance, which evaluates success mostly based on budget, schedule, and requirement goals, becomes an indicator to project success. Benefits evaluation of a project performance through Benefits Realisation Management method shows a positive correlation to project success on the creation strategic value for the business [3]. Project management performance and integration management are linked with 17 components [4]. Integration management is consisted of development of project charter, knowledge integration, staff integration, supply chain integration, and integration of changes. Meanwhile, the project management performance are time, cost, quality, safety, and client satisfaction. The hypothesis regarding the relation reveals the strong interaction between integration and performance.

According to [5], a project success is measured by product and project quality, timeliness, budget compliance, and degree of customer satisfaction. Project management focuses on the ability to deliver the project's product in scope, time, cost and quality known as the traditional iron triangle. Project scheduling and planning research started after the World War 2. Project scheduling provides a detailed plan that represents how and when the project will deliver the products, services, and results [1]. This is a tool for communication, managing stakeholders' expectation, and as a basis for performance reporting. Resources have become a factor contributing to the time and cost uncertainty. A project planning model considering time, budget, and resource constrained activity under uncertainty is developed using Monte Carlo [6]. The Monte Carlo simulation concept is applied to calculate project reliability. Thus, a project manager can evaluate his planning in order to meet the targeted time and cost. Schedule performance needs to be increased related to managerial efficiencies. A benchmarking model of 57 large projects is proposed evolving 5 categories: input upper outlier, optimal input, deficient input, and input lower outlier [7]. These 57 projects are selected and configured as Decision Making Units (DMUs) in a DEA (Data Envelopment Analysis) procedure that is integrated with PCA (Principal Component Analysis). In automobile D&R process in China, several projects developed simultaneously have led to resource conflict problems. It is required a new resource schedule method based on task priority, Critical chain method, and evidence reasoning in order to resolve the problems. This approach studied by [8] bring several benefits to the enterprise such as eliminate multitasking problem by ensuring one task undertaken at only a unit time.

The scheduling method presents the framework for schedule model creation. The most common scheduling methods, supported by all the major scheduling tools, are the precedence diagram method (PDM), Critical Path Method (CPM), and Critical Chain which is based on CPM [1]. Within these methods, there are various techniques such as rolling wave, PERT, Monte Carlo, integrated master scheduling, and agile. CPM and PERT research started in the late 1950s. They are an activity-on-node representation, both assuming unbounded availability of resources. CPM assumes deterministics duration of activities and PERT relates to uncertainty in activity durations. The Critical Path Method (CPM) provides a way for project managers to determine which tasks are critical (zero slack time) which tasks can be delayed as well as the duration delayed.

Production scheduling tools extremely outperform compared to manual scheduling methods. Some organizations standardize on a specific software tool. Project management software (PMS) is required to support organizing, planning and managing resource in project management. They are capable of dealing with estimation, project scheduling, budget management, resource assignment, cooperation, communication, decision-making, quality management and administration system. Spreadsheet program Microsoft Excel 2016 and Microsoft Project 2016 are used for cost optimal time scheduling in construction [9]. Excel's add-in solver was allowed to execute the cost optimization of example project time schedule applying the developed model. Afterwards, Microsoft Project software was employed for further organizing and presentation of optimized time scheduling solution. Using the CPM spreadsheet that implements the CPM/PERT algorithm in Excel can be easily operated than Ms. Project [10].

Regarding to scheduling, an aircraft industry, where this study is undertaken, has dealt with over schedule problem. It is found that the schedule execution does not meet the schedule plan. Based on data gained, the airplane of NC212i-199 version is manufactured for 155 days at station 1 started from March 01st until October 03rd, 2018. One of the main indicators affected the delaying is limited resources allocated. 6 process areas within station 1 have been assigned 2 operators each. Yet, the available resources are smaller than the resources required. Thus, this study aims to rearrange the project schedule under limited resource as well as analyse the critical tasks and critical path. Utilizing the project management software (PMS) Ms. Project, the current schedule has been evaluated and changed into a shorten schedule by leveling method. The critical tasks inform that they cannot be delayed related to accomplish the project on time. The study does not consider the cost, risks, and advanced technology indicators during the process. Our scope is reducing the completion time of the project employing 12 operators instead. The work breakdown structure (WBS) including the duration of each activity do not change either. They remain the same as the original schedule.

2. Method

This study is conducted based on a case study analyzed in an aircraft industry in Indonesia. It produces various type of aircraft to fulfill the needs of civil airlines, military operators and specific missions. Several problem occured while producing the aircraft such as schedule planned missed due to resource-constrained. This study focuses on the NC212i-199 version scheduling particularly at station 1. It comprises 6 processes: landing gear (LG); door (DR); fuselage (FS); center wing (CW); sealant (SL);

and nose (NS). There are 100 activities spreading to 6 processes. 2 operators are assigned into each process. It means that 12 operators are employing in station 1. The research framework is shown in Figure 1 below.

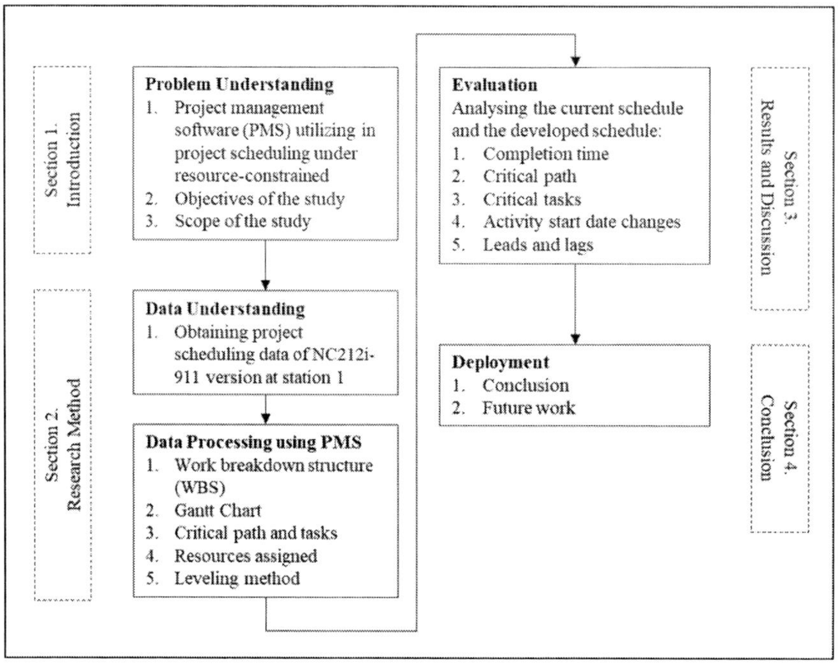

Figure 1. Research Framework

In many cases, the number of resources that are expected to be available to accomplish an activity may determine the activity's duration [1]. A change to a driving resource allocated to the activity will usually have an effect on the duration, yet it is not a simple linear relationship. Increasing the number of resources to twice the original number of the resources does not always reduce the time by half as it may increase extra duration due to risk. Moreover, at some point adding too many resources to the activity may increase duration due to knowledge transfer, learning curve, additional coordination, and other factors involved. Therefore, this study does not consider hiring more resource in terms of finishing the project faster. it is not required any changes to WBS either as the activity of producing the aircraft is a fixed activity based on company regulation. Both manually and auto task mode affect the overallocated tasks. This problem will be resolved by resource optimization. Resource optimization is used to adjust the start and finish date of activities to adjust planned resource use to be equal to or less than resource availability [1]. One technique applied to this study is Resource leveling. It is a technique which start and finish dates are adjusted based on resource constraints with the goal of balancing the demand for resources with the available supply. Resource leveling can be used when shared or critically required resources are available only at certain times or in limited quantities, or are overallocated. Several consideration of data processing are:

- The number of resources available are 12 operators with 2 operators of each task.
- The tasks and duration to produce the aircraft in station 1 are set by company.
- The tasks are sorted from the oldest to the newest date.
- The schedule given is processed on manually task mode, while the new schedule is on auto task mode.
- The overallocated tasks occurred will be resolved by leveling technique.

- Other factors in calculating duration as well as determining start and finish date are excluded such as cost and risk.
- Most activities dependencies are finish-to-start (FS), but 2 activities within fuselage (FS) process area are finish-to-finish (FF) dependencies.

Project constraints will affect the determination of leads and lags. It is a limiting factor that affect the execution of a project, program, portfolio or a process [1]. These includes: scope, schedule, budget, quality, resources, and risks. This study focuses on balancing the schedule and limited resources while others are excluded. These 2 different task modes have changed the completion time as well as the critical tasks and critical path. The results are analysed and evaluated based on several parameters: the completion time, critical tasks, critical path, and leads and lags.

3. Result and Discussion

3.1 The completion time

The original schedule is processed using manually schedule mode while the new schedule is rearranged in auto schedule mode. For both schedules, there are 12 operators assigned with 2 operators each. They work 8 hours a day on weekdays. The processes can be seen in the following Table 1.

Table 1. Station 1 process of NC212i-911 version

Code	Work Specification	Process Area	Resources	Activities
LG	Mechanic	Landing Gear	2	2
DR	Mechanic	Door	2	13
CW	Mechanic	Center Wing	2	7
FS	Mechanic	Fusalage	2	31
SL	Sealant Mechanic	All process	2	6
NS	Mechanic	Nose	2	41
TOTAL		6 processes	12 resources	100

Sealant (SL) activities are part of Nose (NS) and Fuselage (FS) process area. Mixing sealant is conducted once in fuselage, while other 5 mixing sealant activities are done in nose. The original schedule starts on March 01st and finishes on October 03rd, 2018. It takes 155 days to accomplish the project. The auto schedule has reduced the duration to 91 days. Ms. Project automatically calculate the start, finish, and duration values for each task based on dependencies, constraints, calender, and other factors. In the calculating process, one constraint considered is the resources allocation for each activity. It is assumed that 2 operators set in every activity. The example of several activities is shown in Figure 2 and 3.

Based on data, installation of passenger door in door process area and installation of heating system in fuselage area are conducted on the first day of the project, March 01st, 2018. The activities are sorted by the oldest to newest date. Installation of tube for hydraulic system in nose is the last activity accomplished on October 02nd – 03rd, 2018. It is found 4 overallocated activities: installation of supports for hydraulic system in nose [NS] and installation mechanic at transformer and transmitter in nose [NS] started on June 05th; installation of pulley support in center wing [CW] and installation of tube for hydraulic system in nose [NS] are on June 20th.

Task Mode	Task Name	Duration	Start	Finish	Predecessor	Successors
	INST'L OF PASSENGER DOOR [DR2]	2 days	Thu 3/1/18	Fri 3/2/18		42
	INSTALLATION OF HEATING SYSTEM IN FUSELA [FS2]	1 day	Thu 3/1/18	Thu 3/1/18		6
	RECEIVING OF THE FUSELAGE [FS2]	1 day	Fri 3/2/18	Fri 3/2/18		39,8,13,86,38
	ATTACHMENT OF RIGHT RAMP SEAT [DR2]	3 days	Mon 3/5/18	Wed 3/7/18		7
	INST'L OF VENTILATION SYSTEM IN FUSELAGE [FS2]	3 days	Mon 3/5/18	Wed 3/7/18	3	45
	INST. OF SWITCH & LABEL RAMP DOOR AREA [DR2]	1 day	Thu 3/8/18	Thu 3/8/18	5	9
	INSTALLATION OF SUPPORTS FOR LH TAPESTRI [FS2]	2 days	Thu 3/8/18	Fri 3/9/18	4	43
	INST. OF REAR DOOR / PORT DOOR EQUIPPED	2 days	Fri 3/9/18	Mon 3/12/18	7	14
	INST. OF TUBES FOR HYDRAULIC SYSTEM [FS2]	2 days	Mon 3/12/18	Tue 3/13/18		41
	INSTALLATION OF MICROSWITCH IN DOOR [DR2]	1 day	Tue 3/13/18	Tue 3/13/18		16
	INSTL. HINGE AND FITTING OF PILOT DOOR [DR2]	2 days	Wed 3/14/18	Thu 3/15/18		21
	INSTALLATION OF SUPPORTS FOR RH TAPESTRI	2 days	Wed 3/14/18	Thu 3/15/18	4	44
	INSTALLATION OF FORWARD DOOR / RAMP DOOR [DR2]	2 days	Fri 3/16/18	Mon 3/19/18	9	17
	MAKE HOLES & INST'L SPT FOR ROUTING FSLG [FS2]	6 days	Fri 3/16/18	Fri 3/23/18		46

Figure 2. Example of Project scheduling using manually task mode

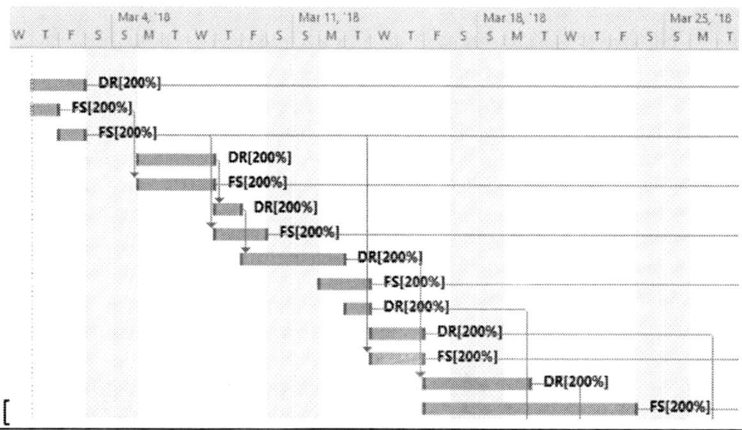

Figure 3. Example of Gantt chart using manually task mode

Due to the leveling technique using manually task mode, installation of pulley support in center wing has changed from 1 day to 2days. Now, the duration total turns to 156 days. Comparing to auto task mode integrated with leveling technique, the new schedule has been reduced 65 days into 91 days. We do not find any task duration changed using this mode. It is proposed to complete the project from March 01st to July 05th, 2018.

3.2 Critical path

The critical path is the sequence of activities that represents the longest path through a project, that determines the shortest possible project duration. It is normally characterized by zero total float or total slack on the critical path. The total slack contains the amount of time a task's finish date can be delayed without delaying the project's finish date. Ms. Project shows the critical task in red. There is only one critical task found among 100 activities in the original schedule that is installation of tubes for hydraulic system in nose process area. This task is the latest conducted within the schedule. Furthermore, around 16 activities have free slack 0. Free slack is the number of days that an activity can have before it starts delaying the next (successor) activity. For example, attachment of right ramp seat in door process area has total slack of 143 days but no free slack. It cannot be delayed otherwise the successor of installation of switch and label ramp door area will be delayed as well.

Based on the result on auto task mode for new schedule, there are 4 critical tasks with the shortest possible project duration of 9 days. 16 activities of 0 free slack have been identified. The following Table 2 and 3 shows the free slack and critical activities of both original and new schedule.

IOP Conf. Series: Materials Science and Engineering **528** (2019) 012037 doi:10.1088/1757-899X/528/1/012037

Table 2. Free slack activities

Original Schedule

Task Name	Duration	Start	Finish	Pred	Successors
ATTACHMENT OF RIGHT RAMP SEAT [DR2]	3 days	Mon 3/5/18	Wed 3/7/18		7
INST. OF SWITCH & LABEL RAMP DOOR AREA [DR2]	1 day	Thu 3/8/18	Thu 3/8/18	5	9
INST. OF LOCKING SYSTEM ON RAMP DOOR [DR2]	1 day	Thu 3/22/18	Thu 3/22/18	14	18
INST'L OF ANTENNA SPT ON FUSELAGE AREA [FS2]	4 days	Mon 4/2/18	Thu 4/5/18		25FF
INST. OF TUBES & HOSES ON MAIN LANDING [LG2]	1 day	Wed 5/30/18	Wed 5/30/18		51
INSTALLATION OF CENTER WING TO FUSELAGE [CW2]	2 days	Wed 5/30/18	Thu 5/31/18		64,53,57,68,66,61
INSTALLATION SUPPORT OF INSTRUMENT PRESS [NS2]	1 day	Wed 5/30/18	Wed 5/30/18		52
INSTALLATION OF SUPPORT ASSY FOR ENGINE [CW2]	2 days	Fri 6/1/18	Mon 6/4/18	48	64,57
INSTALLATION SUPPORT OF INSTRUMENT PANEL [NS2]	1 day	Mon 6/4/18	Mon 6/4/18		62,73,97
INST'L OF JOINING PART ON CTRW WING [CW2]	3 days	Tue 6/5/18	Thu 6/7/18	48,53	68,66,61
INST. OF SUPPORTS FOR HYDRAULIC SYSTEM [NS2]	2 days	Tue 6/5/18	Wed 6/6/18	52	60
INSTALLATION OF TUBES FOR HYDRAULIC SYST [NS2]	1 day	Mon 6/11/18	Mon 6/11/18	58	63
INSTALLATION OF TUBES FOR HYDRAULIC SYST [NS2]	2 days	Thu 7/5/18	Fri 7/6/18	72	76
INSTALLATION OF TUBES FOR HYDRAULIC SYST [NS2]	2 days	Mon 7/23/18	Tue 7/24/18	76	80
INSTALLATION OF WINDSHIELD [NS2]	2 days	Thu 8/2/18	Fri 8/3/18	78	84
INSTALLATION OF TUBES FOR HYDRAULIC SYST [NS2]	2 days	Tue 10/2/18	Wed 10/3/18	52	

New Schedule

Task Name	Duration	Start	Finish	Predecessors	Successors
INST'L OF PASSENGER DOOR [DR2]	2 days	Fri 3/16/18	Mon 3/19/18		42
INSTALLATION OF HEATING SYSTEM IN FUSELA [FS2]	1 day	Thu 3/1/18	Thu 3/1/18		6
RECEIVING OF THE FUSELAGE [FS2]	1 day	Fri 3/2/18	Fri 3/2/18		8,13,38,39,86
ATTACHMENT OF RIGHT RAMP SEAT [DR2]	3 days	Thu 3/1/18	Mon 3/5/18		7
INST'L OF VENTILATION SYSTEM IN FUSELAGE [FS2]	3 days	Mon 3/5/18	Wed 3/7/18	3	45
INST. OF SWITCH & LABEL RAMP DOOR AREA [DR2]	1 day	Tue 3/6/18	Tue 3/6/18	5	9
INSTALLATION OF SUPPORTS FOR LH TAPESTRI [FS2]	2 days	Thu 3/8/18	Fri 3/9/18	4	43
INST. OF REAR DOOR / PORT DOOR EQUIPPED	2 days	Wed 3/7/18	Thu 3/8/18	7	14
INST. OF TUBES FOR HYDRAULIC SYSTEM [FS2]	2 days	Fri 3/16/18	Mon 3/19/18		41
INSTALLATION OF MICROSWITCH IN DOOR [DR2]	1 day	Fri 3/9/18	Fri 3/9/18		16
INSTL. HINGE AND FITTING OF PILOT DOOR [DR2]	2 days	Mon 3/12/18	Tue 3/13/18		21
INSTALLATION OF SUPPORTS FOR RH TAPESTRI	2 days	Tue 3/20/18	Wed 3/21/18	4	44
INSTALLATION OF FORWARD DOOR / RAMP DOOR [DR2]	2 days	Wed 3/14/18	Thu 3/15/18	9	17
MAKE HOLES & INST'L SPT FOR ROUTING FSLG [FS2]	6 days	Thu 3/22/18	Thu 3/29/18		46
INSTALLATION OF EMERGENCY EXIT DOOR [DR2]	2 days	Tue 3/20/18	Wed 3/21/18	11	23
INST. OF LOCKING SYSTEM ON RAMP DOOR [DR2]	1 day	Thu 3/22/18	Thu 3/22/18	14	18

The use of critical path calculation is to determines the earliest possible completion of the project. In other word, it can be called the driving path to project completion. Analysis of the critical path requires several assumptions, including the following: all tasks are known, all links are accurate, all estimates are accurate, and other non-critical paths may have small amount of slack. In this study, all tasks and links are based on the data given by company. The estimation of other factors such as duration are provided as well. In terms of small amount of slack, there are only 2 activities of less than 10 days total slack. The rest, excluded the critical task, is between 10 to 150 days total slack. Most of the non-critical paths have big amount of slack. It can be confusing as the critical task found is only one among 100 tasks. Different results are showed on auto task mode for new schedule. The non-critical paths have vary total slack from 3 to 89 days. This is not small amount of slack, yet it is smaller than the original schedule result. Despite this potential problem, the critical path analysis using auto task mode provides the project manager a good indication of where to focus attention.

Table 3. Critical tasks comparison

Task Name	Duration	Start	Finish	Pred	Success
Original Schedule					
INSTALLATION OF TUBES FOR HYDRAULIC SYST [NS2]	2 days	10/2/18	10/3/18	52	
New Schedule					
INSTALLATION SUPPORT OF INSTRUMENT PANEL [NS2]	1 day	3/2/18	3/2/18		62,73,97
INST'L SUPPORT AT MECH. ELECT. NOSE AREA [NS2]	2 days	3/28/18	3/29/18	56	77
INST. OF ELECTRIC EQUIPMENT ON FRAME 3-5 [NS2]	4 days	4/10/18	4/13/18	73	100
INSTL. OF FUEL CONTROL UNIT ON FR.C4F-5F [NS2]	3 days	7/3/18	7/5/18	77	

3.3 Leads and lags

In project schedule management, sequence activities are the next step after define activities. Sequence activities is the process of identifying and documenting relationship among the project activities. The key benefit is defining the logical sequence of work to obtain the greatest efficiency given all project constraints. One technique used is leads and lags. A lead is the amount of time a successor activity can be advanced with respect to a predecessor activity. Conversely, a lag is the amount of time a successor activity will be delayed with respect to a predecessor activity. Leads and lags are essential for making a good project. They are the basic building blocks of scheduling. We try to identify the leads and lags of each activity based on the start, finish date and free slack information showed in new schedule. Most activity has lags due to limited resources of each area. An example of calculating leads and lags in door area are documented in Table 4 and Figure 4.

Table 4. Activities in door are

Activity	Description	Duration	Start	Finish	Pred	Successors	Free Slack
1	INST'L OF PASSENGER DOOR [DR2]	2 days	Fri 3/16/18	Mon 3/19/18		41	0 days
4	ATTACHMENT OF RIGHT RAMP SEAT [DR2]	3 days	Thu 3/1/18	Mon 3/5/18		6	0 days
6	INST. OF SWITCH & LABEL RAMP DOOR AREA [DR2]	1 day	Tue 3/6/18	Tue 3/6/18	4	8	0 days
8	INST. OF REAR DOOR / PORT DOOR EQUIPPED [DR2]	2 days	Wed 3/7/18	Thu 3/8/18	6	13	0 days
10	INSTALLATION OF MICROSWITCH IN DOOR [DR2]	1 day	Fri 3/9/18	Fri 3/9/18		15	0 days
11	INSTL. HINGE AND FITTING OF PILOT DOOR [DR2]	2 days	Mon 3/12/18	Tue 3/13/18		20	0 days
13	INSTALLATION OF FORWARD DOOR / RAMP DOOR [DR2]	2 days	Wed 3/14/18	Thu 3/15/18	8	16	0 days
15	INSTALLATION OF EMERGENCY EXIT DOOR [DR2]	2 days	Tue 3/20/18	Wed 3/21/18	10	22	0 days
16	INST. OF LOCKING SYSTEM ON RAMP DOOR [DR2]	1 day	Thu 3/22/18	Thu 3/22/18	13	17	0 days
17	INST. OF SUPPORT AND RAILS ON RAMP DOOR [DR2]	1 day	Fri 3/23/18	Fri 3/23/18	16		74 days
18	INSTL.RUBB PROFILE FR. OF EMERGENCY DOOR [DR2]	1 day	Mon 3/26/18	Mon 3/26/18			73 days
20	INSTALLATION OF PILOT DOOR [DR2]	2 days	Tue 3/27/18	Wed 3/28/18	11		71 days
22	INSTALLATION OF LOCK ON LH DOOR [DR2]	1 day	Thu 3/29/18	Thu 3/29/18	15		70 days

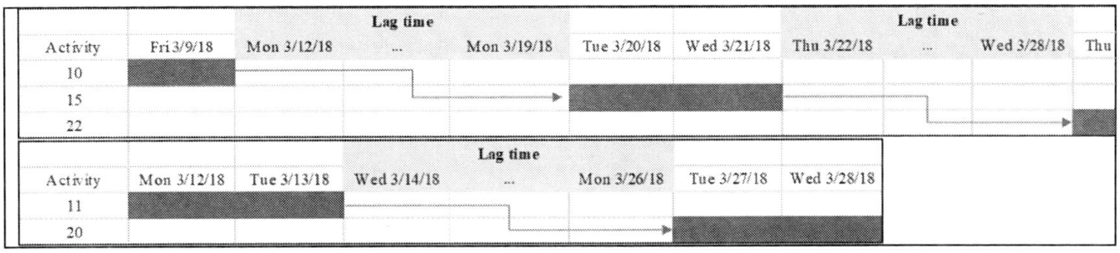

Figure 4. Bar chart representation of lag

The scheduled Start (S) of 15 is 6 days after (+6days) the scheduled Finish (F) of 10. Mathematical representation of lags: 15(S) = 10(F) + 6 days; 22(S) = 15(F) + 5 days; 20(S) = 11(F) + 9 days. Leads and lags can be estimated accurately in the future schedule planning. Many scheduling experts do not use them including this aircraft industry where the study is undertaken.

4. Conclusion

The activities in a project must be completed in a certain period of time, typically the activities in the critical path. This has led the project manager adjusting the resource to ensure the completion time is on schedule. For some projects, the limited resources allocated becomes a factor considered in estimating the schedule. The available resources may be less than resources required to do the tasks. Resources are precious to be spent in the best and most efficient way possible. It is all about the ability to meet customer demand under uncertainties occurred.

Manually and auto task mode will be interchangeable. It depends on the complexity of the project. Since the original schedule has been defined by the company, it is treated as manually task mode in Ms. Project. The new schedule uses the auto task mode. They give different completion of time. For several projects, they are hybrid application. Yet, either manual or auto, each is influenced by project constraints, project assumption, and project dependencies. The concept of leads and lags has many practical applications in Project Scheduling. Sometimes they are mandatory due to the project constraints or environmental reasons. In this case, the project constraints of resource have driven lags determination. For further research, other project management softwares can be applied to get a better result. Cost optimization using mathematical modeling can be integrated with project software application including other constraints such as budget and risks constraints. Understanding project assumptions, project constraints, and project dependencies in terms of project scheduling management particularly resource-constrained project scheduling will create a successfully project completion.

5. References

[1] *A guide to Project Management Body of Knowledge Sixth Edition* 2017 (Pennsylvania: Project Management Institute, Inc.)
[2] The value of project management 2010 *White paper* (Pennsylvania: Project Management Institute, Inc.)
[3] Serra C E M and Kunc M 2015 *Int. Jour. of Proj. Management* **33** pp. 53-66
[4] Demirkesen S and Ozorhon B 2017 *Int. Jour. of Proj. Management* **35** pp. 1639-54
[5] Caccamese A and Bragantini D 2012 Beyond the iron triangle: year zero – *PMI Global Congress 2012-EMEA* (Marsailles, France)
[6] Saputra Y A and Latiffianti E 2015 *Procedia Comp. Science* **72** pp. 561-8
[7] Iyer K C and Banerjee P S 2016 *Int. Jour. of Proj. Management* **34** pp. 219-36
[8] Yang S and Fu L 2014 *Int. Jour. of Proj. Management* **32** pp. 166-77
[9] Valenko T and Klanšek U 2017 *Organization, Technology and Management in Construction* **9** pp.1627-37
[10] Ssempebwa R K 2013 Project Schedule Management Retrieved from: https://www.researchgate.net/publication/273759807_Project_Schedule_Management

Planting System Modeling Of *Chrysanthemum* Seedling Plants Stock for Profit Optimization

A Ilmaniati and D H Taufik

Industrial Engineering Department, Universitas Suryakancana, Cianjur 43216, Indonesia
anitailmaniati@unsur.ac.id

Abstract. Inventories of conventional agricultural cultivation products are closely related to available land area. Often the area of available land is not comparable to meet all demands and cause a profit loss. Appropriate planting management is needed to minimize the profit loss. This study aims to optimize the profit from Chrysanthemum seedling cultivation on Cianjur seedling plantation by developing a model of decision-making system through simulating the planting proportion between varieties. The method used in this research is by using hard method operational research (OR) approaches, while the simulation process is done using spread sheet excel with manual input. The results showed that in one planting period required four times planting with the particular planting proportion. The total profit generated from the best scenario simulation is IDR. 385,779,274.19 with 24.63% profit from first planting, 26.26% from second planting, 27.97% from third planting and 21.13% from fourth planting. The results of this study provide a basis for determining the decision of planting patterns compared with existing conditions that have not used the planning model at all.
Keywords: System modeling; Simulation; Profit optimization; Inventory

1. Introduction

The demands fulfillment to businesses engaged in conventional agricultural cultivation is closely related to land availability as a limitation. Farmers who are used to conventional agricultural technologies are generally not pursued the optimization of land use through application of advanced technologies. Based on that, the determination of the right planting pattern is needed to generate optimal profits; otherwise it will result in profit loss and increasing cost.

Chrysanthemum seedling plant is one of the agricultural commodities that are seeded at Gugenang, Cianjur district, West Java. Flower seeds used in seedling are obtained from Japan and the results of seedling are also exported to Japan, so the supply of raw materials (seeds) as well as the distribution of products (seedling plants) of this nursery business is an export-import process. Business that include import-export processes in material supply and product distribution require proper planning in production process as they generally have a relatively longer lead time than if supply and distribution only within the country, making it particularly vulnerable to profit loss.

The seedling plant produced by this nursery business consists of 16 *Chrysanthemum* varieties *namely Aki, Hibiki, Kougen, Makoto, Hana, Kinka, Kirameki, Tsubasa, Izumi, Otome, Nana, Maiki, Angels, Sheltie, Urizun*, and *Mao*. Each of these varieties has a different level of demand and tends to fluctuate. Diverse types of products and fluctuating demand make nursery managers difficult to make a planting pattern planning.

Previous research has formulated inventory planning for those chrysanthemum seedling businesses using the Q method [1]. The research shows that the amount of seedling plant inventory that must be met exceeds the capacity in one planting, so further research is needed to formulate a system for managing seed planting pattern based on the number of inventory that have been formulated by Q method. Based on the research background, the problems that can be formulated in this study include: (1) how to formulate a planting system model that is in accordance with the quantity of seedling plant inventory; (2) what is the planting proportion of each chrysanthemum seed variety that produces maximum sales profit. The purpose of this study are: (1) to find out the planting system modeling for the supply of chrysanthemum seedlings in accordance with the quantity of calculated inventory; (2) to obtain the planting proportion of each chrysanthemum seed variety which results in maximum sales profit.

This research was conducted with several limitations to simplify the process of mathematical modeling. The limitations of this study are:

a) This research is only focused on planting systems modeling of seedling plants, not discussing delivery systems
b) The quantity of inventory used in this study refers to the results of previous research [1] and no recalculation
c) The decision variables in this study is only the proportion of planting per variety, other factors are considered as parameters with a fixed value

Understanding of a system is a fundamental need to do a simulation or application of analytical methods, because the approach used to solve problems is a system approach, which is a holistic approach to a problem. A holistic approach is needed because the system (where problems arise) can exhibit behaviours or properties that none of its components individually may exhibit [2]. In general, the hard OR system modeling process consists of four stages including: (1) depict a problem situation summary (rich pictures, mind map, etc.); (2) analyse relevant systems; (3) create influence diagrams, and (4) formulate quantitative models or mathematical models [2].

Simulation is an experimental method to explain system behaviour, to construct theories or hypotheses based on the behaviour of the observed system, to predict future system behaviour that will be generated by changes in variables and system parameters or changes in operations [3]. So, simulation is the process of designing a model of an actual system, conducting experiments on the model and evaluating the results of the experiment.

2. Methods

As mention earlier, this study use hard OR system modeling approach and manual simulation to obtain the solution of problem. The following steps are:

1) Depiction of problem situation summary

This stage is a depiction of the problem situation faced, can include the description of the process flow and structure, the components involved, goals and desires, the relationship between components, data and information sources, and hierarchical structure [4]. An effective way to describe the situation perfectly is to draw it into a diagram. The problem situation summary diagram used in this study is a Mind Map diagram.

2) Determination of dominant seed varieties

Determination of dominant seed varieties is carried out to determine which seed varieties are prioritized to be handled based on the highest potential value. The proportion of dominant seed varieties will later become the decision variable to determine the proportion of seedling planting patterns, while other varieties (which are not dominant) will only fill the planting proportion up to 100%. Determination of these dominant varieties is done using the ABC classification. The steps of the ABC classification as described by [5] are as follows:

a. Calculate the potential value for each type of variety per year (Mi) by multiplying the number of usage or demand for each type of variety per year (Di) with the unit price of the variety (pi), mathematically can be expressed as in equation 1.
$$M_i = D_i \, x \, p_i \quad (1)$$
b. Calculate the total potential value for all types of varieties with equation 2.
$$M = \sum M_i \quad (2)$$
c. Calculate the percentage of potential value for each of the varieties (Pi) with the equation 3.
$$P_i = \frac{M_i}{M} x \; 100\% \quad (3)$$
d. Sort the percentage of potential value, starting from the largest percentage to the smallest.
e. Calculate the cumulative value of the percentage of potential value based on the sequence obtained in step (d).

3) Identification of relevant systems

The relevant system identification phase will show that all components and relationships between components are described as having relevance to the problem situation. This will be the basis for forming the right model to study and manipulate problem situations. The output generated from these phase is the Influence Diagram. Influence Diagram is a tool to describe the effect of system input with components in the system, between components in the system, and between components and system output.

4) Formulation of mathematical model

The mathematical model states (in quantitative terms) the relationship between various components, as defined in the relevant system [1]. To facilitate the formulation of mathematical models in this study, researchers used the following assumptions:

a. Chrysanthemum seedling plants produced are assumed to have no damage or failure in production, because the failure rate or defect of this plant is very small and almost non-existent.
b. The price of each chrysanthemum seed variety is assumed to be constant during this study and not affected by the size of the order and exchange rate changes
c. The handling costs for all seed varieties are always the same
d. Supplier always sends chrysanthemum seed on time with a lead time of 10 days
e. Re-ordering all varieties of chrysanthemum seeds are carried out simultaneously when all the seeds supply is used up (after the last planting)
f. Chrysanthemum seedling plants are sent directly to Japan after 2 days storage

5) Verification of mathematical model

Verification of mathematical model aims to ensure that there are no errors when formulating a mathematical model. The verification process needs to be carried out because it will verify that the mathematical model is logical, mathematically correct, and that the mathematical expression is correctly representing the relationship in the system. Verification is done by checking the suitability of all units in each equation.

6) Simulation

The simulation in this study is to conduct experiments on mathematical models that have been formulated, and then evaluate the results of these experiments so that solutions to problems are obtained. The simulation process was carried out in this study by using a spreadsheet with manual input on Microsoft Excel. The columns created in the spreadsheet correspond to the system components in the influence diagram.

7) Sensitivity and error analysis

Error analysis is done to find out the goodness of the model by looking at the difference between the total costs of the system modeling results using historical data with the total cost of the existing data. A small difference shows that the model error is also small and the model is good. Sensitivity analysis is carried out to see how the solution will be affected if there are changes (both single and simultaneous changes) from parameter, and how much the changes will burden the solution. Sensitivity analysis in this study was carried out by making six changes in two parameters, namely inventory cost and handling cost.

3. Results and Discussion

3.1. Data Collection

Data collection in this study was carried out by direct observation and interviews with nursery management. There are several data needed as input in this study, which consists of:

1) Demand per variety per year (Di) and price per unit (Pi)
2) Number of seed stock based on Q method per one period
3) Handling cost (land preparation cost, planting cost, fertilizer cost, watering cost, harvesting cost, and seedling packaging cost) and inventory cost (electricity cost for cold storage, labour cost)

The input data are presented in tables 1, 2, and 3 below.

Table 1. Annual demand per varieties and price of seeds per unit

No.	Varieties	Annual demand (Di) (pcs)	Price per unit (Pi) (IDR)	No.	Varieties	Annual demand (Di) (pcs)	Price per unit (Pi) (IDR)
1	Aki	693,000	570	9	Izumi	299,000	480
2	Hibiki	778,000	570	10	Otome	251,000	480
3	Kougen	316,000	570	11	Nana	1,372,000	480
4	Makoto	2,238,000	570	12	Maiki	155,000	480
5	Hana	536,000	570	13	Malaikat	446,000	480
6	Kinka	1,412,000	480	14	Sheltie	149,000	480
7	Kirameki	924,000	480	15	Urizun	123,000	480
8	Tsubasa	979,000	480	16	Mao	235,000	480
		Total				10,906,000	

Table 2. Number of seed stock based on Q method per one period

No	**Varieties**	***Order qty (Q*)* (pcs)**	***Safety Stock (Ss)* (pcs)**	**Expected seedling plants qty (pcs) ($Q^* + Ss$)**	**No**	**Varieties**	***Order qty (Q*)* (pcs)**	***Safety Stock (Ss)* (pcs)**	**Expected seedling plants qty (pcs) ($Q^* + Ss$)**
1	Makoto	123,387	73,790	197,178	9	Malaikat	52,002	9,190	61,193
2	Kinka	98,812	48,859	147,672	10	Kougen	39,991	7,142	47,133
3	Nana	93,325	33,624	126,949	11	Izumi	42,293	5,924	48,217
4	Tsubasa	78,537	20,364	98,901	12	Otome	38,548	4,711	43,259
5	Kirameki	75,490	18,253	93,743	13	Mao	37,283	4,715	41,998
6	Hibiki	63,583	18,040	81,623	14	Maiki	30,951	4,046	34,997
7	Aki	146,862	7,195	154,057	15	Sheltie	29,736	3,069	32,805
8	Hana	52,760	13,316	66,077	16	Urizun	27,054	2,370	29,423
				Total					1,305,225

Table 3. Handling cost and inventory cost

No.	Component	Cost (IDR)	Unit description
	A. Handling Cost		
1	Land preparation cost (LPC)	12,800,000	per planting
2	Planting cost (PPC)	7,500,000	per planting
3	Fertilizer cost (FPC)	16,900,000	per planting
4	Watering cost (SPC)	24,000,000	per planting
5	Harvesting cost (HPC)	3,750,000	per planting
6	Seedling packaging cost (PKC)	2,375,000	per planting
	Handling cost (HC) = LPC+ PPC+ FPC+ SPC+ HPC+ PKC	67,325,000	per planting
	Unit handling cost (HC1) = HCA/350,000* pcs	**192.36**	per pcs per planting
	B. Inventory Cost		
1	Electricity cost for cold storage (CSC)	20,000,000	per month
2	CSC1 = CSC/30 days	666,667	per day
3	Labor cost (LSC)	280,000	per day
	Inventory cost (IC) = CSC1 + LSC	946,667	per day
	Unit Inventory cost (IC1) = IC/600,000 pcs**	**1.58**	per pcs per day

*capacity per planting
**storage capacity

3.2. Problem Situation Summary

This chrysanthemum nursery business is a business that requires an export-import process in the supply of raw materials and its distribution with relatively long lead times (approximately 10 days). Therefore, proper planning is needed in managing this nursery business, one of which is by calculating the inventory that must be met. The nursery with 35 Ha land area has a planting capacity of 350,000 seeds in one planting, while based on the inventory calculation with the Q method it is known that in one period, the seedling plants that must be provided are 1,305,224 pcs. Based on the gap between land capacity and required inventory, it is necessary to do several plantings in one period. Planting must be done several times and seed varieties (with different seed prices between varieties) make the nursery manager must be able to determine the right planting proportion. The proportion referred to is the proportion (%) of the number of each variety in each planting to produce the most optimal profit in one period. Figure 1 presents a summary of the problem situation of this study.

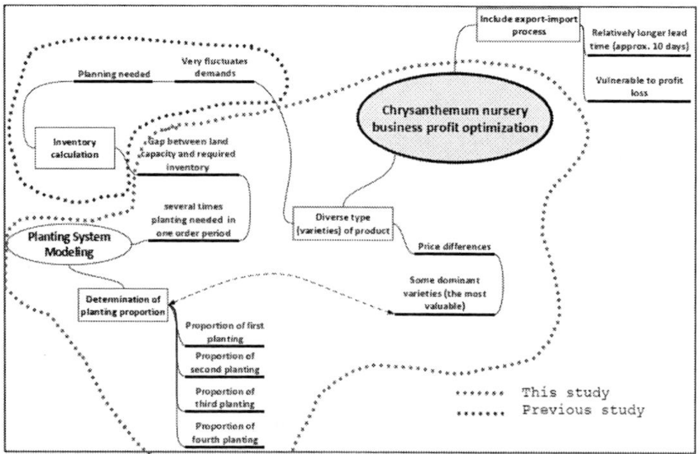

Figure 1. Mind Map of Problem Situation

3.3. Dominant Seed Varieties

Determination of dominant seed varieties is based on the method previously explained. Based on these calculations, it can be known that the seed varieties that fall into the dominant category or category A (% total value ±80%) are 7 varieties, namely Makoto, Kinka, Nana, Tsubasa, Kirameki, Hibiki, and

Aki, while other varieties besides that belong to category B (% total value ±15%) and category C (% total value ±5%). The simulation stage will use the results of this stage because in the simulation stage there will be 2 % reduction or additional proportion from the initial proportion of varieties which fall into category A only. That step was carried out to get some combination of proportion options which will ultimately provide several choices that can generate the maximum profit.

3.4. Relevant System

The relevant system is described based on a summary of predetermined problem situations. Relevant system in this study is described in the influence diagram contained in Figure 2. Elements in circles are components of the system, with the output of the system is profit, uncontrollable inputs are costs, and controllable input or decision variable is planting proportions of dominant varieties.

3.5. Mathematical Model

To measure the modeling performance that has been designed, it can be done by formulating the mathematical model. Table 4 shows the result of the formulation of a mathematical model based on the influence diagram of the relevant system.

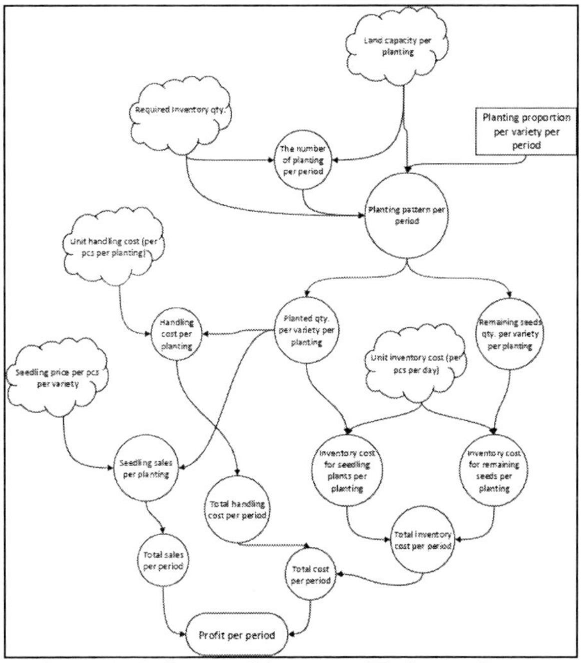

Figure 2. Influence Diagram of Relevant System

Table 4. Mathematical model

No.	Role	Name of Elements	Notation	Equation	Unit
1	Uncontrolable input or parameter	Land capacity per planting	LC	(from data collection)	pcs
2		Required inventory qty.	(Q*+Ss)	(from data collection)	pcs
3		Unit handling cost	HC1	(from data collection)	Rp/pcs/planting
4		Unit inventory cost	IC1	(from data collection)	Rp/pcs/day
5		Seedling price per variety	Pi	(from data collection)	Rp/pcs
6	Controlable input or decision variable	Planting proportion per variety per planting	PF	(obtained from experiments on simulations)	%
7	Components	The number of planting per period	NFP	$\frac{\sum (Q^* + Ss)}{LC}$	-
8		Planting pattern per period	PPF	(based on optimum PF)	
9		Planted qty. per variety per planting	NVF	PF x LC	pcs
10		Remaining seeds qty. per variety per planting	RS	(Q*+Ss) - NVF	pcs
11		Handling cost per planting	HC	HC1 x ΣNVF	Rp
12		Inventory cost for seedling plants per planting	ICFa	ΣNVF x IC(1) x tp*	Rp
13		Inventory cost for remaining seeds per planting	ICFb	ΣRS x IC(1) x tp**	Rp
14		Total handling cost per period	THC	$\sum_{n=\text{planting}} (HC)_n$	Rp
15		Total inventory cost per period	TIC	$\sum_{n=\text{planting}} (ICF_a + ICF_b)_n$	Rp
16		Total cost per periode	TC	THC + TIC	Rp
17		Seedling sales per planting	PSF	ΣNVF x Pi	
18		Total sales per period	TSS	$\sum_{n=\text{planting}} (PSF)_n$	Rp
19	Output	Profit per period	P	TSS - TC	Rp

tp = storage time required in days
tp* = storage time for planted seeds = 4 days
tp** = storage time for remaining seeds to next planting = 12 days

3.6. Simulation

The simulation was carried out to determine some planting proportion options for each variety. The options for this proportion are different for the first, second, third and fourth planting, so the simulation must be done four times according to the number of plantings in one period. Seven varieties belonging to the dominant varieties will be used as a benchmark in determining the proportion option. Each proportion of dominant varieties is added and reduced up to 2% (-1%, -2%, +1%, and +2%) of the initial proportion. The initial proportion is the proportion of each variety obtained from the required inventory quantity per variety divided by the total required inventory quantity then multiplied by 100%. The total option of simulation per planting is 28 options of planting proportion (7 dominant varieties multiplied by 4 changes) plus the initial proportion. Table 5 shows the example of proportion for all varieties in each option. Each proportion option is simulated using an excel spreadsheet with columns according to system elements to obtain output in the form of profit per period.

Table 5. Proportion for all varieties in each option (example)

No	Dominant Varieties	Initial proportion option	1st option	2nd option	3rd option	4th option	(...)	25th option	26th option	27th option	28th option
1	Makoto	15.11%	**14.11%**	**13.11%**	**16.11%**	**17.11%**		15.26%	15.41%	14.96%	14.81%
2	Kinka	11.31%	11.42%	11.54%	11.20%	11.08%		11.42%	11.54%	11.20%	11.08%
3	Nana	9.73%	9.83%	9.92%	9.63%	9.54%		9.83%	9.92%	9.63%	9.54%
4	Tsubasa	7.58%	7.66%	7.73%	7.50%	7.43%		7.66%	7.73%	7.50%	7.43%
5	Kirameki	7.18%	7.25%	7.32%	7.11%	7.04%		7.25%	7.32%	7.11%	7.04%
6	Hibiki	6.25%	6.31%	6.38%	6.19%	6.13%		6.31%	6.38%	6.19%	6.13%
7	Aki	11.80%	11.92%	12.04%	11.68%	11.56%		**10.80%**	**9.80%**	**12.80%**	**13.80%**
8	Hana	5.06%	5.11%	5.16%	5.01%	4.96%		5.11%	5.16%	5.01%	4.96%
9	Malaikat	4.69%	4.74%	4.78%	4.64%	4.60%		4.74%	4.78%	4.64%	4.60%
10	Kougen	3.61%	3.65%	3.68%	3.57%	3.54%		3.65%	3.68%	3.57%	3.54%
11	Izumi	3.69%	3.73%	3.76%	3.65%	3.62%		3.73%	3.76%	3.65%	3.62%
12	Otome	3.31%	3.34%	3.38%	3.28%	3.24%		3.34%	3.38%	3.28%	3.24%
13	Mao	3.22%	3.25%	3.28%	3.19%	3.16%		3.25%	3.28%	3.19%	3.16%
14	Maiki	2.68%	2.71%	2.73%	2.65%	2.63%		2.71%	2.73%	2.65%	2.63%
15	Sheltie	2.51%	2.54%	2.56%	2.48%	2.46%		2.54%	2.56%	2.48%	2.46%
16	Urizun	2.25%	2.27%	2.30%	2.23%	2.21%		2.27%	2.30%	2.23%	2.21%
		100%	**100%**	**100%**	**100%**	**100%**		**100%**	**100%**	**100%**	**100%**

3.7. Sensitivity and Error Analysis

In the error analysis, it can be seen the difference in profit generated by the initial conditions of the company with those generated by the pattern of cropping proportions based on the system modeling that has been made. Overall, the average difference between profit from existing conditions and profit from modeling is 2.43%. The difference can be said to be relatively small, so the model is concluded to have small error. In the sensitivity analysis, changes were made by increase and reduce IC1value (+2%, +4%, +6%, -2%, -4%, -6%), and increase and reduce HC1 value (+2%, +4%, +6%, -2%, -4%, -6%). Based on the sensitivity analysis, it is known that the output of the model will change if the system parameters (IC1 and HC1) change, but the changes are relatively small which is below 6%.

3.8. Discussion

After the simulation calculation for all seed varieties in all options the proportion of planting in the first planting was obtained by the option that gave the highest profit in the first planting, namely the 4th option and the 28th option (as shown at Table 5). The proportion of planting for second, third and fourth planting uses the same calculation as the calculation for the first planting. Table 6 shows the simulation results of the most optimal planting proportions for all plantings.

Table 6. Optimal planting proportion.

No.	Varieties	Planting				No.	Varieties	Planting			
		1st	**2nd**	**3rd**	**4th**			**1st**	**2nd**	**3rd**	**4th**
1	Makoto	17,11%	16,37%	15,22%	10,47%	9	Malaikat	4,55%	4,60%	4,68%	5,00%
2	Kinka	11,08%	11,17%	11,30%	11,85%	10	Kougen	3,49%	3,54%	3,60%	3,88%
3	Nana	9,51%	9,59%	9,71%	10,23%	11	Izumi	3,59%	3,63%	3,69%	3,93%
4	Tsubasa	7,38%	7,45%	7,57%	8,04%	12	Otome	3,23%	3,26%	3,31%	3,51%
5	Kirameki	7,00%	7,07%	7,17%	7,61%	13	Mao	3,15%	3,18%	3,21%	3,37%
6	Hibiki	6,09%	6,15%	6,24%	6,64%	14	Maiki	2,63%	2,65%	2,68%	2,80%
7	Aki	11,55%	11,64%	11,79%	12,38%	15	Sheltie	2,48%	2,49%	2,51%	2,59%
8	Hana	4,91%	4,97%	5,05%	5,41%	16	Urizun	2,24%	2,24%	2,25%	2,29%

Based on the optimal planting proportion of the simulation results, it is known that the optimal proportion is not the same as the initial proportion that can be directly obtained by the formula of ordinary proportions. This shows that this research provides a better solution compared to the calculation of ordinary proportions. Nonetheless, modeling a system with a simpler calculation may make it easier to implement on a real system.

4. Conclusion

System modeling to determine the most appropriate proportion of planting to determine optimum profit can be done by designing a mathematical model and running simulations based on inputs: inventory quantity, costs, and planting capacity. The optimal planting proportion resulted from this study has found in the results section. The optimal proportion yields profit of IDR. 95,036,354.42 for the first planting, IDR. 101,520,387.46 for the second planting, IDR.107,916,538.29 for the third planting, and IDR. 81,505,504.02 for the fourth planting, so the total profit is IDR. 385,779,274.19. The next research can be done by reducing some of the assumptions in this study, so that the model obtained is more representative of the real system.

5. References

[1] Taufik D H and Sutoni A 2018 Perencanaan Persediaan Metode Q untuk Permintaan Probabilistik pada Bibit Bunga Krisan di PT Transplants Indonesia Proc. Industrial Engineering National Conf. (IENACO) 2018 (Surakarta) p 191
[2] Daellenbach H B and McNickle D C 2005 *Management Science: Decision Making through System Thinking* (New York: Palgrave Macmillan)
[3] Sridadi B 2009 *Pemodelan dan Simulasi Sistem* (Bandung: Informatika)
[4] Ristono A 2015 *Pemodelan Sistem* (Yogyakarta: Graha Ilmu)
[5] Senator N B 2006 *Sistem Inventori* (Bandung: Penerbit ITB)

Integer Linear Programming Model and Algorithm to Integrate Heuristics Scheduling EDD, Inventory Control and Distribution Problems in a Modular Production System

P Moengin, E F Harahap, S Adisuwiryo and W A Fransiska

System and Industrial Simulation Laboratory, Department of Industrial Engineering, Trisakti University, Jakarta 11440 Indonesia
parwadi@trisakti.ac.id, elfira@trisakti.ac.id, sc.adisuwiryo@gmail.com, fransiskalefta@yahoo.co.id

Abstract. In this paper, a two-stage algorithm developed for the cutting sequencing problem in a modular manufacturing system. This system has a level of flexibility depends on the cutting stage of raw material. This paper focuses primarily on the work station. In the first stage of the algorithm, an integer linear programming model is used to determine the number of hardboards that will be cut. The model was tested with two different objective function is to minimize waste and to minimize the number of hardboards used. In the second stage, a heuristic developed a scheduling pattern of cutting, where the due date and the number of requests demand a consideration. This algorithm is further implemented in a furniture manufacturer that operates using the make-to-order basis.
Keywords: Manufacturing system, sequential production, algorithm, linear programming, EDD scheduling

1. Introduction

Currently, rapid industrial development leads to high competition. To be able to survive in the competition and the development of this industry, the response of companies to meet consumer demand for products that often change at any time become one of the important goals for each manufacturing company who wants to achieve success. One of them is the timber industry must always deal with consumer demand for wood products size variations produced.

Wood processing industry usually considers the stage of cutting wood in minimizing the residual waste or congestion in the work station. Much research has been done to solve the problems. This can be seen from the formulation of the first model proposed by [1-6].

[1] discussed several basic formulation issues and procedures to solve a cutting problem. Settlement procedures are summarized in a linear programming model. Further sequencing pattern to cut or allocation patterns, analyzed in a study conducted by [2] in order to minimize the maximum queue in the process of cutting. [3-6] proposed a heuristic approach in allocating sawmill.

In this study, a two-stage algorithm was developed for the cutting sequencing problem of a furniture manufacturer XYZ. In the first stage, linear integer programming model is used to determine the number of sheets of wood to be cut by each type of pattern. This cutting pattern produced by a special software program. This cutting pattern is one of the input models. The model was tested by using two different objective functions, which is about minimizing waste, and others minimize the amount of wood to be cut. The second stage consists in a heuristic determined sequence cutting patterns obtained in Phase 1 by considering the maturity date of the order, and taking into consideration the demands of the item.

This paper is organized in the following order: the first part is introduction, the second part presents two-stage algorithm, the third part of the implementation of the algorithm on a furniture company XYZ and the final section presents conclusions and recommendations.

2. Two-phase algorithm

Phase 1: Integer Linear Programming Models

At this first pahse, a linear programming model is developed. This model was used to determine the number of hardboards that will be cut by each cutting pattern. The number of hardboards that will cut the *i*-th cutting pattern given the symbol x_i. There are three types of constraints used in this model, namely:

a) Constraints that ensure that all products demanded by customers will be produced.

$$\sum_{i=1}^{P} \alpha_{ik} x_i \geq d_k, \qquad \forall k \tag{1}$$

where:

α_{ik} = the number of product type k produced using the cutting pattern-i
d_k = total demand of product type k
P = the number of cutting pattern

b) Constraints which considers the capacity of the hardboardss used.

$$\sum_{i=1}^{P} x_i \leq Q \tag{2}$$

where the value of Q is the available capacity of the hardboards used.

c) Constraints that ensure nonnegativity integer number for all variables.

$$x_i \geq 0 \text{ for all integer } i \tag{3}$$

The objective to be achieved in this model is to determine the number of sheets of cut blocks in each cutting pattern. This model was tested by applying two different objective functions, but uses the same constraints for both formulations as a comparison.

The first objective function aims to minimize the total waste (called by Model 1), that is

$$Z = \sum_{i=1}^{P} \left(LWH - \sum_{k=1}^{N} l_k w_k h_k \alpha_{ik} \right) x_i \tag{4}$$

where:
l x w x h = product size (*length x width x height*)
L x W x H = raw material size (*length x width x height*)

The second objective function aims to minimize the number of hardboards that will be cut or used (called by Model 2), that is

$$Z = \sum_{i=1}^{P} x_i \tag{5}$$

Data used in both models are the size of the hardboards, hardboards capacities, demand for products and various combinations of cutting patterns. If the both above mathematical model is rewritten back, then we gained two models as follows.

Model 1: Mathematical model with the aim of minimizing waste

Minimize waste:

$$Z=\sum_{i=1}^{P}\left(LWH-\sum_{k=1}^{N}l_k w_k h_k \alpha_{ik}\right)x_i$$

Subject to:

$$\sum_{i=1}^{P}\alpha_{ik}x_i \geq d_k, \text{for } \forall k$$

$$\sum_{i=1}^{P}x_i \leq Q$$

$x_i \geq 0$, integer for $\forall i$.

Model 2: Mathematical model with the aim of minimizing the number of hardboards used. Minimize the number of hardboards:

$$Z=\sum_{i=1}^{P}x_i$$

Subject to :

$$\sum_{i=1}^{P}\alpha_{ik}x_i \geq d_k \; for \; \forall k$$

$$\sum_{i=1}^{P}x_i \leq Q$$

$x_i \geq 0$ integer for $\forall i$.

Phase 2: Heuristics Scheduling

In the second phase contains about heuristic calculation to determine the scheduling of cutting patterns obtained from the Phase 1. This phase also considers the number of demand for each product on each request. Calculation of the heuristic scheduling is to follow the EDD (Earliest Due Date) rule. This is done in achieving one goal to minimize delays in production scheduling and delivery in order to meet consumer demand and companies are not penalized due to delay of the consumer.

Initialization:

Input the following values:

x_i = the number of hardboards that will cut by the i-th cutting pattern

α_{ik} = the number of product k produced by the i-th cutting pattern

d_{jk} = total demand of product k in order-j

Step 1

Sequence the order according to the EDD rule.

Step 2

Write the vector of demand-j using formula:

$$O_j = (d_{j1}, d_{j2}, \ldots , d_{jk}) \tag{6}$$

Step 3

Compute the weight of the element of demand-j by normalizing the value using formula:

$$w_{jk} = \frac{d_{jk}}{\max_k \{d_{jk}\}} \quad \forall k \tag{7}$$

Step 4

Compute the *matching score* between demand-j and pattern-i using formula:

$$\mathrm{MS_i} = \sum_{k=1}^{N} \alpha_{ik} w_{jk} \qquad \forall \mathrm{i} \tag{8}$$

where N = the number of product type.

Step 5

Select the best pattern θ having the highest machine value MS_i using formula:

$$\mathrm{MS}_\theta = \max_\mathrm{i} \{\mathrm{MS_i}\} \tag{9}$$

Step 6

Determine the required number of hardboards cut by pattern θ for all product *k* using formula:

$$nc_{\theta k} = \begin{cases} d_{jk} / \alpha_{\theta k} & ,\text{if } \alpha_{\theta k} > 0 \text{ and } d_{jk} > 0 \\ 0 & ,elsewhere \end{cases} \tag{10}$$

Step 7

Determine the required maximum number of hardboards cut by pattern θ from among $nc_{\theta k}$ determined in Step 6 using formula:

$$nc_{\theta k*} = \max_k \{nc_{\theta k}\} \tag{11}$$

Step 8

Determine the actual number of hardboards cut by pattern θ using formula:

$$nc_{\theta k*} = \begin{cases} x_\theta & ,\text{if } nc_{\theta k*} > x_\theta , \\ & \text{set } x_\theta \leftarrow 0 \\ nc_{\theta k*} & ,\text{if } x_\theta \geq nc_{\theta k*} , \\ & \text{set } x_\theta \leftarrow x_\theta - nc_{\theta k*} \end{cases} \tag{12}$$

Step 9

Update the vector of demand-*j* using formula:

$$ns_k \leftarrow \alpha_{\theta k} nc_{\theta k*} + ns_k \tag{13}$$

$$d_{jk} \leftarrow d_{jk} - ns_k \qquad \forall k \tag{14}$$

where ns_k = the number of product *k* in stock. Update the vector of pattern θ by:

$$\alpha_{\theta k} \leftarrow \begin{cases} 0 & ,\text{if } x_\theta = 0 \\ \alpha_{\theta k} & ,elsewhere \end{cases} \tag{15}$$

If $d_{jk} \leq 0$, for all product *k*, go to Step 2, and read the next demand-*j*.
If not, go to Step 4, and select another pattern, while renormalize the element of vector demand-*j* (Step 3).

Step 10

Repeat Step 2 – 9 until all order vectors are exhausted.

3. Algorithm Implementation

The proposed two-stage algorithm is implemented in a medium-size furniture manufacturer located in Jakarta Indonesia, named XYZ Company. XYZ Co. is a manufacturing company engaged in timber production. Products produced by the XYZ Co. are wooden sticks that have the length, width and thickness in accordance with consumer demand. Usually the consumer will change the wood products into other products are ready for use. Therefore, XYZ Co. maybe said as a company that produces semi-finished goods for consumers, which will be reprocessed into finished goods.

Raw material used by this companyy is the raw material sheet of wood beams. In this study, the size of the raw materials used in the hardboards has a length of 2000 mm, width 200 mm and 90 mm thick. Whereas the capacity availability of raw material of hardboards sent by the supplier is about 350 m^3 which is equivalent to 9722 units of the hardboards. This study used 5 products derived from the reservation that are product A, B, C, D and E.

Here is a consumer demand for each product that is written in Table 1.

Table 1. Consumers' demand for each product (in mm^3)

Product Type	Demand 1	Demand 2	Demand 3	Total Demand
A	1062	892	937	2891
B	904	867	880	2651
C	672	604	0	1276
D	28	3	33	64
E	43	64	7	114

Where the due date demand 1 <due date demand 2 <due date demand 3.

Variety of combination is obtained by using the help of Visual Basic 6.0 software to produce 13 combinations as shown in Table 2.

Table 2. Variety of combination of cutting

No.	Combination Type	A	B	C	D	E	Remaining Volum (mm3)	Code
1	BA	6	4	0	0	0	5350040	LL
2	AE	6	0	0	0	2	5393336	LL
3	BB	0	8	0	0	0	9445120	LL
4	EB	0	4	0	0	2	9488416	LL
5	E	0	0	0	0	4	9531712	PP
6	CA	6	0	2	0	0	9787880	LL
7	AC	2	0	4	0	0	12529960	PP
8	CB	0	4	2	0	0	13882960	LL
9	EC	0	0	2	0	2	13926256	LL
10	BA	2	4	0	0	0	16931720	PP
11	D	0	0	0	2	0	18092880	LL
12	CC	0	0	4	0	0	18320800	LL
13	AAA	6	0	0	0	0	18627480	PP

Phase 1 Integer Linear Programming Model

Model 1

Using the data shown on Table 1 and Table 2 and considering the avalable capacity, then Model 1 become:

Objective function :

Minimizei $Z = 5350040\,x_1 + 5393336\,x_2 + 9445120\,x_3 + 9488416\,x_4 + 9531712\,x_5 + 9787880\,x_6 + 12529960\,x_7 + 13882960\,x_8 + 13926256\,x_9 + 16931720\,x_{10} + 18092880\,x_{11} + 18320800\,x_{12} + 18627480\,x_{13}$

Subject to:

$$6x_1 + 6x_2 + 6x_6 + 2x_7 + 2x_{10} + 2x_{13} \geq 2891$$
$$4x_1 + 8x_3 + 4x_4 + 4x_8 + 4x_{10} \geq 2651$$
$$2x_6 + 4x_7 + 2x_8 + 2x_9 + 4x_{12} \geq 1276$$
$$2x_{11} \geq 64$$
$$2x_2 + 2x_4 + 4x_5 + 2x_9 \geq 114$$
$$x_1 + x_2 + x_3 + x_4 + x_5 + x_6 + x_7 + x_8 + x_9 + x_{10} + x_{11} + x_{12} + x_{13} \leq 9722$$
$$x_1, x_2, x_3, x_4, x_5, x_6, x_7, x_8, x_9, x_{10}, x_{11}, x_{12}, x_{13} \geq 0 \text{ dan integer}$$

<u>where :</u>

Z = waste

X_i = the number of hardboards that will be cut by cutting pattern i ($i = 1, 2, \ldots, 13$)

By Using software WinQSB, we get the results as follows:

$x_1 = 375$	$x_8 = 0$	$\Sigma x = 899$ units
$x_2 = 1$	$x_9 = 0$	
$x_3 = 142$	$x_{10} = 0$	
$x_4 = 4$	$x_{11} = 32$	
$x_5 = 26$	$x_{12} = 0$	
$x_6 = 0$	$x_{13} = 0$	
$x_7 = 319$	$Z = 8214672000\ mm^3$	

Model 2

Using the data shown on Table 1 and Table 2 and considering the avalable capacity, then Model 2 become:

Objective function :

$$\text{Minimize } Z = x_1 + x_2 + x_3 + x_4 + x_5 + x_6 + x_7 + x_8 + x_9 + x_{10} + x_{11} + x_{12} + x_{13}$$

Subject to :

$$6x_1 + 6x_2 + + 6x_6 + 2x_7 + 2x_{10} + 2x_{13} \geq 2891$$
$$4x_1 + 8x_3 + 4x_4 + 4x_8 + 4x_{10} \geq 2651$$
$$2x_6 + 4x_7 + 2x_8 + 2x_9 + 4x_{12} \geq 1276$$
$$2x_{11} \geq 64$$
$$2x_2 + 2x_4 + 4x_5 + 2x_9 \geq 114$$
$$x_1 + x_2 + x_3 + x_4 + x_5 + x_6 + x_7 + x_8 + x_9 + x_{10} + x_{11} + x_{12} + x_{13} \leq 9722$$
$$x_1, x_2, x_3, x_4, x_5, x_6, x_7, x_8, x_9, x_{10}, x_{11}, x_{12}, x_{13} \geq 0 \text{ dan integer}$$

<u>where :</u>

Z = waste

X_i = the number of hardboards that will be cut by cutting pattern i ($i = 1, 2, \ldots, 13$)

By Using software WinQSB, we get the results as follows:

$x_1 = 376$	$x_8 = 0$
$x_2 = 0$	$x_9 = 0$
$x_3 = 143$	$x_{10} = 0$
$x_4 = 1$	$x_{11} = 32$
$x_5 = 28$	$x_{12} = 0$
$x_6 = 0$	$x_{13} = 0$
$x_7 = 319$	$Z = 899$ units

Phase 2: Heuristics Scheduling

Heuristics Scheduling of Model 1

i = cutting pattern i-th (1, 2, 3, ..., 13)

j = demad j-th (1, 2, 3)

k = product type (A, B, C, D, E)

x_i = the number of hardboards that will be cut by cutting pattern i-th

α_{ik} = the number of product type k produxced by cutting pattern i-th

d_{jk} = the total demand of product type k in demand j-th

Cutting scheduling to minimize waste shown on the following table.

Table 3 Cutting scheduling to minimize waste

Scheduling No.	Cutting pattern (i)	The number of hardboards that will be cut by cutting pattern *i*-th (x_i)	Total demand
1	1	226	1
2	7	168	1
3	5	11	1
4	11	14	1
5	3	109	2
6	7	151	2
7	5	15	2
8	11	2	2
9	4	2	2
10	1	149	3
11	3	33	3
12	11	16	3
13	4	2	3
14	2	1	3

The number of hardboards used = 375 + 1 + 142 + 4 + 26 + 0 + 319 + 0 + 0 + 0 + 32 + 0 + 0 = 899 units
Total waste = 8214672952 mm^3
The percentage of total waste of all the sheets beam used is

$$= \frac{\text{Total Waste}}{\sum \text{total sheet beam used x Volum of the sheet beam}} \text{ x } 100\% = 25.38\%$$

Heuristics Scheduling of Model 2
Cutting acheduling to minimize the number of hardboards shown on the following table.

Table 4 Scheduling cutting patterns for minimizing beam sheets

Scheduling No.	Cutting pattern (i)	The number of hardboards that will be cut by cutting pattern *i*-th (x_i)	Total demand
1	1	226	1
2	7	168	1
3	5	11	1
4	11	14	1
5	3	109	2
6	7	151	2
7	5	16	2
8	11	2	2
9	1	150	3
10	3	34	3
11	11	16	3
12	4	1	3
13	5	1	3

The number of hardboards used =376 + 0 + 143 + 1 + 28 + 0 + 319 + 0 + 0 + 0 + 32 + 0 + 0
= 899 unit beam sheet
Total waste = 8214672952 mm^3
The percentage of total waste of all the sheets beam used is

$$= \frac{\text{Total Waste}}{\sum \text{total sheet beam used x Volum of the sheet beam}} \text{ x } 100\% = 25.38\%$$

From these two model formulations obtained that the total waste of raw materials are the same that is 8214672952 mm^3, and the percentage of total waste of all the sheets beam used is is equal to 25.38%.

Table 5 Comparison results of Model 1 and Model 2

Model	Total number of hardboards used (units)
Model 1	899
Model 2	899

The difference results of both calculations minimizing waste and minimization the hardboards occurs on the order of scheduling based on the EDD, which model of minimize waste obtained 14 cutting pattern scheduling sequence and the model of minimize the total number of hardboards used obtained 13 cutting pattern scheduling sequence. Because the results of both formulations are the same, then the decision to apply one of the two model formulations and cutting scheduling handed to the company.

4. Conclusions and Recommendations

In this study, a two-stage algorithm is developed for the cutting sequencing problem of a furniture manufacturer located in Jakarta. In the first stage, an integer linear programming model is used to determine the number of hardboards that will be cut by each pattern type. The model is tested with two different objective functions for comparison. The cutting patterns, generated by a special software program, are among the inputs of this model. The second stage consists of a heuristic that decides on the sequencing of the cutting patterns that are obtained in stage 1 by considering the due date of the orders, and taking into account the demands of the items. The implementation results were discussed with the planners working in the furniture manufacturer, and they indicated that the proposed algorithm could be employed as a useful tool for the cutting sequencing problem of the company.

5. References

[1] Haessler R W and Sweeney P E 1991 Cutting stock problems and solution procedures. *European Journal of Operational Research* **54**(2) pp 141-50
[2] Yuen B J and Richardson K V 1995 Establishing the optimality of sequencing heuristics for cutting stock problems. *European Journal of Operational Research* **84** pp 590-8
[3] Foronda S U and Carino H F 1991 A heuristic approach to the lumber allocation problem in hardwood dimension and furniture manufacturing. *European Journal of Operational Research* **54** pp 151-62
[4] Yanasse H H 1991 Two-dimensional cutting stock with multiple stock sizes. *Journal of the Operational Research Society* **42**(8) pp 673-83
[5] Carnieri C, Mendoza G A and Gavinho L G 1994 Solution procedures for cutting lumber into furniture parts. *European Journal of Operational Research* **73**(3) pp 495-501
[6] Morabito R and Garcia V 1998 The cutting stock problem in a hardboard industry: a case study. *Computers & Operations Research* **25**(6) pp 469-85
[7] Yazgaç T and Rifat G O 2004 A Cutting Sequencing Approach To Modular Manufacturing. *Journal of Manufacturing Technology Management* **15**(1) 20–28

Optimum Container Network Route in Papua Region

Christine Natalia[1*], Chendrasari Wahyu Oktavia[1], Gabriela Eirene[1]

[1] Department of Industrial Engineering, Faculty of Engineering Atma Jaya Catholic University of Indonesia
*chrisnatalia@atmajaya.ac.id

Abstract. Indonesia is an archipelagic country which around 77% of its area is covered by water. Thus, developing good logistic system is needed especially in water-based route to distribute goods to the entire country. To develop national logistic system, government establishes Sea Highyway concept. This new concept is expected to lower the logistic cost for Eastern side of Indonesia especially Papua. The existing routing system has not yet established to reach all the seaport at Papua. This issue will still force air-based and land-based distribution system to take over most of its distribution activity. The main discussion of this research is to determine the most optimal routing system for container shipment in Papua. The determination of optimal routing system applies Integer Linear Programming for its approach. The scope of this research would be the container shipment originated from Tanjung Perak and Makassar. The seaports taken into account for this research consist of 3 hub ports and 7 feeder ports. Based on the calculation, 54 alternatives routing are determined and 22 of these are the most optimal routing system for container shipment in Papua.
Keywords: Sea highway, Container Routing System Development, Integer Linear Programming.

1.Introduction

Indonesia is an archipelago with 17.000 islands. Total area of Indonesia approximately reach 7,9 million km sq with 1,8 million km sq for lands, 3,2 million km sq for territorial sea, and 2,9 million km sq for Exclusive Economic Zone. According to the facts above, 77% of Indonesia is water or three times larger than lands [1]. By 2016, Logistic Performance Index (LPI) show that Indonesia has dropped from rank 53 to rank 63. Indonesian Institutes of Sciences (LIPI) states that Indonesia is one of country that has the highest logistic cost in ASEAN. It costs 25 to 30% from Gross Domestic Product [2]. Current Indonesian logistic system has not yet efficient because of the high shipping cost. It causes price disparity and development gap between Western side and Eastern side of Indonesia. Government takes effort to give balance for the mentioned issue with the establishment of a programme called Tol-Laut. The idea of Tol-Laut is not to build a highway above the sea, but to increase sea connectivity by routinely order ships to travel from Western side to Eastern side of Indonesia [3]. This developing concept is expected to strengthen shipping route for Eastern side of Indonesia [4]. Currently, container domestic trading stream is still focused on Western side of Indonesia as portrayed below.

According to the Government Regulation Chapter 6 No. 20 Year 2010, sea transportation activities are commenced constantly and regularly, also able to be supplemented with incidental route. The domestic sea transportation activities performed in network route [5]. Currently, government has developed shipment network route through Tol-Laut. Unfortunately, there are many seaports in Papua excluded from the network. The container shipment to Eastern side of Indonesia mostly done from Tanjung Perak port and Makassar port. Tol-Laut programme that supports container shipment to Papua are figured below by the red line.

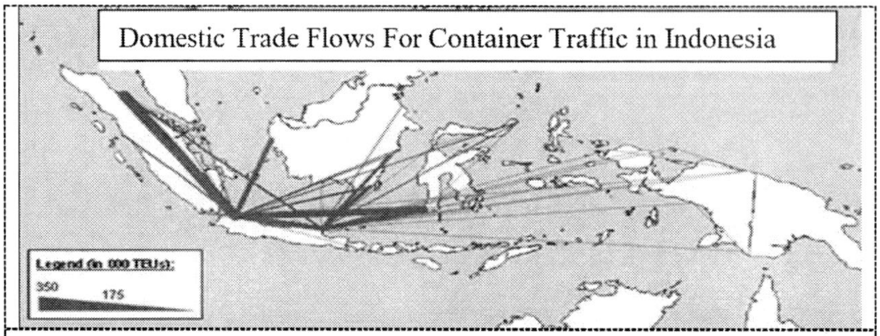

Figure 1. Overview of Trade Flows for Container Traffic in Indonesia.

(Source: Development of Sea highway in IDRJMN 2015-2019 and Implementation 2015)

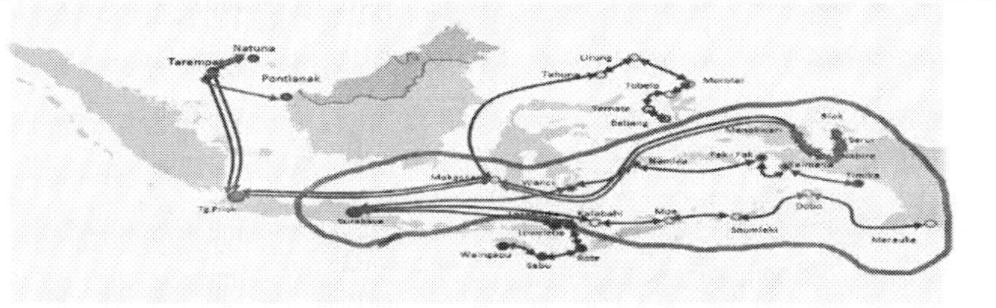

Figure 2. Mapping of Sea highway Route in Papua.

(Source: Director General of Sea Transportation – Ministry of Transportation)

This research is different from any other existing route determination research. In example, [6] did route determination for medical help distribution. [7] did route determination for employee-serving bus. [3] did container route determination using hub and spoke system. [8] constructed sea transportation routes in Sumatra with Sumatran Ring. All of the research mentioned above was done for the same objectives, it is route determination. However, the object of this research differs from them. Little number of research discussing this topic is the main reason for this research. The objective is to develop optimal container shipment routes from 2 major seaports in Indonesia (Tanjung Perak port and Makassar port) to hub ports and feeder ports in Papua. By the result, it is expected to minimize the logistic cost so the commodity price in Papua would lower.

2. Methods

Structured and systematic flow of research should be conducted to gain a more comprehensive result. Collect data is the first step, followed by the formulation of mathematic model and constraints. Choices of alternative routes determination done after that and will be continued to the calculation of total shipping cost. Determination of optimal shipping route will be the last step.

2.1 Data Collection

Source of data for this research comes from primary and secondary data. Primary data collected by conducting interviews and will be a qualitative data. Existing applied sea routes and container shipment system are one of the questions to gain deeper knowledge about current condition. Meanwhile, secondary data is gathered from Ministry of Transportation, Research Center of Ministry of Transporation, official website of PT. Pelabuhan Indonesia, and application to determine distance among ports. Example of secondary data are goods distribution data through seaport in year 2016, seaport cost, distance among ports and seaports name. Table 1 shows the list of seaports taken into account in this research. The location of port members shown by picture 3.

Table 1. The Posts Member in Case Study.

Major Port		Hub Port in Papua		Feeder Port in Papua	
Name	Notation	Name	Notation	Name	Notation
Tanjung Perak	A1	Sorong	B1	Fak-Fak	C1
Makassar	A2	Jayapura	B2	Kaimana	C2
		Merauke	B3	Manokwari	C3
				Biak	C4
				Amamapare	C5
				Nabire	C6
				Serui	C7

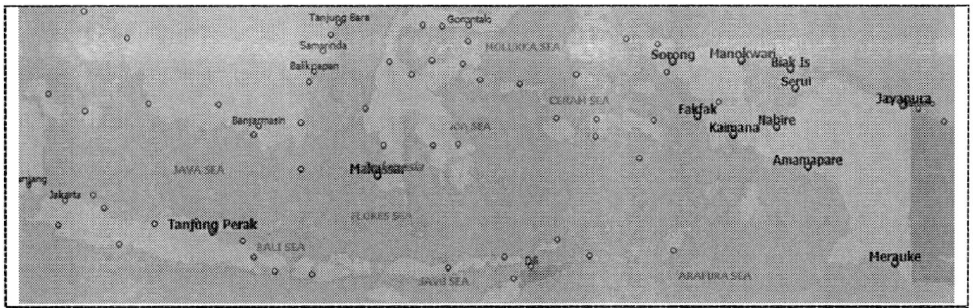

Figure 3. Port Members Location.

2.1.1 Mathematical Model and Constraints Formulation

To process data, adjusted mathematical model is constructed. It is the container shipment system in Indonesia. Next, constraints are constructed to support the mathematical model.

2.1.2 Route Alternatives Determination

Once the mathematical model and constraints has fully constructed, then route alternatives are determined. The determination is based on container shipment flow data and load capacity of the ship especially for the shipping from hub port to the feeder port and also from feeder port to the feeder port. Overload is not allowed, so if one ship has already reached its load limit, then the ship will travel back from the last feeder port to the hub port. Type of ship taken into account for this research is 1000 TEU ship traveling from major port to hub port and 700 TEU ship travelling from hub port to feeder port. Moreover, route alternatives determination also take concern to constraints.

IOP Conf. Series: Materials Science and Engineering **528** (2019) 012040 doi:10.1088/1757-899X/528/1/012040

2.1.3 Calculation of Total Shipping Cost

From the route alternatives, then total shipping cost calculation can be done. Total shipping cost consist of cost components. Hence, calculating all component of the cost should be done first. They are ship operational cost, travelling cost, port cost, main route cost, and hub route cost.

2.1.4 Optimal Shipping Route Decision

This step followed after the previous step has cleared. Determination of optimal shipping route is decided by comparing total cost from each of the alternatives. The chosen route is to have the lowest cost of total shipping cost.

2.2 Mathematical Framework

Optimal shipping route determination would be conducted in this research. The model used in this research developed from [9] to determine objective function and constraints in data processing.

$$St_m = DIS_m / Sp^V \times 24 \quad (1)$$

$$St_f = DIS_f / Sp^v \times 24 \quad (2)$$

Equation (1) explains the function of travelling time from major port to hub port, and equation (2) expresses the function of travelling time from hub port to feeder port and also from one feeder port to another feeder port. This calculation is influenced by distance between ports and speed of the ship.

St_m : Mainline streaming time (day).
DIS_m : The distance between major port and hub port (Nautical Mile).
Sp^V : The speed of ship 1000 TEU (knots).
St_f : Feeder streaming time (day).
DIS_f : The distance between hub port and feeder (Nautical Mile).
Sp^v : The speed of ship 700 TEU (knots).

$$CV_T^V = CV^V \times L \times St_m \quad (3)$$

$$CV_T^v = CV^v \times L \times St_f \quad (4)$$

Equation (3) expresses the function to determine total operational cost applied to the ship travelling from major port to hub port while equation (4) explains the total operational cost for the ship travelling from hub port to feeder port and among other feeder port. This calculation to determine operational costs is influenced by the amount of load and streaming time.

CV_T^V : Total vessel daily operating costs for ship 1000 TEU (IDR)
CV_T^v : Total vessel daily operating cost for ship 700 TEU (IDR)
CV^V : Vessel daily operating cost for ship 1000 TEU (IDR/TEU/day).
CV^v : Vessel daily operating cost for ship 700 TEU (IDR/TEU/day).
L : Load (TEU).
St_m : Mainline streaming time (day).
St_f : Feeder streaming time (day).

$$SC_m = CV_T^V \times St_m \quad (5)$$

$$SC_f = CV_T^v \times St_f \quad (6)$$

Equation (5) is expressing cost expenditure to travel from major port to hub port. Equation (6) is the function to determine cost expenditure to travel from hub port to feeder port and among other feeder port. This calculation to determine sea cost is influenced by the length of time to sail

SC_m : Sailing cost from major port to hub port (IDR/day)
SC_f : Sailing cost from hub port to feeder port or feeder port to feeder port (IDR/day)
CV_T^V : Total vessel daily operating costs for ship 1000 TEU (IDR).
CV_T^v : Total vessel daily operating cost for ship 700 TEU (IDR).
St_m : Mainline streaming time (day).
St_f : Feeder streaming time (day).

$$PC = (PC_a \times L) + (PC_t \times L) \quad (7)$$

IOP Conf. Series: Materials Science and Engineering **528** (2019) 012040 doi:10.1088/1757-899X/528/1/012040

Equation (7) is the function describing the cost of loading activity in origin port and unloading activity in destination port. This calculation to determine port cost is influenced by origin port cost, destination port cost and the amount of load.

PC : Port Cost (IDR)
PC_a : Origin port cost (IDR/TEU)
PC_t : Destination port cost (IDR/TEU)
L : Load (TEU).

$$MC = SC_m + PC \quad (8)$$

Equation (8) expresses the function of container shipping cost from major port to hub port.

MC : Mainline costs from major port to hub port (IDR).
SC_m : Sailing cost from major port to hub port (IDR/day).
PC : Port Cost (IDR).

$$FC = SC_f + PC \quad (9)$$

Equation (9) is the function to determine container shipping cost from hub port to feeder port and also among other feeder port.

FC : Feeder costs from hub port to feeder port or feeder port to feeder port (IDR).
SC_f : Sailing cost from hub port to feeder port or feeder port to feeder port (IDR/day)
PC : Port cost (IDR)

$$TC = MC + \sum FC \quad (10)$$

Equation (10) is the function to calculate the shipping cost for one route consisting of shipment from major port to hub port, hub port to feeder port and also among other feeder port.

TC : Total costs (IDR).

$$\sum FC = Total\ Hub\ Route\ Cost$$

Minimum:

$$\sum_{a=1}^{N}\sum_{b=1}^{N}\sum_{c=1}^{N} TC_{abc}\, X_{abc}$$

Constraints

- Ship will be allowed to travel to each feeder port only once for one container shipping route.

$$\sum_{c=1}^{N} c = 1$$

- Overload is not allowed when shipping container from major port to hub port in one shipping route.

$$\sum_{b=1}^{N} di \sum_{a=1}^{N} X_{abc} \leq L$$

- Overload is not allowed when shipping container from hub port to feeder port in one shipping route.

$$\sum_{c=1}^{N} di \sum_{b=1}^{N} X_{abc} \leq L$$

- Ship from the last feeder port will travel back to hub port if load has reached its limit.

$$\sum_{c=1}^{N} X_{abc} \sum_{b=1}^{N} X_{abc} \geq L$$

- Binary number will be used for decision variable.

$$Y_k \in \{0,1\}$$

Xabc = 1, if travelling the route
Xabc = 0, if not.

Whereas:
N_a : The set of major ports a in the total ports.
N_b : The set of hub ports b in the total ports.
N_c : The set of feeder ports c in the total ports.
TC_{abc} : Total costs from major port a via hub port b to feeder port c (IDR).
X_{abc} : Flow fraction from major port a via hub port b to feeder port c.

3. Analysis and Results

In this research the optimal container shipping route will be designed based on the lowest minum total shipping cost. There are variable costs required for the total amount of shipping costs. Variables-variables cost are operating costs that according with needs, ports costs, major costs, port cost, origin post cost, destination port costs, feeder costs, and sailing costs. In this research, the development of the model from previous research is adding the charge to the mathematical equation in determining operational costs, and ports costs. For determination operating costs obtained by the ship to operate when travelling from one port to another. Operating costs are adjusted to the amount of cargo carried by the ship and also the length of the time to sail. Thus, the more cargo carried and also the longer the time to sail, it will affect the amount of operational costs of the ship needed. Whereas to determine port costs is the handling at the port both from the origin port and destination port. The intended container handling costs is the container loading costs at the port of origin when it wants to ship and the costs of container loading when arriving at the destination port. The results of the calculation port costs are influenced by the amount the cargo loaded and unloaded. To determine the optimum delivery route, alternative routes are first made. Alternatives routes are made based on the flow of container shipping and adjusted to the ships payload capacity. The load cannot be greater than the capacity of the ship. To conduct calculation, alternative routes are made and resulting in 54 route alternatives. Next, total shipping cost is calculated for each of the alternative route. Table below is the recapitulation of total shipping cost for each of the alternative.

Table 2. Recapitulation of Total Shipping Cost.

No	Alternative	Total Delivery Cost (IDR)	No	Alternative	Total Delivery Cost (IDR)
1	a_1-b_1-c_3-c_4-c_6-b_1	2.411.456.146	28	a_1-b_3-c_1-c_2-c_5-b_3	2.122.577.753
2	a_1-b_1-c_3-c_6-b_1	2.481.038.147	29	a_1-b_3-c_5-c_2-c_1-b_3	1.589.406.565
3	a_1-b_1-c_3-c_7-b_1	2.358.813.160	30	a_2-b_3-c_1-c_2-c_5-b_3	1.729.792.795
4	a_1-b_1-c_4-c_6-b_1	2.225.711.531	31	a_2-b_1-c_3-c_4-c_6-b_1	2.316.958.269
5	a_1-b_1-c_4-c_7-b_1	2.396.880.592	32	a_2-b_1-c_3-c_6-b_1	2.386.941.456
6	a_1-b_1-c_4-c_3-b_1	2.398.882.193	33	a_2-b_1-c_3-c_7-b_1	2.264.667.990
7	a_1-b_1-c_6-c_7-c_4 - b_1	1.873.387.057	34	a_2-b_1-c_4-c_6-b_1	2.131.593.609
8	a_1-b_1-c_6-c_4 - b_1	1.875.574.321	35	a_2-b_1-c_4-c_7-b_1	2.302.771.348
9	a_1-b_1-c_6-c_3 - b_1	2.748.544.172	36	a_2-b_1-c_4-c_3-b_1	2.304.694.380
10	a_1-b_1-c_7-c_6 - b_1	2.184.585.290	37	a_2-b_1-c_6-c_7-b_1	2.088.847.707
11	a_1-b_1-c_7-c_4 - b_1	2.134.888.091	38	a_2-b_1-c_6-c_4-b_1	2.092.909.885
12	a_1-b_1-c_7-c_3 - b_1	2.769.819.317	39	a_2-b_1-c_6-c_3-b_1	2.654.486.651
13	a_1-b_2-c_3-c_4-c_6-b_2	3.176.158.589	40	a_2-b_1-c_7-c_6-b_1	2.090.599.824
14	a_1-b_2-c_3-c_7-c_6-b_2	3.218.375.952	41	a_2-b_1-c_7-c_4-b_1	2.090.599.824

15	a_1-b_2-c_3-c_6-b_2	3.246.137.996	42	a_2-b_1-c_7-c_3-b_1	2.675.809.832
16	a_1-b_2-c_4-c_3-b_2	2.294.519.616	43	a_2-b_2-c_3-c_4-c_6-b_2	3.282.627.437
17	a_1-b_2-c_6-c_4-b_2	3.040.769.153	44	a_2-b_2-c_3-c_7-c_6-b_2	3.324.842.075
18	a_1-b_2-c_3-c_4-b_2	3.246.137.996	45	a_2-b_2-c_3-c_6-b_2	3.352.610.624
19	a_1-b_3-c_5-c_7-b_3	1.785.601.880	46	a_2-b_2-c_4-c_3-b_2	2.822.099.299
20	a_1-b_3-c_6-c_7-b_3	3.246.137.996	47	a_2-b_2-c_6-c_4-b_2	3.147.176.432
21	a_1-b_3-c_5-c_7-c_6-b_3	1.809.697.629	48	a_2-b_2-c_3-c_4-b_2	2.801.388.368
22	a_1-b_3-c_7-c_6-b_3	1.384.276.709	49	a_2-b_3-c_5-c_7-b_3	1.481.517.930
23	a_1-b_3-c_7-c_5-b_3	1.728.215.059	50	a_2-b_3-c_6-c_7-b_3	1.370.495.162
24	a_1-b_3-c_6-c_5-b_3	1.988.174.815	51	a_2-b_3-c_5-c_7-c_6-b_3	1.504.857.488
25	a_1-c_5-c_2-c_1-a_1	1.380.133.681	52	a_2-b_3-c_7-c_6-b_3	1.079.935.980
26	a_1-c_1-c_2-c_5-a_1	1.354.741.781	53	a_2-b_3-c_7-c_5-b_3	1.424.685.450
27	a_1-b_3-c_5-c_2-c_1-b_3	1982.191.522	54	a_2-b_3-c_6-c_5-b_3	1.683.942.195

Optimal route determination is decided by comparing the total shipping cost of alternative that share the same route. The chosen route will be the route that yield lowest total shipping cost. After comparing all the total shipping cost, 22 container shipment routes synergized with container shipping stream is obtained. Table 3 is the recapitulation of the optimal container shipment route.

Table 3. Recapitulation of the optimum container network in Papua.

Route 1	Alternative 7	Tanjung Perak-Sorong-Nabire-Serui-Biak-Sorong	Route 12	Alternative 39	Makassar-Sorong-Nabire-Manokwari-Sorong
Route 2	Alternative 8	Tanjung Perak-Sorong-Nabire-Biak-Sorong	Route 13	Alternative 41	Makassar-Sorong-Serui-Biak-Sorong
Route 3	Alternative 14	Tanjung Perak-Jayapura-Manokwari-Serui-Nabire	Route 14	Alternative 42	Makassar-Sorong-Serui-Manokwari-Sorong
Route 4	Alternative 15	Tanjung Perak-Jayapura-Manokwari- Nabire-Jayapura	Route 15	Alternative 49	Makassar-Merauke-Amamapare-Serui-Merauke
Route 5	Alternative 16	Tanjung Perak-Jayapura-Biak- Manokwari-Jayapura	Route 16	Alternative 50	Makassar-Merauke-Nabire-Serui-Merauke
Route 6	Alternative 18	Tanjung Perak- Jayapura-Manokwari-Biak-Jayapura	Route 17	Alternative 51	Makassar-Merauke-Amamapare-Serui-Nabire
Route 7	Alternative 31	Makassar-Sorong-Manokwari-Biak-Jayapura	Route 18	Alternative 52	Makassar-Merauke-Serui-Nabire-Merauke
Route 8	Alternative 32	Makassar-Sorong-Manokwari-Nabire-Sorong	Route 19	Alternative 54	Makassar-Merauke-Serui-Amamapare-Nabire
Route 9	Alternative 33	Makassar-Sorong-Manokwari-Serui-Sorong	Route 20	Alternative 51	Makassar-Merauke-Nabire-

					Amamapare-Merauke
Route 10	Alternative 34	Makassar-Sorong-Biak-Nabire-Sorong	Route 21	Alternative 25	Tanjung Perak-Amamapare-Kaimana-Fak-Fak-Tanjung Perak
Route 11	Alternative 35	Makassar-Sorong-Biak-Serui-Sorong	Route 22	Alternative 26	Tanjung Perak-Fak-Fak-Kaimana-Amamapare

4. Conclusions

1. The total shipping costs in the model are affected by mainline costs, feeder costs and port cost. Port cost for each route can be different because they are affected by ship' load.
2. Based on the flow of container shipping and adjusted to the ship's payload capacity, it is obtained 54 alternatives routes.
3. Based on the minimum total shipping cost and also the shortest shipping route, 22 optimum container shipping routes are selected from these 54 alternatives.

Suggestions for the development of this research in the future are container shipments added from Tanjung Priuk Port and Belawan Port

5. Reference

[1] Suhaidi 2004 *Protection of The Marine Environment from Pollution Sourced by Ship: The Consequences of Implementation on International Shipping Rights Through Indonesian Waters.* (Jakarta: Pustaka Bangsa Press)

[2] Abbas, A. 2012 National Logistics System Not Efficient, [online] (http://lipi.go.id/berita/single/Sistem-Logistik-Nasional-Belum-Efisien, accessed on 8-03-2012)

[3] Natalia, C. and Agus, M.A 2016 Shipping Route Design with Hub and Spoke System (Case Study: Papua Region – Indonesia). *Jurnal Metris* **17** 113-122

[4] Prihartono, Irawan, Bastian, and Setyanto 2015 *Sea Tol Implementation Concept* 2015-2019. (Jakarta: Ministry of National Development Planning, and Ministry of Transportation Republic of Indonesia)

[5] Mappangara, A. S. C., Idrus, M., and Asri, S 2012 Study of National Naval Route Network for Container Cargo in Support of National Connectivity. *Proc. National Conf.On Civil Engineering* (UMS: Central Java) pp 71-80

[6] Prathama, W. A 2016 *Determination of Medical Aid Distribution Routes for Mount Merapi Eruptions in Yogyakarta.* Undergraduate theses. Gadjah Mada University: Yogyakarta.

[7] Arifah, Z 2012 *Determination of Bus Employees by Integer Linear Programming.* Undergraduate theses. Agricultural Institute: Bogor

[8] Yandeza, R. and Razak, R 2016 Design of Marine Transportation Routes for Sumatran Ring with Operational and Simulated Research Approaches. *Proc. National Conf.on Innovation and Technology Applications in The Industry* (ITN Malang: East Java) pp 50-54

[9] Boontaveeyuwat, P. and Hanaoka, S. 2010 Analysing the Optimal Location of a Hub Port in Southeast Asia. *International Journal of Logistic Systems and Management*, **6(4)** 458-75.

Genetic Algorithm for Waste Transportation Route in Eastern Bandung (Case Study: PD. Kebersihan Kota Bandung)

Y Yogaswara[1] and L S M D Saputra

Department of Industrial Engineering, Faculty of Engineering Universitas Pasundan, Bandung, Indonesia
[1]yogiyoga@unpas.ac.id, larasshintamega@yahoo.com

Abstract. VRP highly close related with distance and cost minimization. PD. Kebersihan Kota Bandung has been facing one of the problem related with vehicle routing since this company has not have a fixed route for waste transportation because of their limited vehicle used and it will cost them a lot more if they have to add more vehicle to handle this. Therefore, research on VRP for this case has been done especially in West Bandung Region to find a better solution to reduce total distances and cost in which the company have to pay out. Method used to solve this problem is Genetic Algorithm with Clarke and Wright Saving used first to find initial solution. Genetic Algorithm is problem independent, it means that it can solve a lot of problems in a different field and still have probability to keep continuing to find the best solution including global search. Based on researh that has been done, there are three route with three trucks created with total volume for each vehicle on those route which are 7,053 m^3, 6,753 m^3, and 6,903 m^3 with total distance 65,33 km.
Keywords: Waste transportation, VRP, Clarke and Wright Saving, Genetic Algorithm

1. Introduction

PD. Kebersihan Kota Bandung has been facing one of the problem related with vehicle routing. This company has not have a fixed route for waste transportation because of their limited vehicle used and it will cost them a lot more if they have to add more vehicle to handle this. PD. Kebersihan Bandung for East Bandung region scheduled for pick up a day per week for each TPS. However, the accumulation of trash in a few TPS in most case has reach more than the capacity in only two days. This caused the company to pick it up more than a day per week. There is often that the vehicle which is dump truck with 6 m^3 capacity do the loading right after they unload the trash in SPA without going back to the pool. The company have a lot of consideration to add more vehicle considering for additional cost on maintenance and crew.

VRP (Vehicle Routing Problem) close related with distance and cost minimization. VRP has been applied to a lot of problem related with transportation and logistics. For example, school bus route, trashbin pick up route, retail distribution, expedition service, etc. The shorter the distance is gone through by the dump truck, the bigger the cost which can be reduce therefore operational cost each day can be more thrift. With this model, determining a route for waste transportation vehicle can be more closer to its real situation and a better solution can be proposed to reduce unnecesary cost.

Genetic algorithm is included on a group of metaheuristic problem solving method that could be able to solve a different type of optimization problem such as nonlinier programming, knapsack, TSP, VRP, flowshop sequence, jobshop schedulling, to facility layout design [1]. The data is processed with Genetic Algorithm with initial solution is solve first by heuristic method which is Clarke and Wright. After tour and total distance has been found, a number of population is created to be optimized by Genetic Algorithm [1].

The data used from waste transportation trucks survey in April to May 2018. The data survey was completed with 22 TPS (temporary waste area) in Eastern Bandung.

Based on the background of the problem which has been explained, therefore question research are how is the better waste transportation route based on distance and total amount of waste which is being picked up and how much waste transportation route created in Eastern Bandung?

2. Literature Review

2.1. Vehicle Routing Problem (VRP)

In 1959, Dantzig and Ramser for the first time introduce a theory about Truck Dispatching Problem which has become a basis of Vehicle Routing Problem (VRP) to define fuel delivery route from one terminal to serve a number of gas station.

VRP is one of optimization problem that is related with distribution network in which there are depot with a few vehicles with a certain capacity to serve a number of customer. VRP is not only applied in goods distribution field but also in a lot of field such as trashbin pick up, road clean up route, school bus route, salesman route, or even maintenance unit route.

The purpose of VRP is to minimize total cost on transportation by cutting total distance travelled by the vehicle and number of vehicle used by the company to serve the customer. This way, total cost that the company need to pay out can be reduced. Therefore, objective function for this problem is [1]:

$$\text{minimize } z = \sum_{i\epsilon V}^{K}\sum_{j\epsilon V}^{K} c_{ij} \sum_{k=1}^{K} x_{ijk} \tag{1}$$

constraint subject to:

$$\sum_{k=1}^{K} y_{ijk} = 1\,; i \in V \setminus \{0\} \tag{2}$$

$$\sum_{k=1}^{K} y_{0k} = K \tag{3}$$

$$\sum_{j\in V}^{K} x_{ijk} = \sum_{j\in V}^{K} x_{jik} = y_{ik}\,; \forall i \in V\,, k = 1,2,3,\ldots,K \tag{4}$$

$$\sum_{i\in V}^{K} q_i y_{ik} \leq Q_k\,; k = 1,\ldots,K \tag{5}$$

$$\sum_{i\in S}^{K}\sum_{j\in S}^{K} x_{ijk} \leq |S| - 1\,;\ \forall S \subseteq V \setminus \{0\}, |S| \geq 2, k = 1,2,\ldots,K \tag{6}$$

$$y_{ijk} \in \{0,1\}\ \forall i \in V, k = 1,\ldots,K \tag{7}$$

$$x_{ijk} \in \{0,1\}\ \forall i,j \in V, k = 1,\ldots,K \tag{8}$$

Constraint (2) define that each customer is visited once, (3) define a number of K vehicle that leave the depot, constraint (4) define that the same vehicle will visit and leave the same customer, constraint (5) to (8) define that the load for each k vehicle is not overcapacity with $|S|$ define cardinality of S don't have subroute.

2.2. Problem Solving Method for VRP

VRP can be solved with heuristic or even metaheuristic. The difference between these two methods are that heuristic method depend on the type of problem so it only can be used for some certain case. Meanwhile metaheuristic method is not depend on what kind of problem that want to be solved so it can be used for almost any type problems. A few examples on heuristic methods are Nearest Neighborhood (NN), Saving (Clarke and Wright), insertion method, Gillet-Miller algorithm, and 2-opt method. While on metaheuristic there are Genetic Algorithm (GA), Ant Colony Optimization (ACO), and Tabu Search.

2.2.1. Saving Method (Clarke and Wright)
This method first introduced by Clarke and Wright on 1964, that is why this method is also called as Clarke and Wright method. Saving method is based on saving concept which estimate cost reduction by serving two customer in sequence on the same route instead of different route [3]. This method start with a number of vehicle serving each customer individually in a separate route. There, all those separate route is clustered so that the vehicle can serve two or more customer on the same route considering maximum capacity of the vehicle.

This saving method can be modified in regards of waste transportation. Consider that a truck pool D, while i and j are two TPS's that the waste need to be transported to SPA S, as shown on Figure 1a.

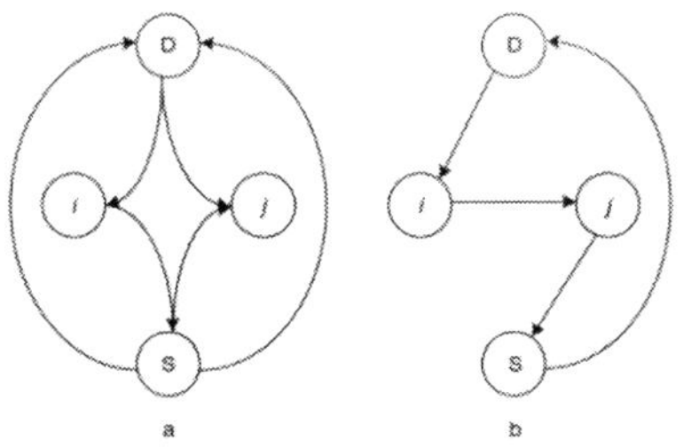

Figure 1. Modified VRP – Saving Method

If a single vehicle is being used to serve two TPS's, say i and j, on a single trip, the total distance traveled is reduced by the amount as shown on Figure 1b.

$$\begin{aligned} s(D, S) &= [d(D, i)+ d(i , S)+ d(D, j)+ d(j, S)+ 2d(j, S)]- [d(D, i)+ d(i, j)+ d(j, S)+ d(S, D) \\ &= d(i, S)+ d(D, j)+ d(S, D)- d(i, j) \end{aligned} \quad (9)$$

The difference of this modified is after collecting waste on i, the truck will not travel back to the pool instead of travel to j to collect the waste and transporting the waste to SPA and after unload the waste in SPA, the vehicle will travel back to the truck pool.

2.2.2. Genetic Algorithm (GA) Operators
Genetic algorithm is a random global search technique which used to solve problem by immitating the process on natural evolution [1]. Genetic algorithm is included on a group of metaheuristic problem solving method that could be able to solve a different type of optimization problem such as nonlinier programming, knapsack, TSP, VRP, flowshop sequence, jobshop schedulling, to facility layout design.

Data representation on genetic algorithm to solve the problem can use binary representation using number 0 and 1, permutation representation which is using integer number, and random keys representation which is using decimal number. The data is represented as a series of number that appoint individual which called chromosome.

Roulette Wheel is one of the most used method in selection process. This method implemented with roulette wheel model [4]. Like a roulette wheel model, the size applied on the wheel of each individuals in a population is based on its fitness value proportion.

In this method, there is choosen a number of position of gene randomly from the first parent, thereafter all those genes that has been choosen in the same position is become part of gene on its offspring. Meanwhile, the rest of position that is still empty on the offspring is filled with genes from the second parent sequentially from left to right [5].

Based on a comparative study about performance of a number of crossover operator in genetic algorithm to solve TSP problem which has been researched by [6], POS has become on of few method that give a good solution after Exchange Recombination Crossover (EX) and Order Crossover (OX) compare to another eight crossover operator.

Reciprocal Exchange Mutation is mutation method in which choose two position of gene randomly and then exchange the position of these two genes.

3. Methods

Research started with literature study that is continued with data collection. Distance matrix is created from existing route. The data is processed with Genetic Algorithm with initial solution is solve first by heuristic method which is Clarke and Wright. After tour and total distance has been found, a number of population is created to be optimized by Genetic Algorithm.
Each steps for Clarke and Wright methods are as mentioned below:

- List volume of demand for each node including vehicle allocated and its maximum capacity to serve the customer.
- Joint the depot/pool to each node wich will be visited.
- Create distance matrix.
- Calculate saving that has been made with $s_{ij} = c_{io} + c_{oj} - c_{ij}$, with :
- s_{ij} = value of saving,
- c_{io} = distance between node *i* to depot,
- c_{oj} = distance between depot to node *j*,
- c_{ij} = distance between node *i* to node *j*.
- List the result of saving from the bigger to the smaller value.
- Create route from a node which have a bigger saving value and the load meet with the capacity in that node.

If the total load is smaller than capacity, therefore widen the route by adding node that have a bigger saving value after the previous node. If the total load meet the capacity, a new route can be created. If all the demands has been loaded, feasible solution has been made.

In general, step by step on genetic algorithm are as mentioned below:

- **Initialization first population**, which is the process of creating a number of individual randomly with a certain chromosome composition.
- **Evaluation** is used to calculate fitness value from each chromosome. The bigger the fitness value obtained, the bigger also the probability of its chromosome to be a better solution. Fitness value usually obtained from objective function. Since VRP is a minimization problem, therefore formulation used for the fitness value = $\frac{1}{f(x)}$.
- **Selection**, the purpose of this step is to choose individual from a population which will be choosen to stay to the next generation. Probability function is used to choose individual. Individual with a better fitness value have a higher probability to be choosen [4].
- **Crossover**, the purpose of this step is to create offspring from individuals which has been choosen from the population and cross them between two parents.
- **Mutation**, the purpose of this step is to create a new individual by modifying its gene.

The step is looping until a few generation. Usually, a better solution will be shown by a constant fitness value even though iteration has been done for a few more time.

4. Result and Discussion

To keep the problem from being explained far too wide on a different field according to the purpose of this research, there are a few constraints which is applied for this problem. Those constraints are as mentioned below :

- Research focus on Eastern Bandung.
- Existing route which is being observed based on route that is used by dump truck with capacity equal to 6 m^3.

- Research only focus on pool to TPS to SPA.
- In this research, maintenance cost is not considered.

A few assumption used in this research are All vehicle used is in a good condition, the road travelled by the vehicle is not in a traffict condition, all road can be travelled by dump truck with capacity equal to 6 m^3, all dump truck do the pick up everyday and there is compaction factor for about 20% which affect on maximum capacity to 7,2 m^3 for dump truck with capacity equal to 6 m^3.

4.1. Initial Solution with Saving Method

In this case, dump truck with capacity equal to 6 m^3 will pick trash up from each TPS based on its route. Each TPS is represented with number from 1 to 22 with 0 represent the pool. The detail of each TPS is explained down below:

1	=	Metro KFC & Superindo	12	=	PT. Mepro
2	=	Pembakaran	13	=	Carrefour
3	=	Santosa Asih RW 05 & 06	14	=	Buana Soetta
4	=	Batu Karang	15	=	Guruminda
5	=	Manisi	16	=	Bebedahan
6	=	Sindang Laya RW 01	17	=	Rancameong
7	=	Cipadung Permai	18	=	Cisaranten Kidul RW 06 & 07
8	=	Cipadung RW 08	19	=	Cipamokolan RW 02
9	=	Komp. Palasari	20	=	Apartemen The Suittes
10	=	Komp. Pasundan	21	=	Kawaluyaan RW 04 dan 05
11	=	Cipagalo RW 07	22	=	Tanabe Abadi

Based on the data processing from each step on Saving method which has been mentioned on the theory overview, the result is explained on Table 1.

Table 1. Route Created with Saving Method

Dump Truck	Route	Amount of Volume Picked/Day	Distance
1	0 – 13 – 20 – 21 – 1 – 2 – 3 – 4 – SPA	7,053 m^3	25,78 km
2	0 – 12 – 17 – 15 – 19 – 18 – 14 – 16 – SPA	6,753 m^3	27.13 km
3	0 – 7 – 8 – 5 – 9 – 22 – 6 – 10 – 11 – SPA	6,903 m^3	28,44 km
Total		20,708 m^3	81,35 km

Based on the result shown on Table 1, total tour created for this case are three route in which first dump truck pick 7,053 m^3 trash with distance travelled is 25,78 km, second dump truck pick 6m753 m^3 trash with distance travelled is 27,13 km, and third dump truck pick 6,903 m^3 trash with distance travelled is 28,44 km.

4.2. Optimization with Genetic Algorithm

First step on Genetic Algorithm is determining parameter and initialize first population. Parameter which is determine are as mentioned below:

Population Size = 20
Crossoover Probability = 0,6
Mutation Probability = 0,1

First chromosome on the population is taken from the tour created with Saving method, meanwhile the rest nineteen chromosomes are determined randomly.

The next step is selection to choose which chromosome that will be choosen for crossover. Method used for this step is roulette wheel where chromosome will be choosen based on random number its get from 0 to 1. If the random number on the said chromosome is less then or equal with fitness value cumulative probabilty, therefore the said chromosome is choosen for crossover.

After selection process, choosen chromosome will be crossed over with each other. The process of choosing chromosome for crossover will be determine with random number from 0 to 1. If the random number on the said chromosome is less then or equal with crossover probability, therefore the said chromosome will be choosen for the next step.

Method used for crossover is Position-Based Crossover (POS) which is a development from Order Crossover (OX). Example on how to do crossover with this method on the first generation is as shown on Figure 2.

Parent 1	13	20	21	1	2	3	4
Offspring	13	20	21	2	1	3	4
Parent 2	13	3	2	20	21	4	1

Figure 2. Crossover in POS with The First Generation

The result of crossover process from all chromosome on the first generation is shown on Table 2.

Table 2. Offspring Created with POS Method

NO	OFFSPRING							DISTANCE
1	13	20	21	2	1	3	4	27.56
2	4	13	2	20	3	21	1	45.5
3	2	13	4	1	3	20	21	37.85
4	20	1	2	21	3	4	13	40.72
5	20	3	21	1	4	13	2	46.16

The next step after crossover is mutation. The process of choosing chromosome to applied mutation is quite similar with the other two genetic operator which is determined with random number from 0 to 1. If the random number on the said chromosome is less then or equal with mutation probability which is 0,1, therefore the said chromosome will be choosen for the next step. Example on how to determine the chromosome on the first generation is shown on Table 3.

Table 3. Random Number for Mutation Process

NO	OFFSPRING							DISTANCE	RANDOM NUMBER
1	13	20	21	2	1	3	4	27.56	0.432
2	4	13	2	20	3	21	1	45.5	0.592
3	2	13	4	1	3	20	21	37.85	0.469
4	20	1	2	21	3	4	13	40.72	0.409
5	20	3	21	1	4	13	2	46.16	0.731

Random number on the first generation are all more than mutation probability as it shown on table 4.6 so the mutation is not applied in this generation. Thereafter, total offspring created on first generation is being put back to the population and the position of all individual will be ranked from the biggest fitness value to the smallest. Then, chromosome located on the twenty first position to the rest next chromosome which is left will be removed from the population and this new population will be used as initial population on the second generation.

The same step from selection to putting the offspring back to the population will be repeated until there is no change on fitness value. In this case, iteration has been done until thirtieth generation for each route and fitness value has been in a constant condition in average on the twentieth generation. Final result on this VRP which is solved by Genetic Algorithm is as shown on Table 4.

Table 4. Route Created with Genetic Algorithm

Dump Truck	Route	Amount of Volume Picked/Day	Distance
1	0 – 4 – 3 – 2 – 1 – 13 – 20 – 21 – SPA	7,053 m^3	24,30 km
2	0 – 14 – 18 – 16 – 15 – 19 – 17 – 12 – SPA	6,753 m^3	19.22 km
3	0 – 11 – 10 – 6 – 22 – 9 – 8 – 7 – 5 – SPA	6,903 m^3	21,81 km
	Total	20,708 m^3	65,33 km

Based on Table 4, there is a few change on the position of TPS for each route. With the same amount of volume that is being picked each day, distance travelled by first dump truck reduce from 25,78 km to 24,30 km, distance travelled by second dump truck reduce from 27,13 km to 19,22 km, and distance travelled by the third dump truck reduce from 28,44 km to 21,81 km, therefore total distance travelled is 65,33 km.

5. Conclusion and Further Research

Based on data processing that has been done to determine the shortest path for trash bin pick up route in East Bandung Region, there are a few conclusion that can be made to answer problem formulation which previously has been determine. The conclusion of this research are Total route created are three in which each vehicle that will serve these route is constrained by maximum capacity equal to 7,2 m^3. So that based on this constraint total amount of volume of trash that can be picked up for first route is 7,053 m^3 with total number of TPS that is being visited are 7, on the second route is 6,753 m^3 with total number of TPS that is being visited are 7, and on the third route is 6,903 m^3 with total number of TPS that is being visited are 8.

In this case, metaheuristic method can give a better solution than heuristic method in which shown by the result created from Clarke and Wright Saving with total distance travelled is 81,53 km, meanwhile completion by Genetic Algorithm resulted a significantly total distance travelled which is 65,33 km. First route created is 0 – 4 – 3 – 2 – 1 – 13 – 20 – 21 – SPA with distance travelled 24,30 km, second route created is 0 – 14 – 18 – 16 – 15 – 19 – 17 – 12 – SPA with distance travelled 19,22 km, third route created is 0 – 11 – 10 – 6 – 22 – 9 – 8 – 7 – 5 – SPA with distance travelled 21,81 km.

Based on the conclusion that has been made, a few of recommendation that can be proposed for the next research are Traffict and maintenance cost can be considered for the next research. Hybrid Genetic Algorithm can be used so that a better solution can be found instead of using Standard Genetic Algorithm. A comparative study can be considered for genetic operator since each operator has a lot of method that can be applied, for example on crossover itself there are a few method such as PMX, OX, POS, *One Cut Point, Heuristic Crossover*, etc and each of them have different performance.

Acknowledgements

The author would like to express gratitude to the Ministry of Research, Technology and Higher Education of The Republic of Indonesia for the financial support of this research.

6. References

[1] Toth P and Vigo D 2002 *The Vehicle Routing Problem* (Philadelphia: Society for Industrial and Applied Mathemathics)

[2] Schrage L 1981 Formulation and Structure of More Complex/Realistic Routing and Scheduling Problems. *Networks An International Journal* **11** pp 229–32

[3] Cordeau J F, Laporte G, Savelsbergh M W P and Vigo D 2007 Vehicle Routing *in* Banhart C and Laporte G (*Eds,*) *Handbook in Operation Research and Management Science* 14 375 pg. (North Holland: Elsevier)

IOP Conf. Series: Materials Science and Engineering **528** (2019) 012041 doi:10.1088/1757-899X/528/1/012041

[4] Gen M and Cheng R 1997 *Genetic Algorithms and Engineering Design*. (New York: John Wiley & Sons Inc.)
[5] Zukhri Z 2014 *Algoritma Genetika Metode Komputasi Evolusioner untuk Menyelesaikan Masalah Optimasi*. (Yogyakarta: Andi Publisher)
[6] Larranaga P, et.al. 1999 Genetic Algorithms for the Travelling Salesman Problem: A Review of Representations and Operators, *Artificial Intelligence Review* **13** pp 129-170

Improvement Route for Distribution Solutions MDVRP (Multi Depot Vehicle Routing Problem) using Genetic Algorithm

R Fitriana, P Moengin and U Kusumaningrum

Simulation and System Industry Laboratorium, Industrial Engineering Department
Industrial Technology Faculty, Universitas Trisakti
rinaf@trisakti.ac.id, parwadi@trisakti.ac.id, utami88@hotmail.com

Abstract. The Supply Chain Management Division (SCM) of KM Rubber Company is responsible for delivering the sole product, crumb rubber, from its various factories in Sumatera to the destination ports for further delivery to customers or export destinations. The aim of this study is to provide the improved route for the depot distribution to each destination ports with the shortest time and distance. Writing stage begins with preliminary studies, followed by the identification of the problem, determine the purpose of research, data collection, data processing that contains the initial distribution system optimization solved by methods of transportation and distribution of a new route that is resolved by Genetic Algorithm, and closed by the analysis and conclusions. The manual calculation method to obtain the optimal route gives a total transport distance of 8832.12 km; while the distribution of the proposed route using Genetic Algorithms generate total distance routes as far as 5188.06 km. As the distance is shorter the new route will generate minimal costs and shorter travel time.
Keywords: Multi Depot Vehicle Routing Problem, Genetic Algorithm, Transportation.

1. Introduction

PT. KM Factory Operation department is the department that responsible for managing the operations of production, SCM (Supply Chain Management) and planned production needs. This study focused on the part of SCM companies. The main product produced in the form of crumb rubber (rubber crumb) with quality standards based on SIR (Standard Indonesian Rubber). Factories are the focus in this thesis is the entire plant KM group located on the island of Sumatra.

The factory which is located on the island of Sumatera Indonesia, there are 12 factories, PT. N in Medan, PT. PS in Simelungun, PT. TSS in Riau, PT DJ 1 in Jambi, PT. KM in Bangka, PT. KS Padang, PT. DS 2 in Jujuhan, PT. ALB in Muara Bungo, PT. KW in Windu, PT KMP in Sekayu, PT. KP in Prabumulih, and PT. KJP in Tulang Bawang.

The whole factory in Sumatra sends the result of the products to 6 different port is the port of Jambi, the port of Belawan, the port of Padang, the post of Palembang, the port of Perawang, and the port of Tanjung Priok. Distribution costs incurred by the company for each plant is very diverse and depend on the destination port. In this research will model the distribution of the most optimal route for each plant on the island of Sumatra.

Problems Appears PT KM distribution system is the produce includes distribution costs are very diverse and have a route that is ineffective. It initials caused because distribution system the previously used process less efficient currently where hearts one plant can send prayers or more to ports. Mileage and travel time distribution is very dependent to port the destination, travel time different in each ports as well as the distance from factory to ports that also different.

One way is to determine the most efficient distribution of the products of time and distance as well as the maximum in transportation. Here is the distribution that is used today by the company. The table below shows the factory which will deliver the products to the ports available.

Table 1. Distribution System Company Now

No.	Factory	Port	No.	Factory	Port
1	ABL1	Belawan	10	KPT1	Palembang
2	ABL1	Padang	11	KSP1	Belawan
3	DWJ1	Jambi	12	KUT1	Jakarta
4	DWJ1	Palembang	13	KWI1	Jambi
5	DWJ2	Belawan	14	KWI1	Palembang
6	DWJ2	Padang	15	NSI1	Belawan
7	KJP1	Palembang	16	PSUI1	Belawan
8	KMP1	Palembang	17	TSS1	Belawan
9	KPT1	Jakarta	18	TSS1	Perawang

The aim of this study is to provide the proposed improvements to the depot distribution of each destination and determine the distribution of the depot to the most optimal destination resulting in distance and time most efficiently.

There many research about vehicle routing problem. Okude proposed an approximate solution calculation method for vehicle routing problems (VRPs) that obtains a better solution in a shorter time. The proposed method generates an approximate solution by using a hierarchical traffic network composed on the basis of a vehicle's behavior, which is the frequency of using roads [1].

At MDVRP, number and location of the depot has been determined in advance. Each depot has certainly large enough to accommodate orders from customers. Every vehicle departs and returns to the depot respectively. Location and demand of each customer are also known in advance and every customer will be served exactly once. Customers are served from the same depot will be allocated passes through several routes with methods of saving [2]. The goal of routing is to minimize the number of lanes without neglecting the capacity. The better the process of routing and scheduling , it will be closer spacing will be pursued vehicle , the shorter the time of customer service , more efficient , and more affordable price of its delivery. Generally MDVRP used to shorten the distance and shorten the time of service to all customers. The less the vehicle will make the operating costs are also getting cheaper, can be said MDVRP also an effort to reduce vehicle.

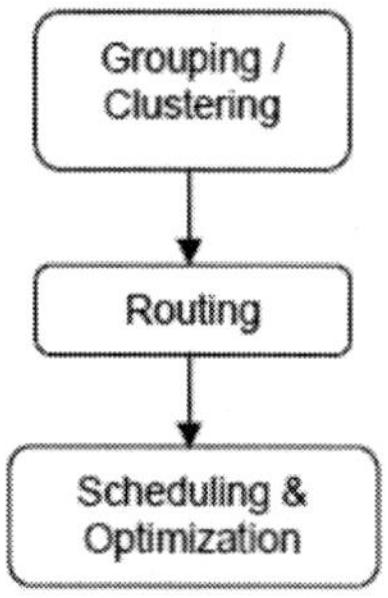

Figure 1. The decision making process on MDVRP

IOP Conf. Series: Materials Science and Engineering **528** (2019) 012042 doi:10.1088/1757-899X/528/1/012042

Formulation for MDVRP with the objective of modeling the customer order contained in the path of the vehicle. In this model it is assumed that all vehicles have a capacity of uniforms, and each vehicle will depart and return to their respective depots. Notation is used for mathematical models as follows [3]:

I : set the whole depot
J :set the whole consumer
K : set the entire vehicle
I: indexes depot
j : indexes consumer
k : indexes track
Parameter :
N :number of vehicles
C_{ij} : distance of points i to j
V_i: The maximum number of products on the depot i
d_j : demand consumer j
Decision variable :

$$X_{ijk} = \begin{cases} 1, if\ i\ immediately\ preceeds\ j\ on\ route\ k \\ 0, otherwise \end{cases}$$

$$Z_{ij} = \begin{cases} 1, if\ customer\ j\ is\ allotted\ to\ depot\ i \\ 0, otherwise \end{cases}$$

U_{ik} = auxiliary variable to subtour elimination constrain in route k.

Mathematical model:

The function objective to minimize the total distance of all vehicles

$$Min \sum_{i \in I \cup J} \sum_{j \in I \cup J} \sum_{k \in K} C_{ij} X_{ijk} \quad (1)$$

Every consumer should be allocated to a track.

$$\sum_{k \in K} \sum_{i \in I \cup J} X_{ijk} = 1, j \in J \quad (2)$$

Every track will be served one time.

$$\sum_{i \in I} \sum_{j \in J} x_{ijk} \leq 1, k \in K \quad (3)$$

Below is the capacity for each depot.

$$\sum_{j \in J} \text{dizij} \leq \text{Vi}, \text{i} \in I \quad (4)$$

Below is the equation that specifies that a consumer can be allocated to a depot only if there is a path from the depot to the consumer.

$$-Z_{ij} + \sum_{U \in I \cup J} (X_{iuk} + X_{ujk}) \leq 1, i \in I, j \in J, k \in K \quad (5)$$

Equationn (6) dan (7) give binary requirements to decision variable.

$$X_{ijk} \in \{0,1\}, i \in I, j \in J, k \in K \quad (6)$$

$$Z_{ij} \in \{0,1\}, i \in I, j \in J \quad (7)$$

Variable score auxiliary has to be positive.

$$U_{lk} \geq 0, l \in J, k \in K \quad (8)$$

Genetic Algorithm

Genetic Algorithm (GA) work based on parallel mechanisms, and therefore more efficient than other optimization techniques such as branch and bound, tabu search and simulated annealing. Basically GA will keep the population of the solutions involved based on existing boundaries. By using GA (genetic operators, crossover, dan mutation) will prevent the processing of data trapped in a local optimum solution. Because it is possible to produce global optimization, then genetic algorithm is suitable to apply to the MDVRP. GA applying natural selection work like genetics [3]. Research about determination of base-stock levels in a ten member serial supply chain with multiple products produced by factories using Uniform Crossover Genetic Algorithms. The complexity of the problem increases when more distribution centers and agents and multiple products were involved. These considerations leading to very complex inventory management process has been resolved in this work[4]. Research about Genetic algorithm developed a mathematical model for a distribution network in a three echelon supply chain that minimizes the system-wide costs and delays on delivery of products. The mathematical model is designed as a multi-objective optimization problem taking into account the two conflicting objective functions[5]. Research about Node duplication Genetic Algorithm (NGA) based technique by comparing against some of the existing deterministic scheduling techniques for minimizing inter processor traffic communication. A comparative study of the results obtained from simulations shows that genetic algorithm can be used to schedule tasks to meet deadlines, in turn to obtain high processor utilization[6]. Research about a method for optimizing software testing efficiency by identifying the most critical path clusters in a program had done by developing variable length Genetic Algorithms that optimize and select the software path clusters which are weighted in accordance with the criticality of the path[8].

Cromosom Representation

Cromosom in MDVRP is a list of consumer that have been visited. There are 6 customer, customer 1-6. Route that representative is first route (02410) and second route (0650). In the first route move from depot to customer 2, then 4, 1, and back to the first place. In the second route is done the same thing. GA for MDVRP is done with three step :grouping, routing, and scheduling.

Grouping- in this step customer will be connected into every n links. The purpose of it is costumer will be divided into depot that have small destination with this rules:

- If $D(ci,A) < D(ci,B)$, then customer ci will be transferred from the depot A
- If $D(ci,A) > D(ci,B)$, then customer ci will be transferred from the depot B
- If $D(ci,A) = D(ci,B)$, then customer ci will be transferred from the depot A or B

From that cases, this is the distance between customer ci and depot k [3].

$$D(c_i,k)=\sqrt{(x_{ci}-x_k)^2+(y_{ci}-y_k)^2} \quad (9)$$

Routing- customer in one link will be alocated into several route using saving matrix [2]. A saving matrix $S(c_i,c_j)$ divided into two customer i and j in the same link. At the end costomer that have the biggest saving will be join into one route without. Saving matrix formulated as follows.

$$S(c_i,c_j)=D(k,c_i)+D(k,c_j)-D(c_i,c_j) \quad (10)$$

Scheduling- from the first costumer, delivery step by step until the distance of next customer is not too far from the previous consumer. This process will be done until all customer can get delivery line. In the final step, the most visible solution will be done.

Fitness Evaluation

One of the objective MDVRP is to minimaze time that needed to process delivery. Process delivery is started at the same time, but can be finished at different time depend on how many costumer can be served.

Time to spend for delivery process is counted from vehicle that have longest time in finished his job to send their product to customer. Dt is total time delivery that needed in depot k and min(Dt) is delivery time minimum that spent in all depot [3].

$$D_t=\sum_{k=1}^{mk} d[c(m_c),c(0)]+\sum_{i=1}^{mc} d[c(i-1),c(i)] \quad (11)$$

Whish d(a,b) is time travel a vehicle from customer a ke b, V is the speed of vehicle, c(i) is location from customer to i, c(0) is from adalah point start or depot position, m_c is customer number in r rute, m_k is route number for depot k. Equation function is define as $F = \sum_{p=1}^{psize} \min(Dt)$

Selection

A common method of selection that used is roulette-wheel. As the name, this methode copy games roulette-wheel where each cromosoms occupied circle pieces on roulette wheel proportionally according to fitness value. Cromosom that have the biggest fitness score occupying a larger circle than a chromosome value is low fitness chromosom. This method is very easy to use. The first step is to make cummulative interval number (in interval [0,1]) from fitness score every cromosoms is divided by total nilai fitness score from every cromosom. The cromosom will be chosen if random number will be raised by in interval accumulation [7].

Crossover

IOP Conf. Series: Materials Science and Engineering **528** (2019) 012042 doi:10.1088/1757-899X/528/1/012042

Crossover technik Best Cost Route Crossover (BCRC) develop to solve route problem with time windows (VRPTW), now it can be used to solve problem route multi depot (MDVRP) with modification.

Mutation

Mutation is modification process of child cromosom randomly. [9]. In the TSP, mutation carriers typically implemented by exchanging genes with the other genes mutated randomly selected. For example, chromosome { 2 , 3 , 4 , 1 , 5 } can be mutated into a chromosome { 4 , 3 , 2 , 1 , 5 }. In this case the gene 1 and gene 3 can be exchanged. This mutation scheme is known as swapping mutation.

There are many research about MDVRP for instance like research about a new VRP solution based on a bi-directional road network, which aims to minimizing fuel consumption for gasoline powered delivery vehicle[10],

2. Research Method

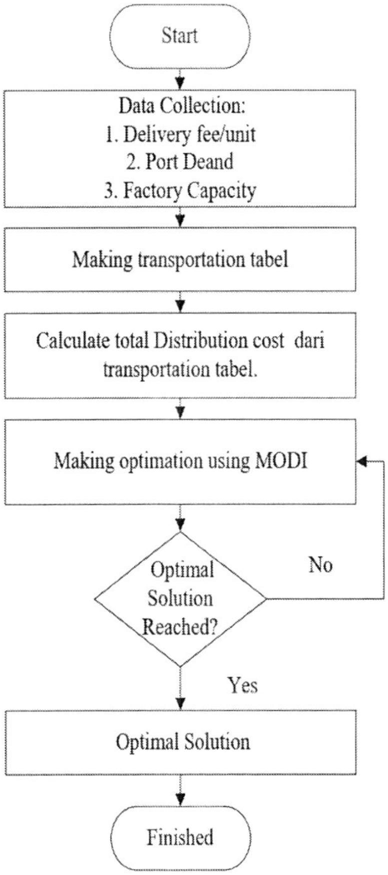

Figure 2. Flowchart Transportation Methode

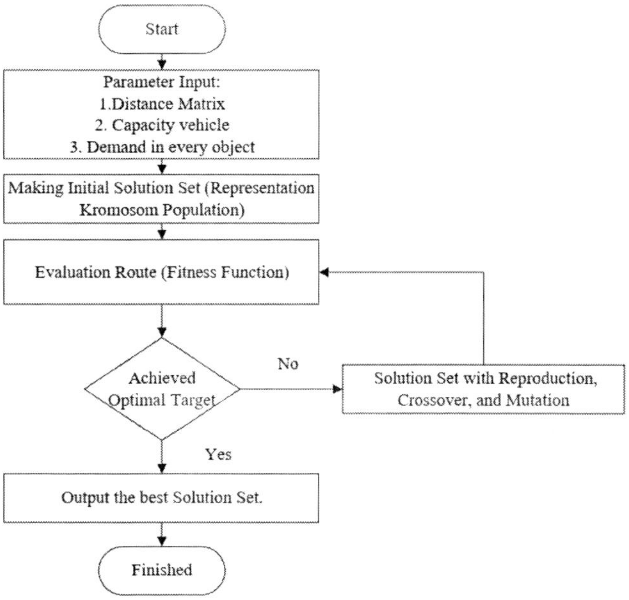

Figure 3. Flowchart Genetic Algorithm

3. Result And Discussion

Results Distribution Company Rute Optimization

Optimization of the distribution system applied by the company nowadays, can be settled with the transportation table and MODI. Here is the result of the calculation using the transport table is least cost way of charging cell that has the lowest price. Previously it had to known in advance the cost of distribution per unit which each unit is used pallet (Rp/Pallet).

The request data is also needed to determine the demand for each of the objectives and the capacity of the plant. By way of a column filled multiplied by the cost it will obtain the total cost of least cost for distribution of products for Rp 98.724.817.800,00 in one year.

After obtaining least cost charges can proceed to the stage of optimization using MODI. Basically MODI performed to optimize the results of the table transportation so can result the most optimal transport routes.

Table 2. Table Description

	Name of the Port	No	Code	No	Code
A	Belawan Port	1	ABL1	7	KSP1
B	Teluk Bayur Padang Port	2	DWJ1	8	KUT1
C	Talang Duku Jambi Port	3	DWJ2	9	KWI1
D	Boom Baru Palembang Port	4	KJP1	10	NSI1
E	Tanjung Priok Jakarta Port	5	KMP1	11	PSU1
F	Perawang Port	6	KPT1	12	TSS1

Process completion using MODI is by giving the row and column values for all variables in the formula base with bi + kj = c ij, where bi is a line to i, kj a is column to j, an cij is a charge on the cell. After all the columns and rows had filled, then calculated non-base variables with the formula $zij - cij = bi + kj - cij$ where $zij \leq 0$ for the optimal result. The result calculation The result of the calculation of non-variable basis as follows [11].

$zij - cij = bi + kj - cij$

$z12 - c12 = 0 + 357840 - 379260 = -21420$

$z24 - c24 = 157500 + 52920 - 359100 = 0$

$z44 - c44 = 365400 + 201600 - 567000 = 0$

$z65 - c65 = 45360 + 201600 - 246960 = 0$

The transport table is optimal for a total cost Rp 98.724.817.800,00, and the total distance of the route as far 8832,12 km.

The result optimation route proposal with Genetika Algoritma

The making of cromosom

The making of cromosom starts with grouping process is grouping or allocation of interest to the nearest depot. Furthermore, the routing process is making of these groups based on the allocation of the value of saving [10]. And then scheduling process namely the manufacture of a complete service to all destinations across the depot that allows for no change in population size. This is the result of the making cromosom prosess.

Table 3. Grouping Result

Group 1	Group 2	Group 3	Group 4	Group 5	Group 7
A	B	C	D	E	F
11	3	2	5	8	1
10		9	6		12
			4		7

Fitness Evaluation

Indication that a route had optimal distance route is to know time schedule in the travel route. And then will calculate using persamaan evaluation fitness from population [3].

Selection

Selection process can be done to choose population the most fit of the others based on their fitness values. From that selection had been choosen *Parents* and then will be crossover. This is the result of selection. From the selection had been choosen P1 dan P4 as a *Parents*.

Table 4. Selection

	Dt	Sum of Dt	Score of Fitness	Sum of Fitness	Interval	Rn	Chosen
P1	2478		0,2496		0,2496	0,2269	0,2496
P2	2481	9926	0,2499	1,0	0,2499	0,9909	0,2499
P3	2485		0,2503		0,2502	0,2546	0,2503
P4	2482		0,2500		0,2500	0,2290	0,2500

Crossover

Crossover process is done between between two population parents and then resulting new generation. Change P1 and P4 into biner form before crossover with give score1 if point i have passed point j, and 0 if not.

Mutation

This is the result from sum of biner score then it become weight and then resulting new allocation route.

Table 5. New Allocation Route

Step

10	**11**	**A**	**3**	**B**	**2**	**9**	**C**	**6**	**4**	**5**	**D**	**8**	**E**	**12**	**7**	**1**	**F**
1	2	3	1	2	1	2	3	1	2	3	4	1	2	1	2	3	4

Allocation

Route	**A**	**10**	**11**	**A**	**B**	**3**	**B**	**C**	**2**	**9**	**C**	**D**	**6**	**4**	**5**	**D**	**E**	**8**	**E**	**F**	**12**	**7**	**1**	**F**
Distance		35	106	25		239	239		575	226	247		155	245	283	17		704	704		208	513	620	48

The total distance route with that route is 5188,06 km.

From the above calculation with optimization of routes used by the company using the transport table and optimization MODI obtained total mileage is 8832,12 km. Where the distance is the total distance of all route from starting the vehicle departs from the place of origin to destination and back again to destination.

If compared with calculation using Algoritma Genetika, total mileage of the vehicle is shorter 5188,06 km. This can be happened because the repair service is opened all lanes of the depot to the destination that has the shortest distance to the depot, then the product distribution area may include the area aound the depot area. Different with distribution route that used by the company now where in one delivery, vehicle only delivery to one place and then back to the first place.

4. Conclusion

1. The new more efficient distribution routes formed after making repairs are A-10-11-A, B-3-B, C-2-9-C, D-6-4-5-D, E-8-E, F-12-7-1-F.
2. These new distribution routes produce a shorter total distance of 5,188.06 km which gives a saving of 3,644.06 km from the initial distribution distance of 8,832.12 km.

5. References

[1] Okude M and Taniguchi E 2014 *Hierarchical Traffic Network For Heuristic Approximation Method Of Vehicle Routing Problems The 8th International Conference on City Logistics. Procedia - Social and Behavioral Sciences* **125** pp 262–274

IOP Conf. Series: Materials Science and Engineering **528** (2019) 012042 doi:10.1088/1757-899X/528/1/012042

[2] Clarke G and Wright J W 1964 Scheduling of Vehicles from a Central Depot to a Number of Delivery Points. *Operations Research* **12** pp 568-581

[3] Surekha P and Sumathi S 2011 Solution To Multi-Depot Vehicle Routing Problem Using Genetic Algorithms. *Journal of Algorithm Optimization-World Applied Programming* **1**(3) pp 118-131

[4] Narmadha S, Selladurai V and Sathish G 2010 Multi-Product Inventory Optimization using Uniform Crossover Genetic Algorithm. *(IJCSIS) International Journal of Computer Science and Information Security* **7**(1)

[5] Dzupire N C and Nkansah-Gyekye1 Y 2014 A Multi-Stage Supply Chain Network Optimization Using Genetic Algorithms. *Mathematical Theory and Modeling* **4**(8)

[6] Valentina A 2012 Penyelesaian *Traveling Salesman Problem* (TSP) Asimetris Dengan Algoritma Genetik *Commonality Program Studi Matematika* Skripsi Jurusan Matematika Fakultas Matematika Dan Ilmu Pengetahuan Alam Universitas Jember

[7] Heidari H and Chalechale A 2012 Scheduling in Multiprocessor System Using Genetic Algorithm. *International Journal of Advanced Science and Technology* **43**

[8] Srivastava P. R. and T.Kim 2009 *Application of Genetic Algorithm in Software Testing*. International Journal of Software Engineering and Its Applications Vol. 3, No.4.

[9] Kusumadewi S and Purnomo H 2005 *Penyelesaian Masalah Optimasi dengan Teknik-Teknik Heuristik*. (Yogyakarta: Graha Ilmu)

[10] Lang Z, Yao E, Hu W and Pan Z 2014 A Vehicle Routing Problem Solution Considering Alternative Stop Points. *The 9th International Conference on Traffic & Transportation Studies (ICTTS'2014). Procedia - Social and Behavioral Sciences* **138** pp 584–591

[11] Taha H A 2007 *Operation Research: An Introduction.* 8th Ed. (New Jersey: Pearson Prentice Hall)

IOP Conf. Series: Materials Science and Engineering **528** (2019) 012043 doi:10.1088/1757-899X/528/1/012043

Domino algorithm: a novel constructive heuristics for traveling salesman problem

Asrul Harun Ismail

[1]The Department of Mechanical Engineering, University of Birmingham, Edgbaston, Birmingham B15 2TT, United Kingdom
AHI729@student.bham.ac.uk

Abstract. Developing algorithms for solving complex optimisation problems has become a challenging topic recently. This study has applied a novel constructive heuristics algorithm named Domino Algorithm for the Traveling Salesman Problem (TSP) case which is aimed to efficiently reduce the calculation complexity and to find the optimal results of TSP best solution of tour lengths. As the study is a basic version of Domino Algorithm, it is decided to use the small TSP data sets consisting of 100 cities or less, such as Eil51, Berlin52, St70, Eil76, Pr76, and Rat99. This study also applied Nearest Neighbor approach to compare its result with that of Domino Algorithm. The results have shown that better tour lengths were achieved by Domino Algorithm (using 1 or 2 player (s)) for Eil51, Eil76, St70, Rat 99; and were resulted by Nearest Neighbor for Berlin52, and Pr76. For further research, the algorithm should be intensively combined with applying the improvement heuristic and should also be hybridized with meta-heuristics algorithm focusing on finding the optimal results by which the application will be moreover quite simple and easy.
Keywords: Constructive heuristic algorithm, domino algorithm, nearest neighbor algorithm, traveling salesman problem.

1. Introduction

Undoubtedly, the most prominent member of the rich set of combinatorial optimization problems is the traveling salesman problem (TSP). TSP is comprised of finding the shortest closed tour, Hamilton cycle, or visiting N cities. It is an NP-hard problem in combinatorial optimization whose computation will naturally rise exponentially by expanding the quantity of cities [1], that is important in operations research and theoretical computer science.

TSP is a complicated or a NP-complete problem that cannot be solved by using traditional algorithms. This problem consists of N cities and between every two cities there can exist a connections (paths) which every path has specified distance or cost. The salesman wants to start from one city and then proceed with other cities in a way that he will pass through each city for only once and finally, he will return to the origin point. The aim of TSP is finding the sequence of the cities to be traveled through by the traveling salesman that will result in a reduced total distance or cost.

TSP can be solved either with exact or heuristic algorithm. Although it is easy to understand traveling salesman problem, but if the size is getting larger, its solution using exact algorithm will become very difficult. In term of quality and speed, solving the problem using heuristic or meta-heuristic algorithms is more feasible. Discovering a suboptimal solution with a sensible time or cost might be more beneficial rather than finding an optimal solution with time wasted. Various heuristic and meta-heuristic algorithms resulting in quick and good solutions have been devised. Modern heuristic or meta-heuristic methods can find solutions for extremely large problems (millions of cities) in a reasonable time with only a gap of 2–3% from the optimal solution.[2]

Some of the meta-heuristic algorithms which are used to solve such problems are genetic algorithm, simulated annealing algorithm, bee algorithm, etc. Furthermore, most of the searching step, global and local search of the meta-heuristic algorithms are using heuristic algorithms either the constructive heuristics or improvement ones. There are several popular constructive heuristic algorithms in combinatorial domain such as nearest neighbor, nearest arbitrary city, nearest insertion, farthest insertion, and cheapest insertion.

These algorithms are also already used widely and some of its application are like Machine scheduling problem, Cellular manufacturing, Arc routing problem, Frequency assignment problem, Structuring of matrices, Printed-circuit-boards manufacturing, data transmission in computer networks, power-distribution networks, image processing and pattern recognition, robot navigation, and data partitioning [3],[4].

To date, a few approaches based on deterministic or probabilistic heuristics have been proposed for optimizing TSP and all of these algorithms have to use constructive heuristics algorithms to generate prospective solution(s) as the initial good solution(s). Later, the improvement heuristics will enhance those solutions to get closer to the optimal solution. For a large-size TSP case, it is almost impossible to find an optimal or near-optimal solution using constructive heuristic alone. However, for several small-size TSP cases (less than 50 cities), the constructive heuristics mentioned-above might provide an optimal solution.

In constructive heuristics research, reaching an optimal solution becomes less important compare to resulting a great seed leading to near-optimal solution by providing the randomness of prospective solutions. In term of generating initial solutions, the constructive heuristics also give a faster calculation rather than using random permutation. Population-based heuristics such as ant colony optimization (ACO) algorithms [5],[6],[7],[8], neural networks [3],[9], evolutionary algorithms [10],[11], simulated annealing [12],[13],[14],[15], tabu search [16],[15], artificial bee colony algorithm [17],[18], particle swarm optimization [19],[18], and Bee Algorithm [20],[21],[22],[23] are also using constructive heuristics to solve combinatorial problems by employing tour construction. Hence, building up a constructive heuristics algorithm that generates good random solutions to become as good seed and efficient complex computation remains a challenging topic, especially when it can be used for developing a meta-heuristic algorithm for generating sub-optimum or optimum solution.

In this paper, a novel constructive heuristic "Domino Algorithm" is proposed to provide better seed to be improved. Most of current constructive heuristics are using a "greed" strategy. It makes them provide limited number of prospective solutions and hard to reach optimum solution by improving heuristics on larger-size TSP [24]. The main idea of proposed approach is derived from the domino game. The traditional Sino-European domino game is featuring all combinations of spot counts between zero and six (see figure 1), which generates a sequence of the tiles. Each tile represents a path consisting 2 points as original point and destination one. The game will be over when the sequence gets formed or the player turns down all the tiles. This basic version of proposed algorithm will order each player to make turn and each tile selected from a player's hand should become the minimum distance and be inserted in the matching edge of the tiles sequence. Although the selection and insertion strategy could be exchanged to the others, but this paper will focus on the basic version strategy.

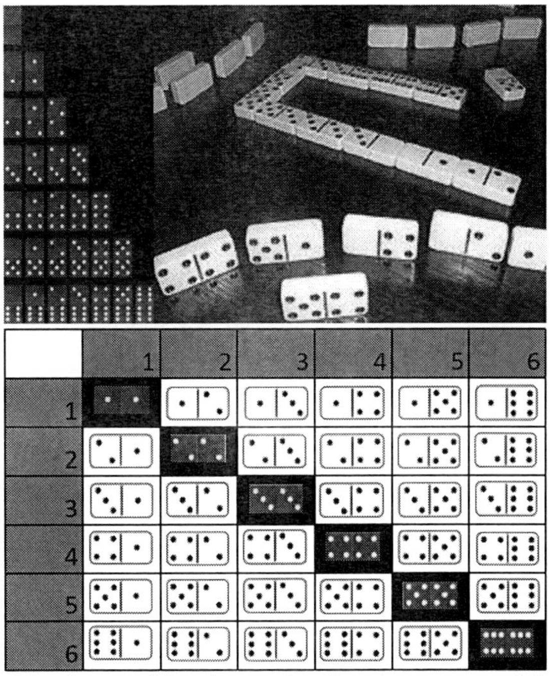

Figure 1. A complete double-six set of dominoes, the tiles sequence, and dominoes in TSP matrix cell (clockwise).

The main part of this paper will consist of the literature review and the result of experiment that will compare the proposed method with the nearest neighbor since it is very popular and widely used as constructive heuristics. The TSPLIB's data sets are used as problem sets to compare their performances. The comparison between proposed approach and nearest neighbor seems fair enough to provide comprehensive experiment result – compare to other constructive methods – since both of them use the same selection and insertion strategy. As it is a basic version of Domino Algorithm so we choose to use the small TSP data set, 100 cities or less, such as Eil51, Berlin52, St70, Eil76, Pr76, and Rat99 [25],[26]. This paper will be organized as follows. In next section, we will present the literature review. Proposed Domino Algorithm as a basic version will be presented in section 3. Section 4 will focus on the evaluation of the performance of Domino Algorithm compare to its counterparts using six sets of TSP problem (TSPLIB) as mentioned above [25]. The 2-opt, the improvement heuristics, is used in this experiment to prove that the proposed algorithm has generated a better initial solution as a seed. Finally, conclusions will be presented in section 5.

2. Literature review

2.1. Traveling salesman problem

Combinatorial problems has arisen in many areas of computer science and other disciplines in which computational methods are applied, such as artificial intelligence, operations research, bioinformatics and electronic commerce. Some popular problems are like finding shortest tour also known as the travelling salesman problem, finding models of propositional formulae and determining the 3D-structure of proteins. Other well-known combinatorial problems are found in planning, scheduling, time-tabling, resource allocation, code design, hardware design and genome sequencing [27]. Some of combinatorial problems are N-P Hard type. It is generally believed that N-P hard problems cannot be solved to optimality within polynomial bounded computation times. Consequently, it has increased an interest in approximation algorithm that can find near-optimal solution with reasonable running times.

TSP can be symmetric or asymmetric. In the symmetric one, the distance between the two cities does not depend on the distance of travel. For example, if we show the distance between two nodes called i and j with d_{ij} and if we have $d_{ji}=d_{ij}$, then TSP is symmetric or otherwise it is asymmetric.

2.2. Nearest neighbor

There are three types of heuristics method i.e. constructive, improvement, and compound. The constructive starts from building a solution with step by step, based on a set of rules defined beforehand. This approach is also used by meta-heuristics in their global search calculation. The improvement heuristics starts from a feasible solution and improve it by applying successive small changes. There are several popular algorithms such as 2-opt, 3-opt, k-opt, etc. Meanwhile, the compound heuristics has both constructive and improvement phases. Normally, the improvement phase will be conducted following the constructive phase. Hence, the compound is usually built from a combination of constructive and improvement, such as nearest neighbor-2Optimal, nearest neighbor-3Optimal, etc.

Nearest neighbor belongs to constructive heuristics which starts from nothing to then build a prospective solutions. Today, there are several popular constructive heuristic algorithm in combinatorial domain such as nearest neighbor, nearest arbitrary city, nearest insertion, farthest insertion, and cheapest insertion. The nearest neighbor algorithm is one of the first algorithms used to determine a solution to the travelling salesman problem. In such the algorithm, the salesman starts at a random city and proceeds with visiting the nearest city until all cities are finished to visit. It quickly yields a short tour, but usually not the optimal one. The nearest neighbor algorithm is easy to implement and execute quickly, but it can sometimes miss shorter routes which are easily noticed with human eyesight, due to its "greed" nature [24]. The procedure follows is the nearest neighbor algorithm [1] where $V = \{1, 2, . . . , n\}$ is a set of nodes representing the cities and $E = \{ (i, j) \mid i,j \in V\}$.

(1) Select an arbitrary node k,
(2) Find city k+1 that is not yet visited and closer to k (min d_{ij}). Let T= {1, 2, . . ., k} be the current partial tour (k<n).
(3) Insert k + 1 at the end of the partial tour; and k = k+1.
(4) If all nodes or cities are already inserted then STOP, else go back to step (2).

3. Proposed Algorithm

Basically this proposed method is using the idea of domino game. This game is a traditional tile-based strategy game usually played by 2 to 4 players. Each domino tile is a rectangular shape with a line parting its face in half into two equal square-shaped: front half and end one. Each half is marked with either a blank, one spot, or a number of spots. The game is played this way. A first player will select a tile, then the next player will proceed with selecting a matching tile having a similar value in its front or end half with the tile of the first player to then make a sequential set of matching tiles. They will in turn play this way until the end of the game. A domino set is a generic gaming device, similar to playing cards or dice, in which a variety of games can be played with a set. Generally this game has six main steps: Shuffle the dominoes, Draw open hand, Decide the order of play, Lay first domino, Take turn adding domino, and end the round and award points. This rule will be implemented to solve a TSP case.

Domino Algorithm generates a sequence from a set of matching tiles which each tile represents a path consisting 2 points as original and destination cities. The tiles used in the game are the combinations of spot (s) represent a number of cities between 1 to N-cities. Fig 2 shows the combination of tiles (excluding the zero value represented by the blank square) that is used in 6 cities' TSP case. The algorithm consists of 4 common steps of constructive heuristics: initialization, selection, insertion, and termination. The first step is distributing the tiles equally to each player in a random. Players make turns to add tiles into the front half or into the end half of the initial tile. The game will be over after a sequence gets formed or the players turns down all the tiles. This algorithm will order each player to make turn and each tile selected from player's hand should become the minimum distance. The Fig 3 below is showing the flow chart of Domino algorithm procedure.

4. Methodology

4.1. Problem Representation

TSP can be represented as a graph of $G = (V, E)$, where $V = \{1, 2, \ldots, n\}$ is a set of nodes representing the cities, and $E = \{ (i, j) \mid i,j \in V\}$ is the set of all connecting edges between them. Each edge means the possible path between two connected cities. The distance d_{ij} or $d(i,j)$ is associated with edge (i, j) and represents the Euclidean distance from city i to city j as equation (1). The information is calculated before performing the main algorithm. The distances of all edges were calculated and saved as a distance matrix.

$$d_{ij} = \sqrt{(x_i - x_j)^2 - (y_i - y_j)^2} \tag{1}$$

The TSP is the problem of finding a minimum length of Hamiltonian tour on the undirected graph $G = (V, E)$, where an Hamiltonian tour of graph G is a close-distance tour that needs visiting all the nodes of G one time only, and its length is given by the sum of the lengths of all the edges by which it is composed. The aim of solving TSP is to minimize the total close tour length L which is defined as equation (2).

$$L = \sum_{i=1}^{n} d(i, i+1) + d(n, 1) \tag{2}$$

4.2. Initialization

To solve TSP using Domino algorithm we should conduct some following steps. First, we generate some distributed initial random tiles by firstly setting up the number of players (p), followed by shuffling the tiles up, then distributing the initial set of tiles to each player in equal number or n/p or it will lead to a little adjustment when the number of nodes is odd number.

4.3. Selection and Insertion

In this manner a random city is selected as initial of the tour (e. g. city i). Then the next cities k consecutively are added to the tour which corresponds to their distance from city i. Let (1, … , k) be the current partial tour (k < n). The new city k will be added in the end of the tour if d(k,k+1) < d(k+1,k), otherwise it will be added or inserted at the beginning of partial tour. The players have to extend the partial tour with their remaining tiles (nodes) in turn.

4.4. Termination

When the next city k is equal to number of nodes (cities) then it will be the last insertion on the algorithm. The algorithm will calculate the final close tour or Hamilton cycle using Eq. (2). The figure 2 below depicts the flowchart of Domino algorithm where $V = \{1, 2, \ldots, n\}$ is a set of nodes representing the cities, $P = \{1,\ldots,p\}$ is a number of players, $Vp = \{1,\ldots,n/p\}$ and $E = \{ (i, j) \mid i,j \in V\}$.

5. Experimental result

In order to evaluate the performance of Domino algorithm, we tested it in the benchmark problems of TSPLIB. The experimental results are summarized in table 1. The first and 2nd column is the problem names and optimum tour lengths (best known solution), 3th column is the values of the compared algorithms, and 4th and 5th columns are the achieved values by proposed method. Each 3rd, 4th, 5th columns have sub-columns which respectively are the best tour length generated by those methods, the average value of tours length possibly generated, and the number of possibly generated tours. The possible generated tour achieved by Nearest Neighbor is always as much as the-N (number of nodes) since it uses the minimum distance. The Domino Algorithm with 1 player always produces possible tours as much as those of Nearest Neighbor since there is no other players in turn generating sequence. We could also say that every player's combination of nodes generates n possible constructed tours and in this case, a player only has $C(n, n) = 1$. However Domino Algorithm with 2 players generates more tours since each of them has the same number of nodes (n/2) or $V_p=\{1,\ldots,n/2\}$ and having their own turn to build the sequence. The number of possible constructed tours of proposed algorithm with two players is $\frac{n}{2}.C(n,\frac{n}{2})$ where $C(n,\frac{n}{2})$ is a number of nodes combination of 2 players. We assume that the larger number of constructed tours, the better solution will be produced when it is enhanced with the improvement heuristic methods. The experiment uses 1000 iterations, which neither efficient nor effective for Nearest Neighbor and Domino with 1 player, and random number as a seed (initial city) for each method.

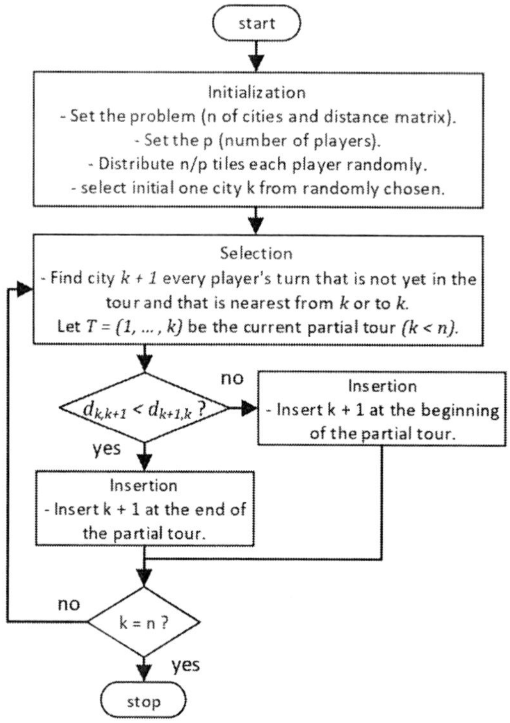

Figure 2. The flowchart of Domino algorithm.

As comparison to the performance shown with a graph, the best solution every approach has made is summarized in table 2. All the best tours still have crossing lines or unsmooth shapes because of the prior research and the exclusion of the improvement heuristics. We believed in further research the problems will be eliminated by using improvement heuristic methods. A better constructed tour length has been achieved by Domino Algorithm (using 1 or 2 player (s)) for Eil51, Eil76, St70, Rat 99; and Nearest Neighbor for Berlin52, and Pr76 results can be seen respectively in table 1.

Table 1. Comparison value between best solution tour length of Nearest Neighbor and Domino Algorithm

Problem	Optimal Solution	Nearest Neighbor (NN)		Domino-2 players (Dom2pl)		Domino-1 players (Dom1pl)	
		Best	Average	Best	Average	Best	Average
Eil51	426	482	525.90	507	622.63	469	540.34
Berlin52	7542	8181	9375.58	9343	11293.75	8541	9226.44
St70	675	796	841.26	854	1067.32	748	822.52
Eil76	538	608	665.75	654	796.44	583	654.07
Pr76	108159	130921	147178.41	143351	181450.47	137788	143343.23
Rat99	1211	1437	1535.83	1509	1946.40	1372	1513.58

ISIEM 2018 IOP Publishing
IOP Conf. Series: Materials Science and Engineering **528** (2019) 012043 doi:10.1088/1757-899X/528/1/012043

Table 2. Comparison graph between best solution of tour lengths between Nearest Neighbor and Domino Algorithm

Problem	Nearest Neighbor (NN)	Domino-2 players (Dom2pl)	Domino-1 players (Dom1pl)
Eil51 (Opt. Solution=426)	482 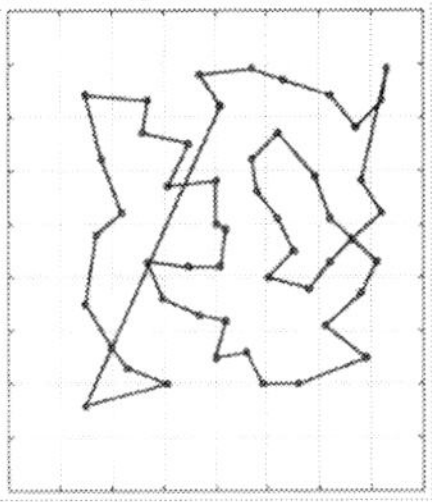	507 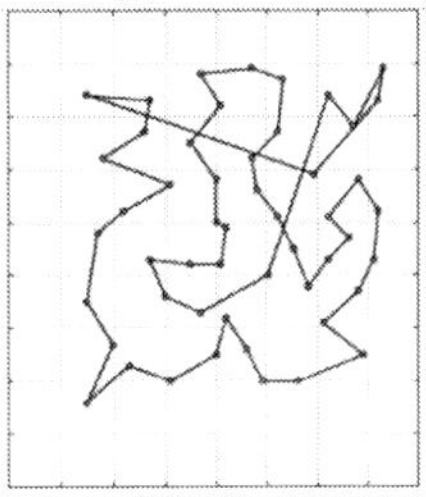	469 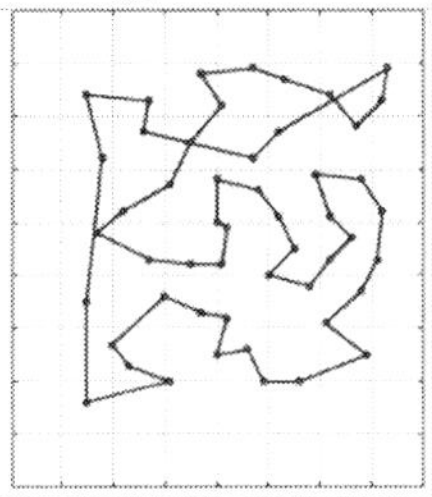
Berlin52 (Opt. Solution=7542)	8181 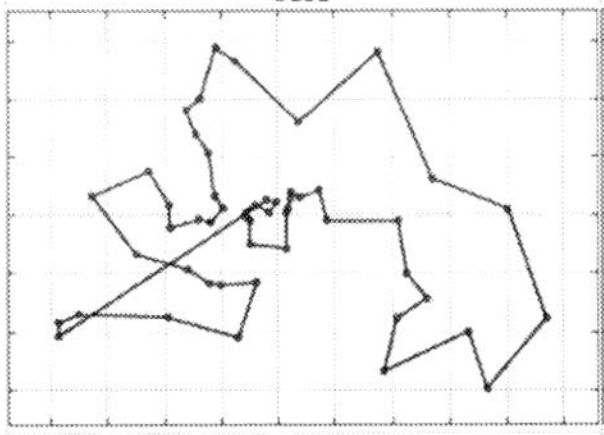	9343 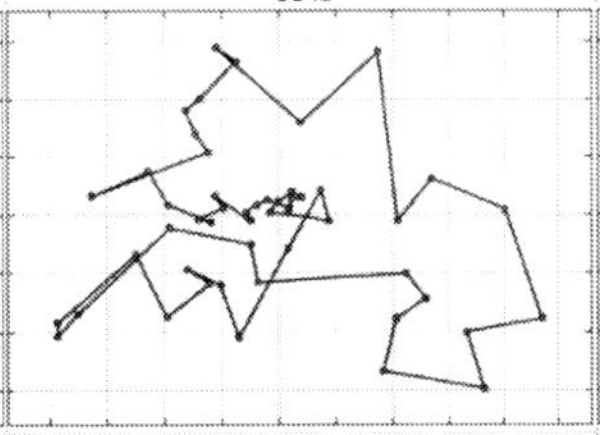	8541 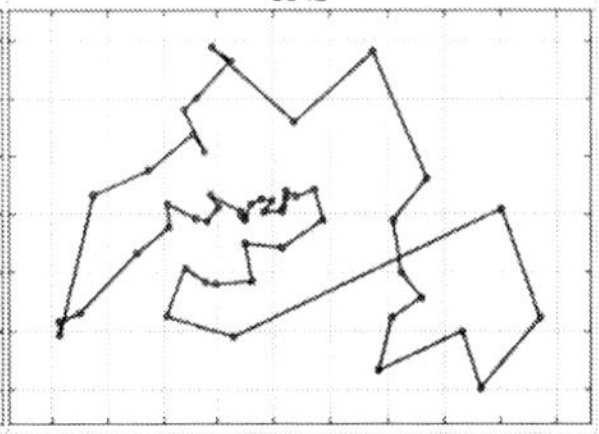
St70 (Opt. Solution=675)	796 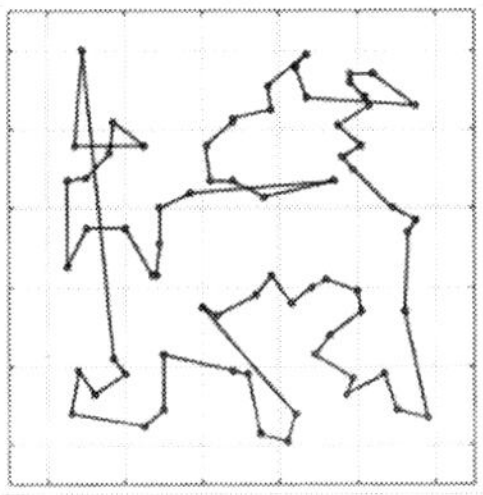	854 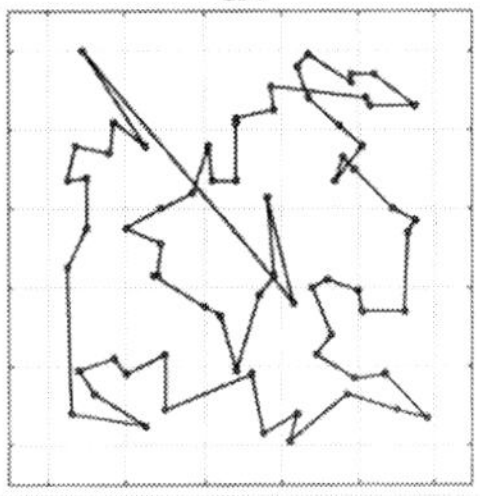	748 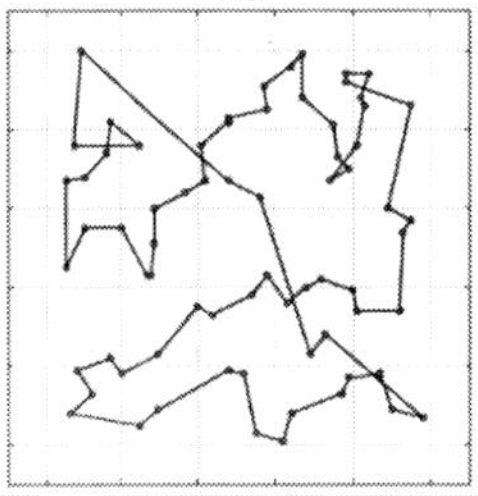
Eil76 (Opt. Solution=538)	608 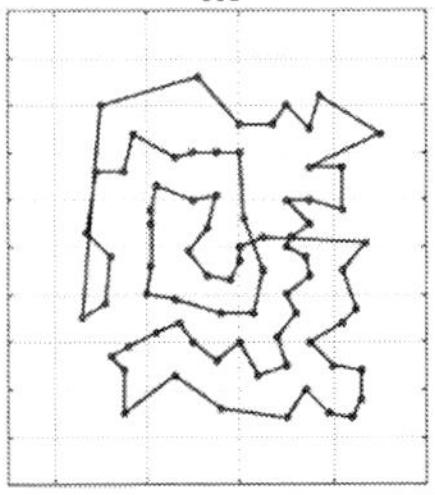	654 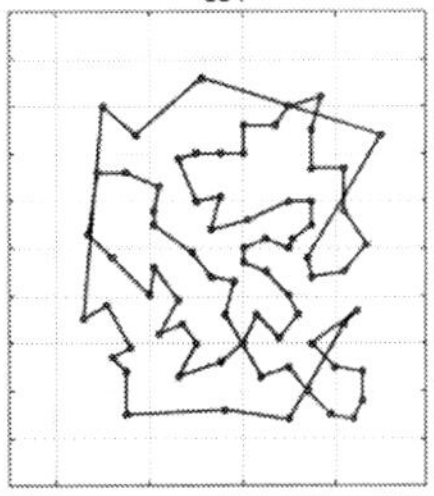	583 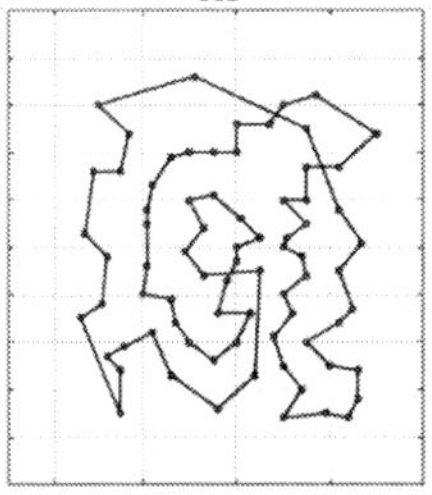
Pr76 (Opt. Solution=108159)	130921 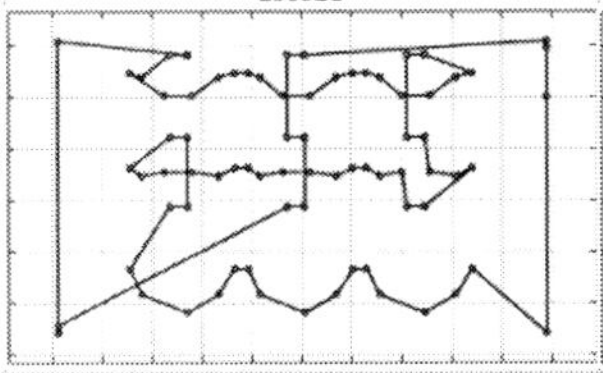	143351 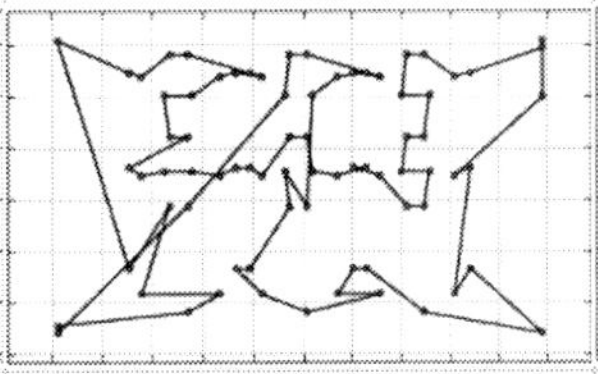	137788 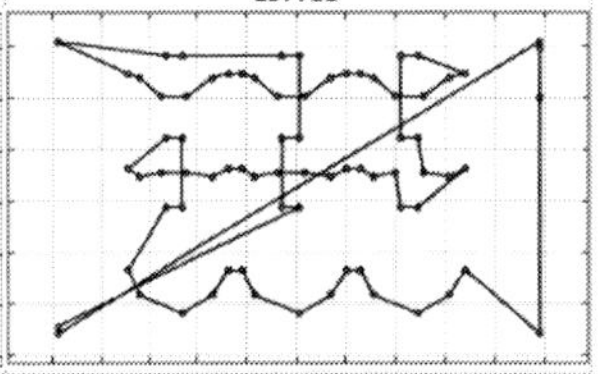
Rat99 (Opt. Solution=1211)	1437 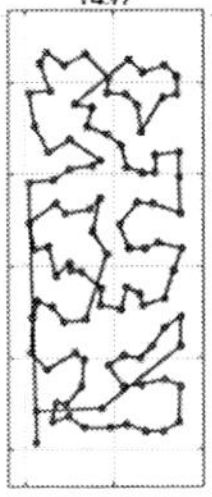	1509 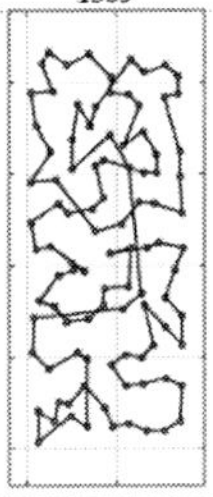	1372 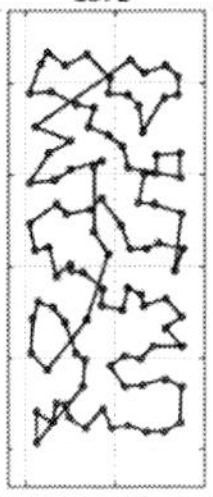

6. Conclusion

In this paper, a novel constructive heuristics algorithm named Domino algorithm has been introduced for TSP, which is aimed to efficiently reduce the calculation complexity of meta-heuristics algorithms which can generate $\frac{n}{2}.C(n,\frac{n}{2})$ constructed tours as good prospective solutions. Using random permutation in global search of meta-heuristics will waste the time since it generates $n!$ of prospective solutions.

For further research, the algorithm should be intensively combined with applying the improvement heuristic and hybridize it with meta-heuristics algorithm focusing on finding the optimal results which will be moreover quite simple and easy to apply.

Acknowledgement

The authors would like to acknowledge the generous support from the Newton Fund - Ristekdikti.

7. References

[1] Reinelt G 1994 *Traveling Salesman: Computational solution The TSP applications*. (Springer-Verlag, Berlin).

[2] Rego C, Gamboa D, Glover F and Osterman C 2011 Traveling salesman problem heuristics: Leading methods, implementations and latest advances. *Eur. J. Oper. Res.* vol 211 no 3 pp 427–41

[3] Saadatmand-Tarzjan M, Khademi M, Akbarzadeh-T M R and Abrishami Moghaddam H 2007 A novel constructive-optimizer neural network for the traveling salesman problem. *IEEE Trans. Syst. Man, Cybern. Part B Cybern.* vol 37 no 4 pp 754–70

[4] Gutin A P, Punnen G 2006 *The Traveling Salesman Problem and Its Variations*. (Springer).

[5] Dorigo M and Gambardella L M 1997 Ant colonies for the travelling salesman problem. *Biosystems*. vol 43 no 2 pp 73–81

[6] Dorigo M and Gambardella L M 1997 Ant Colony System : A Cooperative Learning Approach to the Traveling Salesman Problem. *IEEE Trans. Evol. Comput* vol 1 no 1 pp 1–24

[7] Gambardella L and Dorigo M 1997 Ant-Q: Reinforcement Learning Approach to the Traveling Salesman Problem. *IEEE Trans. Evol. Comput.* vol 1 no 1 pp 252–60

[8] Liao E and Liu C 2018 A Hierarchical Algorithm Based on Density Peaks Clustering and Ant Colony Optimization for Traveling Salesman Problem. *IEEE Access*. vol 6 pp 38921–33

[9] Creput J and Koukam A 2009 A memetic neural network for the Euclidean traveling salesman problem. *Neurocomputing*. vol 72 pp 1250–64

[10] Vaishnav P, Choudhary N and Jain K 2017 Traveling Salesman Problem Using Genetic Algorithm : A Survey. *Int. J. of Scientific Research in Computer Science, Engineering and Information Technology*. vol 2 no 3 pp 105–8

[11] Hussain A, Muhammad Y S, Sajid M N, Hussain I, Shoukry A M and Gani S 2017 Genetic Algorithm for Traveling Salesman Problem with Modified Cycle Crossover Operator. *Hindawi Computational Intelligence and Neuroscience*. vol 2017 pp 1–7

[12] Mafarja M M and Mirjalili S 2017 Hybrid Whale Optimization Algorithm with simulated annealing for feature selection. *Neurocomputing* vol 260 pp 302–12

[13] Wang C, Lin M and Zhong Y 2016 Swarm simulated annealing algorithm with knowledge-based sampling for travelling salesman problem. *Int. J. Intelligent Systems Technologies and Applications*. vol 15 no 1 pp 74–94

[14] Helshani L 2016 Solving the Traveling Salesman Problem using Google Services and Simulated Annealing Algorithm. *European Academic Research*. vol IV no 3 pp 2321–30

[15] Lin Y, Bian Z and Liu X 2016 Developing a dynamic neighborhood structure for an adaptive hybrid simulated annealing – tabu search algorithm to solve the symmetrical traveling salesman problem. *Appl. Soft Comput. J.* vol 49 pp 937–52

[16] Ali Z A 2016 Concentric Tabu Search Algorithm for Solving Traveling Salesman Problem (Eastern Mediterranean University January-North Cyprus)

[17] Sabet S, Farokhi F and Shokouhifar M 2013 A Hybrid Mutation-based Artificial Bee Colony for Traveling Salesman Problem. *Lect. Notes Inf. Theory*. vol 1 no 3 pp 99–103
[18] Neelima S, Satyanarayana N and Murthy P K 2016 A Comprehensive Survey on Variants in Artificial Bee Colony. *Int. J.of Computer Science and Information Technologies*. vol 7 no 4 pp 1684–89
[19] Shi X H, Liang Y C, Lee H P, Lu C and Wang Q X 2007 Particle swarm optimization-based algorithms for TSP and generalized TSP. *Inf. Process. Lett.* vol 103 no 5 pp 169–76
[20] Pham D T, Otri S and Darwish A 2007 Application of the Bees Algorithm to PCB assembly optimisation. *3rd Virtual Int. Conf. Innov. Prod. Mach. Syst.* pp. 511-16
[21] Ang M C, Pham D T, Soroka A J and Ng K W 2010 PCB assembly optimisation using the bees algorithm enhanced with TRIZ operators. *IECON Proc. Industrial Electron. Conf.* pp 2708–13
[22] Ahmad S A, Pham D T, Ng K W and Ang M C 2012 TRIZ-inspired asymmetrical search neighborhood in the bees algorithm *Proc. - 6th Asia Int. Conf. Math. Model. Comput. Simulation, AMS 2012*. pp 29–33
[23] Otri S 2011 Improving The Bees Algorithm for Complex Optimisation Problems. (Cardiff-United Kingdom)
[24] Gutin G, Yeo A and Zverovich A 2002 Traveling salesman should not be greedy : domination analysis of greedy-type heuristics for the TSP. *Discrete Applied Mathematics*. vol 117 pp 81–6
[25] Reinelt G 1991 TSPLIB-- A traveling salesman problem library. *ORSA Journal on Computing*. vol 3 no 4 pp 376–84
[26] Reinelt G 1995 *Tsplib 95* (Universit¨at Heidelberg- Heidelberg) pp. 1–16
[27] Hoos H and Stutzle T 2005 *Stochastic Local Search: Foundations and Applications*. (Morgan Kaufmann Publishers)

IOP Conf. Series: Materials Science and Engineering **528** (2019) 012044 doi:10.1088/1757-899X/528/1/012044

A Comparison of Forecasting Building Material Inventory between Backpropagation Neural Network And Arima

I A Soenandi[1] and C Hayat[2]

[1]Department of Industrial Engineering Faculty of Engineering & Computer Science Krida Wacana Christian University, Tanjung Duren Raya No. 4 Jakarta 11470, Indonesia
[2]Department of Information System Faculty of Engineering & Computer Science, Krida Wacana Christian University Tanjung Duren Raya No. 4 Jakarta 11470, Indonesia
[1] iwan.as@ukrida.ac.id

Abstract. The demand for forecasting task is very important to determine the number of stocks efficiently. This process should accommodate the demand for a company's product or service and control the inventory level. Especially for products such as building materials that needed capitals to buy and wide space to keep it safe. This research has objective to minimize the excessive amount of product in inventory and minimize loss in sales. This study was compared between a method named Back Propagation Neural Network (BPNN) that known as one of the most accurate and widely used forecasting model and ARIMA as a time series model to find the most accurate in forecasting of inventory. In this case, the model of BPNN used 6 input neurons as a monthly period of sale, the price of the product, amount of historical selling, an approximation of project renovation, an approximation of new project building and number of a competitor. And for Arima method we have three trials of tentative models. To compare the accuracy between them, we used the performance criteria such as MAD, MAE, RMSE, RRSE and RAE. In this research, we obtained that forecasting with BPNN is more accurate than ARIMA with error prediction of 19.6, 19.6, 30.4, 0.6, 0.5 for those performance criteria consecutively.
Keywords: inventory; forecasting; building material; backpropagation neural network; Arima;

1. Introduction

The demand forecasting is a critical task to improve the efficiency of a supply chain system [1]. Since each part in the supply chain will process the order in response to the demand signal, the accuracy of demand forecasts will significantly improve the production scheduling, capacity planning, and material requirement planning and inventory management [2]. A qualitative method, time series method, and causal method are 3 important forecasting techniques. Qualitative methods are based on the opinion of subject matter expert and are therefore subjective. Time series methods forecast the future demand based on historical data. Causal methods are based on the assumption that demand forecasting is based on certain factors and explore the correlation between these factors. This prediction of forecast predicts the company sales at any time horizon. Overall, the price of goods becomes a major reference in determining inventory.

For building material inventory, the price of goods affects the effectiveness of venture capital management, while for consumer price the goods become benchmark and adjustment in purchasing the building materials needed.

In the previous literature, many articles stated that the ability to forecast of the future based on past data is a key tool to support individual and organizational decision making. In particular, the goal of Time Series Forecasting (TSF) is to predict the behavior of complex systems by looking only at past patterns of the same phenomenon [3]. Nowadays, forecasting is an integral part of supply chain management and traditional forecasting methods suffer from serious limitations which affect the forecasting accuracy. New methods such as neural network algorithms have been found to be useful techniques for demand forecasting due to their ability to accommodate non- linear data, to capture subtle functional relationships among empirical data, even where the underlying relationships are unknown or hard to describe[4][5]. Recently, the implementation for inventory forecasting is a method that was named back propagation neural networks (BPNNs) have been extensively studied and used in time series forecasting presented a recent review in this area [6]. The major advantage of this method is their flexible nonlinear modeling capability. With BPNNs, there is no need to specify a particular model form.

In this study, we compared the BPNN method with another forecasting method that one of the most important and widely used time series models is the Autoregressive Integrated Moving Average (ARIMA). The popularity of the ARIMA model is due to its statistical properties as well as the well-known Box–Jenkins methodology [7]. In the model building process. ARIMA model is user-friendly in practice and sensitive to the trend of time series data. Its high accuracy of prediction earns the fame as one of best models in time series and highly demand of data abundances for training [8].

This paper presents a time series approach for determining building material stocks amount. We used a comparative approach for modeling the prediction. First, we modeled the forecasting using standard ARIMA [9] models. Then we used nonlinear neural network model trained with "BFGS quasi-Newton Method" based on Back-propagation [10] for modeling the clone evolution series.

We also proposed a forecasting techniques that are suitable to predict the amount of procurement of material supplies accurately. This chosen forecasting method to be used as a supportive qualitative decision, which provides the additional possibility of accommodate certainty demand and minimizes inventory that has a good impact on business sustainability.

2. Methods

2.1. Forecasting with Backpropagation Neural Network

Back Propagation (BP) is a supervised learning algorithm and it often used to train the perceptron. The neural network model which is trained using the BP algorithm is referred to as BPNN [11]. This study has applied the standard procedure of back -propagation neural network as 3 phases. Phase one: Forward Propagation Each Input Unit receives the signal and passes it to the hidden unit on it. In phase two: Backpropagation. This phase has 2 steps. First: count of factor units output based on an error of unit output. Next, we count of hidden factor unit based on an error in hidden units. In phase 3: Putting all the values together and calculating the updated weight value. Next, update of line weights to hidden units. Then this algorithm is done.

The network architecture used in this research was the multi-layer network. In this network, other than the input and output units, there can be one or more other units called the hidden layer which is determined in the training data. The multi-layer neural network can solve the more complex problems compared to the single layer one, although it sometimes requires the more complex and longer training process. The design of artificial neural network architecture was made by determining the numbers of input layer neurons, hidden layer neurons, and output layer neurons.

The dataset was built with several steps. First, we collected data using a questionnaire and interview with several managers of building material shops based on correlation level of procurement and several variables. Those variables are a monthly period of sale (X1), the price of the product (X2), amount of historical selling (X3), an approximation of project renovation (X4), an approximation of new project building (X5) and a number of a competitor (X6). By analyzing all those variables we want to determine the amount of inventory stock (Y) as shown in Figure 1.

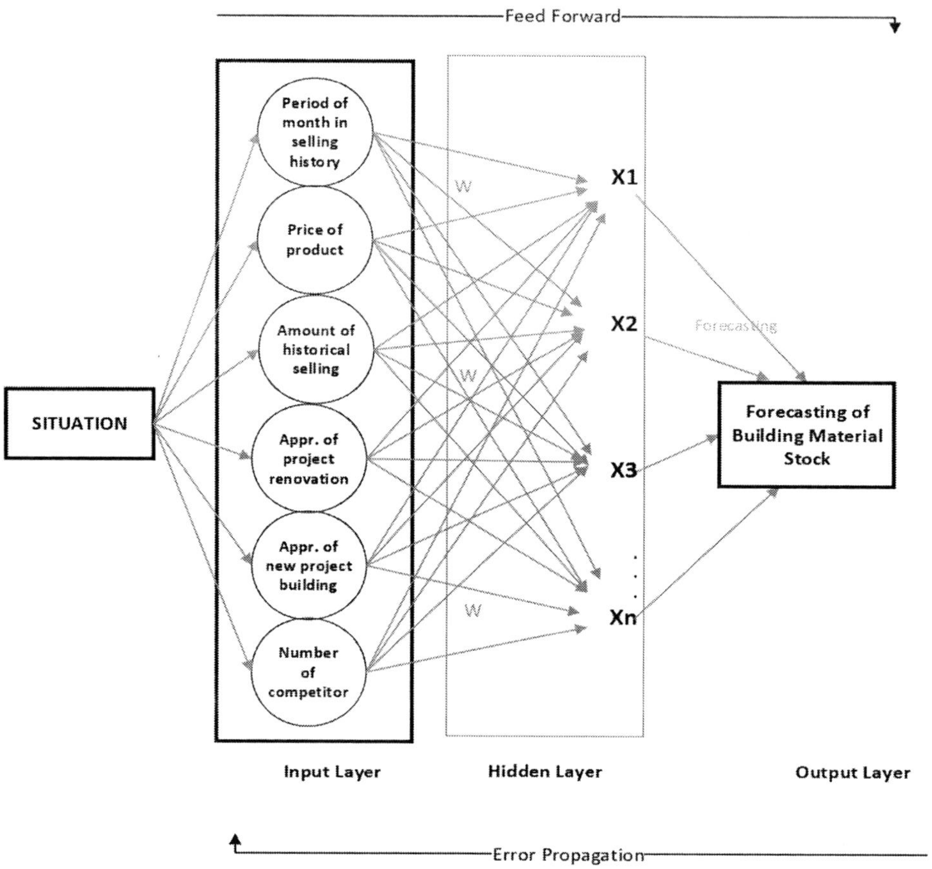

Figure 1. A forecasting model for BPNN

2.2. Forecasting with ARIMA

Autoregressive integrated moving average model (ARIMA [12]) is capable of identifying complex patterns in temporal dataset and thus widely applied for short-term forecasts. The ARIMA model consists of three basic steps, namely identification, assessment and testing, and diagnostic examination. Furthermore, the ARIMA model can be used for forecasting if the model obtained is sufficient. It should be noted that most periodic sequences are non-stationary and that AR and MA aspects of the ARIMA model are only related to stationary periodic sequences. Stationary means having no growth or decline in the data. The data must roughly be horizontal along the time axis. In other words, the fluctuation of the data around a constant average value is independent of the time and variance of the fluctuation which essentially remains constant at all times.

A non-stationary time series must be converted into stationary data by performing differencing. Differencing is to calculate the change or the difference in the value of observation. The difference value obtained is checked further whether it is stationary or not. If it is not stationary, then there will be another differencing performed. If the variance is not stationary, then there will be a transformation of the algorithm.

In this study, we applied ARIMA for testing variable in time series is inventory level. It's necessary to accommodate the stationary of variants, it should be noted that most periodicals are non-stationary and aspects AR and MA aspects of the ARIMA model are only relevant to periodic series of stations.

IOP Conf. Series: Materials Science and Engineering **528** (2019) 012044 doi:10.1088/1757-899X/528/1/012044

Distrust means no growth or decrease in data. Non-stationary time series must be converted to stationary data by making a difference. To find the best model to use in this forecasting problem we stated several options. We were modeled ARIMA as conducted a trial against the tentatively model (1-1-0 or 0-1-1 or 1-1-1),

3. Result and discussion

3.1. Data normalization

(i) As a first step, to define the best of architectural network, the data was divided into two sections. The two sections of data are: trained and tested data. Totally, we get 22 data from shop managers than we divided two groups of data as trained data 15 value and the rest of 7 data using the testing, this total of data gathering is concerned as our limitation of this research.

(ii) After we collected all the data that related to inventory of this building material, we need to normalize the data as from data input (X1, X2, X3,... Xn) and the output (Y) with formula from Jain et.al [13].

$$x' = \frac{0.8(x-b)}{(a-b)} + 0.1 \tag{1}$$

Which:

X'= normalized data; X= existing data; a= maximum data; b= minimum data

The secondary datasets were collected from shop managers and showed in Table 1.

Table 1. A sample of secondary data from shops managers

X_1	X_2	X_3	X_4	X_5	X_6	Y
Feb 2015	80.000	25	8	1	18	50
Mar 2015	80.000	42	14	2	18	20
Apr 2015	85.000	128	2	5	18	100
Aug 2016	83.000	53	3	1	19	50
Sep 2016	83.000	16	2	0	19	50
Oct 2016	83.000	131	5	3	19	100
Nov 2016	83.000	27	8	1	19	50

(iii) The data for input and target after normalized with formula in Equation 1 were shown in Table 2.

Table 2. A sample of normalization of data input and target

X_1	X_2	X_3	X_4	X_5	X_6	Y
0.17	0.10	0.17	0.40	0.26	0.50	0.23
0.25	0.10	0.23	0.70	0.42	0.90	0.10
0.32	0.90	0.53	0.10	0.90	0.90	0.46
0.39	0.90	0.18	0.60	0.26	0.90	0.23
0.46	0.26	0.27	0.90	0.42	0.90	0.23
0.54	0.26	0.40	0.55	0.58	0.90	0.23
0.61	0.26	0.19	0.15	0.26	0.90	0.23
0.68	0.26	0.23	0.35	0.26	0.85	0.23
0.75	0.26	0.10	0.15	0.10	0.85	0.23
0.83	0.42	0.26	0.45	0.26	0.85	0.23
0.90	0.42	0.90	0.15	0.42	0.90	0.90

The data training is a confirmation of the network model, which is doing multiple times of the attempt that count error to get the best network by determining the number of neurons manually. And this step finally obtained the best network with the smallest errors.

Table 3. MSE comparison for each training with various neurons

o	hl	Neuron	lr	mc	Activation Function	MSE
1	1	5-1	0.1	0.95	logsig-purelin	0.00403
2	1	6-1	0.1	0.95	logsig-purelin	0.00160
3	1	7-1	0.1	0.95	logsig-purelin	0.00170
4	1	8-1	0.1	0.95	logsig-purelin	0.00125
5	1	10-1	0.1	0.95	logsig-purelin	0.00099

In Table 3 we showed each MSE in training for 10 neurons and 1 hidden layer has the smallest MSE as 0.00099. Based on those result, we can define the best of architecture for neural network in this research are 6-10-1 it means has 6 inputs, 10 hidden layer neurons, and 1 output that represent the level of inventory.

Next, we used several of trained parameters which selected to find an optimal result as learning rate, constant momentum, Epoch, and Goal. In this research, we obtained best training performance is 0.00099057 with 265 epochs. All these results showed in Figure 2.

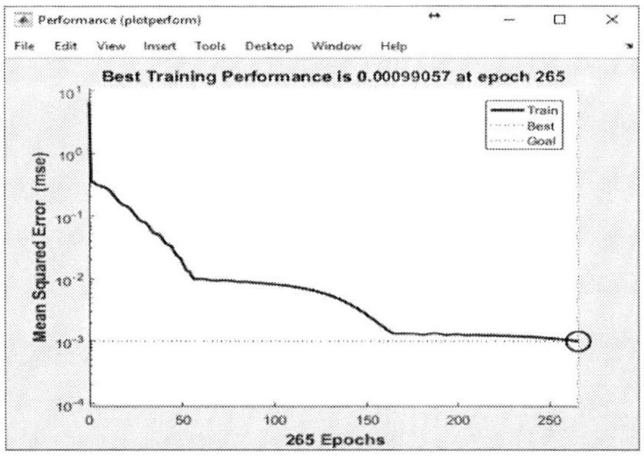

Figure 2. Graphics of network performance

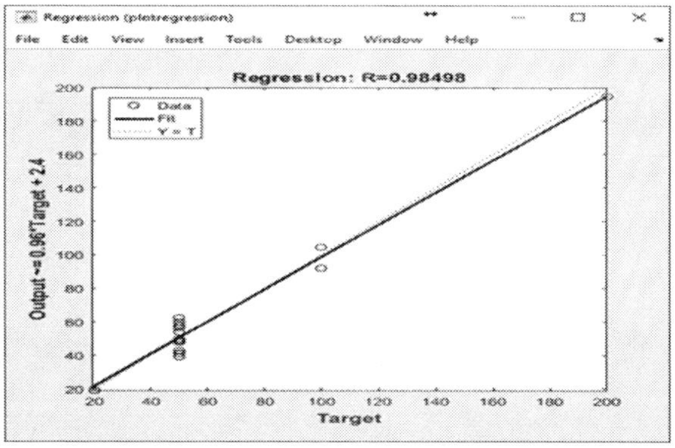

Figure 3. Graphics of trained data regression

This test is done by using the best architectural designs obtained from training, which uses a network structure that consisting of a layer with 6 input neurons. The first hidden layer consists of 10 neurons and the output layer consists of 1 neuron. For the activation function is sigmoid binary (logsig) function and identity function (purelin).

In Figure 3 shows the regression number is 0.98498 which means between the real variables and BPNN on the test has a good correlation. The correlation size shows a good level of equality between the relevant variable variables i.e.: a monthly period of sale, the price of product, amount of historical selling, an approximation of project renovation, an approximation of new project building and number of competitors

3.2 Autocorrelation and Partial Autocorrelation

It should be noted that most periodic sequences are non-stationary and that AR and MA aspects of the ARIMA model are only related to stationary periodic sequences. Stationary means having no growth or decline in the data. A non-stationary time series must be converted into stationary data by performing *differencing*. A scaled sequence is said to be stationary or showing a random error if the autocorrelation coefficient for all lags, the numbers are shown at each interval, are statistically not different from zero or different from zero for just several lags ahead. An autocorrelation coefficient is said to be not different from zero if it is in the interval. There was a Lambda (Rounded Value) = -0.5 obtained in the stationary step.

The autocorrelation coefficient for all lag, i.e. the figures shown at each interval is statistically no different from zero or different from zero to just a little lag behind. The autocorrelation coefficient is said to be no different from zero if in interval. At Stationery step to the various results obtained by Lambda (round values) = -0.5.

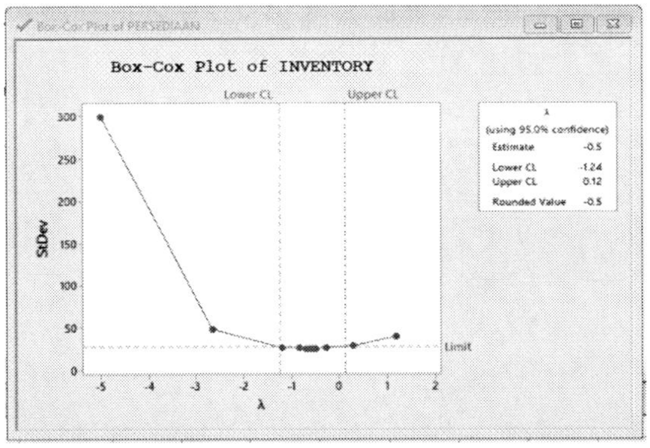

Figure 4. Plot graphics of previous stationary data

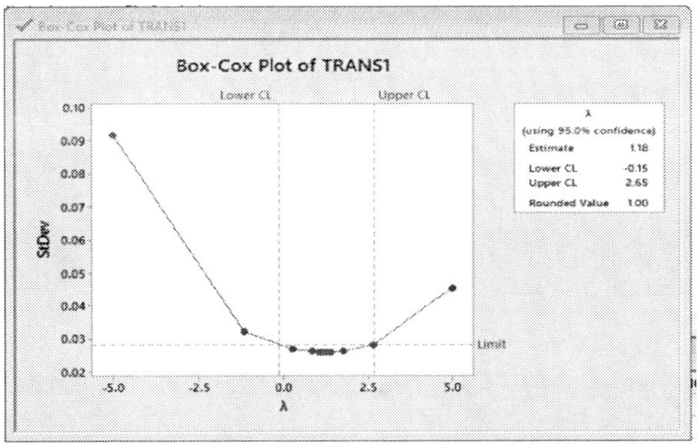

Figure 5. Plot graphics of stationary data after transformation

A data was said to be stationary for the variety as shown in Figure 4 and the mean, if the rounded value or lambda is worth = 1, then the existing data must be transformed first so that the Lambda value becomes optimal, by the formula: X'= X-2, or multiplied by 1 / λ.

TABLE 4. List of sample stationary data after transformation

Month	Inventory	Trans1
January 2016	100	0.1
February 2016	50	0.141421
March 2016	50	0.141421
April 2016	50	0.141421
May 2016	50	0.141421
June 2016	200	0.070711
July 2016	100	0.1
Augustus 2016	50	0.141421
September 2016	50	0.141421
October 2016	100	0.1
November 2016	50	0.141421

The results obtained from the initial data are still stationary as shown in Table 4, therefore must perform the initial data transformation, to accommodate the stationary of means in step to find stationary of the means, it is not acceptable if the amount of Lag is more than 3 [14] and the result of transformation data was shown in Figure 5.

From the result that showed in graph Autocorrelation and Partial correlation using MINITAB 17 there is no lag outside the confident interval that means the data was stationary and no need differencing process. Then we have three trials of tentative models and for the third Arima 1-1-1 that means order p = 1, d = 1 and order q =1 has significant parameters that be obtained. Finally, we obtained the comparison of forecast result from BPNN and ARIMA in period May 2016 until November 2016 as shown in Table 5.

Table 5. Forecasting comparison with BPNN and Arima

Month	Target (Qty)	Forecast	
		BPNN (Qty)	ARIMA(Qty)
May 2016	50	60.5	49.732
June 2016	200	134.94	101,153
July 2016	100	55.93	24.79
August 2016	50	62.71	49.228
September 2016	50	51.61	49.594
October 2016	100	100.68	109.533
November 2016	50	52.57	50.322

We followed the method of performance criteria measurement for forecasting model from Tay and Cao [15] and Lu.et.al [16]. We evaluated and used respectively mean absolute difference (MAD), mean absolute error (MAE), root mean square error (RMSE), root relative squared error (RRSE), and relative absolute error (RAE).All these results were listed in Table 6.

Table 6. Error prediction result of BPNN and Arima

	MAD	MAE	RMSE	RRSE	RAE
BPNN	19.6	19.6	30.4	0.6	0.5
ARIMA	26.5	26.5	47.0	0.9	0.6

In improving the efficiency of the supply chain system, there is a need for demand forecasting because each stakeholder involved makes a demand order in response to a demand signal. A good demand forecasting is which can have accurate estimation so as to facilitate the operations of the company such as in planning the material needs and inventory management.

This paper presents a time series approach for modeling in the building materials demand forecasting. The approach used is a linear ARIMA model and non-linear BPNN model. The level of accuracy can be concluded into a good category because it shows the data verification between the actual data and BPNN which most of the data do not have significant differences. As discussed with the store manager that final result is expected. This forecast with BPNN will enable the companies to maximize sales and profits, so they can have a good impact on their business continuity.

4. Conclusion

Based on our case study, we concluded the back propagation neural network architecture can be used for inventory predictions in the company is a multi-feed forward layer network with 6-10-1 neuronal structures, or 6 input neurons, 10 neurons, and 1 hidden layer. In this study, we defined the six input neurons that consist of: a monthly period of sale, the price of the product, amount of historical selling, an approximation of project renovation, an approximation of new project building and number of competitor. As an output of neurons is an inventory forecast value. The activation function was used is binary function sigmoid (logsig) and identity function (purelin) and this model was selected for the best forecast by using parameter of learning rate value 0.1 and momentum constant value 0.95. As a limitation, in this research, we only get 30 data.

As accuracy of prediction is concerned, we compared the error of prediction with performance criteria such as MAD, MAE, RMSE, RRSE and RAE. From the experimental result, this research found that the forecasting with BPNN algorithm has more accurate than ARIMA. This research also suggested in predictions that obtained considerable variation in results, this may be due to limited training data variables. It is expected to continue investigations by adding product variant data, as well as adding or converting with variables that may affect the procurement of goods, for example by reviewing the number of building projects or building renovations that are the major factors affecting the sale of building materials.

5. References

[1] Albarune A R B and Habib M M 2015 A study of forecasting practices in supply chain management **4** pp 55-61

[2] Mishra B K, Raghunathan S and Yue X H 2009 Demand forecast sharing in supply chains. *Production & Operations Management* **18** (2) pp 152-166

[3] Friedman J H 1991 Multivariate adaptive regression splines. The Annals of Statistics **19** (1) ,pp. 1–141

[4] Chang P C and Wang Y W 2006 Fuzzy Delphi and backpropagation model for sales forecasting in PCB industry. *Expert Systems with Applications* **30** (4) pp 715–726

[5] Fildes R, Nikolopoulos K, Crone S F, Syntetos A A 2008 Forecasting and operational research: a review. *Journal of the Operational Research Society* **59** (9) pp 1150–1172

[6] Lican H, Yuhong Z, Xin X and Fan F 2010 Prediction of investment on inventory clearance based on improved BP neural network in *Proceedings of the 1st International Conference on Networking and Distributed Computing* pp 73–75

[7] Khasel M, Bijari M and Ardali G A R 2009 Improvement of Auto-Regressive Integrated Moving Average models using Fuzzy logic and control. pp 956-967

[8] Mai T, Ghosh B and Wilson S 2012 Multivariate short-term traffic flow forecasting using Bayesian vector autoregressive moving average model. *Transportation Research Board 91st Annual Meeting* **12** 3728

[9] Wu W, Zhang W, Yang Y and Wang Q 2010 Time series analysis for bug number prediction," in *Proc. 2nd Int. Conf. Softw. Eng. Data Mining (SEDM)* pp 589-596

[10] Ghosh A and Chakraborty M 2012 Hybrid optimized back propagation learning algorithm for multi-layer perceptron, *Int. J. Comput. Appl* **60** (13)

[11] Wang X P, Shi Y, Ruan J B and Shang H Y 2010 Study on the inventory forecasting in supply chains based on rough set theory and improved BP neural network. *Advances in Intelligent Decision Technologies Smart Innovation, Systems and Technologies* **4** pp 215–225

[12] Box G E, Jenkins M and Reinsel G C 1994 Time Series Analysis: Forecasting and Control. 3rd ed. (Englewood Cliffs, NJ: Prentice Hal)

[13] Jain A K, Mao J and Mohuddin K M 1996 Artificial Neural Networks: A Tutorial. *IEEE Computer* **29** (3) pp 31-44

[14] Wu W, Zhang W, Yang Y and Wang Q 2010 Time series analysis for bug number prediction, in *Proc. 2nd Int. Conf. Softw. Eng. Data Mining (SEDM)* pp 589-596

[15] Tay F E H and Cao L 2001 Application of support vector machines in financial time series forecasting Omega **29**(4) pp 309-317

[16] Lu C J, Lee T S and Chiu C C 2009 Financial time series forecasting using independent component analysis and support vector machine. Decision Support Systems 47(2) pp 115-125

Desin of Job Scheduling System and Software for Packaging Process with SPT, EDD, LPT, CDS and NEH algorithm at PT. ACP

L Gozali[1] , V Kurniawan[2], S R Nasution[3]

[1]Lecturer of Industrial Engineering Department of Tarumanagara University, Jakarta
[2]Graduated Student of Industrial Engineering Department of Tarumanagara University, Jakarta
[3]Lecturer of Industrial Engineering Department of Pancasila University, Jakarta
ligoz@ymail.com

Abstract. PT. Avesta Continental Pack is a manufacturer of flexible packaging for pharmaceutical products. The company faces ineffective production scheduling and repeatedly failed to finish orders at the expected time limit. Customers requesting orders in different finishing time contributes in causing delays in the delivery process. Consequently, customers ended up claiming the fee because of the late delivery. Therefore, a more advanced scheduling system will assist the company in managing the delivery time. The purpose of this research is to establish the most effective job scheduling arrangement to minimize mean tardiness as well as mean lateness job. As the production process uses flow shop method, this research applies heuristic methods, such as Earliest Due Date (EDD) method, Short Processing Time (SPT), Nawaz, Enscore, and Ham (NEH) algorithm and Campbell, Dudek, and Smith (CDS) algorithm. The result shows that SPT method provides the most optimum outcome, specifically by reducing mean tardiness by 801.81 minutes or 85.57% and decreasing mean lateness job by 2 or 66.67%. The best job sequence is subsequently job 8 - job 4 - job 1 - job 6 - job 7 - job 5 - job 2 - job 3, which minimizes the average delay time from 937.05 minutes to 135.24 minutes.
Keywords: Flow Shop Scheduling; Mean Tardiness; Mean Lateness Job, Heuristic Scheduling Method

1. Introduction

Rapid industrialization has caused companies to be highly competitive. To fulfil market demands, manufacturing companies are inquired to achieve a certain level of effectivity and efficiency in their production process. One of the ways to reach the target of production is by arranging meticulous scheduling and job management. Inefficient job scheduling causes massive pile up in the product processing stage, increases the total of production time, and inflicts lateness of the product delivery time. PT. Avesta Continental Pack is a company that manufactures flexible packaging for pharmaceutical products. PT. Avesta Continental Pack is a large and reputable company that regularly acquires numerous demands of production as shown in the increasing number of their machines, employees, and operators. Its production process includes printing, inspection, laminating, slitting, and packaging.

The company has been having difficulties in meeting the expected delivery time. Based on the last 3 months data in 2017, there was 28.92% delay in October, 25.5% in November, and 23.83% in December. One of the main causes of the delay is ineffective production scheduling, which was mostly arranged manually. Companies must pay for a penalty claim for the delay which also reduces consumers' trust and satisfaction. The company serves customers' orders using First Come First Serve (FCFS) method. Unfortunately, this method failed to solve the delay delivery time problem.

2. Method

2.1. Terminology of scheduling

Scheduling method is included in the body of science, particularly related to frameworks, techniques, and understandings of the systems (Baker, 1974). In manufacturing companies, types of scheduling have similar structures. It contains a set of jobs to be calculated and a set of machines available to execute the jobs. General problem regarding scheduling revolves around the failure to arrange the timing of jobs based on the capability of the machines. This problem frequently comes up during the decision-making process as the result of using ineffective scheduling method.

While Krajewski and Ritzman (2001) stated that scheduling is the allocation of resources from time to time to support the implementation and completion of a specific work activity. Decisions regarding companies' resources allocation (human resources, capacity resources, production equipment or machines, and time) should realize the efficient and the effective use of resources in the process of producing the right amount of output at the expected time and in a certain expected quality (Subagyo, 2000).

2.2. Nawaz, Enscore, dan Ham (NEH) Algorithm Scheduling Method

Nawaz, Enscore, dan Ham (NEH) method was first coined by Muhammad Nawaz, E. Emory Enscore Jr, and Inyong Ham in year 1983. In a general flow shop, where all the jobs must be produced by the machines in the same order, certain heuristic algorithms propose that the jobs with higher total processing time should be given a higher priority than the jobs with less total processing time (Nawaz et. al., 1983).

2.3. Campbell, Dudek, dan Smith (CDS) Scheduling Method

This method was introduced by H.G. Campbell, R.A. Dudek, and M.L. Smith (1970), and it is based on the Johnson algorithm scheduling method. This method solves the problem of n jobs in m flow shop machines. Campbell, Dudek, and Smith (CDS) scheduling method is one of the many methods of production scheduling that can minimize the number of late jobs. The CDS method looks for the best priority sequence by combining the existing work stations into two groups of machines. Machines in both groups are sorted by priority based on the shortest processing time. Hence, the shortest processing machine is sorted to be the first. Meanwhile, the longest processing machine is sorted to be the last (Herjanto, 2007).

3. Results and Discussion

3.1. Data Collection

The collected time data refers to the setup time data from each main process. The setup time data was collected on March 21, 2018. The 30 set up time data were recorded for each process.

3.2. Data Testing

Data testing in this research includes normality test, uniformity test, and adequacy test. If the data is tested and declared feasible, then it is feasible for next data processing. All processed data involve some adjustment factors. For example, the standard time for 1 colour of printing process can be obtained as below:

$$\text{Standard time} = 74{,}13 * (1 + 52\%) = 112{,}68 \text{ menit}$$ (Karanjkar,2008)

3.3. Data Calculation

The data for this research includes date of order, product data, order quantity, and delivery deadline time. The following table is the production data taken from the period of 8 January 2018 to 12 January 2018. Research Production data shows in Table 1.

After obtaining the research production data in table 1, the number of orders in the roll units will be converted into meters to find out the length of raw material needed to make the roll.

Table 1. Research Production Data

Job	Product Name	Date of Order	Quantity (Roll)	Measurement	Colours	Due Date
1	MICROTINA CAPSULE	1/8/2018	1620	130 x 500 Meter	3	1/21/2018
2	GRANTUSIF KAPLET	1/8/2018	1450	261 x 500 meter	2	1/20/2018
3	MIXAGRIP FLU & BATUK	1/9/2018	815	350 x 500 meter	5	1/24/2018
4	VIFERRON EXP KAPLET	1/9/2018	670	292 x 500 meter	1	1/20/2018
5	GLIQUIDONE TABLET	1/10/2018	1760	187 x 500 Meter	3	1/21/2018
6	CEFADROXIL KAPS	1/10/2018	900	260 x 500 Meter	5	1/20/2018
7	ORNAMENT	1/11/2018	1000	260 x 500 meter	1	1/21/2018
8	ANTIMO ANAK RASA JERUK	1/12/2018	850	180 x 500 Meter	7	1/22/2018

Data of customers' orders from 8 January 2018 to 12 January 2018 which was collected in each process and each job for each machine is shown in Table 2.

Table 2. Processing Time in each Machine

Processing and Setup Time (minutes)					
Job	Printing	Inspection	Laminating	Slitting	Total Time
1	2938	1856	2316	1893	9004
2	5114	3323	4076	3360	15873
3	5779	3735	4571	3772	17858
4	2416	1535	1931	1572	7455
5	4187	2689	3315	2726	12917
6	3270	2063	2564	2100	9995
7	3550	2292	2839	2329	11009
8	2157	1299	1647	1336	6438

The data of processing time for each machine was recorded by monitoring the speed of each production machine. For example, the calculations for obtaining processing time on a printing machine for job 1 are as follows.

Process time= $\frac{measurement\ of\ raw\ material\ needed}{speed\ of\ the\ machine} + set\ up\ time = \frac{222750\ m}{80\ m/minutes} + 154{,}1\ minutes =$ $2938{,}48\ minutes.$ (Herjanto, 2007)

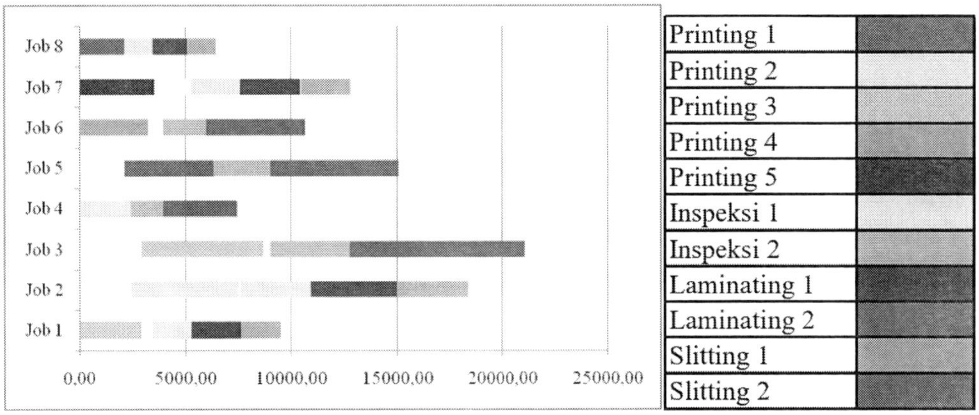

Figure 1. Gantt Chart of SPT Scheduling Method

The Shortest Processing Time (SPT) scheduling method (Heizer, 2005) sorts 8 jobs based on the shortest to the longest total processing time. In this scheduling method, the sequence of the job preparation results in 8-4-1-6-7-5-2-3 order with 1 late job and the mean tardiness of 135.24 minutes. Gantt Chart Scheduling with SPT Method shows in Figure 1. Sorting the 8 jobs starting from the longest to the shortest total processing time (LPT) as shown in Table 3.

Table 3 Sorted Job with LPT Scheduling Method

	Processing and Setup Time (minutes)				
Job	Printing	Inspection	Laminating	Slitting	Total Time
3	5779	3735	4571	3772	17858
2	5114	3323	4076	3360	15873
5	4187	2689	3315	2726	12917
7	3550	2292	2839	2329	11009
6	3270	2063	2564	2100	9995
1	2938	1856	2316	1893	9004
4	2416	1535	1931	1572	7455
8	2157	1299	1647	1336	6438

In NEH scheduling algorithm, the best job sequence is 3-2-5-7-8-6-1-4, which resulted in 3 late jobs and mean tardiness of 355.04 minutes. The Gantt Scheduling Chart with the NEH Method is shown in Figure 2.

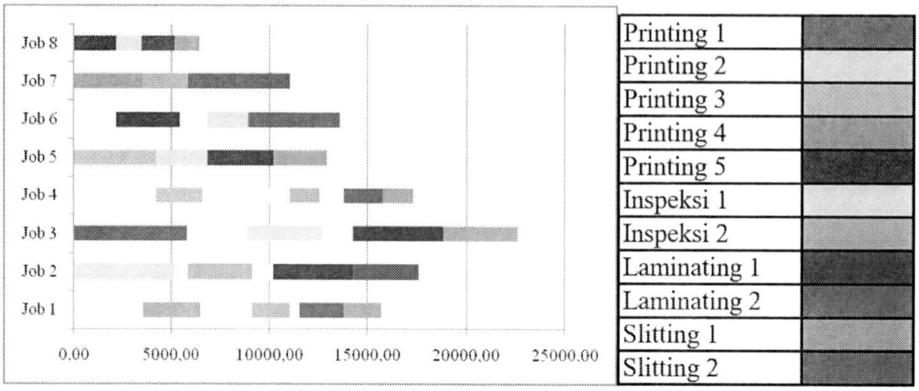

Figure 2. Gantt Chart of NEH Scheduling Method

Table 4 Sorted Job First Iteration with CDS algorithm

	Printing	Slitting
Job 1	2938	1893
Job 2	5114	3360
Job 3	5779	3772
Job 4	2416	1572
Job 5	4187	2726
Job 6	3270	2100
Job 7	3550	2329
Job 8	2157	1336

The First Job Iteration Ordering Table with the CDS Algorithm is shown in Table 4. As shown at Table 5, the shortest processing time is at job 8. Therefore, based on Johnson's method, the job 8 will be placed in the back order and will be removed from this job order in the next calculation. In CDS scheduling algorithm acquires a job sequence scheduling from 3 iterations.

Table 5. Calculation Result of Each Iteration with CDS Algorithm

Iteration	Job Sorted	Mean Tardiness (minutes)
K1	3-2-5-7-6-1-4-8	520.73
K2	3-2-5-7-6-1-4-8	520.73
K3	3-2-5-7-6-1-4-8	520.73

Table 5 exhibits three iterations which resulted the same mean tardiness. The job sequence with CDS algorithm generates 3-2-5-7-6-1-4-8 which caused 1 late job and mean tardiness of 520.73 minutes. Gantt Chart Scheduling with the CDS Method exhibit in Figure 3.

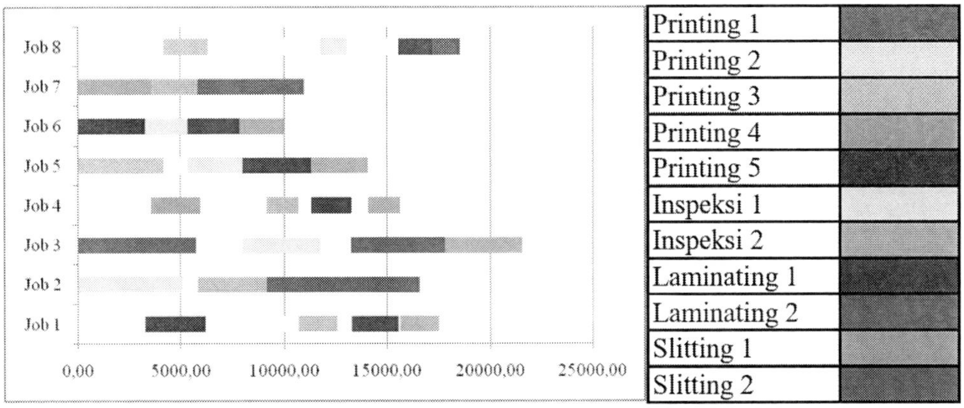

Figure 3. Gantt Chart CDS Scheduling Method

After all the calculation with EDD, SPT, NEH algorithms, and CDS scheduling algorithms and methods (Gozali, 2014), the next step is comparing the best results. Comparison of all scheduling methods is shown in Table 6.

Table 6. Comparison of the Result of Each Method

Method	Mean Tardiness (mnt)	Reducing (mnt)	% reducing	No Lateness jobs
Initial	937.05	-	-	3
EDD	355.63	581.42	62	3
SPT	135.24	801.81	86	1
NEH	355.04	582.01	62	3
CDS	520.73	416.32	44	1

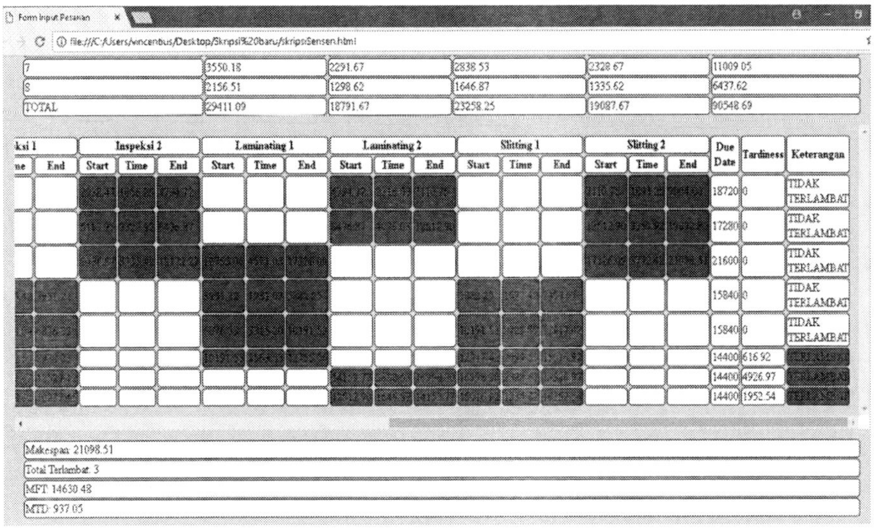

Figure 4 Input screen from scheduling method for this company

IOP Conf. Series: Materials Science and Engineering **528** (2019) 012045 doi:10.1088/1757-899X/528/1/012045

After finding the best proposed method that can minimize the delay time (mean tardiness) and the number of late jobs, a decision-making system is designed to conduct production scheduling using the SPT method which is the best proposed method. This system provides the best solution for scheduling method and capable of producing the best output. The application created is a web-based application, which means that this application can be used in web browsers, such as Google Chrome, Mozilla Firefox, and Microsoft Edge. This application has two main screens, including the scheduling input screen that contains order input and capacity, and the other screen to generate output from the scheduling or the process screen. Figure 4 exhibits the results of scheduling calculations by the FCFS method which resulted in 3 late jobs and an average delay time (MTD) of 937.05 minutes.

To minimize these conditions (lateness and tardiness), the SPT scheduling method produces the best solution. The output display of SPT Scheduling method shows in Figure 5.

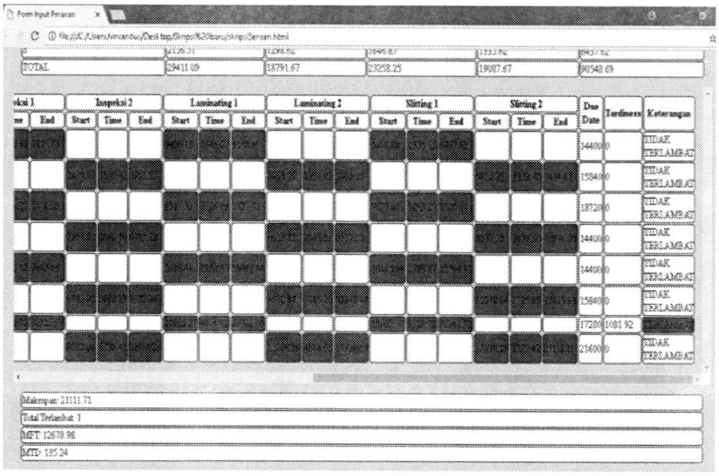

Figure 5. Output screen of SPT scheduling method for this company

4. Discussion and conclusion

Tyagi (2017) stated that eight Dispatching rules as SPT, LPT, EDD, MDD, FCFS, LCFS, MST and CR, an exact algorithm as Branch and Bound algorithm, Johnson's algorithm for two and three machines flow shop scheduling problems are described. In addition, four constructed and four improved heuristic algorithms are proposed for more than three machines scheduling problems. Palmer, Gupta, CDS, and NEH heuristic algorithm are proposed as constructed heuristic algorithm. Metaheuristic algorithms works as an Improved heuristic algorithm. Genetic Algorithm, Ant Colony Optimization, Simulated Annealing and Tabu Search are proposed as Improved Heuristic algorithms. All the proposed techniques are effective for solving the scheduling problems in various environments for obtaining the optimum or near optimum sequences.

Gozali (2014) found that NEH is the best solution compared to CDS and Gupta Scheduling Method. Febianti (2017) stated that CDS, NEH and Heuristic Pour scheduling method achieved all the best same result. Jungwattanakit (2006a) declared that the simple dispatching rules the SPT, LPT, ERD, and HSE rules are good algorithms whereas for the flow shop makespan heuristics, the NEH algorithm is most superior to the other constructive algorithms. Jungwattanakit (2006b) also found that the NEH algorithm is an excellent constructive algorithm for minimizing the objective functions. The NEH algorithm is the most superior compared to the other constructive algorithms.

The results in this research show that from the proposed methods using EDD, SPT, NEH, and CDS scheduling methods, the best scheduling method is the Shortest Processing Time (SPT) method. This research also found that the best job sequence is job 8 - job 4 - job 1 - job 6 - job 7 - job 5 - job 2 - job 3, which minimizes the average delay time from 937.05 minutes to 135.24 minutes, meaning this calculation reduces 801.81 minutes or 85.57% tardiness. Moreover, this calculation also reduces late jobs from 3 jobs to 1 job. The scheduling application system is web-based; hence it is easy to use. It also contains comparisons of all proposed methods, so the scheduling processes will calculate quickly and accurately.

5. References

[1] Baker, K. R. 1974. *Introduction to Sequencing and Scheduling*. John Wiley & Sons.

[2] Campbell, H. G., Dudek, R. A., & Smith, M. L. 1970. A heuristic algorithm for the n job, m machine sequencing problem. Management science, **16** 10, B-630.

[3] Febianti, E., & Zikry, Z. 2018. PENJADWALAN PRODUKSI FLOWSHOP DENGAN METODE CAMPBELL DUDEK SMITH, NAWAZ ENSCORE HAM DAN HEURISTIC POUR. In Prosiding Seminar Nasional Teknologi, Inovasi dan Aplikasi di Lingkungan Tropis 1 1, pp. 25-33).

[4] Gozali, L., Ariyanti, S., & Natali, F. L 2014 Machine Scheduling Proposed for Minimizing Makespan at PT Adiperkasa Anugrah Pratama, *7th International Seminar on Industrial Engineering and Management*, Bali, Indonesia pp PS1-PS6

[5] Heizer, Jay and, B. Render 2005. *Manajemen Operasi* Ed 7 Jilid 2. Jakarta : Salemba Empat, Indonesia

[6] Harjanto, E. 2007. *Manajemen Operasi* Ed 3. Grasindo, Indonesia

[7] Jungwattanakit, J., Reodecha, M., Chaovalitwongse, P., & Werner, F. 2006a. Sequencing heuristics for flexible flow shop scheduling problems with unrelated parallel machines and setup times. In *Proceedings of the 2006 IE network national conference*, Bangkok, Thailand (pp. 1-8).

[8] Jungwattanakit, J., Reodecha, M., Chaovalitwongse, P., & Werner, F. 2006b. *Sequencing algorithms for flexible flow shop problems with unrelated parallel machines, setup times, and dual criteria*. Univ., Fak. für Mathematik.

[9] Karanjkar, A. 2008. *Manufacturing and Operations Management*. Nırali Prakashan.

[10] Krajewski, L. J., & Ritzman, L. P. 2001. *Operations management: strategy and an*alysis. Pearson College Division.

[11] Subagyo, P. 2000. *Manajemen Operasi*. BPFE Yogyakarta, Indonesia

[12] Nawaz, M., Enscore Jr, E. E., & Ham, I. 1983. *A heuristic algorithm for the m-machine, n-job flow-shop sequencing problem*. Omega, **11**(1), 91-95.

[13] Tyagi, N., Tripathi, R. P., & Chandramouli, A. B. 2017. SEQUENCING AND SCHEDULING METHODOLOGIES. In Proceedings of National Conference on Future Innovations & Research in Science and Technology **22**, p. 23.

IOP Conf. Series: Materials Science and Engineering **528** (2019) 012046 doi:10.1088/1757-899X/528/1/012046

Analysis of Double indian Ballbreaker Net Sorter Machine Based on Overall Equipment Effectiveness Method Cases in Tea Plantation Plants

J Alhilman and A F Abdillah

School of Industrial and Systems Engineering
Telkom University
Bandung, West Java, Indonesia
alhilman@telkomuniversity.ac.id, afabdilah96@gmail.comIndonesia

Abstract. In a factory, the position of the production machine plays an important role because if the engine is damaged it will disrupt the production target. This study analyzes the performance of Double Indian Ballbreaker Net Sorter (DIBN) machines, one of the machines used to produce orthodox tea which in the fact, this machine often damaged and has high downtime. This leads to a low level of machine availability in the Production Department. For that, we need an Overall Equipment Effectiveness (OEE) method to measure the performance and the level of effectiveness of the machine and six big losses factor analysis to find out what factors cause low of OEE value. Based on the OEE calculation results, the value of OEE DIBN machine is 53.98%. The result is still far from the standard set by the Japanese Institute of Plant Maintenance of 85%. From six big losses, it is known that the most influencing factors to decrease the effectiveness of DIBN machines are rework losses (23.33%), reduced yield losses (20.17%) and reduced speed of 19.49%.
Keywords: overall equipment effectiveness, six big losses, performance

1. Introduction

1.1 Background
Nowadays, the major daily problems that are encountered by many manufacturing companies are equipment breakdown, repair, and quality defects. These problems have a great impact on quality and delivery time [2]. Machine maintenance is inseparable from the problems of efficiency and effectiveness of machinery. Therefore, steps are needed to prevent or overcome the problem of machine downtime. The purpose of this study is to measure the performance of production equipment, find the root cause of the problem and provide suggestions for improvement. There have been many previous studies using OEE methods, including the OEE method used to measure the effectiveness of a machine in a sugar factory and to find the cause of the low effectiveness of the machine [3] and according to [4] to measure the effectiveness of the production machine in the fibre cement plant, the OEE method is used and to determine the critical downtime, Pareto diagram is used. Furthermore, OEE is used to measure the effectiveness of rotary car dumper engines and to find the cause of the greatest contribution to decreasing efficiency, the cause and effect diagrams are used. The result is reduced speed losses factor of 68.25% and breakdown losses of 21.06%. [5] and according to [6] OEE is used to measure the effectiveness of

an 80 ton Press machine on a line of P3C03 3 & 4 steel wheel factories. The results of the fishbone diagram found the biggest cause of low efficiency is the less optimal nursing time and less effective cutting machine factors. The maintenance propose is carried out every 1900 minutes.

This research is implemented in PT. X. PT. X is one of the companies engaged in agro-business and agro-industry. The products produced are various kinds of tea both for domestic consumption and export quality. With high demand, companies always produce tea on a large scale. To pursue the target requires the production department to run non-stop tea production for 24 hours, this causes frequent engine damage that results in not meeting the number of production targets. For this reason, a method is needed to determine the effectiveness of the overall use of factory facilities. The method used is the OEE method. Furthermore, to find out what factors are most influential in reducing the effectiveness of the machine will be analyzed by the six big losses method.

In the tea production process, it is divided into 3 production rooms, namely Laying & Milling Room, Drying Room, and Sorting & Packing Room. This research is a focus to the highest downtime occurred that in the Laying & Milling Room on the Double Indian Ballbreaker Net Sorter machine.

2. Methods

2.1 Overall Equipment Effectiveness

OEE is a product of six big losses that can measure the overall effectiveness of the equipment by multiplying availability, performance rate, and rate of the quality product [7]. This effectiveness measurement combines the factors of time, speed, and quality of operation of the equipment and measures how these factors can increase added value. OEE tell us whether our equipment is under-utilization or over utilization[8]. According to [9] to facilitate OEE analysis in determining treatment policies, the OEE application was made.

2.2 Availability

According to Moubray [7], availability (A) is the ratio between the operating time of the tool minus the downtime, compared to the operating time of the tool. In its calculation, availability is formulated in the following equation [3]:

$$Availability = \frac{Operation\ Time}{Loading\ Time} \qquad (1)$$

$$A = \frac{Loading\ Time - Downtime}{Loading\ Time} \qquad (2)$$

2.3 Performance Efficiency

Performance Efficiency (P) is the ratio of the ratio of the actual production level to the expected level of production or the ratio of machine capability that is realized by multiplying the number of goods produced with the ideal time which is then divided by the operating time stated in the percentage [9], or the product times operating speed rate and net operating rate. For this reason, the mathematical formula is:

$$\text{Performance Efficiency} = = \frac{Theoritical\ Cycle\ Time \times Processed\ Amount}{Operating\ Time} \times 100\% \qquad (3)$$

Where :
Theoretical Cycle Time = the operating time performed by the machine per day or per hours based on theoretical / standard.
Processed Amount = number of products processed by the machine
Operation Time = the operating time of a machine in one day

2.4 Rate Of Quality Product

The third element of OEE is quality (Q). The rate of quality is the ratio of the number of products that is better to the total number of products processed. In the calculation, the rate of quality is formulated in the following equation:

$$Q = \frac{Input - volume\ of\ quality\ defects}{Input} \tag{4}$$

2.5 OEE Calculation Procedure

Steps involved to calculate OEE based on [8] are as follows

- Input the value of Total shift time
- Input the value of Planned downtime
- Calculate Loading time = Total time - Planned downtime
- Input the value of all downtime and all stop time
- Calculate operating time = Loading time (all downtime + all stop time)
- Availability = Operating time / Loading time
- Input value of actual cycle time and theoretical cycle time
- Calculate operating speed rate = Theoretical cycle time / Actual cycle time
- The input value of the Actual processing time and the total amount produced
- Calculate net operating rate = Actual processing time / operating time.
- Calculate performance efficiency = net operating rate X operating speed rate
- Input the value of defect amount
- Calculate Quality rate = (total amount produced - defect amount) / total amount Produced.
- OEE = Availability X Performance Efficiency X Quality

2.6 Six Big Losses

Furthermore, after the OEE calculation is carried out, to analyze the causes of losses and to measure the effectiveness of machinery or equipment can be identified through the six big losses [3], including:

2.6.1. Equipment Failures

Equipment failures are caused by equipment malfunction that requires upgrading and a breakdown time. Major losses include product opportunity loss, spare parts loss, and sporadic losses. Equipment failures can be calculated by the following formula [4]:

$$\text{Breakdown Loss} = \frac{\text{Total } Breakdown\ Time}{Loading\ Time} \times 100\ \% \tag{5}$$

2.6.2. Setup and Adjustment

Setup and adjustment are caused by changes that occur when the engine is operating, such as changes in the type of product that is made, changes in work shifts, and adjustments to operating conditions that make the machine stop working. Setup and adjustment can be calculated by the following formula [4]:

$$\text{Set-up and Adjustment} = \frac{\text{Total } Set-up\ and\ Adjustment}{Loading\ Time} \times 100\% \tag{6}$$

2.6.3. Idling and Minor Stoppages

Idling and minor stoppages are sensor errors or activities waiting for material/parts to come or be processed. Idling and minor stoppages can be caused by the engine being blocked or having stopped for a while, or the engine that stops working because it has to wait (idling). Idling and minor stoppages can be calculated by the following formula [4]

$$\text{Idling and Minor Stoppages} = \frac{Nonproductive\ Time}{Loading\ Time} \times 100\% \tag{7}$$

$$\text{Nonproductive Time} = \text{Ideal Cycle Time} \times \text{Production Difference} \quad (8)$$

2.6.4. Reduce Speed Losses

Reduced speed losses are caused by a decrease in engine speed when operating, ie when the engine does not work at its normal speed. Reduced speed losses can be calculated by the following formula [4]:

$$\text{Reduced Speed Losses} = \frac{(Actual - Ideal\ Operation\ Time)}{Loading\ Time} \times 100\% \quad (9)$$

2.6.5. Defect Losses

The defect is caused by the product being produced outside the specified specifications, or defective during the normal production process. The quality of the resulting product is bad. Products must be reworked or redesigned so that they can be used or sold. Defect losses can be calculated by the following formula [4]:

$$\text{Defect Losses} = \frac{(\text{Total } Reject\ Product \times Ideal\ Cycle\ Time)}{Loading\ Time} \times 100\% \quad (10)$$

2.6.6. Reduce Yield

Reduced yield is caused by the length of the machine to adjust conditions to normal conditions, causing the number of products to be rejected. The loss is caused by a situation where the product produced is not in accordance with the standard, because there is a difference in quality between the time the machine was first turned on and when the machine was operating stably. Reduced yield can be calculated by the following formula [4]:

$$\text{Reduce Yield} = \frac{(Ideal\ Cycle\ Time \times Reject\ Product)}{Loading\ Time} \times 100\% \quad (11)$$

The relationship between equipment, OEE, and Six Big Losses can be seen in table 1.

Table 1. OEE metrics [2]

OEE Loss Classifications	Six Big Loss Category	Computation of OEE
	Equipment failure	
Availability Rate	Setup and adjustment	Availability=Operating time/loading time
Performance Rate	Idling and minor stoppage	Performance rate = Net operating time/operating time
	Reduced speed	
Quality Rate	Defects in process	Quality rate = (processed amount – defect amount)/processed amount
	Reduced yield	

3. Results and Discussion

3.1 Calculation of Overall Equipment Effectiveness

The DIBN engine OEE calculation based on damage data for 2016-2017 is shown in Table 2.

Table 2. Overall Equipment Effectiveness Value of the DIBN Machine

Year	Month	A	P	Q	OEE
	1	98.45	75.90	80.20	59.93
	2	97.99	89.39	65.89	57.72
	3	95.81	99.65	78.88	75.31
	4	88.45	87.55	65.67	50.85
	5	86.80	85.55	75.45	56.03
2016	6	96.95	98.65	55.80	53.37
	7	97.85	90.90	80.53	71.63
	8	99.65	78.90	72.20	56.77
	9	89.75	89.75	65.58	52.83
	10	79.90	90.75	55.51	40.25
	11	85.78	90.25	57.78	44.73
	12	89.86	89.50	76.87	61.82
	1	90.79	88.78	45.80	36.92
	2	89.55	90.98	54.68	44.55
	3	87.88	79.88	67.89	47.66
	4	90.42	78.43	78.76	55.85
	5	92.35	87.80	68.90	55.87
2017	6	94.21	86.56	78.65	64.14
	7	90.35	88.66	55.65	44.58
	8	89.57	87.56	75.65	59.33
	9	94.25	89.89	60.65	51.38
	10	91.12	87.76	65.67	52.51
	11	97.66	88.99	78.65	68.35
	12	81.85	78.45	55.60	35.70
Average		91.55	87.52	67.37	53.98

The OEE value for DIBN machines in 2016-2017 was obtained from the results multiplication of the Availability(A), Performance(P) and Quality(Q) rate.
The final OEE result (53.98) is the average of OEE values from year 2016 to 2017.

3.2 Six Big Losses calculations

After obtaining the results of the OEE, an analysis of the six big losses is conducted to find out the factors which resulted losses. The results of the six big losses are expressed as a percentage of total losses so that it can be seen what factors influence the effectiveness of the machine which can be seen in Table 3.

From table 3 it can be seen that the biggest factor that causes loss are the rework factor of 23.33% , the reduced yield of 20.17% and reduced speed of 19.49%.

3.3. Overall Equipment Effectiveness Total

Based on the calculations above, according to Jim's Japanese Institute of Plant Maintenance[10], OEE values can be said to meet the criteria standards if they get a minimum value of 85%. Explanation of the classification of the fulfillment of JIPM criteria for DIBN machines in 2016 and 2017 can be seen in Table 4.

Table 3. Value of the Six Big DIBN Machine Losses

No	Six Big Losses	Persentase Losses(%)	Total Losses (%)
1	Equipment Failures	1.45	3.99
2	Setup and Adjustment	5.55	15.24
3	Idle and Minor Stoppages	6.48	17.78
4	Reduce Speed	7.10	19.49
5	Rework	8.5	23.33
6	Reduced Yield	7.35	20.17
	Total	36.43	100

Table 4. Classification of Fulfillment of 2016 DIBN JIPM Engine Criteria

Year	Availability	Performance Efficiency	The rate of Quality Product	OEE
2016	Fulfillment	Fulfillment	NO	NO
2017	Fulfillment	Fulfillment	NO	NO

Based on Table 4 it can be seen that the Availability and Performance Efficiency factors in 2016 and 2017 meet JIPM criteria standards because the average score is above 85%. The causes that make up the Availability value meet the standards, one of which is the implementation of preventive maintenance activities that are quite good so that the machine is always ready to be produced. Furthermore, the Performance Efficiency factor also meets the standard, one of which can be seen in terms of the total amount of products produced is around the value of the average production target, this indicates the performance of the engine is still pretty good. The last factor is the Rate of Quality which the value 67.37 % does not meet the standard. This can be due to the fact that there are still many products with defects that do not meet the established criteria. Then for the overall OEE value of 2016-2017 is 53.98% does not meet the JIPM criteria standards this is due to the unfavorable quality rate factor which is only around 67.37%, so that in the future it is expected to need preventive maintenance especially an increase in the quality control section than defective products can be reduced and the average good product will increase.

3.4. Analysis of Six Big Losses

Based on the results of the analysis of the six losses factors that have been done, it can be seen that the factors of Equipment Failures, Setup and Adjustment, and Idle and Minor Stoppages, contributes relatively low contribution to the OEE value, while the Rework (23.33%), Reduced yield (20.17%) and Reduced Speed of 19.49% contributes significantly to OEE values. this can be due to poor quality control or due to poor raw materials because the raw materials are affected by weather conditions such as the rainy or dry season, to minimize these losses, the machine needs more serious handling that related to reduced speed of machine and also attention to the quality of raw materials
so it is expected defective production will decrease.

4. Conclusion

The value of OEE at DBIN machines in 2017 based on the results of calculations that have been done is 53.98% this results indicate that the OEE value is below the JIPM standard of 85%. To over come the problem, the machine needs better preventive maintenance with task and certain schedule.

Based on the calculation on the six big losses factor approach, the most influential cause of the problem are the Rework Losses, Reduced Yield Losses and Reduced Speed factors in which the amount of production is not standardized and required rework too much, the percentage of rework losses are

23.33% , reduced yield of 20.17% and reduced speed of19.49%. So to minimize losses, the company must increase in quality control activities in production line.

5. References

[1] J. B. Mapokgole 2014 "To What Degree Is Sustainability Practiced In South African Paper Manufacturing Companies ?," *Int. Conf. Ind. Eng. Oper. Manag.* (January 2014) p 2030
[2] A. Azizi 2015 "Evaluation Improvement of Production Productivity Performance using Statistical Process Control, Overall Equipment Efficiency, and Autonomous Maintenance," *Procedia Manuf.*
[3] Rahmad, Pratikto, and S. Wahyudi 2012 "Penerapan Overall Equipment Effectiveness (OEE) Dalam Implementasi Total Productive Maintenance (TPM) (Studi Kasus di Pabrik Gula PT. 'Y')," *J. Rekayasa Mesin* 3(3) pp 431–437
[4] O. E. Effectiveness 2008 "Overall equipment effectiveness," *Control Eng.* 55(2) p 64
[5] M. Anggraini, M. W. W, and K. E. F. B 2017 "Implementasi Metode Overall Equipment Effectiveness Dalam Menentukan Produktivitas Mesin Rotary Car Dumper," pp. 8–9
[6] Sanjaya and Mutmainah 2016 "Analisis Perawatan Mesin Press 80 Ton Pada Lini P3C03 3 & 4 Denganmetode TPM (Total Productive Maintenance) Di Pt. Xyz," 2(November) pp. 1–12
[7] J. Moubray 1997 *Reliability-centred maintenance.*
[8] R. Singh, D. B. Shah, A. M. Gohil, and M. H. Shah 2013 "Overall equipment effectiveness (OEE) calculation - Automation through hardware & software development," in *Procedia Engineering.*
[9] J. Alhilman 2017 "(ICoICT 2017)_Software Application for Maintenance System."
[10] J. Moubray 2001 "Introduction to Reliability-centered maintenance," in *Reliability-centered maintenance.*

Inventory Level Optimization of Raw Materials for Ready-Made Garment Industry XYZ Pty Ltd using Mamdani Method of Fuzzy Interference System

N Nurhasanah[1], SW Fauzia[1], B Aribowo[1], R Safitri[2], B Samiono[3], CF Lutfia[1], M Devana[1], P Kalifa[1], A Supriyanto[2]

[1]Industrial Engineering Department, Faculty of Science and Technology, Universitas Al Azhar Indonesia, Sisingamangaraja Street, Jakarta 12110, Indonesia
[2]Informatic Engineering Department, Faculty of Science and Technology, Universitas Al Azhar Indonesia, Sisingamangaraja Street, Jakarta 12110, Indonesia
[3]Management Department, Faculty of Economy, Universitas Al Azhar Indonesia, Sisingamangaraja Street, Jakarta 12110, Indonesia
nunungnurhasanah74@gmail.com, syitawf@gmail.com

Abstract. Every company who performs production activities needs inventory of raw materials. The most important thing which is mandatory for all companies when performing their production activities is management of inventory, as inventory is an asset for a company. Availability of raw materials in the industry is expected to facilitate production/service activities to meet consumer needs. Ready-made garment industry, XYZ Pty Ltd is a family business established in 2004 located in Jakarta which engages in textile manufacture and has shown a relatively high demand. While XYZ Pty Ltd attempted to meet the demand, the inventory level was not able to balance with it, resulting in after-hour work of employees. In order to provide appropriate solutions, we proposed some strategies using Mamdani method and TFN (Triangular Fuzzy Number) range which selected appropriate rules to optimize inventory level of the company. Average of range score Fuzzy Interference System obtained for both production level and demand level were 36,500, while average of range score Fuzzy Interference System obtained for Inventory Cost and Inventory Level were 2,250,000,000 and 13,900, respectively. Rules resulting in a high inventory level are very good for use either together with other levels or not. Rules resulting in an intermediate inventory level are also good for use, in particular together with low and intermediate demand levels. Rules resulting in a high demand level are still good for use. On the other hand, rules resulting in a low inventory level are not good for use.
Keywords: Inventory, Demand, Production, Optimization, Ready Made Garment, Fuzzy Inference System

1. Introduction

In order to obtain a quality production, a company pays attention to details of manufacturing activities. Management of inventory is essential for a company, considering that inventory is a company's asset. Availability of raw materials allows production process running in accordance with consumer needs. In addition, an adequate availability of raw materials in warehouse is expected to facilitate production/service activities to meet consumer needs. This importance was also recognized by a ready-made garment industry, XYZ Pty Ltd. The problem dealt with by XYZ Pty Ltd currently was inexistence of production plan as production still relied on market demands which were fluctuating. Due to inexistence of a production planning system, demands might sometimes be higher which was not balanced with the inventory level. Using Fuzzy Logic Mamdani method, a number of rules, such as low, intermediate, and high production levels and costs, were determined that ensured inventory level to meet demand level **[1]**.

2. Methods

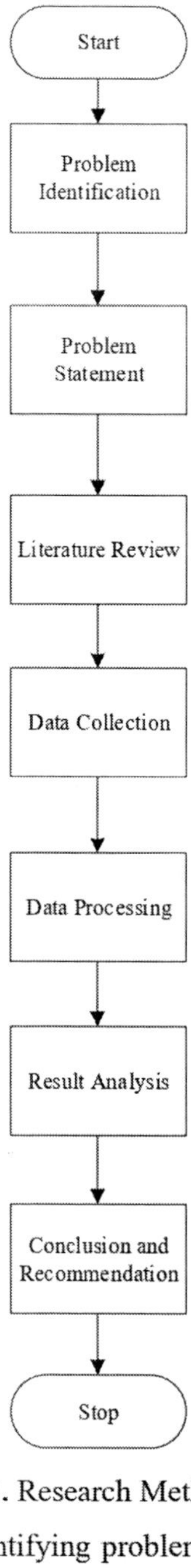

Figure 1. Research Methodology

First step carried out in this study was identifying problems in production activities of XYZ Pty Ltd. Problem statement was composed based on the existing problems found. Type of literature study used was determined subsequently, which was Mamdani method and TFN (Triangular Fuzzy Number) range. Data collection was performed from production level, demand level, inventory costs, and inventory level. Data processing was carried out to obtain rules which were applied in Mamdani method. The results obtained from data processing were analyzed, and conclusion was drawn while recommendation was given to XYZ Pty Ltd.

3. Results and Discussion

3.1 Fuzzy Analytic Hierarchy Process

F-AHP (Fuzzy Analytic Hierarchy Process) is a ranking method in decision making which incorporates different criteria. In order to determine variable weights in F-AHP, a function rule in the form of Triangular Fuzzy Number (TFN) which was arranged according to a linguistic set was employed [2].

In defining values of AHP intensity into Triangular Fuzzy Number (TFN) scale, each fuzzy set was divided by two, except that intensity of importance was divided by 1, according to [3] in his definition on F-AHP. Triangle Fuzzy scale used by Chang is given in Table 1 [4].

Table 1. Fuzzification of Comparison between 2 Criteria [3]

Definition	Likert Scale	TFN	Invers Fuzzy Scale
Equally Important	1	(1, 1, 1) if diagonal (1, 1, 3) if others	(1/1, 1/1, 1/1) if diagonal (1/3, 1/1, 1/1) if the opposite
Slightly More Important	2	(1, 2, 4)	(1/4, 1/2, 1/1)
	3	(1, 3, 5)	(1/5, 1/3, 1/1)
	4	(2, 4, 6)	(1/6, 1/4, 1/2)
More Important	5	(3, 5, 7)	(1/7, 1/5, 1/3)
	6	(4, 6, 8)	(1/8, 1/6, 1/4)
Extremely More Important	7	(5, 7, 9)	(1/9, 1/7, 1/5)
	8	(6, 8, 9)	(1/9, 1/8, 1/6)
Absolutely More Important	9	(7, 9, 9)	(1/9, 1/9, 1/7)

Fuzzification scale of comparison between two criteria was described in a graph, as given in Figure 2 [4].

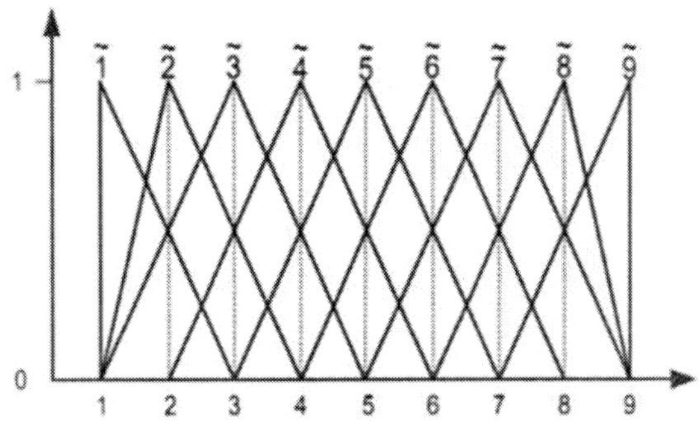

Figure 2. Fuzzification graph of F-AHP scale [3]

3.2 TFN (Triangular Fuzzy Number) Range

TFN (Triangular Fuzzy Number) range is quantitative data which shows [4] highest and lowest values. This study used range for production level, demand level, inventory cost, and inventory level at X Pty

Ltd. The range of values obtained from interviews with an expert, where the results for the production level and the lowest demand level were 26,000 dozen and the highest was 47,000 dozen, the value range for the lowest inventory cost is 1.5 billion and the highest is 3 billion, and value range for the lowest inventory level of 9,288 dozen and the highest of 18,577 dozen. TFN (Triangular Fuzzy Number) range is presented in Table 2.

Table 2. TFN (Triangular Fuzzy Number) range

TFN (Triangular Fuzzy Number) Range			
No	**Description**	**Lowest**	**Highest**
1	PL (Production Level)	26,000 dozens	47,000 dozens
2	DL (Demand Level)	26,000 dozens	47,000 dozens
3	IC (Inventory Cost)	1.5 billions	3 billions
4	IL (Inventory Level)	9,288 dozens	18,577 dozens

3.3 Rules of Fuzzy Interference System
In this step, determination of rules of Fuzzy Interference System was made from opinion of experts, resulting in 27 rules which would be processed using Fuzzy Logic Mamdani on Matlab software [5]. The obtained rules are presented in Table 3.

Table 3. Rules of Fuzzy Interference System

No	**PL**	**DL**	**IC**	**IL**
1	H	H	H	H
2	H	H	M	H
3	H	H	L	H
4	H	M	H	M
5	H	M	M	M
6	H	M	L	M
7	H	L	H	H
8	H	L	M	H
9	H	L	L	L
10	M	H	H	M
11	M	H	M	H
12	M	H	L	M
13	M	M	H	M
14	M	M	M	M
15	M	M	L	M
16	M	L	H	L
17	M	L	M	L
18	M	L	L	L
19	L	H	H	H
20	L	H	M	H
21	L	H	L	H

22	L	M	H	M
23	L	M	M	M
24	L	M	L	L
25	L	L	H	L
26	L	L	M	L
27	L	L	L	L

3.3.1 Matlab Software for Fuzzy Logic Mamdani

In performing Matlab Fuzzy Logic Mamdani, data TFN (Triangular Fuzzy Number) range and the obtained rules had to be inputted. Steps of Matlab software use in this study is as follows [6].

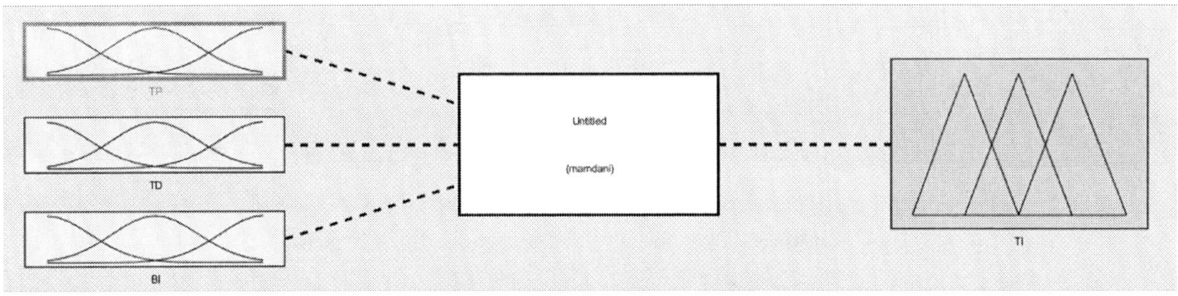

Figure 3. Determination of Input

The first step to perform was selection of Fuzzy Logic Mamdani, followed by completing Input and Output. In this research, we had 3 inputs, including Production Level, Demand Level, and Inventory Cost, as well as 1 output, namely Inventory Level.

IOP Conf. Series: Materials Science and Engineering **528** (2019) 012047 doi:10.1088/1757-899X/528/1/012047

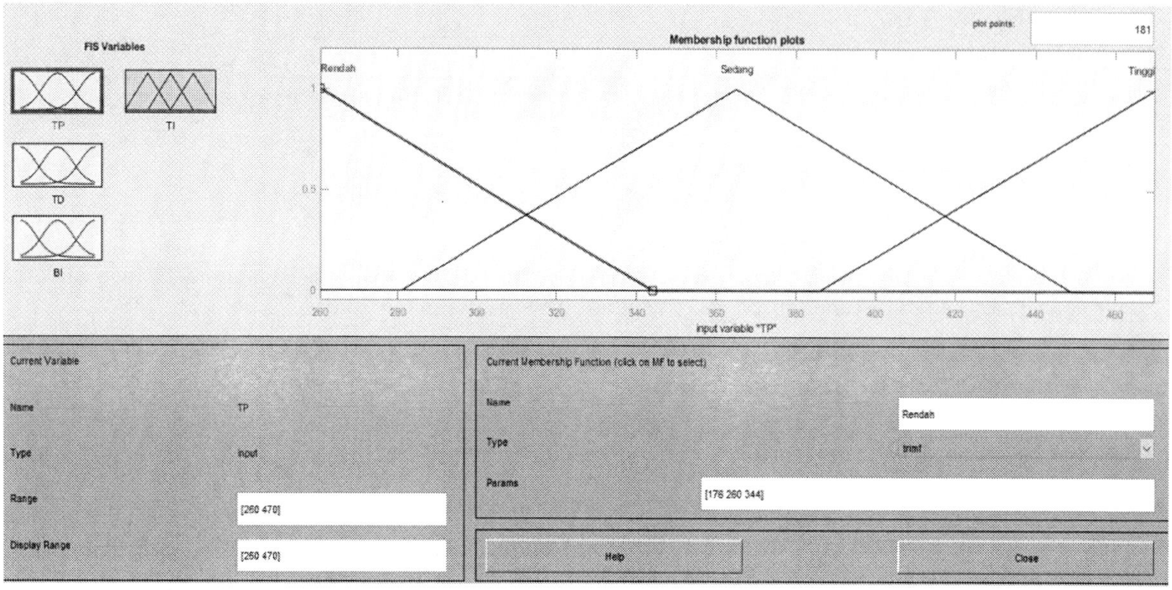

Figure 4. Determination of Range for Each Criteria

The next step was completion range of each criteria by filling out low, intermediate, and high rules of each criteria.

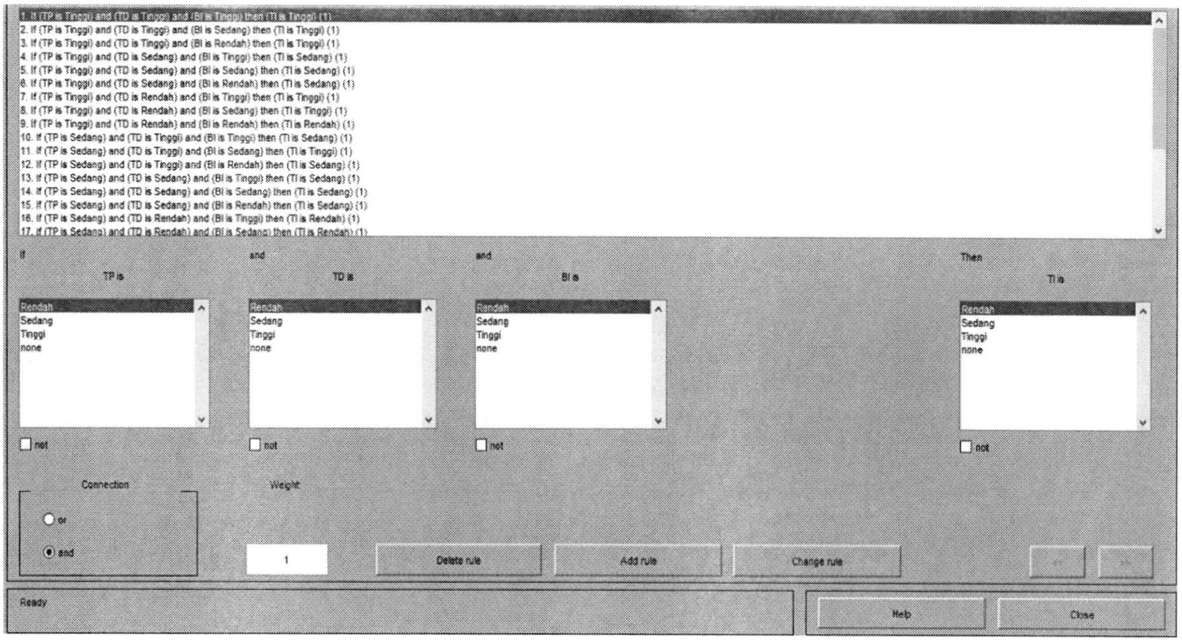

Figure 5. Range Input

After filling out range and criteria, we moved to the third step, in which the 27 rules were inputted into Matlab.

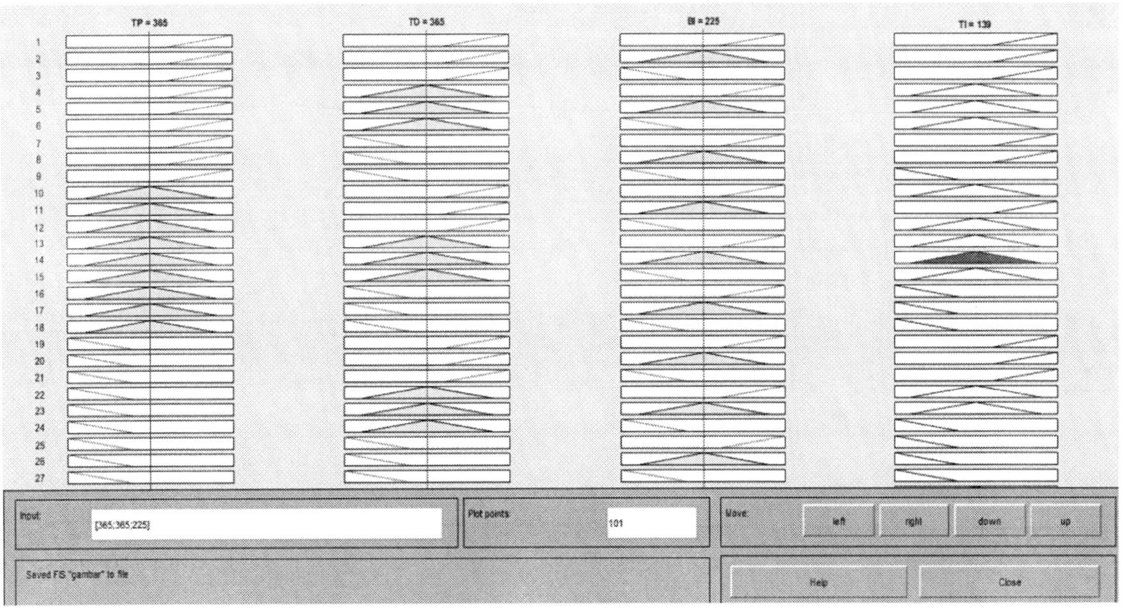

Figure 6. Results from Fuzzy Interference System

Figure 6 shows that rules number 9, 16, 17, 18, 24, 25, 26, and 27 had a low Inventory Level (IL), and therefore they were not able to be used. Rules number 4, 5, 6, 10, 12, 13, 14, 15, 22, and 23 had an intermediate Inventory Level (IL), and therefore they were still considerable for use. Moreover, a high inventory level as shown by rules number 1, 2, 3, 7, 8, 11, 19, 20, and 21 was good for use as through a high inventory level, demand level (DL) is able to be met at different production levels and cost levels. Average of range score Fuzzy Interference System obtained for PL, DL, IC, and IL were obtained for both production level and demand level were 36,500, while average of range score Fuzzy Interference System obtained for Inventory Cost and Inventory Level were 2,250,000,000 and 13,900, respectively.

4. Conclusion

Based on the experiment on Fuzzy Interference System using Matlab Software in order to investigate Inventory Level of XYZ Pty Ltd, it was concluded that:

1. The average value of the Fuzzy Inference System range is the Production Level of 36500, the Demand Level of 36500, the Inventory Cost of 2250000000, and the Inventory Level of 13900. The value is obtained using the matlab software.
2. Rules with high inventory levels are very good to use, because they can meet the demand level with low, medium and high conditions.
3. Rules with good inventory levels are used, especially with small and medium demand levels. While with a high level of demand, these rules can still be used.
4. Rules with low inventory levels are not good to use, because they cannot meet medium and high demand levels.

5. Acknowledgement
Thanks to Directorate of Research and Community Service of the Ministry of Research and Technology of the Republic of Indonesia which funded this research activity in 2018 with an Applied Product Research scheme.

6. References
[1] E. Mamdani, Application of Fuzzy Logic To Approximate Reasoning Using Linguistic Synthesis, London: Queen Mary College, London University, 1975.

[2] L. A. Zadeh, "Fuzzy Sets," *Information And Control,* vol. VIII, pp. 338-353, 1965.

[3] D. Y. Chang, "Application of The extent Analysis Method On fuzzy AHP," *European Journal of Operational Research,* pp. 649-655, 1996.

[4] Y. M. Wang and K. S. Chin, "Fuzzy Analytic Hierarchy Process: A Logarithmic Fuzzy Preference Programming Methodology," *International Journal of Approximate Reasoning,* p. 13, 2010.

[5] A. H. Prihamayu, "Mamdani Method Fuzzy Logic Application in Deciding the Amount of Production Decision," p. 10, 2015.

[6] A. Naba, Belajar Cepat Fuzzy Logic Menggunakan MATLAB, Yogyakarta: ANDI, 2009.

Advanced ERP Application for Marine Transportation Industry in the South Asia Pacific Country; a Case Study

Lalu Tri Wijaya Nata Kusuma[1], Jun-Der Leu[2] and Fu-Shiang Tseng[2]

[1]Faculty of Engineer, University of Brawijaya, Indonesia
[2]National Central University, Taiwan
eltriwijaya@ub.ac.id, leujunder@mgt.ncu.edu.tw, fushiang@mgt.ncu.edu.tw

Abstract. Marine transportation industry is one of the most vital activities in the world. Demand to provide effective and efficient inter-regional connectivity within the marine transportation framework will be a common agenda in some areas. South Asia Pacific region has a member country of more than 14 countries. With the region's economic growth of about 5.3 percent, compared to the world average of 3.2 percent, because it is the most safe area compared to other regions (Antara, 2016). Almost 70% of total world trade takes place among countries in Asia Pacific. In this study, we try to focus on comparative analysis of IT support system to support the operation of marine transportation services in 3 countries namely Indonesia, Malaysia and USA. Some port systems in the archipelagic country particularly the South Asia Pacific region have adopted approaches such as the ERP system, but are still partial. Each part of the port management creates its own information system, and links it to every organization's website. Utilizing existing ERP systems, will be key in designing new frameworks with IOT concepts. With the development of IOT techniques and major investments in marine transportation systems in the Asia Pacific region, intelligent transportation will be able to improve the effectiveness and efficiency of marine and port transportation services processes. The technology of the Internet of Things will enable the development of transportation systems to be more intelligent, safe, harmonious, and energy efficient.
Keywords: Marine Transportation System, ERP, IoT

1. Introduction

Based on The Review of Maritime Transport 2017 published by the United Nations Conference of Trade and Development [1], the demand for world marine freight increased slightly. The increase in world trade volume by marine was 2.6 percent, up slightly from 1.8 percent in 2015, but still below the average growth in the last four decades of 3 percent. With a total volume of 10.3 billion tons, an increase of more than 260 million tons of cargo was added. In UNCTAD calculations, world trade by marine in 2017 will reach 2.8 percent, with a total volume of 10.6 billion tons. The medium-term projection also shows a tendency to increase cargo flows. Still according to UNCTAD, the seaborne trade will increase by an average of 3.2 percent between the period from 2017 to 2022. Surely the value has a profound effect on the design and operation of the marine transportation system, especially the marine port.

Several global issues related to the marine transportation system between countries in recent years have been widely discussed, especially related to marine information systems. The marine information system becomes a standard of analytical resources for public and private institutions in planning development, managing resources, and supporting Geographic Information Systems (GIS) marine transportation management decisions. Nartey and Shi in their study has developed the concept of marine information system [2]. Marine Information System is limited to providing access to information primarily for ship owners, company ships, and port authorities. However, the information cannot be accessed by external users due to the design of marine information systems.

The development of integrated marine information system has not been fully able to work optimally if implemented in an archipelagic country region, as in several southern and southeastern pacific countries. Each archipelago has its own authority to manage its service system under the control of the local government. Particularly the marine port service system, several governments in each region tend to have their own rules and policies in decision making. Service systems that are still conventional and not yet integrated among stakeholders, also make the marine information system can not be implemented effectively and efficiently.

Some study investigated the relationship between information systems, especially ERP implementation and transportation or logistics systems (Rondinelli et al., 2000; Ince, 2013; Kandananond, 2014). As transport and logistics systems continue to integrate, their impact on the physical environment (air, water, and land resources) will become more complex. Economic globalization, agile manufacturing, fast delivery to markets, and supply chain management create greater demand for intermodal transport services and multi-modal transport infrastructure [3].

Implementation of logistics and ERP systems has revealed that logistics practices and ERP systems have a positive effect on company performance and competitive advantage [4]. On the other study that the successful implementation of ERP relies on five important factors: defining business cases, setting up systems and users, stabilizing for normal operation and maintenance, and improvement [5].

However, some previous studies have not discussed the effect of the IoT (Internet of Things) application trend on the implementation of the ERP system framework to support the marine transportation system, particularly with regard to the geographical conditions of the pacific region. Implementation of ERP and IOT as a new generation of marine port system especially in south pacific region of south pacific island most of the country become the key factor to optimize the performance of marine port.

2. Methodology

In accordance with our previous explanation, some previous studies have not discussed the influence of IoT (Internet of Things) for the implementation of the ERP system framework to support the marine transportation system. In this study, we reviewed the data qualitatively, related to the comparison of port operational systems in the USA, Malaysia and Indonesia. The process of determining the variables as the basis of data comparing the needs of ERP and IOT service users is done by interviewing some port authority authorities and related parties, as well as field observations, major ports in Port Klang Malaysia and Tanjung Perak in Indonesia. While our USA uses secondary data released by Maritime Administrator in 2017. As the basis of comparison data, especially in the pacific asia region we chose to observe the port of Klang Malaysia which has tried to implement this IoT system and has the same geographical characteristics as Indonesia as an archipelagic country.

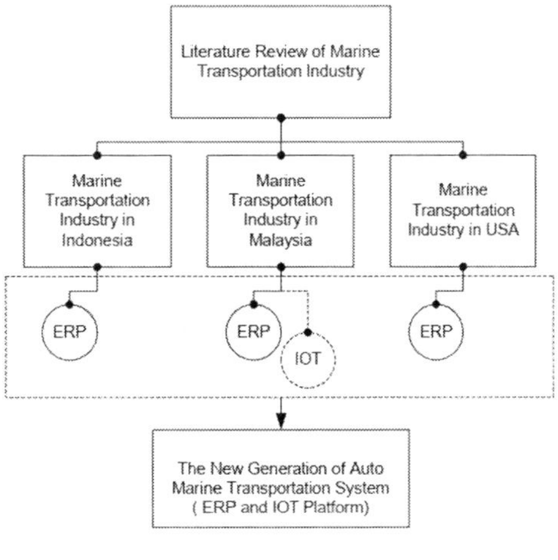

Fig. 1 Research Methodology

It is also important to note that the current operating system is still conventional rather than integrated. Local government, shipping companies, and port authorities as service suppliers work individually. On the other hand, there are many additional costs without permission to facilitate the process to be paid by the customer.

As stated in the purpose of this study, it is important to develop a new system to optimize the performance of logistics systems and marine highway programs in Indonesia using the IoT approach, while only using ERP in each stakeholder system for the operating system has not been able to work optimally and fully integrated.

3. The Analysis of Marine Transportation System between Indonesia, Malaysia and USA Country

Utilization of information systems technology in several ports of the world has become one of the key solutions to optimize the performance of integrated systems. The US marine port system for example, has implemented the American Marine Highway program. The program run by the Maritime Administration (MARAD) since 2007 involves dozens of ports along the inland rivers in the United States, by implementing a highway system, diverting the cargo transportation process on land to the water sector [6]. The program has also integrated the port service information system with stakeholders to make marine transfers more effective and efficient.

In the Asia Pacific region especially in southern and southeastern parts, we take the example of archipelagic countries such as Malaysia and Indonesia. One of the main ports in Malaysia is Port Klang. The port management has even implemented the IOT concept, with the release of e-PAN for the benefit of early arrival of ships related to the ship's security process which will be anchored. And also the e-DCFZ system, which aims to facilitate the processing of dangerous cargo documents and free zones. Implementation of e-DCFZ will help streamline the trading process in accordance with the vision of the port authority of Klang (PKA) and anticipate the government's vision to improve the ease of doing business in Malaysia [7]. Another case in Indonesia, which recently implemented marine toll programs. The marine transportation system is trying to develop a marine toll strategy that has been done US through the American Marine Highway program.

However, the results of our observations and analysis, especially in Malaysia and Indonesia, found that the implementation of the marine information system still tends to be partial, where each field makes its own application and not yet integrated as a whole. Then also still tend to be technically focused only in certain fields only. In conclusion, not all information systems from each stakeholder can be optimally integrated, and from the user side of the port service is still not satisfied

Table 1. Comparison of Marine Transportation Industry Conditions

	Indonesian	Malaysia	USA
Operation and Policy of Marine Transportation Industry	- There are approximately 70 major ports spread across 34 provinces. - Implement marine toll programs. - The main port is managed by the government as a state-owned enterprise.	- There are about 7 major ports of cargo. - Some ports are managed by private companies.	- There are about 100 major cargo ports. - Implement the American Marine Highway program.

Table 1. Comparison of Marine Transportation Industry Conditions

	Indonesian	Malaysia	USA
ERP System Perspectivenes	- Implement marine information systems for internal port authority authorities. - Implementation of ERP is still not optimal, and only in some main ports only.	- Implement marine information systems for each stakeholders. - Implementation of new ERP is done in some ports only.	- Each port authority has its own information system.
ERP and IOT Platform	- Has not implemented integration between ERP and IOT. - Has not integrated the entire service process from each stakeholders in all major ports.	- Has applied the e-PAN application for the benefit of early arrival of vessel related to vessel security process which will be anchored. - Implementation of e-DCFZ application, which aims to facilitate processing of dangerous cargo documents and free zones.	- Has not implemented integration between ERP and IOT especially for integration of all service process.

3.1 Marine Transportation Industry in the South Asia Pacific Country

South Asia Pacific region has a member country of more than 14 countries. With the region's economic growth of about 5.3 percent, compared to the world average of 3.2 percent, because it is the most safe area compared to other regions [8]. Almost 70% of total world trade takes place among countries in Asia Pacific.

Based on the description in Table 1, we tried to use data logistic performance index (LPI) which was released by World Bank in 2106 to compare the growth condition of logistics performance index between Indonesia, Malaysia and USA as shown in Fig. 2 [9]. The results show significant differences from the three countries, from scale 1 to 5 the highest value of USA about 4.0, followed by Malaysia 3.4 and last Indonesia 3.0. Of course with these conditions the opportunity to improve infrastructure and service processes, especially integrated information system tools such as ERP system wide open. Countries with archipelago areas have a good opportunity, if coupled with the approach of marine toll programs such as those run by the US government.

IOP Conf. Series: Materials Science and Engineering **528** (2019) 012048 doi:10.1088/1757-899X/528/1/012048

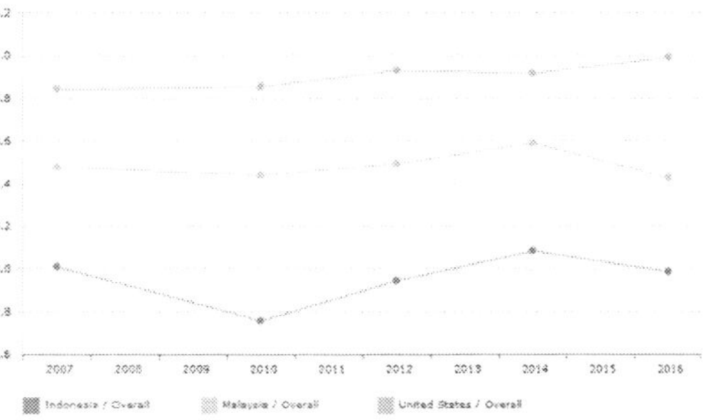

Fig. 2 Comparison of LPI Index between Indonesia, Malaysia and USA

However, before we build an integration system with ERP, it is necessary to map and clarify in advance about the business processes of each stakeholder. By understanding the existing conditions of the stakeholder business processes, it will help build the initial foundation in the ERP integrated network system later.

3.2 ERP System

The development of ERP system was initially more implemented in the manufacturing world, but along with the development of technology in the world of marine transportation system was now widely used. Especially in the marine transport system, based on our case studies on the three countries representing the Asia Pacific and American regions, the ERP system has been implemented in each country, starting with the marine transportation management information system in its region.

Overall, however, the issue of sectoral ego of each stakeholder remains a constraint in the implementation of the system. ERP system that is built is still partial which means each stakeholders create its own system for internal organization only. Has not optimally integrated ERP, also caused by each government in every archipelago or province has its own authority to manage the marine transportation system. So many policies that are made precisely complicate the process of system integration in the field.

3.3 IOT in Marine Port System

The development of IoT platform in the world of industry and especially marine transportation system has not been widely applied. If anyone has started, just an online application that works for one of the processes in a support organization of the marine port system. As we have explained in table 1, the Malaysian port authority has tried to implement an IOT platform to assist the cargoship service process that will carry out loading and unloading process in its territory. However, the application is only specialized to meet the security needs and process of dangerous cargo documents.

While in fact many opportunities that can be taken by utilizing the IOT platform, especially to integrate the entire process of service from each stakeholders of marine ports in their respective regions.

4. The New Generation of Auto Marine Transportation System

Based on our literature study, accompanied by secondary data that we obtained both observation and interviews directly with some authorized officials in Indonesia and Malaysia, there are some fundamental variables that must be met in designing the marine port system to be optimal. All stakeholders' interests in the port service system must be effectively and efficiently selected.

The fundamental variables in the service system include, among others, the administrative process of the port authority, the shipping company service system, the local government service system including the customs inspection process, the freighter operator information system and of course access as the external service users.

The new generation design of marine transport system, all fundamental variables of each stakeholder need to be integrated in one application service. The use of IOT platforms to be one of the main objectives at this time, to optimize the performance of marine transport service system.

5. Conclusion

With the rapid development of the pacific south Asia region, together with uneven infrastructure support throughout the region, the demands of a fast and efficient intelligent transportation system is inevitable. Information systems technology, especially in the transportation sector with the IOT concept, is required today.

However, the current perception of the marine transportation system to service users, both shipping companies, governments and some stakeholders is passive, incomplete information is partial and inaccessible, and networking and monitoring functions are not strong enough, all of which result in integration low and low intelligence levels.

Currently, the development of IOT in the early stages and key technology is still in the exploration stage, but has a very strong momentum. With the development of IOT techniques and major investments in marine transportation systems in the Asia Pacific region, intelligent transportation will be able to improve the effectiveness and efficiency of marine and port transportation services processes. The technology of the Internet of Things will enable the development of transportation systems to be more intelligent, safe, harmonious, and energy efficient.

6. References

[1] *The Review of Maritime Transport, (2017),* Nations Conference of Trade And Development (UNCTAD), Access : May, 2018

[2] Nartey E., Shi J. C., (2012), Analysis of Future Marine Information System Based on Cloud Computing. International Conference on Fuzzy Systems and Knowledge Discovery, pp. 2590-2593

[3] Rondinelli, (2000), Multimodal Transportation, Logistics, and the Environment: Managing Interactions in a Global Econom. *European Management Journal* Vol. 18, No. 4, pp. 398–410.

[4] Ince H., (2013), The Impact of ERP Systems and Supply Chain Management Practices on Firm Performance: Case of Turkish Companies. Procedia - Social and Behavioral Sciences Vol. 99 ,1124 – 1133

[5] Kandananond K., (2014), A Roadmap to Green Supply Chain System Through Enterprise Resource Planning (ERP) Implementation. Procedia Engineering 69 (2014) 377 – 382

[6] America's Marine Highway Report to Congress, (2011). MARAD AMH Report to Congress. https://www.marad.dot.gov. Acces : March,2017

[7] Port Klang Authority (PKA), (2018). http://www.pka.gov.my/. Acces : Januari,2018

[8] Ekonomi Dunia, (2016). https://www.antaranews.com/ekonomi. Acees : Janurai, 2018.

[9] LPI World Bank. (2016) https://lpi.worldbank.org/international/global. Access : May, 2018.

[10] Kementerian Perhubungan RI. (2013) http://www.dephub.go.id/. Acces : March, 2017.

[11]Pelindo Company, (2012). Annual Report. http://www.indonesiaport.co.id/sub/annual-report.html. Access : March, 2017.

Forecasting Methods and Implementation of DRP (Distribution Requirement Planning) Methods in Determining the Master Production Schedule

Riana Magdalena[1,2] and Theresia Suli[1]

[1]Industrial Engineering, Engineering Faculty,Atma Jaya Indonesian Catholic University, Jln. Raya Cisauk Lapan No. 10, Sampora, Cisauk, Tangerang, Banten 15345, Indonesia.
[2]riana.magdalena@atmajaya.ac.id

Abstract. PT ABC President Indonesia is a manufacturing company engaged in the food sector by producing ABC noodles and Nu Greentea. At PT ABC President Indonesia, the production scheduling process is one of the main processes carried out to determine the number of products that must be produced. Based on the problems that occur regarding stock, it is proposed that the best forecasting method is based on the time series of each Distribution Centre (DC) and processing Distribution Centre (DC) requests using the Distribution Requirement Planning (DRP) method. Where the DRP method is used, so that the company can meet the demand of every DC so that there is no shortage of products and the production process continues to run without the company experiencing losses. In addition, with the use of DRP method there will be additional stock so that if there is a defect, the company still has stock remaining so that it can still meet the customer demands for each DC and the product distribution process runs more regularly.
Keywords: Forecasting Method, Distribution Requirement Planning Method, Stock.

1. Introduction

In this era, the industrial world is growing rapidly. This can be seen, where many manufacturing companies are engaged in various industrial fields. To be able to survive and compete with other manufacturing companies, a company must be able to produce products that have good quality and are able to maintain the level of satisfaction of customers towards the products produced by the company.

One way to maintain and compete with other companies is to create a good production system and one which is mutually integrated by considering many things. One of them considers the production scheduling carried out and the distribution flow carried out in a company. Production scheduling is one important aspect of a company where a good production scheduling strategy can minimize losses and increase customer satisfaction with the product being produced. One of the losses that can happen is that there are several requests from customers that cannot be fulfilled due to delay in the distribution process to customers. So, if there is an unfulfilled demand, it will reduce the level of corporate service to customers and can reduce the level of customer satisfaction with the product being produced. This could cause the possibility that customers will switch to competing products.

PT ABC President Indonesia is a manufacturing company that produces noodles and beverages such as Nu Greentea. In the world of the food industry, there are often uncertainties in the time and number of products ordered by customers. This results in an excess or lack of stock which was sometimes experienced by PT ABC President Indonesia where PT ABC President Indonesia implemented the make to stock system. The existence of excess or lack of stock sometimes results in losses such as the existence of slow moving (products that are not distributed), non-fulfillment of customer demand, or the existence of products that cannot be consumed again. The importance of strategies in production scheduling can be used as a solution to overcome the problems that occur.

Therefore, in this study, the topic raised was about production scheduling planning by implementing the DRP method (Distribution Requirement Planning). The DRP method is a distribution planning method using several control parameters to calculate product requirements at each distribution point. The advantage of the DRP method is that the DRP method is centered on the demand of each depot or agent so that the production process will be scheduled based on requests from each depot or agent. Where using the DRP method can provide recommendations on the production master schedule and can overcome the stock issues at PT ABC President Indonesia.

2. Methods

In this study, DRP (Distribution Requirement Planning) was used as a research model. DRM is Distribution resource planning (DRP) is a method used in business administration for planning orders within a supply chain. DRP enables the user to set certain inventory control parameters (like a safety stock) and calculate the time-phased inventory requirements. The method in this study consisted of: (1) preparation, (2) research design, and (3) data collection (4) data processing.

2.1. Preparation

In the preparation phase in this study, identification of real conditions was conducted at PT ABC President Indonesia through observation (direct research in the field) as well as the interview process with parties from several. In addition, in this phase literature studies from books and research journals are carried out to discuss method used in this study, namely the forecasting techniques, DRP method, and MPS.

2.2. Research Design

DRP is carried out to identify problems in the questions related to the topic of the issue to be discussed which focused on the problems regarding the existence of undistributed products experienced by PT ABC President Indonesia. By using DRP, product production is regulated to meet the demand until the next product production arrives for each distribution center. In making MPS using the DRP method, a forecast is needed from each DC to estimate the number of products that must be produced. With the determination of good forecast methods, the forecast has a lesser error value so there is no excess stock.

2.3. Data Collection

This research was conducted at PT ABC President Indonesia by means of observation (direct research in the field) as well as the interview process with parties from several divisions on questions related to the topic of the issue to be discussed which focused on the problems regarding the existence of undistributed products experienced by PT ABC President Indonesia. The data collected consists of historical data requests, lead times of each distribution center, stock from each DC and available distribution centers.

2.4. Data Processing

The first data processing is to determine the time series plot of each DC. The time series plot determination aims to find out the exact demand forecasting method in accordance with the available data. Then the forecasting process is carried out using the selected method. After forecasting process, the process of determining the proposed lot size for each DC is appropriate and continued with the calculation of the Master Production Schedule with the DRP method by considering the safety stock and the remaining stock. Where the results of the data processing are the MPS proposal for the 2018 period for the production of Alfamart private label noodles.

3. Result and Discussion

The method used in the study is the DRP method but is closely related to the DRP method with forecasting. Because the data obtained from the company is only in the form of historical data requests per DC so that forecasting needs to be done first so that the existing data can be used for subsequent DRP calculations.The data collected was the data from distribution centre at PT ABC President Indonesia for Alfamart ABC private label noodles, and lead time data and inventory for each DC spread on the island of Java.

Table 1. Data Demand from each DC at Java Island

Bulan	DC 1	DC 2	DC 3	DC 4	DC 5	DC 6	DC 7	DC 8	DC 9	DC 10	Total
January	1636	795	392	840	448	400	336	1060	1976	736	8619
February	1674	1078	672	756	728	220	560	860	2228	1060	9836
March	2215	404	336	504	112	414	0	1260	2264	732	8241
April	1400	514	504	196	616	612	168	780	1274	898	6962
May	1588	1404	322	728	728	580	952	100	2382	2170	10954
June	3450	846	868	1008	504	0	168	1940	2072	420	11276
July	2276	1010	224	84	700	220	784	1080	1803	916	9097
August	1872	1538	560	468	476	220	504	760	1760	620	8778
September	1160	794	210	264	434	384	420	625	1386	328	6005
October	1244	630	336	416	0	212	56	251	2218	534	5897
November	908	1102	126	340	42	280	196	500	2575	1156	7225
December	1334	772	252	290	156	360	796	880	1348	202	6390

The above data was obtained from requests for each DC Alfamart for 2017.

Table 2. Lead Time Data

DC	Lead Time (month)	DC	Lead Time (month)
1	1	5	2
2	2	6	1
3	2	7	1
4	2	8	1
		9	1

The data above was the ordering time data from each DC to the time the product arrived at each DC.

Table 3. Project Data on Hand for Every DC

DC	Project On Hand	DC	Project On Hand
1	160	6	160
2	200	7	160
3	80	8	200
4	200	9	360
5	160	10	360

The data above shows the project on hand data that is owned by each DC where the data above is obtained from the product remaining in each DC at the end of 2017 (December 2017). Besides that, the other data needed to calculate the DRP method is that the lot used is Fixed Order Quantity (FOQ) of 40 pcs or multiples thereof for each order.

3.1. Forecasting

3.1.1. Time Series Plot

Time Series is used to find out the graphic form of each DC to determine the correct forecasting method.

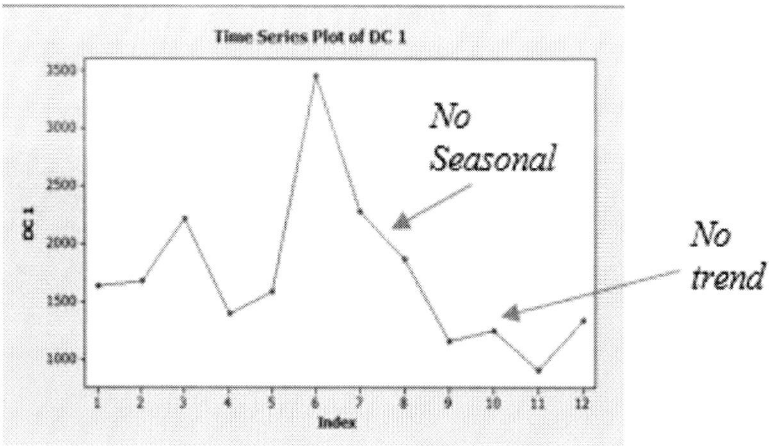

Figure 1. Time Series Plot DC 1

Based on the picture above, it can be seen that the form of the DC 1 time series plot graph shows the form of non-trendy data due to the absence of rising or falling and non-seasonal demand trends due to the absence of a repetitive pattern in the request data. This means forecasting methods that must be used are Moving Average and Single Exponential.

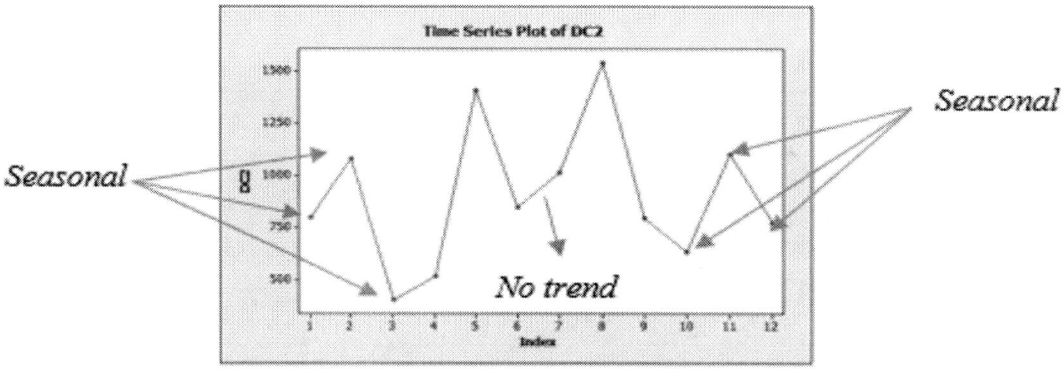

Figure 2. Time Series Plot DC 2

Based on the picture above, it can be seen that the form of the DC 2 time series plot graph shows the form of non-trendy and seasonal data with a period of 3 periods as in the 1-2-3 period and 10-11-12 period. Seasonal said despite having the same pattern at the end of the data, due to limited data taken only 12 periods so that there is a possibility of a repeat pattern that is not visible on this graph. This means that the forecasting method that must be used is Decomposition of Additive and Multiplicative and Winter Additive and Multiplicative.

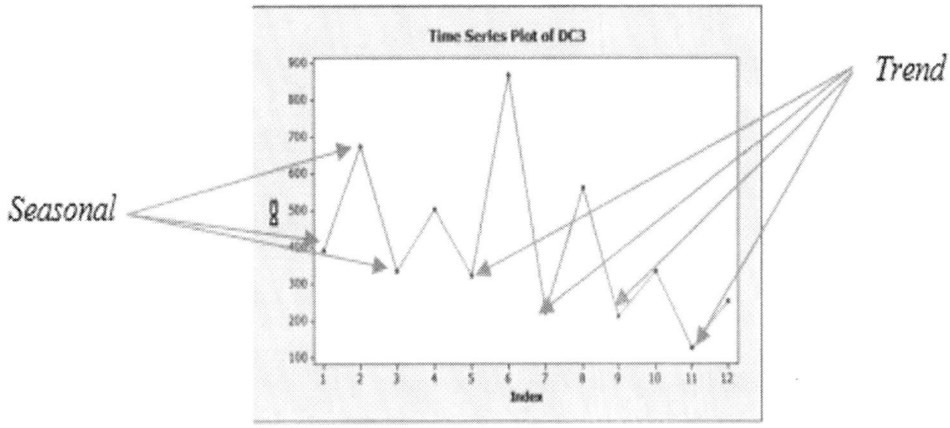

Figure 3. Time Series Plot DC 3

Based on the picture above, it can be seen that the shape of the DC 3 time series plot graph shows the form of trend (declining) data which can be seen in odd periods (1-3-5-7-9-11) which tends to decrease and seasonal with interval 2 period seen in the pattern of periods 1-2-3, 3-4-5, 5-6-7, 7-8-9, 9-10-11. This means that the forecasting method that must be used is Decomposition of Additive and Multiplicative and Winter Additive and Multiplicative.

3.1.2. Forecasting
Forecasting is done using MINITAB 17 software. Here are the results of forecasting:

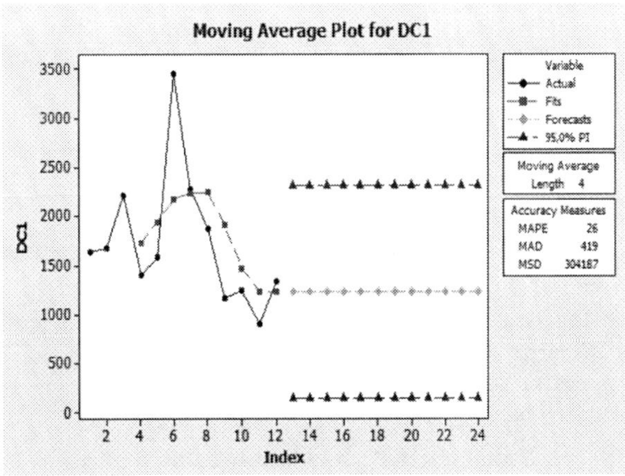

Figure 4. Double Moving Average for DC 1

Forecasting results above, which must be considered, is the magnitude of the error value listed where the error values listed are MAPE, MAD, and MSD.

3.1.3. Accuracy
After forecasting, all the resulting error values are recapitulated and compared to what method has the smallest error value.

IOP Conf. Series: Materials Science and Engineering **528** (2019) 012049 doi:10.1088/1757-899X/528/1/012049

Table 4. Accuracy Table from DC 1

Distribution Center 1			
	Moving Average		*Single Exponential Smoothing*
	Double Moving Average	*Single Moving Average*	
MAPE	26	42	31
MAD	419	643	489
MSD	304187	730327	462683

Based on the recapitulation of DC 1 accuracy level it can be seen that the best forecasting method used for DC 1 is the Double Moving Average method because the error value of the Double Moving Average method is the smallest error value indicating that the error of the forecasting method is smaller than the 2 methods others.

Table 5. Results of Forecasting Every DC

Period	DC 1	DC 2	DC 3	DC 4	DC 5	DC 6	DC 7	DC 8	DC 9	DC 10
13	1229	723	115	350	198	292	331	549	1934	608
14	1229	1306	192	350	198	292	331	549	1934	608
15	1229	694	69	350	198	292	331	549	1934	608
16	1229	723	97	350	198	292	331	549	1934	608
17	1229	1306	24	350	198	292	331	549	1934	608
18	1229	694	2	350	198	292	331	549	1934	608
19	1229	723	0	350	198	292	331	549	1934	608
20	1229	1306	0	350	198	292	331	549	1934	608
21	1229	694	0	350	198	292	331	549	1934	608
22	1229	723	0	350	198	292	331	549	1934	608
23	1229	1306	0	350	198	292	331	549	1934	608
24	1229	694	0	350	198	292	331	549	1934	608

The table above is a request recapitulation table obtained from the MINITAB software according to the chosen forecasting method. For the period, it starts with the period 13-24 which indicates that the forecasting is forecasting for the period January - December 2018.

3.1.4. Distribution Requirement Planning (DRP)

Table 6. DRP Calculation Table for DC 1

DRP DC 1													
FOQ = 40		*Lead Time = 1*		*Safety Stock = 0*		*On Hand = 160*							
Periode	0	1	2	3	4	5	6	7	8	9	10	11	12
Gross Requirement		1229	1229	1229	1229	1229	1229	1229	1229	1229	1229	1229	1229
Schedule Receipt		0	0	0	0	0	0	0	0	0	0	0	0
Project on Hand	160	11	22	33	4	15	26	37	8	19	30	1	12
Net Requirement		1069	1218	1207	1196	1225	1214	1203	1192	1221	1210	1199	1228
Planned Order Receipt		1080	1240	1240	1200	1240	1240	1240	1200	1240	1240	1200	1240
Planned Order Release	1080	1240	1240	1200	1240	1240	1240	1200	1240	1240	1200	1240	0

Calculation example:

- Gross Requirement: obtained from DC forecasting results 1
- Project on Hand: obtained from production at the end of 2017 for DC 1 for period 1
- Planned Order Receipt - Net Requirements = 1080 - 1069 = 11
- Net Requirement: Gross Requirement - Project on Hand = 1229 - 160 = 1069
- Planned Order Receipt: FOQ = 40 then multiples of 40 where 40 x 27 = 1080
- Planned Order Release: Based on lead time (1 month).

Data in Table 6 describes the calculation of Distribution Requirement Planning for DC 1, where the data needed in this calculation is the gross requirement data, receipt schedule, and project on hand. Gross requirement is the number of requests from each DC obtained from the prediction that has been done previously with the chosen forecasting method. For the receipt schedule, a product that has been scheduled will be ordered by the company. In this case the receipt schedule is 0 because the company produces its own product without the presence of other companies to produce the noodle product. As for the project on hand, the stock available to the company is taken from the remaining production in the previous month. The results from Table 6 are scheduling of each distribution for each month by considering things like gross requirement data, receipt schedule, and project on hand and lot size of each distribution.

3.1.5. Master Production Schedule (MPS)

Table 7. MPS Table

	MPS											
Quantity	1	2	3	4	5	6	7	8	9	10	11	12
	6320	6240	6840	6200	6280	6760	6160	6240	6760	6200	4960	0

MPS calculation is obtained from recapitulation of period 1 to period 12 by adding up each DC.
Data in Table 7 is an MPS table where the MPS value is obtained from the combination of each DC where the merge is obtained by planned order release. The relation from tables 6 and 7 is that after the calculation of table 6, it is obtained that the planned order release from each DC is combined so that it becomes the total that must be produced in one month.

3.1.6. Analysis

Based on observations made at PT ABC President Indonesia, it can be seen that the current production scheduling process is still not optimal. It is said that it is not optimal because there are still several distribution centres that place orders simultaneously so that there is a bottleneck in the product distribution process and there are some DCs that are not fulfilled; the demand caused by the existing process has not been able to collect in line with existing requests.

Based on the problems at PT ABC President Indonesia above, the authors recommend Proposed Forecasting Methods and Implementation of the DRP Method (Distribution Requirement Planning) in Determining Production Master Schedules. In the forecasting method, the time series plot is carried out in advance on the request data to determine the trend pattern of demand data where there are 2 types of data plots, namely seasonal trend and data. For trend data are time series data where there is a tendency for demand to rise or fall. As for seasonal data, the data that is formed has a tendency to repeat over a period of time.

After determining the demand pattern using a time series plot, it can be known that the forecasting method is in accordance with the time series plot of each DC. For DC 1,4,5,6,7,8,9 and 10 because it has a non-trendy and non-seasonal pattern, the Moving Average and Single Exponential Smoothing forecasting methods are used. For DC 2 and 3 using the Decomposition and Winter methods.

Furthermore, data processing is performed using forecasting methods that are in accordance with the time series.

In the forecasting process, MINITAB software was used. In processing data by using this software, in addition to obtaining forecasting results, an error value from the forecasting is also obtained. This error value indicates the magnitude of the forecasting error or non-conformity that is carried out with the conditions that will occur in the future. Therefore, in choosing the best forecasting method, we need to pay attention to the error value where from all DCs look for a method that has the smallest error value where the error value used is MAPE, MAD and MSD which will show a more accurate level of data accuracy than forecasting data with use other methods. After forecasting it will proceed to the calculation using the DRP method.

The use of the DRP method is intended to collect all requests from each DC by considering the lead time, lot and the number of requests from each DC so that the company can meet the existing demand without stock out or stock shortage. Forecasting data on data processing is used as a gross requirement (GR) because the DRP method is assessed for the number of products to be produced for the next 1 year period where this assessment uses historical data one year earlier.

The period in DRP starts from period 0-12 but in reality the period starts in period 1 while the use of period 0 in the DRP method is due to the existence of project on hand which a stock before the commencement of production is. In DRP there is also a safety stock where the stock is a stock of deposits. However, in this case, the safety stock is assumed to be zero because the product produced is a food product whose storage period cannot be too long so that if used as a safety stock can cause risks such as expired products, product damage, and large storage costs.

The lot sizing method used for each DC is a fixed order quantity of 40 pcs. Where this determination is based on the agreement of the company with Alfamart where the order is at least 1 box where 1 box contains 40 pcs of ABC noodles. So the ordering process can be done by ordering 40 pcs of noodles or multiples. The results of this DRP are Production Master Schedule or MPS which is obtained from the sum of planned order releases for each DC in the same period.

Based on the research and analysis that has been done, the evaluation that must be done is the importance of the accuracy of forecasting methods and the use of the DRP method can affect the amount of product production each period. With the forecasting method and the proposed DRP method, it can minimize unmet demand for each DC.

4. Conclusion

The best forecasting method to do on DC 1, 4, and 5,6,7,8,9,10 is Double Moving Average, DC 2 is Decomposition Additive and DC 3 is Multiplicative Decomposition. In the calculation of MPS with the DRP method, the results show that the number of production in period 1 is 6320 pcs, period 2 is 6240 pcs, period 3 is 6840 pcs, period 4 is 6200 pcs, period 5 is 6280 pcs, period 6 is 6760 pcs, period 7 is 6160 pcs, period 8 is 6240 pcs, period 9 is 6760, period 10 is 6200, and period 11 is 4960 pcs. By scheduling production using the DRP method, there is a stock due to the use of FOQ lot size so as to minimize the risk of not fulfilling the request if a defect occurs. Companies should forecast each DC so that more optimal forecasting is done. Production scheduling should be done per DC so that there is no miscalculation of product production. The limitations of the data in this study are due to historical data collection that was carried out only for 12 months. Suggestions for further research, it is better to take historical data for at least 2 years in order to see more clearly the time series plots available so that the results will be more optimal.

5. References

[1] Abdillah, Adib Fahrozi 2009 *Perencanaan dan Penjadwalan Aktivitas Distribusi Hasil Perikanan dengan Menggunakan Distribution Requirement Planning (DRP)*. (Jawa Timur: Universitas Pembangunan Nasional "Veteran")

IOP Conf. Series: Materials Science and Engineering **528** (2019) 012049 doi:10.1088/1757-899X/528/1/012049

[2] Ong, Johan Oscar 2013 *Implementasi Distribusi Requirement Planning dan Saving Matrix untuk Minimimasi Total Biaya Distribusi di Industri Bahan Kimia*. (Cikarang: President University)
[3] Pasaribu, Theresia Oshin Rosmaria 2014 *Penentuan Metode Peramalan Sebagai Dasar Penentuan Tingkat Kebutuhan Persediaan Pengaman Pada Produk Karet Remah SIR 20.* (Depok: Universitas Gunadarma)
[4] Setyawan, Eko 2016 *Analisis Peramalan (Forecasting) Produksi Karet (Hevea Brasiliensis) di PT Perkebunan Nusantara IX Kebun Sukamangli Kabupaten Kendal.* (Semarang: Universitas Wahid Hasyim)
[5] Supriyadi 2016 *Penjadwalan Produksi IKS-Filler pada Proses Ground Calcium Carbonate menggunakan Metode MPS di Perusahaan Kertas.* (Serang: Universitas Serang Raya)
[6] Surya, Regina Steven. 2013 *Implementasi Metode Distribution Requirements Planning (DRP) pada CV Karya Mandiri Sejahtera di Surabaya.* (Surabaya: Universitas Surabaya)
[7] Wardah, Siti 2016 *Analisis Peramalan Penjualan Produk Keripik Pisang Kemasan Bungkus*. (Tembilahan: Universitas Islam Indragi)

IOP Conf. Series: Materials Science and Engineering **528** (2019) 012050 doi:10.1088/1757-899X/528/1/012050

Applying value stream mapping tools and kanban system for waste identification and reduction (case study: a basic chemical company)

Wilson Kosasih[1], I K Sriwana[2], E C Sari[1], C O Doaly[1]

[1] Department of Industrial Engineering, Universitas Tarumanagara, Jakarta, Indonesia
[2] Department of Industrial Engineering, Universitas Esa Unggul, Jakarta Indonesia
wilsonk@ft.untar.ac.id

Abstract. This paper describes how to apply both of value stream mapping tools and kanban system for identifying and reducing wastes that happened on shop-floor. This research was conducted at a basic chemical company which produces the thinner. This study aims to reduce even to eliminate the types of waste that occur on the production line of jerrycan product type. The plot of VSM is used to picture the entire of value stream. The value stream mapping tools are used to analyze the certain waste after it has been figured out by using waste assessment model. Hence in this study two selected tools were applied and discussed further, i.e. process activity mapping, and supply chain response matrix. The Current-VSM shown that value-added time of 472.33 s compared with total lead time of 132,568.76 s, consequently obtained process cycle efficiency (PCE) approximately 0.36%. Kanban system will be implemented to eliminate waste of unnecessary inventory throughout the value stream. At last, the result of improvements shown that value of PCE probably will increase by 67.25%, non-value activity will reduce 6.74%, as well as order fulfillment time of the whole process will be 13 days with 3.59 days of physical stock.
Keywords: waste, value stream mapping tools, kanban system.

1. Introduction

An old paradigm states that "creating a quality product, expensive." The era of mass customization, all manufacturing industry players strive to produce products that have a high level of conformity to the standards but still able to fulfill the speed of customer demand. If Stalk & Hout's proposition [1] is correct, that time is a powerful source of competitive advantage, then focus for improvement in the process must include time-based method. Lately, many manufacturing industries in Indonesia have tried to apply the lean concept in transforming their business.

The word "lean" was first introduced by Krafcik [2] to explain the Toyota Production System (TPS). Lean is a set of tools or a set of methods or systems to reduce the time between customer orders and delivery of goods by eliminating waste that does not add value. Five lean principles as an implementation framework in an organization, among others; 1) Value: identify the value from customers perspective, 2) Value stream: identify "specific activities required to design, order, and provide a specific product, from concept to launch, order to delivery, and raw materials into the hands of the customer." 3) Flow: "progressive achievement of tasks along the stream so that the product goes from design to launch, order to delivery and raw materials into the hands of the customer with no stoppages, scrap or backflows." 4) Pull: only make what is pulled by the customer signal of need 5) Perfection: by continually removing successive waste from value stream [3].

The lean production concept introduces a highly effective and efficient production system, which uses smaller resources to produce higher quality at lower costs as a result. Lean production is also known as Just-in-Time (JIT) [4]. Management based on lean production principles enables a company to obtain higher levels of efficiency, competitiveness with the lowest costs, high levels of productivity, speed of delivery, minimum inventory levels, and optimal quality [5].

Lean manufacturing means [3]"... a way to do more and more with less and less ...." Based on the TPS, lean manufacturing is a systematic approach to identifying and eliminating waste of non-value added activities, through continuous improvement activities and optimizing value stream [6]. The application of lean manufacturing will create a synchronized production that is pulled according to the "pulse" of customer demand. The lean approach focuses on efficiency without reducing the effectiveness of processes including increasing value-added operations, reducing waste, and meeting customer needs [7]. In manufacturing industry, there are seven types of waste, among others [6,8,9,10]; a) overproduction, b) defects, c) waiting, d) unnecessary inventory, e) inappropriate processing, f) excessive transportation, g) unnecessary motion [3].

This paper discusses an application of value stream mapping tools and kanban system to identify and reduce the wastes. This research was conducted on a basic chemical manufacturing company. First of all, we need to figure out which type of waste is most dominant in influencing the system or the other types of waste. In discussion section, all activities both value adding and non-value adding would be visualized to compare between current state and future state so that improvements could be designed well.

2. Methods

2.1. The value stream mapping tools

Value Stream Mapping (VSM) is used as an underlying rationale by using a set of tools that can help in identifying waste along the value stream. VSM visualizes the process or activity in the form of a flowchart that is useful for mapping activities that provide added value in realizing lean transformation [4,11]. VSM tools is able to identify value added and non-value added activities in the manufacturing industry, making it easier to find the roots of problems in the entire processes. VSM maps not only the flow of material, but also the flow of information that signifies and controls the flow of material. The path of material flow from a product is traced back from the last operation and its journey to the location of raw material storage as a starting point in implementing lean manufacturing. In its development, the typology of seven new tools [12,13,14,15,16] was presented to further analyze the seven types of waste, including: a) process activity mapping (PAM); b) supply chain response matrix (SCRM); c) production variety funnel (PVF); d) quality filter mapping (QFM); e) demand amplification mapping (DAM); f) decision point analysis (DPA); g) physical structure (PS) as shown in Table 1. Further details of the selection process can be found in Hines and Rich [11].

2.2 The seven waste relationships

According to Rawabdeh [17], all types of waste are inter-dependent, and affect other types of waste. The inter-relationships of wastes are very complex because the influence of each type of waste on other types may arise directly or indirectly (O: overproduction, P: processing, I: inventory, T: transportation, D: defects, W: waiting, and M: motion). The relationships between waste consists of; types of waste as O, D and T affect all other wastes, except P; while the type of waste P affects all other waste, except T; and so on until the type of waste W only affects O, I and D. In total there are 31 relationships where type of waste *i* affects type of waste *j* [17].

2.2.1 Waste relationship matrix. A waste relationship matrix (WRM) questionnaire [17] was adopted to measure the inter-relationships of wastes that occurred. The WRM questionnaire consists of 6 questions with rating from 0 to 4, which includes all relationships of wastes . Furthermore, the steps to analyze the survey results of WRM are as follows; 1) add all scores of each waste relationship; 2) convert a total score into a relationship symbol refers to a range of scores of each inter-relationship of waste; 3) then, convert the results into numbers according to the applicable provisions, where the symbols A = 10, E = 8, I = 6, O = 4, U = 2, and X = 0 [17]; 4) last, calculated the total score and percentage of each waste to create a waste matrix value.

Table 1. Selection matrix for the seven value stream mapping tools [11]

Wastes/Structure	Process Activity Mapping	Supply Chain Response Matrix	Production Variety Funnel	Quality Filter Mapping	Demand Amplication Mapping	Decision Point Analysis	Physical Structure (a)Volume (b)Value
Overproduction	L	M		L	M	M	
Waiting	H	H	L		M	M	
Transport	H						L
Inappropriate processing	H		M	L		L	
Unnecessary inventory	M	H	M		H	M	L
Unnecessary motion	H	L					
Defects	L			H			
Overall Structure	L	L	M	L	H	M	H

Notes : H = 9 – High correlation and usefulness
M = 3 – Medium correlation and usefulness
L = 1 – Low correlation and usefulness

2.2.2 Waste assessment questionnaire. WAQ consists of 68 types of questions which are divided into two categories, namely: 1) the "from" category, if the answer is "yes" means that there is a waste, where the answer score for this category is 1 if "yes", 0.5 if "medium", and 0 if "no"; 2) category "to", if the answer is "yes" means there is no waste that occurs, where the answer score for this category is 0 if "yes", 0.5 if "medium" , and 1 if "no". The steps are used in analyzing WAQ [17], among others; 1) group and calculate the number of questionnaire questions both of categories "from" and "to" for each type of waste; 2) weighting for each type of waste from each type of WAQ question based on the weight of WRM; 3) eliminate the effect of variation for each type of question by dividing the weight of each line with the number of questions grouped (Ni) for each question, as formulated by the following equation:

$$S_j = \sum_{k=1}^{K} \frac{W_{j.k}}{N_i} \text{ ; for each type of waste j} \tag{1}$$

4) sum up the score (Sj) for each column of type of waste; 5) calculate the frequency (Fj) of the appearance of the values in each waste column by ignoring the value of 0 (zero); 6) input the value of WAQ results (1, 0.5, or 0) into each weight of the waste value by using the following mathematical equation:

$$s_j = \sum_{k=1}^{K} X_k \times \frac{W_{j.k}}{N_i} \text{; for each type of waste j} \tag{2}$$

7) sum up the score (sj) for each weight value in each waste column; 8) calculate the frequency (fj) for each weight value in each waste column by ignoring the value of 0 (zero); 9) calculate the initial indicator for each waste (Yj) by using the following equation:

$$Yj = \frac{sj}{Sj} \times \frac{fj}{Fj} \text{ ; for each type of waste j} \tag{3}$$

10) Calculate the final value of each waste ($Yj\ final$), where previously determining the probability factor of the effect of each waste Pj, by using the mathematical equation below:

$$Yj_{final} = Yj\ x\ Pj = \frac{sj}{Sj} x \frac{fj}{Fj} xPj \text{ ; for each type of waste j} \tag{4}$$

The next step is to determine the rank of each waste by calculating the percentage of a final result. Furthermore the WAQ result is used as a reference in selecting value stream mapping tools (see Table 3, Table 4) that will be used for further analysis.

After the steps mentioned earlier, the whole picture of current process was mapped to calculate the process cycle efficiency (PCE) before improvement. The kanban system is developed and designed so that it will eliminate waste of unnecessary inventory throughout the value stream in a future. Eventually, a comparative analysis is used to explain the results between before and after improvement.

3. Results and discussion

3.1 The selection of value stream mapping tools

Firstly, a survey was conducted by using instruments of waste assessment model. Those questionnares were distributed to respondents who are eligible. Table 2 shows that the "form" value of waste of excessive transportation has the highest percentage approximately 19.08%, which means that if this waste has considerable influence to cause other wastes. The "to" value of waste of unnecessary motion has the highest percentage, ie 19.08%. This indicates that the waste of unnecessary motion is the most widely waste caused by the others. Table 3 shows that the following types of waste which are sorted from three highest occurrences respectively are unnecessary motion, then excessive transportation and unnecessary inventory.

After obtaining the results of WRM and WAQ, next step is a selection of appropriate mapping tools against certain type of waste that occurs in the company refers to the result of weighting of each tool. Table 4 shows that two tools are chosen, i.e. process mapping activity and supply chain response matrix. Therefore both are used to identify the wastes and would be discussed further.

Table 2. The result of waste relationship value

F/T	O	I	D	M	T	P	W	Score	%
O	10	6	4	6	6	0	6	38	14.50
I	4	10	4	8	6	0	0	32	12.21
D	4	2	10	6	6	0	4	32	12.21
M	0	6	6	10	0	10	8	40	15.27
T	4	10	6	10	10	0	10	50	19.08
P	4	6	8	10	0	10	10	48	18.32
W	2	4	6	0	0	0	10	22	8.40
Score	28	44	44	50	28	20	48	262	100
%	10.69	16.79	16.79	19.08	10.69	7.63	18.32	100	

Table 3. The result of waste assessment questionnaire

	O	I	D	M	T	P	W	Total
Score (Y_j)	0.25	0.23	0.23	0.26	0.25	0.27	0.24	1.74
P_j factor	155.00	205.12	205.12	291.36	203.95	139.85	153.84	1354.23
Y_j final	39.33	47.87	47.09	74.31	51.48	38.15	37.42	335.65
Final result (%)	11.72	14.26	14.03	22.14	15.34	11.36	11.15	100.00
Ranking	5	3	4	1	2	6	7	

Table 4. The selected result for mapping tools

Waste	Weight	Mapping tools						
		PAM	SCRM	PVF	QFM	DAM	DPA	PS
Overproduction	11.72	11.72	35.16	0.00	11.72	35.16	35.16	0.00
Unnecessary inventory	14.26	42.79	128.36	42.79	0.00	128.36	42.79	14.26
Defect/reject	14.03	14.03	0.00	0.00	126.26	0.00	0.00	0.00
Unnecessary motion	22.14	199.26	22.14	0.00	0.00	0.00	0.00	0.00
Excessive transportation	15.34	138.03	0.00	0.00	0.00	0.00	0.00	15.34
Innapropriate processing	11.36	102.28	0.00	34.09	11.36	0.00	11.36	0.00
Waiting idle	11.15	100.34	100.34	11.15	0.00	33.45	33.45	0.00
Total		608.45	286.00	88.30	149.34	196.97	122.76	29.60
Ranking		**1**	**2**	6	4	3	5	7

3.2 Waste reduction by using the selected mapping tools

Current-VSM is conducted to understand a big picture that currently occurs on the information flow and physical flow throughout the value stream, as illustrated in Figure 1. CVSM result shows that the production process is still implementing a push system, where there is a relatively high inventory on the production floor. Current state indicates that the process runs inefficiency with PCE approximately 0.36% and the number of manpower is 14 workers.

Process activity mapping describes the proportion between value added activities, non value added activities and necessary but non value added activities in an internal manufacturing context [18]. This mapping tool can also be used to identify the waste that occurs in the value stream and optimize the process to be more effective and efficient by designing it more simply. The result of PAM could be seen in Table 5. Further analysis of 5W-1H (asking: What activity does occur? Why does an activity occur? Who does it? On which machine? Where? When? And How?) is used to identify the root causes of non-value-added activities that are unnecessary, so that it can be eliminated.

The SCRM result illustrates that the average time to fulfill an order is 13 days, where the cumulative inventory is 5.75 days. Cumulative inventory or days of physical stock describes the average time of a material throughout the value stream. Table 6 shows that the longest days of physical stock are in the thinner area. This happens because the average production per day is 18,866.32 liters while the absorption of raw materials on average per day is 67,75 liters, resulting occur in a pile of items that will be processed in the thinner area (noted as "work in process"). Even though the raw material warehouse has the longest lead time which is 5 days since ordering, the physical stock circulation in the raw material is relatively normal as well as finished goods area.

Table 5. The result of process activity mapping

Category	Number of Activities	Time (seconds)	Percentage (%)
Value adding	12	472.23	9.58
Non value adding	1	332.47	6.74
Necessary non-value adding	24	4,125.30	83.68
Total	37	4,930.10	100

Table 6. The detail result of SCRM

Area	Days Physical Stock	Lead Time	Cumulative Inventory (Days Physical Stock)	Cumulative Lead Time
Raw material warehouse	1.13	5	1.13	5
Thinner area	3.59	1	4.73	6
Finished goods area	1.02	1	5.75	7
Total				12.75

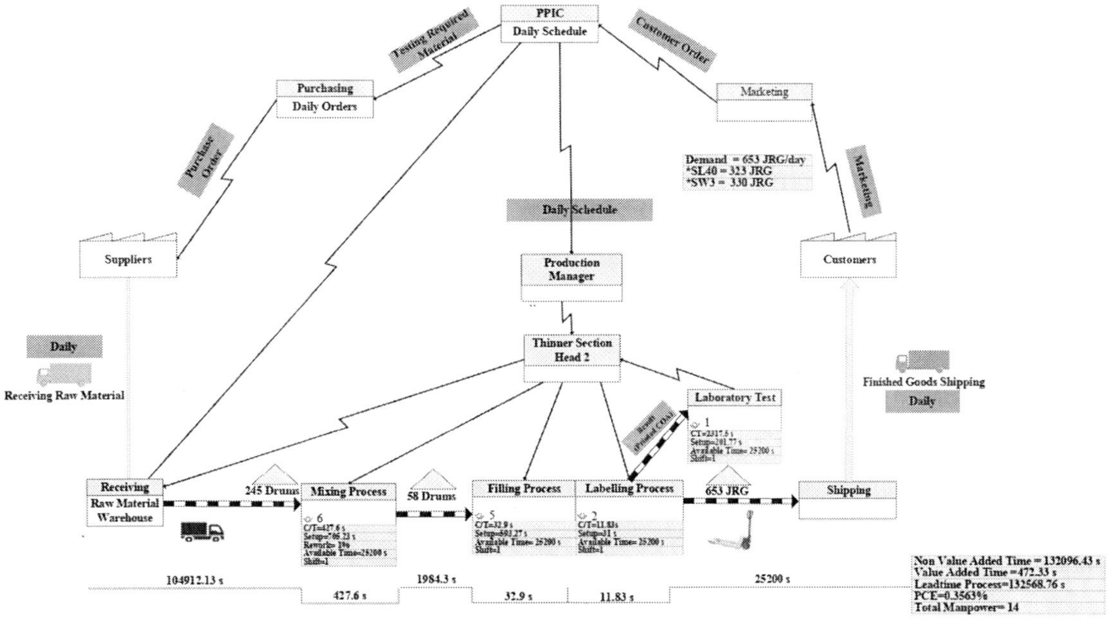

Figure 1. Current value stream mapping

3.3 Waste of inventory removal by using kanban system

Future-VSM as illustrated in Figure 2, shows that the kanban system is developed and implemented to eliminate waste of inventory that accumulates throughout the value stream. In shipping, it is known that numbers of daily demand are 653 jerry cans, the takt time can be calculated as follows:

$$Takt\ time = \frac{\text{available working time}}{\text{daily customer demand}} \tag{5}$$

$$Takt\ time = \frac{25200}{653} = 38.59 \text{ seconds}$$

Therefore, number of transport kanban is needed to move goods from labelling process to shipping which can be calculated as follows:

$$\text{Number of kanban } = \frac{\text{Daily demand} \times (1 + \%\ \text{Safety Stock})}{\text{Container Capacity}} \tag{6}$$

$$Number\ of\ kanban = \frac{653 + (653 \times 0.1)}{50} = 14.40\ \approx 15 \text{ Kanban}$$

and, pack-out quantity and pitch can be formulated and calculated as follows:

$$Packout\ quantity = \frac{daily\ customer\ demand}{amount\ of\ kanban} \tag{7}$$

$$Packout\ quantity = \frac{653}{15} = 44 \text{ jerry cans/kanban}$$

$$Pitch = takt\ time \times packout\ quantity \tag{8}$$

$$Pitch = 38.59 \times 44 = 1{,}697.96\ seconds$$

Meanwhile, required the number of production kanban in filling process as well as the number of transport kanban to move goods from mixing process to filling process as follows:

$$Number\ of\ kanban = \frac{65.3 + (65.3 \times 0.1)}{5} = 14.37\ \approx 15 \text{ kanban}$$

Where assume that one drum could produce ten jerry cans and takt time is 32.9 s hence pack-out quantity and pitch are respectively 5 drums/kanban and 1,645 s. Similarly, needed one production kanban in mixing process, as well as one transport kanban to move goods from receiving to this process. Consequently, here has pack-out quantity and pitch are respectively 66 drums/kanban and 28,221.6 s.

Table 7 describes the results of a comparison between CVSM and FVSM, where non-value added time and total lead time decreased which impact to the increase of PCE by 67.25% with total workers unchanged. After eliminating a non-value added activity which is identified by using PAM, it will decrease setup time of filling process to be 260.8 s. Hence the lead time of filling process decreased by 332.47 s, could be seen in Figure 2.

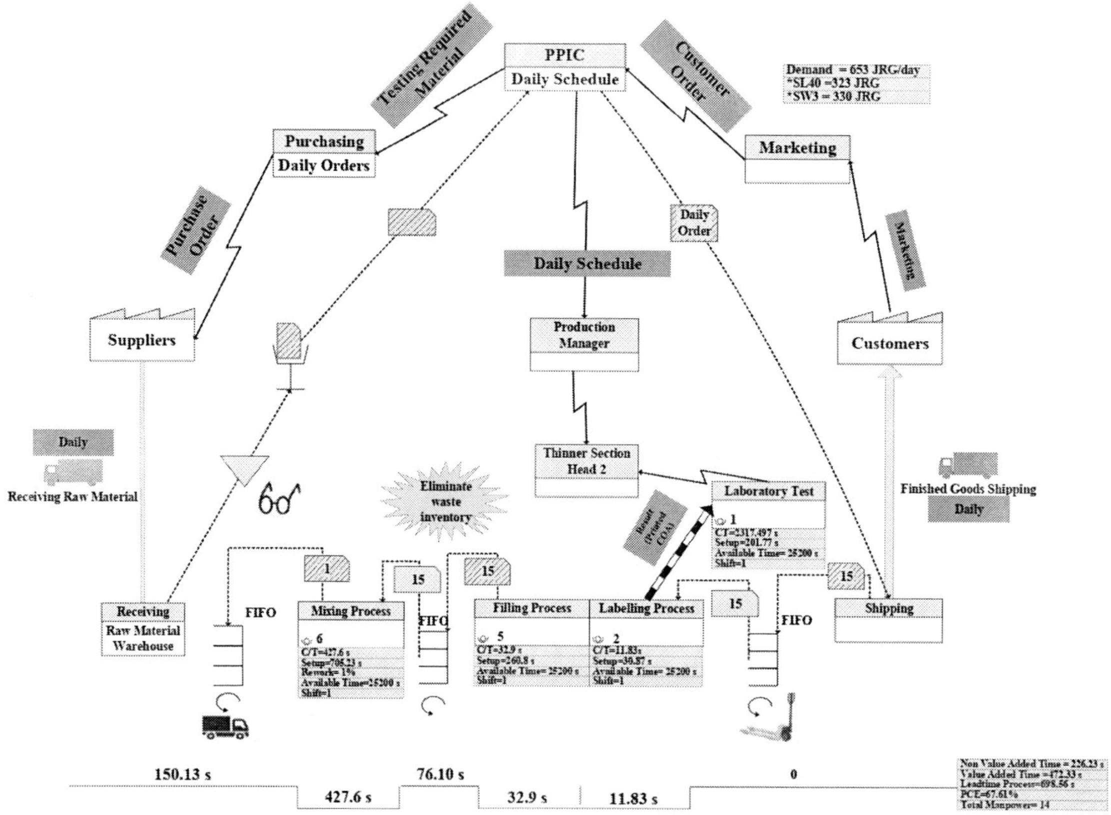

Figure 2. Future value stream mapping

Table 7. A comparison result between CVSM and FVSM

Indicator(s)	CVSM	FVSM	Result
Non value added time	132,096.43 s	226.23 s	Decrease 99.83%
Value added time	47.33 s	472.33 s	
Total Lead time	132,568.76 s	698.56 s	Decrease 99.47%
Process cycle efficiency	0.36 %	67.61 %	Increase 67.25 %
Total workers	14	14	

4. Conclusion

In this study, the wastes could be identified and further analyzed using value stream mapping tools where previously conducted a survey using a waste assessment model. In order to obtain the detailed description of each process at current state, a big picture of value stream was mapped. Kanban system is also applied on this improvement. Finally, some type of identified wastes can be reduced or even eliminated. Thus, process cycle efficiency will increase significantly by 67.25%. However, various studies still need to be developed and elaborated to transform this organization into a fully lean enterprise.

Acknowledgements
This research recieved a grant from the Directorate of Research and Society Service, Universitas Tarumanagara, Jakarta, Indonesia. Thus on this occasion, Authors would like to thank to this institution and also to everyone who assisted Authors to complete this research.

5. References

[1] Stalk G and Hout T 1990 *Competing Against Time.* (New York: The Free Press)
[2] Krafcik J F 1988 Triumph of the Lean Production System. *Sloan Management Review* **30**(1) pp 41–52
[3] Womack J P and Jones D T 1996 *Lean Thinking: Banish Waste and Create Wealth in Your Corporation.* (New York: Simon & Scuster)
[4] Taylor D and Brunt D 2001 *Manufacturing Operations and Supply Chain Management.* (London: Thomson Learning)
[5] Arbós L C 2002 Design of a Rapid Response and High Efficiency Service by Lean Production Principles: Methodology and Evaluation of Variability of Performance. *International Journal of Production Economics* **80**(2) pp 169–183
[6] Ohno T 1988 *Toyota Production System beyond Large-scale Production.* (Portland: Productivity Press)
[7] Hines P and Taylor D 2000 *Going Lean.* (United Kingdom: Lean Enterprise Research Center Cardiff)
[8] Shingo S 1989 *A Study of the Toyota Production System from an Industrial Engineering Viewpoint.* (Cambridge, MA: Productivity Press)
[9] Bicheno J 1994 *The Quality 50.* (Buckingham, UK: Picsie Books)
[10] Japan Management Association 1985 *Kanban: Just-in-Time at Toyota.* (Cambridge, MA: Productivity Press)
[11] Hines P and Rich N 1997 The Seven Value Stream Mapping Tools. *International Journal of Operations & Production Management* **17**(1) pp 46-64
[12] Ishiwata J 1991 *Productivity Through Process Analysis.* (Cambridge, MA: Productivity Press)
[13] Practical Management Research Group 1994 *Seven Tools for Industrial Engineering.* (Tokyo, Japan: PHP Institute)
[14] New C 1993 The Use of Throughput Efficiency as a Key Performance Measure for the New Manufacturing Era. *The International Journal of Logistics Management* **4**(2) pp 95-104
[15] Forrester J 1958 Industrial Dynamics: A Major Breakthrough for Decision Makers. *Harvard Business Review* (July-August) pp 37-66
[16] Hoekstra S and Romme S 1992 *Towards Integral Logistics Structure – Developing Customer Oriented Goods Flows.* (New York: McGraw-Hill)
[17] Rawabdeh I 2005 A Model For The Assessment of Waste in Job Shop Enviroments. *International Journal of Operations & Production Management* **25**(8) pp 800-822
[18] Monden Y 1993 Toyota Production System: An Integrated Approach to Just in Time. (Georgia: Industrial Engineering and Management Press)

IOP Conf. Series: Materials Science and Engineering **528** (2019) 012051 doi:10.1088/1757-899X/528/1/012051

Waste assessment using lean manufacturing in rubber production

Elita Amrina, Nilda Tri Putri, Dwara Mitha Anjani

Department of Industrial Engineering, Andalas University, Padang, Indonesia
elita@eng.unand.ac.id

Abstract. During the rubber production process, there are several wastes caused by breakdown machines, damaged equipments, and production delays. Those wastes have negative impact to companies and thus, need to be eliminated. This study aims to assess wastes in rubber production process using lean manufacturing approach. Waste Relationship Matrix (WRM), Waste Assessment Questionnaire (WAQ), and Value Stream Analysis Tools (VALSAT) are applied to identify the wastes. Fishbone Diagram is then used to determine the root cause of problems. Finally, Failure Mode and Effect Analysis (FMEA) method is applied to evaluate the source of wastes and develop the action plans. It indicated that over production, inventory, and defects as the three highest wastes in the rubber production. Process Activity Mapping (PAM) is selected as the analysis tools in VALSAT and it obtained the activities in the rubber production are mostly categorized into value added activities. The necessary but non value added activities are also identified at a fairly high frequent. The major risks during the rubber production is related to dynamo damage and machine clogged in both hammer mill and mangle machine. It needs high attention from the production manager of rubber industry. Then, action plans are suggested to improve the maintenance program.
Keywords: waste assessment, lean manufacturing, rubber industry

1. Introduction

Lean manufacturing has been recognized as a powerful tool in finding and eliminating the wastes during the production processes. It is defined as a systematic and structured methodology that used to find, solve, and prevent the performance problems through tracking-back approaches to obtain the main hidden roots of wastes [1]. Lean manufacturing is known as a philosophy of Toyota Production System that attempts to reduce the time line between the customer order and the product delivery by constantly eliminate the wastes [2]. The basic ideas of lean manufacturing are waste elimination, cost reduction, and employee empowerment [3].

In the concept of lean manufacturing, the activities in production processes are divided into three types, namely value added activities, non-value added activities, and necessary but non-value added activities [4]. Eliminating wastes can reduce the non-value added activities and increase the value added activities. The implementation of lean manufacturing is expected to improve the flexibility of production process, overcome the changes in customer needs, and reduce the inventory level [2]. Elimination of wastes provides a high impact to companies that can increase the added value of final product. Therefore, the activities that do not provide added value to the product must be eliminated because it is categorized as a waste in the production process.

It is no doubt that the elimination of waste is a critical part for companies in order to survive in today's manufacturing world. Wastes will have a negative impact to companies such as inventory that increase the production costs due to require the additional handling and space, delay times that create the work in process between workstations, and excessive processing that will increase the rate of redundancy operations [1]. In other hand, companies today must strive to produce high-quality and low-cost products that can reach the customers in the shortest possible time. Thus, it is necessary to improve the efficiency of the production line in order to reduce the downtime rates and fulfill the market demand [1].

Based on the preliminary study conducted in a rubber manufacturing company in Padang, Indonesia, it can be seen there are several wastes during the rubber production processes. It caused by the breakdown machines, damaged equipments, and production delays due to low maintenance on machines and equipments. The most frequently wastes occured in the rubber production processes are over production, delays, and defects. Therefore, it is needed to eliminate those wastes. This paper attempts to assess the wastes during the rubber production processes using lean manufacturing approach.

2. Methodology

The methodology has four main stages.

2.1. Waste Identification

Waste identification is conducted using Waste Relationship Matrix (WRM) and Waste Assessment Questionnaire (WAQ). WRM is applied to determine the relationship amongst wastes, while WAQ is used to assess the type of wastes and rank the wastes. In this study, seven wastes are considered in the assessment consist of over production (O), inventory (I), defects (D), motion (M), transportation (T), over processing (P), and waiting (W). The waste relationships are assessed using a questionnaire consisting of six questions and the weight of answers ranging from zero to four adopted from [5]. The results are presented in a WRM that show the relationship value amongst the wastes. WAQ consists of 68 different questions for the purpose of allocating the wastes. Each of the questions represents an activity, a condition or a behaviour that may lead to a specific type of waste. Some questions are assigned a "From" note, which means the question represents an existing type of waste that may lead to other wastes, with reference to the WRM. Other questions are assigned a "To" note, which means the question represents any existing type of waste that may have been caused by other types of waste. All the questions have three answers and each answer is assigned a weight of 0, 0.5 or 1. The questions then categorized into four groups consist of man, machine, material and method. The results show the value of each waste and ranked in descending order.

2.2. Value Stream Analysis Tools (VALSAT)

VALSAT developed by Hines and Rich in 1997 is aimed to understand the current value stream and aid in developing the improvements to eliminate wastes in the value stream. It is a decision heuristic for selection of value stream mapping techniques. VALSAT develops the correlation matrix between the various types of waste and the seven value stream mapping tools. Implementation of improper mapping tools can result in wasting the additional resources such as time and money, and decreasing the employee confidence to the lean philosophy [6]. In this stage, an analysis tool with the highest value is selected to obtain a detail mapping analysis.

2.3. Fishbone Diagram

The Fishbone diagram is then developed to identify the root causes of waste. In this stage, the three highest wastes identified from the waste identification stage are selected to obtain the root causes of those wastes. The root causes are analyzed in which categories of process, product, machine, material, marketing, and environment. Identification of the root causes is conducted using brainstorming approach with industry experts.

2.4. Failure Mode and Effect Analysis (FMEA)

In this stage, the failure modes caused by the effects of wastes are identified using Failure Mode and Effect Analysis (FMEA) method. For that purpose, a FMEA questionnaire is then designed to determine the severity, occurrence, and detection of failure modes. Next, the Risk Priority Number (RPN) is calculated to obtain the highest failure modes. Finally, the action plans are developed to minimize the failure modes.

3. Results and discussions

This study is conducted in a rubber manufacturing company located in Padang, Indonesia. Established in 1972, the company produces the crumb rubber. The production flow is shown in Figure 1.

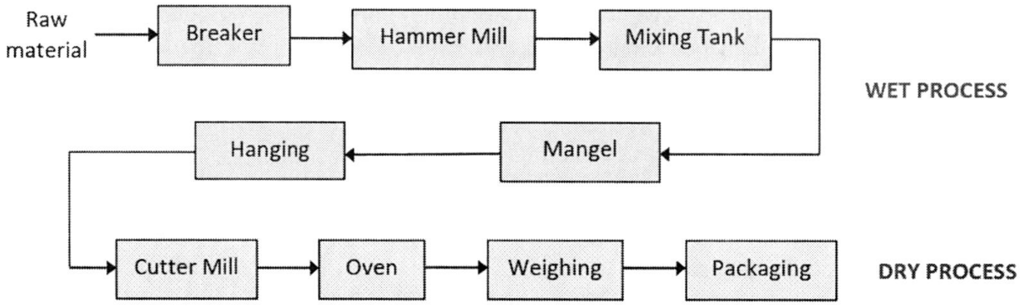

Figure 1. The crumb rubber production flow.

3.1. Waste Identification

A total of 14 production managers and staffs are consulted to determine the waste relationship. The results are shown in Table 1.

Table 1. Waste relationship score.

Question	Score	Relationship	Question	Score	Relationship
O_I	13.50	E	M_P	10.50	I
O_D	9.67	I	M_W	11.75	I
O_M	12.33	I	P_O	11.50	I
O_T	14.67	E	P_I	10.33	I
O_W	12.33	I	P_D	10.50	I
I_O	14.17	E	P_M	11.17	I
I_D	9.83	I	P_W	10.17	I
I_M	12.50	I	T_O	13.50	E
I_T	13.33	E	T_I	14.17	E
D_O	9.25	I	T_D	10.83	I
D_I	10.75	I	T_M	12.83	I
D_M	10.25	I	T_W	13.17	E
D_T	7.75	O	W_O	7.50	O
D_W	12.25	I	W_I	11.00	I
M_I	13.00	E	W_D	9.50	I
M_D	9.75	I			

The Waste Relationship Matrix of rubber production is then constructed based on the table above as presented in Table 2. The Waste Relationship Matrix is then converted into a value where A= 10, E= 8, I= 6, O= 4, U = 2, and X= 0 adopted from [5]. The percentage value is calculated and the results are shown in Table 3. It can be seen from the table, transportation (T), over production (O), and over processing (P) are the most influencing wastes, while inventory (I), defects (D), and over production

(O) are the most influenced by other wastes. The next step is validating the wastes using Waste Assessment Questionnaire (WAQ). The results are shown in Table 4.

Table 2. Waste relationship matrix.

F/T	O	I	D	M	T	P	W
O	A	E	I	I	E	X	I
I	E	A	I	I	E	X	X
D	I	I	A	I	O	X	I
M	X	E	I	A	X	I	I
T	E	E	I	I	A	X	E
P	I	I	I	I	X	A	I
W	O	I	I	X	X	X	A

Table 3. Waste relationship value.

F/T	O	I	D	M	T	P	W	Score	%
O	10	8	6	6	8	0	6	44	16.42
I	8	10	6	6	8	0	0	38	14.18
D	6	6	10	6	4	0	6	38	14.18
M	0	8	6	10	0	6	6	36	13.43
T	8	8	6	6	10	0	8	46	17.16
P	6	6	6	6	0	10	6	40	14.93
W	4	6	6	0	0	0	10	26	9.70
Score	42	52	46	40	30	16	42	268	100
%	5.67	9.40	7.16	4.93	1.19	.97	5.67	100	

Table 4. Waste relationship matrix.

Wastes	Score (Y_j)	P_j Factor	Final Result (Y_{jfinal})	Final Result (%)	Rank
O	0.195	257.296	50.149	20.244	1
I	0.179	275.117	49.323	19.911	2
D	0.181	243.373	44.124	17.812	3
M	0.155	200.490	31.012	12.519	5
T	0.173	192.136	33.178	13.393	4
P	0.126	89.107	11.261	4.546	7
W	0.189	152.038	28.673	11.575	6

The results show over production (O) indicated as the major type of waste. It is followed by inventory (I) and defects (D). Over production has considered as the most important waste because it has a high influence on the occurrence of other wastes [7]. Producing more items than required by the next station or the market will create an excess inventory. It clearly shows the strong relationship between over production and inventory. Over production also tends to lead to increased lead time and increased storage. As a result, defects may not be detected at early stage.

An investigation was carried out and shows that the production is often incompatible with the predetermined schedules. It can be caused by a sudden downtime on the machine that cause delays in the production process and loss of production day. This leads to over production and increased inventory on the crumb rubber production. In addition, it might also be due to the push-based production process that causes the high inventory levels on raw material and finished product [8].

3.2. Value Stream Analysis Tools (VALSAT)

VALSAT approach is applied to select an analysis tool by multiplying the final results (% of Y_{jfinal}) from the waste identification stage with the scale of VALSAT. In this stage, the correlation value of VALSAT consisting of high (9), medium (3), and low (1) is determined for each of the wastes and the mapping tools. The results are presented in Table 5.

Table 5. The results of VALSAT.

		Mapping Tools						
Wastes	Weight	Process Activity Mapping	Supply Chain Response Matrix	Production Variety Funnel	Quality Filter Mapping	Demand Amplification Mapping	Decision Point Analysis	Physical Structure Mapping
O	20.244	20.244	60.733	0	20.244	60.733	60.733	0
I	19.911	59.732	179.197	59.732	0	179.197	59.732	19.911
D	17.812	17.812	0	0	160.308	0	0	0
M	12.519	112.672	12.519	0	0	0	0	0
T	13.393	120.540	0	0	0	0	0	13.393
P	4.546	40.912	0	13.637	4.546	0	4.546	0
W	11.575	104.172	104.172	11.575	0	34.724	34.724	0
Total	476.085	356.621	84.944	185.098	274.654	159.735	33.304	268

The results are identified the Process Activity Mapping (PAM) as the highest value of mapping tools with a score of 476.085. The PAM is then used to describe the crumb rubber production process in detail using symbols that represent the activities such as operation, storage, transportation, inspection, and delay. PAM is a tool used to describe in details the order fulfilment process and find out the activities which include Value Added (VA), Non Value Added (NVA), and Necessary but Non Value Added (NNVA) activities. This tool is also used to identify wastes in the value stream and optimize the process to be more effective and efficient by eliminating unnecessary activities, simplifying processes, and combining processes. The results are presented in Table 6.

Table 6. Results of process activity mapping.

Classification	Quantity	Percentage (%)
Value Added	32	66.67
Non Value Added	2	4.17
Necessary but Non Value Added	14	29.17
Total	48	100.00

The value added activities have identified as the most frequent activities in the rubber production processes with a value of 66.67%. The necessary but non value added activities also have a fairly high value of 29.17%. It mostly caused by the activities of transportation and inspection. The transportation in the rubber production process highly affects the material flow. It might be caused by the plant layout and the continuous production flow of rubber production. In term of inspection, a high frequent of inspections are conducted to maintain the quality of products in the company.

3.3. Fishbone diagram

The Fishbone diagram is used to identify any possible causes of a problem [9]. A series of discussions were conducted to production managers and staffs of the rubber company to develop the Fishbone diagram. The three highest wastes identified in the waste identification stage consisting of over production, inventory, and defects are described. The causes of problem are defined into some categories. The Fishbone diagrams of three wastes are presented in Figure 2–4.

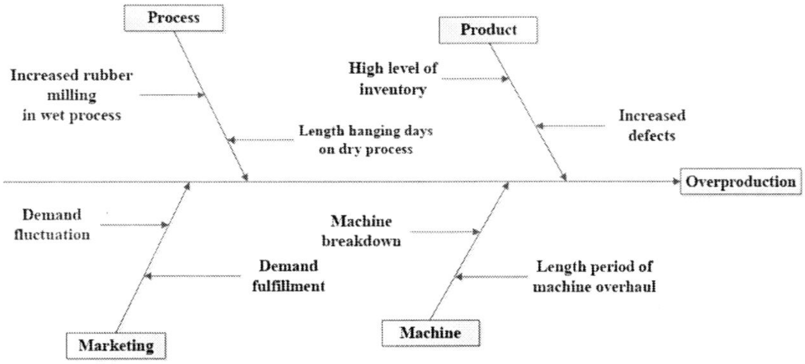

Figure 2. Fishbone diagram of over production.

The root causes of over production are categorized into process, products, marketing, and machine. Over production in the rubber production is mostly caused by the increased milling process in the wet process and the long hanging days. This causes the excess inventory. Over production is usually caused by the high level of inventory and increased defects [8]. In addition, the demand fluctuation from customer makes the company difficult to predict the future demand and thus causing the over production.

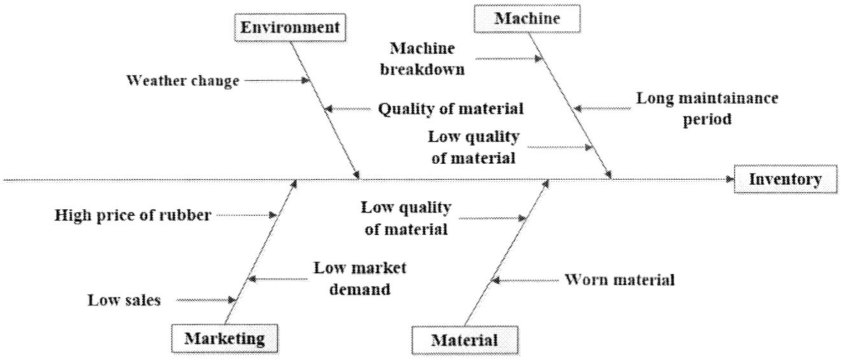

Figure 3. Fishbone diagram of inventory.

Identification of the root causes of inventory consist of four categorizes of environment, machine, marketing, and material. When the sunny, raw material of rubber at a high quality and a high quantity. Company usually buys the raw material at this time. The inventory level also increases during the machine breakdown. In term of marketing, the inventory is caused by high price of material, low sales, and low market demand. Improper production planning causes the excess inventory.

The root causes of defects are divided into four categories of process, machine, product, and material. Defects in the rubber production processes caused by nonconformance process, duration of hanging days, operator error. Nonconformance process will result the product produced not conform to the quality standard [8]. Nonconformance process can be caused by operator and machine. Raw material contaminated with other substances such as iron, stone, and plastic can cause the defects. Besides, low quality of material and high speed of rotation will also affect the product defect.

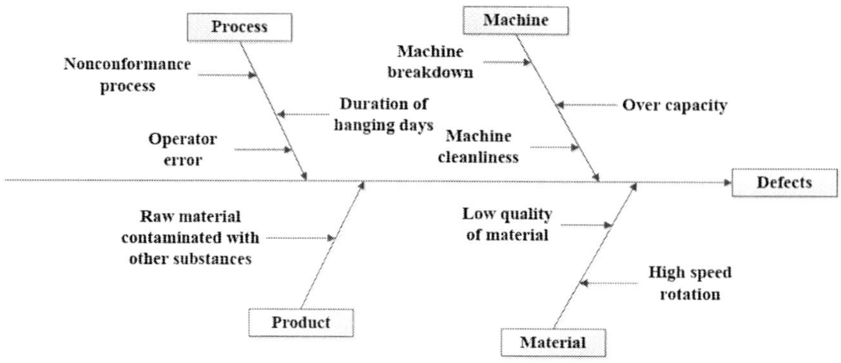

Figure 4. Fishbone diagram of defects.

3.4. Failure Mode and Effect Analysis (FMEA)

Failure Mode and Effect Analysis (FMEA) method is applied to identify the failure modes caused by the affected wastes. Failure modes are the error conditions which affecting waste in the system [9]. The failure modes are assessed in terms of severity, occurrence, and detection with a scale ranging from 1 to 10 adopted from [4]. A total of 34 failure modes are identified from three main machines in the rubber production based on the results of Fishbone diagram. The Risk Priority Number (RPN) is computed to each of the failure modes. The failure mode with a high RPN means the activity needs an urgent attention and a proper improvement. The results of Risk Priority Number of failure modes in the rubber production are shown in Figure 5.

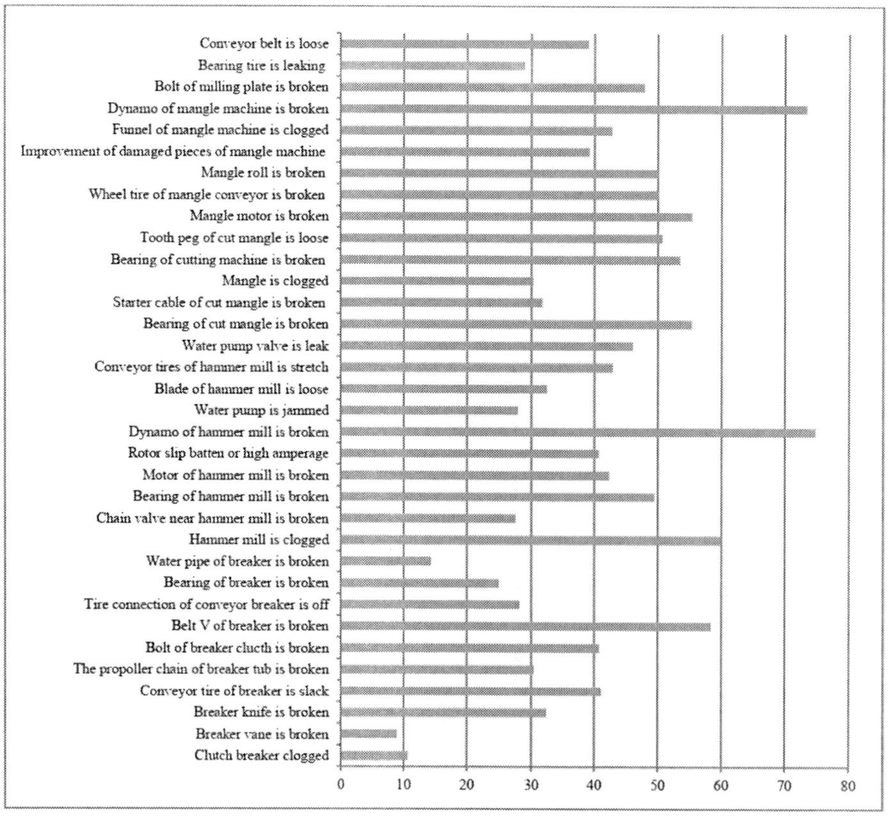

Figure 5. Risk priority number.

It can be seen that the dynamo of hammer mill is broken identified as the major failure in the rubber production with the highest RPN of 74.813. It is followed by the dynamo of mangle machine is broken with RPN of 73.813 and the hammer mill is clogged with RPN of 73.486. Based on the results, it can be concluded that major failures are mostly caused by the production equipment. Thus, it needs urgent attention from production department to improve the maintenance program. The preventive maintenance should be conducted to production equipments. The quality control on the raw material is also suggested since the failure can cause the customer dissatisfaction, major system problems, and rework. It is required the appropriate inspection for material supplied. The comprehensive inspection is needed rather than regular inspection. Besides, the relationships with the supplier need to be improved in order to provide a good quality of material.

Conclusions

This paper has applied lean manufacturing to assess wastes in the rubber production processes. Waste Relationship Matrix (WRM) and Waste Assessment Questionnaire (WAQ) are used to identify the wastes. The results show over production is indicated as the major type of waste in the rubber production processes, followed by inventory and defects. Through Value Stream Analysis Tools (VALSAT), the Process Activity Mapping (PAM) is selected as the analysis tools. It identified the activities in the rubber production processes are mostly categorized as value added activities with a value of 66.67%. However, necessary but non value added activities also have a fairly high value of 29.17%. It mostly caused by the activities of transportation and inspection. Then Fishbone Diagram is applied to determine the root cause of the three highest wastes identified. Finally, Failure Mode and Effect Analysis (FMEA) is used to evaluate the source of wastes and develop the action plans. FMEA indicates the highest failure was caused by the production equipments. Thus, the action plans suggested to the rubber companies are to improve the maintenance program. Besides, the quality control of raw material and the relationships with suppliers should also need to be improved. It is hoped the rubber company can improve their production processes to be more efficient.

Acknowledgments

The authors would like to thanks to Faculty of Engineering, Andalas University and Directorate General of Higher Education, Ministry of Research, Technology, and Higher Education, Indonesia.

References

[1] Khalil, R A, Stockton, D J, Tourki, T and Mukhongo, L M. 2013. Implementation of lean in continuous process-based industries. *Int. J. of Scientific & Engineering Research.* 4(10). p 723.

[2] Singh, S K, Sharma, K, Kumar, D and Gupta, T. 2014. Role & importance of lean manufacturing in manufacturing industry. *Int. J. Engineering and Science.* 3(6). p 1.

[3] Ramesh B K, Jayachitra R and Abinath M. 2016. Implementation of lean in automotive component manufacturing process. *Int. Research J. of Engineering and Technology.* 3(8). p 49.

[4] Gasperz, V. 2007. *Lean six sigma for manufacturing and service industries.* PT Gramedia Pustaka Utama Jakarta.

[5] Rawabdeh, I A. 2005. *A model for the assessment of waste in job shop environments.* University of Jordan.

[6] Ramesh, V and Kodali, R. 2012. *A decision framework for maximizing lean manufacturing performance.* Birla Institute of Technology and Science: India.

[7] Kobayashi, I. 1995. *Twenty Keys to Workplace Improvement.* Productivity Press: Cambridge, MA.

[8] Liker, J K. 2010. *The Toyota Way.* McGraw-Hill Publication: United States.

[9] Amrina E and Lubis A A A. 2017. Minimizing waste using lean manufacturing: a case in cement production. *Proc. Int. Conf. On Industrial Engineering & Application.* IEEE. p 71.

IOP Conf. Series: Materials Science and Engineering **528** (2019) 012052 doi:10.1088/1757-899X/528/1/012052

Evaluating Storage Tank Cap 10000L Manufacturer by Using Lean Project Management

T G Amran, D Saraswati and E F Harahap

University of Trisakti, Jl. Kyai Tapa No. 1, Jakarta, Indonesia
tiena@trisakti.ac.id, devisaraswati@gmail.com; elfira@trisakti.ac.id

Abstract. Business competitions between manufacturing fabrication industries increase fiercely. Companies were demanded to produce high quality products with competitive price that can be delivered to the consumer on time. Storage tank (10000 liter capacity) is one of the most ordered item but most of the time the delivery is unpunctual. During the initial planning, buffer time was unavailable, which caused 32% cost increase by waste. In this study, the concept between lean manufacture and lean project management was bridged to obtain minimal delay. The first research methodology was to identify and analyze waste, second method was risk analysis by FMEA, and final method was cost and time estimation by critical chain project management. S Curve depicted the decrease of delay, and value stream mapping was used to map optimum result from lean project management. Potential waste was "waiting", and the highest FMEA risk value was material delay caused by weather conditions. Project buffer result shows 39% decrease in engineering task and the actual cost was still within the budget. The company need to integrate HR cost in the project and measure the time buffer to predict delay gap.
Keywords: CCPM, delay project, lean project management, value stream mapping

1. Introduction

Lean philosophy was first introduced by Toyota as "Lean Manufacture". This concept believes that through waste identification and elimination, efficient process can be reached and customer value can be fulfilled [5]. *Lean Management* is a systematic and integrative method that is implemented continuously to minimalize waste. Separation of non-value added process through continuos improvement. In value stream, all employees starting from operational field to top management were mapped. This concept is applied in service companies, government agencies, health, and education services to provide more effective and efficient process, faster service, lower cost, and better quality [6].

Project management is often used to handle design-to-order demand. Products related to the design-to-order demand type has total uniqueness in every unit. A few of product example that is produces with project management approach such as: plane, ship, building, bridge, etc. [4]. Unique characteristic that differentiate project from other manufacturing project is that there is no product flow. Materials, tools and workers are brought into the project location to be done on site. Nevertheless, a project still has order of activities. Project also has a tendency to spend a high cost and relatively difficult to plan and control compared to other manufacturing process [4].

This paper objective is to decrease the project manufacturing fabrication delay that is ordered by the consumers with product variety and the single or repeated order. Each company project new design has to make a new customized product plan [1] [2]. Work divisions are classified into a few project, high product variety, risk complexity in the project, dense time with high cost, competent worker skill and high work standard [1] [2] [3]. Lean Project Management approach with VSM emphasize added value, decrease waste and unproductivity Non Added Value and Non Value Added [4] and help in implementing the system [7] Assessment evaluation of project risk, project cost, project difficulty, HR quality and project completion time [5] [2] is subjective and determined by the boards. The use of S-Curve [4] is to estimate project total cost if it is according to project value that has been determined by project owner or might be exceeding. Cost estimation is conducted by detailing required material and human resource in every type of work. Critical Chain Project Management (CCPM) approach [3] [4] change safety time with buffer time. Project buffer is a buffer time that is placed in the end of critical chain of a project as a backup time of the whole project [6]. The determination of buffer size can be conducted by Square Root of the Sum of Square (SSQ) method [4]. The obstacles encountered by company are the unknown of process difficulty and production of products, scheduling suitability, human resource and target achievement that has been determined earlier [2] In this research, manufacturing fabrication for storage tank L10000 is conducted by (1) Fishbone diagram and If then formulation to understand the root that cause waste, [9] [10] (2) Right Solution [4] [3] [10]. (3) Lean Project Management approach with VSM (4) Critical Chain Project management (CCPM) approach [3] [4] changing safety time into buffer time.

Recent project management researches [12] [13] show its efficiency of CCPM in construction management, telecomunication [14] and ganggar construction. Comparative analysis between CPM and CCPM [21] showed that CCPM is more efficient and should be considered as the new paradigm [15]. Several studies on project risk management [16] combined with lean project management and CCPM for safety during waste arrival combined with scheduling [17][18] revealed a decrease in working hour, cost saving, and worker. Value Stream Mapping (VSM) dan Value Stream Analysis Tools (VALSAT) [18] [19] also showed the decrease of project waiting time 9% to 1.09%.

2. Methodology

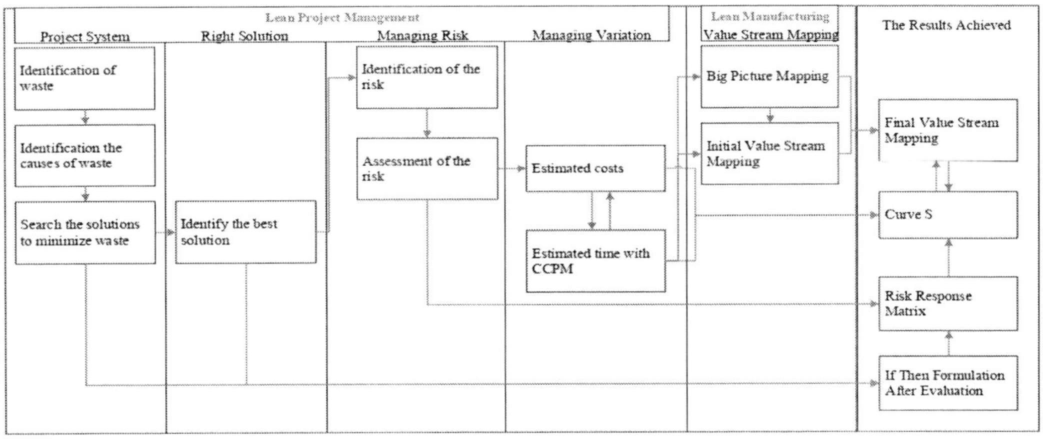

Figure 1. The Research Framwork

3. Results And Discussion

3.1. Cause of Waste Identification

The project system work consists of the following activities: material description, consumable, shrearing, bending, bending roll, welding filter, TIG welding, polishing, fabrication manpower, lathe, milling, packaging and delivery, Quality Control, and engineering services. Occurred wastes were identified by questionnaire which caused delays in 1000L Storage Tank Cap projects. The cause of waste were analyzed with fishbone approach.

Table 1. Summary of Waste Identifcation for Storage Tank 10000l

No	Type of Waste	Ranking	Weight	No	Type of Waste	Ranking	Weight
1	Defect	21	0.15	5	Unnecessary Motion	7	0.05
2	Overproduction	9	0.064	6	Excessive Transportation	15	0.107
3	Waiting	35	0.25	7	Unnecessary Inventory	12	0.086
4	Unappropriate Processing	18	0.129	8	Unsatisfied Design	23	0.164

The waste in Storage Tank 10000L production shows the highest delay in waiting activity (35), unsuitable design (23), and defect (21).

3.2. *Fishbone Diagram*

The fishbone diagram explains the highest cause of waste: waiting time, unsuitable design, and defect as described:

- Cause of delay: waiting factor are length of drawing approval waiting from the customer and sub contractor work;
- Cause delay: man factor are lack of engineering skill in drawing and drawing reading. Method factor,: mixed up drawing and BOM before and after revision.
- Cause of delay method factor are: SOP violation and tight deadline. Material factor: argon level below standar & welding wire not suitable to the welding machine.

3.3. *If Then Formulation*

The If Then formulation is a follow-up of some action-solutions for each cause of waste. The causes that have more than one proposed solution will be processed into the evaluation matrix to get the best solution based on the criteria and conditions that have been set. Hence, each event that causes waste has only one best solution.

Table 2. If Then Formulation as Solution for Every Cause of Waste

	Controling Waste	
If	Then	When
Waiting for sub contractor work 5 days late production work becomes 41 days late	Provide a reminder letter to the sub contractor 2 times a week to control progress stor work Do other work that does not use part of sub contractor work	implementation sub work contractor
Waiting for drawing approval from The customer results in a withdrawal engineering work for 64 days	Provide approval drawing drawing reminders to customers intensively that is 2 times a week Give deadline drawing approval for 7 the day after the customer order process	Implementation work engineering
Lack of skills in engineering in drafting drawing / P17155 / VIII so there was a revision on the drawing	Provide in house training and seminars for as long as less than 3 days Give sanctions to workers with poor performance based on assessment results workers according to SOP-HR-03 standards	Implementation work engineering
Less careful in reading drawings and specifications BOM / P17155 / VIII by the parties production and PPIC	Checking drawing / P17155/ VII for reference or for working before the distribution of data to production side and PPIC Hold a project kick of meeting with the division related before the project implementation process	Implementation work production and PPIC

Table 2. If Then Formulation as Solution for Every Cause of Waste (cont.)

Controling Waste		
If	Then	When
Project work is 43 days late so the deadline for completion urgent project	Hold overtime with maximum limits working hours until 21.00 WIB Adding worker personnel	Implementation work production
There is a violation of MS / SOP SOP HR/ SOP QC/ SOP PC	Give sanctions to violators accordingly provisions of SOP MS / SOP HR / SOP QC / SOP PC Provide training and evaluation regarding understanding SOP regularly, which is 3 months	Implementation work production and engineering
The quality of argon is less than 99.999% standard and wrong use welding wire type or welding wire does not match ampere voltage on the welding machine	Provide a warning letter to the supplier if the material does not match the order Check the quality of the material periodically, every 3 months	Implementation work production

3.4. *Right Solution*

aims to choose solution based on criterias such as cost, time, impact towards output and project risk by using weighing and it is obtained a score for "Go" or "Not Go". The ranking scale given is 1-10 (highest rank) based on expert judgement. Total value can be obtained by formula:
Weighted Score = Weight Factor x Ranking

Table 3. Evaluation Matrix on the Cause of Sub Contractor

Criteria	Weight factor	Waiting for a sub-contractor job that is 5 days late so that the production work become 41 days late			
		Provide a reminder letter to the sub contractor 2 times a week for control the work progress		Do other jobs that don't use part of sub work contractor	
		Ranking	Weighted Score (Weight factor x Ranking)	Ranking	Weighted Score (Weight factor x Ranking)
Cost	7	9	63	8	56
Time	7	8	56	7	49
Impact of results	5	7	35	9	45
Risk	8	10	80	10	80
Total Weight Score		234		230	
GO/ NOT GO		GO		NOT GO (GO II)	

The action recommendation table that is done by project executor of the company after conducting evaluation with evaluation matrix.

3.5. *Managing Risk*

Risk Identification arrange list of unexpected events that cause failure in the making of *Storage Tank Cap*. 10000 L. Source of risk are from external (unpredictable), internal (predictable) and internal non-technical. Table 4 result of assessment each risk indocators. Equation determines FMEA Risk Value = Impact x Possibility x Detection (4)

Table 4. Risk Assessment Form in Making Storage Tank Cap. 10000 L

Risk Indicators	Possibility	Impact	Difficulty Detection	FMEA	Time
Natural disaster	9	8	9	648	All the time
Argon material and welding provision do not meet standard.	4	7	2	56	Before & on execution
Worse project cost condition	3	5	3	45	Before & on execution
Worse project time condition	3	5	2	30	On execution
Violation in Work Health and Safety	1	8	2	16	On execution
Stealing, Negligence and Dishonesty	3	9	6	162	All the time
Damage in measuring tools, welding and grinding machine	3	5	3	45	On execution

3.6. Risk analysis

Responses to risks are grouped as mitigation, avoiding, transferring, sharing and retaining and also contingency planning, risk response matrix tools in Table 5.

Table 5. Risk Response Matrix

Risk Indicator	Possibility	Contigency Plan	Trigger
Natural disaster	Reduction	Submitting project delay letter to the customers	Bad weather or natural disasters
Argon material and welding provision do not meet standard.	Prevention	Planning a standar material and worker requirement before executing project	Lack of preparation and material and worker control by the executors
Worse project cost condition	Reduction	Planning cost and provides contingency fund through investors	Overloaded project in the same time
Worse project time condition	Reduction	Planning schedule properly using buffer time as backup time	Worse schedule planning and lack of preparation by the project executor
Violation in Work Health and Safety	Prevention	Increase Work Health and Safety training, supervision implementing SOP	Negligence in supervision
Stealing, Negligence and Dishonesty	Prevention	Give sanction to raise supervision and security	Negligence in supervision
Damage in measuring tools, welding & grinding	Reduction	Increace maintenance periodically	Error in utilization

3.7. Managing Variation:

To estimate at before project execution in terms of cost, time, and resource that is used previously. From the cost details Rp. Rp 200,530,548.03 has not included by human resource cost, and that is still below project value Rp. 385,486,000.00. Therefore, the project cost can be accepted, worker cost is counted based on 5% yearly omzet and do not impact in execution of only one project.

Table 6. Project Cost

No	Job Description	Cost (IDR)		Total (IDR)	
		PPIC Works			
1	Material Descriptions	Rp	130,047,960.65	Rp	140,958,288.03
	Consumable	Rp	10,910,327.38		
		Production Works			
2	Shearing	Rp	4,183,200.00	Rp	48,772,260.00
	Bending	Rp	1,738,800.00		
	Bending Roll	Rp	5,796,000.00		
	Fitter WELDING	Rp	6,350,400.00		
	Welding TIG (316)	Rp	8,424,000.00		
	Polishing	Rp	10,395,540.00		
	Fabrication Manpower	Rp	2,747,520.00		
	Lathe	Rp	5,832,000.00		
	Milling	Rp	3,304,800.00		
		Delivery			
3	Packing and Delivery	Rp	4,050,000.00	Rp	4,050,000.00
		Sub Contractor Works			
4	Dieshing, Forming, Flanging Head	Rp	3,150,000.00	Rp	4,050,000.00
	Transport to Vendor	Rp	900,000.00		
		QC Works			
5	Inspection and Documentation	Rp	900,000.00	Rp	1,800,000.00
	Test and Commisioning	Rp	900,000.00		
		Engineering Works			
6	Drawing Engineering	Rp	900,000.00	Rp	900,000.00
	Total	**Rp**	**200,530,548.03**	**Rp**	**200,530,548.03**

3.8. Buffer Calculation

Critical Chain Project Management (CCPM) prioritize more in the whole project success, and focused in critical chain project completion. Below is the project plan and actual project in making Storage Tank Cap. 10000 L. Based on previous project buffer, it is obtained an optimum time of project execution by giving suggestion in using buffer time as a backup time in plan scheduling. To determine buffer:

$$B = 2\sqrt{\left(\frac{S_1 - A_1}{2}\right)^2 + \left(\frac{S_2 - A_2}{2}\right)^2 + \cdots + \left(\frac{S_n - A_n}{2}\right)^2} \quad (2)$$

Description: B : *Buffer time*; S : *Most Likely* (average standard time that is assumed as time that still save backup time); A: Optimistic (fastest time that is assumed without backup time)

Table 7 Suggestion Result of Project Execution

Work	Plan (days)	Actual (days)	Buffer (days)	Proposed (days) (Plan + Buffer)	Description
Engineering	7	71	64	71	Maximum 71 days
PPIC (Material)	15	15	-	15	The same as the actual
Production	21	62	17	38	Maximum 38 days
Sub Contractor	3	8	5	8	Maximum 8 days
QC	1	1	-	1	The same as the actual
Delivery	3	1	-	1	The same as the actual

PPIC (material) work remains 15 working days, however the production work, project execution within 38 days and sub contractor work, it is suggested 8 days. In QC and delivery work it is suggested 1 day where there is no change from the actual project execution.

3.9. Buffer Cost calculation

Executors can save cost in resource which is in production work, because production worker is paid based on working hours per day. In engineering work and sub contractor, saving can not be counted because the human resource cost calculation is calculated based on company's project revenue per year and the number of unit done, even if the execution experienced acceleration.

Table 8 Recapitulation of Project Total Cost

Actual Project Costs	Rp	200,530,548.03	**Savings**
Project Plan Costs	Rp	168,642,288.03	
Project Buffer Costs	Rp	20,303,100.00	
Proposed Project Costs (Plan+ Buffer)	Rp	188,945,388.03	
Cost Savings Using Buffers (Actual Costs - (Plan + Buffer))	Rp	11,585,160.00	6 %

3.10. Value Stream Mapping

All value stream that existed in the company by making an end of Big Picture Mapping, that describe optimum lead time as an improvement to the company. In Value Stream Mapping in the early production work (Table 7) and QC actually executed in 63 days, and in the end of Value Stream Mapping, project execution suggestion by using buffer time in the production work and QC is 39 days (Figure 3).

Based on CCPM scheduling estimation, it is obtained an optimal timeline condition in the making of *Storage Tank Cap* 10000 L that is illustrated in the form of Value Stream Mapping ending. Project execution suggestion with buffer time in the production work and QC is 39 days, and for production and QC work is actually done in 63 days

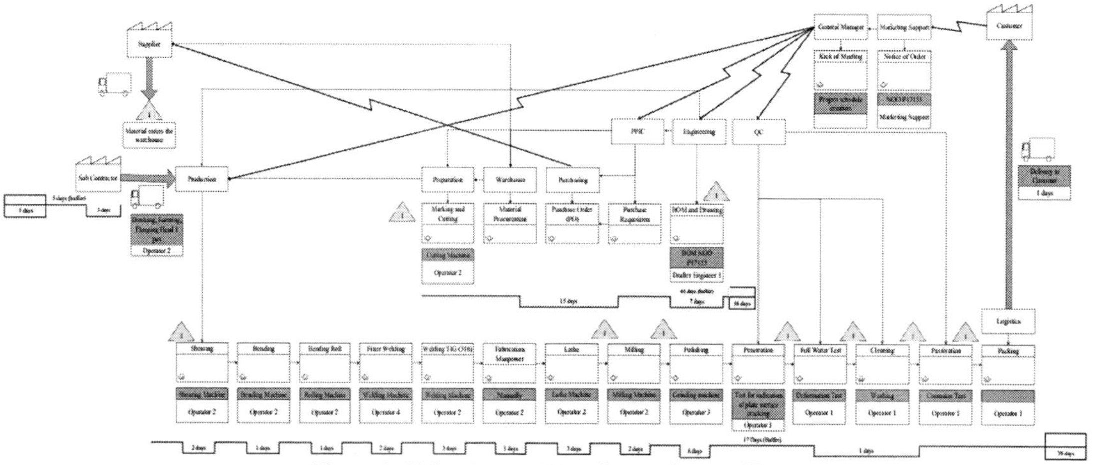

Figure 2. Value Stream Mapping Project Buffer

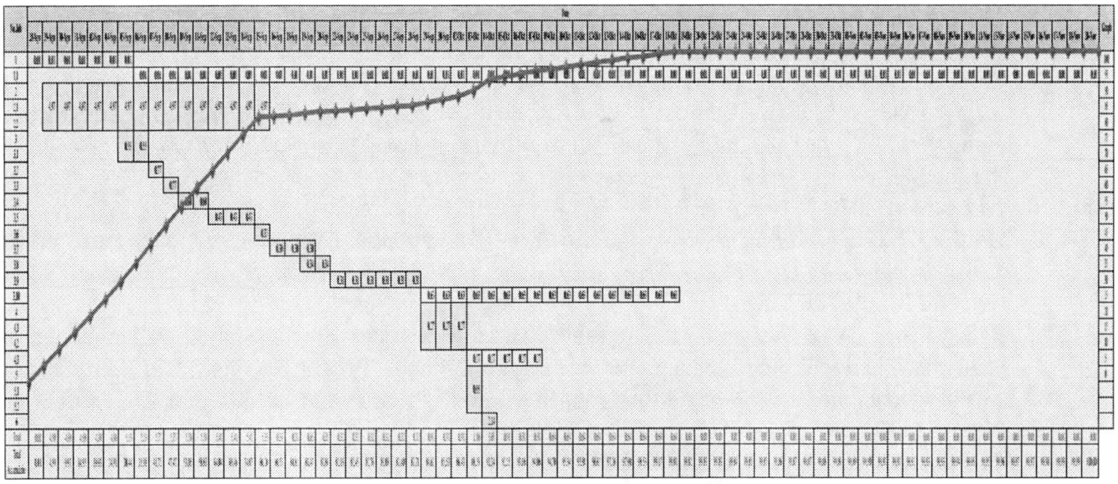

Figure 3. S-Curve Project Buffer

4. Conclusion

Strorage Tank Cap 10000L evaluation by the Lean Project Management approach shows the cause of the biggest project implementation delay is waiting and unsatisfied design. Some indicators of risks that occur in the project are natural disaster and natural hazard, if both risk can be handled, work finalization can be accelerated. Application of buffer time save 0.06% cost and the work schedule is reduced by 39%. The CCPM approach to lean management programs helps companies reduce risk and reduce financial inputs. The S curve shows the comparison of the actual conditions and the use of buffer time, especially engineering tasks.

5. References

[1] C. F. Gray and E. W. Larson, 2006 *Manajemen Proyek Proses Manajerial*, 3rd ed.

[2] PMBOK Guide, 2000 *Project Managemnt Body of Knowledge*, 2000th ed. Newtown Square, Pennsylvania, USA: Project Management Institute Four Campus Boulevard.

[3] V. Gaspersz, 1998 *Production Planning and Inventory Control Berdasarkan Pendekatan Sistem Terintegrasi MRP II dan JIT Menuju Manufakturing 21*. Jakarta: PT Gramedia Pustaka Umum.

[4] B. Yuliandra, 2015 *Manajemen Proyek : Sebuah Perspektif Teknik Industri*. Andalas University.

[5] D. B. Baskara and B. A. Noer, 2012 "Perencanaan dan Pengendalian Proyek Periklanan menggunakan Lean Critical Chain Project Management dan S-Curve Monitoring," *Tek. Pomits*, vol. 1, pp. 1–6.

[6] S. H. Stavania Untu, A. K. T. Dundu, and R. J. . Mandagi, 2014 "Penerapan Metode Lean Project Management dalam Perencanaan Proyek Konstruksi (Studi Kasus : Pembangunana Gedung Mantos Tahap III)," *Sipil Statik*, vol. 2, pp. 1–10.

[7] D. Artika, 2014 "Penerapan Metode Lean Project Management dalam Proyek Konstruksi pada Pembangunan Gedung DPRD Kabupaten Ogan Ilir," *Tek. Sipil dan Lingkungan.*, Vol. 2, pp. 1–9.

[8] R. I. Hapsari, 2011 "Penerapan Metode Lean Project Management dalam Perencanaan Proyek Konstruksi Pada Pembangunan Gedung SDN Bektiharjo II Semanding Tuban," pp. 1–15.

[9] M. Aulady and C. Orleans, 2016 "Perbandingan Durasi Waktu Proyek Konstruksi Antara Metode Critical Path Method (CPM) dengan Metode Critical Chain Project Management (Studi Kasus : Proyek Pembangunan Apartemen Menara Rungkut)," *IPTEK*, vol. 20, pp. 1–13.

[10] Y. C. Fernando and S. Noya, 2014 "Optimasi Lini Produksi dengan Value Stream Mapping dan Value Stream Analysis Tools," pp. 1–9.

[11] D. Kasidi, 2008 "Penerapan Metode Critical Chain Project Management Pada Penjadwalan

Proyek konstruksi (Studi Kasus : Proyek Sudirman Tower)," *J. Tek. Ind.*

[12] U. Dhanu Saputra, 2008 "Komparasi Metode Critical Chain Dengan Metode Critical Path Pada Manajemen Proyek Pembangunan Menara Telekomunikasi," *J. Tek. Ind.*.

[13] D. Supriadi, 2017 "Evaluasi Penjadwalan Pembangunan Hanggar PT. Gudang Garam, Tbk Menggunakan Metode Critical Chain Project Management (Studi Kasus Pembangunan Hanggar PT. Gudang Garam, Tbk di Waru, Sidoarjo, Jawa Timur)," *J. Info Manaj. Proy.*, vol. 8.2, p. 30.

[14] T. G. Lechler, 2010 "Critical Chain: A New Project Management Paradigm or Old Wine in New Bottles?," *Eng. Manag. J.*, vol. 7, No. 4, p. 58, 2005.R. Wibowo, W. Siswanto, and P. Samosir, *Standar Konstruksi Jaringan Tegangan Menengah Tenaga Listrik*, 5th ed. Jakarta Selatan: PT. PLN (PERSERO).

[15] C H Suyansen, W S Martanto, P Nugraha, 2014" Penentuan Buffer Sizing Dalam Penerapan Critical Chain Project Management Pada Sebuah Proyek Konstruksi., http://www. usti.blogspot.com/2012/03/sistem-tenaga-litarik-pusat pembangkit.html. diakses 14 february 2014

[16] Ryan Ramanda and N Ary Arvianto. 2014 Penerapan Critical Chain Project Management Untuk Mengatasi Masalah Multi Proyek Dengan Keterbatasan Resources di PT Berkat Manunggal Jaya., http://www. usti.blogspot.com/2012/03/sistem-tenaga-litarik-pusat pembangkit.html. diakses 14 february 2014

[17] Luh Nyoman Widyatuti 2014 Analisis Gangguan Sistem Transmisi Listrik Menggnakan Metode Root Cause Anlyasis (RCA). http://www. usti.blogspot.com/2012/03/sistem-tenaga-litarik-pusat pembangkit.html. diakses 14 february 2014

[18] G. S. Intifada and Witantyo, 2012 "Minimasi Waste (Pemborosan) Menggunakan Value Stream Analysis Tools untuk Meningkatkan Efisiensi Waktu Produksi (Studi Kasus PT. Barata Indonesia, Gresik)," *Jurnal Teknik. Pomits*, vol. 1, pp. 1–6.

[19] Daonil 2012 "Implementasi Lean Manufacturing Untuk Eliminasi Waste Pada Lini Produksi Machining Cast Wheel Dengan Menggunakan Metode WAM dan VALSAT," Jurusan Fakultas Teknik. Universitas Indonesia, Jakarta.

The Effect of The Processing Time's Variance to The Performance of Sequencing Rule

Arum Sari [1] and Y C Yulia[2]

Pasundan University, Bandung, Indonesia
[1]aarum310355@yahoo.com, [2]yolandayuliacitra@gmail.com

Abstract. SPT was the most studied rule in sequencing. Most of the research concluded that SPT was the best rule but other research concluded that SPT did not achieve the best performance. Studies shown that performance varies due to a great different environments because there were so many factors affected the performance. One of the factors that lead to differences in the conclusions was varying of processing time. The objective of the paper was to evaluate the effect of processing time's variance to the performance of sequencing rules. Only processing time variance was being investigated while number of machine was treated as nuisance factor. To achieve this objective, an experiment was designed and the experiment result was analyzed using anova. Performance measure used in the experiment was lateness. Only five sequencing rules were used. The rules were First In First Out (FIFO), EDD, SPT and Slack Time. The experiment found that for a small number of job, the effect of processing time's variance was not significant to the performance of sequencing rule. For large number of jobs, processing time's variance had a significant effect on the performance of sequencing rule. For further research, more sequencing rule to be consider while some assumptions is relaxed.
Keywords: Sequencing rule, FCFS,SPT,LPT,EDD,Slack Time, DOE

1. Introduction

The production scheduling problem is one of the main areas in operation management. Generally, the problem can be described to find best way to assign the available resources to the tasks, within the existing constraints that would maximize the desired performance of the system (Nadia Bhuiyan et. al.,2011) Thus, the problem is how to decide the priority of the tasks and create the sequence of the activities in the shop floor. The decision include the selection of the next part to be processed by the machines or the next part to be released into the system.

In many situations, sequencing rule have been used to address scheduling problems due to their simplicity and ease of use. Sequencing rules are rules that prioritize the jobs waiting for processing. Panwalker (1977) states that the most studied sequencing rule is SPT compare to many other sequencing rule. Baker (1965) conclude that SPT is the best rule when the objective is to minimize the mean flow time (Baker&Dzielinski) but Conway (1965) when comparing with 31 rules conclude that SPT dominate the rules despite the fact that it did not achive the best performance in any experiment. Montazeri and Van Wassenhove (1990) have also studied the effectiveness of the scheduling rules for various system performance measures using a discrete event simulator. They concluded that the SPT priority rule was the second best priority rule for the system under study in terms of average waiting time. Many studies have shown that performance varies due to a great different environments and thus no dispatching or sequencing rule has been found to be optimal for all planning and scheduling problems. The difference in the conclusion occurs because there are many factors that affect the performance of the sequencing rule. One of the factors that lead to differences in the conclusions is varying of processing time which is measured by variance of the processing time. For this reason a research is extended to evaluate the effect of variance's processing time into the sequencing rules performance. The object to be studied is a dummy system representing a production system to process N job on a single machine with different processing time.

2. Methodology

The effect of variance's processing time into the sequencing rules performance is studied through an experiment. Design of Experiments (D.O.E.) is used to investigate it's effect. Design of Experiments is a structured approach in experimentation used to identify the significance of selected variables (factors) on the performance of the system under study. In this approach the required set of experimental runs is created based on the combinations of factors and the levels of these factors. The experiments are performed by measuring the performance measures of the system under study, while changing the values of other factors in the system. The result is to evaluate the impact of these changes.

To evaluate the effect of factor on the performance of the system, it is necessary to define the factors to be studied. The factor to be studied is then classified into controllable factor and uncontrollable factor. The controllable factors is the main design factor. In this research the main design factor is the variance of the processing time. The uncontrollable factor considered in this study are the number of machine and the priority rule. The other factor is not included as controllable factor because the effect of them is already clear. While variance's processing factor is treated as main factor, the number of job and priority rule used is treated as a nuisance factor because these factors cannot be eliminated. Other factors that can affect the performance is eliminated by setting them at the same condition. Number of job cannot be eliminated because the job is the subject under study. So do the priority rule used in the experiment. The design of experiment for one main factor with 2 nuisance factor is Latin Square Design. Variance of processing time is set at 5 level. The variance is measured by the range of processing time. There are 5 level of variance of the processing time expressed in range (R). These levels are (R) = 1;5;10;20; and 40. R=1 represent a homogenous system. R=5 and 10 represent a low variance and R= 20 and R=40 represent a high variance. As a consequences of a Latin Square Design, the level factor of number of job (N) is also set at 5 namely N= 4; 8; 12; 20 and 28 job. N=4 and N=8 represent a small number of job, N=12 represent a medium number, while N= 20 and N=28 represent a large number of job. The priority rule is selected among many sequencing rule. The rules used in this study are SPT, FIFO EDD, LPT, and SLACK TIME. These rules are selected because it is well known algorithm and it has been shown effective used as a reference. All the factor and its level factor are presented in Table 1. The total number of treatment to be carried for the defined factor is presented in Table 1 are 25.

Table 1. Factor and Level Factor of Experiment

Level	Main Design Factor	Nuisance Factor	
	Processing time	Number of job	Priority rule
1	Range = 1	4	SPT
2	Range = 5	8	FIFO
3	Range = 10	12	EDD
4	Range = 20	20	LPT
5	Range = 40	28	SLACK

To perform high reliability, each experiment is replicated 10 times. Thus, all the factor-level combinations of the experiment run is replicated 10 times. Hence there will be 250 (25 conditions * 10 experiment replications) experiment treatments is created. In this paper a test on adequacy of data is carried out to prove that the data required is eligible

3. Planning the Experiment

The experiment require processing time, due date, assumption and response variable. Each of them are explained in the follow

3.1 The Processing Time

The main data required to run the experiment is the processing time. The data is a dummy data of N job which is processed in a single production process. The processing time is generated to simulate a single production process of N job with a certain processing time and different variances for a certain due date. The processing time of each job follows a uniform distribution having mean 30. The variance is expressed in term of range. The range is set at 5 different level of range and the number of job is set at 5 level as presented in table 1. So there will be 25 set data of processing time required to run the experiment for each replication. All the dummy data is generated using Microsoft excel software.

3.2 The Due Date Setting

EDD is one of the rule to be studied. To run the EDD rule, it is essential to have a due date for each job. The due date is set at the time the processing time is generated. In this research, the due date is set so that there will be 25% of the job will be delayed. Table 2 perform the arrangement of the due date setting for each set number of job

Table 2. Due Date Setting For Each Set Number of Job

Number of job	Due date Range
4	90-120
8	180-240
12	270-360
20	450-600
28	630-840

3.3 The Assumption

The assumption used in this study is the normative assumption as follow:

a. The machine can only perform one job at a time and 100% reliable
b. Preemption and alternate routing are not allowed
c. Due date are fixed
d. Set up included in processing time

3.4 The Response Variable Of The Experiment

The objective of the research is to study the effect of processing time's variance to the sequencing rule performance. ANOVA is used to analyze the experiment result. The performance measure used to analyse the experiment result is lateness.

4. The Experiment Run

The design of the experiment used is Latin Square Design. The effect of processing time's variance to the performance of sequencing rule under a set of the number of job as a nuisance factor .To randomized the experiment the treatment is set as expressed in Table 3 so that the setup of the experiment design run for 5 level factor mention above, are presented in Table 4

Table 3. The Randomization Of the Experiment Run

Sequencing rule	Number of job				
	4 job	8 job	12 job	20 job	28 job
SPT	R = 1	R = 5	R = 10	R = 20	R = 40
LPT	R = 5	R = 1	R = 40	R = 10	R = 20
FCFS	R = 10	R = 20	R = 1	R = 40	R = 5
EDD	R = 20	R = 40	R = 5	R = 1	R = 10
SLACK	R = 40	R = 10	R = 20	R = 5	R = 1

Table 4. The Treatment of The Experiment Run

Run no	Variance of Processing Time	Number of job	Sequencing rule
1	R=1	4	SPT
2	R=5	8	SPT
3	R=10	12	SPT
4	R=20	20	SPT
5	R=40	28	SPT
6	R=1	4	LPT
7	R=5	8	LPT
8	R=10	12	LPT
9	R=20	20	LPT
10	R=40	28	LPT
11	R=1	4	FCFS
12	R=5	8	FCFS
13	R=10	12	FCFS
14	R=20	20	FCFS
15	R=40	28	FCFS
16	R=1	4	EDD
17	R=5	8	EDD
18	R=10	12	EDD
19	R=20	20	EDD
20	R=40	28	EDD
21	R=1	4	Slack
22	R=5	8	Slack
23	R=10	12	Slack
24	R=20	20	Slack
25	R=40	28	Slack

5. Result

After the data have been generated, there will be 5 set of the processing time and their due date for each level of the number of job. The experiment is ready to run as designed in Table 4
The 1st experiment is running for data set of processing time of 4 jobs having Range =1 using SPT rule. The sequence is then determined and the lateness as the experiment response is computed. Rerun the experiment for the same data set for the rest 9 replication. Continue for the second experiment for the second data set of processing time for 8 jobs having Range =5. Use SPT to find the sequence and its lateness. Re run for all the replication. Continue to run all the design treatment presented in Table 4. The response variable is then averaged for analysis its variances. The experience result is then analysed using ANOVA. The hypothesis is set as follows:

$H_0 : \tau_i = 0$ for all j=1,2,3,4, and 5

$H_a : \tau_i \neq 0$ at least for one j

Table 5 shows the result of ANOVA on average lateness for all number of job (N) used in the experiment including the F-test. Overall experiment result is presented in appendices. The Anova is presented in Table 5.

Table 5. ANOVA For Average Lateness

Source	Dof	Sum of square	Mean square	Ftest	Ftable
Treatment	4	474122	118530.6	34.65	3.6
Sequencing rule	4	780112.5	195028.1	57.01	
Number of job	4	922358.4	230589.6	67.41	
Replication	9	10100.3	1122.3		
Error	228	779975.1	3420.9		
Total	249	2966668.8			

The result shows that the effect of processing time's variance is significant at 99% level of significant. The *F* value at 3.6 indicate that there was a significant effect of variance of the processing time to the use of sequencing rule.

Some fact findings are obtained from this study:

1. Performance of sequencing rule for small number of jobs was not significantly different For a small number of jobs, all sequencing rule almost performed the same result
2. The greater the number of job, the greater its variance. The study showed that, the use of sequencing rule for greater number of jobs perform a different result significantly
3. For a small number of job, the performance of EDD and Slack in term of lateness was not significantly different compare to another rule.

6. Conclusion

The conclusion of the research are explained below.

1. Variance of the processing time had a significant effect to the performance of sequencing rule selected especially for large number of job.
2. The greater the number of job the more significant the effect to the performance of sequencing rule selected
3. For a small number of jobs, all sequencing rule almost performed the same result

Appendices: Experiment Result

	Number of Jobs(N)									
Sequencing rule	N=4		N=8		N=12		N=20		N=24	
SPT	Range=1		Range=5		Range=10		Range=20		Range=40	
	12.0	16.2	8.3	41.5	19.1	74.3	139.8	50.0	369.1	353.6
	12.0	11.1	56.3	31.4	69.0	54.1	64.5	107.2	203.1	255.0
	11.0	27.1	32.8	62.1	135.9	135.0	136.5	164.3	222.0	153.7
	16.0	27.6	1.5	31.0	61.1	123.2	172.4	53.0	333.4	354.8
	12.3	27.0	0.3	53.5	103.3	75.3	130.5	33.4	310.5	308.7
LPT	Range=5		Range=1		Range=40		Range=10		Range=20	
	8.0	17.5	1.3	1.4	87.6	81.7	159.3	127.1	373.5	507.4
	27.7	24.8	50.9	17.5	102.5	61.1	47.0	183.0	328.5	236.2
	28.8	12.4	46.4	40.3	104.0	93.7	187.4	173.1	505.1	426.8

	5.3 7.6	46.1 33.4	124.6 43.3	221.6 134.3	524.2 314.7
	29.0 7.2	32.4 15.5	76.7 122,5	154.7 126.7	331.9 250.0
FCFS	Range=10	Range=20	Range=1	Range=40	Range=5
	3.0 4.0	38.5 30.2	35.9 35.5	316.2 21.0	159.0 168.7
	2.1 6.0	43.6 41.3	36.0 36.5	293.8 318.7	162.2 161.0
	2.0 4.7	42.6 34.2	36.7 36.9	292.5 263.8	170.0 159.0
	2.8 5.0	34.0 37.1	37.9 38.7	254.1 271.0	189.4 153.0
	5.6 2.6	38.2 38.6	37.1 38.5	278.8 364.0	156.3 255.6
EDD	Range=20	Range=40	Range=5	Range=1	Range=10
	7.0 3.2	3,7 8.5	0.0 0.0	0.0 0.0	5.0 3.9
	4.7 4.6	3.4 12.7	0.0 3.3	0.0 0.0	16.3 16.4
	4.8 4.5	8.4 6.0	4.7 0.0	0.0 0.0	13.1 17.3
	5.4 6.5	5.6 9.8	0.0 0.0	0.0 0.0	3.2 1.0
	6.0 4.5	2.2 5.7	0.0 0.0	0.0 0.0	17.0 17.1
SLACK	Range=40	Range=10	Range=20	Range=5	Range=1
	3.8 10.8	0.0 0.2	3.5 9.2	8.5 4.5	0.0 0.0
	3.4 1.2	0.0 0.8	0.5 8.2	4.8 5.1	0.0 0.0
	4.0 1.2	4.8 1.3	5.5 11.7	5.1 9.7	0.0 0.0
	13.0 4.2	6.0 4.2	4.0 1.0	4.5 4.4	0.0 0.0
	2.0 1.2	0.0 2.3	0.8 5.2	6.4 8.4	0.0 0.0

7. References

[1] Baker K R, 1974, *Introducing to sequencing and scheduling*, John Willey & Sons, Inc, Canada

[2] Baker, K.R, 2009, Tiertse, Dan, *Principle of sequencing and scheduling*, John Willey & Sons, Inc, Hooken, New York

[3] Bhuiyan, N, Gouw, G, and Yazdi, D, (2011). *Scheduling of a computer integrated manufacturing system: A simulation study. Journal of Industrial Engineering and Management*, 4(4), pp 576-609. http://dx.doi.org/10.3926/jiem.331

[4] Conway, R W (1965a), *Priority Dispatching and Work-In-Process Inventory In a Job Shop. Journal of Industrial Engineering*, 16(2), pp 123-130.

[5] Conway, R W (1965b). *Priority Dispatching and Job Lateness in a Job Shop. Journal of Industrial Engineering*, 16(4), pp 228-236.

[6] Kelton, W D & Law, A M (1984). *An Analytical Evaluation of Alternative Strategies in Steady-state Simulation. Operational Research*, 32(1), p 69 http://dx.doi.org/10.1287/opre.32.1.169

[7] Montgomery, D C (2001). *Design and Analysis of Experiments* (5th Ed.). John Willey & Sons Inc.

[8] Montazeri, M, & Van Wassenhove, L.N. (1990). *Analysis of Scheduling Rules For An FMS. International Journal of Production Research*, 28(4), pp 785-802 http://dx.doi.org/10.1080/00207549008942754

Maintenance management improvement based on reliability centered maintenance II in energy generating industries

M L Singgih[1] , Y Prasetyawan[1] ,Sutikno[2], D Hartanto[1], F R Kurniawan[1] ,W T Wicaksana[1]

[1] Industrial Engineering Department, Institut Teknologi Sepuluh Nopember
[2] Mechanical Engineering Department, Institut Teknologi Sepuluh Nopember, Surabaya, Indonesia
moseslsinggih@gmail.com, yudhaprase@gmail.com, sutikno.its@gmail.com, dody_hartanto@ie.its.ac.id, kurniawanrindy@gmail.com, winahyutyas@gmail.com

Abstract: The *X* power plant is one of an electric company subsidiaries. Engaged in providing electricity for the certain regions. In general, the type of treatment from *X* power plant is short and long term maintenance. The application of these two systems helps *X* power plant to maintain the machines. The author argues that the type of maintenance from *X* power plant is not or is not yet optimal, the author wants to test whether reliability centered maintenance II treatment method can be applied in *X* power plant so that the efficiency and maintenance of the machine is in accordance with their individual needs. After implementing the RCM II information worksheet and RCM II decision worksheet, it is found that the existing maintenance management system in *X* power plant can be. Several incompatibility are detected, and should be matched with the corresponding maintenance method. For further research, the author recommend that the object of research are broadened so that the RCM II method can be used on other energy generating industries.
Keywords: Reliability Centered Maintenance II, RCM II information worksheet, RCM II decision worksheet, Maintenance Management

1. Introduction
In this part, the research background, problem formulation, and research objective will be shown.

1.1 Research Background
The availability of electricity supply is one of the important needs in society. Electricity development is an alternative that can be done to guarantee the availability of electricity. This will ensure the availability of electric energy supply so the eletricity will be in sufficient quantities, of good quality and prices that can be reached by the wider community. A sufficient amount means that existing supply can accommodate the needs of the wider community. While good quality means the community will be guaranteed to receive the supply as it should and prices that do not burden the public. The main objective of this eletric energy development focuses on improving the welfare and prosperity of the community fairly and equitably and realizing sustainable development.

Existing data shows that the existing electrification ratio increases continuously every year. For example, in 2013 with a target of 76.8%, it experienced an increase of 3.71% so that the realization in

2013 was 80.51%. Likewise in 2014 with a target of 80%, an increase of 3.45% also happen so that the realization was 84.35%. The data shows that the electrification ratio has increased positively.
Compared to the ASEAN countries, the electrification ratio of Indonesia with 84,35% in 2015 is far behind. All the ASEAN countries has been able to get electrification ratio above 98%. Based on this condition, the Indonesian government through *national employment general plan* (RKUN) target to reach electrification ratio of 97.35% in 2019. To support this target, a 35.000 MW electrical program is designed.

The 35.000 MW electricity program is one of the RUKN programs. The program then carry out the construction of new plants from Sumatra to Papua. In addition to building new plants, there are other alterations, namely maintaining and carrying out maintenance on existing plants. The purpose of doing maintenance is to maintain national electricity needs. Existing programs certainly have very expensive costs. The program that is carried out must still pay attention to the fact that there are maintenance costs attached to it. The maintenance costs arise from maintenance for old and new electrical energy providers. If the problem is not taken care of, there will be overrun costs. To prevent the occurrence of large costs, appropriate treatment methods are needed. Proper maintenance according to the type of treatment will reduce costs so that efficiency is created which is certainly very profitable for the industry.

1.2 Problem Formulation
Based on the finding above, the problem that will be solved in this research is how to improve and maintain the maintenance management activity in the energy generating industries using reliability centered maintenance II method.

1.3 Research Objective
Based on the problem formulation, the objective for the research are (1) to identify functions, malfunctions, causes of malfunctions, and the impact of failures of components, facilities and tools found in energy-producing industries and make RCM II information worksheet, (2) to determine the most appropriate maintenance activities for each energy-producing industry using RCM II Decision Worksheet, and (3) Determine the appropriate maintenance activity interval for the components, facilities and tools found in the energy-producing industry

2. Methods
In this part, first the research procedure will be discused. Then the method of collecting data will be presented. After that an example of te result from data collection will be presented. This includes FMEA, RCM II information worksheet, RCM II decision worksheet.

2.1 Research Procedure
The procedure of conducting the research is pictured in the figure 1 below:

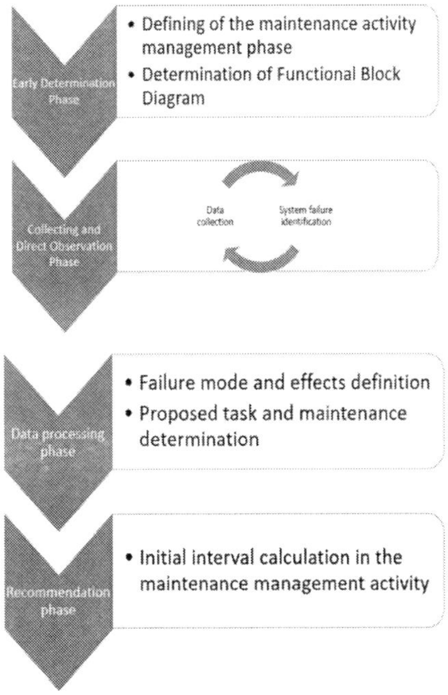

Figure 1 Research Procedure

The research begins with the initial determination phase. The initial determination phase includes defining maintenance management activities. Defining is done specifically for maintenance management activities in the energy industry. After the maintenance management activities have been defined, then the construction of functional block diagrams is carried out. In this diagram an explanation of the relationship between the functions of a part of a system, and other parts of a system will be given. After that, it will proceed with the stage of direct collection and observation. This stage is a cycle of data collection and identifying system failures. The cyclic nature is because the activity is carried out on every object observed and each activity adds data collected. System failure identification will focus on identifying the functions of existing systems in the energy industry. The function that has been defined is then used and analyzed to determine the failure that will occur.

The next stage is the data processing stage. The data processing stage consists of defining failure modules and their effects as well as determining proposed tasks and maintenance activities. Defining failure mode will be based on the function defined in the previous stage. After that, the proposed task and technical maintenance will be carried out based on RCM II.

The final stage of this study is the preparation of an improvement in the efficiency of care management by calculating the initial interval. The calculation is carried out based on RCM II and has recommendations for carrying out appropriate maintenance activities both in terms of method, time and cost for each facility.

3. Result and Discussion

In this part, the result of the data collection will be presented. The, RCM II information worksheet, and RCM II decision worksheet will be shown and discussed. At the end of the part, ther will be a thorough discussion about the finding.

3.1 RCM II Information Worksheet

Reliability centered maintenance II information worksheet is the development result from the FMEA of the observation object. FMEA aims to identify the flow of a failure may happened in product or process, and then estimate the risk that may happen. The result from FMEA can be categorized into function, function failure, failure mode, and failure effect.

In the function part, the function of the machine in the observation object will be explained. In the function failure part, the possibility of failure while the machine is operating is explained. In the failure mode part, after the possibility of failure is listed, the possible cause of the function failure will be listed. The failure mode can be obtained from the failure record in the observation object and through interview with the supervisor and the field operator. The last part, the failure effect contain the possible impact of the failure mode. After the FMEA identification is finished, the result will be described into the RCM II information worksheet.

The research conducted in *X* power plant involved 31 different machines. In this part an example of the RCM II information worksheet will be presented in table 1

3.2 RCM II Decision Worksheet

RCM II decision worksheet is the next phase after the FMEA or the RCM II information worksheet is done. Decision worksheet is a summary of every question listed in the RCM II decision diagram. The RCM II decision diagram is presented in figure 2. There are some condition that become the main attention of the decision worksheet, those are the hidden failure, safety, environment, and operational. The result of the decision worksheet will be the a list of proposed task that contains scheduled on condition, scheduled restoration task, schedule discard task and finding failure.

The RCM II decision worksheet will identify all 31 machines observed in *X* power plant. An example of RCM II decision worksheet will be presented in table 2. The result or the proposed task will determine the best time for each machine to be maintained by related department.

Table 1 An example of RCM II information worksheet in superheater machine

<table>
<tr><td colspan="2" rowspan="2">RCM II Information Worksheet</td><td colspan="2" rowspan="2">Sistem : UNIT PEMBANGKIT LISTRIK TENAGA UAP</td><td colspan="3">Unit 4 UP Gresik</td></tr>
<tr><td>1</td><td colspan="2">SUPERHEATER</td></tr>
<tr><td colspan="2">Function</td><td colspan="2">Function Failure</td><td colspan="2">Failure Mode</td><td>Failure Effect</td></tr>
<tr><td rowspan="4">1</td><td rowspan="4">Transform saturated steam to super heated steam that is exhausted from the high pressure steam drum</td><td rowspan="4">1</td><td rowspan="4">Fail to transform saturated steam to super heated steam in the primary superheated, secondary superheated, and final superheated part.</td><td>1</td><td>Cracked drain inlet primary superheated boiler pipe</td><td>Cracked pipe caused by corrosion and unable to withstand the heat when the combustion process is on. If the pipe is cracked can cause shutdown and trip.</td></tr>
<tr><td>2</td><td>Leaked secondary superheated inlet header drain valve</td><td>Super heated steam formation process cannot be done properly and may cause shutdown</td></tr>
<tr><td>3</td><td>Leaked final superheater inlet header valve</td><td>Super heated steam formation process cannot be done properly and may cause shutdown</td></tr>
<tr><td>4</td><td>Obsolete and degrading pipes</td><td>Corrotion and deposit happen because overload of several compound into the pipes, which are Na, Mg, Cl. The pipes become easily broken. Aside from that the mixing of the compound and the superheated steam can cause trip.</td></tr>
</table>

Table 2 An example of RCM II decision worksheet in superheater machine

RCM II Decision Worksheet			System : ELECTRIC STEAM POWER PLANT											
			Sub System : SUPERHEATER											
Information Reference			Consequences Evaluation				H1 / S1 / O1	H2 / S2 / O2	H3 / S3 / O3	Default Task			Proposed Task	Can be Done by
F	FF	FM	H	S	O	N	N1	N2	N3	H4	H5	S4		
1	1	1	N	N	N	Y	N	N	Y	Y			Do Scheduled Discard task by replacing inlet drain primary superheated pipe. Also by doing finding failure task by checking the material composition by using spectrometer as a predictive maintenance effort	Engineering mechanic
		2	Y	N	N	N	N	Y	N	Y			Do Scheduled Restoration Task by fixing the inler header drain valve in the secondary superheated by closing leaked plug. Do finding falure buy checking the condition of the valve.	Engineering mechanic
		3	Y	N	N	N	N	Y	N	Y			Do Scheduled Restoration Task by fixing the inler header drain valve in the final superheated by closing the leaked plug. Do finding failure activity by checking the valve condition.	Engineering mechanic
		4	Y	N	Y	N	N	Y	N	N			Do scheduled restoration task by cleaning using cathodic protection	Har Mekanik, Engineering

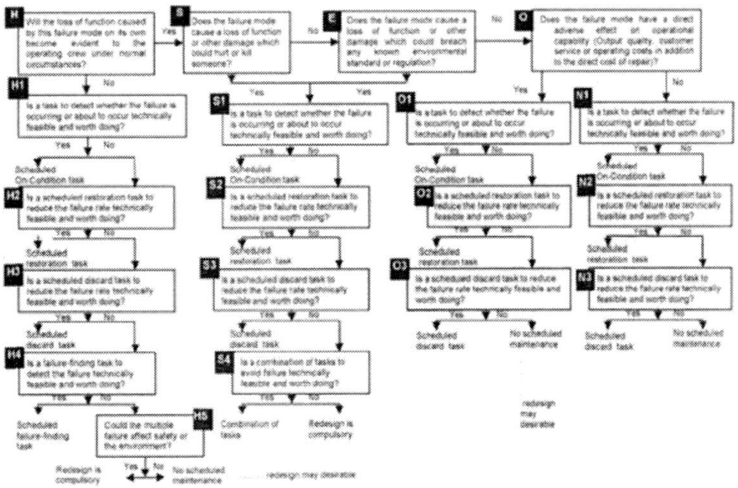

Figure 2 RCM II decision diagram

3.3 Discussion

The work flow of making the RCM II information worksheet and then the RCM II decision worksheet is done on whole 31 machine in the *X* power plant. The 31 machine are related in the machines that functioned in the rankine cycle in the power plant from the boiler to the condensor. From the RCM II information worksheet developed, the failure modes are identified. The failure modes in the heating related machines are mostly about the potency of leaking and unpurified compund mixed in the burner chamber. The RCM II decision worksheet for the machines mostly recommend a finding failure activity such as checking the leak potency in the pipes. It is highly critical to keep the heating related machines functioning due to shutdown possibility if the machines are broken down.

Meanwhile, in the aerator and condensing related machines, the potensial failure modes are related to corrotion and leaking in the cooling pipes. The corrotion may happen becuase of the altering temperature happen in the condensing related machines. While the leaking may happen because of the corrotion weaken the pipes structure. One of the main propsed task is by scheduling a regular calibrating task in the display unit so that the displayed condition are accurate. This will lead to early prevention of leaking caused by extreme temperature change.

4. Conclusion

Based on the fiding in the result and the discussion above, the conclusion for this research are:

- The result from FMEA identification of the machines shows that there are a few improvement that can be made because in the existing condition is not yet affordable
- The result from the identification of information worksheet indicates that there are a few machines that is assigned maintenance activity not compatible with the machines. Through RCM II, a suitable maintenance activity has been proposed so that the maintenance activity can be improved

5. References

[1] Administration, N. A. A. S., 2008. *Reliability Centered Maintenance Guide.* s.l.:National Aeronautics And Space Administration

[2] Assauri, D. S., 1980. *Management Produksi.* (Jakarta: Fakultas Ekonomi Universitas Indonesia)

[3] B, Y. & A, A., 2015. Rational Reliability Centered Maintenance Optimation for Power Distribution Systems. *Electrical Power And Energy Systems,* 23 May.pp. 350-360.

[4] Corder, A., 1996. *Teknik Manajemen Pemeliharaan.* 3rd edition. Jakarta: Erlangga. Ebeling, C. E., 1996. *An Introduction to Reliability and Maintainability Engineering.* 1st edition. (United State of America: McGraw-Hill Companies, Inc.)

[5] Elsayed, E. A., 2012. *Reliability Engineering.* 2nd. (Hoboken: John Wiley and Sons Ltd.)

[6] Groover, M. P., 2001. *Automation, Production System, and Computer-Integrated Manufacturing.* 2nd. (Upper Saddle River, New Jersey 07458: Prentice-Hall, Inc.)

[7] Harvard, T., 2000. Determination of a Cost Optimal, Predetermined Maintenance Schedule. *Determination of a Cost Optimal, Predetermined Maintenance Schedule.*

[8] Jardine, A. K. and Tsang, A. H., 2013. *Maintenance, Replacement, and Reliability.* (New York: CRC Press)

[9] Moubray, J., 1992. *Reliability Centered Maintenance (RCM).* 1st. (Oxford: Butterwort Heinemann)

[10] Moubray, J., 1997. *Reliability Centered Maintenance (RCM) II.* 2nd. (New York: Industrial Press Inc.)

[11] Priyanta, D., 2000. *Keandalan dan Perawatan.* (Surabaya: Jurusan Teknik Sistem Perkapalan)

IOP Conf. Series: Materials Science and Engineering **528** (2019) 012054 doi:10.1088/1757-899X/528/1/012054

[12] Yssaad, B., Khiat, M. & Chaker, A., 2013. Reliability Centered Maintenance Optimazation for Power Distribution Systems. *Electrical Power and Energy Systems 55,* pp. 108-115.

The applications of Cobb-Douglas Production Function in remanufacturing industry

D Saraswati[1], D K Sari[2] and D Hapsari[3]

Industrial Engineering Department, Trisakti University, Jakarta, Indonesia
[1]docki_saraswati@trisakti.ac.id, [2]debbie.kemala@trisakti.ac.id.
[3]dinahapsari731@gmail.com

Abstract. The study was conducted in a heavy equipment remanufacturing industry. The problem faced by the industry is in order to make the right decisions about the number of products that must be produced and the labor and parts needed to carry out the remanufacturing process. All decisions regarding this matter consider the cost of the new and reprocessed parts and the workforce used, since it depends on the condition of the product returned. In this regard, it will be useful to provide recommendations for developing the production process framework for remanufacturing. The main purpose of this paper is to measure the appropriate production quantity with an existing remanufacturing alternative, during the 3 months observations. This study based on the Cobb-Douglas production function with labor and parts as the input. The remanufacturing process itself consists of three stages; disassembly, reconditioning and reassembly. There were two cases that occurred in reassembly process during 3 months of observation. The input for reassembly process uses new parts and reconditioned parts. In Case I all parts from disassembly process are forwarded to reconditioning process without any constraints. In Case II the parts are binding by the constraints of the return parts. As a result, the existing production quantity from reassembly process is below the maximum output, except for month 2. This indicates that production framework needs to pay attention to the proportion of input parameters. It means that the maximum quantity of the new products could get from a given combination of labor and parts at the reassembly process.
Keywords: remanufacturing, Cobb-Douglas Production Function, returns to scale.

1. INTRODUCTION

Inventory management issues have been widely discussed, especially in terms of deterministic models. The development of inventory management tends towards a global scale. Just-in-time philosophy, vendor managed inventory and consignment are some of the development concepts of the inventory model [1]. In recent years, inventory management is also developed for raw materials such as components or spare parts that come from a used return product and then to be reused to produce a new product after going through a process called remanufacturing [2]. In remanufacturing, inventory management is more complex due to the unknown condition of the used return products [3]. The inventory management framework for components obtained from used products is different from the conditions for new components purchased from suppliers [4]. This study was carried out in the heavy equipment remanufacturing industry. The purpose of this study is to design the production planning of remanufacturing process using Cobb-Douglas production function for a certain engine component.

IOP Conf. Series: Materials Science and Engineering **528** (2019) 012055 doi:10.1088/1757-899X/528/1/012055

Cobb-Douglas Production Function is widely used in economics and productivity for representing the relationship between input and output. The production function states the physical relationship between production input and output. The analysis of production function is often used in research, hence with the limited resources can provide the maximum results. In his thesis Hassani [5] has developed the use of Cobb-Douglas production function for a cost function model that shows the relationship between labor and equipment to the costs required if the construction project schedule is shortened from the desired time. This paper discusses the combination from a given labor and material/parts to achieve the maximum output of production in a heavy equipment remanufacturing industry.

2. METHODS

The Cobb-Douglas production function is a function or equation that involves two or more variables, dependent variable / Y (which is explained) and independent variable / X (which explains). This approach is used as a production function to show the effect of the input used on the desired output. In this case the data is processed into a natural logarithm (Ln) and converted to the regression equation as $Y = a + bX$ or $Y = Ln.Q$ and $X = Ln.I$. Therefore, the the regression equation can be written as; $Ln.Q = a + b(Ln.I)$; $Ln.Q = a + Ln.I^b$; $Ln.Q - Ln.\mathrm{I}^{\mathrm{b}} = a$ → $Q = e^a I^b$

The general form of the equation is as follows [2]:

$$Q = \delta.I^{\theta} \tag{1}$$

where, Q = output; I = the type of input used; δ = efficiency index of input use in generating output; θ = production elasticity of the input used. In remanufacturing the production system will pass through disassembly, cleaning, reprocessing, testing and reassembly processes. With Cobb-Douglas production function the decisions will focus on the influence of cost of labor and material. For this purpose, the equation (1) in production system can be written as follows;

$$Q = \delta L^a M^b \tag{2}$$

where, Q = total output from manufacturing production, L = labor input, M = material input, a = production elasticity of labor, b = production elasticity of material and δ as the efficiency index or total factor productivity. The value of a and b are constant determined by the state of technology in the remanufacturing processes. There are 3 alternatives; 1) if a+b = 1, the production function has constant returns to scale (RTS), 2) If a+b<1, the RTS are decreasing and 3) if a+b>1, the RTS are increasing. For example in case of L and M are increased by 20% respectively, 1) if Q increases exactly by 20%, it becomes constant RTS, 2) if Q increases less than 20%, RTS will decreasing, and 3) if Q increases more than 20%, RTS will increasing. Returns to scale (RTS) examines the changes in total production related to a proportional changes in all inputs.

2.1. Model Formulation

Production process flow in the remanufacturing industry begins with the arrival of used products from consumers. Inspection of used products is conducted by the collection unit. Remanufacturing begins with disassembly and reconditioning process of the return products. The result of the dissasembly process is to group the parts into 3 categories; 1) can not be used and must be replaced with a new one, 2) can still be used through the machining process, and 3) can be used without going through the machining process. In addition, judgment is carried out in the disassembly process to determine the level of damage. The low, medium and high damage level categories are labeled with A, B and C respectively. The next step is to do the washing process of all parts that can be reused. Parts that cannot be reused are disposed of. In order to meet the requirements at the production floor, the company should make an order to the supplier. After the machining process, the reconditioned parts will be re-inspected, and at the last process new and reconditioned parts will be reassembled into new products.

2.2. Notation and Terminology

The notation and terminology used refer to [2] ;

Decision variables:

L_m	: labor input at reassembly process
L_r	: labor input at disassembly and reconditioning process
M_m	: new parts input at the reassembly process
R	: return products into disassembly process
λ	: Lagrange multiplier for the budget constraint
μ	: Lagrange multiplier for the return products constraints
M_r	: used parts after reconditioning as input at the reassembly process
M	: total parts input at reassembly process ($M = M_m + M_r$)
Q	: output from reassembly process

Parameters

γ	: return ratio for the return products, $\gamma < 1$, γQ the maximum number of return products to be disassembled
P_l	: price of input labor
P_m	: price of input material
P_r	: price of used product
C	: budget limit
α, a, b	: positive values of Q
β, c, d	: positive values of M_r

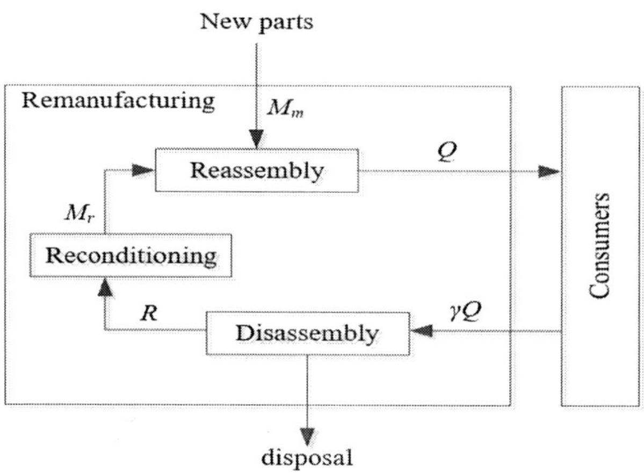

Figure 1 Remanufacturing System

The equations for the output from reassembly process and the ouput from the reconditioning process are represented by the Cobb-Douglas function. It is assumed that returns to scale (RTS) are decreasing, i.e. $a+b<1$ and $c+d<1$.

$$Q = \alpha L_m^a M^b \tag{3}$$

$$M_r = \beta L_r^c R^d \tag{4}$$

The total parts input at reassembly process are consisted of used parts after reconditioning and new purchased parts. It substitutes equation 4) into 3) for equation $Q = \alpha L_m^a (M_m + M_r)^{b.}$. It obtains as follows;

$$Q = \alpha L_m^a \left(M_m + \beta L_r^c R^d\right)^b \tag{5}$$

The total cost (TC) consists of labour cost plus material cost plus returned product cost and it cannot exceed the budget limit, C. Furthermore, the returned parts, R, is limited by the maximum number of used products to be diassembled, γQ.

$$TC = P_l(L_r + L_m) + P_m M_m + P_r.R \leq C \tag{6}$$

$$R \leq \gamma Q \tag{7}$$

In this case, the problem is to maximize Q, the output from reassembly process, subject to the budget limit C. Lagrangian function is used in order to maximize the problem, where λ and μ are the Lagrange multipliers for equation (6) and (7), respectively. The Lagrangian function is the sum of equation (5), (6) and (7) as follows [2];

$$L = \alpha L_m^a \left(M_m + \beta L_r^c R^d\right)^b + \lambda\left(C - P_l(L_r + L_m) - P_m M_m - P_r R\right) + \mu(\gamma Q - R) \tag{8}$$

2.3. *Reconditioning process*

The return parts from reconditioning and disassembly process should be maximized, subject to the budget limit for these processes, namely C_R. The Lagrangian function is the sum of equation (4), (6) for reconditioning process only and equation (7);

$$L_R = \beta L_r^c R^d + \lambda_R\left(C_R - P_l L_r - P_r R\right) + \mu_R(Q - \gamma R) \tag{9}$$

The limited budget for the reconditioning process will bind the labor at reconditioning and disassembly, reconditioning parts and Lagrange multipliers for λ_R, L_R, and $M_R > 0$. Since there are inequality constraints, the non-linear programming problem needs to apply Kuhn Tucker conditions to find the optimal value [6]. The first order Kuhn Tucker conditions are:

$\frac{\partial L_R}{\partial L_r}; \frac{\partial L_R}{\partial R}; \frac{\partial L_R}{\partial \lambda_R}$ and $\frac{\partial L_R}{\partial \mu_R}$.

The solution for the decision variables are as follows;

$$L_r^* = C_R c P_l^{-1}\left(c + d y_R\right)^{-1} \tag{10}$$

$$R^* = C_R d P_r^{-1} y_R \left(c + d y_R\right)^{-1} \tag{11}$$

$$\text{where,} \quad y_R = P_r\left(P_r + \frac{\mu_R}{\lambda_R}\right)^{-1} \tag{12}$$

The output of the decision variable is;

$$M_r^* = C_R^{c+d} \beta c^c d^d P_l^{-c} P_r^{-d} y_R^d \left(c + d y_R\right)^{-(c+d)} \tag{13}$$

2.4. *Reassembly process*

The consequences of a limited budget will also have an impact on the process of reassembly. As the final process of remanufacturing, for M_r, L_r, and $R > 0$. The first order Kuhn Tucker conditions are [2]; $\frac{\partial L}{\partial L_r}=0$; $\frac{\partial L}{\partial L_m}=0$; $\frac{\partial L}{\partial M_m}\leq 0$; $\frac{\partial L}{\partial R}=0$; $\frac{\partial L}{\partial \lambda}=0$ and $\frac{\partial L}{\partial \mu}\mu=0$

The inequality is in the condition of $\frac{\partial L}{\partial M_m}\leq 0$. In Case I, all parts from disassembly process are forwarded to reconditioning ($R^* < yQ^*$ with $\mu = 0$ and $y = 1$). The input for reassembly process consists of reconditioned parts ($M_r > 0$) and new parts ($M_m > 0$). The formulation for this condition becomes, as follows [2];

$$L_m^* = \frac{aP_l^{-1}C}{a+b}\left(1+hy^{d/(1-c-d)}(1-c-dy)\right) \tag{14}$$

$$M_m^* = \frac{bP_m^{-1}C}{a+b}\left(1-hy^{d/(1-c-d)}\left(\frac{a+bc}{b}+dy\right)\right) \tag{15}$$

$$L_r^* = cP_l^{-1}Chy^{d/(1-c-d} \tag{16}$$

$$R^* = dP_r^{-1}Chy^{1-c/(1-c-d)} \tag{17}$$

$$Q^* = \frac{\alpha a^a b^b P_l^{-1} P_m^{-b} C^{a+b}}{(a+b)^{a+b}}\left(1+hy^{d/(1-c-d)}(1-c-d)\right)^{a+b} \tag{18}$$

$$\lambda^* = \alpha a^a b^b P_l^{-a} P_m^{-b}\left(\frac{a+b}{C(1+h(1-c-d))}\right)^{1-a-b} \tag{19}$$

where,
$$h = \beta^{1/(1-c-d)} c^{c/(1-c-d)} d^{d/(1-c-d)} P_l^{-c/(1-c-d)} P_m^{1/(1-c-d)} P_r^{-d/(1-c-d)} C^{-1} \tag{20}$$

According to [2], if the value of h is less than $b/(a+bc+bd)$, then the solution is optimal. For Case II the value of $y = y_{II}$ and it can be calculated using the following formula;

$$h = \frac{by_{II}^{-d/(1-c-d)}}{a+bc+bdy_{II}} \tag{21}$$

3. Result & Discussion

In this paper, the Cobb Douglas production function is implemented in industries that remanufacture heavy equipment. This study focuses on observing one type of product that is part of excavator equipment, namely core engine type XX. For example; the core engine requires 210 items that must be replaced. These items consist of 6325 parts. Inspections are carried out by workers to group between parts that require reconditioning process and parts that are directly used in the disassembly process. There are 2974 parts need to be reconditioned and 3351 parts do not require a reconditioned process. Return ratio for the core engine's parts is, $\gamma = 2974/6325 = 0.47$.

Technical efficiency at reassembly process (α) and disassembly process (β) are 4 and 2, respectively. These values indicate that the use of technology is more dominant in the reassembly process than the disassembly process. Parameters a, b, c, and d indicate the elasticity of the rate of change in output as a consequence of the input used. The parameter values a, b, c, and d are 0.1, 0.6, 0.1 and 0.3, respectively. The elasticity value of a = 0.1 means that the reassembly process requires only a few workers because the use of technology is more dominant. The value of b = 0.6, means the use of new parts is greater than the reconditioned parts. Obviously, the number of reconditioned parts used as input in the reassembly process is less than the new parts, it is shown by c = 0.1. The value of R (input return) is denoted by $d = 0.3$, since the input parts used in the reassembly process consists of new parts, reconditioned parts and returned parts without machining. The total labor cost for the core engine is USD 3523.24 (P_l). The price of returned core engine is USD 6427 (P_r). The total price for all new components (P_m) is USD 27767.93.

3.1. Case I

In Case I, the input for reassembly process uses new parts ($M_m > 0$) and reconditioned parts ($M_r > 0$). All parts from disassembly process are forwarded to reconditioning without any constraints ($\mu = 0$ and $y = 1$). The budget limit for producing a new core engine (C) is USD 90,000. Referring to equation (20), it obtains $h = 1.0718$. The results show the number of workers in the disassembly process (L_m) is 3 workers and reassembly (L_r) is 6 workers. The number of return parts to the reconditioning process (R) is 5. The number of new materials (M_m) is equal to 2. The total reconditioning material (M_r) is 4, and the optimal amount of production (Q) is 12 units. In summary, the result are $L_r^*= 3$, $R^* = 5$, $L_m^*= 6$, $M_m^* = 2$, $M_r^* = 4$, and $Q^* = 12$ units.

3.2. Case II

Similar to Case I, the new parts ($M_m > 0$) and the reconditioned parts ($M_r > 0$) are the input to the reassembly process. In Case II, the Lagrange multiplier value for the constraints of return parts is greater than 0 ($\mu > 0$) and the value of y is less than 1 ($y < 1$). The budget limit for producing the new core engine for Case II is USD 100,000. Using equation (20) with C = 100,000, the value of h is 0.9646. Case II optimal if the value of $h < b/(a + bc + bd)$ = 0.6/(0.1+ (0.6 x 0.1) + (0.6 x 0.3)) = 0.6/0.34 = 1.765. It shows h (= 0.9646) less than 1.765. Using equation (21) for Case II with h equals to 0.9646, then

$$0.9646 = \frac{0.6 y_{II}^{\,0.3/(1-0.1-0.3)}}{0.1 + (0.6 \text{ x } 0.1) + (0.6 \text{ x } 0.3 \text{ x } y_{II})},$$ it obtains y_{II} equals to 0.5942, where $y = y_{II}$.

Table 1 shows the resources allocation for Case I and II during 3 month observations.

Table 1. Resources allocation for 3 month observations

	Decision Variables					
Case	R^*	L_r^*	L_m^*	M_m^*	M_r^*	Q^*
I	5	3	6	2	4	12
II	2	3	7	3	3	13

The existing production quantity that produced by reassembly process in month 1, 2 and 3 were 5, 12 and 2 units, respectively. It is shown that only in month 2 has the same production quantity with Cobb-Douglas production function approaches. The existing production quantity in month 1 and 3 are below the maximum output of the reassembly process.

4. Conclusion

The main purpose of this paper is to measure the appropriate production quantity with an existing remanufacturing alternative, during the 3 months observations. The existing production quantity for the last 3 months in month 1, 2 and 3 are 5 units, 12 units and 2 units, respectively. The final result of the Cobb-Douglas production function approach shows that at Case I optimal production has been achieved in month 2, while in month 1 and 3 the production are not optimal. Similar with Case 1, the optimal conditions in Case II is achieved at period 2. For further research the application of the Cobb Douglas production function can be tried again for different parameter values. It is shown that production framework needs to pay attention to the proportion of input parameters so that the influence of production inputs can have a positive influence on production results. The recommendations are to induce the industry to search a way to get as much output as possible from the existing combination of inputs.

5. References

[1] Saraswati D, Cakravastia A, Iskandar B P and Halim A H A Single manufacturer – A Single buyer partnership with vendorm managed inventory and consignment *Proceeding the 4th IEOM* Bali (2014) 1020-1029

[2] Grubbstrom R W and Tang O Optimal production opportunities in a remanufacturing system Int *J of Prod Res* 44 18-19 (2006) 3953-3966

[3] Daniel R and Guide Jr Production planning and control for remanufacturing industry practice and research needs *J of Op Mgmt* 18 (2000) 467-483

[4] Fleischmann M Jacqueline M Ruwaard B Dekker R Laan E Nunen J A E Wassenhove L N *Quantitative model for reverse logistics: A review*, J of Op Res 103 (1997) 1-17

[5] Hassani A Applications of Cobb-Douglas Production Function in Construction Time-Cost *Thesis* Univ of Nebraska USA 2012

[6] Rao S S *Engineering Optimization – Theory and Practice*, Third Edition, New Age International Limited, New Delhi, 2000

IOP Conf. Series: Materials Science and Engineering **528** (2019) 012056 doi:10.1088/1757-899X/528/1/012056

Redesign of office layout using activity relationship chart (ARC) at the "X" department administration office of a "Y" university

Wahyukaton[1], Ghina Affifah[2]

[1]Industrial Engineering Department, Engineering Faculty, Universitas Pasundan
Jl. Setiabudi 193 Bandung, West Java, Indonesia, wahyukaton@unpas.ac.id
[2]Industrial Engineering Department, Engineering Faculty, Universitas Pasundan
Jl. Setiabudi 193 Bandung, West Java, Indonesia, ghnaff04@gmail.com
wahyukaton@unpas.ac.id, ghnaff04@gmail.com

Abstract. Administration office at "X" Department is a room for managing papers and letters administration, documentation, archiving, and also a room for communication to lecturers and students and office administrators. This room needs to have good layout in order to support the activity for all elements within the department. Redesign the office layout will reduce the time to wait for completing the letters which are needed by the students and the lecturers. This paper's objectives are to improve the information flow and data flow, to reduce the documents stack, and to create a comfortable office to meet the facility layout principles. Activity Relation Chart (ARC) is been used to redesign the office layout by distributing questionnaires to the students and officers and lecturers to obtain the preference of closeness to construct the chart. The result is to redesign the desk into "L" shape to reduce the movements. Also, this result was explained later by a simulation using ARENA.
Keywords: ARC, layout, office, facility, simulation

1. Introduction

Facility layout is an activity to analyze, to concept, to design, and to realize the system for goods producing or services. Generally, facility layout is described as floor plan, a physical facility to optimize the relationship of officers, workers, materials flow, information flow, and a needed procedure to obtain effective and efficient effort [1]. The scoop of facility layout is not only for manufacture problems, also it is been applied to office, transportation, warehousing, packaging, and so on, and those fields corresponding to input, process, and output. In manufacturing fields, the input, process, and output could be tangible, raw materials and components, but in office layout, the input, process, and output would be intangible like information flow and data flow.

The administration office is dealing with documents, papers, letters to run the administration for the department. This paper's objectives are to improve the information flow and data flow, to reduce the documents stack, and to create a comfortable office to meet the facility layout principles, in this paper, Activity Relationship Chart (ARC) is been used.

2. Method

Facility layout is an activity to analyze, to concept, to design, and to realize the system for goods producing or services. Generally, facility layout is described as floor plan, a physical facility to optimize the relationship of officers, workers, materials flow, information flow, and a needed procedure to obtain effective and efficient effort [1]. The scoop of facility layout is not only for manufacture problems, also it is been applied to office, transportation, warehousing, packaging, and so on, and those fields corresponding to input, process, and output. In manufacturing fields, the input, process, and output could be tangible, raw materials and components, but in office layout, the input, process, and output would be intangible like information flow and data flow.
The administration office is dealing with documents, papers, letters to run the administration for the department. This paper's goals are to improve the information flow and data flow, to reduce the documents stack, and to create a comfortable office to meet the facility layout principles.
Facilities in an office must be well planned and arranged in a layout such a manner that the objective of increase in productivity and efficiency of an office operations achieved. There are different techniques and approaches for facility layout design; among this systematic layout planning approach is adopted for design office facility layout in this paper is chosen Activity Relationship Chart (ARC) to meet the objectives. A simplified framework for the application of systematic layout planning for the design of office facility layout is proposed by making enough modifications in the basic approach.
This technique can be successfully applied for both production and service facilities. The systematic layout planning starts with data collection regarding the activities, closeness, rooms, equipment. Based on these input data an activity relationship diagram is drawn. Then a space relationship diagram is prepared by considering the availability and requirement of space constraints. A number of alternative layouts are generated by incorporating the results from relationship diagrams. These layouts are evaluated and select the best one that meets the objectives [2]. Simplified Systematic Layout Planning is especially useful in job shops or process layouts, where there are many different products that are produced in limited quantities. Where there is no consistent process flow, the development of a relationship chart may be the best data you have to determine relative placement of separate work areas. The value of a relationship chart should not be ignored in product-oriented layouts or production lines. The six-step process will aid you in locating all the necessary support functions. [5][6]

2.1 Variables for Questionnaire

The questionnaires were distributed to three groups of respondents, which were lecturers, officers, and students. For those three groups, the questionnaire's variables are as follows,

Table 1. Closeness

	AD	LR1	LR2	MDB	IB	SIB	DC	PM	OR	T
Administration Desk (AD)										
Lecturer Room 1 (LR1)										
Lecturer Room 2 (LR2)										
Mark Display Board (MDB)										
Information Board (IB)										
Seminar Information Board (SIB)										
Documents Cabinet (DC)										
Photocopy/Printer Machine (PM)										
Officer Room (OR)										
Toilet (T)										

Table 2. Layout Variables, Ordinal Scale 1 – 5 for agreement

Sub Variable	Indicators
Desk layout	U shape L Shape
Cabinet layout	U shape L Shape
Orientation	In the middle of the room At the corner of the room

Table 3. Waiting room Variables, Ordinal Scale 1 – 5 for agreement

Sub Variable	Indicators
Bench	Long bench Chair
Orientation	Outside the room Inside the room

2.2 Activity Relationship Chart (ARC)

The objective of ARC is to obtain a closeness relationship from every activity within office or factory organization. To explain the degree of closeness among activities in ARC, symbols are used, which are A, E, I, O, U, and X.

A : *Absolutely necessary*
E : *Especially important*
I : *Important*
O : *Ordinary closeness*
U : *Unimportant*
X : *Undesirable*

2.3 Layout

Layout is a decision to determine a long term for an operation efficiency. There are many effects based on layout decision, such as capacity, process, flexibility, cost, work environment quality, customer contact, and enterprise image. The effective layout will help the enterprises to support they business's strategy for differentiation strategy, low cost, and quick respond.

In order to planning a layout, it needs to,

1. High benefit of rooms, facilities, and labors.
2. Improvement of information flow, material flow, labor flow
3. Increase work morale and a better safety
4. Increase enterprise and customer interaction
5. Increase flexibility

To describe layout, a block layout is used regarding relationship among activities.

2.4 Simulation

Plant layout design requires diverse field of knowledge. Among others the application and use of computers become an advantage. Computer can perform tedious computations and generate several alternative solutions much more rapidly and effectively than manual procedures. [3]

Simulation is a method to duplicate or to describe a real system characteristic. The early idea of simulation is to imitate the real system as mathematically, then to learn the characteristics and operational characters, and decision is made based on a simulation eventually. With this method, the real system will remain untouched until a benefit or loss is found based on a simulation. A simulation software ARENA is used to modelling a redesign layout.

3. Result and Discussion

After questionnaires been distributed, collected, and summarized, all questions are valid and the questionnaire is reliable.
The ARC can be formed as follows,

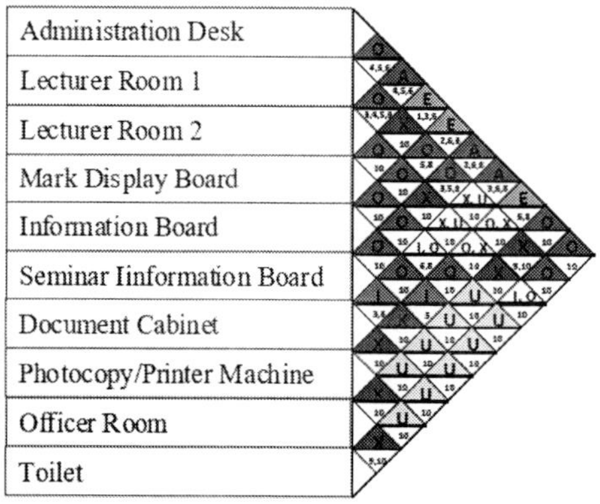

Figure 1. Activity Relationship Chart

Table 4. Closeness of ARC

Code	Closeness
1	Using the same record
2	Using the same personal
3	Using the same room
4	Personal relationship level
5	Paperwork relationship level
6	Paper flow sequence
7	Using the same work flow
8	Using the same equipment and facility
9	Noisy, dirty, dusty, etc.
10	Anything might need

Based on the processed questionnaires, the chosen administration desk layout is "L" shape, cabinet layout is "U" shape, and long bench for the waiting room, with regards to ARC, then the layout redesign of the administration office will be as follows,

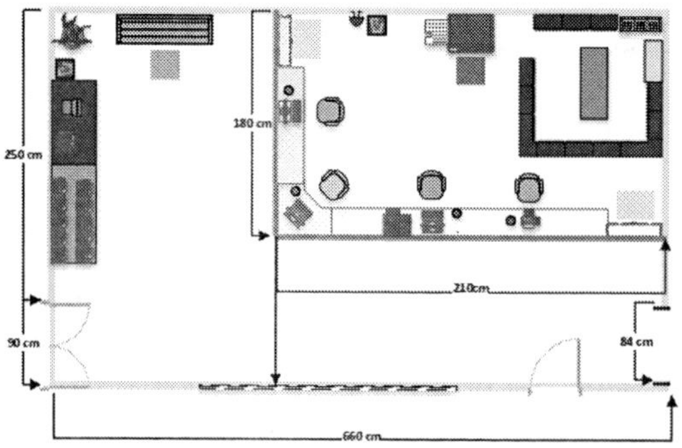

Figure 2. Layout Redesign

The simulation using ARENA software, the result as follows,

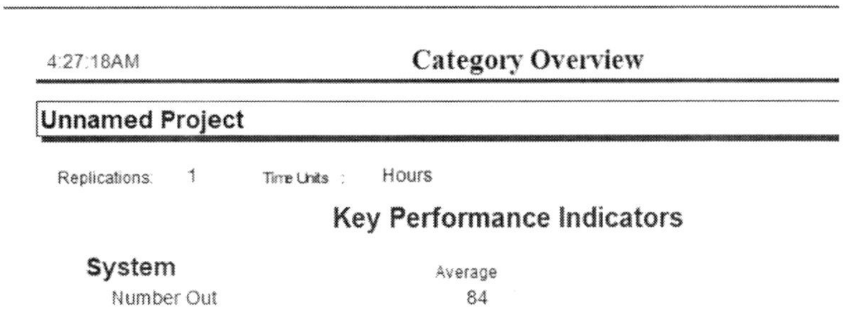

Figure 3. Key Performance Indicators (KPI)

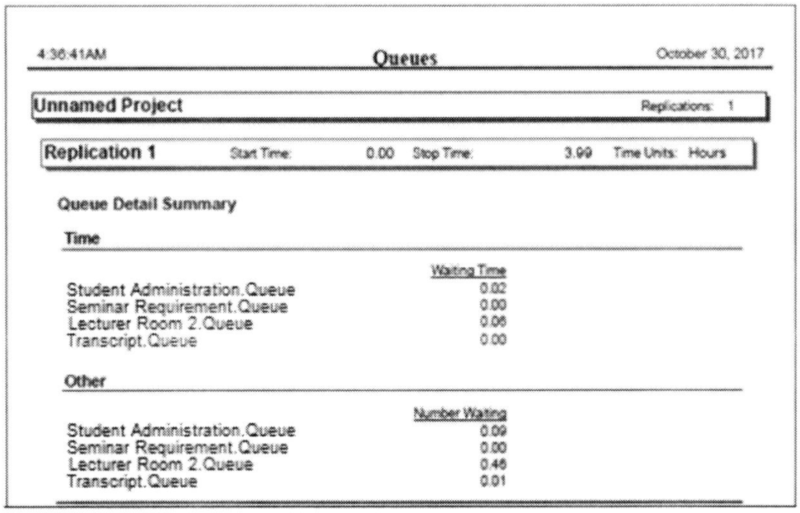

Figure 4. Queue Detail Summary

The result of the simulation shows that the queue on managing documents are minimum, indicated in the Table 4 that the waiting time is 0.06, and number of waiting is under 1 person, it means that the work flow runs well.

4. Conclusion

Based on an ARC and layout redesign, the conclusion can be drawn that to improve activity flow, work flow, and information flow, the layout is using "L" shape desk, cabinet layout is "U" shape will minimize the movement and increase the efficiency in regard of managing paper work and administration work in administration office, and it has been proved by the simulation.

5. References

[1] James M. Apple., 1990, Plant Layout and Material Handling, John Willey & Son, New Jersey

[2] Anish I, Arish Ibrahim, Facility Layout Design of Library Using Systematic Layout Planning, *International Journal of Library and Information Studies*, Vol.4 (3) Jul-Sep, 2014, pp. 23-27.

[3] Ameha Mulugeta, Birhanu Beshah and Daniel Kitaw, Computerized Facilities Layout Design, *Journal of EEA,* Vol. 30, 2013, pp. 27-32.

[4] Marni Astuti, Eko Poerwanto, Agus Trianingsih, Analisis Tata Letak Fasilitas Dengan Menggunakan Metode Activity Relationship Chart Pada Industri Mebel Bambu Karya

Manunggal Yogyakarta, *Seminar Nasional Teknologi Informasi dan Kedirgantaraan (SENATIK)*, Vol. III, 21 Desember 2017, pp. 112-116.

[5] Jeff Mohr, and Mike Willett, Simplified Systematic Plant Layout, *Center for Industrial Research and Service*, Vol. 34, No. 1, Fall 1999, pp. 5-8.

[6] Shah, C. R., A. M. Joshi, Increased Productivity in Factory Layout by Using Systematic Layout Planning (SLP), *International Journal of Advanced Engineering Technology*, Oct-Dec, 2013, pp. 61-63.

IOP Conf. Series: Materials Science and Engineering **528** (2019) 012057 doi:10.1088/1757-899X/528/1/012057

Line Balancing Application Analysis of Generator Manufacturing Process in DPG Inc.

S. Alif[1] and B. Aribowo[2]

Industrial engineering department, Faculty of science and technology, Universitas Al Azhar Indonesia, Jakarta 12110, Indonesia
salsabilaa.alif@gmail.com, budiaribowo@uai.ac.id

Abstract. Every company always try to fulfill their customers' demand in order to gain maximum profit with minimum cost. To achieve that, company needs to pay attention to all aspects of the industry from upstream to downstream and create an integrated system so all processes can run effectively and efficiently. One important aspect that needs to be considered closely is production time. Company have to ensure that the production activity can be done within the time target so customers can get the right product in the right time. In the production floor, the balance of production lines is effecting the efficiency of production process. This research measures the efficiency of a manufacturing process in a generator company (DPG Inc.) based on its production time. Then, line balancing is applied to the generator manufacture line using RPW, J-Wagon, and LCR method to know the comparison of efficiency before and after line balancing applied. Based on the line efficiency, smoothness index and balance delay obtained, the best line balancing method to use in the generator manufacture line is LCR. The result is the generator manufacture line is more efficient after line balancing applied, especially using LCR method.
Keywords: Line Balancing, Line Efficiency, Smoothness Index, Balance Delay, Ranked Positional Weight Method(RPW), J-Wagon Method, Largest Candidate Rule Method(LCR)

1. Introduction

Electricity is one of human needs that is very important to Indonesian society, whether in urban or rural areas. People's needs toward electricity can be categorized as a primary need because this time, almost every household appliances uses electricity as its energy source. Even so, along with the growth of population rate and economic rate in each region, the electricity provided by the State Electricity Company (PLN) sometimes cannot meet the needs of the community. This results in frequent power outages that are quite disruptive to community activities. To address this, people can use electricity generators. Electric generators convert mechanical energy into electrical energy that is suitable for lighting and electricity reserve use.

Just like any other industries, DPG Inc. always try to fulfill their customers' demand on time in order to get maximum profit with minimum cost. But in reality, often production target that has been scheduled cannot be done on time. This might happen because of some factors that occur during the manufacturing process. One of those factors came from the company's production floor, that the manufacturing process is not yet efficient.

Line balancing is a way that can be applied in production floor, in order to balance the work element weight on each workstation. These balanced work element weight will prevent the production line from 'bottleneck effect' or in other words, too many stacked product or semi-finished product in one workstation. Moreover, line balancing can also be an alternative solution that can be applied to smoothen out the production stream and to increase line production efficiency. Therefore, in this research, researcher applied line balancing to know how much efficiency can DPG Inc. get after the line balancing applied in its production line. The line balancing itself, in this research, is done using three different methods; ranked positional weight (RPW) method, J-Wagon Method, and Largest Candidate Rule Method (LCR).

RPW Method prioritizes work element with the largest weight, followed by work elements with smaller weight. Different from RPW, J-Wagon Method prioritizes work element with most work included followed by fewer work included. While LCR method sorts work element with the longest time to the shortest time. For this research, researcher uses three indicators to measure a production line efficiency for each method. Those indicators are; line efficiency, smoothness index, and balance delay.

2. Methods

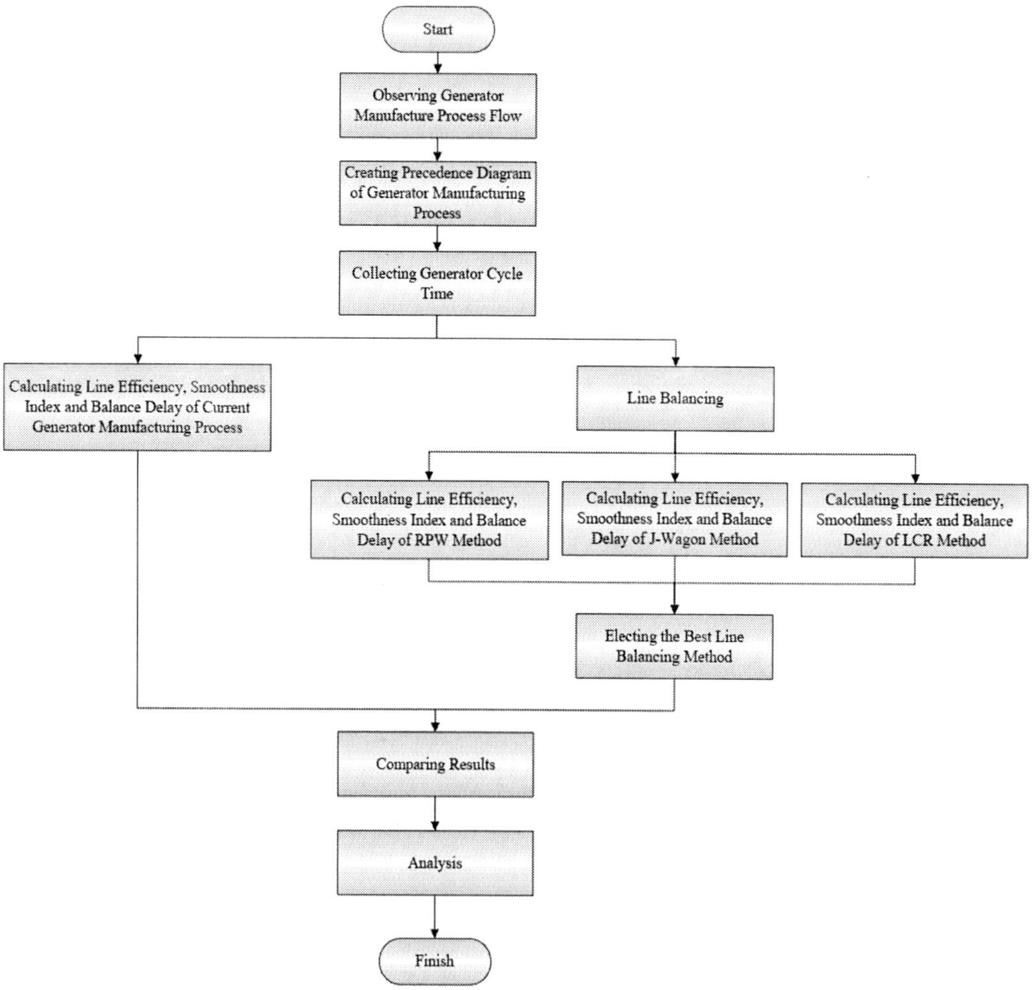

Figure 1. Flowchart

The first step taken in this research is to observe generator production process flow. This step is important to understand step-by-step of creating a unit of generator. After that, researcher created a precedence diagram to visualize the production process flow. The third step is to collect generator cycle time which is a sum of time needed in every work elements presented in minutes. This cycle time is obtained from the sum of the work elements average time. The next step is to calculate line efficiency, smoothness index, and balance delay of current generator production process. Meanwhile, line balancing approach is applicated using three methods; RPW, J-Wagon, and LCR so that line efficiency, smoothness index, and balance delay of every methods can be calculated. After line balancing using the three methods been calculated, researcher compare and elect one method that has the best result among the others. Then, the results of the current generator manufacturing process indicators are being compared with the results of the chosen line balancing method indicators. Finally, researcher analyze those results to know which one of those calculation shows the most efficient among the other.

3. Result and Discussion

3.1. Precedence Diagram

Precedence diagram is a graphic visualization of work operation order and relationship between one operation to others in order to support control and planning activity for each operation [1]. DPG Inc. has 12 workstations in its generator production line which consist 24 work elements. The generator manufacturing flow is being visualized into a precedence diagram to to describe the sequence of work elements, each nodes represents work elements.

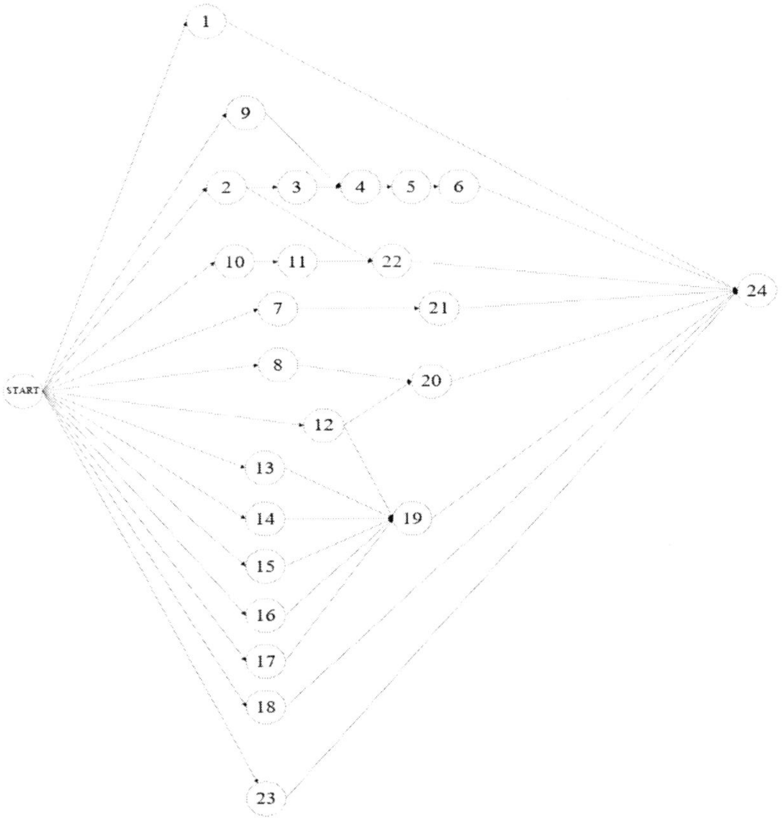

Figure 2. Precedence Diagram

Precedence diagram is supposed to be read from left to right. Each node represents work element according to the job number shown in the table 1. While the line or branch shows relationship between one or more work elements to another. Nodes that comes from more than one branch can only be done after the related work elements have been completed.

3.2. Generator Cycle Time

The cycle obtained from the sum of the work elements average time. The cycle time is 520,95 minutes or 8,7 hour.

Table 1. Generator Cycle Time

Work Element Number	Workstation	Process	Output Product	Operation Time (min)
1	1	Winding Wire	Arm Coil Winding	23.20
2		Winding Wire	Field Coil Winding	49.13
7	2	Wire Harness	Wire Harness Rocker	5.23
8		Wire Harness	Wire Harness Box	4.08
9	3	Stamping	Armature Core FA 5	7.50
10		Stamping	Pole Core FA 5	9.49
11	4	Welding	Pole Block FA - 5	72.00
12	5	Plate Bending	Control Box FA - 5	15.60
13		Plate Bending	Control Box Cover	5.60
14		Plate Bending	Protection Cover (S)	7.50
15		Plate Bending	Protection Cover (F)	6.90
16		Plate Bending	Bearing Cover S	4.00
17		Plate Bending	Bearing Cover F	3.00
18		Plate Bending	Clamper Wire	0.50
3	6	Armature Varnish	Varnished Field Coil	8.00
4	7	Rotor Varnish	Varnished Rotor Assy	9.88
5	8	Rotor Process	Rotor Assy	116.86
6	9	Lathe Process	Balanced Rotor Assy	13.57
20	10	Assembly Process	Control Box Assy	8.10
21		Assembly Process	Rocker Assy	10.53
22		Assembly Process	Pole Assy	14.77
23		Assembly Process	Output Term Bord Assy	2.57
19	11	Painting	Painting	65.80
24	12	Setting	Generator	57.13
			Cycle Time	520.95

3.3. Current Generator Manufacturing Process Efficiency

To know and to compare manufacturing process efficiency before and after line balancing applicated, measurement of current generator manufacturing process efficiency is done using three indicators; line efficiency, smoothness index and balance delay.

3.3.1. Line Efficiency. Current generator manufacturing process line efficiency (LE) is calculated using the equation:

$$LE = \frac{\sum T}{(K \; x \; C)} x \; 100\% \quad (1)$$

Explanation:

LE : Line Efficiency
$\sum$T : Cycle Time
K : Largest Work Station Time
C : Number of Workstations

Cycle time of current generator manufacturing process as shown in table 1 is 520,95 minutes. And for the largest workstation time is at the 5th workstation (rotor process) which needs 116.86 minutes to finish the activity. Therefore, the equation becomes:

$$\text{LE} = \frac{520{,}95 \text{ (min)}}{(116{,}86 \text{ (min)} x \; 12)} \text{x}100\% = 37{,}15\%$$

3.3.2. Smoothness Index. This is the formulation to get the smoothness index:

$$SI = \sqrt{(\sum STi\ max - \sum STi)^2} \tag{2}$$

Explanation:
SI : Smoothness Index
STi max : Maximum Workstation Time
STi : i-Workstation Time

Smoothness index (SI) is an index that shows a relative smoothness of any assembly or manufacturing lines. The closer the index to zero, the smoother a production line gets [2]. A process line will get the perfect balance state when the smoothness index is equal to zero. In this equation, Maximum workstation time (STi-max) has the same meaning as 'K' in line efficiency's equation. Therefore, the equation for smoothness index for current generator manufacturing process becomes:

$$\mathrm{SI} = \sqrt{\begin{matrix}(72{,}33(\mathrm{min}) - 116{,}86(\mathrm{min}))^2 + (9{,}31(\mathrm{min}) - 116{,}86(\mathrm{min}))^2 + \cdots + \\ (57{,}13(\mathrm{min}) - 116{,}86(\mathrm{min}))^2\end{matrix}} = 278{,}672$$

3.3.3. Balance Delay. The formulation of balance delay is:

$$BD = \frac{(K\ x\ C) - \sum ti}{(K\ x\ C)}\ x\ 100\% \tag{3}$$

Explanation:
BD : Balance Delay
K : Number of Workstations
C : Largest Workstation Time
∑ti : Sum of Operation Time

Balance delay (BD) is the ratio between waiting time and time available in the process line. So, the smaller the balance delay value is, the less waiting time occur in the process line. Based on table 1, the balance delay equation of current generator manufacturing process is:

$$BD = \frac{(12x116{,}86(min)) - 520{,}96(min)}{(12\ x\ 116{,}86(min))} x100 = 62{,}85\%$$

3.4. Line Balancing

Line balancing is a task division of interrelated workstations in a production line [3]. The ultimate goal of line balancing is to maximize the speed of each workstation so that a high work efficiency can be achieved in every station [4]. There are some line balancing methods, three of them that are used in this research is ranked positional weight, J-Wagon, and largest candidate rule.

3.4.1. Ranked Positional Weight Method. This method prioritizes work element with the largest weight, followed by work elements with smaller weight [5]. The way to determine the weight of each work element is based on a precedence diagram that starts from the end of the process to the beginning of the process

Table 2 shows the last iteration data of line balancing using RPW method. Based on the result, the number of modifed workstation is five and the longest operation time is 116,86 minutes. By those data, LE, SI and BD of generator manufacturing process using RPW method can be calculated. The results are 89,16%, 48,14 and 10,24%.

3.4.2. J-Wagon Method. J-Wagon method prioritizes work process based on the number of operations it contains. In this method, the number of work element called number of wagon [6].

IOP Conf. Series: Materials Science and Engineering **528** (2019) 012057 doi:10.1088/1757-899X/528/1/012057

Table 2. Modified Workstation Using RPW Method

Work Element Number	Process	Operation Time (min)	Modified Operation Time (min)	Modified Workstation
24	Setting	57.13	116.13	I
3	Armature Varnish	8.00		
9	Stamping	7.50		
10	Stamping	9.49		
12	Plate Bending	15.60		
14	Plate Bending	7.50		
15	Plate Bending	6.90		
16	Plate Bending	4.00		
2	Winding Wire	49.13	115.78	II
8	Wire Harness	4.08		
21	Assembly Process	10.53		
18	Plate Bending	0.50		
1	Winding Wire	23.20		
22	Assembly Process	14.77		
6	Lathe Process	13.57		
4	Rotor Varnish	116.86	116.86	III
20	Assembly Process	8.10	100.18	IV
23	Assembly Process	2.57		
19	Painting	65.80		
17	Plate Bending	3.00		
5	Rotor Process	9.88		
7	Wire Harness	5.23		
13	Plate Bending	5.60		
11	Welding	72.00	72.00	V

Table 3. Modified Workstation Using J-Wagon Method

Work Element Number	Process	Operation Time (min)	Number of Wagon	Modified Operation Time (min)	Modified Workstation
2	Winding Wire	49.13	6	106.54	I
9	Stamping	7.50	4		
3	Armature Varnish	8.00	4		
10	Stamping	9.49	3		
12	Plate Bending	15.60	3		
7	Wire Harness	5.23	2		
8	Wire Harness	4.08	2		
14	Plate Bending	7.50	2		
4	Rotor Varnish	116.86	3	116.86	II
11	Welding	72.00	2	115.45	III
13	Plate Bending	5.60	2		
15	Plate Bending	6.90	2		
16	Plate Bending	4.00	2		
17	Plate Bending	3.00	2		
18	Plate Bending	0.50	2		
5	Rotor Process	9.88	2		
6	Lathe Process	13.57	1		
1	Winding Wire	23.20	1	116.30	IV
20	Assembly Process	8.10	1		
21	Assembly Process	10.53	1		
22	Assembly Process	14.77	1		
23	Assembly Process	2.57	1		
24	Setting	57.13	0		
19	Painting	65.80	1	65.80	V

Table 3 shows that number of workstation modification for line balancing using J-Wagon method is five and the 2nd modified workstation has the longest operation time. Based on the line balancing result shown in table 3, the three efficiency indicators results are: LE = 89,16% , SI = 52,11 and BD = 10,84%.

3.4.3. Largest Candidate Rule Method. The last line balancing method used for this research is LCR, which sorts operation times from the longest to the shortest [5]. The last iteration result shown on table 4.

Table 4. Modified Workstation Using LCR Method

Work Element Number	Process	Operation Time (min)	Modified Operation Time (min)	Modified Workstation
4	Rotor Varnish	116.86	116.86	I
11	Welding	72.00	116.40	II
1	Winding Wire	23.20		
12	Plate Bending	15.60		
13	Plate Bending	5.60		
19	Painting	65.80	115.43	III
18	Plate Bending	0.50		
24	Setting	57.13	116.80	IV
2	Winding Wire	49.13		
21	Assembly Process	10.53		
22	Assembly Process	14.77	104.59	V
6	Lathe Process	13.57		
5	Rotor Process	9.88		
10	Stamping	9.49		
20	Assembly Process	8.10		
3	Armature Varnish	8.00		
9	Stamping	7.50		
14	Plate Bending	7.50		
15	Plate Bending	6.90		
7	Wire Harness	5.23		
8	Wire Harness	4.08		
16	Plate Bending	4.00		
17	Plate Bending	3.00		
23	Assembly Process	2.57		

Table 4 shows that number of workstation modification for line balancing using LCR method is five and the 1st modified workstation has the longest operation time (116,86 minutes). Based on the line balancing result shown in table 4, the three efficiency indicators results are: LE = 89,16%, SI = 12,36 and BD = 10,84%.

3.4.4. Best Line Balancing Method. According to the line balancing results using RPW, J-Wagon, and LCR method, the value of LE and BD using those methods are the same. They are 89,16% and 10,84%. Meanwhile the SI's are different. For RPW it is 47,51; for J-Wagon it is 52,11; and for LCR it is 12,36. Since the smallest SI is by the LCR method, then it can be concluded that the best line balancing method to be applicated in the generator manufacturing process between RPW, J-Wagon, and LCR is LCR.

3.5. Before and After Line Balancing Result Comparison

3.5.1. Line Efficiency. Before: 37,15%; After: 89,16%
There is a line efficiency value difference of 52,01% after line balancing using LCR method applicated.

3.5.2. Smoothness Index. Before: 278,67; After: 12,36.
Smoothes index after line balancing applicated using LCR method reduce (it gets closer to perfect balance state) with value difference of 266,31.

3.5.3. Balance Delay. Before: 62,85%; After: 10,84%
The percentage of balance delay after line balancing using LCR method applicated decrease with 52% value difference.

4. Conclusion

Based on the result of line efficiency (LE), smoothness index (SI), and balance delay (BD) values of generator manufacturing process in DPG Inc., it can be concluded that the best line balancing to be applicated is LCR method. According to the result comparison of before and after line balancing applicated especially using LCR, the efficiency increased 52,01% , the smoothness index also increased to 12,36 point closer to perfect balance state and idle time decreased 52% from the current condition. Briefly, the proposed workstation setting modified using LCR method is more efficient than the current generator manufacturing process setting.

Acknowledgement

Thank you for DPG inc. that gave the opportunity for the researcher to do research in their generator production line.

5. References

[1] Gozali L, Widodo L and Bernhard M 2012 *Analisa Keseimbangan Lini Pada Departemen Chasis PT. TOYOTA MOTOR Manufacturing Indonesia dengan Algoritma Ant Colony, Rank Positional Weight, dan Algoritma Genetika* (Jakarta: Jurnal Teknik Industri Universitas Tarumanagara)
[2] Azwir H H and Pratomo H W 2017 Implementasi Line Balancing untuk Peningkatan Efisiensi di Line Welding Studi Kasus: PT X *Jurnal Rekayasa Sistem Industri* **6**(1) p 61 (Bandung: Universitas Parahyangan)
[3] Baroto T 2002 *Perencanaan dan Pengendalian Produksi* (Jakarta: Ghalia Indonesia)
[4] Kusuma H 1999 *Manajemen Produksi Perencanaan dan Pengendalian Produksi* (Yogyakarta: Andi Offset)
[5] Groover M P 2008 *Automation Production Systems, and Computer-Integrated Manufacturing, 3rd edition* (New Jersey: Prentice-Hall Inc.)
[6] Chase B R and Aquilano J N 1995 *Production and Operations Management* (USA: Richard D Irwin Inc.)

Integration of Lean Manufacturing and Group Technology Layout to increase production speed in the Manufacture of Furniture

U Tarigan[1], U P P Tarigan and V Sukirman

Faculty of Engineering, Industrial Engineering Department, University of Sumatera Utara, Indonesia
[1]ukurta.tarigan@yahoo.com

Abstract. PT. ABC is a wood processing company which manufacture household appliances such as chair, table, wagon, shelf, and clothen hanger using rubber wood as the base material. Based on the results of observations made on the company, it was known that the production process in the company still having many non-value added activities such as delay and long distance transportation activities, causing longer lead time of the production process. From the problems that exist in the company, lean manufacturing approach is used to reduce non value added activities by using Value Stream Mapping to identify activities which are value added and which are the non-value added then to reduce non-value added activities 5W and 1H principles is used. After non-value added activities have been reduced, followed by Group Technology Layout approach to design three alternative layouts with the purpose of minimize the distance between station. This study aims to eliminate non-value added activities through simplification of the production process and redesign of the layout to reduce production time. The results of this study concluded that using this method the manufacturing lead time is decreased from the beginning of 19237.4 seconds to 14340.3 seconds. In addition, process cycle efficiency also increased by 34.15% and the completion of the process also increased by 20 units or by 25.64%.
Keywords: Facility Layout, Lean Manufacturing, Group Technology Layout, Value Stream Mapping

1. Introduction

The importance of export activities in an effort to increase the country's foreign exchange causes the importance of efforts needed to create work productivity and reliable product quality. Increased productivity is important because the quality of service of a company has a very important role in maintaining a smooth business [1]. One of the leading exports in Indonesia from the manufacturing industry is furniture. But for business development, in the long run, the company always faces the challenge of high levels of competition in furniture sales in the global market. [2]

Manufacturing companies are companies that need a process with a considerable amount of material usage. In an effort to increase productivity, companies must know activities that can increase value added (goods and / services) and eliminate (waste), therefore a lean approach is needed [3]. Lean Manufacturing distinguishes which adds value, by eliminating other things that have no added value [4]

The design of production facilities is one of the factors that greatly influence the performance of a company. This is due to poor layout of the facility which will cause a pattern of unfavorable material flow and the movement of materials, products, information, equipment, and labor to be relatively high which causes delays product completion and increase production costs [5].

Group Technology Layout methods classify products that are not identical based on processing steps, forms, machines, or equipment used, etc. [6]. The GT principle is to realize problems that have in common by grouping problems based on similarity in order to obtain a single solution for a group of problems so that it can save time and effort [7-8].

In recent years the level of productivity at this company decreased significantly. From the preliminary study conducted on this company, it can be seen that the state of production still has weaknesses in the flow of production and the arrangement of laying work stations. Based on the layout of the company which can be seen in Figure 1, So far the location of the workstations at PT. ABC is irregular and has a long distance between stations which results in production that is not optimal.

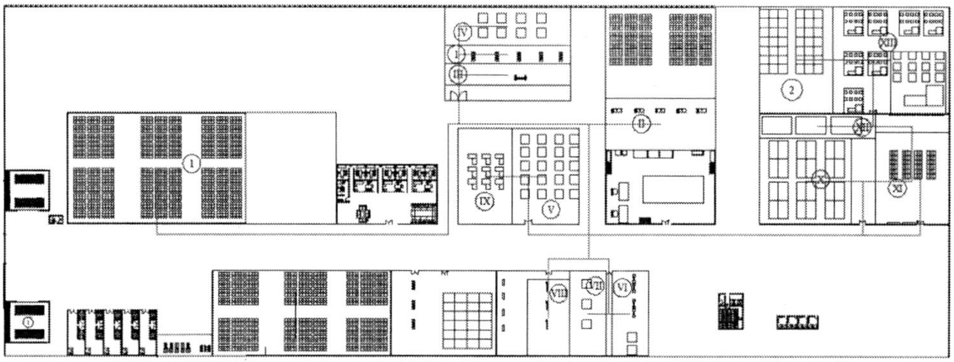

Symbol	Remark	Symbol	Remark
1	Raw Material Storage	VII	Drilling Station
2	Product Storage	VIII	Lathe Station
I	Tapping Station	IX	Sanding Station
II	Cutting Station	X	Painting Station
III	F-Joint Station	XI	Vernishing Station
IV	Laminating Station	XII	Drying Station
V	Pressing Station	XIII	Packing Station
VI	Tenon Station		

Figure 1. Initial Layout

This study tries to solve the problem by integrating the Lean Manufacturing approach which aims to reduce non-value added activity and with Group Technology Layout as a method to improve the location of production facilities on the production floor of PT. ABC in order to increase production and reduce lead times so that it has an impact on increasing the productivity of the company.

2. Research Method

Research begins with a review and collection of data. Data collected from company documents is production volume, and description of the production process, conducting direct observations and direct interviews in the field. The dependent variable in this research design is the proposed production floor layout which has a minimum material handling displacement.

Data processing in this research uses the following steps:

1. Current State Map [9-10]
   a. Value Stream Manager
   b. SIPOC Diagram
   c. Standard Time Calculation
   d. Waste Identification

e. Map Creation for Each Process Category Throughout the Value Stream
f. Overall Factory Process Chart
g. Process Cycle Efficiency Calculation
h. Process Added Mapping Analysis

2. Improvement using Lean Manufacturing and Relayout [11-12]
   a. Initial Layout
   b. Distance Calculation
   c. Transportation Frequency Calculation
   d. Moment of Displacement
   e. Alternative Layout construction using ROC method (Group Technology Layout)
   f. Proposed Process Activity Mapping
   g. Future State Map

3. Result and Discussion

3.1. Deciding Value Stream Manager
In this research, the selected value stream manager is the production manager.

3.2. Product Choosing
In this research, the selected product is wooden chair because it got the highest production volume.

3.3. SIPOC Diagram (Supplier-Input-Prcoess-Output-Customer)
Sipoc Diagram (Supplier-Input-Process-Output-Customer) of wooden chair production process can be seen in Figure 2.

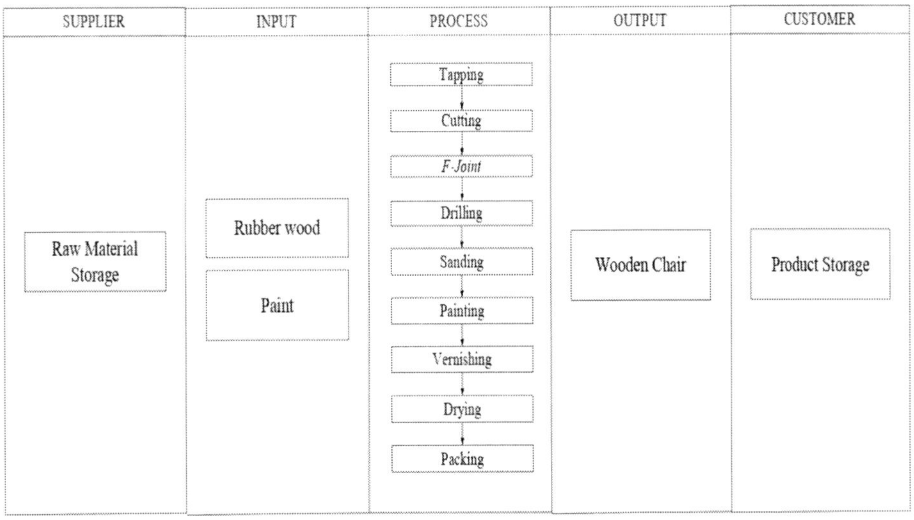

Figure 2. SIPOC Diagram of Wooden Chair Production

3.4. Calculation of Standard Time
The example of standard time calculation for WC 1 is as follow [13]:

$$\text{Standard Time} = \text{Normal Time} \times \frac{100\%}{100\% - Allowance\%} = 12.6 \times \frac{100\%}{100\% - 16\%} = 15{,}0 \text{ seconds} \quad (1)$$

3.5. Waste Identification with Process Added Mapping
Process Activity Mapping (PAM) in identifying necessary activities of non value added and non value Added for company, conducted direct observation [10].

Through process activity mapping obtained the number of operation process, transportation, inspection, storage and delay along with its time can be seen in Table 1.

Table 1. Process Added Mapping Recapitulation

Symbol	Amount	Time (sec)	Percentage (%)
Operation	8	11054,4	57,46
Transportation	10	4688,9	24,37
Inspection	2	1333,1	6,93
Delay	4	2161,0	11,23

3.6. Map Creation for Each Process Category Throughout the Value Stream

Mapmaking for each process category along the value stream uses the standard time data of each process plus other data such as processing time, and the number of operators. a preliminary process category map for tapping as shown in Figure 3 is obtained.

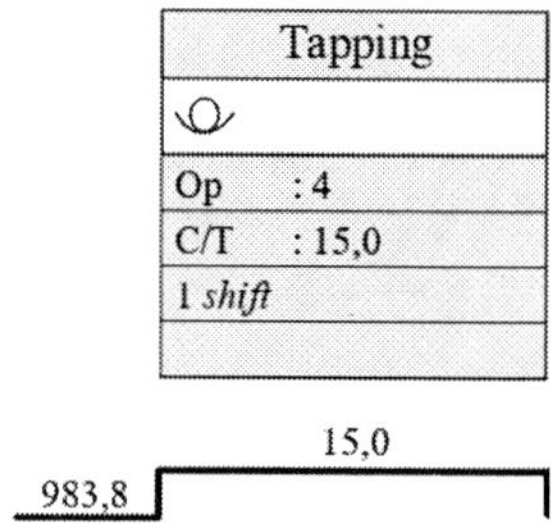

Figure 3. Map of Tapping Category

3.7. Forming the Overall Factory Flow Chart

Each process along the value stream is combined with material flow and information flow so that it becomes a single flow in the factory. After all, information is obtained, thus the current state map can be formed by placing all material and information flows into the folder. Current state map of wooden chair products can be seen in Figure 4.

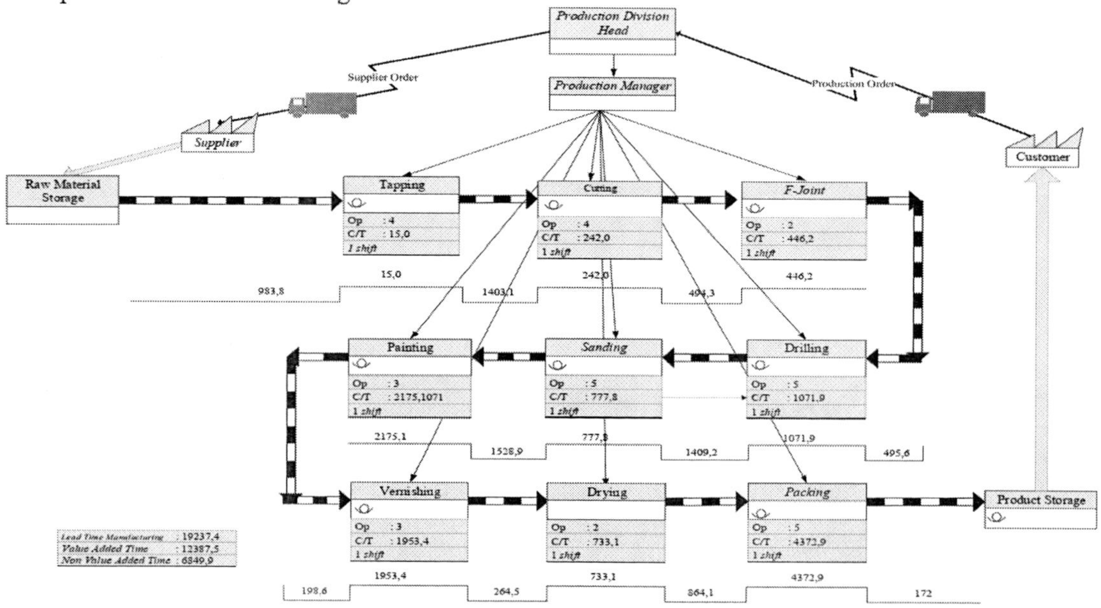

Figure 4. Current State Map of Wooden Chair Product

3.8. Calculating Process Cycle Efficency

The calculation of process cycle efficiency is as follows [14]:

$$\text{Process Cycle Efficiency} = \frac{\text{Value Added Time}}{\text{Manufacturing Lead Time}} = \frac{12387{,}5}{19237{,}4} = 64{,}39\ \% \tag{2}$$

$$\text{Average Finishing Time} = \frac{\text{Total Production (year)}}{\text{Amount of Work Days}} = \frac{19500}{250} = 78 \text{ units/day} \tag{3}$$

3.9. Improvement using Lean Manufacturing

Activities included in NVA and can be improved can be seen in Table 2.

Table 2. Imporvable Non-Value Added Activity

Number	Activity Name
A	The raw material is carried to the tapping station
B	Stacking the results of the tapping before being transported
C	The wood from the impacting is carried to the cutting station
D	Wood measuring 40 cm is carried to the finger joint station
E	Wood is brought to the drilling station
F	Piling up all the drilling parts
G	Brought to the sanding station
H	Accumulate sanding results
I	Brought to the painting station
J	Piling up dry parts
K	Taken to the packing station

Activity included above will be analyzed using 5W and 1H method.

3.10. Improvement Using Group Technology Layout

The results of grouping machines using the Rank Order Clustering method can be seen in Figure 5 [11].

i/j			Component																						DE
			C2	C3	C4	A4	A1	A2	A3	A5	A6	C1	B2	D1	D2	D3	D4	E1	E2	E3	B1	B3	B4	C5	
			2^{21}	2^{20}	2^{19}	2^{18}	2^{17}	2^{16}	2^{15}	2^{14}	2^{13}	2^{12}	2^{11}	2^{10}	2^{9}	2^{8}	2^{7}	2^{6}	2^{5}	2^{4}	2^{3}	2^{2}	2^{1}	2^{0}	
	XIII	2^{12}	1	1	1	1	1	1	1	1	1	1	1	1	1	1	1	1	1	1	1	1	1	1	4194303
	I	2^{11}	1	1	1	1	1	1	1	1	1	1	1	1	1	1	1	1	1	1	1	1	1	0	4194175
	II	2^{10}	1	1	1	1	1	1	1	1	1	1	1	1	1	1	1	1	1	1	1	1	1	0	4194175
	IX	2^{9}	1	1	1	1	1	1	1	1	1	1	1	1	1	1	1	1	1	1	1	1	1	0	4194175
	X	2^{8}	1	1	1	1	1	1	1	1	1	1	1	1	1	1	1	1	1	1	1	1	1	0	4194175
	XI	2^{7}	1	1	1	1	1	1	1	1	1	1	1	1	1	1	1	1	1	1	1	1	1	0	4194175
Machine	XII	2^{6}	1	1	1	1	1	1	1	1	1	1	1	1	1	1	1	1	1	1	1	1	1	0	4194175
	VII	2^{5}	1	1	1	1	1	1	1	1	1	1	0	0	0	0	0	0	0	0	0	0	0	0	4132608
	III	2^{4}	1	1	1	1	0	0	0	0	0	0	1	1	1	1	1	0	0	0	0	0	0	0	280440
	VI	2^{3}	0	0	0	0	0	0	0	0	0	0	1	1	1	1	1	1	1	1	1	1	1	0	61567
	IV	2^{2}	1	1	1	0	0	0	0	0	0	0	1	1	1	1	1	0	0	0	0	0	0	0	18296
	V	2^{1}	1	1	1	0	0	0	0	0	0	0	1	1	1	1	1	0	0	0	0	0	0	0	18296
	VIII	2^{0}	0	0	0	0	0	0	0	0	0	0	0	0	0	0	0	1	1	1	0	0	0	0	7
DE			8182	8182	8182	8176	8160	8160	8160	8160	8160	8160	8158	8158	8158	8158	8158	8137	8137	8137	8136	8136	8136	4096	

Figure 5. Machine Grouping

Based on the results of the grouping above, it can be seen that there are several components that are not colored, because that the machine used in the construction of the component already exists in another group. This can be overcome by dividing more than one machine into the group of machines that are formed. The results of the improvement of machine grouping can be seen in Table 3.

Table 3. Grouping Correction

Group	Component	Machine
I	C2,C3,C4,A4,A1,A2,A3,A5,A6,C1,B2	XIII(3), I(2), II(2), IX(3), X(2), XI(2), XII(1), VII(5), III(1), VI(1), IV(1), V(1)
II	D1,D2,D3,D4,E1,E2,E3,B1,B2,B3,B4	XIII(2), I(2), II(2), IX(2), X(1), XI(1), XII(1), III(1), VI(4), IV(1), V(1), VIII(2)

3.11. Alternative Layout Construction

After grouping machines, 3 types of alternative layouts are designed which will then be selected based on the moment of displacement and line efficiency rate[12].

The chosen layout is the second alternative with the lowest moment of displacement 831051,9 meter/month and line efficiency rate 82,77% [13]. The chosen alternative can be seen in Figure 6.

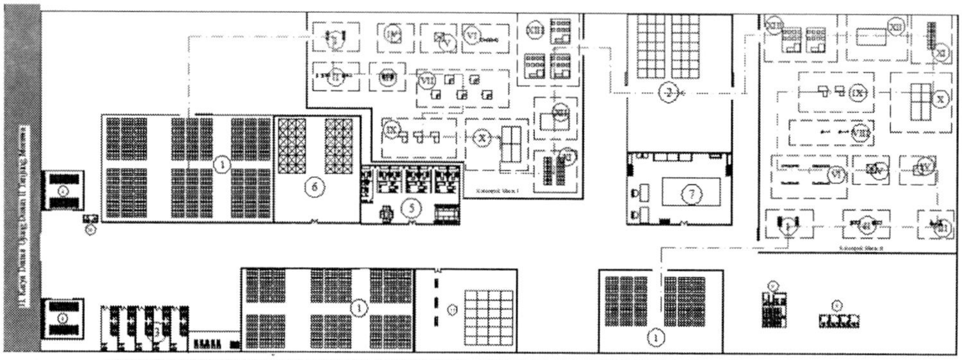

Symbol	Remark	Symbol	Remark
1	Raw Material Storage	VII	Drilling Station
2	Product Storage	VIII	Lathe Station
I	Tapping Station	IX	Sanding Station
II	Cutting Station	X	Painting Station
III	F-Joint Station	XI	Vernishing Station
IV	Laminating Station	XII	Drying Station
V	Pressing Station	XIII	Packing Station
VI	Tenon Station		

Figure 6. Chosen Layout

3.12. Proposed Process Activity Mapping

Through the proposed Process Activity Mapping (PAM). The number of operations, transportation, delays, and inspections can be seen as shown in Table 4.

Table 4. Proposed PAM Recapitulation

Symbol	Amount	Time (sec)	Percentage (%)
Operation	8	11054,4	77,09
Transportation	10	1952,8	13,62
Inspection	2	1333,1	9,29
Delay	0	0	0,00

3.13. Future State Map Depiction

Future State Map of wooden chair production can be seen in Figure 7.

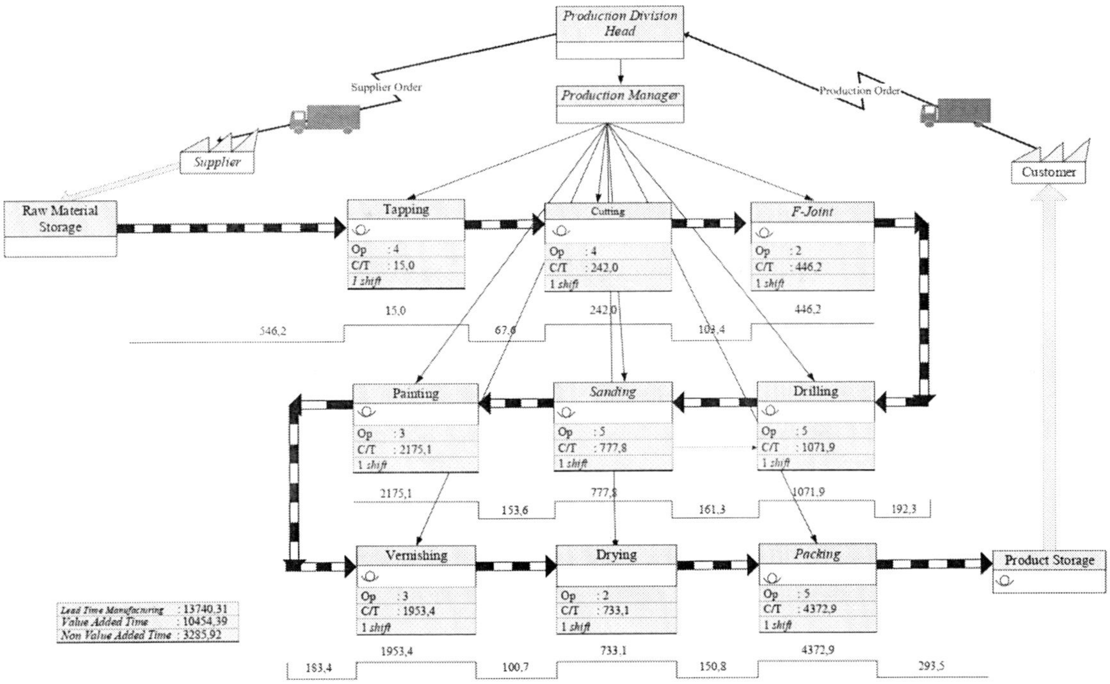

Figure 7. Proposed Future State Map

3.14. Calculating Improved Process Cycle Efficiency

The calculation of imporved process cycle efficiency is as follows:

$$\text{Process Cycle Efficiency} = \frac{\text{Value Added Time}}{\text{Manufacturing Lead Time}} = \frac{12387{,}5}{14340{,}3} = 86.38\ \% \tag{4}$$

$$\text{Production Improvement} = \frac{\text{Initial Lead Time-Proposed Lead Time}}{\text{Innitial Lead Time}} = \frac{19237{,}4\text{-}12387{,}5}{19237{,}4}\ \text{x Average Finishing Time}$$

$$= 0{,}2546 \text{ x } 78 \text{ units/day} = 20 \text{ units} \tag{5}$$

After repaired using the lean manufacturing method, increase in production of 20 units to 98 units / day.

4. Conclusion

Based on the results of the analysis using the lean manufacturing approach, delay or non-value added activities that occur in the production process affect the productivity of the company. It can be seen that after improvements with the lean manufacturing method and layout redesign with the Group Technology Layout method there is a change in the manufacturing lead time from the beginning of 19237.4 seconds to 14340.3 seconds. Improvements using the Lean Manufacturing and Group Technology Layout approaches also increased the process cycle efficiency by 34.15% and increased production per day by 20 units.

Acknowledgments

In this research, the authors would like to thank profusely to Universitas Sumatera Utara that gives chances to publish and develop author in the process.

5. References

[1] Koesmono T 2006 *Pengaruh Budaya Organisasi Terhadap Motivasi, Kepuasan Kerja dan Kinerja Karyawan (Studi Pada Perusahaan Pengolahan Kayu Skala Besar)*. (Surabaya: Unika Widya Mandala)

[2] Kurniawati D and Rohmah A 2018 *Tantangan Eksportir Furnitur di Yogyakarta Studi Kasus CV Dbest Furniture*. (Yogyakarta: Universitas Ahmad Dahlan)
[3] Womacks P J 2003 *Lean Thinking*. (New York: Free Press)
[4] Jadhav P K 2018 Implementing Lean Manufacturing Principle in Fabrication Process-A Case Study. *Int. Research J. of Eng. and Tech. (IRJET)* **5**(6) pp 1843-47
[5] Sofyan D 2015 *Perancangan Ulang Tata Letak Fasilitas Menggunakan Metode Konvensional Berbasis 5S*. (Aceh: Universitas Malikussaleh)
[6] Hutabarat J 2017 *Re-Layout dengan Metode Group Technology*. (Malang: Institut Teknologi Malang)
[7] Hasibuan A 2017 *Aplikasi Metode Group Technology dalam Memperbaiki Tata Letak untuk Meminimalkan Jarak Perpindahan Bahan* (Medan: Universitas Islam Sumatera Utara)
[8] Apple J 1997 *Tata Letak Pabrik dan Pemindahan Bahan*. (Bandung: ITB)
[9] Standnicka and Dorota 2015 Application of Value Stream Mapping and Possibilities of Manufacturing Processes Simulations in Automotive Industry. *FME Transactions* **43**(4)
[10] Wijayant B 2018 *Rancangan Proses Produksi Untuk Mengurangi Pemborosan Dengan Penggunaan Konsep Lean Manufacturing PT Mizan Grafika Sarana*. (Bandung: Institut Teknologi Nasional)
[11] Vanany I 2005 *Aplikasi Pemetaan Aliran Nilai Di Industri Kemasan Semen*. (Surabaya: Institut Teknologi Sepuluh Nopember)
[12] Grepcka and Alketa 2017 *Production Flow Analysis Based on GT Concepts in Wood Production Factory – Case Study*. (Albania: Agriculture University of Tirana)
[13] Sutalaksana I Z 2006 *Teknik Perancangan Sistem Kerja*. (Bandung: ITB)
[14] Heragu S S 1997 *Facilities Design*. (Boston: PWS Publishing Company)

Flowshop Scheduling with Drum-Buffer-Rope and CDS Algorithm to Minimize Lateness and Work in Process at PT. AKS

A S Viady[1]**, P P Suryadhini**[2] **and M Rendra**[3]

School of Industrial and System Engineering, Telkom University, 40257 Bandung, West Java, Indonesia
[1]akmaseptia@gmail.com, [2]poeripps@gmail.com, [3]meldirendra@telkomuniversity.ac.id

Abstract. In many years, the problem of lateness remains a problem for companies. This problem also occurs in textile company. At PT. AKS, one of the textile company, the lateness causes by bottleneck on one of their work station, which is Split Workstation, and the other caused by large unit load. The objective of this research is to minimize lateness by reducing the bottleneck and unit load. This research purposed a Drum-Buffer-Rope method and CDS algorithm to solve the problem. The constraint work station which is Split Workstation as the drum which is the control point of the whole system. The rope systematic which is backward scheduling applied at work stations before Split Workstation to minimize the queue time and to control the work in process. CDS Algorithm used in the interest of jobs sequencing to be processes after Split Workstation by using a forward scheduling. To solve the large unit load, we determine the unit load by trial and error. The result of this research manufacturing lead time decrease is by 61.88 percent from the current condition, queue time decrease is by 82.45 percent from current condition and the lateness decrease is by 35.71 percent from current condition.
Keywords: Flowshop scheduling, Drum-Buffer-Rope, CDS Algorithm

1. Introduction

PT. AKS as textile company produce a half-finished fabric i.e. polyester and cotton and produce a leather as material for clothes and shoes, their using a make to order system to fulfil the customer orders. The last four months, they faced a high demand of polyester, based on that condition, the company sustain a late delivery to customer. Based on survey and interviews, the root cause of late delivery are defect product and improper schedule which result in idle and bottleneck. Built upon the root cause, this research focus on improper schedule which inflict a high quantity of work in process.

This company produces various types of fabric, but the most common orders are polyester and cotton. Polyester and cotton production processes are carried out in different divisions, but at the end of the process, both types of fabric will be processed in the dyeing and finishing division. In the dyeing and finishing division there are eleven workstations, that is Partition Inspection, Mercerizing, Split, Dyeing, Calator, Centrifugal, Dryer, Callender, Setting, and Finishing.

The bottleneck occurs in Split Workstation, the input for four hours is 6000 kilograms or 237 rolls, meanwhile the output for four hours is 2400 kilograms or 96 rolls. From the current condition, there's no rule of sequencing the job, they process randomly, and it is resulting in late delivery to customers. According to the problem occur, this research proposes a scheduling and sequencing jobs to minimize a lateness and work in process.

There are several studies that discuss a job scheduling with Drum-Buffer-Rope Technique such as research conducted by [1] which applied Drum-Buffer-Rope for re-entrant flowshop and [2] which applied Drum-Buffer-Rope in aircraft industry.

Drum-Buffer-Rope method was chosen in this research because this method focuses on improvements in the workstation that has problems or the workstation that affect throughput. In this matter the workstation that is having problems is a workstation that has a bottleneck.

2. Methods

According to [3], Dr. Eliyahu Goldratt, the founder of Theory of Constraint, defined the constraint resource, the role of constraint in determining the output of organization, method to identify and exploit the constraint. Goldratt also gave the steps for ongoing improvement. The proposed steps for improvement are follows:

1. Identify the constraint system
2. Decide how to exploit these constraint
3. Sub-ordinate everything else to the above decision
4. Elevate the constraint system.
5. If in the above process generate other constraint, go back to the first step.

In this research, the research methodology was based on a practice-based research, the step of improvements adopts as describe on conceptual model as follow:

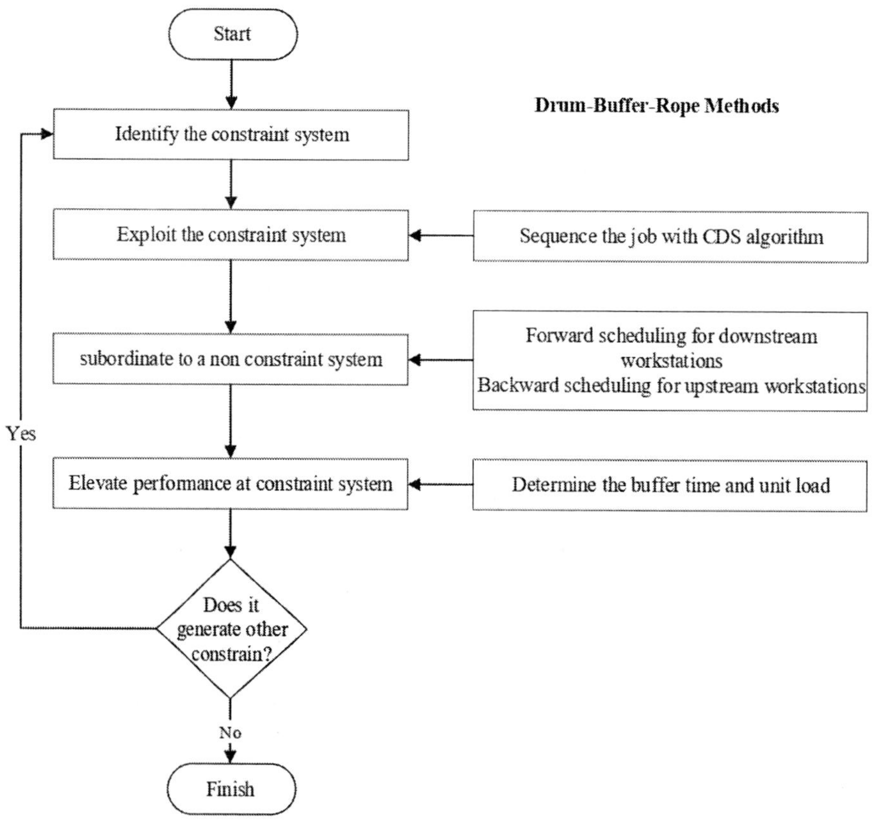

Figure 1. The Conceptual Model

The production control technique to implement the step of Theory of Constraint (TOC) which is exploiting, subordinating and elevating is Drum-Buffer-Rope [4]. Drum is bottleneck resource that drive the system, buffer placed before bottleneck operation, the purpose of the buffer is to preserve feeding rate in the bottleneck resource, the buffers usually consist of three types: capacity buffer [5], time buffer [6] and stock buffer [7]. The rope is feedback loop which connecting the buffer to the raw material dispatching point.

The steps of Drum-Buffer-Rope technique as follows:

1. Scheduled the bottleneck resource.
2. Do the backward scheduling from bottleneck resource to first operation.
3. Do the forward scheduling from bottleneck resource to last operation.
4. Determine the buffer time in bottleneck resource.

From the step of exploit the constraint system, mostly, the sequencing in Drum-Buffer-Rope using SPT, EDD or random rule, but in this research, the jobs sequence is done using CDS Algorithm, this is because the processing time for all jobs in the constraint workstation is the same, so there is no difference with the actual conditions.

CDS algorithm is a scheduling method to get the optimal job sequence from n jobs to m machines. This scheduling is the development of Johnson's algorithm with heuristic approach and making several alternative job sequences to get the optimal scheduling. According to [8] the CDS algorithm applied the Johnson Rules in a heuristic manner and create several schedules from which the "best" schedule can be selected. The algorithm appropriate to a multistage use of Johnson's Rule applied to a pseudo-problem, reviled from the original, with processing times aj and bj. At stage 1, $aj = p1\,j$ and $bj = pmj$. In other term, the first and last processing times for each job involve the pseudo-problem. At stage 2, $aj = p1\,j + p2\,j$ and $bj = pm-1, j + pmj$. Here, the first two and last two processing times, accumulated for each job, involve the pseudo-problem. In general, at stage i; $a_j = \sum_{k=1}^{i} P_{kj}$ and $b_j = \sum_{k=m-i+1}^{m} P_{kj}$. The procedure be composed of $(m - 1)$ stages, some of which may resulting the same sequence, after which the algorithm chooses the best makespan calculated.

The data used in this research were obtained from interviews and field observations. The data collected from Dyeing and Finishing Division is the cycle time data, the machine's productive time data which is the number of hours / shifts / working days, the machine capacity and routing data of the machine.

3. Result and Discussion

By following the Drum-Buffer-Rope steps, the result as follows:

1. Identify the constraint resource as the Drum.
   In this case, the constrain resource is Split Workstation. The input in Split Workstation for four hours is 6000 kilograms or 237 rolls, meanwhile the output for four hours is 2400 kilograms or 96 rolls, which is the input bigger than output so that it can cause bottleneck and high value of work in process.
2. Determined the total time.
   In this step calculate the total time for each job in each workstation, herewith the example of total time for several jobs.

Table 1. The example of total time for several jobs.

Job Code	Roll	Workstation	Processing Time	Setup Time	Move Time	Total Time
01	12	Inspection	11.06	1.17	1.56	12.23
	12	Split	29.30	0.41	3.27	29.71
	12	Jet Dyeing	24.21	1.93	6.54	26.14
	12	Centrifugal	16.09	4.85	5.60	20.95
	12	Setting	26.62	0.72	1.18	27.34
	12	Stenter	29.90	1.07	10.39	30.97
03	12	Inspection	11.02	1.17	1.56	12.19
	12	Split	29.30	0.41	3.27	29.71
	12	Jet Dyeing	27.21	1.93	6.54	29.14
	12	Centrifugal	16.09	4.85	5.60	20,95
	12	Setting	36.62	0.72	1.18	37.34
	12	Stenter	29.90	1.07	10.39	30.97

Table 1. The example of total time for several jobs (cont.).

07	12	Inspection	11.02	1.17	1.56	12.19
	12	Split	29.30	0.41	3.27	29.71
	12	Jet Dyeing	24.21	1.94	6.54	26.15
	12	Centrifugal	16,09	4,85	5,60	20,95
	12	Setting	26.62	0.72	1.18	27.34
	12	Stenter	27.40	1.07	10.39	28.47

3. Sequence the job with EDD, SPT and random rule in constraint resource.
The result of sequencing with EDD, SPT and random rule can be seen in Table 2. The sequence result with these three rules have the same results with current condition because all jobs has same processing time, therefore we use the CDS algorithm to sequences the jobs.

Table 2. The sequences of the job using EDD, SPT and random rule.

Job Code	Earliest Start	Latest Finish	Processing Time	Job Sequence
J01	0	1440	29.71	1
J02	0	1440	29.71	2
J03	0	1440	29.71	3
J04	0	1440	29.71	4
J05	0	1440	29.71	5
J06	0	1440	29.71	6
J07	0	1440	29.71	7
J08	0	1440	29.71	8
J09	0	1440	29.71	9
J10	0	1440	29.71	10
J11	0	1440	29.71	11
J12	0	1440	29.71	12
J13	0	1440	29.71	13
J14	0	1440	29.71	14
J15	0	1440	29.71	15
J16	0	1440	29.71	16

4. Sequence the job with CDS algorithm.
By using the CDS algorithm, there are five alternative job sequences. The result of CDS algorithm for all alternatives can be seen in Table 3. The best result of sequencing is alternative 4 and alternative 5, because both alternatives have the shortest makespan, which is equal to 2118.18. Both alternatives also have the same sequences.

Table 3. The alternatives of job sequence with CDS algorithm.

Alternative 1 (K = 1)	**Makespan**
J01-J02-J03-J04-J05-J06-J07-J08-J09-J10-J11-J12-J13-J14-J15-J16-J17-J18-J19-J20-J21-J22-J23-J24-J25-J26-J27-J28-J29-J30-J31-J32-J33-J34-J35-J36-J37-J38-J39-J40-J41-J42-J43-J44-J45-J46-J47-J48-J49-J50	2121.27
Alternative 2 (K = 2)	
J01-J02-J03-J04-J05-J06-J07-J08-J09-J10-J11-J12-J13-J14-J15-J16-J17-J18-J19-J20-J21-J22-J23-J24-J25-J26-J27-J28-J29-J30-J31-J32-J33-J34-J35-J36-J37-J38-J39-J40-J41-J42-J43-J44-J45-J46-J47-J48-J49-J50	2121.27
Alternative 3 (K = 3)	
J06-J15-J24-J33-J42-J05-J14-J23-J32-J41-J50-J03-J12-J21-J30-J39-J48-J04-J09-J13-J18-J22-J27-J31-J36-J40-J45-J49-J02-J11-J20-J29-J38-J47-J08-J17-J26-J35-J44-J01-J10-J19-J28-J37-J46-J07-J16-J25-J34-J43	2121.27

Table 3. The alternatives of job sequence with CDS algorithm (cont.).

Alternative 4 (K = 4)	
J01-J04-J07-J10-J13-J16-J19-J22-J25-J28-J31-J34-J37-J40-J43-J46-J49-J02-J05-J08-J11-J14-J17-J20-J23-J26-J29-J32-J35-J38-J41-J44-J47-J50-J03-J06-J09-J12-J15-J18-J21-J24-J27-J30-J33-J36-J39-J42-J45-J48	2118.18
Alternative 5 (K = 5)	
J01-J04-J07-J10-J13-J16-J19-J22-J25-J28-J31-J34-J37-J40-J43-J46-J49-J02-J05-J08-J11-J14-J17-J20-J23-J26-J29-J32-J35-J38-J41-J44-J47-J50-J03-J06-J09-J12-J15-J18-J21-J24-J27-J30-J33-J36-J39-J42-J45-J48	2118.18

5. Scheduling in Split Workstation.
The Split workstation is the point to be controlled for scheduled due to a bottleneck. Two kind scheduling approach used to solve problems, which is forward approach from Split workstation to the last work station and backward approach from Split workstation to first workstation.
According to research done by [2], following is the step of scheduling in Split workstation:
- Calculate the manufacturing lead time from Split workstation to last workstation
- Calculate the latest finish in Split workstation using backward approach.
- Calculate the earliest finish for all jobs in Split workstation.
- Calculate the earliest start in Split workstation.

The following are some of the results of scheduling based on constraint resource using alternatives 4.

Table 4. The example of result of scheduling

Job Code	Operation	Workstation	Start	Finish
J01	1	Inspection 1	0.00	24.36
	2	Split	25.92	55.63
	3	Jet Dyeing 1	58.90	320.30
	4	Centrifugal	326.85	347.80
	5	Setting 1	353.40	408.06
	6	Stenter 1	409.24	471.17
J04	1	Inspection 2	0.00	54.07
	2	Split	55.63	85.34
	3	Jet Dyeing 2	88.61	350.01
	4	Centrifugal	356.56	377.50
	5	Setting 2	383.10	437.77
	6	Stenter 2	438.95	515.88
J07	1	Inspection 1	24.36	83.78
	2	Split	85.34	115.05
	3	Jet Dyeing 3	118.32	379.72
	4	Centrifugal	386.27	407.21
	5	Setting 1	412.81	467.48
	6	Stenter 1	471.17	528.09
J10	1	Inspection 2	54.07	113.49
	2	Split	115.05	144.76
	3	Jet Dyeing 4	148.03	409.43
	4	Centrifugal	415.97	436.92
	5	Setting 2	442.52	497.19
	6	Stenter 2	515.88	577.80

Table 4. The example of result of scheduling (cont.).

Job Code	Operation	Workstation	Start	Finish
J13	1	Inspection 1	83.78	143.20
	2	Split	144.76	174.47
	3	Jet Dyeing 5	177.74	439.14
	4	Centrifugal	445.68	466.63
	5	Setting 1	472.23	526.90
	6	Stenter 1	528.09	605.02

6. Calculate the buffer time in Drum.
   Recommended size of buffer time is ≤10% from total lead time in actual condition, however the optimum size is appropriate with the requirement of shop floor [3]. In this research, the buffer time be appointed as 5% from total lead time.
7. Calculate the unit load with trial and error.
   Unit load aims to get the optimal amount to be moved and processed from one work station to another work station. In actual conditions the company uses a unit load of 12 rolls or fit to maximum capacity of MHE. The exploitation of the constraints is done by delivering the fabric and processing it in a Split Workstation in the amount of 3, 4, or 6 based on lot packaging of fabric and minimal fabric capacity in jet dyeing machines. Following is the result of unit load:

Table 5. The result of unit load determination.

Unit Load	Makespan
12 rolls (actual condition)	2121.27
6 rolls	1920.11
4 rolls	1903.47
3 rolls	1901.47

Based on calculation of unit load, we have the best unit load with the shortest makespan which three rolls per delivery.

Overall, the following is the comparison between the actual condition and proposed condition:

Table 6. The comparison between actual and proposed condition.

Performance	Actual Condition	Proposed Condition	Decrease
Manufacturing lead time average	1024.04 minutes	390.31 minutes	61.88%
Queue time average	435.55minutes	76.41 minutes	82.45%
Lateness	252 rolls	162 rolls	35.71%
Makespan	2121.27 minutes	1901.47 minutes	10.36%

4. Conclusion

In the actual condition, a set of jobs that must be processed do not have a clear job sequence, so that the sequence of the job is done randomly. This has resulted in delays in producing fabric due to a bottleneck, delays in completing orders at one of the workstation impacting on the completion of orders at other work stations.

The use of drum buffer rope method in the proposed condition with the initial step of making the workstation constrained as a drum that functions as a control point, can improve the overall performance of the system. In the proposed conditions, improvements are made by exploiting the constraint system, by scheduling jobs at the workstation constraints first, then doing rope or backward scheduling to determine the right time so that the job arrives at the Split Workstation according to the machine's ability. To anticipate delays that occur at workstations before the Split Workstation, buffer time is applied so that the job arrives faster as if there is an optimum buffer stock in the Split Workstation. Finally, forward scheduling from Split Workstation to the last workstation and continue to improve system performance by calculating the unit load so that the best quantity can be obtained that minimizes system constraints and improves the whole system. The result of this research is the manufacturing lead time decrease is by 61.88 percent from the current condition, queue time decrease is by 82.45 percent from current condition and the lateness decrease is by 35.71 percent from current condition.

5. References

[1] Chang Y C and Huang W T 2011 Using simplified drum-buffer-rope for re-entrant flowshop scheduling in a random environment *African Journal of Business Management* **5**(26) pp 10796-10810

[2] Rieswien R, Suryadhini P P and Juliani W 2014 Perancangan Sistem Scheduling Job Menggunakan Drum Buffer Rope untuk Meminimasi Keterlambatan Order dan Manufacturing Lead Time pada Bagian Machining MPM di PT. Dirgantara Indonesia *Jurnal Rekayasa Sistem dan Industri* **1** pp 14–21

[3] Bhardwarj A, Gupta A and Kanda A 2010 Drum-Buffer-Rope: The technique to plan and control the production using theory of constraints. *Int. J. of Industrial and Manufacturing Engineering* **4**(9) pp 784-787

[4] Sipper D and Bulfin R L Jr 1997 *Production Planning, Control and Integration* (New York: The McGraw-Hill) pp 577

[5] Schmitt T G 1984 Resolving uncertainty in production systems *Journal of Operations Management* **4**(4) pp 331–345

[6] Chang C A, Brown L G and Johnson L R 1983 Analysis of buffering techniques in MRP systems with a matrix loading model. *IIE Trans* **15** pp 305–312

[7] New C and Mapes J 1984 MRP with high uncertain yield losses *Journal of Operations Management* **4** pp 315–330

[8] Baker K R and Trietsch D 2009 *Principles of Sequencing and Scheduling* (New Jersey: John Wiley and Son)

Line Balancing with Reduced Number of Operator: A Productivity Improvement

A Hasta and Harwati[1]

Industrial Engineering Department, Universitas Islam Indonesia, Yogyakarta, Indonesia
[1]harwati@uii.ac.id

Abstract. The Objective of this research was to improve the productivity and efficiency in a line process of a case study. The observation showed that current efficiency was about 87% and operator productivity was about 13.3 pcs/hour/man power. Productivity improvement can be reached by three step. First, identification and reduction of non-added value activity. There are 13 non value added activities which are prioritized to be eliminated based on Pareto principles. Second, the reduction number of operators while doing the third step: reallocate the work of each operator after reduction. Operator reduction is done from 3 to 2 operators only. Then the job that is already efficient from the stepped operator is moved to two other operators. After repairing and balancing tasks for each operator, the efficiency value becomes 96% and productivity increases to 20.5 pcs/hour/man power and man power savings to two men power.
Keywords: Productivity, Line Balancing, Operator, Reduction, Reallocation

1. Introduction

PT.X is one of the automotive companies that has been established in 1982, which located in North Jakarta. As the development of automotive industry in Indonesia increases, PT X has started to improve its automotive business by establishing supplementary company. PT. X Group that located in Karawang has 2 plants, namely Plant PT. X Karawang Plant 1 A that produces Machining Fly Wheel and Hub Front for Light Passenger Car, Front Axle Parts for Light Duty Truck, Differential Case and Differential Carrier, while the PT X Karawang plant 1B produces Machining of Rear Axle and Propeller Saft for SUV and MPV. With various of products, PT. X has a vision to be able to compete with other companies to become worldwide Rear Axle & Propeller Shaft supplier and to become primary partners in Indonesia. Hence, to accommodate that vision, PT. X should have annual activity plan.

In 2018, the company has set several targeted activities, such as: focus to man power's competencies, Improved working environment and Strengthen Core Value, level up people to level up performance, dojo level up knowledge and skill, productivity improvement, improvement of management and target that correlated with financial, internal process, QCD excellence as well as learning and growth. From above several targets, PT. X is focused to productivity improvement that aims to improve the productivity since it becomes one of the aspects that determines the success of a company in the fiercer competition in the industrial world. The level of productivity that achieved by a company is the indicator of company's efficiency in combining its economics resources.

Productivity as a traditional formulation on the entire productivity, the ratio of output towards entire input [1]. Therefore, beside optimal output, productivity also depends on optimal input, which is, among others, man power productivity. Operator's productivity could be calculated by multiplying the total numbers of operator with working hour of each operator. In this case, to measure the productivity, numbers of operators could influence numbers of products that could be manufactured in one hour.

Later, to increase its productivity, PT. X keeps on persisting to create a target to establish its resources effectively and efficiently by reducing numbers of operators on each production process.

Target on operator's optimization to improve productivity in PT. X, is implemented also in fly wheel production that located in Blok B Plant 1 A. Initial condition recorded that there were 3 operators involved in production process with 40 outputs every hour. Numbers of operators will be reduced to 2 operators in accordance with productivity improvement in line of fly wheel 3. The cycle time will be improved with the reduction and yet the output will be diminished. Hence, a certain method is required to maintain stable output of 40 products/hour and stable cycle time with only 2 man powers. It could be achieved by decreasing the idle time, improving the efficiency and smoothing the production line by using line balancing. Line balancing is a methodology to increase the availability of the existing line. Line balancing results in maximum effectiveness of the equipment [2]. The operator reduction is one from five ways to solve line balancing problems [3]. The improvement of efficiency and productivity could be improved by reducing the waste activities.

The approach that can be used to eliminate waste and increase the productivity is Toyota Production System (TPS). TPS is a system that developed by Taichi Ohno in 1950 that has successfully applied in Toyota company. This system is one of the key factors that have made Toyota as a successful company. The main idea of Toyota Production System is to eliminate the waste activities to identify the potential issues by reducing the supplies. In order to explore real case and regulate it in organizational culture, TPS performs sustainable improvement. Two important factors on TPS is Just In Time (JIT) and Jidoka [4]. Toyota Production System has main goal to reduce cost or productivity improvement. The cost reducing and productivity improvement could be established by omitting the waste activities, such as excessive man powers. Later, Toyota Production System is an approach that will be applied as basic steps to improve productivity at the fly wheel 3 production.

This research is focused on the productivity improvement by performing resources efficiency at the fly wheel 3 production process in PT. X. The researcher applies line balancing method using tools of Yamazumi chart and waste subtraction as well as work standardization with Toyota Production System approach. Those methods are expected to be able to assist PT. X in achieving one of its targets that described in activity plan 2018.

2. Research Question

The questions that should be answered in this research is how much improvements on productivity and line efficiency at the fly wheel 3 production process in PT. X by using line balancing method and Toyota Production System. Later, to response the question of the research, it is conducted to achieve below three objectives:

1. To identify actual condition at the fly wheel 3 production process in PT. X
2. To suggest the improvement to reduce *waste* and process improvement at the fly wheel 3 production process in PT.X
3. To perform efficiency at the fly wheel 3 production line as well as to identify the productivity condition and future line efficiency based on fly wheel 3 production process improvement in PT. X

3. Method

Line balancing is an adjustment towards work elements assignments from one assembly line to work station that addressed to minimize numbers of work stations and to minimize the idle time on all stations in the certain output level. Line balancing could enhance the process efficiency by minimizing work stations, work cycle and maximize working load as well as improving flexibility among work stations [5]. Efficiency of production line could be managed by using following formula [6].

$$\textit{Line Efficiency}: \frac{\textit{Total Work Station CT}}{\textit{Takt time X Numbers of Workers}} x100\% \tag{1}$$

Efficiency covers two basic concepts, which are cost reduction followed by the elimination of waste activities and the full empowerment on man powers' capability [7]. Efficiency involves concept or cycle in reducing numbers of workers. The first step suggests the omission of useless operation, followed by operation reallocation. Final step advices the reduction of man powers.

IOP Conf. Series: Materials Science and Engineering **528** (2019) 012060 doi:10.1088/1757-899X/528/1/012060

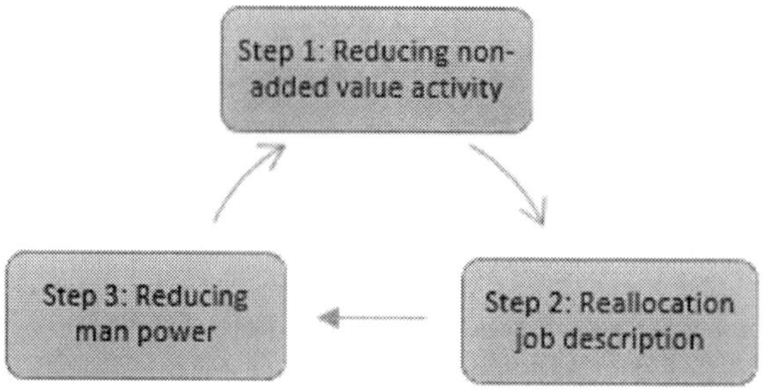

Figure 1. Cycle of Man Powers' Reduction on Toyota Production System [8]

The productivity improvement in this research is carried out through several stages, which are measurement on line efficiency and initial productivity, identification on value – non value activities towards each operator (man power), operator reducing and job reallocation. Eventually, final stage covers the measurement of line efficiency and productivity after treatment. Therefore, final condition could be compared with the initial condition.

4. Result and Discussion

4.1. Line efficiency and productivity in initial condition

Before calculating the line efficiency and initial productivity, initial production cycle time and operators' cycle time should be previously measured. Cycle time is derived from bottleneck time or the longest time at the fly wheel 3 production process. It is identified that the bottleneck at the fly wheel 3 production process occurred on the operator 2, with machine cycle time of 90 seconds. Thus, it can be concluded that the cycle time is 90 seconds.

The measurement of cycle time is conducted on each total 3 operators. The observation is carried out five times on that production process. Following is the result of observations on total cycle time for each operator:

Table 1. Cycle Time for Each Operator

Number Operator	Mode	Average
Operator 1	65	67.8
Operator 2	90	92.2
Operator 3	80	83.6
Total	235	

After initial cycle time is obtained from previous calculation, initial line efficiency and productivity could be formulated to be compared with targeted line efficiency and productivity after the implementation of proposed improvement.

$$1. \frac{Output}{hour} = \frac{3600s}{90s} = 40\ unit/hour$$

$$2. Operator\ Productivity = \frac{Output}{Operator\ Input} = \frac{40}{3} = 13.33$$

$$3. Line\ Efficiency\ (\%) = \frac{Total\ CT\ operator}{Longest\ CT\ xnumber\ operator} x100\% = \frac{235}{(90)x2} x100\% = 87\%$$

4.2. Identify Non value added activity

The efforts to enhance the productivity is began with the identification of non-value added activity on every operator. By eliminating non-value added activity will reduce cycle time. The omission of *waste* is conducted by identifying work elements to be value work, non-value work and walking. The examples of non-value added activities are part wiping, part picking up, component's marking, etc. Below is the identification result of value work, non-value work and walking on every operator.

Table 2. Recapitulation of *value work, non-value work* and *walking operator*

	Operator 1		Operator 2		Operator 3	
	Total Time in second	Total Time in %	Total Time in second	Total Time in %	Total Time in second	Total Time in %
Value Work	123	65.43%	228	71.70%	228	71.70%
Non-value work	60	31.91%	82	25.79%	82	25.79%
Walking	5	2.66%	8	2.52%	8	2.52%

Later, activities on every operator's work elements will be described in pareto diagram. It will be employed to recognize the activity that experiences the longest time. It will be applied as the basic of *waste* elimination and the improvement on designated activity. Below is the pareto diagram represented for each operator.

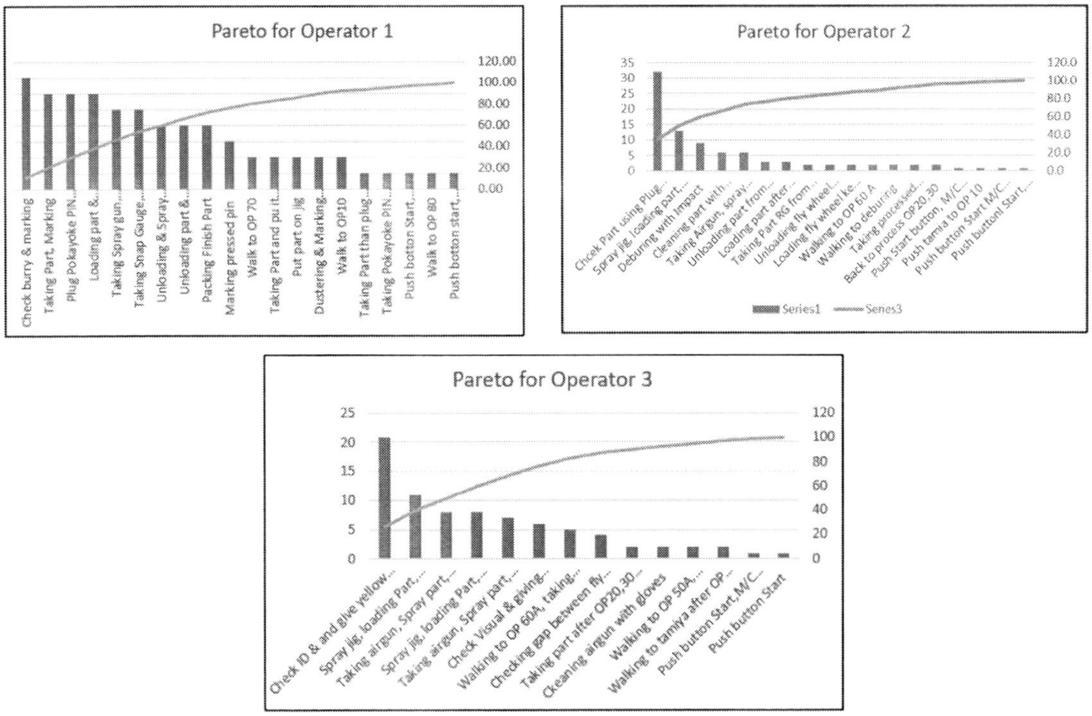

Figure 2. Pareto Diagram for Operator Non Added Value Activity

Productivity improvement is made by eliminating the non-value added activities using Pareto principles. For instance, operator 1 performs the biggest non value added activity on part marking that carried out manually during the time. Therefore, to eliminate it, stamp is suggested to be used in reducing marking time up to 2 seconds.

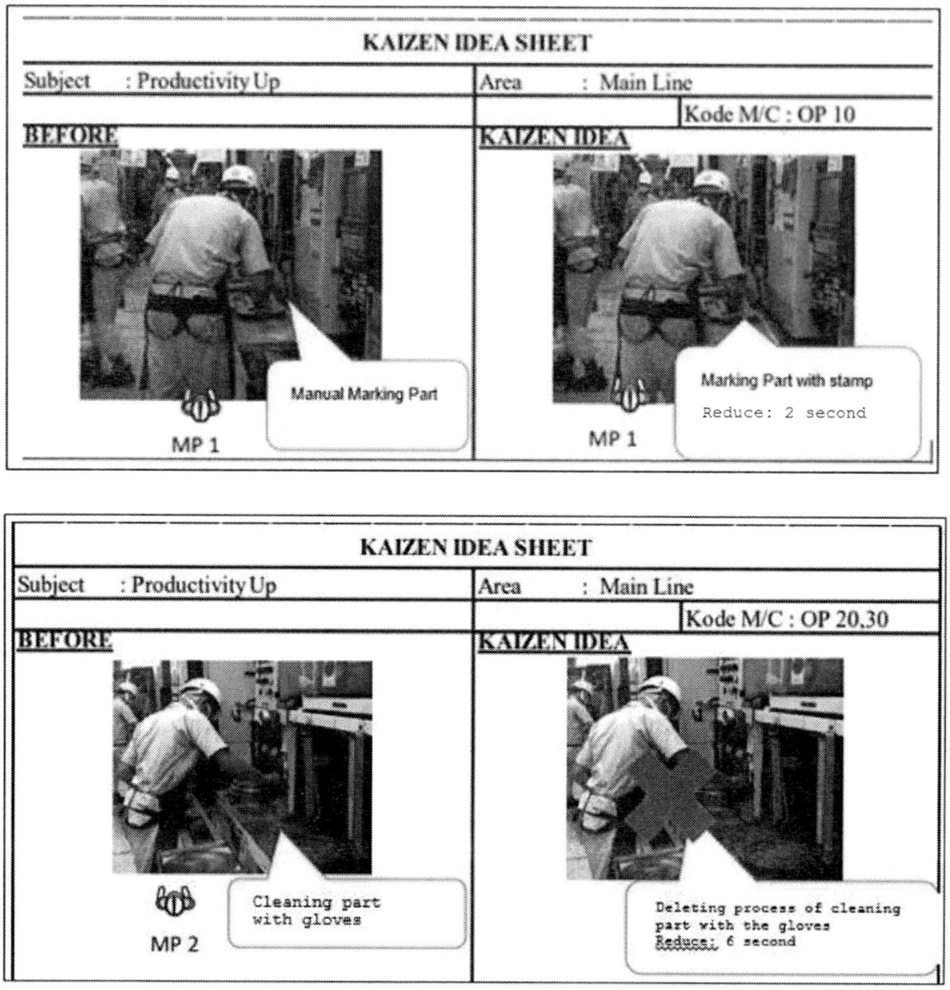

Figure 3. Sample of Improvement Activity

4.3. Operator Reduction

4.3.1. Yamazumi chart Operator current condition

Following is the *yamazumi* chart on the actual condition before improvement:

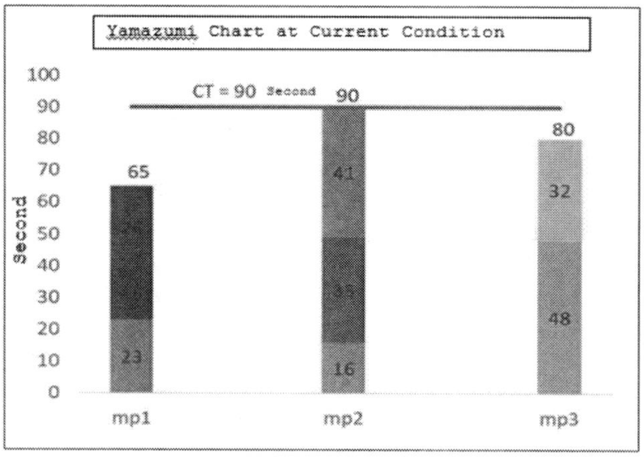

Figure 4. *Yamazumi chart Man power*

From the calculation of cycle time after *waste* elimination and process improvement, *yamazumi* chart could be implemented, the chart after *waste* elimination and process improvement are employed as tool for line balancing process. It could be performed by reallocating one of the operators on the next step.

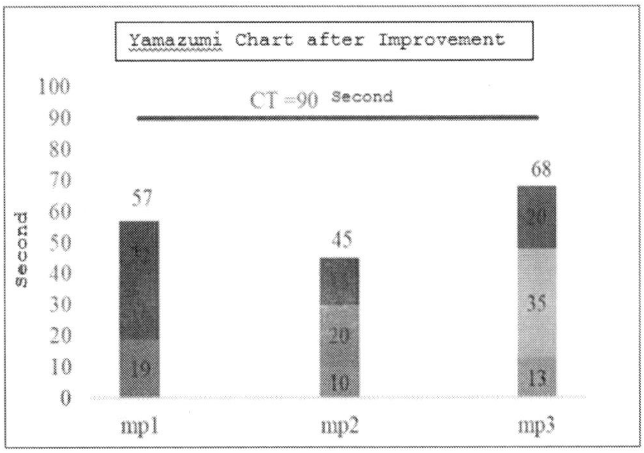

Figure 5. *Yamazumi chart after kaizen*

After *yamazumi* chart operator is identified once after the waste elimination and process improvement are carried out, later, operation reallocation could be performed from one operator to another operator. It could be seen from remaining time of initial targeted cycle time on each man power. The operator reallocation is conducted to balance the inter operator line in order to improve line efficiency. In this case, operation that conducted by operator 2 will be relocated to operator 1 and operator 3. Below is the operator reallocation plan described by *yamazumi* chart operator after improvement:

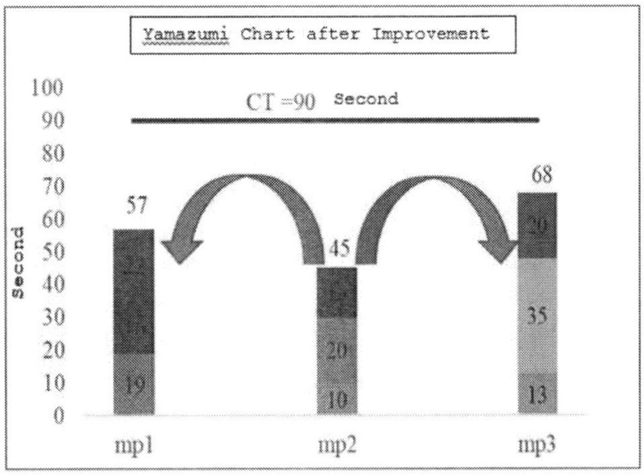

Figure 6. *Yamazumi* chart after reallocation

Work element and cycle time that already calculated will be later implemented in *yamazumi* chart to understand each operator's cycle time, whether it reaches targeted cycle time or not. Below is *Yamazumi* chart operator after improvement:

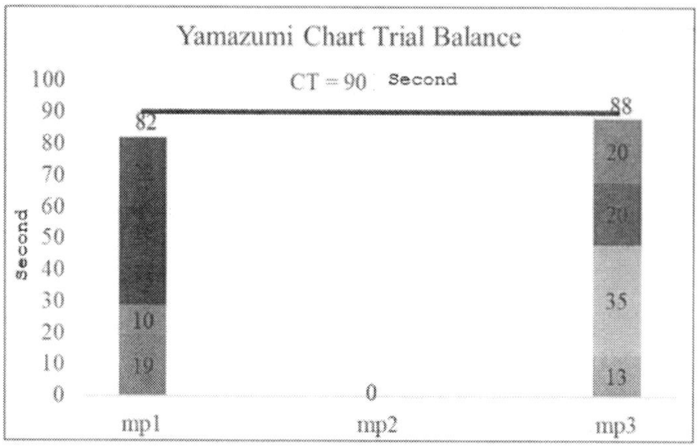

Figure 7. *Yamazumi chart Man power* after improvement

4.4. Line efficiency and productivity after improvement

Below is the calculation on efficiency line and operator productivity after improvement

1. $\frac{\text{Output}}{\text{hour}} = \frac{3600\text{s}}{88\text{s}} = 41 \text{ unit/hour}$
2. $\text{Operator Productivity} = \frac{\text{Output}}{\text{Operator Input}} = \frac{41}{3} = 20.5$
3. $\text{Line Efficiency } (\%) = \frac{\text{Total CT operator}}{\text{Longest CT xnumber operator}} \text{x}100\% = \frac{170}{(88)\text{x}2}\text{x}100\% = 96\%$

Below is the comparison graph between line efficiency initial condition and after improvement

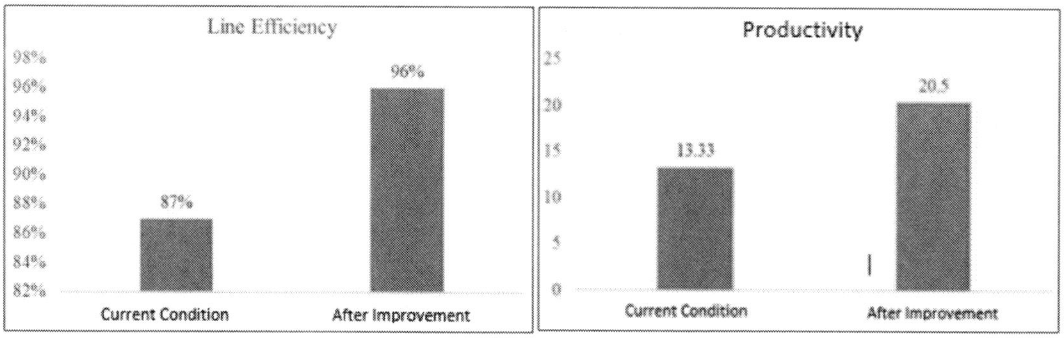

Figure 8. The comparison of line efficiency and productivity before and after improvement

From above figure, it can be notified that initial condition of line efficiency is 87% with 3 man powers employed. It experiences a raise to 96% with 2 man powers, it raises up to 9%. The improvement of line efficiency value is caused by the work elements balancing on every man power.

5. Conclusion

Hence, the conclusion could be drawn as follows:

1. The level of productivity before improvement is 13.33pcs/hour on every operator at the fly wheel 3 production process in PT. X and becomes 20,5 pcs/hour on every operator, after improvement
2. The level of efficiency at fly wheel 3 production process before improvement is 87% and raises to 96% after improvement.

6. References

[1] Kapyla J, Jaaskelainen A and Lonnqvist A 2009 Identifying future challenges for productivity research: evidence from Finland. *International Journal of Productivity and Performance Management* **59**(7) pp 607-623

[2] Reddy K D, Bhupal M, and Naidu G C 2018 Productivity Improvement by Using Line Balancing and Automation Strategies in order to Improve Overall Equipment Effectiveness. *International Journal of Engineering Science Invention (IJESI)* **7**(1) pp 21-27
[3] Tasan S O and Tunali S 2008 A Review of the Current Applications of Genetic Algorithms In Assembly Line Balancing. *Journal of Intelligent Manufacturing* **19**(1) pp 49–69
[4] Liker J K and Meier D 2006 *The Toyota Way Fieldbook: A Practical Guide for Implementing Toyota's 4Ps.* (New York: McGraw-Hill)
[5] Adeppa A 2015 A Study on Basics of Assembly Line Balancing. *International Journal on Emerging Technologies* pp 294-297
[6] Adnan A N, Arbaai N A and Ismail A 2016 Improvement of Overall Efficiency of Production *Line* by using Line Balancing. *ARPN Journal of Engineering and Applied Sciences* pp 7752-7758
[7] Sugimori Y, Kusunoki K, Cho F and Uchikawa S 1977 Toyota Production System and Kanban system Materialization of just-in-time and respect for human system. *The International Journal of Production Research* pp 553-564
[8] Monden Y 1995 *Sistem Produksi Toyota.* (Jakarta: Pustaka Binaman Pressindo)

Design of Computer Aided Process Planning System for Holster Mold at PT. Carnegie Universal Industries

Amal Witonohadi[1], Nanang Ali Sutisna[2] and Martulan Suryanto Naibaho[3]

Department Industrial Engineering, Trisakti University
[1]amal@trisakti.ac.id [2]snanang@yahoo.com, [3]martulanmsn@gmail.com

Abstract. PT. Carnegie Universal Industries manufactures holster products and magazine carrier. Holsters have the most variants and the highest demand. To produce a holster, a mold is needed for each type of handgun. A mold uses wooden mold and blue gun. Mold making takes the longest time compared to other stages in the holster production process. The mold making process does not yet have standards, therefore it takes a long time and high cost. The purpose of this study is to develop a process planning to facilitate the making of molds with automated systems using the Computer Aided Process Planning (CAPP) method, speed up production time, and reduce production costs. The CAPP generative approach is generated through decision logic, formulation, search, algorithm and geometric based data. Database programs are created to store detailed information on each print that has been created. This approach design proved that the mold making procedures are simplified and speeded up and have standards, so that the time and costs are used more effectively and efficiently.
Keywords: CAPP (Computer Aided Process Planning), Holster Mold, Geometrics Database, Generative method.

1. Introduction

PT. Carnegie Universal Industries have 2 brand namely Pax Militari for weapon accessories and Pax Dynamic for gear equipment. Pax militari consisting of two the kind of product that produced the holster and magazine carrier with some variants that uses a material of KYDEX®, Holstex® and Boltaron®. The material of KYDEX® for holster has highest customer demand. A type of product a holsters containing KYDEX® material produced by several phase in the process of production. The phase that done in making molds, heating raw materials, forming, cooling, cutting, the assembling of, and finishing. The phase that was done among manufacture of mold, warming up, forming, cooling, cutting, assembling, and finishing. Now, mold used are a mold manual with timber and blue gun which have developed in accordance with the desired pattern.

A manual pattern from blue gun and timber set formed to the process of making mold phases. Making manual mold need around 2 hours, due to complexity and the number of cavity that is covered. Making patterns in process of a holster conducted twice, because a holster needs a side front and back of a weapon to load details. Holsters with manual mold used two sheets of KYDEX® material and have a mistake while cavity covered and improper pattern will result in defect of product. A defect which a manual mold is holster not conforming so that gun loose and details weapons are not formed.

The use of a manual mold have a really effect on the effectiveness and efficiency of the production. Time for used pattern is relatively longer and the use of two pieces of KYDEX® material considered wasteful because of KYDEX® material thrown out when cutting process. Holster that produced by PT. Carnegie Universal Industries maximum 4th product a day. This process is really had an influence on the quality and the speed of holster production PT. Carnegie Universal Industries. The company wanted increases the quantity and quality of production manual mold.

2. Problem Case

At the moment blue gun mold is made manually and takes long time and high cost, so that the efficiency is low. This study aims to improve the manual production of mold by automation system using Computer Aided Design (CAD), Computer Aided Process Planning (CAPP), and Computer Aided Manufacturing (CAM). The goal is to produce a better and more flexible planning system. The product focus of the study is the holster XD-9M.

2.1. Process Engineering

The planning process is a function in setting in the manufacturing process and determined process or parameters which are used to change early part to the end one, which was preceded by mechanical drawing [1]. The planning process of constituting the determination of the process of assembling and the manufacture and the ranking of in which this process should be completed to finish the product until the final form [4]. The steps of the planning process covering; (1) interpretation picture design, (2) the process and the order (3) the selection of equipment (4) the selection of tools, dies, mold, and gages, (4) the method of analysis, (5) Standard, and (6) Cutting tools and cutting condition

2.2. Process Planning

The planning process also called manufacturing planning, material process, process engineering, dan machine routing. Process engineering consisting of the process machinery and parameters used to convert early form part into the end of part based on mechanical drawing. The whole process needed preparation to declare an order to producing part and how to realize product design [2] Figure 1.

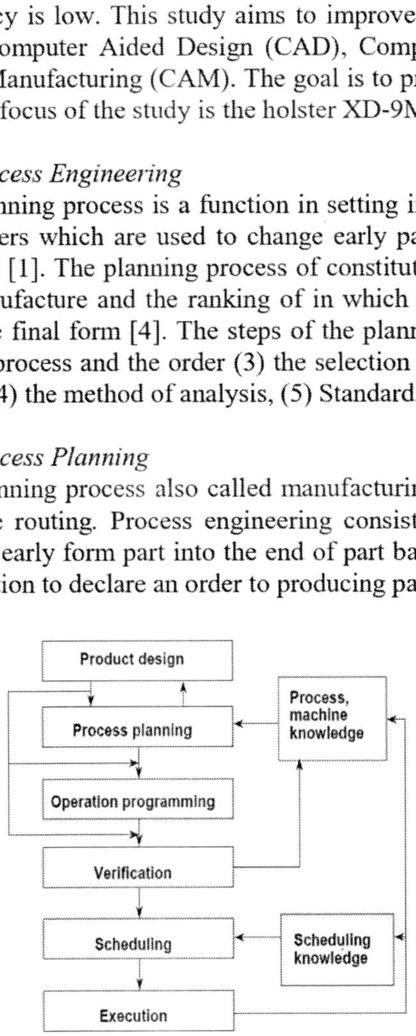

Figure 1. Production Flow Chat (Chang, 1998)

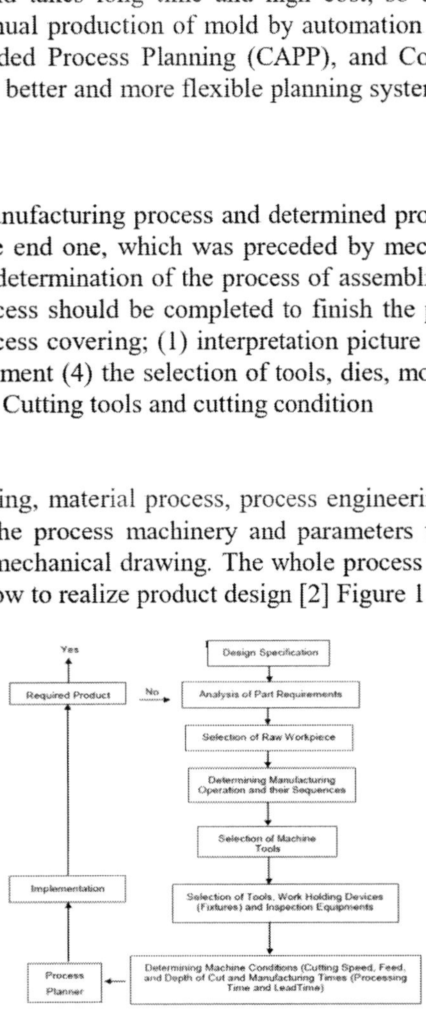

Figure 2. Process Planning Flow chat (IGNOU, 2017)

Process planning in manufacturing production is the phase that use to product design and processed to be product. Process planning of using the engine selection parameters, the selection of tools, scheduling and control the production as a Figure 2. A part of process planning [2] consisting of: Route, Processes, Process Parameters, Machine and Tool Selections, Fixtures. The form of process planning used namely operation planning table that containing operation description, tools, machine, process parameter, machinery time, cost etc. Detail of process planning is presented in Figure 3.

2.3. Computer-Aided Process Planning (CAPP)

Process planning defined as a function in a series of manufacturing activity which it was decided that production process and the parameters further used to transform a material from originally into a shape in accordance with the desirable design [2]. Process planning of constituting liaison between the design and production. Process planning should be optimal, in which production based on this process, the last part should be can produced at the right time and lower cost of production. All will lead to the need for a computer based system that can function planning this process or can be called Computer-Aided Process Planning (Figure 4)

PROCESS PLAN ACE Inc.

Part No. S0125-F Material: steel 4340Si
Part Name: Housing
Original: S.D. Smart Date: 1/1/89 Changes: ______ Date: ______
Checked: C.S. Good Date: 2/1/89 Approved: T.C. Chang Date: 2/14/89

No.	Operation Description	Workstation	Setup	Tool	Time (Min)
10	Mill bottom surface1	MILL01	see attach#1 for illustration	Face mill 6 teeth/4" dia	3 setup 5 machining
20	Mill top surface	MILL01	see attach#1	Face mill 6 teeth/4" dia	2 setup 6 machining
30	Drill 4 holes	DRL02	set on surface1	twist drill 1/2" dia 2" long	2 setup 3 machining

Figure 3. Process Plan (Chang, 1998)

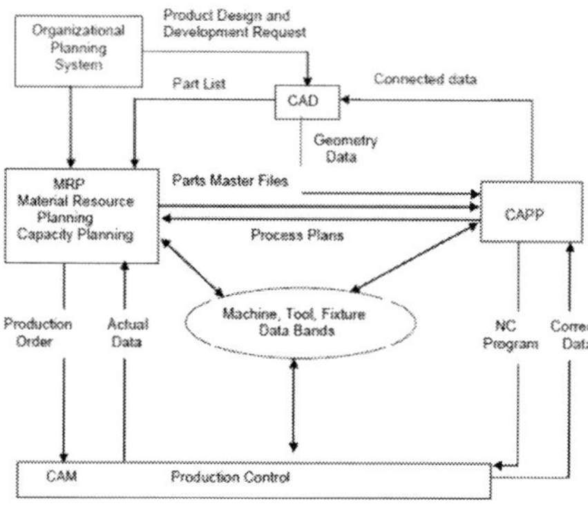

Figure 4. *Computer-Aided Process Planning* Flow chat (IGNOU, 2017)

a. *Metode Varian*

A method of variant than any manual with the methods, where new plan the process of parts arranged with identifying and looking for a plan that already exists in a database to part the same (is often called master part) categorized based the concept of group technology and make the modification which required to a new part. One of the weakness of a method of variant is the quality of the process plan are still highly dependent on planner the process. Computer is just be used as a device to help as manual processes planning activities. But, this method is still many be used, a number of reasons that form the basis of such as [3].

1. Hardware investment and small software. And there are many vendors to supports then the generative system.
2. Faster time to develop and unneeded many power. Installation easier than generative system
3. In a few instances, variant are more reliable system to be applied to the environment of production that is real, that to low class especially for corporations.
4. The quality of the plan process that is extremely dependent planner with knowledge and experience of the process

b. *Method of Generative*

A method the process of generative planning result through logic a decision, formulation, search algorithms, and data base of geometric designs. On the whole, the format of the input of a CAPP system can be divided into two categories all over the place, text input and graphic input. The text input is defined as interactive, and the graphic where data part is drawn in through cad module (is defined as the input interface) [3] .

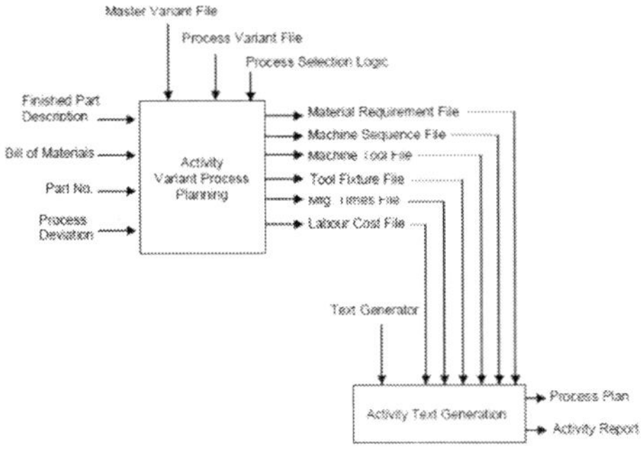

Figure 5. Varian of CAPP Flow chat (IGNOU, 2017)

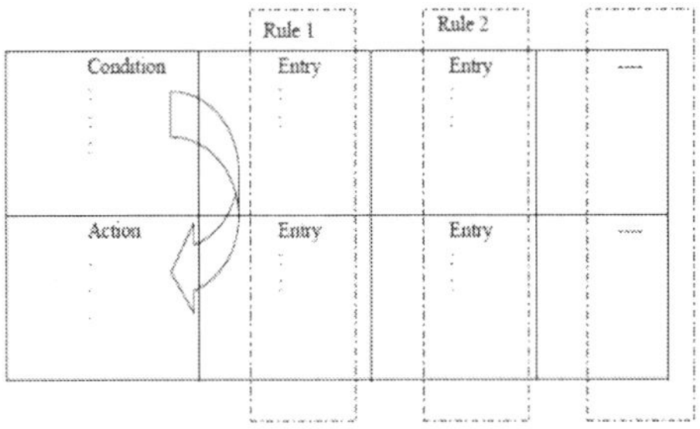

Figure 6. Generative of CAPP Flow chat (IGNOU, 2017)

Planning process with the methods generative do not used plan a standard for planning new components. This approach used manufacturing to produce a process planning. Its method try to imitate a planner the process who make plans the process of new part based on knowledge possessed [6]. In this method, plan a process not are kept on database, but database consisting of information on part, machine, tools and well as the rules process planning. Approach planning application process generative is a fit of a system based on knowledge [7].

c. ***Method of Semi – Generative***

This method is approach while when there was the problem in system development planning process by using the method generative. A method of semi-generative can define as an application latest of technology variant by using type features generative. A system that uses a method of spring generative must be more being together with planner, where having knowledge technology. Responsibility planner is interpretation of data decision or image work.

3. Methods

A stage of methodology is;

1. Design Product by CAD
2. Coding design with fixture of product design
3. Part programme design made G-Code model with CAM tools
4. CAPP plan model
5. Result Operation Process Chat (OPC)
6. Compare production time between manual and automation of mold making.
7. Compare product cost between manual and automation of mold making.

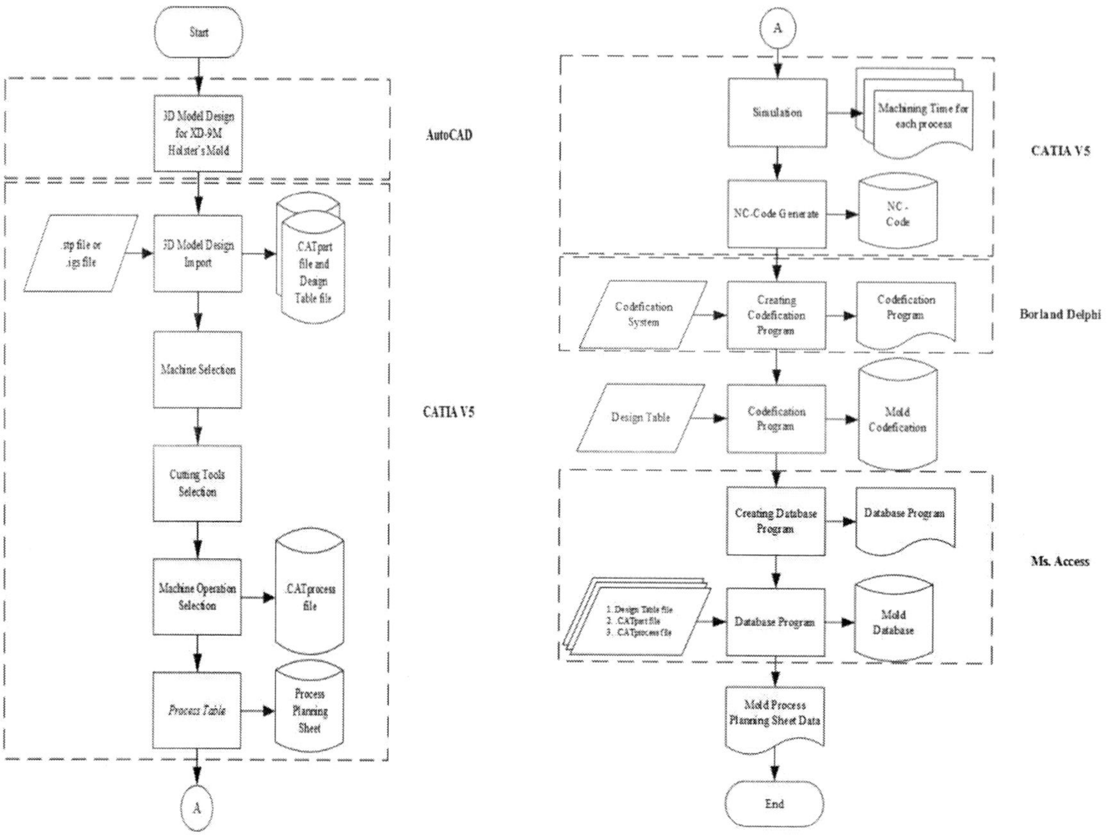

Figure 7. CAPP System Design Flow chat

4. Result and Discussion

4.1. Product Design

Figure 8. Holster Springfield XD-9M

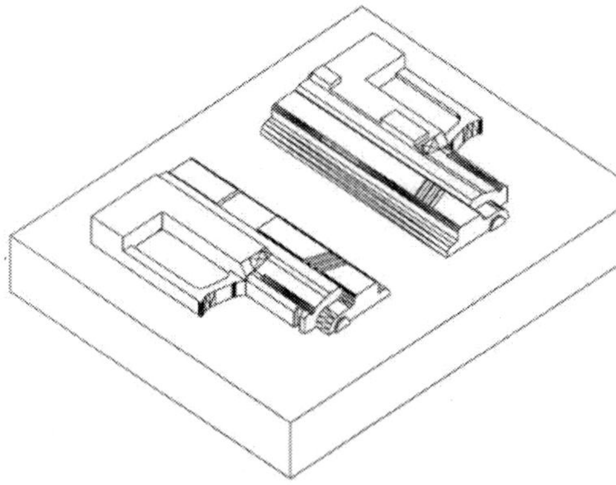

Figure 9. Holster Mold XD-9M

Activity	Value
Hole 1 Diameter (mm)	4
Hole 1 Hole Limit 1 Depth (mm)	36
Hole 1 Tap depth (mm)	10
Hole 2 Diameter (mm)	4
Hole 2 Hole Limit 2 Depth (mm)	36
Hole 2 Tap depth (mm)	10
Pad 1 First Limit Length (mm)	2.5
Pad 1 Second Limit Length (mm)	0
Pad 1 Thick Thin1 (mm)	1
Pad 1 Thick Thin2 (mm)	0
Pad 2 First Limit Length (mm)	2.5
Pad 2 Second Limit Length (mm)	0
Pad 2 Thick Thin1 (mm)	1
Pad 2 Thick Thin2 (mm)	0
Pocket 1 First Limit Depth (mm)	1.5
Pocket 1 Second Limit Depth (mm)	0
Pocket 1 Thick Thin1 (mm)	1
Pocket 1 Thick Thin2 (mm)	0
Pocket 2 First Limit Depth (mm)	1
Pocket 2 Second Limit Depth (mm)	0
Pocket 2 Thick Thin1 (mm)	1
Pocket 2 Thick Thin2 (mm)	0
Pocket 3 First Limit Depth (mm)	1
Pocket 3 Second Limit Depth (mm)	0
Pocket 3 Thick Thin1 (mm)	1
Pocket 3 Thick Thin2 (mm)	0
Shaft 1 Activity	true
Shaft 2 Activity	true

Figure 10. Design Mold Holster at Excel Format

4.2. Coding

A Step to design coding is access design table with format.xls than used parameter; pad, pocket, hole, shaft, chamfer, and fillet. Steps of coding process resulted table about operation information, coding for holster mold XD-9M. Then the final model of coding result is 0100000933011.

4.3. Part Programming (G-Code)

Holster XD-9M mold made by CNC machining process that the CNC Siemens and the Sinumerik840D to control system. Machining process arrange by CatiaV5 and G-Code program. The reason G-Code used is simple to purpose. The steps G-Code generation at CatiaV5 are process, relief, parameter used, simulation and generator of G-Code. For presentation a plan process needed a route sheet process. A route sheet collected number, name, material, tread machine, process, tools, and time each product.

IOP Conf. Series: Materials Science and Engineering **528** (2019) 012061 doi:10.1088/1757-899X/528/1/012061

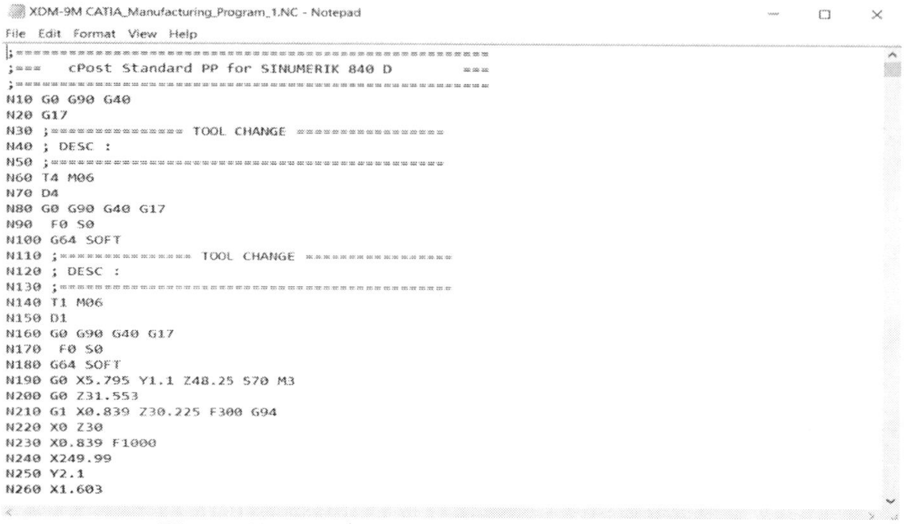

```
XDM-9M CATIA_Manufacturing_Program_1.NC - Notepad
File Edit Format View Help
;=========================================================
;===   cPost Standard PP for SINUMERIK 840 D          ===
;=========================================================
N10 G0 G90 G40
N20 G17
N30 ;=============== TOOL CHANGE ================
N40 ; DESC :
N50 ;==========================================
N60 T4 M06
N70 D4
N80 G0 G90 G40 G17
N90  F0 S0
N100 G64 SOFT
N110 ;=============== TOOL CHANGE ================
N120 ; DESC :
N130 ;==========================================
N140 T1 M06
N150 D1
N160 G0 G90 G40 G17
N170  F0 S0
N180 G64 SOFT
N190 G0 X5.795 Y1.1 Z48.25 S70 M3
N200 G0 Z31.553
N210 G1 X0.839 Z30.225 F300 G94
N220 X0 Z30
N230 X0.839 F1000
N240 X249.99
N250 Y2.1
N260 X1.603
```

Figure 11. G-Code Programme Holster Mold XD-9M

4.4. CAPP Programs

Thera are two steps to plan CAPP program that the preparation with analysis of information system and generative CAPP program. The information system planned from customer needs until to final product. Steps of identification and analysis are PIECES Framework, Use Case Diagram, Data Flow Diagram, Entity Resources Diagram, and Physical Data Model. CAPP method with the generative has been implementation at the company that facilities are computer, HPE material, CNC Milling machine, Pressing machine, vacuum, oven etc.

Figure 12. Coding Programme Screen

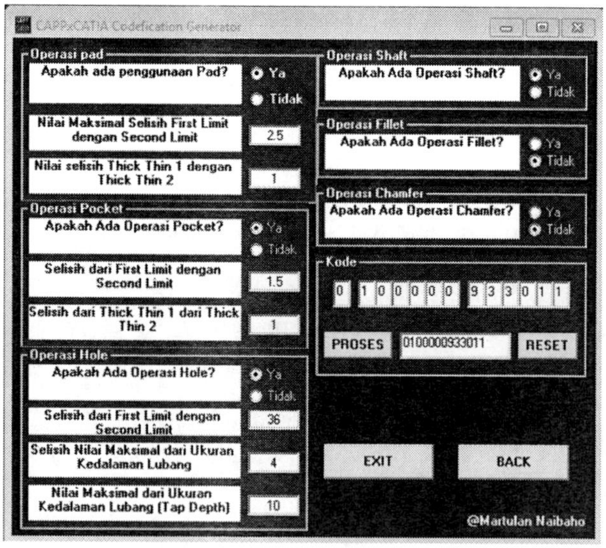

Figure 13. *Input* Data Screen

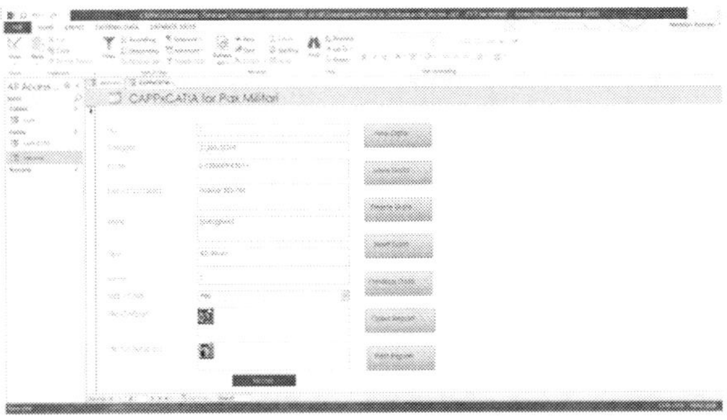

Figure 14. Screen of Input Data at Database Programme

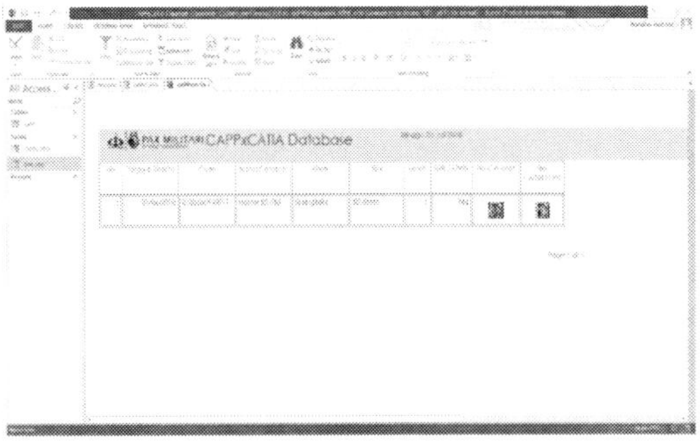

Figure 15. Screen of Database Programme Result

Result from implementation CAPP suggestion system at Table 1 and Table 2.

Table 1. Times Process Ratio Actual and Suggestion of Holster

Times Ratio Actual and Suggestion		
Operation Name	Actual (Minute)	Suggestion (Minute)
Mold Preparation	120	2
Melting and Casting Process	10	10
Setting and Rough Cutting	60	2
Finishing	60	1
Total	250	15

Table 2. Cost Ratio Holster Production

Cost Ratio Actual and Suggestion		
Operation Name	Actual	Suggestion
Material	Blue Gun	HPE
Description	Repetition Blue Gun yield mirror defect of mold could be product defect.	Used HPE hold up impact pressure and high temperature than restrain product defect
Cost	US $ 60 – 80	US $ 250-300
Capacity	1-3 product each of mold	10-30 product each of mold

5. Conclusion

1. The output produced by this study is a program code and database design parameter with Computer-Aided Process Planning (CAPP) system. The Database software systems were Borland Delphi and Microsoft Access.
2. The time to process a plan by CAPP decreases. With CAPP System the process to access the database is simple.
3. The time to produce a holster with HPE Mold using the standard process planning is increase from 262 minute to 169 minutes.
4. The time to produce a Holster Mold using CNC Milling Machine with G-Code programme is 125.2 minute. HPE mold by CNC Milling is the best choice to use. This is because it can be used repetitively, with lower risk defect compared to the blue gun mold method. The timing comparison between HPE and Blue Gun mold is 1:10.

IOP Conf. Series: Materials Science and Engineering **528** (2019) 012061 doi:10.1088/1757-899X/528/1/012061

6. References

[1] Chang, Tien-Chien. 1990 *Expert Process Planning for Manufacturing*. (United States of America: Addison Wisley Publishing Company)

[2] Chang, Tien-Chien., Wysk, Richard A., and Wang, Hsu-Pin. 1998 *Computer Aided Manufacturing*. Second Edition. (United States of America: Prentice Hall International)

[3] Gawlik, Edward. 2002 The Analysis of Methods for Computer Aided Process Planning. *Journal of Cracow* (Poland: University of Technology Cracow).

[4] Groover, Mikell P. 2008 *Automation Production Systems and Computer Aided Manufacturing*. Third Edition. Englewood Cliffs (New Jersey: Prentice Hall Inc.)

[5] Gu, P., and Norrie, DH. 1995 *Intelligent Manufacturing Planning* (United Kingdom: Chapman & Hall)

[6] Indira Gandhi National Open University. 2016 *Computer-Aided Process Planning*. IGNOU. India. Available: www.ignou.ac.in/upload/Unit-9-CRC.pdf [Accessed, August 7, 2018].

Design and improvement layout of a production floor using automated layout design program (ALDEP) and CRAFT algorithm at CV. Aji Jaya Mandiri

D Suhardini[1] and S D Rahmawati[2]

[1,2]Industrial Engineering Department, Universitas Trisakti, Jakarta 11440, Indonesia
[1]didien@trisakti.ac.id, [2]srhdn28@gmail.com

Abstract. Different computerized algorithms are used to assist the layout planner in generating alternate layout. CRAFT (Computerized Relative Allocation of Facilities Technique) has been developed to improve an existing layout, while ALDEP (Automated Layout Design Program) is a construction layout. The improvement given by ALDEP over CRAFT technique was found to reduce annual material handling cost by 23% [1]. This study hypothesize that better result could be gained by combined two algorithms rather than their independent application. The hypothesis is verified by taking layout problems of CV Aji Jaya Mandiri. The objective of the study is to get the optimal layout of CV Aji Jaya Mandiri using ALDEP and followed by improving it with CRAFT. ALDEP based on the value of proximity (total closeness rating). The layout resulted from ALDEP is improved by exchanging machines positions. The newly layout obtained by ALDEP is compared with the result layout improved by CRAFT. There are two criteria for comparing the layout: material handling cost taken from MHES (material handling evaluation sheet), and the manufacturing lead time or processing time using Promodel simulation. The material handling cost of the layout resulted by CRAFT was decreased by 6,24%, that is mean that CRAFT was improved the layout effectively. The result of simulation showed the manufacturing lead time of the improved layout decreased 23 minutes. Layout evaluation using Adjacency Based Scoring for the CRAFT layout increase 15% compare with the initial layout so it was selected as the best layout. This study concluded that combined use of computerized layout design algorithms provides better result,
Keywords: ALDEP, CRAFT, Facility layout Design, ProModel, Simulation

1. Introduction

Facility layout design is the field of selecting the most effective arrangement of physical facilities to allow the greater efficiency. It reduces the total cost of manufacturing activity provides optimum space to give maximum output. The plant layout or facilities layout can be defined as an arrangement of the factory facilities to support the smooth running of the production process. These settings will increase the utilization of the area for the placement of the machinery or other production support facilities, smooth movement transfer of material, material storage which is either temporary or permanent, operator and so on other personnel. Setting a good and planned layout will determine the efficiency of the production process. According to [2] a good layout will give more output although the cost is the same or less, less man hours, and or machine hours.

2. Method

The hypothesis is developing a new layout using construction routine followed by improvement routine is more fruitful than constructing new layout with a construction routine alone or improving a working existing layout using improved routine [1]. The hypothesis is verified by taking CV Aji Jaya Mandiri layout problems. The initial layout of production floor of the CV Aji Jaya Mandiri was not well-ordered. The flows of material are not efficient. The first step evaluated of current layout using adjacency based scoring. The optimal layout of CV Aji Jaya Mandiri was obtained using ALDEP and followed by improving it with CRAFT. Data collection for designing layout, such as the initial layout, characteristic of the product, operation process chart (OPC), standard times, capacities and size of machines. These data was processed to develop routing sheet, material handling planning sheet (MHPS), from to chart, total closeness rating, and material handling evaluation sheet.

The first method used to design the layout of the production floor is ALDEP (Automated Layout Design Program). This method is one of many construction algorithms for designing layout. The use of ALDEP method is expected to make a better layout proposal. After getting the results of ALDEP layout then calculate material handling cost based on material handling evaluation sheet (MHES). The proposed layout was improved by CRAFT and compare the cost of material handling of improved layout than the previous layout. Next, running a simulation to obtain the processing time of the proposed layout by ALDEP and the improved layout as well, which further evaluated the optimal layout using Adjacency Based Scoring to make sure the optimal layout the better layout.

3. Result and Discussion

The important things in facility layout design is reducing cost by maximizing adjacency of highly interacting components of a system or reducing material handling cost or distance between work stations. The first step is to evaluate the initial layout CV. Aji Jaya Mandiri using Adjacency Based Scoring method resulted the value of the relationship of proximity calculation. The priority scale is from FTC (from to chart) and MHPS (material handling planning sheet). Which for calculation using the following formula:

$$\text{Max } Z = \sum_{i=1}^{m} \sum_{j=1}^{m} fij\ xij \quad (1)$$

Table 1. The Value of Adjacency Based Scoring for Initial Layout

From		To	Inflow Coefficient	Rank	Rating	Score	Value of ABS
Finishing area		Circular Saw Machine	1.17	1	A	16	16
Radial Drilling Machine	-	Circular Saw Machine	1.04	2	A	16	0
Painting Area	-	drying area	1.04	3	A	16	16
Packing Area	-	Warehouse	1	4	A	16	16
Welding Machine		Finishing Area	0.96	5	E	8	8
Finishing area	-	Packing Area	0.94	6	E	8	0
Storage	-	Cutting Area	0.89	7	E	8	0
Cutting Area	-	Puttying Area	0.74	8	E	8	8
Patching area	-	drying area	0.69	9	E	8	8
Patching area	-	Puttying Area	0.51	10	I	4	0
Auxiliary storage	-	Painting Area	0.39	11	I	4	0
Finishing Area	-	Radial Drilling Machine	0.33	12	I	4	0
Puttying area	-	Painting Area	0.3	13	I	4	4
Auxiliary storage	-	Puttying area	0.26	14	I	4	4
Warehouse	-	Circular Saw Machine	0.2	15	O	2	2
Welding Machine		Radial Drilling Machine	0.15	16	O	2	2
Cutting Area	-	drying area	0.11	17	O	2	2
Auxiliary storage	-	Patching area Puttying area	0.07	18	O	2	2
Auxiliary storage	-	Packing Area	0.06	19	O	2	2
Auxiliary storage	-	Welding Machine	0.05	20	O	2	2
Auxiliary storage	-	Finishing Area	0.04	21	U	1	1
Patching area	-	Finishing Area	0.04	22	U	1	1
Finishing area		drying area	0.03	23	U	1	1
		Total Score				139	95
		Percentage of Adjacency Based Scoring				68%	

The value of the Adjacency Based Scoring for initial layout is 68% that is mean the layout must to be improved to reduce cost by maximizing adjacency of highly interacting machines or work station or reducing material handling cost or distance between work stations. Facility layout design requires diverse field of knowledge. Among other the application and use of computers become an advantage. The computerized layout methods, either construction or improvement-type routines are heuristics. Construction-type layout routine generates a block layout based on the relationship between different departments. Automated Layout Program design (ALDEP) is commonly use software for construction-type layout. A newly constructed layout alone will not be nearly optimal, the new construction must be evaluated by an improvement routine. The most popular improvement-type methods are Computerized Relative Allocation of Facilities Technique (CRAFT).

Before designing a new layout, the number of machines needed must be checked using Routing Sheet. From the Routing sheet of CV. Aji Jaya Mandiri the number of machines which need to be increase are 3 machines there are Grinding, adjustable work benches 1, adjustable work benches as seen below.

Table 2. Number of Machines

No	Machine	Number of machine/work station	Number of machine	Number of machine/Station	Number of machine (round-up)	Number of machine available	+/-
		Two Way	Bench				
1	Circular Saw Machine	0.087	0.286	0.373	1	1	0
2	Drying Area	0.295	0.128	0.423	1	1	0
3	Radial Drilling Machine	0.105	0.667	0.772	1	1	0
4	Spray Gun	0.071	0	0.071	1	1	0
5	Welding Machine	0.48	0.501	0.981	1	2	-1
6	Grinding Machine	0.929	1.462	2.391	3	2	+1
7	Adjustable Work benches 1	0	0.258	0.258	1	0	+1
8	Adjustable Work Benches 2	0	0.198	0.198	1	0	+1
9	Packing Area	0.037	0.084	0.121	1	1	0
10	Work Table	0.108	0.114	0.222	1	0	+1
11	Cutting Table	0.04	0.103	0.143	1	1	0

From to Chart made after the number of machines will be arranged, are obtained from the routing sheet and compared to the number of machines available. The space of Department of specified machine was defined and then calculated the distance between the departments multiplied by the number of products being transferred and the frequency of removal to get the cost of material handling between Department.

To determine the rating of closeness used coefficient of inflow that calculate from the chart above (formula 2), which can be seen below. The coefficient then used as closeness rating A,E,I,U and X. The quantitative methods used to design the layout, particularly concerning the transfer of material with the minimal possible distance. Furthermore, the results of calculations used as input from FTC Inflow, which for calculation using the following formula:

$$\text{FTC Inflow} = \frac{\text{FTC Cost}}{\text{total column per machine}} \quad (2)$$

Legend:

X	Storage	6	Grinding Machine
Y	Storage of Auxiliary Materials	7	Adjustable Work Benches 1
1	Circular Saw Machine	8	Adjustable Work Benches 2
2	Drying Area	9	Packing Area
3	Radial Drilling Machine	10	Work Table
4	Spray Gun	11	Cutting Table
5	Las Machine	Z	Warehouse

Table 3. From to Chart Between Department

From \To	X	Y	1	2	3	4	5	6	7	8	9	10	11	Z	Total
X			112087.1										47935.32		160022.4
Y						35063.2	19284.76	2713.32	12324.2	37772.92	12792.22	17606			137556.62
1					192643.4			264551.1							457194.51
2						21777.08		20770.56	30286.8				11033.49		83867.93
3			25127.4				54684.5	124288.5							204100.35
4				224471.4											224471.44
5								108412.5							108412.5
6			267998				208152					22597.9			498747.93
7				37017.2								11830.14			48847.34
8						22664.88			37098.6						59763.48
9														84842.52	84842.52
10											85208.68				85208.68
11										60391.18					60391.18
Z															0
Total	0	0	405212.5	261488.6	192643.4	79505.16	282121.3	520735.9	79709.6	98164.1	98000.9	52034.04	58868.81	84842.52	2213426.88

Table 4. Coefficient of Inflow

	X	Y	1	2	3	4	5	6	7	8	9	10	11	Z	Total
X			0.28										0.81		1.09
Y						0.44	0.24	0.69	0.16	0.48	0.16	0.22			2.39
1					1.06	0.86		0.51							2.43
2						0.27		0.04	0.38	0.14			0.19		1.02
3							0.19	0.24							0.43
4										0.29					0.29
5								0.95							0.95
6												0.43			0.43
7										0.47		0.23			0.7
8													0.62		0.62
9												0.87		1	1.87
10															0
11															0
Z															0
Total	0	0	0.28	0	1.06	1.57	0.43	2.43	0.54	1.38	0.16	1.75	1.62	1	12.22

ALDEP allows the department to provide specific placed and can enter a dummy. Design with Algorithm ALDEP is subdivided into 2 procedures i.e. procedures of selection and placement procedures [3]. Input is needed to run the software of ALDEP it is the maximum value of TCR (Total Closeness Rating), the number and area of the Department that will be iterated, spacious layout dimensions and the minimum degree of closeness of relationship that is used.

The first elected Department to get done at random, for the next selected departments taken from departments that have a relationship "A" with the Department's first elected, if there is more than one Department, which is worth A then the Department will be selected randomly, but when there is no Department of value A, then Department of equal value or the value is equal to the lower limit of the level of closeness that is already put before by the user will be elected. The next selected departments in a manner that is similar to the previous one, this process is continued until the entire Department goes into the layout.

Procedure for placement on the first Department of ALDEP i.e. placing selected at the top left corner of the layout and continued towards the bottom. The length and width layout is determined by the user who entered earlier. This is also known as placement process with vertical sweep pattern (vertical road pattern). Total Closeness Rating (TCR) score become input in using software ALDEP can be seen at Table 1 below. Space available that is 447.11 m2, but as an input of ALDEP it is inserted in the template that is the number of 117 unit with a length of 13 and a width of 9. The number of departments that will be as much iteration 14 departments. Minimum degree of relationship of the TCR already calculated earlier U (unimportant).

Tabel 5. Total Closeness Rating

From\To	X	Y	1	2	3	4	5	6	7	8	9	10	11	Z	Total
X		U	O	U	U	U	U	U	U	U	U	E	U	U	21
Y			U	U	U	I	O	E	U	E	U	O	U	U	31
1				U	A	A	U	E	I	U	U	U	U	U	51
2					U	O	U	U	U	U	U	U	O	U	12
3						U	O	O	U	U	U	U	U	U	11
4							U	U	U	I	U	U	U	U	11
5								A	U	U	U	U	U	U	22
6									U	I	U	I	U	U	12
7										U	U	O	U	U	6
8											U	U	E	U	11
9												A	U	A	33
10													U	U	2
11														U	1
Z															0
Total	0	1	2	1	1	1	1	1	1	1	1	8	1	1	224

The data processing software of ALDEP found optimum layout design of a product, with closeness rating 630, can be seen in Figure 6 as follows.

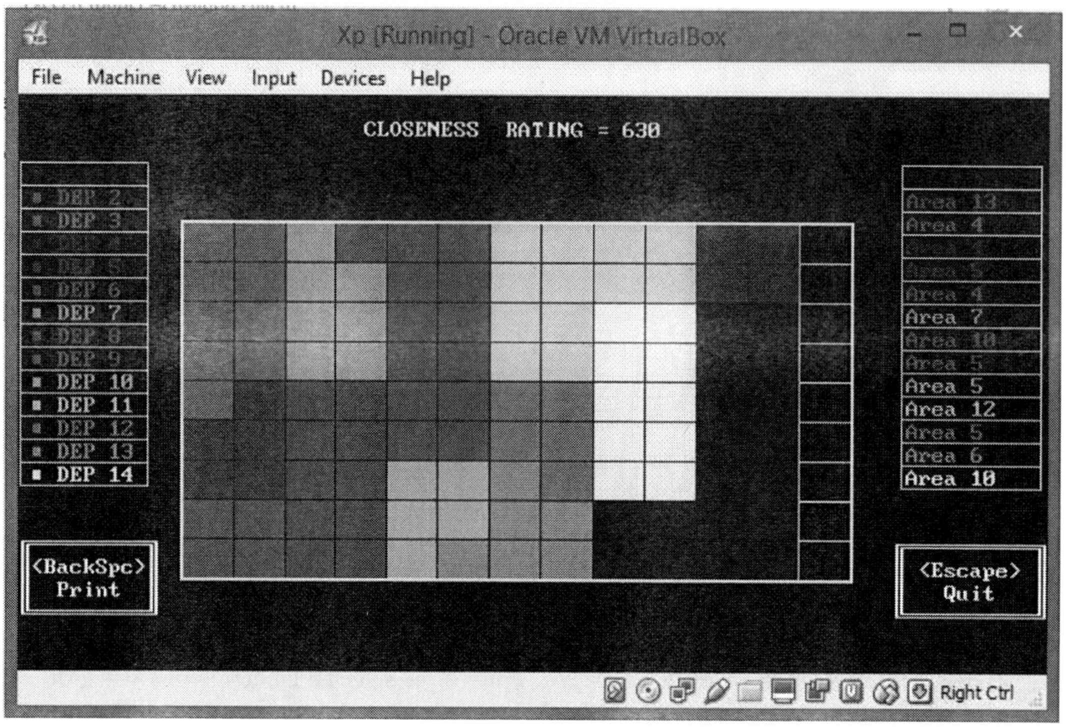

Figure 1. The optimum layout from ALDEP software

Improvement-type routines require an input from ALDEP block layout and aim to reduce internal transport cost by attempting simultaneous pair-wise position exchanging among the departments. The basic concept of the CRAFT is the exchange location/department activities on initial layout in order to find a better solution based on the flow of materials. Exchange will further bring to facility layout direction approaching the optimum costs. Algorithm of CRAFT can exchange 2 or 3 departments, so it can be used to generate a better layout. [4]. The data required as input CRAFT:

1. Initial layout from ALDEP block plan
2. Flow of material (frequency displacement)
3. Cost (material handling cost per meter)
4. The number of departments that have not changed (fixed)

CRAFT is a useful for improve the previous layout. The result layout obtained from CRAFT software as follows can be seen in Figure 2. MHES is a table used to calculate the cost of materials handling. MHES optimize the flow of materials by planning a sequence of operations and equipment settings. Material handling used here is manual, where the operator performs the transfer of materials from one Department to another Department. The result of the calculation of the MHPS will be as input for the FTC. Unlike the Material Handling Planning Sheet (MHPS), the Material Handling Evaluation Sheet (MHES) using real distance from the layout. The distance in the study using aisle distance. The purpose of doing an evaluation of the design of the layout using material handling that is to know the difference in the cost of material handling. Evaluation performed towards the cost of material handling on the layout of ALDEP and layout of the CRAFT, The Material Handling cost of proposed layout from ALDEP was Rp. 2,562,331, which when compared with the improved layout of the results of CRAFT

IOP Conf. Series: Materials Science and Engineering **528** (2019) 012062 doi:10.1088/1757-899X/528/1/012062

was Rp. 2,402,358, the results of CRAFT showed a decrease in the cost of material handling Rp. 159,973 (6,24%)

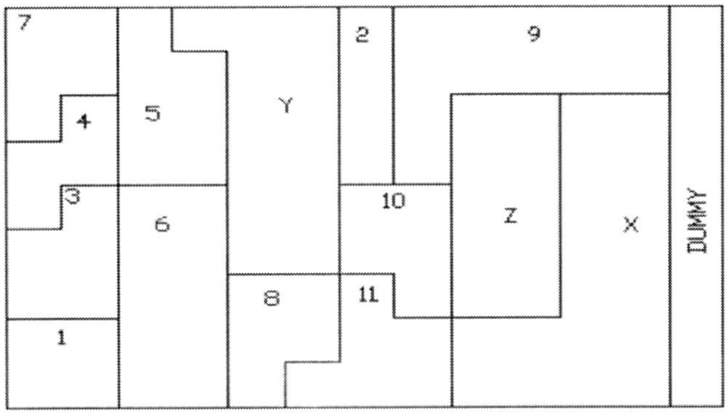

Figure 2. Output CRAFT Layout

Simulation is a tool that can be used to predict performance. The simulation could not design a system by itself. The user must provide various design scenarios and evaluate each scenario simulation tool to determine which scenario has the best performance. The simulation can also be defined as a technique that mimics the operation of the computer manufacturing real or system services from time to time. [5]. In the simulation model development, the basic elements are required are location, entities, arrival, processing and resource.

Table 6. The Result of Evaluation Layout with Adjacency Based Scoring

From	To		Rank	Rating	Score	ALDEP	CRAFT
Circular Saw Machine	Radial Drilling Machine	1.06	1	A	16	16	16
Packing Area	Warehouse	1	2	A	16	16	16
Las Machine	Grinding Machine	0.95	3	A	16	16	16
Packing Area	Work Table	0.87	4	A	16	16	16
Circular Saw Machine	Spray Gun	0.86	5	A	16	0	0
Warehouse	Cutting Table	0.81	6	E	8	8	8
Auxiliary storage	Grinding Machine	0.69	7	E	8	8	8
Awb2	Cutting Table	0.62	8	E	8	8	8
Circular Saw Machine	Grinding Machine	0.51	9	E	8	8	8
Auxiliary storage	Adjustable Work Benches 2	0.48	10	E	8	8	8
Adjustable Work Benches 1	Adjustable Work Benches 2	0.47	11	I	4	0	0
Auxiliary storage	Spray Gun	0.44	12	I	4	4	4
Grinding Machine	Work Table	0.43	13	I	4	4	4
Drying Area	Adjustable Work Benches 1	0.38	14	I	4	0	0
Spray Gun	Adjustable Work Benches 2	0.29	15	I	4	0	4
Warehouse	Circular Saw Machine	0.28	16	I	4	0	0
Drying Area	Spray Gun	0.27	17	O	2	2	2
Auxiliary storage	Las Machine	0.24	18	O	2	2	2
Radial Drilling Machine	Grinding Machine	0.24	19	O	2	2	2
Adjustable Work Benches 1	Work Table	0.23	20	O	2	2	2
Auxiliary storage	Work Table	0.22	21	O	2	2	2
Drying Area	Cutting Table	0.19	22	O	2	2	2
Radial Drilling Machine	Las Machine	0.19	23	O	2	2	2
Auxiliary storage	Adjustable Work Benches 1	0.16	24	U	1	1	1
Auxiliary storage	Packing Area	0.16	25	U	1	1	1
Drying Area	Adjustable Work Benches 2	0.14	26	U	1	1	1
Dying Area	Grinding Machine	0.04	27	U	1	1	1
	Total Score				162	130	134
	Percentage of Adjacency Based Score					80%	83%

IOP Conf. Series: Materials Science and Engineering **528** (2019) 012062 doi:10.1088/1757-899X/528/1/012062

This percentage of adjacency based score by applying ALDEP algorithm followed by improvement routine methodology CRAFT was better than that of ALDEP construction routine alone. Evaluation of improved layout resulted from CRAFT generate Adjacency Based Score percentage of 83% than 80% of ALDEP. There are also significant consistency between the cost of Material handling, and manufacturing lead time. There are also significant consistency between the cost of Material handling and manufacturing lead time.

4. Conclusion

This paper has shown federated usage of construction and improvement routine results better performance of factory layouts. The improved layout of CRAFT was the best layout according to the cost of material handling, manufacturing lead time and adjacency based score. Further research can be apply in other industries and other method of construction as well improvement type algorithmic.

5. References

[1] Deshpande V, Patil N D, Baviskar V and Gandhi J 2016 Plant Layout Optimization using CRAFT and ALDEP Methodology. *Productivity Journal by National Productivity Council* ISSN: 0032-9924 **57**(1) pp 32-42

[2] Wignjosoebroto S 2009 *Tata Letak Pabrik dan Pemindahan Bahan.* 3rd Ed. (Surabaya: Guna Widya)

[3] Tompkins, White, Bozer, Frazelle, Tanchoco and Trevino 2010 *Facilities Planning.* 3rd Ed. (Canada: John Wiley & Sons)

[4] Heragu S S 2016 *Facilities Planning.* 4th Ed. (New York: CRC Press)

[5] Apple J M 1990 *Tata Letak Pabrik dan Pemindahan Bahan,* 3rd Ed. (Bandung: Penerbit ITB)

[6] Pamularsih T, Mustofa F H and Susanty S 2015 Usulan Rancangan Tata Letak Fasilitas Dengan Menggunakan Metode Automated Layout Design Program (ALDEP) di Edem Ceramic. *Reka Integra* (Bandung: Jurusan Teknik Industri Itenas) **3**(2)

IOP Conf. Series: Materials Science and Engineering **528** (2019) 012063 doi:10.1088/1757-899X/528/1/012063

Job Scheduling for Hybrid Assembly Differentiation Flow Shop to Minimize Total Actual Flow Time considering Multi-Due-Dates

R Maulidya, Suprayogi, R Wangsaputra and AH Halim

Industrial Engineering Faculty, Institut Teknologi Bandung, Bandung 40132, Indonesia
rahmimaulidya@gmail.com, yogi@mail.ti.itb.ac.id, rwangsap@mail.ti.itb.ac.id, ahakimhalim@ti.itb.ac.id

Abstract. The hybrid assembly differentiation flow shop is a three-stage flow shop system with machining, assembly and differentiation stages. The machining stage produces *k* parts in independently unrelated dedicated machines. After *k* parts of a job completed, they are assembled at the assembly stage and assigned to dedicated machines for a final process. At the end, several types of finished products are resulted. The finished products should be delivered at their respective different due dates. The problem is to find a schedule considering multi due dates, and the criterion of minimizing total actual flow time. This paper proposes a model for job scheduling in a hybrid assembly differentiation flow shops and its algorithm for solving the problems. The initial solution is defined using SPT-based heuristic and it is optimized using variable neighbourhood descend method. The proposed model is tested using a set of hypothetic data. The solution shows that the algorithm can solve the problems effectively.
Keywords: job scheduling, hybrid assembly differentiation flow shop, total actual flow time, multi due dates

1. Introduction

This paper deals with a three-stage hybrid flow shop that consists of machining parts, an assembly operation and processing different end products called Hybrid Assembly Differentiation Flow Shop. The first stage is the machining stage where several dedicated machines produce different parts in parallel manner. The second stage is the stage where all parts of a product will be assembled in a single machine after they completed. The final stage is the differentiation stage consists of dedicated machines where an assembled part is processed into a certain type of products. The application of this system is illustrated from the electronic manufacturing where the system consists of a three-stage production system producing different types of products.

The recent pieces of research [1-4] developed the three-stage production system for job scheduling problem. According to [1], the three stage of production system consists of machining, transportation and assembly stages. Transportation is seen as an important aspect in job scheduling problem. The difference of [1], [2] define a three-stage production system as the machining processes in three stages. Other pieces of research [3-4] developed a three-stage production system that consists of machining, assembly and differentiation stages. In this paper, we extend the research proposed in [4] to cover a condition of multi-due-date.

The literature of [1-3] and [5] for job scheduling usually solve the problem using forward approach with the assumption of parts that should be available at time zero. The Model proposed in [3] uses the total flow time as the criteria to minimize the time spending by parts in the shop from time zero until the completion time. Other criteria developed by [5] using the performance measure of total tardiness to minimize the penalty between the job's due date and its completion time. In this paper, we use backward approach and the criterion of minimizing total actual flow time to minimize the interval time of parts spend in the shop from the arrival time until its due date. Total actual flow time is defined for some cases where parts are not available simultaneously in time zero, but will arrive when its needed, and other cases that consider the delivery time from due date of parts [6].

This paper deals with a problem of job scheduling model for a hybrid assembly differentiation flow shop to minimize the total actual flow time. We extend the problem in [4] to cover a condition where the finished products are demanded at several due dates meaning that there are several delivery times within a production planning. This paper is organized as section 2 for formulating the model, section 3 for the proposed algorithm, section 4 for the illustration and section 5 for the conclusion and the future research.

2. Method

2.1. Problem Formulation

Consider the problem of scheduling J jobs at a hybrid assembly differentiation flow shop to make H different types of finished products. We can denote the jobs belong to a certain type as JH_h. An illustration in Figure 1 show that there are four jobs in the system that consist of product type 1, JH_1=J_1,J_2, and product type 2, JH_2=J_3,J_4. Each job consists of part#1, part#2...part#*k* produced by unrelated parallel machines in the machining stage. Parts are processed on each machine in the same order. After all parts of a job are completed, they are assembled in a single assembly machine. The differentiation stage processes the assembled parts into a certain types of products in particular dedicated machines. There is a condition where the products must be delivered in different due dates. It means that each job must fulfil their own due date.

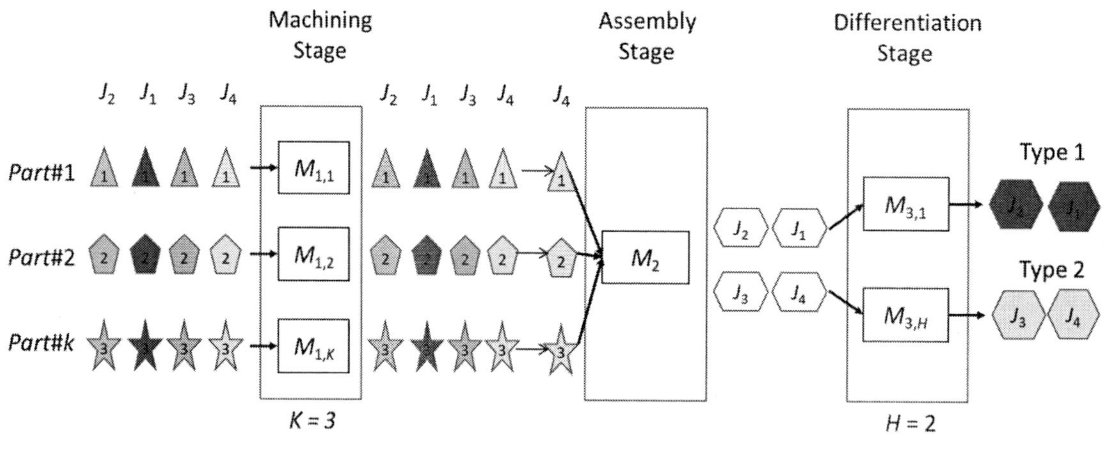

Figure 1. The three stages production system

The assumptions in this paper are:

- Production are finished at their corresponding due dates
- Each machine can process at most one job at a time
- Each job can be processed on at most one machine at a time
- Machines are always available

- Travel time between stages are neglected
- Job processing cannot be interrupted, until the completion of job.
- Jobs are available for processing at a stage immediately after processing completion at the previous stage.

2.2. Model

The actual flow time of a job ($F^a_{[i]}$), defined by [7], is the time spend by a job from its arrival time in the shop until its due date. The formulation for $F^a_{[i]}$ is as follows.

$$F^a_{[i]} = d - B_{[i]}, \quad i=1,2...J \quad (1)$$

Where d is a due date and $B_{[i]}$ is the starting time for processing J_j. The schedule is started backwardly and the position of $[i]$ is started from its due date.

The formulation of the actual flow time of a job with a due date is as follows.

$$F^a_{[i]} = d - \min_{k \le K}\left(B^{(1)}_{[i],k}\right), \ \forall i, \forall k \quad (2)$$

Figure 2 shows the actual flow time of several due dates d_1 and d_2, respectively.

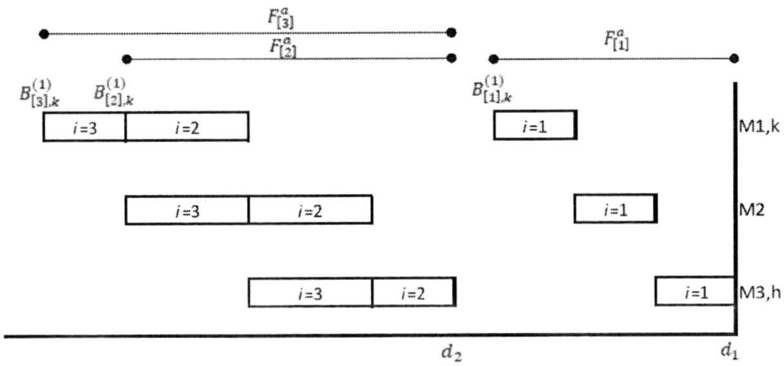

Figure 2. Actual Flow Time for multi due dates

From Figure 2, we have the actual flow time for position i=1,2,3.

$$F^a_{[1]} = d_1 - \min_{k \le K}\left(B^{(1)}_{[1],k}\right)$$

$$F^a_{[2]} = d_2 - \min_{k \le K}\left(B^{(1)}_{[2],k}\right)$$

$$F^a_{[3]} = d_2 - \min_{k \le K}\left(B^{(1)}_{[3],k}\right)$$

$$TAFT = F^a_{[1]} + F^a_{[2]} + F^a_{[3]}$$

$$TAFT = \sum_{i=1}^{J}\left(d_j - \min_{k \le K}\left(B^{(1)}_{[i],k}\right)\right) \quad (3)$$

Notation

Index

j = index of jobs, j=1, 2...J
i = index of positions, i=1, 2...J
k = index of machines at the machining stage, k=1, 2...K
h = index of machines and the product types at the differentiation stage, h=1, 2...H

Parameter

K = number of machines at the machining stage
H = number of machines at the differentiation stage
J = number of processed job
JH_h = number of job processed in machine h
d_j = due date for job j
G_g = interval of two consecutive due dates

M_{1k} = machine for processing part k of all the jobs at the machining stage, $k = 1, 2...K$
M_2 = assembly machine in the assembly stage
M_{3h} = differentiation machine for processing types of products in the differentiation stage, $h = 1, 2...H$
$p_{j,k}^{(1)}$ = the processing time of part k job j in machine M_{1k} at the machining stage
$p_j^{(2)}$ = the processing time of assembly operation of job j on machine M_2 at the assembly stage
$p_{j,h}^{(3)}$ = the processing time of job j type h in machine M_{3h} at the differentiation stage. $p_{j,h}^{(3)} > 0$ if $j \in N_h$; 0, otherwise.
$A_{j,h}$ = the binary variable equals 1 if *job j* is assigned in machine h with the processing time $p_{j,h}^{(3)} > 0$; otherwise 0
Decision variables
$X_{j[i]}$ = the binary variable equals 1 if job j is assigned to position i; otherwise 0
$Y_{[i],h}$ = the binary variable for the differentiation stage equals 1 if job in position i is processed in machine h; otherwise 0;
$B_{[i],k}^{(1)}$ = the starting time of job in position i in machine M_{1k} at the machining stage
$B_{[i]}^{(2)}$ = the starting time of assembly job in position i in machine M_2 at the assembly stage
$B_{[i],h}^{(3)}$ = the starting time of job in position i in machine M_{3h} at the differentiation stage
Objective function
$TAFT$= total actual flow time

The problem can be formulated as follows.

$$\text{Min } TAFT = \sum_{i=1}^{n}\left(\left(X_{j[i]} * d_j\right) - \min_{k \le K}\left(B_{[i],k}^{(1)}\right)\right) \tag{4}$$

Subject to:

$$B_{[i],h}^{(3)} = Y_{[i],h}\left(X_{j[i]}\left(d_j - p_{j,h}^{(3)}\right) - G_{[z],h}\left(\left(d_j * X_{j[z]}\right) - B_{[z],h}^{(3)}\right)\right), \quad z \le i, \ \forall i, \forall h \tag{5}$$

$$B_{[i]}^{(2)} + \sum_{j=1}^{n}\left(X_{j[i]} \cdot p_j^{(2)}\right) \le d_j - \left(Y_{[i],h} * \sum_{j=1}^{n}\sum_{i^*=1}^{i}\left(X_{j[z]} \cdot p_{j,h}^{(3)}\right)\right), z \le i, \forall i, \forall h \tag{6}$$

$$B_{[i+1]}^{(2)} \le B_{[i]}^{(2)} - \sum_{j=1}^{n}\left(X_{j[i+1]} \cdot p_j^{(2)}\right), \quad i < n, \forall i \tag{7}$$

$$B_{[i],k}^{(1)} + \sum_{j=1}^{n}\left(X_{j[i]} \cdot p_{j,k}^{(1)}\right) \le B_{[i]}^{(2)} \quad \forall i, \forall k \tag{8}$$

$$B_{[i+1],k}^{(1)} \le B_{[i],k}^{(1)} - \sum_{j=1}^{n}\left(X_{j[i+1]} \cdot p_{j,k}^{(1)}\right), i < n, \forall i, \forall k \tag{9}$$

$$A_{j,h} = \begin{cases} 1, & \text{if } p_{j,h}^{(3)} > 0 \\ 0, & \text{if } p_{j,h}^{(3)} = 0 \end{cases} \quad j = 1,2..n, \forall h \tag{10}$$

$$\sum_{j=1}^{n}\left(X_{j[i]} \cdot A_{j,h}\right) = Y_{[i],h} \qquad \forall i, \ \forall h \tag{11}$$

$$\sum_{j=1}^{n} X_{j[i]} = 1, \qquad i = 1,2..n \tag{12}$$

$$\sum_{i=1}^{n} X_{j[i]} = 1, \qquad j = 1,2..n \tag{13}$$

$$G_{[z],h} = \begin{cases} 1, & \text{if } B_{[z],h}^{(3)} < \left(d_j * Y_{[z],h} * X_{j[z]}\right) \\ 0, & \text{if } B_{[z],h}^{(3)} \ge \left(d_j * Y_{[z],h} * X_{j[z]}\right) \end{cases} \quad z < i, j = 1,2..n, \forall h \tag{14}$$

$$X_{j[i]} \in \{0,1\} \tag{15}$$

$$G_{[z],h} \in \{0,1\} \tag{16}$$

$$B_{[i],k}^{(1)}, B_{[i]}^{(2)}, B_{[i],h}^{(3)} \ge 0, \text{ and integer} \tag{17}$$

The objective function (4) shows the total actual flow time that is calculated from the actual flow time of all position where the actual flow time of each position is the longest interval between the part being processed in the first stage until its common due date. Constraint (5) defines the starting time of job in position i in machine h at the third stage from the due date of the job considering the decision that position i is assigned in machine h and there is no starting time of other job before the due date d_j. Constraint (6) defines the starting time of job in position i at the second stage from backward based on job in position i at the third stage, with condition where $i^* \leq i$. Constraint (7) and (9) ensure that each position can only be processed at the same stage after the next position is finished. Constraint (8) defines the starting time of job in position i on machine k at the first stage based on the starting time of the same job in position i at the second stage. Constraint (10) defines the job with the processing time at the third stage greater than zero, is processed in machine h at the third stage. Constraint (11) decides the job in position i that is assigned on machine h at the third stage only if the value is 1. Constraint (12) and (13) ensure that each position in the sequence of job is assigned to only one job, and each job is assigned only to one position in the sequence of job. Constraint (14) defines that there is no other job of position z processed until the due date. Constraint (15) and (16) define the domain of the decision variables. Constraint (17) ensures that the starting time is feasible from time zero.

2.3. *Algorithm*

We develop a solution using SPT based heuristic and optimized by Variable Neighbourhood Descent (VND) methods. The multi-due-date has condition where the jobs must be scheduled to meet their due dates. Figure 3 show that there are several due dates and between two consecutive due dates the jobs must be processed. The proposed algorithm divided the jobs according to the interval they belong.

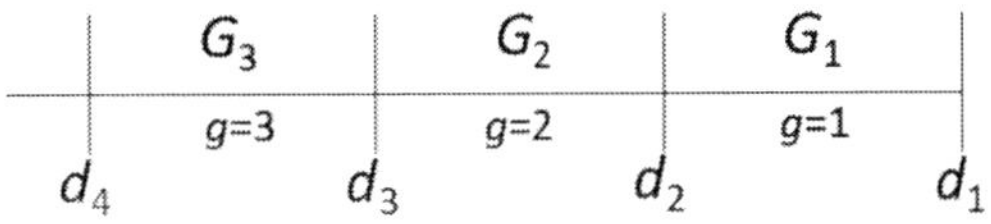

Figure 3. Interval of two consecutive due dates

SPT-based heuristic generates the job sequences based on the processing time of the three stages in increasing orders. The jobs are scheduled from their corresponding due dates and the value of TAFT is defined. The best initial solution from SPT-based heuristic is optimized using VND method. VND with the insert move determined the job sequence that minimizes the total actual flow time. In VND-insert move, a job is moved to a position so that the current positions are shifted one position forward along the solution [8]. The stopping rule is the condition where the iterations give no new minimum total actual flow time after a set of iterations.

The proposed algorithm

Step 0: Initialize the problem

Step 1: Sequence the due dates in increasing order, the greatest due dates will be g=1.

Step 2: Determine G_g as the interval between two consecutive due dates using:

$$G_g = \begin{cases} d_g - d_{g+1}; & g = 1, \dots, r-1 \\ d_r; & g = r \end{cases}$$

Step 3: Set g=1

Step 4: Generate the job sequences in increasing orders for all jobs in interval G_g based on the SPT-based heuristic for $\max_{1\leq k\leq 3}\left(p_{j,k}^{(1)}\right), p_j^{(2)}, p_{j,h}^{(3)}$, $\max_{1\leq k\leq 3}\left(p_{j,k}^{(1)}\right) + p_j^{(2)}, p_j^{(2)} + p_{j,h}^{(3)}$ and $\max_{1\leq k\leq 3}\left(p_{j,k}^{(1)}\right) + p_j^{(2)} + p_{j,h}^{(3)}$, respectively. Schedule the jobs from theirs corresponding due dates.

Step 5: Set the best total actual flow time as the best initial solution.

Step 6: Perform procedure Variable Neighbourhood Descent (insert move) as follows.

- For all the jobs, i=1,2…J, in vn_{seq}, repeat
- Set i^{th} job in vn_{seq} sequence as J'_{vn}
- Insert J'_{vn} in all possible positions in vn_{seq} to get $N^1(x)$, a subset of $N(x)$.
- Find *TAFT* of all sequences in $N^1(x)$
- Set sequence with minimum *TAFT* in vn_{seq}.
- Output the final sequence vn_{best}.

Step 6: Is the solution better than the best initial solution?
If Yes, output the best solution
If No, then stop.

3. Result and Discussion

Consider an illustration of eight jobs (J=8) to analyse the model. There are two product types (Type 1 and Type 2). Table 1 shows the data that Type 1 consists of jobs J_1, J_2, J_3 and J_4, written as $JH_1=\{J_1,J_2,J_3,J_4\}$ and type 2 consists of jobs J_5, J_6 ,J_7 and J_8 written as $JH_2=\{J_5,J_6,J_7,J_8\}$. The number of machine in the machining stage (K)=3, and the number of machine in the differentiation stage (H)=2. This illustration is run using the processor of Intel Core i7-6500U CPU, 2,50 GHz and 12 GB RAM.

Table 1. Processing time and due date of jobs

Job	Type	$p_{j,1}^{(1)}$	$p_{j,2}^{(1)}$	$p_{j,3}^{(1)}$	$p_j^{(2)}$	$p_{j,1}^{(3)}$	$p_{j,2}^{(3)}$	d_j
J_1	1	5	4	3	3	6	0	990
J_2	1	6	3	4	9	10	0	990
J_3	1	6	4	10	3	9	0	980
J_4	1	3	5	5	4	5	0	1000
J_5	2	3	4	5	6	0	8	980
J_6	2	3	4	6	4	0	7	990
J_7	2	10	4	6	5	0	6	980
J_8	2	5	4	3	10	0	4	1000

We can divide the due dates into three sections (g=3) and the jobs are put from their corresponding due dates backwardly. Table 2 shows the initial solution using SPT-based heuristic. There are two sequences with the best TAFT, J_4-J_8-J_1-J_6-J_2-J_5-J_3-J_7 and J_4-J_8-J_1-J_6-J_2-J_5-J_7-J_3.

Table 2. SPT-based heuristic for multi-due-date

$\max_{k\le K}(p_{j,k}^{(1)})$	J_8	J_4	J_1	J_2	J_6	J_5	J_3	J_7	TAFT
	5	5	5	6	6	5	10	10	241
$p_{j,2}^{(1)}$	J_4	J_8	J_1	J_6	J_2	J_3	J_7	J_5	
	4	10	3	4	9	3	5	6	216
$p_{j,h}^{(3)}$	J_8	J_4	J_1	J_1	J_2	J_7	J_5	J_3	
	4	5	6	7	10	6	8	9	218
$\max_{k\le K}(p_{j,k}^{(1)}) + p_{j,2}^{(1)}$	**J_4**	**J_8**	**J_1**	**J_6**	**J_2**	**J_5**	**J_3**	**J_7**	
	9	**15**	**8**	**10**	**15**	**11**	**13**	**15**	**214**
$p_{j,2}^{(1)} + p_{j,h}^{(3)}$	J_4	J_8	J_1	J_6	J_2	J_7	J_3	J_5	
	9	14	9	11	19	11	12	14	219
$\max_{k\le K}(p_{j,k}^{(1)}) + p_{j,2}^{(1)} + p_{j,h}^{(3)}$	**J_4**	**J_8**	**J_1**	**J_6**	**J_2**	**J_5**	**J_7**	**J_3**	
	14	**19**	**14**	**17**	**25**	**19**	**21**	**22**	**214**

The proposed algorithm optimized the initial solution using variable neighbourhood descent with the insert moves that can be seen in Table 3. From the two best initial solution, we select the first sequence J_4-J_8-J_1-J_6-J_2-J_5-J_3-J_7. The iteration of Variable Neighbourhood Descent started with the first job of J_4 to be inserted into vn_{seq} (sequence 1 until sequence 7) with different positions. The *TAFT* of each sequences are defined. In iteration 1, there are no improvement for the vn_{best}. The iteration continued with the next job from the sequence selected. In iteration 5 and 8, we get the new vn_{best} for the same TAFT value of 211. Those sequences are J_4-J_8-J_1-J_6-J_5-J_3-J_7-J_2 and J_4-J_8-J_1-J_6-J_7-J_2-J_5-J_3. We can see that J_2 with the longest processing time in both assembly and differentiation stages is scheduled at the third section (the respective due date). It is because the interval between the second and the third due date are smaller than the total three job's processing time (J_1, J_2, J_6). Figure 4 is the Gantt chart for the best sequence after the optimization. Figure 4 also shows the cluster of jobs according to the interval between two due dates and the position of J_2.

Table 3. Variable Neighbourhood Descent-insert moves

Iteration 1 (J_4)									Iteration 2 (J_8)								
vn_{best}	**J_4**	**J_8**	**J_1**	**J_6**	**J_2**	**J_5**	**J3**	**J_7**	**214**	vn_{best}	**J_4**	**J_8**	**J_1**	**J_6**	**J_2**	**J_5**	**J_3**	**J_7**	**214**
1	J_8	J_4	J_1	J_6	J_2	J_5	J3	J_7	216	1	J_8	J_4	J_1	J_6	J_2	J_5	J_3	J_7	216
2	J_8	J_1	J_4	J_6	J_2	J_5	J3	J_7	244	2	J_4	J_1	J_8	J_6	J_2	J_5	J_3	J_7	254
3	J_8	J_1	J_6	J_4	J_2	J_5	J3	J_7	241	3	J_4	J_1	J_6	J_8	J_2	J_5	J_3	J_7	248
4	J_8	J_1	J_6	J_2	J_4	J_5	J3	J_7	243	4	J_4	J_1	J_6	J_2	J_8	J_5	J_3	J_7	262
5	J_8	J_1	J_6	J_2	J_5	J_4	J3	J_7	248	5	J_4	J_1	J_6	J_2	J_5	J_8	J_3	J_7	256
6	J_8	J_1	J_6	J_2	J_5	J3	J_4	J_7	253	6	J_4	J_1	J_6	J_2	J_5	J_3	J_8	J_7	248
7	J_8	J_1	J_6	J_2	J_5	J3	J_7	J_4	254	7	J_4	J_1	J_6	J_2	J_5	J_3	J_7	J_8	247
Iteration 3 (J_1)									Iteration 4 (J_6)								
vn_{best}	**J_4**	**J_8**	**J_1**	**J_6**	**J_2**	**J_5**	**J_3**	**J_7**	**214**	vn_{best}	**J_4**	**J_8**	**J_1**	**J_6**	**J_2**	**J_5**	**J_3**	**J_7**	**214**
1	J_1	J_4	J_8	J_6	J_2	J_5	J_3	J_7	306	1	J_6	J_4	J_8	J_1	J_2	J_5	J_3	J_7	307
2	J_4	J_1	J_8	J_6	J_2	J_5	J_3	J_7	254	2	J_4	J_6	J_8	J_1	J_2	J_5	J_3	J_7	265
3	J_4	J_8	J_6	J_1	J_2	J_5	J_3	J_7	214	3	J_4	J_8	J_1	J_6	J_2	J_5	J_3	J_7	214
4	J_4	J_8	J_6	J_2	J_1	J_5	J_3	J_7	237	4	J_4	J_8	J_1	J_2	J_6	J_5	J_3	J_7	239
5	J_4	J_8	J_6	J_2	J_5	J_1	J_3	J_7	222	5	J_4	J_8	J_1	J_2	J_5	J_6	J_3	J_7	246
6	J_4	J_8	J_6	J_2	J_5	J_3	J_1	J_7	229	6	J_4	J_8	J_1	J_2	J_5	J_3	J_6	J_7	250
7	J_4	J_8	J_6	J_2	J_5	J_3	J_7	J_1	232	7	J_4	J_8	J_1	J_2	J_5	J_3	J_7	J_6	250
Iteration 5 (J_2)									Iteration 6 (J_5)								
vn_{best}	**J_4**	**J_8**	**J_1**	**J_6**	**J_2**	**J_5**	**J_3**	**J_7**	**214**	vn_{best}	**J_4**	**J_8**	**J_1**	**J_6**	**J_5**	**J_3**	**J_7**	**J_2**	**211**
1	J_2	J_4	J_8	J_1	J_6	J_5	J_3	J_7	349	1	J_5	J_4	J_8	J_1	J_6	J_2	J_3	J_7	399
2	J_4	J_2	J_8	J_1	J_6	J_5	J_3	J_7	301	2	J_4	J_5	J_8	J_1	J_6	J_2	J_3	J_7	229
3	J_4	J_8	J_2	J_1	J_6	J_5	J_3	J_7	233	3	J_4	J_8	J_5	J_1	J_6	J_2	J_3	J_7	269
4	J_4	J_8	J_1	J_2	J_6	J_5	J_3	J_7	239	4	J_4	J_8	J_1	J_5	J_6	J_2	J_3	J_7	247
5	J_4	J_8	J_1	J_6	J_5	J_2	J_3	J_7	219	5	J_4	J_8	J_1	J_6	J_2	J_5	J_3	J_7	214
6	J_4	J_8	J_1	J_6	J_5	J_3	J_2	J_7	215	6	J_4	J_8	J_1	J_6	J_2	J_3	J_5	J_7	213
7	**J_4**	**J_8**	**J_1**	**J_6**	**J_5**	**J_3**	**J_7**	**J_2**	**211**	7	J_4	J_8	J_1	J_6	J_2	J_3	J_7	J_5	230
Iteration 7 (J_3)									Iteration 8 (J_7)								
vn_{best}	**J_4**	**J_8**	**J_1**	**J_6**	**J_5**	**J_3**	**J_7**	**J_2**	**211**	vn_{best}	**J_4**	**J_8**	**J_1**	**J_6**	**J_5**	**J_3**	**J_7**	**J_2**	**211**
1	J_3	J_4	J_8	J_1	J_6	J_2	J_5	J_7	401	1	J_7	J_4	J_8	J_1	J_6	J_2	J_5	J_3	379
2	J_4	J_3	J_8	J_1	J_6	J_2	J_5	J_7	273	2	J_4	J_7	J_8	J_1	J_6	J_2	J_5	J_3	325
3	J_4	J_8	J_3	J_1	J_6	J_2	J_5	J_7	231	3	J_4	J_8	J_7	J_1	J_6	J_2	J_5	J_3	265
4	J_4	J_8	J_1	J_3	J_6	J_2	J_5	J_7	249	4	J_4	J_8	J_1	J_7	J_6	J_2	J_5	J_3	238
5	J_4	J_8	J_1	J_6	J_3	J_2	J_5	J_7	233	5	**J_4**	**J_8**	**J_1**	**J_6**	**J_7**	**J_2**	**J_5**	**J_3**	**211**
6	J_4	J_8	J_1	J_6	J_2	J_3	J_5	J_7	213	6	J_4	J_8	J_1	J_6	J_2	J_7	J_5	J_3	216
7	J_4	J_8	J_1	J_6	J_2	J_5	J_7	J_3	230	7	J_4	J_8	J_1	J_6	J_2	J_5	J_7	J_3	230

The optimal model has run for five hours using a programming software, but the result is still local optimal with the best TAFT=346. The performance from this illustrative example is 14.31%. For this illustration, the result is better than the optimal model. Though it still need more illustrative data to make the best conclusion.

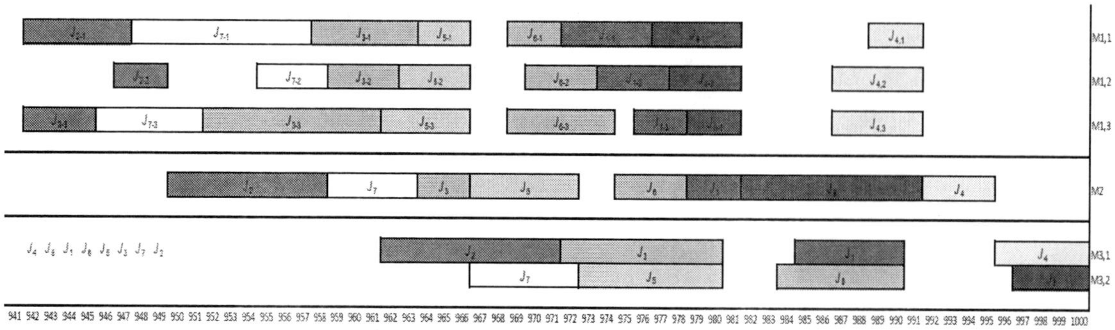

Figure 4. Gantt Chart for *TAFT*=211

4. Conclusions

This paper deals with job scheduling problems for hybrid assembly differentiation flow shop to minimize total actual flow time with the condition of multi-due-date. The results of this paper are job scheduling model for multi-due-date condition and the proposed algorithm consists of SPT based heuristic for initial solution and variable neighbourhood descend method for optimization. The illustration is only for eight jobs and there must be more experiments to have a better conclusion. The proposed algorithm can give the better solution and shows the improvement from the initial solution. The future research is to find the proposed algorithm that can give optimal solution in short time compare to the current approach.

Acknowledgments

This research was fully funded by Doctoral Dissertation Research Funding from Directorate General Higher Education Indonesia (2018), Ministry of Research, Technology and Higher Education under contract No.011/KM/PNT/2018 with Universitas Trisakti.

5. References

[1] Koulamas, C and Kyparisis, G J 2001 The Three-stage Assembly Flowshop Scheduling Problem. *Computers & Operations Research* **28** pp 689-704

[2] Babayan, A and He, D 2004 Solving The n-job 3-stage Flexible Flowshop Scheduling Problem using An Agent-based Approach. *International Journal of Production Research* **42**(4) pp 777-799.

[3] Xiong F, Xing K and Wang F 2015 Scheduling a Hybrid Assembly-Differentiation Flow Shop to Minimize Total Flow Time. *European Journal of Operation Research* **240** pp 338-54

[4] Maulidya R, Suprayogi, Wangsaputra R and Halim A H 2016 Job Scheduling for Hybrid Assembly Differentiation Flowshop to Minimize Total Actual Flow Time. *Proc. of 17th Asia Pacific Industrial Engineering and Management System Conference (APIEMS) Conf.* December 7-10, 2016 (Taipei) pp 1198-1204

[5] Allahverdi A and Aydilek H 2015 The two-stage assembly flowshop scheduling problem to minimize total tardiness. *J Intell Manuf* **26** pp 225–237

[6] Halim A H Miyazaki S and Ohta H 1994a Batch-Scheduling Problems to Minimize Actual Flow Times of Parts through the Shop under JIT Environment. *European Journal of Operation Research* **72** pp 529-44

[7] Halim AH, Miyazaki S and Ohta H 1994b Lot Scheduling Problems of Multiple Items in the shop with Both Receiving and Delivery Just in Times. *Production Planning and Control* **5**(2) pp 175-184

[8] Vanchipura R, Sridharan R and Babu A S 2014 Improvement of constructive heuristics using variable neighbourhood descent for scheduling a flow shop with sequence dependent setup time. *Journal of Manufacturing Systems* **33** pp 65–75

Implementation of Material Requirement Planning (MRP) on Raw Material Order Planning System for Garment Industry

Nidaul Hasanati[1], Effrizka Permatasari[2], Nunung Nurhasanah[3], and Syarif Hidayat[4]

UIN Syarif Hidayatullah[1], Universitas Al Azhar Indonesia[2,3,4]
[1]nidaul.hasanati@uinjkt.ac.id, [2]frizkapisca@gmail.com,
[3]nunungnurhasanah@uai.ac.id, [4]syarif_hidayat@uai.ac.id

Abstract. The increasingly competitive business environment has encouraged many actors to use information and communication technology as a competitive advantage, including the garment industry. So far, difficulties in controlling raw material inventories which causes problems for the production process. If there is a shortage of raw materials, the production process cannot be carried out, on the contrary if the raw material ordered is too excessive it will cause excess and the high cost of raw material inventory. This research develops an information system that can plan the needs of raw materials needed for production. This system is web-based, applying the MRP (Material Requirement Planning) method to develop raw material ordering planning systems, case studies at garment companies, using PHP programming and FIREBIRD databases. This system gets input from the production schedule resulting from production scheduling information system, then provides output in the form of recommendations for scheduling raw material orders. This output can later be used as a reference for the company's operational decisions in planning raw material requirements.
Keywords: information system, raw materials, order planning system, MRP, material requirement planning

1. Introduction

The development of the current global era influences the way companies do business. To be able to survive, an organization or company innovates to beat competitors, and strives to advance in utilizing Information Technology (IT). One of the keys to success, companies must have the right mission and strategy in their business activities and improve their performance.

As a case study, XYZ company is engaged in Moslem fashion industry in Indonesia, which provides various models of Moslem clothing for sale (make to stock). During this time, XYZ companies have difficulty in planning and controlling the supply of raw materials that will lead to obstacles in the production process. There are many suppliers as business partners who supply raw materials in the production process, but on the other hand consumer demand cannot be met. If the raw materials ordered are lacking, the production process cannot be carried out, so that some consumer requests cannot be met. However, if the raw material is ordered too much, it will cause excess and accumulate raw materials, as well as the high cost of raw material inventory. If this continues to happen, it will result in a waste of working capital and could be a loss for the company.

IOP Conf. Series: Materials Science and Engineering 528 (2019) 012064 doi:10.1088/1757-899X/528/1/012064

Therefore, to enhance the existing business strategy, needed an information system that can do the planning needs of the raw materials needed for production. This system refers to the production schedule resulting from demand forecasting, then output the form of a recommendation of scheduling reservations raw materials [1].

This paper discusses how to analyze, design, and develop a computer based and web based application system, for raw material order planning system, using Material Requirement Planning (MRP) method. Input of this system is in the form of Master Production Schedule (MPS) resulting from Production Scheduling Information System [2]. MPS are generated from the demand forecasting results of sales transaction history data in Online Transaction Processing (OLTP) [3]. The system will produce output in the form of raw materials on the booking schedule, using Material Requirement Planning (MRP) per week (period), and not using safety stock with the assumption that goods arrive on time.

Materials Requirement Planning (MRP) is a technique for determining the quantity and the time to purchase dependent demand items necessary to meet the needs of the Master Production Schedule (MPS) [3]. Materials Requirement Planning (MRP) as a technique that uses a list of bill of material, inventory, receipts were estimated, and the master production schedule to determine material requirements [4]. The implementation of MRP would create competition and efficiency, which would lead to a better quality of life for customers at lower costs [5].

This system plays an important role for maintaining the stability of the amount of raw material inventory that is no less and no exaggeration, according to production requirements. In other words, this system will control the supply of raw materials effectively, reducing the risk of a less amount of raw materials or excessive, so that the production activity of company XYZ can run optimally.

2. METHODS

2.1 Data collection

The data collection is the procedure for obtaining the necessary data in this study. After knowing the main problem faced by company, then collecting the data needed in order to find solutions to these problems. Literature studies are also conducted and appropriate reference about MRP are looked to support and obtain information that is true and accurate.

The technique used by the author collection data for raw materials is a subsystem of data analysis results of Muhammad Aulia Taqwa [6] of raw material inventory control which is currently running at company XZY. The data includes the number of raw materials needed to produce certain products, as well as data lead Time or length of time to send raw materials from its suppliers respectively.

Bill of Material (BOM) contains information on all materials, components or sub-assembly that is used in making a product [7]. The data amount of the raw material needs as well as bill of material from other models can be seen recapitulation bill of material collected from five products.

Table 1. Summary of Bill Of Material 5 Model

Model	Type Material	Total Reuirements (m^2)
X-101	Kaos Broken White	9.21
	Shirt Dark Brown	0.21
	Shirt Brown Young	0.65
	Label	2
	Rubber	0.3
	Tape	0.3
X-68	Pink Kaos Kids	8.44
	White shirt	0.42
	Pink Shirt Young	0.38

Model	Type Material	Total Reuirements (m^2)
	Label	2
	Metal studs	3
X-95	Kaos Mosaline	9.06
	Shirt Magenta strit	3.664
	Label	2
	Rubber	0.4
X-100	Shirt Brown Young	7.926
	Shirt Dark Brown	0.945
	Label	2
	Rubber	0.2
	Metal studs	2
X-102	Kaos Batik	7.64
	Shirt Dark Brown	0.8
	Label	2
	Rubber	0.2
	Tape	0.8

Table 2 describes the results of the calculation data forecasting demand for each model is in April and May 2013[4]. The results of forecasting (pieces) explains the forecast amount of clothing that must be produced or forecasts of the number of requests by the prediction clothes.

Table 2. Summary of Demand Forecasting for 5 Moslem Clothes Models

Model	Year	Month	Results Forecasting (Pieces)
X 68	2013	April	571.00
		May	579.00
X 95	2013	April	377.43
		May	374.93
X 100	2013	April	88.49
		May	1.11
X 101	2013	April	467.84
		May	451.58
X 102	2013	April	1,025.38
		May	1,038.01

2.2 Data Analysis

Improper scheduling of production activities, resulting in PT. XYZ also has difficulty in determining the time and number of orders for raw materials that must be ordered to suppliers. For example, the delivery time can reach 4 weeks to arrive at the raw material warehouse. This will certainly slow down the production process of finished goods if the raw material delivery time does not match the initial estimate. While the number of orders for excessive raw materials can make a buildup in a limited warehouse of raw materials.

3. RESULT AND DISCUSSION

3.1 Analysis and Design

Workflow system procedures that would apply to the proposed system Booking Raw Materials Planning System in this scenario:

1. Historical data from polling stations processed into forecast demand.
2. Information System Demand Forecasting make forecasting to produce a model with a certain amount to be produced at a certain time.
3. From the results it will bring demand forecasting production or MPS (Master Production schedules) which will be input Booking Planning Information System Raw Materials.
4. Planning system for ordering raw materials will process production demand data by looking at the BOM (Bill of Material) of each model to be produced; estimate the amount of raw material ordered and when it will arrive at the warehouse at the specified production time (lead time).
5. Output from the planning system in the form of Scheduling of Raw Materials / Materials based on suppliers and the date to be ordered.

   The focus of the planning system ordering raw materials are at No. 4 and 5.

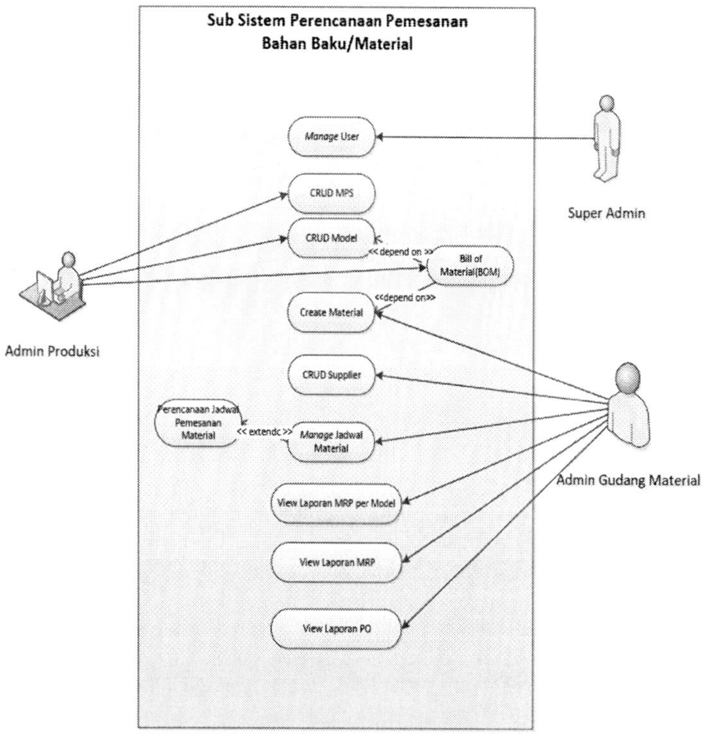

Figure 1. Usecase Diagram of Raw Materials Order Planning System

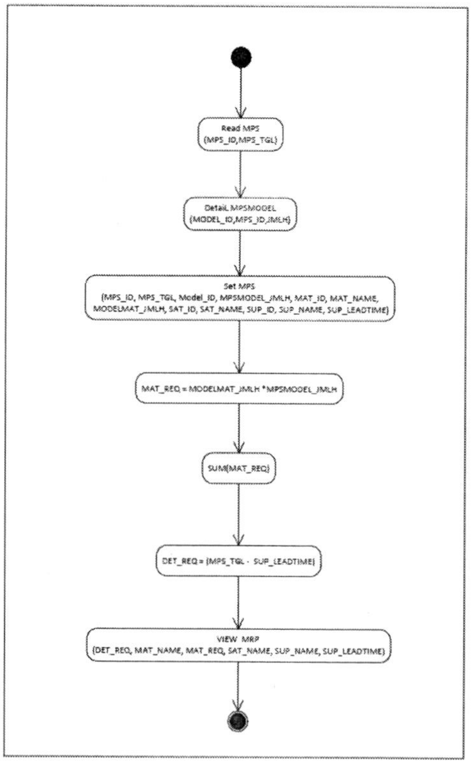

Figure 2. Algorithm for Ordering Raw Materials (for each Material)

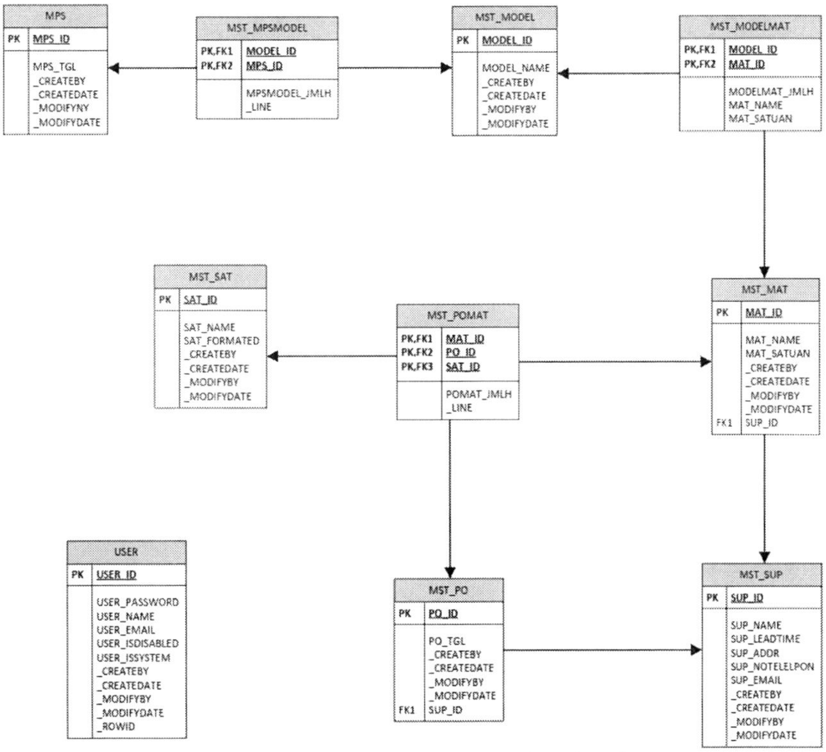

Figure 3. Databases of Raw Materials Order Planning System

3.2 Algorithm

Here is an overview of the activities that occur in the system . Figure 2 is the algorithm of ordering each raw materials calculated based on a request from the production form of MPS. MPS consists of MPS_ID and MPS_TGL. MPS has MPSMODEL detail consisting MPS_ID, MODEL_ID and MPSMODEL_JMLH. Then the request by the specified date of production will produce a model with a certain amount.

For admin of raw materials can know what materials will be ordered system will calculate it automatically. MAT_REQ is the material to be booked calculated by MODELMAT_JMLH multiplied by MPSMODEL_JMLH to obtain the required material for each model, after it was totalled by SUM (MAT_REQ), after getting the material needed to determine the time of booking material supplier. MPS_TGL decreased by SUP_LEADTIME it will get on how to be ordered material for each supplier.

The menu structure is designed based login to allow a user to use this system. The system is simple not too many designs that create confusion.

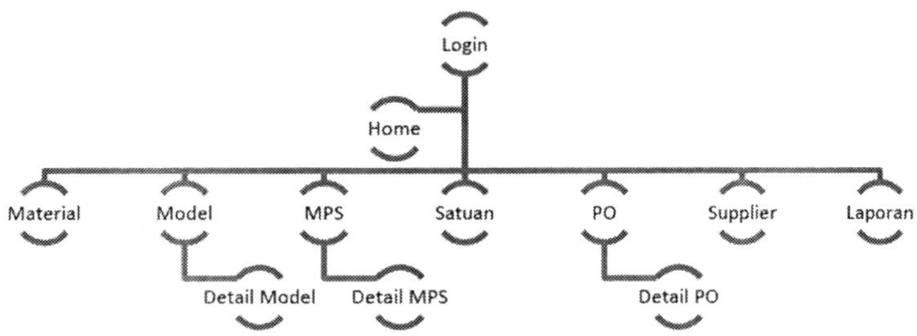

Figure 4. Structure Menu System

3.3 Implementation of User Interface

User Interface the design of the raw materials order scheduling system for company XYZ will be indicated by these figure:

1. Pages Management Bill of Material (BOM)

Bill of Materials is located in the Model menu. Due to indicate the model requires the name of the raw material and how much material.

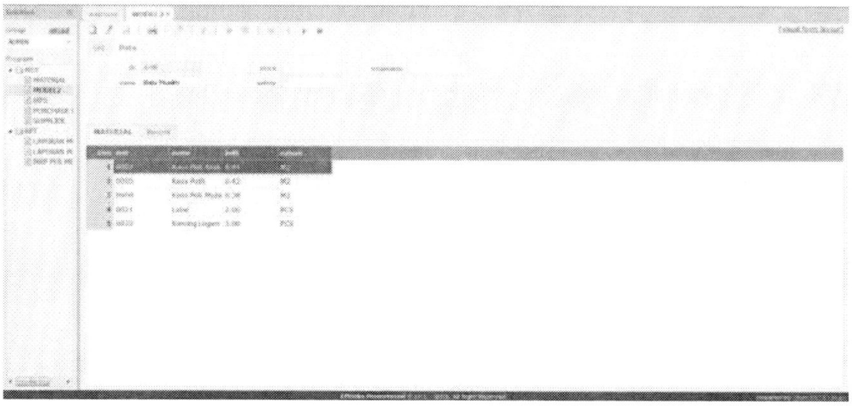

Figure 5. Page Management Bill of Material (BOM)

2. Page MPS

This page is a request from the production department to ask for the necessary raw materials warehouse section. Which will be the input of a reservation system Raw Materials

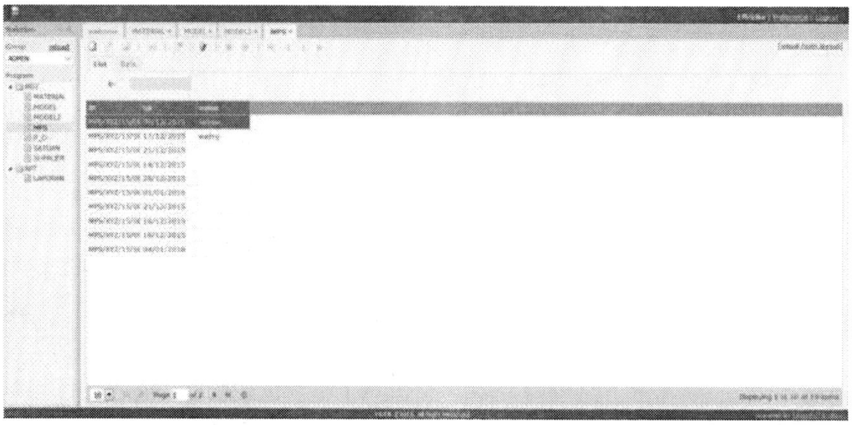

Figure 6. Page MPS

3. Reports MRP for each Model

Weather is the output of the planning system in the form of raw material ordering schedule. It is a calculation of raw materials in a dress model.

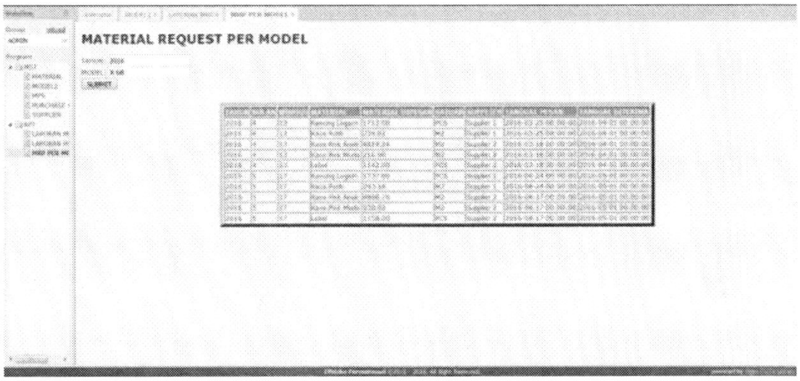

Figure 7. Report of MRP per Model

4. Planning Reports for Booking Raw Material (MRP)

This is a scheduling page that can be done on the date calculated by the system. The calculation of all MPS that have been prepared by the production department which generates the whole or recapitulation of the required raw materials and their order schedule.

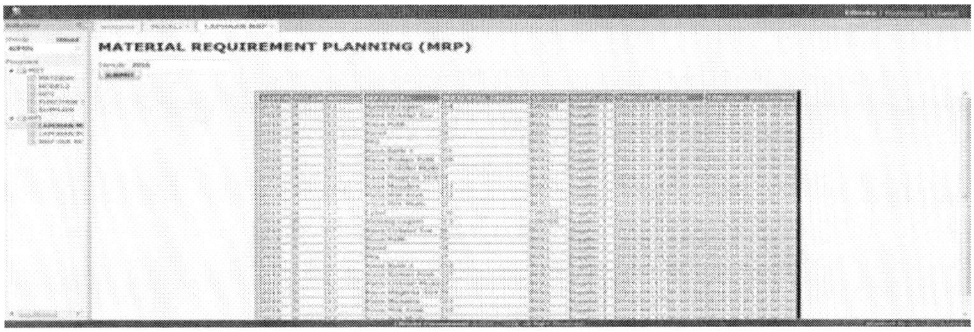

Figure 8. Report of Raw Materials Planning Order

5. The Reports page PO

This is the PO to supplier page report that will be made by the warehouse to order raw materials. Laporan shown on this page is based on a PO to a single supplier.

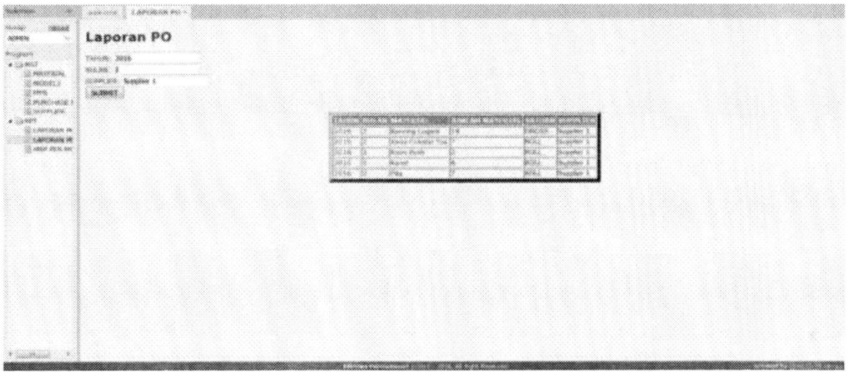

Figure 9. Report of Purchased Order (PO)

4. CONCLUSION

4.1 Conclusions

Raw Materials Planning Order System can be used in Garment Industry and has services and benefits:

1. This system provides solutions and convenience for users to order raw materials. The system produce output in the form of raw materials booking schedule, using Material Requirement Planning (MRP) method with the assumption that goods arrive on time.
2. Master Production Schedule (MPS) is a requirement to be entered into this system. MPS are resulted from Production Scheduling Information System and generated from the demand forecasting results of sales transaction history data in Online Transaction Processing (OLTP).
3. With the construction of this system, it can reduce errors in ordering raw materials up to 80%, which includes the number of orders and order time. Errors can occur if there is a change in market demand from what has been predicted or there are supplier time constraints in sending goods outside of previous planning.
4. The development of this system results in the planning of ordering raw materials in the form of a schedule for ordering raw materials according to the previously estimated, so that the risk of delays in the delivery of raw materials from suppliers can be handled properly.
5. Based on the results of research conducted by the company, the suggestion of acquisition of raw materials must be 3 days to anticipate things that are not desirable. This can actually be done by this system.

4.2 Recommendations

Suggestions for further research authors who want to develop or continue research of this system can be integrated with the sub-systems of production, marketing and storage of raw goods so that data can be integrated and usefulness of this system can be stronger.

Acknowledgments

Author want to acknowledge the help or encouragement of co-workers, special work by lecturers and students from UAI, who have provided support and opportunities for joint research.

5. REFERENCES

[1] Shih-Chih Chen, et al. 2012 Advanced planning and scheduling (APS) and enterprise planning system (ERP) planning reference models for Personal Computer assembly industry: A process perspective. *African Journal of Business Management* **6**(14), pp. 4784-4794. [online]. Available at http://www.academicjournals.org/AJBM.

[2] Biki, Munawir 2013 Sistem Informasi Penjadwalan Produksi. (Jakarta: Universitas Al-Azhar Indonesia)

[3] Nuraini, Cut 2013 Sistem Informasi Sales Order dan Distribution Requirements Planning (DRP). (Jakarta: Unversitas Al-Azhar Indonesia)

[4] Arnold Tony, et al. 2008 Introduction to Materials Management. (Pearson Prentice Hall)

[5] Heizer and Render 2014 Principles of Operations Management, Ninth Edition. (Pearson)

[6] Ismail Salaheldin 2005 An Investigation of MRP Benefit-Determinant Relationships: ACE Model. *Problems and Perspectives in Management* **2**.

[7] Taqwa, Muhammad Aulia 2013 Optimalisasi Perencanaan Dan Pengendalian Persediaan Bahan Baku Produk Baju Muslim Di Company XYZ. (Jakarta: Universitas Al-Azhar Indonesia)

[8] Kurbel, K. E. 2013 Enterprise Resource Planning and Supply Chain Management Functions, Business Processes and Software for Manufacturing Companies. Springer.

IOP Conf. Series: Materials Science and Engineering **528** (2019) 012065 doi:10.1088/1757-899X/528/1/012065

Simulation Model Development for Determination of Components Production Quantity and Lead Time Reduction in Mass Customisation of Single Production Stage

M R A Purnomo[1] and R C Shinta[2]

[1,2]Department of Industrial Engineering, Faculty of Industrial Technology, Universitas Islam Indonesia, Yogyakarta, Indonesia
ridwan_64@yahoo.com, riadho.clarashinta@gmail.com

Abstract. Market rivalry forces manufacturing companies to be customer focus and have an agile system. In manufacturing companies that produce semi-common products based on customer order, the pressure from customers are usually faster delivery, reasonable price and higher product variability. This paper presents the use of Mass Customisation (MC) concept in order to cope with faster delivery and higher product variability. The MC concept is realised by determining Customer Order Decoupling Point (CODP). Besides MC, Theory of Constraint (TOC) concept is also applied in order to address problem that may occurred after MC implementation. Based on a simulation analysis, it shows that CODP implementation could improve delivery lead time and product variability performance dramatically up to 5.15 times compared to existing system. An optimisation algorithm is also proposed for further study in order to improve cost performance of the products.
Keywords: Mass customization, simulation model, lead time reduction, CODP, TOC.

1. Introduction

Market rivalry has forced manufacturing systems to focus on customer satisfaction as one of their important objectives. It has been realised that several big manufacturing companies are not able to survive due to their fail in satisfying the customers. With globalization and pressure from customers for faster delivery, reasonable price, and higher product variability, it is expected that manufacturing companies strive for a more effective and efficient production. Thus, there are often happened a tradeoff between cost and other performances when efficiency is strived for [1]

In a production system, there are production planning and control activities arranged based on customer demand. This is one of operational strategic that applied by company to satisfying the customer. One of the outstanding results in production planning that could satisfy the customer is by improving the inventory control system especially in Make to Stock (MTS) to Make to Order (MTO). According to [2] inventory control is the act of maintaining the inventory at a reasonable level that could fulfill customer's demand in terms of date and amount, which leads for minimizing total costs and maximizing profit.

[3] stated that MTS production system is based on forecasts of product demands and production is triggered without taking into account customer orders. Hence, considerable holding costs or stock-out costs are inevitable in a context of highly fluctuating demands. According to [4], in MTO

environment, the production process begins only after orders are received from the customers. In some cases, the company used this method to reduce inventory because the product orders are known in advance. However, the MTO company requires longer delivery lead time due to materials preparation and late production starting time. Hence, based on the complexity which happened in MTO and MTS company, a potential paradigm called Mass Customization (MC) could be proposed to address the complexity.

Elaborated company in this study is a company produces especially hospital equipments and other minor products. This company is still struggling to fulfil customer satisfaction in term of shorter delivery lead time and optimum production quantity. The problem was occurred due to there is a shared production machine to produce several components and that machine then become a bottleneck machine. Objective of this study is to develop a simulation model to determine optimum production quantity with shorter delivery lead time. A simulation model is proposed due to very complicated process in the production system.

Problem identification was carried out by implementing well known theory called Theory of Constraint (TOC). The shared production machine causes blocked process, and after the blocked process could be identified, then Mass Customisation (MC) concept was applied by determining Customer Order Decoupling Point (CODP). Identifying CODP is to know the position of stock and order which has purpose to reduce the delivery lead time to customers and increase the manufacturing efficiency by optimising the bottleneck machine [5].

2. Related Works

There are several previous studies that are much related to Mass Customisation (MC) study. [6] stated that in an electronics industry company, the concept of MC could be used to gain higher improvements in the efficiency and performance. In the concept of MC which concern on a module-based product range and the use of product configuration systems for sales and order processing, the company record shown that American Power Conversion (APC) did a mass production of the standard components and fulfil the customer orders based on the final assembly. By implemented the concept of MC, the whole delivery time for a complete system could be reduced from 400 to 16 days. It also influences the production cost that significantly reduced.

[7] explained about timing games and Flexible Manufacturing System (FMS) using barbell model with continuous time and make a repetition for purchasing to identify the product differentiation and production flexibility affect firms' timing of order entry. That study focuses on the maximum differentiation in product location and minimum timing differentiation as the outcomes. The result of that research is there is no market failure in the product location choices but there were lateness and the earlier entry could be classified as flexible production system. Otherwise, there also lateness entry which can be classified as less flexible production system.

TOC usually concern in high restricted contribution margin, reducing inventories and operating expenses. Hence, profitability could be improved efficiently. A study showed that TOC could be implemented in a furniture company by identifying the bottleneck or capacity constraint. Then, the constraint could be explored effectively and it increased up to 42% of company throughput [8].

Another study has been carried out in a garment industry that concerns on the MC that integrated with custom-made goods on a mass scale with good quality, low price and high productivity. In that study, there is a need to respond the customers' demand quickly and could be categorised in which product that could be customised and could be reached in very short time. Because of the fluctuating demand, there is a need quick respond to market that leads to the lead time to be shorter. In that industry distribution network, it has three tiers that consist of customers, three logistics service providers and factories. Based on those tiers, customers were categorised as end customer and the third logistic service providers and the company were categorised as the upstream for manufacturers. Thus, it could be concluded that the half-finished products are produced with mass production method with very low cost. Besides that, the product modules to be produced for customised products are based on the customised data [9].

[10] explained that the variety of customer needs and wants forced the manufacture to be agile. The variety of product must be high with short delivery time. Thus, this situation forces the company to combine among the order based and stock based with high flexibility and short manufacturing lead time using an MC concept. According to that study, the company was able to reduce the manufacturing lead time from 43 days to 24 days when producing higher product variety. In the analysis, that study was using product picture analysis to classified the basic components of every product, customised components and required manufacturing process. Operation process chart (OPC) was also required for MC analysis. That study also proposed for future study to apply an optimisation algorithm to control the inventory amount and set up time before CODP manufacturing process.

Thus, based on the previous studies, this study would conduct the analysis of MC among the components that should be produce first. After MC analysis by identifying the CODP, then a simulation model was developed to do further analysis to optimise the production quantity. A simulation software called FlexSim6 was used to develop the simulation model.

3. Analysis Method

Data collection was carried out using observation and interview. The observation applied direct observation to know the production process flow and to collect quantitative secondary data. The interview was also carried out by interviewed the production manager and Production Planning and Inventory Control (PPIC) staffs. Data processing was carried out in several stages which are:

3.1. Product structure analysis

There are 4 main products considered in this study. The 4 main products have higher demand from customer and therefore could be used as the basis of the product analysis. The first step in this study was analysing the products designs in detail in order to find components that actually could be considered as common components.

3.2. Product process analysis

This stage was carried out based on operation and routing sheet. It resulted that the required machines and their processing time to make every component could be identified. Besides that, production layout was also need to be identified in order to know exactly position of every machine in the production shop floor. Rich Picture Diagram (RPD) was used to depict the production process flow on the real machines position.

3.3. TOC

In this study, TOC was used to identify existence of blocked or bottleneck machine which can cause the longest processing time during components production. TOC was also used as the basis of buffer size determination to keep the bottleneck machine works efficiently.

3.4. Simulation model development

Due to complexity of the manufacturing system, it will be more effective and efficient when analysis is carried out based on a simulation model. Development of the manufacturing system simulation was carried out based on the production layout and Operation Process Chart (OPC). The simulation model would be confirmed with the real manufacturing system based on two parameters which are the production quantity and the production lead time. The valid simulation model would then be used to do manipulation to improve performance of the both parameters.

3.5. Identification of CODP

In this stage, an analysis based on MC was carried out in order to determine the CODP. First, all of the products would be disassembled in order to analyse the uniformity of the required components based on their production process. In determining the CODP, there is a trade off between MTS concept that is started from raw materials and MTO concept that is close to end products. The CODP is determined by pushing it forward to be closer to end product to increase mass production level and pushing it backward from end product to be closer to raw materials to increase product flexibility. After the CODP is determined, then the bottleneck machine must be exploited effectively by applying TOC. Bottleneck machine protection could be carried out by providing buffer in front of it to keep it works effectively. The number of buffer was determined based on several simulation scenarios.

3.6. Improvement Verification
CODP would be used as the reference point to manage the production system. MTS concept would be implemented from raw materials to CODP and MTO concept would be implemented from CODP to end products. However, before implementation, verification is required and the simulation model could be used as the experiment object.

3.7. Identification of Lead Time Reduction
After improvement verification and improvements could be seen visually from the simulation model, and then new production lead time based on CODP implementation could be estimated based on the new production quantity.

4. Result and Discussion

4.1. CODP Implementation
In this study, according to product structure and product process analysis which be used as the basis to identify CODP position, there are 62 components that should be analysed. Analysing all of the components lead to ineffective analysis and the result might be infeasible. Hence, a critical component that has highest uniformity and longest processing time was considered as the basis of analysis. The component is called RawMat17. CODP position based on the RawMat17 is shown in Figure 1.

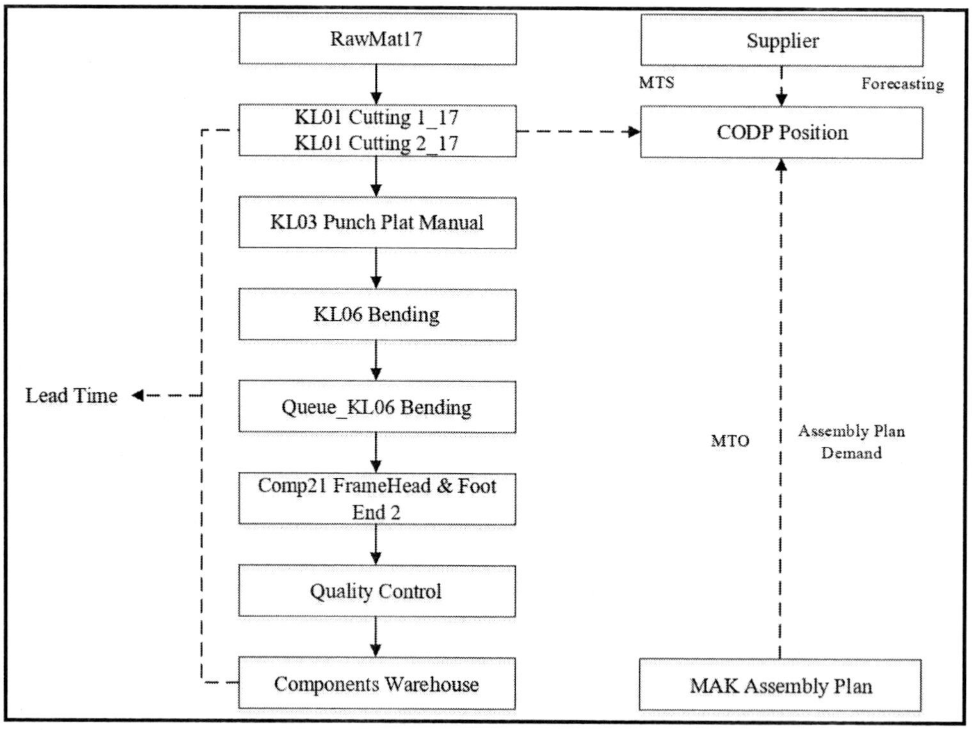

Figure 1. CODP Position based on RawMat17

There are 2 common sub-components required by the investigated component, which are Comp20 and Comp21. In this study, based on production process analysis, there are two potential CODP in the production system, which are at the cutting process carried out by machine KL01 (cutting CODP) and at the Quality Control (QC CODP) process. In order to know the best result, then the simulation system was executed based on cutting CODP and QC CODP. The simulation result shows that QC CODP has superior performance in production quantity. Table 1 shows the comparison of cutting and QC CODP.

Table 1. Summary Simulation Report

No	Simulation Model Development for	CODP Position (Cutting Section)	Improvement (pieces)		Components Warehouse
			Comp20	Comp21	
1	Existing System	-	43	11	336
2	Cutting CODP	KL01 Cutting	36	112	334
3	QC CODP	Quality Control	36	112	1732

Based on the simulation study, it shows that CODP implementation could increase manufacturing lead time and product variety dramatically. However, there is some potentiality to have high amount of components inventory in before-CODP manufacturing processes. Besides that, reduction of set-up time is also considerable to improve performance of manufacturing lead time. For further study, it is recommended to apply an optimisation algorithm to control the inventory amount and set-up time reduction analysis in before-CODP manufacturing processes.

4.2. Lead Time Reduction

It is very hard calculate production lead time in a simulation due to dynamics that have been simulated in the system. However, with referring to Table 1, the component warehouse could be used as the basis to estimate the production lead time. When QC CODP is proposed then the output could increase dramatically 5.15 times than the existing production system. It is reasonable because when the CODP is at the QC point, means that the company could produce the components even though customer order still has not received. The analysis is also show that the difference between products in the investigated company is not much. Customisation could be conducted after QC point and it takes very short time.

5. Conclusion

Based on explanation above, it could be concluded that MC concept is very potential to be implemented to improve delivery lead time and production quantity performance. By implementing CODP, the production system could be managed based on MTS and MTO environment. However, in the MTS sub-system, process block could occur due to a bottleneck machine. TOC could be implemented in order to exploit the bottleneck machine effectively. An optimisation algorithm to control inventory units is also could be implemented to improve cost performance of the production system.

6. References

[1] Freiheit T, Wang W, Spicer P, Wang W and Spicer P 2007 A case study in productivity-cost trade-off in the design of paced parallel production systems. *Int. J. of Production Research* **45** pp 3263–3288

[2] Nagib A N M, Adnan A N, Ismail A, Halim N H A and Khusaini N S 2016 The Role of Hybrid Make-to-Stock (MTS) - Make-to-Order (MTO) and Economic Order Quantity (EOQ) Inventory Control Models in Food and Beverage Processing Industry. *IOP Conference Series: Materials Science and Engineering* **160** 012003

[3] Rafiei H and Rabbani M 2014 Hybrid MTS/MTO order partitioning framework based upon fuzzy analytic network process. *Applied Soft Computing* **19** pp 312–321
[4] Tadeuz and Maruf A 2016 Scheduling Method for MTS/MTO Production System Scheduling Method for MTS/MTO Production System. *IOP Conf. Series: Materials Science and Engineering* **114**.
[5] Olhager J 2003 Strategic positioning of the order penetration point. *Int. J. of Production Economics* **85** pp 319–329
[6] Hvam L 2006 Mass customisation in the electronics industry: based on modular products and product configuration. *Int. J. of Mass Customisation* **1**(4) p 410.
[7] Sun C H 2017 Timing of entry and product location in a linear barbell model: Application to flexible manufacturing systems. *Int. Review of Economics and Finance* pp 1–23
[8] Okutmuş E, Kahveci A and Kartašova J 2015 Using theory of constraints for reaching optimal product mix: An application in the furniture sector. *Intellectual Economics* **9** pp 138–149
[9] Dong B, Jia H, Li Z and Dong K 2012 Implementing Mass Customization in Garment Industry. *Systems Engineering Procedia* **3** pp 372–380
[10] Purnomo M R A and Sufa M F 2015 Simulation-based Performance Improvement Towards Mass Customization in Make to Order Repetitive Company. *Procedia Manufacturing* **2** pp 408–412

Mapping of noise levels made by drilling machines on project x using contour zone method

B Cahyadi[1] and G A Timang[2]

Industrial Engineering, Faculty of Engineering, Pancasila University, Jakarta, Indonesia
[1]bambang.cahyadi@univpancasila.ac.id, [2]gitatimang20@gmail.com

Abstract. Project x is an apartment development activities in Tangerang which in the process of construction resulted in the occurrence of noise. Noise is one of the incidents that can be lead to health problems for the surrounding community. Therefore, this research aims to know the spread of noise, the amount of noise reduction that can be reduced by using a hearing protective device and workers' perceptions of noise against hearing loss. The method used is the contour zone. Data used is the result of noise measurement Project x in the daytime and nighttime conditions. The result obtained is the noise level in the project work area during the daytime in area 1, 2 and 3 with a distance of 2 meters are 97.1 dB, 94.4 dB, 93.9 dB, and length of 4 meters are 94.4 dB, 91.5 dB, 93.0 dB and night measurements at a distance of 2 meters are 102.2 dB, 100.1 dB, 96.5 dB, and length of 4 meters are 94.7 dB, 94.7 dB, 97.4 dB. The calculation of day and night above exceeds the noise threshold value set by the Ministry of Labor which is 85 dB. The perception of the worker with the noise disturbances that occur at The Yukata Project. Using hearing protection devices will reduce the noises, but at nighttime conditions in areas 1 and 2 at a distance of 4 meters, the noise level still exceeds the noise threshold value which has been determined by the ministry of labor.
Keywords: Likert, Noise, Mapping, Contour Zone.

1. Introduction

In this part, the research background, and literature review will be shown.

1.1 Research background

The current state of basic human needs especially in living areas experiencing an increase, this can be seen from the increasing number of apartment buildings. One of the developing residential projects is the x project, which produces a lot of noise in its construction process.

Noise is all unwanted sounds that are sourced from the tools of the production process and or work tools that are at a certain point can cause hearing loss [5]. Noise is one type of pollution that needs to be considered at this time because it has a direct impact on the level of comfort and human health. The impact of the noise itself varies, namely hearing loss, health problems such as psychological disorders, physiological disorders, and balance disorders [5]. The damaging effects that often occur if a person is exposed to noise for too long are the effect of hearing, but using the personal protective equipment provided by the company can be attempted.

In 2012, Waskita Realty was established and used as a division of PT. Waskita Karya (Persero), Tbk. This company focuses on providing reliable services in the property sector in Indonesia. In 2017, Waskita Realty in collaboration with Trinity Group is in the process of constructing project x. The construction process of this project resulted in the creation of a noise due to the collision between a mini crane pile drilling machine and building material from the construction of the Yukata apartment.

An effort to control noise in the project work and housing environment to be more effective, it is necessary to measure noise levels using noise mapping or contour maps to determine the pattern of noise distribution that occurs in the project work and housing environment. To create a noise contour map due to the activity can use a software called Surfer. It is also necessary to analyze the workers' perceptions of the noise to hearing loss whether it is influential or not, using a Likert scale in processing questionnaire data and calculating the value of noise reduction that can be dimmed by the hearing protection devices used by workers.

1.2 Literature Review

1.2.1. Effects of Noise on Health. Noise is an event that can cause a variety of health problems, such as physiological disorders, psychological disorders, communication disorders and hearing loss (deafness) [6].

The most common effects of noise on worker health are effects on hearing, the main effect of noise on health is damage to the sense of hearing, which can cause temporary or permanent deafness.

1.2.2. Noise Measurement Method. Based on the Decree of the Ministry of Environment No: KEP-48/MenLH/11/1996, Regarding the Noise Standard Level, 25 November 1996. The calculation of noise level can be done in two ways [4], as follows:

1. Simple Way
   With a sound level meter, the dB sound pressure level is measured for 10 minutes for each measurement and readings are done every 5 seconds.
2. Direct Way
   With an integrating sound level meter that has a measuring facility of Lm_5, which is a leq with a measuring time of every 5 seconds, it is measured for 10 minutes.

1.2.3. Noise Level Calculation. Here are some ways to calculate the noise standard level according to the ministry of environment No:Kep-48/MenLH/1/1996 [4], as follows :

1. Noise level during the daytime (L_d). The level of noise that occurs during the day with a sound pressure level for 16 hours between 06.00 - 22.00. Daytime noise levels can be denoted by the symbol Ls. Can be formulated as follows :

   $$Leq_{daytime} = L_d = 10Log^{1}/_{16}\{\sum t_i . 10^{0,1.l_1}\} dB \quad (1)$$

   L_s is daytime noise level (dB), t_i is measurement time interval, and l_i is Leq averages at a certain time interval.
2. Noise Level at Night (L_n). The level of noise that occurs at night with the sound pressure level for 8 hours between 22.00 - 06.00 with a minimum of data collection for 3 measurements with a certain frequency range. Can be formulated as follows :

   $$Leq_{nighttime} = L_n = 10Log^{1}/_{8}\{\sum t_i . 10^{0,1.l_1}\} dB \quad (2)$$

   L_m is night noise level (dB), t_i is interval of measurement, and l_i is Leq averages at a certain time interval.
3. Noise Levels During Day and Night (L_{sm}). Daytime to nighttime noise levels is used to find out whether noise has exceeded the noise level from field measurements or not. Can be formulated as follows :

$$L_{sm}=10Log^{1}/_{24}\{16.\,10^{l_s/10} + 6.\,10^{(l_m+5)/10}\}\text{dB} \quad (3)$$

L_{sm} is value of noise level during day and night, L_s is value of noise level during daytime, and L_m is value of noise level during the night.

1.2.4. Hearing Protection Equipment for Noise. The using of hearing protection equipment for noise is something that must be done, especially for a worker who is exposed to more than 85 dB per day.

Personal protective equipment used must be able to reduce noise to less than 85 dB. There are three types of hearing protection devices [4], namely:

1. Earplugs : can reduce noise 8 - 30 dB, usually used for protection up to 100 dB.
2. Earmuff : can reduce noise from 25 to 40 dB, used for protection up to 110 dB.
3. Helmet : can reduce noise 40 - 50 dB.

1.2.5. Threshold Value and Noise Zone. Threshold Value The noise limit permitted by the Ministry of Manpower is 85 dB because it is considered safe for most workers when working 8 hours/day or 40 hours/week.

Each noise has an area that is divided according to the noise points that are permitted by the Ministry of Environment as regulated in KEP-48/MENLH/11/1996 [4], namely :

1. Zone A: Intensity 35 - 45 dB.
   Zones designated for research sites, health / social care places.
2. Zone B: Intensity 45 - 55 dB.
   Zones designated for housing and public settlements, schools, and places of worship.
3. Zone C: Intensity 55-65 dB.
   Zones designated for offices, cultural heritage, public facilities, and markets.
4. Zone D: intensity 65 - 75 dB.
   Zones designated for industries, recreation areas, terminals, and factories.
5. Other zones (transportation): An intensity of more than 75 dB. Zones adapted to the Ministry of Transportation such as railway stations, seaports, and airports.

1.2.6. Noise Measuring Tool. Sound Level Meter is a device used for measuring a sound intensity, sound level meters used to measure sound intensity levels can use sound sources and used to measure sound levels in decibels (dB) [3].

Measurements using sound level meters are usually used in the study of noise pollution for noise quantification, but especially for industry, environment and aircraft noise.

1.2.7. Surfer Software. A surfer is one of the software used for making contour maps and three-dimensional modeling, surfer does not require high hardware or operating systems. Therefore, surfers are comparatively easy in their applications. Surfer can be used on Windows 9x and Windows NT operating systems. Provide convenience in loading various types of contour maps or 3-dimensional (3D) spatial models [2].

1.2.8. Likert scale. Likert scale is a scale that can be used to measure attitudes, opinions, and perceptions of a person or group of people about a phenomenon or phenomenon of education [1]. With a Likert scale, the variables to be measured are translated into variable indicators.

The indicators used for questions/statements are strongly agreed (5), agree (4), quite agree (3), disagree (2), strongly disagree (1). In determining the total score of each statement using the following formula [1].

$$T\, x\, Pn \quad (4)$$

T is total number of respondents who choose, and Pn is choice of Likert score.

1.2.9. Noise Reduction Rate. Noise Reduction Rate is a measure of the ability of a hearing protector to reduce noise levels [7]. The following are the calculation methods used to calculate NRR according to The United States Occupational Safety and Health Administration (OSHA) [7], namely:

1. Single Use

   Single-use is the use of one ear protector only, can use earplugs or earmuffs. The following formula is used:

$$NRR_{actual} = (NRR_{label} - 7) \times 50\% \quad (5)$$

NRRactual is actual noise reduction level, NRRlabel is the level of noise reduction seen on the packaging tool label, value of 7dB is determination of noise reduction using hearing protection devices by the National Institute for Occupational Safety and Health (NIOSH), and 50% is average safety provisions used based on the National Institute for Occupational Safety and Health (NIOSH) methods.

2. Methods

The method used in this study is divided into 3 stages, namely:

1. Perform noise mapping, the steps are:
   a. Calculate noise equivalent during the daytime (L_d). Calculates nighttime equivalent noise (L_n) and calculates noise day-night equivalent (L_{dn}). The data used is noise level measurement data.
   b. Performing noise mapping with the contour zone method using Surfer 15.0 to determine the spread of noise that occurs. The data used is the result of the calculation of the equivalent noise level.
2. Analyzing worker perceptions of hearing loss due to noise with the following steps:
   a. Distributing questionnaires.
   b. Analyze worker perceptions using a Likert scale.
3. Calculating the value of noise reduction from the hearing protection equipment used by workers.

3. Result and Discussion

3.1. Noise Mapping

After calculating noise level data, the data is then processed by the surfer for noise/contour noise mapping. The results of the pattern of noise distribution are as follows.

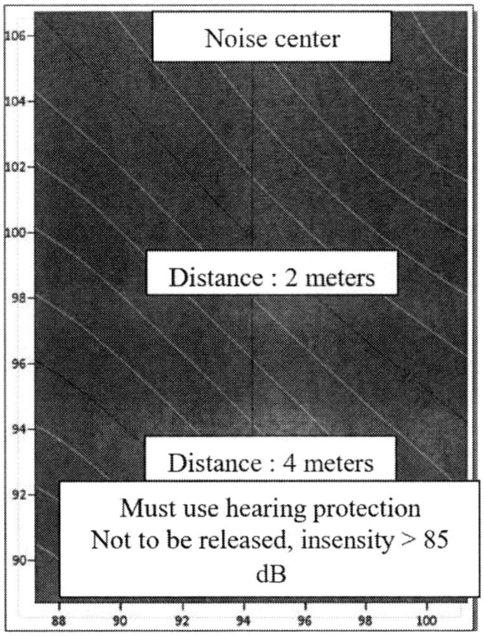

Figure 1. Work Area Contour Map 1

Based on the result in **Figure 1** in the project work area 1, the noise threshold value is 85 dB. All areas in the countor map are red which means the noise value is more than 85 dB. Region 1 has a maximum average noise intensity of 106 dB and a minimum of 88 dB. So, workers are required to use hearing protection equipment.

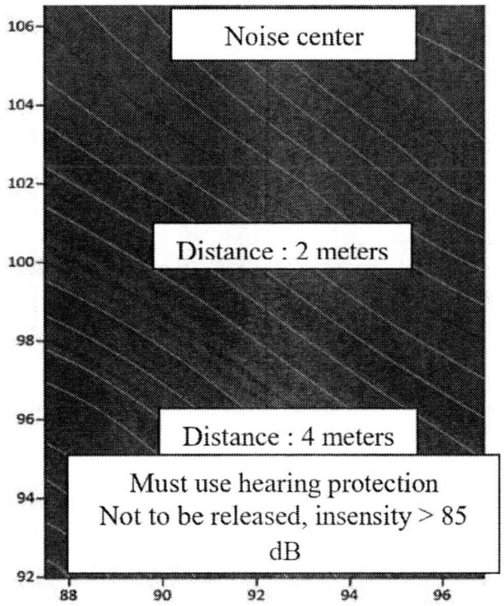

Figure 2. Work Area Contour Map 2

Based on the result in **Figure 2** in the project work area 2, it is obtained to exceed the noise threshold value of 85 dB. All areas in the countor map are red which means the noise value is more than 85 dB, region 2 has a maximum average noise intensity of 106 dB and a minimum of 88 dB. So, workers are required to use hearing protection equipment.

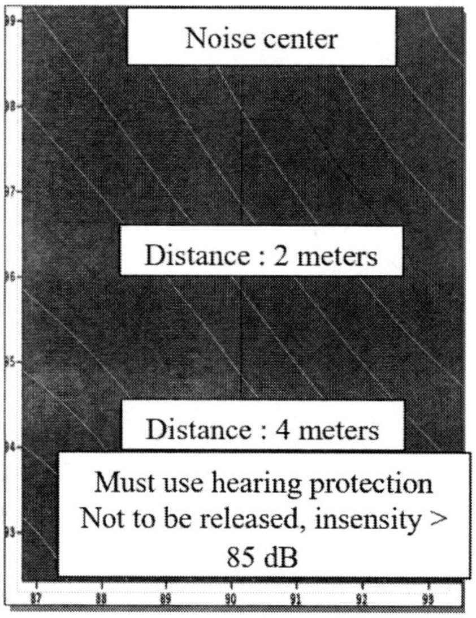

Figure 3. Work Area Contour Map 3

Based on the result in **Figure 3** in the project work area 3, it is obtained to exceed the noise threshold value of 85 dB. All areas in the countor map are red which means the noise value is more than 85 dB, region 3 has a maximum average noise intensity of 99 dB and a minimum of 87 dB. So, workers are required to use hearing protection equipment.

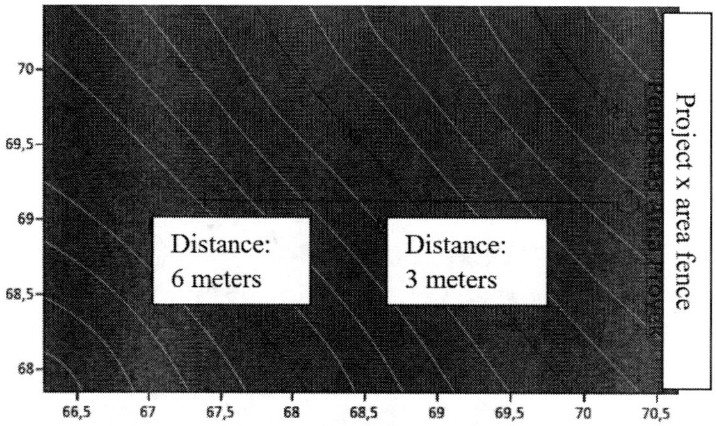

Figure 4. Residential Area Contour Map 1

Based on the result in **Figure 4** in the housing area 1 was obtained, exceeding the noise threshold value of 55 dB for the housing zone. All areas in the countor map are red which means the noise value is more than 55 dB. Area 1 has a maximum average noise intensity of 70.5 dB and a minimum of 66.5 dB.

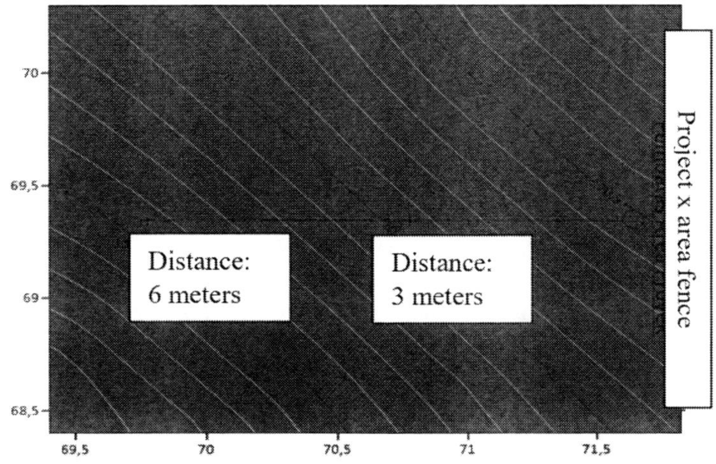

Figure 5. Residential Area Contour Map 2

Based on the result in **Figure 5** in the housing area 2 was obtained, exceeding the noise threshold value of 55 dB for the housing zone. All areas in the countor map are red which means the noise value is more than 55 dB, region 2 has a maximum average noise intensity of 71.5 dB and a minimum of 69.5 dB.

3.2. Noise Reduction Rate

To reduce the level of noise that occurs, the actual noise reduction level of the hearing protection equipment used by workers is carried out. Hearing equipment (earplug and earmuff) used by workers has the same Noise Reduction Rating (NRR) is 33 dB. So, the results after using the formula (5) the results of NRR actual is 13 dB.

3.3. Workers' Perceptions of Noise against Hearing Loss

Analyze worker perceptions about noise against hearing loss with the following steps:

a. Conducting questionnaires.

Questionnaires were distributed to workers aged between 25-40 years, male sex and previously worked in a place that has a noise of 1-year duration. From this limit, 12 respondents were obtained.

b. Calculate the questionnaire using a Likert scale.

1. Calculate Likert total score.

   The score parameter used to calculate the resulting value is strongly agree (5 point), agree (4 point), quite agree (3 point), disagree (2 point) and strongly disagree (1 point). The result of likert total score by using the formula (4) is question 1 36 points, question 2 40 points, question 3 is 36 points, question 4 is 37 and question 5 is 44 points.

2. Determine the highest score (Y) and the lowest score (X).

   Y = the highest score of likert x number of respondents
   = 5 x 12 = 60

   X = lowest score x number of respondents
   = 1 x 12 = 12

3. Determine the interpretation of Likert scores with intervals.

   I = 100/score value (likert)
   = 100/5 = 20

   The distance interval from the lowest 0% to the highest 100% is 20. The following are the interpretation criteria based on the interval:

- 0% - 19.99% = Strongly Disagree
- 20% - 39.99% = Disagree
- 40% - 59.99% = Simply Agree
- 60% - 79.99% = Agree
- 80% - 100% = Strongly Agree

4. Calculate the index (%) of each question.
   Index (%) = (Total Score / Y) x 100
   a. Question 1, Index = (36/60) x 100 = 60% = Agree
   b. Question 2, Index = (40/60) x 100 = 66.67% = Agree
   c. Question 3, Index = (36/60) x 100 = 60% = Agree
   d. Question 4, Index = (37/60) x 100 = 61.67% = Agree
   e. Question 5, Index = (44/60) x 100 = 73.33% = Agree
5. Average index calculation for all questions :
   $X = \Sigma Xi / n$
   $X = (\Sigma 321.67\%) / 5 = 64.3\%$
   Based on the index value obtained from the calculation of all questions into the score interpretation of 64.3%, it can be concluded that the workers feel disturbed by the noise that occurs in the Yukata Project.

4. Conclusions

Based on the results and discussion, the following conclusions are obtained:

1. From the results of contour noise mapping, noise distribution patterns that occur in the construction of project x in area 1, 2 and 3 on project's area with a distance of 2 meters during day time the noise level that occurs in the area exceeds the noise threshold value there are 97.1 dB; 94.4 dB and 93.9 dB and at a distance of 4 meters the noise level are 94.4 dB; 91.5 dB and 93.0 dB. While at night time measurements in area 1, 2 and 3 with a distance of 2 meters, the noise level that occurs in the area exceeds the noise threshold value are 102.2 dB; 100.1 dB and 96.5 dB and at a distance of 4 meters the noise level are 94.7 dB; 94.7 dB and 97.4 dB.
2. Workers' perceptions of noise disturbances that occur at agreed intervals, workers feel disturbed by the noise that occurs at the project x.
3. The value of reduction of the hearing protection device used by workers, either earplug or earmuff during daytime conditions at a distance of 2 meters and 4 meters in area 1, 2 and 3 is 13 dB, the same thing is also obtained at night conditions.

5. References

[1] Djaali A 2008 *Skala Likert* (Jakarta: Pustaka Utama)
[2] Graham S 2007 *Mapping a Global Business Model.* Colorado.
[3] Harris C M 1991 *Handbook of Acoustical Measurements and Noise Control.* 3rd Ed. (New York: McGraw-Hill, Inc.)
[4] Kementerian Lingkungan Hidup 1996 *Baku Tingkat Kebisingan Jakarta.*
[5] Peraturan Menteri Tenaga Kerja dan Transmigrasi RI No: 25/2008 2008 Pedoman. Diagnosis dan Penilaian Cacat karena Kecelakaan dan Penyakit Akibat Kerja Kementerian Tenaga Kerja dan Transmigrasi Jakarta.
[6] Peraturan Menteri Kesehatan RI Nomor 718/MENKES/Per/IX/1987 tentang kebisingan yang berhubungan dengan kesehatan Departemen Kesehatan RI Jakarta.
[7] The United States Occupational Safety and Health Administration (OSHA) Noise Reduction Rate.

Competency Measurement Instrument Design for Maintenance Staff of Electronic Expertise with SECI Method

Agisni[1], R P Soesanto[2], A Kurniawati[3], N Ambarsari[4] and L Andrawina[5]

School of Industrial Engineering, Telkom University, Bandung, Indonesia
gisagisni@gmail.com[1],raysoesanto@telkomuniversity.ac.id[2],
amelia.kurniawati@telkomuniversity.ac.id[3],niaambarsari@telkomuniversity.ac.id[4],
luciana@telkomuniversity.ac.id[5]

Abstract. To gain competitive advantages, companies must combine and collaborate with the tangible and intangible assets. The most important intangible assets in companies are people because people have the knowledge to run the companies. Competencies of each individual in a company is a basic characteristic that must be taken into account. The purpose of this research is to improve the quality of human resources of maintenance staff in the manufacturing company by designing competency measurement instrument that aims to assess the competency of employees. The focus of this research is the mechanical expertise of maintenance staff. SECI method is used in this research for managing knowledge that is held by senior employees regarding employee competence in electronic expertise. The SECI method converts the knowledge of a person's tacit knowledge into an explicit knowledge so that the knowledge can be used by others. The knowledge that is gathered from SECI method is converted into a list of competence and break down into the detailed competency. The result of this paper is the competency measurement instrument that aims to assess the competency of employees. Future research can be done to test the instrument and add more specific criteria.
Keywords: Competency Measurement, Maintenance, SECI Method

1. Introduction

Nowadays, companies are a set of the collaboration of tangible and intangible assets to achieve companies objective. The example of tangible assets in companies are land, labor, and capital. Tangible assets are easy to develop by increasing the number of such assets. The intangible assets in companies usually embedded in people, which come in a form of knowledge. People as an asset is considered as the important factors to gain a competitive advantage [1]. Type of knowledge in people can be divided into two category, which is explicit and tacit knowledge [2]. Explicit knowledge is defined as the knowledge that is easy to articulate and in form of formal and systematic language, where tacit knowledge is defined as the knowledge within the people that forge through experience in form of the mental model. The competitiveness of industry comes with the concept of the needs of bringing the availability of knowledge worker into an industry. The knowledge worker in the industry will bring value-added and increase the productivity of companies to gain a competitive advantage.

Knowledge lies within a people, there is a risk of losing the resource when the people in the organization is retired or resign from the organization. As stated by Alwis, Hartmann, and Gemunden [3], the most important knowledge in people is tacit knowledge and it is hard to be completely transferred into formal language. Company Z is a manufacturing company that focused on manufacturing part of automotive and aircraft. In order to maintain the productivity in optimum level, there is maintenance department which focused on maintain the machine in the company. There is different education background within the maintenance department staff, the difference can be used by the supervisor to establish the work assignment by understanding each competency needed for each specific job. There are three specific competencies that are set by the companies, which is mechanical, electrical, and mechanical. Figure 1 shows the number of an employee from the maintenance department in company X.

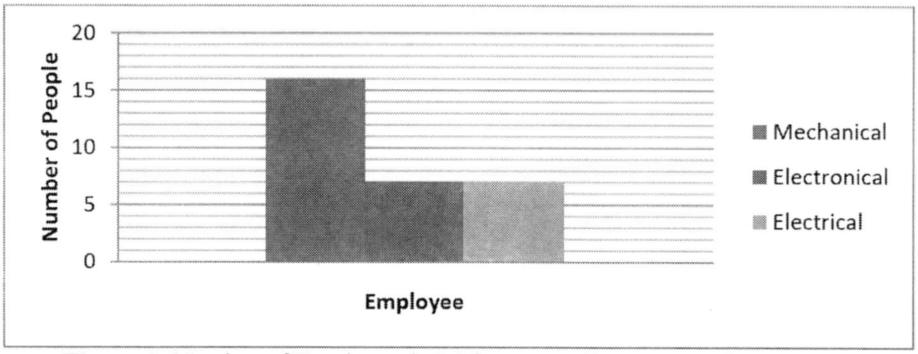

Figure 1. Number of Employee in Maintenance Department Company X,

From figure 1, it is known that the mechanical expertise employee is the largest in the maintenance department and the others are a smaller size than the mechanical expertise employee. It is important to have good competencies and skill from the maintenance department employee because it is needed a specific set of skill to maintain a machine. According to the head of the maintenance department, it is known that there is enough employee to do the maintenance job, but there is no standardization about how to assess the employee according to the current skill and competencies. The assessment of the maintenance operator usually just intuition from the operator supervisor. The standardized assessment and assignment process is considered important because in 5 years from now a lot of maintenance operator retired. The needs of measurement for assessing employee competencies is high. The purpose of this research is to design a competency measurement instrument that aims to assess the competency of employees. Previous research about the knowledge conversion practice is the research from Mahatmavidya, Soesanto, Kurniawati, and Andrawina [1] that focused on converting the expertise of maintenance staff. The study from Lis [4] focused on knowledge conversion in military organizations. Another research of knowledge conversion application is the research from Nugraha, Kurniawati, and Hediyanto [5] that focused on knowledge conversion in a research institution. Research from Gao and Nee [6] focused on converted knowledge for the product development process. Research from Kalogeraki, Apostolou, Polemi, and Papastergiou [7] focused on converting knowledge for maritime and logistics supply chain using SECI and semantic web technologies.

2. Methods

The purpose of this research is to improve the quality of human resources of maintenance staff in the manufacturing company by designing competency measurement instrument that aims to assess the competency of employees. Nonaka and Takeuchi [2] describe the four phase of converting knowledge which is socialization, externalization, combination, and internalization. Socialization consists of sharing process and creation of tacit knowledge through interaction and direct observation. Externalization consists of an articulation of tacit knowledge into explicit knowledge through dialogue and reflexion. Combination consists of the conversion process of explicit knowledge into new explicit knowledge through systemization and information. Internalization consists of the learning process and knowledge acquisition by organization member towards explicit knowledge that spread into the organization through self-experience so it can be tacit knowledge of organization member. Figure 2 shows the SECI Phase.

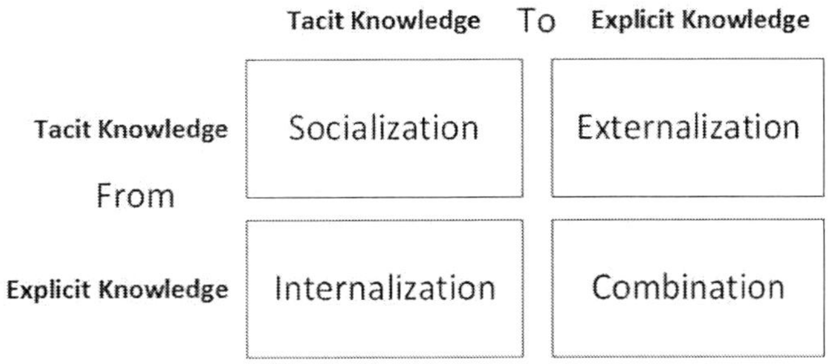

Figure 2. SECI Phase

In the socialization phase, tacit knowledge is transformed into another tacit knowledge, knowledge about job title, expertise, and etc. General competencies for electronic competencies such as specific skills, indicator, how to measure such an indicator and the result is gained from the interview with the supervisor and senior staff. In the externalization phase, the previous knowledge about the competencies and the interview result is documented. Previous tacit knowledge then converted into explicit knowledge in form of review result. In the combination phase, explicit knowledge is converted into new explicit knowledge through a combination of knowledge. From the interview of three respondent, the information will merge into one complete information. All breakdown competencies from the socialization result are merged into one best competencies measurement. In internalization phase, the competencies measurement design is shared through the entire department. Focus group discussion is conducted to share the new explicit knowledge.

3. Result and Discussions

3.1. Socialization

Socialization is a sharing process and the creation of tacit knowledge through interaction and live experience. In the socialization phase, there is knowledge capturing process from the respondent. Figure 3 shows the socialization phase scheme. The result of this phase is tacit knowledge about job title, employee expertise, specific skills, indicators, how to measure indicators, and the result needed to asses and evaluate the competencies for electronic expertise in the maintenance department. Figure 3 shows the scheme of the socialization phase.

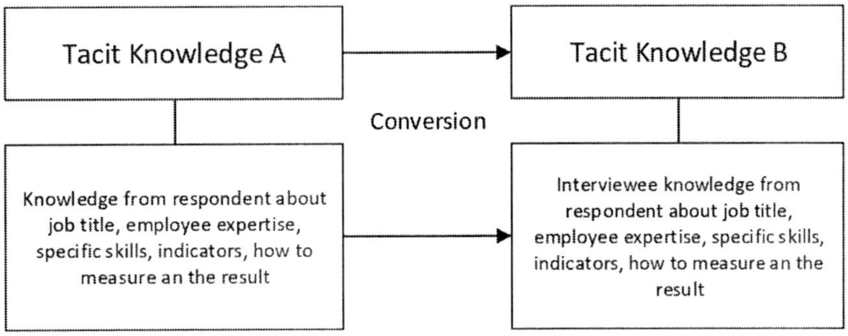

Figure 3. Socialization Phase Scheme

The electronic expertise is a skill that needed for repairing the component of the machine. According to the supervisor, all electronic expertise staff must have all general competencies for repairing the electronics part of the machine. Table 1 shows the competencies needed for electronic competencies,

Table 1. Electronic Expertise Competencies List

No.	Competencies	Specific Competencies
1	Able to use measurement tools for electric devices	Able to use a digital multimeter
2	Have basic mathematical knowledge	Able to understand the number system
3	Have basic mathematical knowledge	Able to understand the number conversion system
4	Understand the maintenance procedure philosophy	Able to understand the preventive maintenance procedure
5	Understand the maintenance procedure philosophy	Able to understand the corrective maintenance procedure
6	Understand the machine diagram	Able to understand block diagram of a machine
7	Have basic English skill	Able to understand the content of the manual book
8	Have basic electronic skills	Able to understand the type of electricity and electric lines.
9	Have machine calibrating system skills	Able to understand tools for calibrating a machine
10	Have machine calibrating system skills	Able to calibrate a machine
11	Have machine work system knowledge	Able to understand how machine work
12	Have the ability to use CMMS application	Able to understand general terms in CMMS
13	Have the ability to use CMMS application	Able to operate the CMMS
14	Have Human Safety and Environment knowledge	Understand basic HSE
15	Have the ability to operate Microsoft office application	Able to use all kind of Microsoft office software

After understanding the job title, employee expertise, specific skills, indicators, how to measure indicators, and the result needed to asses and evaluate the competencies then the next step is to design the scale for assessing the competencies. The socialization process is done by interviewing and explore previous knowledge from the supervisor and the senior staff.

3.2. Externalization

The externalization phase articulate the previous tacit knowledge gain in the socialization phase into explicit knowledge in the form of a business process. In the externalization phase, knowledge will be documented informal ways. Figure 4 shows the externalization phase scheme

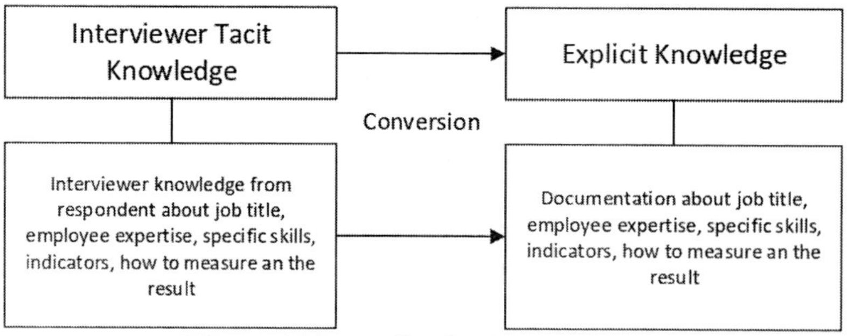

Figure 4. Externalization Phase Scheme

The job title is a brief description of what is the main task of the job and is used to categorize the position of the employee in the organization. A job title can describe the responsibilities. Every staff in the maintenance department have different basic skills. With specific categorization for each job, the line of work for each staff is clear. Table 2 shows the example of externalized knowledge from the respondent.

Table 2. Example of Externalization Process

General Competencies	Specific Competencies	Indicators	How to Measure	Measurement Result
Have basic electronic skill	Know how to use a digital multimeter	Knows part of a digital multimeter	The employee is asked to show each part and the function of a digital multimeter	In general, there are a digital display, pushbuttons, rotary switch, and input terminals in a digital multimeter, 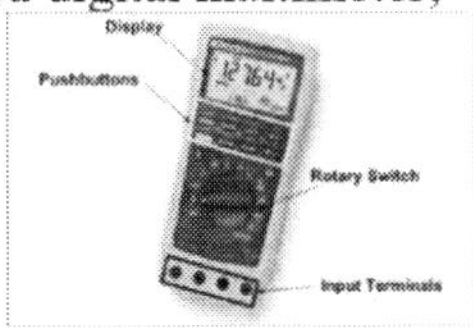

3.3. Combination

In the combination phase, the previous explicit knowledge from each respondent is combined and merged into one new knowledge. In the externalization phase, the knowledge form respondent is gathered, but there is a possibility that there is a difference between respondent in deciding what is competencies needed for electronic expertise. The externalized knowledge from 3 respondent merges into one new externalized knowledge. If 2 out of 3 respondent include the item then it will be used in the new competencies measurement breakdown. Figure 5 shows the combination phase scheme.

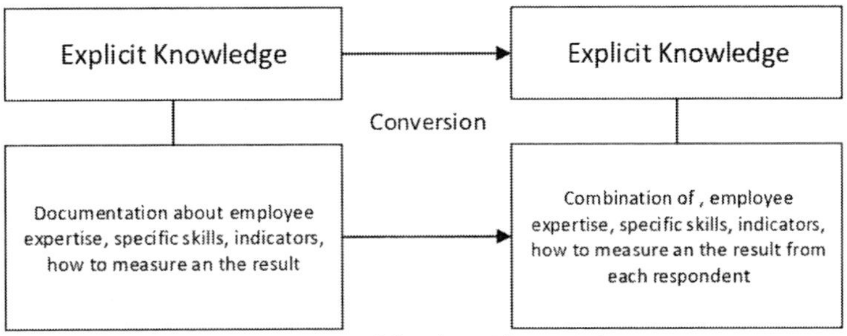

Figure 5. Combination Phase Scheme

The merging process of specific competencies is done based on the knowledge of each respondent. Specific competencies are the detail explanation about the general competencies, therefore all electronically expertise staff must have all the competencies. In deciding which competencies are needed, knowledge exploration is conducted to gain the brief knowledge about the job of maintaining the machine in electronic point of view. Indicators represent the variable that indicates or as a guide to measure about an event or activity. The indicator is used as a base for assessing the staff competencies. How to measure explain how to measure the competencies according to each indicator. The measurement result is a measurement standard that is used for measuring competencies. Table 3 shows the example of the combination process.

Table 3. Example of Combination Process

Competencies	Respondent A	Respondent B	Respondent C	Combined Spesific Competencies	Notes
Able to use the machine's accuracy gauge	Able to use dial indicator	Able to use dial indicator	Able to use dial indicator	Able to use dial indicator	Abilities to use dial indicator and spirit level to measure machine accuracy.
	Able to use spirit level	Able to use spirit level	Able to use spirit level	Able to use spirit level	
Basic Electronic skills	Able to understand the type of electricity and electric lines.	Able to understand the type of electricity and electric components	Able to understand the type of electricity and electric components	Able to understand the type of electricity, electric lines, and electric components	Basic competency for electronic expertise that focused on the knowledge of basic electric skills.

3.4. Internalization
Internalization phase is the last step in knowledge conversion. Internalization is a learning process and knowledge acquisition that is done by the member of the organization. In this phase, the explicit knowledge which is the best practice is to disseminate into tacit knowledge. Focus group discussion (FGD) is conducted in this phase to share the new knowledge into all department employee. FGD is done by discussing the competencies measurement instrument. The purpose of the FGD are (1) to convert the knowledge from explicit to tacit about the competencies, indicators, how to measure and the measurement result, (2) to discuss the competencies and etc. to the employee for sharing process, (3) to understand the willingness of the stakeholder implementing the competencies measurement. From the FGD result, it is known that with the proposed competencies measurement design all electronic expertise staff competencies can be measured so the assignment, evaluation, and recruitment activities can be exactly as needed.

4. Conclusions
This research tends to improve the quality of human resources of maintenance staff in the manufacturing company by designing competency measurement instrument that aims to assess the competency of employees. In the socialization phase, there is knowledge capturing process from the respondent. In the externalization phase, knowledge will be documented informal ways. In the combination phase, the previous explicit knowledge from each respondent is combined and merged into one new knowledge. In the Internalization phase, the explicit knowledge which is the best practice is to disseminate into tacit knowledge. The knowledge about the competencies measurement breakdown is well documented in the form of explicit knowledge. The measurement of employee competencies can be done by using the proposed instrument. Future research can be done to test the instrument and add more specific criteria.

5. References
[1] Mahatmavidya P A, Soesanto R P, Kurniawati A and Andrawina L 2017 Characteristic and Competency Measurement Instrument Development for Maintenance Staff of Mechanical Expertise with SECI Method: A Case of Manufacturing Company in *APCOMS 2017*, Jogjakarta.

[2] Nonaka I and H Takeuchi 1995 *The Knowledge-Creating Company*. (New York: Oxford University Press)

[3] Alwis R S-d, Hartmann E and Gemunden H G 2004 *The role of tacit knowledge in innovation management*. Copenhagen.

[4] Lis A 2014 Knoweldge Creation and Conversion in Military Organizations: How the SECI Model is Applied Within Armed Forces," *Journal of Entrepreneurship Management and Innovation (JEMI)* 10(1) pp 57-78

[5] Nugraha N W, Kurniawati A and Hediyanto U Y K S 2015 The Design Of Best Practice On The Media Transfer Activities And Preservation Based On Knowledge Conversion With SECI Method," in *Proceeding of Industrial Engineering and Service Science*, Yogyakarta.

[6] J. Gao and Nee A Y 2018 An overview of manufacturing knowledge sharing in the product development process. *Journal of Engineering Manufacture* **1**(11) pp 1-11

[7] Kalogeraki E-M, Apostolou D, Polemi N and Papastergiou S 2018 Knowledge management methodology for identifying threats in maritime/logistics supply chains. *Knowledge Management Research & Practice* pp 1-16

[8] Finley D and Sathe V 2013 Nonaka's SECI Framework: Case Study Evidence and an Extension. *Kindai Management Review* **1** pp 59-68

[9] Nisaa A S, Kurniawati A and Pratami D 2013 Knowledge Conversion Pada Proses Perencanaan Proyek Di PT. Len Railway System untuk Standardisasi Proses Dengan Metode SECI," *J@TI* **VIII**(1) pp 27-36

Taguchi Experimental Design to Optimize the Sugar Content of Candied Carrot

Maria S I Notowidjaja, Yurida Ekawati[1] and Sunday Noya

Industrial Engineering Department, Ma Chung University
[1]yurida.ekawati@machung.ac.id

Abstract. Robust parameter design is used to design product that is robust to noise factors so the product's performance fits the target and delivers a better quality. In the process of designing and developing the innovation product of candied carrot, robust parameter design is done using Taguchi Method. The performance of the candied carrot is evaluated in terms of its percentage of sugar. Orthogonal Array (OA) of Taguchi, Analysis of Variance (ANOVA), signal-to-noise (S/N) ratio, and main effect were employed to analyze the effect of parameters on the percentage of sugar in the candied carrot. The optimal level is determined based on the highest signal-to-noise (S/N) ratio. The analysis of the results shows that the optimal composition for candied carrot are the amount of sugar of 25%, 40%, 50%, 60%, the soaking duration for 8 hours, the amount of cinnamon of 10 grams and the age of carrot of 5 months.
Keywords: robust parameter design, Taguchi method, candied carrot

1. Introduction

Carrot is an agricultural product that can be easily found in Indonesia. Although it has abundant harvest, the price of carrots is cheap because carrots are usually sold as a raw commodity. To increase the price of carrot, carrot can be made into processed products which desired by the customers. One of the product that was preferred by customers was candied carrot [1]. The design of candied carrot product had been developed to determine the critical parts while producing candied carrot [2]. Based on the design of candied carrot product, further research was carried out to determine the optimal production process and composition of candied carrot using Taguchi method. This method was used because it can determine the best combination through design of experiment [3]. Additionally, the researches using Taguchi experimental design had been conducted for many kind of food products [4], [5], [6] and [7].

This study attempts to determine the optimization of the production process and composition parameters for optimal percentage of sugar in candied carrot. Parameters have to be optimized to make the candied carrot with highest percentage of sugar.

2. Methods

Robust parameter design (RPD) is a method to product realization activities which focuses on choosing the controllable factors or parameters and their levels in a product. The objectives of RPD are to ensure that the mean of the product output is at target and to ensure that the variability around this target value is small. The general RPD problem was developed by Genichi Taguchi. Taguchi proposed an approach to solve the RPD problem based on designed experiments and some novel methods for analysis of the resulting data [8].

2.1 Experimental design
The experimental design of Taguchi is conducted by stating the problems of concern, stating the objectives of the experiment, selecting the quality characteristics and measurement systems, selecting the factor which may influence the selected quality characteristics, identifying control factors and noise factors, selecting levels for the factors, and selecting the appropriate orthogonal array (OA) [3].

2.2 Perform the experiment and Anthrone test
The experiment was done by making the combination of factor and level as described in OA. In this study, the percentage of sugar in candied carrot is measured using Anthrone test.

2.3 Data processing
Data processing is done by calculating the Signal-to-Noise (S/N) ratio and Analysis of Variance (ANOVA). S/N ratio is a transformation of the repetition data to another value which is a measure of the variation present in Taguchi method. The S/N ratio consolidates several repetitions into one value which reflects the variation of the result. The S/N ratio for larger the better quality characteristic can be defined as [3]:

$$S/N_{LTB} = -10\log\left(\frac{1}{r}\sum_{i=1}^{r}\frac{1}{y_i^2}\right) \tag{1}$$

where r is the number of repetitions or observations i and y is the observed data.
ANOVA is a predominant statistical method that is used to interpret experimental data and make the necessary decisions. The determination of influential factors and the relative strengths is based on their levels. ANOVA is an objective decision making tool for detecting any differences in performance of a product. The decision on ANOVA calculation takes variation into account [3]. The purpose of ANOVA is to find how much variation for each factor affecting the total variation in the final result [9].

3. Result and Discussion

3.1 Experimental design planning
Experimental design planning started by determining the production process of candied carrot which had already conducted [10]. The planning stage was continued by selecting the parameters and their levels. Percentage of sugar in the candied carrot parameters and their levels were discussed with some experts in food technology. The experts said that there are three control parameters affecting the percentage of sugar. The parameters are amount of sugar, soaking duration in sugar solution, and amount of cinnamon. The experts also said that age of carrot also affecting the percentage of sugar. But the age of carrot in the market can't be controlled. So in this study, the age of carrot is set as a noise factor that is tried to be controlled. The level of each parameters is shown in Table 1.

Table 1. The Level of Each Parameters

Factor	Parameter	Unit	Level 1	Level 2	Level 3
A	Amount of Sugar	Percentage	- 15 (1st soaking) - 25 (2nd soaking) - 40 (3rd soaking) - 50 (4th soaking)	-25 (1st soaking) -40 (2nd soaking) -50 (3rd soaking) -60 (4th soaking)	- 35 (1st soaking) - 50 (2nd soaking) - 60 (3rd soaking) - 70 (4th soaking)
B	Soaking Duration	Hour	7	8	9
C	Amount of Cinnamon	Gram	10	30	
D	Age of Carrot	Months	3.5	5	

The next step was selection of orthogonal array (OA). The selection of OA depends on the parameters and their levels. In this section, there are one parameter that has two levels and two parameters that have three levels. Therefore, the OA that is appropriate is L_{18} ($2^1 \times 3^7$).

3.2 Data processing for each age of carrot

The percentage of sugar data resulted from Anthrone test then transformed into S/N ratio. To obtain the optimal percentage of sugar in candied carrot, larger the better quality characteristic must be taken. Table 2. shows the results for percentage of sugar and the S/N ratio for each age of carrot.

Table 2. Percentage of Sugar and S/N Ratio for Each Age of Carrot

Experiment	Parameters			Percentage of Sugar		S/N Ratio	
	A	B	C	3.5 months	5 months	3.5 months	5 months
1	1	1	1	26.19	53.81	28.3627	34.6173
2	2	1	1	35.52	54.97	27.7014	33.6284
3	3	1	1	34.70	51.66	31.0095	34.8025
4	1	2	1	24.27	48.02	26.7092	34.8608
5	2	2	1	21.65	55.34	30.8066	34.2631
6	3	2	1	25.02	53.17	27.9657	34.5133
7	1	3	1	39.71	40.98	31.9780	32.2514
8	2	3	1	29.51	44.29	29.3994	32.9261
9	3	3	1	31.42	49.08	29.9441	33.8181
10	1	1	2	16.73	37.52	24.4699	31.4853
11	2	1	2	19.58	41.20	25.8363	32.2979
12	3	1	2	18.88	40.27	25.5200	32.0996
13	1	2	2	33.95	46.35	30.6168	33.3210
14	2	2	2	30.16	50.84	29.5886	34.1241
15	3	2	2	28.82	50.77	29.1939	34.1121
16	1	3	2	21.21	44.55	26.5308	32.9770
17	2	3	2	27.62	43.03	28.8245	32.6754
18	3	3	2	27.47	39.15	28.7772	31.8546

The result of the S/N ratio is converted into a main effect plot by calculating the mean of S/N ratio. The main effect plot shows the effect of each parameter at different levels. The main effect plot was conducted using Minitab 17 software. The main effect plot for each age of carrot is shown as follows:

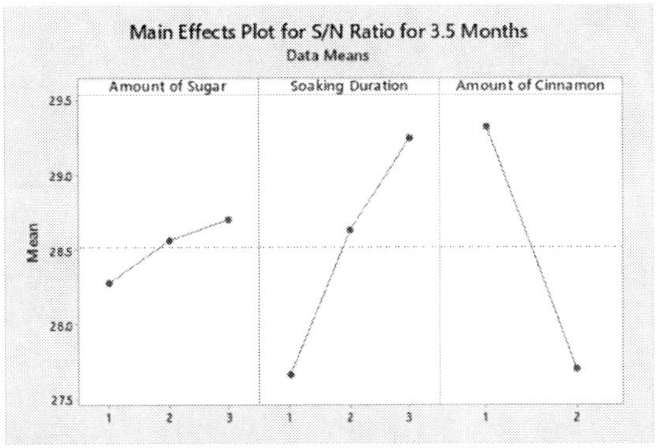

Figure 1. Main Effects Plot for 3.5 Months Carrot

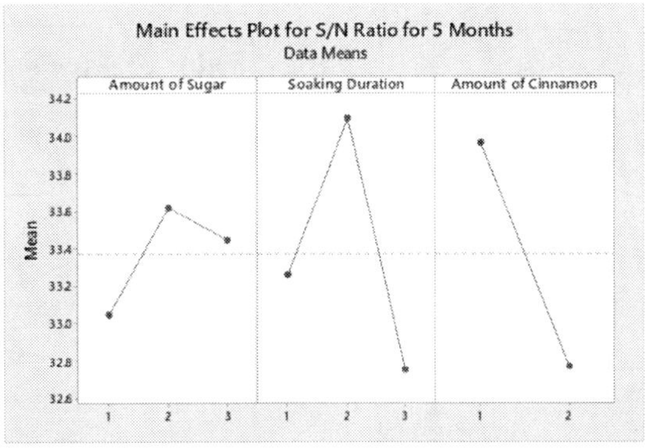

Figure 2. Main Effects Plot for 5 Months Carrot

To determine the optimal combinations of the parameter levels, the relative importance amongst the parameters for percentage of sugar had to be known using ANOVA. The test of hypothesis was conducted for all factors for 95% confidence level. The hypothesis for each factor is described as follows:

1. Factor A
H_0: there is no effect of factor A to percentage of sugar
H_1: there is an effect of factor A to percentage of sugar
2. Factor B
H_0: there is no effect of factor B to percentage of sugar
H_1: there is an effect of factor B to percentage of sugar
3. Factor C
H_0: there is no effect of factor C to percentage of sugar
H_1: there is an effect of factor C to percentage of sugar

The ANOVA calculation was conducted using Minitab 17 software. ANOVA for each age of carrot is shown as follows:

Analysis of Variance

Source	DF	Adj SS	Adj MS	F-Value	P-Value
Amount of Sugar	2	1.508	0.754	0.02	0.983
Soaking Duration	2	53.527	26.764	0.62	0.555
Amount of Cinnamon	1	105.464	105.464	2.44	0.144
Error	12	519.325	43.277		
Total	17	679.824			

Figure 3. ANOVA for 3.5 Months Carrot

Analysis of Variance

Source	DF	Adj SS	Adj MS	F-Value	P-Value
Amount of Sugar	2	29.82	14.91	0.91	0.429
Soaking Duration	2	158.29	79.14	4.82	0.029
Amount of Cinnamon	1	184.58	184.58	11.24	0.006
Error	12	197.02	16.42		
Total	17	569.70			

Figure 4. ANOVA for 5 Months Carrot

Based on the hypothesis test, the factor A, factor B, factor C for the 3.5 months carrot and factor A for the 5 months carrot accept the H_0, that means these factors didn't affect the percentage of sugar. But the factor B and factor C for the 5 months carrot reject the H_0, that means these factors affect the percentage of sugar. Because the result of ANOVA for each age of carrot is different, so the calculation using age of carrot factor is conducted to find the effect of age of carrot.

3.3 Data processing for all factors

In this section, there are two parameters that have two levels and two parameters that have three levels. The OA that is suitable for this section is L_{36} (2^3x3^{13}). The next step is to calculate the S/N ratio and to conduct the main effect plot by calculating the mean of S/N ratio. Table 3. shows the results for percentage of sugar and the S/N ratio.

Table 3. Percentage of Sugar and S/N Ratio

Exp.	A	B	C	D	Percentage of Sugar	S/N Ratio	Exp.	A	B	C	D	Percentage of Sugar	S/N Ratio
1	1	1	1	1	26.19	28.3627	19	2	1	1	2	54.97	34.8025
2	1	2	1	1	24.27	27.7014	20	2	2	1	2	55.34	34.8608
3	1	3	1	1	39.71	31.9780	21	2	3	1	2	44.29	32.9261
4	1	1	2	1	16.73	24.4699	22	2	1	2	2	41.20	32.2979
5	1	2	2	1	33.95	30.6168	23	2	2	2	2	50.84	34.1241
6	1	3	2	1	21.21	26.5308	24	2	3	2	2	43.03	32.6754
7	1	1	1	2	53.81	34.6173	25	3	1	1	1	34.70	30.8066
8	1	2	1	2	48.02	33.6284	26	3	2	1	1	25.02	27.9657
9	1	3	1	2	40.98	32.2514	27	3	3	1	1	31.42	29.9441
10	1	1	2	2	37.52	31.4853	28	3	1	2	1	18.88	25.5200
11	1	2	2	2	46.35	33.3210	29	3	2	2	1	28.82	29.1939
12	1	3	2	2	44.55	32.9770	30	3	3	2	1	27.47	28.7772
13	2	1	1	1	35.52	31.0095	31	3	1	1	2	51.66	34.2631
14	2	2	1	1	21.65	26.7092	32	3	2	1	2	53.17	34.5133
15	2	3	1	1	29.51	29.3994	33	3	3	1	2	49.08	33.8181
16	2	1	2	1	19.58	25.8363	34	3	1	2	2	40.27	32.0996
17	2	2	2	1	30.16	29.5886	35	3	2	2	2	50.77	34.1121
18	2	3	2	1	27.62	28.8245	36	3	3	2	2	39.15	31.8546

The result of the S/N ratio is converted into a main effect plot. The main effect plot was conducted using Minitab 17 software. The main effect plot for S/N ratio for all factors are used is shown in Figure 5.

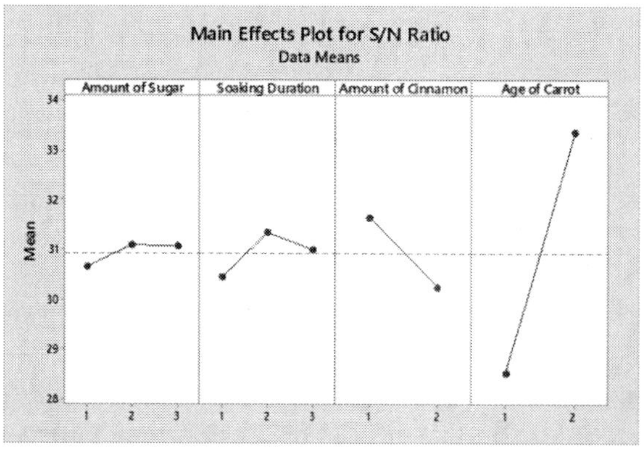

Figure 5. Main Effects Plot

IOP Conf. Series: Materials Science and Engineering **528** (2019) 012068 doi:10.1088/1757-899X/528/1/012068

To determine the optimal combinations of the parameter levels, the relative importance amongst the parameters for percentage of sugar had to be known using ANOVA. The test of hypothesis was conducted for all factors for 95% confidence level. The hypothesis for each factor is described as follows:

1. Factor A
H_0: there is no effect of factor A to percentage of sugar
H_1: there is an effect of factor A to percentage of sugar
2. Factor B
H_0: there is no effect of factor B to percentage of sugar
H_1: there is an effect of factor B to percentage of sugar
3. Factor C
H_0: there is no effect of factor C to percentage of sugar
H_1: there is an effect of factor C to percentage of sugar
4. Factor D
H_0: there is no effect of factor D to percentage of sugar
H_1: there is an effect of factor D to percentage of sugar

The ANOVA calculation was conducted using Minitab 17 software. ANOVA is shown as follows:

Analysis of Variance

Source	DF	Adj SS	Adj MS	F-Value	P-Value
Amount of Sugar	2	20.03	10.01	0.33	0.721
Soaking Duration	2	65.64	32.82	1.08	0.352
Amount of Cinnamon	1	284.54	284.54	9.38	0.005
Age of Carrot	1	3453.33	3453.33	113.89	0.000
Error	29	879.32	30.32		
Total	35	4702.85			

Figure 6. ANOVA

Based on the hypothesis test, the factor A and factor B accept the H_0, that means these factors didn't affect the percentage of sugar. But the factor C and factor D reject the H_0, that means these factors affect the percentage of sugar. In addition, percentage contribution of each parameter was also calculated. The percentage contribution of each parameter is shown as follows:

Table 4. Percentage Contribution of Each Parameter

Parameter	Percentage Contribution
Amount of Sugar	0.86%
Soaking Duration	0.11%
Amount of Cinnamon	5.41%
Age of Carrot	72.79%

3.4 Comparison and analysis of the result
Optimal composition was determined by choosing the highest mean of S/N ratio from each parameter which are shown in the main effects plot. The optimal composition for each age of carrot and for all factors is shown in Table 5.

Table 5. Optimal Composition of Each Parameter

Factor	Optimal Level		
	3.5 Months Carrot	5 Months Carrot	All Factors
A	3	2	2
B	3	2	2
C	1	1	1
D	-	-	2

Based on Table 5 the optimal composition for 3.5 months carrot was the amount of sugar is 35%, 50%, 60%, 70%, the soaking duration is 9 hours and the amount of cinnamon is 10 grams. The optimal composition for 5 months carrot was the amount of sugar is 25%, 40%, 50%, 60%, the soaking duration is 8 hours and the amount of cinnamon is 10 grams. Based on Table 4. the age of carrot has a huge contribution for percentage of sugar in candied carrot (72.79%).

Therefore, it is better to use the 5 months carrot with less amount of sugar and less soaking duration, but has a higher percentage of sugar. Also, the percentage of sugar in the optimal composition for 5 months carrot is higher percentage of sugar (55.34%) than in the optimal composition for 3.5 months carrot (31.42%). Therefore, the optimal composition for candied carrot is shown in Table 6.

Table 6. The Level of Each Parameters

Factor	Parameter	Level 2
A	Amount of Sugar	-25% (1st soaking) -40% (2nd soaking) -50% (3rd soaking) -60% (4th soaking)
B	Soaking Duration	8 hours
C	Amount of Cinnamon	10 grams
D	Age of Carrot	5 months

4. Conclusion

Percentage of sugar in the candied carrot was affected by the amount of sugar (factor A), the soaking duration in sugar solution (factor B), the amount of cinnamon (factor C) and the age of carrot (factor D). The experiment was done by making the combination of factor and level as described in OA. The percentage of sugar data that was tested using Anthrone test then transformed into S/N ratio. The result of the S/N ratio is converted into a main effect plot. Next, the hypothesis test was conducted to determine the optimal combinations of the parameter levels using ANOVA. Because the result of ANOVA for each age of carrot is different, so the calculation using age of carrot factor is conducted to find the effect of age of carrot. The result showed that the age of carrot has a huge contribution for percentage of sugar in candied carrot. The optimal composition for candied carrot are the amount of sugar of 25%, 40%, 50%, 60%, the soaking duration for 8 hours, the amount of cinnamon of 10 grams and the age of carrot of 5 months.

Acknowledgment

This work would not have been possible without the financial support from Directorate of Research and Community Service, Directorate General of Research and Development Strengthening, Ministry of Research, Technology and Higher Education of the Republic Of Indonesia

5. References

[1] Halim V and Ekawati Y 2014 Perencanaan Produk Olahan Wortel Menggunakan Metode Quality Function Deployment *Symbol* **1** pp 57-67

[2] Ekawati Y and Bazarado M 2016 Designing Food Products Based on Carrots Using the Product Design Phase of Quality Function Deployment *ARPN Journal of Engineering and Applied Science* **11** pp 3109-16

[3] Ross, P J 1996 *Taguchi Techniques for Quality Engineering* (New York: Mc Graw-Hill)

[4] Sihombing H, Hafiz M K, Yuhazri M Y and Kannan R 2012 Taguchi's Quality Improvement Analysis of the SME Bread Manufacturing *Global Engineers & Technologists Review* **2** pp 6-13

[5] Habibi N, Soleimanian-Zad S and Zeinoddin M S 2011 Optimization of Kefir Grains Production by Using Taguchi Technique and Mini-Fermentation *World Applied Sciences Journal* **12** pp 613-18

[6] Noorwali A 2013 Apply Lean and Taguchi in Different Level of Variability of Food Flow Processing System *Procedia Engineering* **63** pp 728-34

[7] Naik E K M and Bala E K L 2013 Study on Optimization of Microwave Frying of Potato Slices by Using Taguchi Design *International Journal of Engineering Research & Technology* **2** pp 141-8

[8] Montgomery D.C 2013 *Design and Analysis of Experiments* (New York: John Wesley and Sons)

[9] Roy R K 2001 *Design of Experiments Using The Taguchi Approach* (New York: John Wiley & Sons, Inc)

[10] Noya S, Ekawati Y and Utami N 2017 Quality Function Deployment for Candied Carrots Process Planning *International Journal of Engineering and Advanced Technology Studies* **5** pp 1-7

The Pattern Failure Analysis of Sulfuric Acid Production Process with the Association Rules Algorithm Apriori

W Septiani[1], I A Marie[2], D Sugiarto[3] and L Hakim

Faculty of Industrial Technology, Kampus A Universitas Trisakti, Jakarta 11440, Indonesia
[1]winnie_septiani@trisakti.ac.id, [2]iveline.annemarie@trisakti.ac.id, [3]dedy@trisakti.ac.id

Abstract. PT. Asam Sulfat is industrial Sulfuric Acid that was able to meet the needs of most battery manufacturers in Indonesia. Problems in production operations were mostly caused by production equipment, such as engine stop suddenly, length of set up and adjustment time and decrease of machine production speed. These results caused a decrease level in the efficiency and effectiveness of the engine. This study aims to identify and analyze the machine and components that dominate the sulfuric acid production process damage. The method of analysis used is the Association Rules Algorithm Apriori. Based on the results of initial identification obtained 23 machines and 9 components are often damaged. The variables analyzed include unit, machine type, component, time to repair and turbidity. Using a minimum value of support 0.1 and confidence 0.9 resulted in 17 rules dominated by low turbidity, especially in AT Pump engines and Gasket components. The High turbidity occurred in patterns with the Ash Pump component, The Time to Repair was high and it happened on the Sulfur Pump engine. By knowing the patterns that occurred absolutely the failure in the production process can be more quickly handled.
Keywords: association rules, machine, component, turbidity, sulphuric acid

1. Introduction

PT Asam Sulfat is the first Sulfuric Acid Industry in Indonesia. One of the products produced by PT Asam Sulfat is sulfuric acid supplied to the automotive battery industry that meets standards prescribed by the Indonesian National Standards (Indonesian: *Standar Nasional Indonesia,* abbreviated to SNI) as can be seen in Table 1. The capacity for sulfuric acid production is 82,500 ton per annum. The company has obtained the ISO 9001 certificate and meet standards prescribed by SNI for sulfuric acid.

The quality of sulfuric acid products produced by PT Asam Sulfat is able to meet the needs of most battery plants in Indonesia and the sulfuric acid offers good translucence and iron content that other sulfuric acid plants might find it hard to produce, giving it a big advantage over other plants. This is because of the highly corrosive nature of sulfuric acid itself, thus requiring that the operation of production machinery and supervision of the processes be arranged carefully. The production of sulfuric acid is carried out continuously. Cessation of a production process often results from problems with the production equipment, for example in the event of downtime, time-consuming set-up and adjustment, and decreased speed of production machinery. Those cause losses to the company as they reduce the efficiency and effectiveness of the machines and result in repair costs. In anticipation of those problems, an analysis needs to be performed using a method called association rules in order to reduce production failure.

Table 1. Specifications Prescribed by SNI 0030: 2011

Specification	Content	Specification	Content
Sulfuric Acid Content	Min 98.20%	Sulfuric Acid	Min 98.20%
Residue on Ignition	Max 0.02%	Residue on	Max 0.02%
Turbidity	Max 60 NTU	Turbidity	Max 60 NTU
Chloride (CI) Content	Max 5 ppm	Chloride (CI)	Max 5 ppm
Iron (Fe) Content	Max 40 ppm	Iron (Fe)	Max 40 ppm
Lead (Pb) Content	Max 9 ppm	Lead (Pb)	Max 9 ppm
Arsenic (As) Content	Max 0.2 ppm	Arsenic (As)	Max 0.2 ppm
Copper (Cu) Content	Max 1 ppm	Copper (Cu)	Max 1 ppm
Selenium (Se) Content	Max 0.5 ppm	Selenium (Se)	Max 0.5 ppm
Zinc (Zn) Content	Max 2.0 ppm	Zinc (Zn)	Max 2.0 ppm

Association rules are a useful data mining technique to find out a correlation or the most important/ interesting pattern of a large data set [1]. There are a great number of previous studies undertaken based on scientific works that use the method *association rules*, for example those conducted by: [2], which studied the implementation of data mining on the sale of plane tickets using the apriori algorithm; [3], which conducted research into mining and visualization of association rules over relational dbmss; [4], which studied data mining to analyze the rate of street crimes using the algorithm of the association rules and the apriori method; [5], which studied the application of association rules using the apriori algorithm to analyze fish catch data patterns; and [6], which applied association rules to determine traffic accident patterns.

Poor maintenance at PT Asam Sulfat greatly affect the resulting downtime level and frequent engine breakdowns as a result of the engine condition that is less suitable and the absence of production machine effectiveness calculation. This study aims to identify and analyze the machine and components that dominate the sulfuric acid production process damage using *association rules* with the apriori algorithm.

2. Literature Review

According to [7], association rules are defined as a data mining technique used to determine the rules of association between particular combinations of an item. Association rules are one of the methods that aim to find out patterns that often appear or occur between many transactions, where each transaction consists of several items, and, thus, this method will support the recommendation system through the discovery of patterns between items in the existing transactions [4].

Whether particular association is important or not can be determined using two benchmarks, namely: support and confidence. Support (the support value) refers to the percentage of combinations of the items in the database, while confidence (the value of certainty) is the extent to which the relationship between items in association rules is deemed strong [2]. Support in this research is defined as the probability of occurrence of several item sets in the variable *plane crash* resulting from the association rules of the overall variable *crash*.

3. Method

In this research, the data used were the ones on sulfuric acid production obtained from PT Asam Sulfat in 2017. The stage of research carried out as follows:

1. Preliminary data analysis
   The analysis used represents the overall description of each variable, especially the variables *machine type* and *component* using the Pareto chart. The next step was to take several data that run into problems most often.

2. Determine the analysis variable
   There were a total of five variables, namely: *unit, machine type, component, time to repair,* and *turbidity*. The categories for each variable are determined as follows:
   - Unit : Two, Four
   - Machine Type : AT pump, DT pump, Sulfur pum
   - Component : Ash pump, bearing, breaker, bushing pump, electro motor, gasket, impeller, pipa steam, hulley pump
   - Time to repair : high, low, medium
   - Turbidity : high, low
3. Compile association rules
   The basic methodology of the association analysis is divided into three stage :
   a. *High-frequency pattern analysis*
   This stage aims to discover the combination of items that meets the requirement of the minimum support value in the predetermined database. Support refers to a percentage of occurrence in the data wherein both Events A and B occur simultaneously. Thus, the support value of an item is obtained using the following formula [8]:

$$\text{Support (A} \rightarrow \text{B)} = (A \cup B) = \frac{n(A \cup B)}{n(S)} \tag{1}$$

$$\text{Support (A} \rightarrow \text{B)} = \frac{\text{number of transactions containing A and B}}{\text{total number of transactions}} \tag{2}$$

where:
$(A \cup B)$: Probability of simultaneous occurrence of Events A and B
$(A \cup B)$: Rate of simultaneous occurrence of Events A and B
(S) : Number of S members

   b. *Establishment of the association rules*
   After all high-frequency patterns have been discovered, the next step is to find out the association rules that meet the minimum confidence value requirement. Confidence refers to the percentage of occurrence where if the Event A occurs, then the Event B will occur as well. A combination of items is obtained using the following formula [8]:

$$\text{Confidence (A} \rightarrow \text{B)} = P(B|A) = \frac{P(A \cup B)}{P(A)} \tag{3}$$

$$\text{Confidence (A} \rightarrow \text{B)} = \frac{\text{number of transactions containing A and B}}{\text{number of transactions of A}} \tag{4}$$

where:
P(B|A) : Conditional probability of Event B if Event A has occurred
$P(A \cup B)$: Probability of simultaneous occurrence of Events A and B
P(A) : Probability of Event A

In addition to the two parameters above, which are the support value and the confidence value, there is another parameter known as the lift ratio. The lift ratio indicates a particular level of power that rules have over the resulting association rules. [8] defines the lift ratio as the confidence of the association rules divided by the probability of Event A and the probability of Event B independent of each other, which is expressed in the following formula:

$$\text{Lift ratio (A} \rightarrow \text{B)} = \frac{P(A \cup B)}{P(A)P(B)} \tag{5}$$

$$\text{Lift ratio} = \frac{\text{confidence (A=>B)}}{P(B)} \tag{6}$$

$$= \frac{\text{confidence (A=>B)}}{\text{support B}} \tag{7}$$

where:
P(A∪B) : Probability of simultaneous occurrence of Events A and B
P(A) : Probability of Event A
P(B) : Probability of Event B

c. *Apriori algorithm*
The apriori algorithm is a basic algorithm proposed by Agrawal & Srikant in 1994 to determine frequent item sets for association rules. The rules expressing the association between several attributes are often called *market basket analysis*.

4. The descriptive picture of the rules established using software *R*
Afterwards, association rules were established, followed by the descriptive picture of the rules established using software *R* with the assistance of several packages such as: *arules* and *arulesViz*.

4. Result and Discussion

4.1 Preliminary Data Analysis
Figure 1 below provides information about machine types that commonly run into problems.

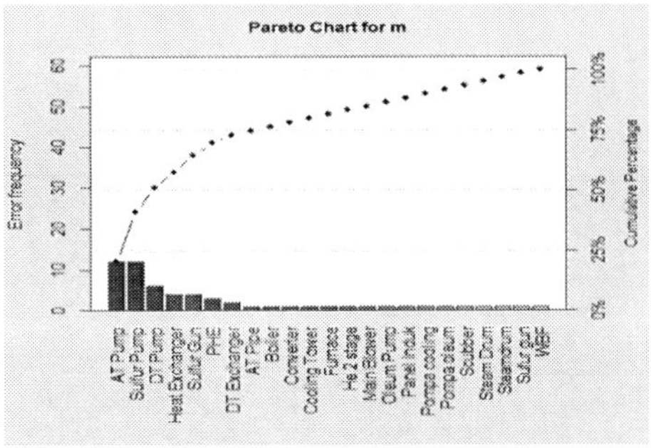

Figure 1. The type of machine with the highest level of problem

If looking closely at Figure 1, machine types *AT Pump, Sulfur Pump,* and *DT Pump* are the three machine types that run into problems most often compared to the other machine types. If there are no efforts made to solve this issue, it will result in unstable product quality and, in turn, will lead to customer dissatisfaction. In addition, components that often run into problems were also examined using the same diagram.

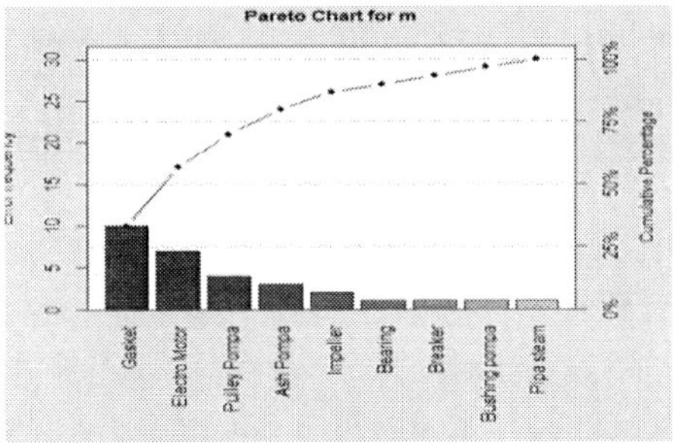

Figure 2. The type of machine with the highest level of problem

In Figure 2, it is revealed that *Gasket, Electro Motor,* and *Pulley Pump* are components that run into problems most often. It is not only damage to the engine that causes a high turbidity value, but damage to the components will also affect the value of turbidity.

IOP Conf. Series: Materials Science and Engineering **528** (2019) 012069 doi:10.1088/1757-899X/528/1/012069

After finding out the dominant damage taking place during the production process, the next step was to perform an analysis using the association rules of the apriori algorithm using only the machine types that often run into problems, namely AT Pump, Sulfur Pump, and DT Pump.

4.2 Determine the analysis of category variable

There were a total of five variables, namely: *unit, machine type, component, time to repair,* and *turbidity*. The categories for each variable are determined as follows:

- Unit : Two, Four
- Machine Type : AT pump, DT pump, Sulfur pum
- Component : Ash pump, bearing, breaker, bushing pump, electro motor, gasket, impeller, pipe steam, hulley pump
- Time to repair : high, low, medium
- Turbidity : high, low

4.3 Compile association rules

4.3.1 High-frequency pattern analysis. In this research, to determine the relation between the resulting patterns or rules of association, the researcher tried to set limits on the support value and the confidence value, which were equal to 0.03 and 0.9, respectively. Using those limits, there are a total of 104 rules established.

4.3.2 Establishment of the association rules. To illustrate how association rules work using the apriori algorithm using five sample, refer to the results below:

1. Data association rules

Table 2. Example data Association Rules of machine type, component and Turbidity

No	Machine Type	Component	Turbidity
1	AT Pump	Gasket	Low
2	Sulfur Pump	Impeller	High
3	AT Pump	Impeller	High
4	AT Pump	Gasket	Low
5	Sulfur Pump	Gasket	High

2. To convert data presented in the table above into a matrix of binary numbers, where 1 indicates that the process exists while 0 indicates that the process does not exist (Table 3).

Table 3. Convert association rules into a binary numbers

No	AT Pump	Sulfur Pump	Gasket	Impeller	High	Low
1	1	0	1	0	0	1
2	0	1	0	1	1	0
3	1	0	0	1	1	0
4	1	0	1	0	0	1
5	0	1	1	0	1	0
Total	3	2	3	2	3	2

3. To determine the total frequency of item sets and the number of item sets used (Table 4 – Table 11). The total frequency of item sets is equal to 2 with a total of 3 item sets.

Table 4. Combination 1 for Turbidity Low

AT Pump	Gasket	Low	x
1	1	1	1
0	0	0	0
1	0	0	0
1	1	1	1
0	1	0	0
Σ			2

Table 5. Combination 2 for Turbidity Low

AT Pump	Impeller	Low	x
1	0	1	0
0	1	0	0
1	1	0	0
1	0	1	0
0	0	0	0
Σ			0

Table 6. Combination 3 for Turbidity High

AT Pump	Gasket	High	x
1	1	0	0
0	0	1	0
1	0	1	0
1	1	0	0
0	1	1	0
Σ			0

Table 7. Combination 4 for Turbidity High

AT Pump	Impeller	High	x
1	0	0	0
0	1	1	0
1	1	1	1
1	0	0	0
0	0	1	0
Σ			1

Table 8. Combination 5 for Turbidity Low

Sulfur Pump	Gasket	Low	x
0	1	1	0
1	0	0	0
0	0	0	0
0	1	1	1
1	1	0	0
Σ			1

Table 9. Combination 6 for Turbidity Low

Sulfur Pump	Impeller	Low	x
0	0	1	0
1	1	0	0
0	1	0	0
0	0	1	0
1	0	0	0
Σ			0

Table 10. Combination 5 for Turbidity High

Sulfur Pump	Gasket	High	x
0	1	0	0
1	0	1	0
0	0	1	0
0	1	0	0
1	1	1	1
Σ	1		

Table 11. Combination 6 for Turbidity High

Sulfur Pump	Impeller	High	x
0	0	0	0
1	1	1	1
0	1	1	0
0	0	0	0
1	0	1	0
Σ	1		

In Column X, Number 1 indicates items that occur simultaneously, while Number 0 means that no items occur simultaneously or there are no transactions. The symbol Σ represents total frequency of item sets. The total frequency of item sets must be greater than or equal to the predetermined total frequency of set items. The predetermined frequency of item sets above is as many as 2 item sets, if there are less than 2 item sets, then the rule is not used. From the tables above, there is one combination with value of the frequency of item sets equal to the predetermined total frequency of item sets, which is Combination 1.

4. The next step is to take the combination that has met the predetermined requirement, i.e. the one with total frequency of item sets of 2 and a total of 3 item sets used. The combination that met the predetermined requirement is Combination 1 with total frequency of item sets of 2 (Table 12).
   Based on the table 12, it is revealed that the resulting rules are listed below: (AT Pump, Gasket, Low)
5. The rule used is "if x, then y", where x is the antecedent while y is the consequent. Based on the rule, there must be 2 item sets, one of which serves as the antecedent and the other serves as the consequent (Tabel 13).

Table 12. Combination AT PUM and Gasket for Turbidity Low

AT Pump	Gasket	Low	x
1	1	1	1
0	0	0	0
1	0	0	0
1	1	1	1
0	1	0	0
Total	2		

Table 13. If Then Rule for Machine Type, Component and Turbidity

No	If antecedent, then consequent	Support	Confidence	Lift
1	If there is damage to the machine type *AT Pump* and to the component *Gasket*, then such a condition will result in *Low Turbidity*.	{Total simultaneous occurrence/ Total data} = {2/5=0.4}	{ Total simultaneous occurrence/ Antecedent (X)}= {2/2=1)	{Total data/ Consequent (Y)}={12/2=6}

Rule No. 1. If there is damage to the machine type *AT Pump* and to the component *Gasket*, then such a condition will result in *Low Turbidity* with support value (domination) of 0.4 or 40% and confidence value (level of confidence) of 1 or 100%.

4.3.3 Apriori algorithm. Based on the limits set, it is revealed that there are a total of 17 rules established. The following is information obtained from Table 14. In Rule No. 1. In the event of failure to Unit 4, especially to component Ash Pump, the resulting turbidity is usually low with support value (domination) of 10% and confidence value (the level of confidence) of 100%. In Rule No. 2. In the event of failure to Unit 2 and it occurs to machine type DT Pump, the resulting turbidity is usually low with support value (domination) of 10% and confidence value (the level of confidence) of 100%. In Rule No. 16. In the event of failure to Unit 4 and it occurs to machine type AT Pump, especially component Gasket, the resulting turbidity is usually low with support value (domination) of 16% and confidence value (the level of confidence) of 100%. In Rule No. 17. In the event of failure to Unit 4 and it occurs to machine type AT Pump, especially component Gasket, the resulting turbidity is high with support value (domination) of 10% and confidence value (the level of confidence) of 100%. Table 14 presents a summary of the 104 rules established.

Table 14. Association Rules

No	Sup	Conf	Lift	If (Antecedent)				Then (Consequent)
				1st Condition	2nd Condition	3rd Condition	4rd Condition	Response
1	0.10	1	2.73	Unit=Four	Component= Ash Pump			Turbidity=High
2	0.10	1	1.57	Unit=Two	Machine.Type= DT Pump			Turbidity=Low
--	----	---	---	-----------------	--------------	------------		------------
--	----	---	---	-----------------	--------------	------------		------------
16	0.16	1	1.57	Unit=Four	Machine.Type= AT Pump	Component= Gasket		Turbidity=Low
17	0.10	1	1.57	Unit=Four	Machine.Type= AT Pump	Component= Gasket	Time.to.Repair= Low	Turbidity=Low

4.4 Graph of items of the association Rules

Then, those association rules are visualized in the form of a Graph of Items of the Association Rules illustrated in Figure 3.

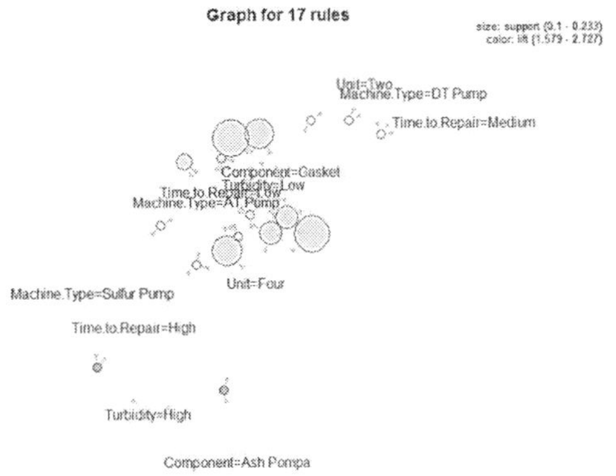

Figure 3. Visualization Graph

Based on Figure 3, it is revealed that the measures used in the graph of items were based on the support value and the lift ratio. Based on the figure above, it can be concluded that:

- A larger circle size indicates a higher support value (domination in the association rules). This can be seen from the association rules where machine type AT Pump, component Gasket, Unit 4 have a larger circle size than the other circles, thus it is revealed that the rule is the rule with the highest support value as can be seen in rules No. 4 and 14 with support value of 23%.
- A darker circle color indicates a higher lift value (the power of association to generate patterns). This can be seen from the association rules where machine type Sulfur Pump, High Time to Repair, High Turbidity, and component Ash Pump have a darker circle color than the other circles, thus it is revealed that the rule is the rule with the highest lift value as can be seen in rules No. 1 and 3. According to the rules, the existing patterns describe events that must happen or whose probability of occurrence is higher than that of other events.

5. Conclusions

Engine failure during the production process will certainly affect the sulfuric acid produced. Failure is more common in machine types AT Pump, Sulfur Pump, and DT Pump as well as components Gasket and Electro Motor. If there are no efforts made to solve this issue, it will affect the resulting turbidity, leading to the unstable production process. In addition, using a minimum support value of 0.1 and a minimum confidence value of 0.9, there are a total of 17 rules established, most of which generate low turbidity, especially in machine type AT Pump and component Gasket. On the other hand, high turbidity is generated by the pattern with the component Ash Pump, High Time to Repair, and machine type Sulfur Pump. Knowing the resulting patterns, failure during the production process can be solved. Thus, production outcomes will be stable and it in turn will affect customer satisfaction.

Acknowledgement

This paper is the result of a research grant funded by the Ministry of Research, Technology and Higher Education of the Republic of Indonesia in accordance with the Research Contract: No. 011/KM/PNT/2018. The author is very grateful for all the data assistance and support from the leadership and staff of PT GS Battery and the Trisakti University Research Institute for the implementation of this research.

6. References

[1] Kuswardani D, Widyanto MR, Trihandini I. 2011 Metode association rules untuk analisis citra CT organ pasien kanker ovarium. *Kursor* **6**(2) pp 111-20
[2] Siregar S R 2014 Implementasi data mining pada penjualan tiket pesawat menggunakan algoritma apriori. *Pelita Informatika Budi Dharma* **7**(1) pp 152-6

IOP Conf. Series: Materials Science and Engineering **528** (2019) 012069 doi:10.1088/1757-899X/528/1/012069

[3] Zhang H 2000 *Mining and visualization of association rules over relational Dbmss*. (Florida USA: University Of Florida In Partial Fulfillment Of The Requirements For The Degree Of Master Of Science)
[4] Fadlina 2014 Data mining untuk analisa tingkat kejahatan jalanan dengan algoritma association rules metode apriori. *Informasi dan Teknologi Ilmiah (INTI)* **3**(1) pp 144-54
[5] Tyas E W 2008 Penerapan metode association rule menggunakan algoritma apriori untuk analisa pola data hasil tangkapan ikan. *e-Indonesia Initiative 2008* pp 1-4
[6] Hakim L and Fauzy A 2015 Penentuan pola hubungan kecelakaan lalu lintas menggunakan metode association rules dengan algoritma apriori. *The 1st University Research Colloquium (URECOL)*
[7] Kusrini and Luthfi E T 2009 *Algoritma Data Mining* (Yogyakarta: Andi Offset)
[8] Zhao Y and Cen Y 2013 *Data Mining Applications with R* (Waltham: Elsevier)

Application of Taguchi Method for Optimization of Parameter in Improving Soybean Cracking Process on Dry Process of tempeh Production

R A Anugraha[1], M Y Wiraditya, M Iqbal and N M Darmawan

[1] Industrial Engineering, School of Industrial Engineering, Telkom University, Jl. Telekomunikasi No. 01, Bandung, Indonesia
rinoandias@telkomuniversity.ac.id

Abstract. Taguchi method is a Design of Experiment approach for parameter optimization of a process. The purpose of this study is to determine the optimum parameters and the most influential parameters in the process of soybean cracking in the dry process of tempeh production. In the process of cracking the dry soybeans found a defect of 14.5%. after applying the Taguchi method and obtaining the optimum parameters, the defect on the dry process of cracking the dry soybeans is reduced to 6.6%. Determination of parameters and levels, objective functions, orthogonal array, signal-to-noise ratio, and analysis of variance (ANOVA) will be done in this study to improve the performance of the soy-cracking process. In this study there are three parameters used in conducting experiments with Taguchi method, three parameters namely distance between stones, hopper outlet diameter, and sun-drying duration. The results of this study are: distance between stones with setting optimum 1mm and percentage contribution 13%; hopper outlet diameter with setting optimum 25mm and percentage contribution 85%; and sun-drying duration with drying optimum 1 day and percentage contribution 2%. the most influential parameter is the hopper outlet diameter in the soybean cracking process, as it determines the soybean discharge that goes into the cracking phase.
Keywords: optimization parameter, Taguchi method, Dry process, Tempeh

1 Introduction

Design of experiments (DOE) is one of the powerful tools used to investigate deeply hidden causes of process variation. DOE techniques are useful for surfacing the effects of hidden variables, and studying possible effects of variables during process design and development. Experiments range from uncontrollable factors introduced randomly to carefully controlled factors [1]. The Taguchi experimental design reduces cost, improve quality, and provides robust design solutions. The advantages of Taguchi method over the other methods are that numerous factors can be simultaneously optimized and more quantitative information can be extracted from fewer experimental trials. [2]. Taguchi proposed using the signal-to-noise (S/N) ratio to measure the quality characteristics that deviated from the designed parameters [3]. Initially developed to improve the quality of manufactured goods (development of manufacturing process), then the application is extended to many other fields in Engineering, such as Biotechnology [4]. The Taguchi method involves identifying the appropriate control factors to obtain optimal results from the process. Orthogonal Arrays (OA) are used to perform a series of experiments.

The results of this experiment were used to analysis the data and predict the quality of the resulting components [5].

This study will apply the Taguchi method to improve the quality of the soybean cracking process in the dry process of soybean production. In the dry process of soybean production, which was initially broken down in wet conditions, using a dry soybean crusher, the soy bean was split dry. Soaking and boiling of soybeans is done after the cracking process using this machine. Therefore, the soybeans are split into 2 and the epidermis are separated in a dry state. The following figure 1 is a flowchart of the dry process of soybean production.

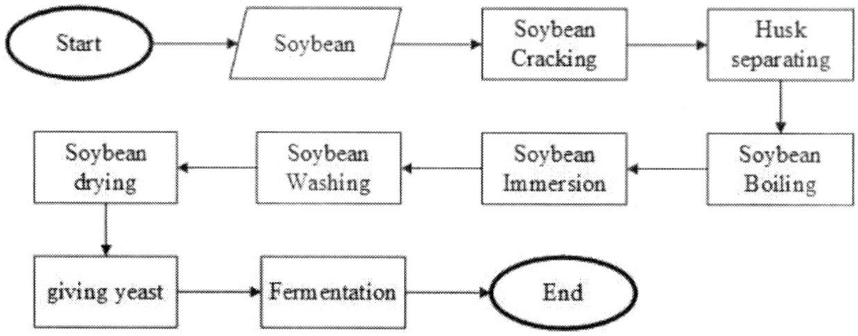

Figure 1. Dry Process of Tempeh Production.

Production of tempeh by using dry process known there is defect at result. Defect on the dry process of tempeh production occurs due to the nonconformity of soybean processing with Critical to Quality (CTQ). The CTQ analysis is done to achieve a high level of understanding of what is important to the success of the process [6]. Critical to Quality (CTQ) elements that directly relate to the needs and satisfaction of the exporter [7]. Tempe production is a field of great interest by SMEs residing in Indonesia. In Indonesia SMEs are scattered in various regions, the existence of SMEs is very dominant, especially in Java, especially in Bandung [8].

The result of application of Taguchi method is the optimum parameter on soybean cracking process. After obtaining the S/N ratio, this study will calculate the Analysis of Variance (ANOVA). ANOVA is a statistical model for comparing differences among means over two populations [9]. The nature of ANOVA is the sum of the squares SS_t (total variation) equal to the sum of the squares of the deviations of all parameters [5]. In this study ANOVA calculation aims to get percentage contribution of the most influential parameters.

2 Methods

2.1 Taguchi Method

Design of Experiment (DOE) is an approach often used in product development. Design of Experiment (DOE) is based on the principles of experimental design, equations or mathematical models and the result of these factors [10]. One method that uses the DOE approach is the Taguchi method. The Taguchi method was developed by Genichi Taguchi to analysis the effect of the experimental parameters and determine the optimal process parameters [11]. Taguchi suggested engineers to use orthogonal arrays of control and level factors to design experiments [12]. This method can not only reduce the experimental cost but also improve the reliability of the results of the analysis. In this method, the signal-to-noise ratio (S/N ratio) is used to analysis the contribution of each experimental data. The S/N ratio can be calculated in accordance with nominal-the-best, smaller-the-better, and larger-the-better [12].

2.2 Step of Taguchi Method

There are the steps involved of the Taguchi method [5]:

a. Identify main functions & Quality Characteristics

b. Identification of Objective Function
c. Identify parameters & levels
d. Selection of Orthogonal Array design
e. Conduct Experiment
f. Calculation of Signal-to-Noise

2.3 *Analysis of Variance*

Analysis of Variance (ANOVA) is the most commonly used method for GR & R analysis. ANOVA for GR & R is performed with operator N number and M number of random parts. The sections are labeled first for purpose and the operator will perform the measurement of the parts with the L number of replication. For a certain degree of trust, Minitab can calculate from operator, from part-to-part, from measurement equipment (repetition), and from interaction between operator and spare parts. R & R is the number of repetitions and reproducibility [6]. In search of various parameters by calculating the total of the sum of squares (SSt). Here is a solution for finding SS. Here is the equation of calculating SSt [7].

$$SSt = \sum_{i=1}^{n}(\bar{x}_i - \bar{\bar{x}}) \quad (1)$$

2.4 *Cracking Soybean Process*

In the process of cracking soybeans there are 5 stages passed for soybeans to break. 5 stages are soybeans entering the hopper, drainage to screw conveyor, drainage toward cracking, soy cracking, and product streaming. The following figure 2 is a transparent box of soybean breakers.

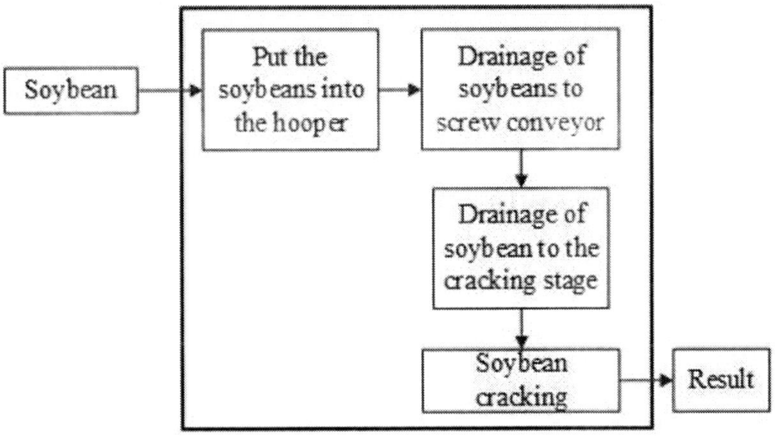

Figure 2. Transparent Box of Soybean-Cracking Machine.

In the dry process of tempeh production, soybeans are broken down using soy-cracking machine. The machine produced by Rumah Tempe Indonesia is made of Stainless steel. Machines in general have 3 main parts, namely prime mover, power transition, and rotation motion. In a soybean-cracking machine, the originator is an AC dynamo with a power of 220 V. The power transition used is a belt pulley. And the movement is a screw conveyor and grinding wheels. Screw conveyor serves to push soy into the cracking part. Grinding stone serves to break the soybeans into 2 parts. Here figure 3 is a schematic image of a soy breaker machine.

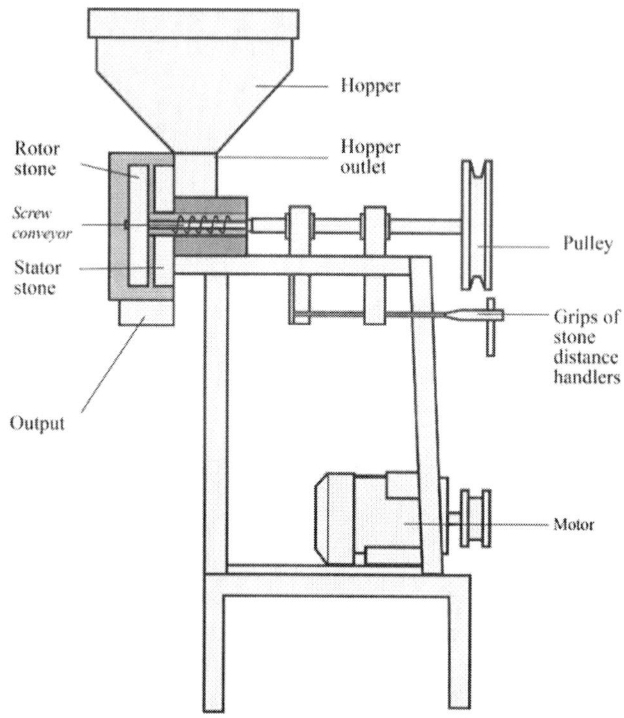

Figure 3. Schematic of Soybean-cracking Machine.

In the soy-cracking process, soybeans will be fed through the hopper. Then the soybean will fall to the screw conveyor through the hopper hole. Screw conveyor will push the soybean to the grinding stone. The grinding wheel will break the soybeans into 2 parts. Then the split result will fall to the resulted splice channel. The distance between the grinders can be adjusted using the grinding wheel handle.

3 Result and Discussion

Identify Main Function & Characteristics Quality

Main Function	: Soybean-cracking process
Characteristic Quality	: Defect
Testing Tools	: Scales

The main function is the process done in the application of Taguchi method in conducting experiments. Characteristic quality is the response to be calculated. Testing tools are tools that are calculated to measure response.

3.1 Identify Objective Function

The purpose of this study is to minimize the defect which means the smaller the defect is identified the better the quality of the splitting process. Therefore, the objective function in this study is smaller-the-better. Here is the formula used to calculate S/N ratio smaller-the-better [5].

$$Signal\ to\ noise = -10\log_{10}\left(\frac{1}{n}\sum_{i=1}^{n} y_i^2\right) \quad (2)$$

3.2 Identify Parameter & Level

The parameters and levels are determined to experiment with the Taguchi method. Based on brainstorming with researchers and expert operators of soybean cracking machine, obtained parameters and levels presented in table 1.

Table 1. Parameter and Level.

Parameter	Level 1	Level 2	Level 3
Distance between stone	1 mm	0,9 mm	0,8 mm
Hopper outlet diameter	21 mm	19 mm	25 mm
Sun-drying duration	0 day	1 day	3 day

3.3 Selection of orthogonal array design

Selection of orthogonal array design using Minitab software. The number of parameters and the number of each level entered to obtain the choice of orthogonal array design. The orthogonal array design to be used is orthogonal array L9 which will be simulated 3 times experiment on soybean cracking process. The following table 2 is an orthogonal array design used.

Table 2. Orthogonal Array.

Experiment	Parameter 1	Parameter 2	Parameter 3
1	1	1	1
2	1	2	2
3	1	3	3
4	2	1	2
5	2	2	3
6	2	3	1
7	3	1	3
8	3	2	1
9	3	3	2

3.4 Experiment

At this stage experiments will be conducted based on selected orthogonal arrays. Each experiment on an orthogonal array will be performed three times to calculate the noise factor interference. The following Table 3 is the result obtained from conducting the experiments performed. Response or result of defect that happened from soybean cracking process.

Table 3. Response of Experiment Result.

Experiment	Trial 1	Trial 2	Trial 3	Mean
1	85	70	70	75
2	205	165	170	180
3	60	90	60	70
4	105	95	90	96.67
5	240	250	245	245
6	110	100	95	101.67
7	120	115	120	118.33
8	330	340	360	343.33
9	110	70	80	86.67

Below figure 4 is a Design of Experiment from the experiments conducted on this study. Design of Experiment is a stage that is done when doing experiments. DOE is performed 9 times according to the number of combinations of orthogonal array tables that have been designed.

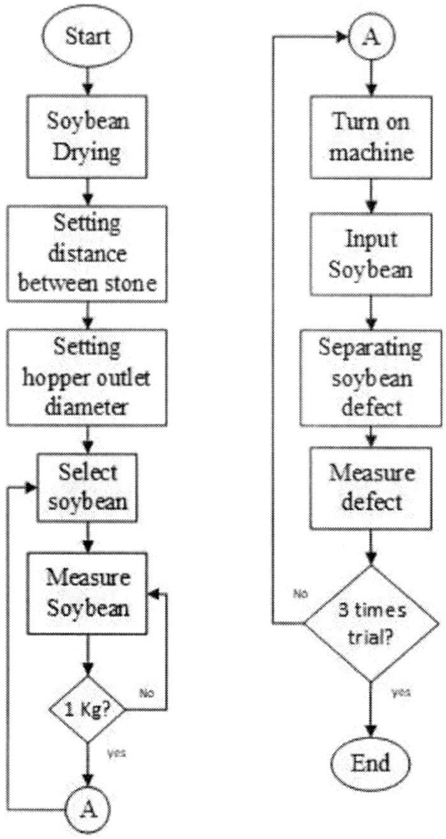

Figure 4. Design of Experiment.

3.5 Examine Data; predict the optimum parameter levels and its percentage contribution

Since the objective function of the soybean-cracking process is smaller-the-better, the S/N ratio calculation is performed with the smaller-the-better equation. The following table 5 is the calculation of the S/N ratio of each experiment.

Table 4. S/N Ratio Experiment.

Experiment	S/N Ratio
1	-37.540
2	-45.148
3	-37.076
4	-39.724
5	-47.785
6	-40.160
7	-41.464
8	-50.720
9	-38.921

After obtaining S/N ratio from each experiment, then calculate S/N ratio from each level of each factor.This study use Minitab software to calculate S/N ratio which the data to be process is response of experiment. The following table 6 is the calculation of the S / N ratio and figure 5 is the graphic of each S/N ratio parameter.

Table 5. Result of S/N Ratio.

	Distance between stones		Hopper outlet diameter		Sun-drying duration	
Level	Total	Average of S/N Ratio	Total	Average of S/N Ratio	Total	Average of S/N Ratio
1	-119.76	-39.92	-118.73	-39.58	-128.42	-42.81
2	-127.67	-42.56	-143.65	-47.88	-123.79	-41.26
3	-131.10	-43.70	-116.16	-38.72	-126.32	-42.11

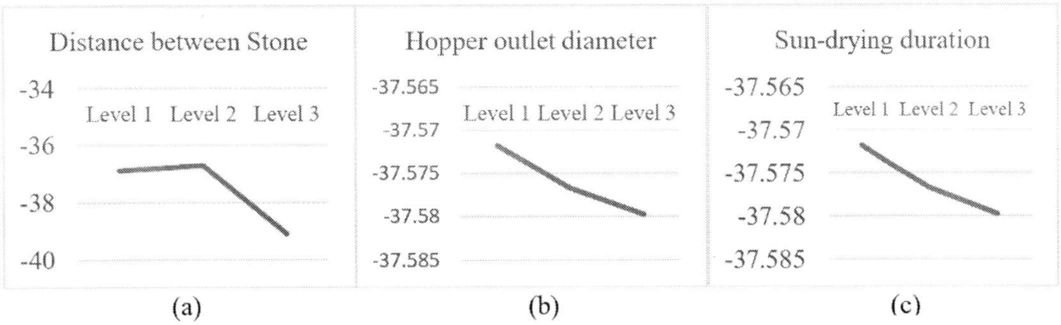

Figure 5 Graphic S/N ratio of (a) Distance between stone, (b) hopper outlet diameter, and (c) Sun-drying duration.

The optimum parameters are selected based on the value of the highest S / N ratio. Based on the above graph, the optimum parameter in soybean cracking process is shown in table 7.

Table 6. Parameter Optimum.

Parameter	Optimum Value
Distance between stones	1 mm
Hopper outlet diameter	25 mm
Sun-drying duration	1 day

The calculation of the analysis of variance is done to obtain the percentage of contribution parameters to obtain the most influential parameters of the parameters undertaken trials. Here is the result of SSt calculation shown by table 8.

Table 7. Result of ANOVA.

Parameter	SS	SS%
Distance between stones	7.51	13%
Hopper outlet diameter	51.25	85%
Sun-drying duration	1.19	2%
Total	59.96	100%

From the table above can be seen in the process of solving soybean parameters that affect the occurrence of defect is the percentage of 13% grinding distance between grinding, hole hopper diameter 85%, and duration of soybean 2%.

4 Conclusion

This journal applies Taguchi method for parameter optimization in soybean cracking process on dried production process of tempeh. Here are the conclusions based on the experimental results that have been done in this journal: In the soybean cracking process the most influential parameter is the hopper outlet diameter as it determines the soybean discharge that goes into the cracking phase; The application of the Taguchi method can perform experiments more efficiently for parameter optimization of a process.

5 References

[1] Konda, R., 1999 *Design of experiments to study and optimize process performance* (Australia: International Journal of Quality & Reliability Management) p. 2.

[2] Reena P., G.H.V.C.Chary and M.G.Dastidar 2016 *Application of Taguchi Method for Optimizing the Process Parameters for the Removal of Copper and Nickel by Growing Aspergillus sp* (India: Water Resources and Industry) p. 2.

[3] Liang L., Jiang L., Wang C., Wei W., Xing Z. and Guangming X. 2016 *Application of Taguchi Method in the Optimization of Swimming Capability for Robotic Fish* (International Journal of Advanced Robotic Systems) p. 6.

[4] Rao, R. Sreenivas, C. Ganesh Kumar, R. Shetty Prakasham, and Phil J. Hobbs *The Taguchi Methodology as a statistical tool for biotechnological applications: A critical appraisal* (Biotechnology Journal 3) p. 510–523.

[5] Athreya, S. and Venkatesh, Y. D. 2012 *Application of Taguchi Method for Optimization of Process Parameters in Improving the Surface Roughness of Lathe Facing Operation* (International Refereed Journal of Engineering and Science) p. 13–19.

[6] Zhan, W. and Ding, X. 2016 *Collection and Statistical Tools for Engineers and Engineering.*

[7] Sofan, A. T., Anugraha, R. A. and Suryadhini, P. P. 2014 *Meminimasi Defect Pada Proses (STUDI KASUS DI PT ABO FARM)* vol 1 no 1 (Bandung: Jurnal Rekayasa Sistem & Industri) p. 1–7 https://doi.org/10.25124/jrsi.v1i01.3.

[8] Iqbal, M. 2015 *usulan perbaikan desain kemasan stick strawberry kencana mas menggunakan metode quality function deployment* (Bandung: Jurnal Rekayasa Sistem & Industri).

[9] Moore, D.S., McCabe, G.P. and Craig, B.A. 2009 *Introduction to the practice of statistics, 7th Edition* (New York: W.H. Freeman and Company).

[10] Kumar, L. 2015 *Full factorial design for optimization, development and validation of HPLC method to determine valsartan in nanoparticles* (King Saud University: Saudi Pharmaceutical Journal) p. 549–555.

[11] G. Taguchi and Y. Yokoyama 1993 *Taguchi methods: design of experiments* (ASI Press)

[12] Hsu C.F. 2015 *Optimization of the Copper Plating Process Using the Taguchi Experimental Design Method* (China: Journal of The Electrochemical Society).

IOP Conf. Series: Materials Science and Engineering **528** (2019) 012071 doi:10.1088/1757-899X/528/1/012071

Analyze of mitigation waste in reconditioning process of Iron Drum with Lean Six Sigma (Case study at PT Mulya Adhi Paramita)

Ahmad[1], Lilyana[2], L Widodo[3], L Gozali[4] and A Maryadi[5]

Industrial Engineering Study Program Tarumanagara, Jakarta, Indonesia
ahmad@ft.teknik.untar.ac.id[1], lilyanajap@yahoo.com[2], lamtow@yahoo.com[3], ligoz@ymail.com[4], angaldian@gmail.com[5]

Abstract. PT. Mulya Adhi Paramita is a chemical company that focuses on the distribution of chemicals, but also performs the process of reconditioning used iron drums. In the process of reconditioning drum is still found some waste, for that this research using lean six sigma concept which aims to reduce waste on drum recondition process. This research uses DMAIC stage and using Waste Assessment Model method with the result of 2 biggest waste identification is waste motion and defect which will become priority improvement, using Defect Per Million Opportunity to know sigma level that is equal to 3,77 sigma, Value Stream Mapping with percentage result PCE on CVSM of 16.31%; Process Activity Mapping result is value added time of 60.22 minutes, non value added time of 9.74 minutes, and necessary non value added time of 21.88 minutes. After performing the improvement implementation can be seen the result is the increase in PCE on FVSM of 21.5%, decreased percentage defect by 0.72% and DPMO with sigma level of 3.81 sigma
Keywords: Iron drum, Waste Assessment Model, Value Stream Mapping, Process Activity Mapping, Failure Mode and Effect Analysis

1. Introduction

The development of the chemical industry in Indonesia is rapidly grown. It can be seen from the increasing distribution of chemicals by the small to large business sector in mostly all areas of Indonesia which use chemicals as main core business. In the distribution of the chemicals, occasionally requires special packaging, depends on the characteristics of chemicals. One of packaging tools that has been used often is the drum. Drum is unreactable material which perfectly fit to load the chemicals for a long time.

PT. Mulya Adhi Paramita (MAP) is a company focusing on the distribution of liquid or solvent chemicals. In the process of packaging the product, MAP uses the reconditioned iron drum that made by company itself in order to save production costs.

In the real conditions in the reconditioning process drum has still found problems with high occurancy, which leads to be wasted. The bunches of waste can be caused by various factors such as humans, machines, methods, and others. The impact of these problems could impact to the high cost of the company. Based on these problems, it will be necessary to conduct research with a special study to reduce waste in MAP. One of the methode that been using in this research with oncoming lean six sigma include DMAIC phase, waste assessment model method, and value stream mapping.

The research goals are about to identify the waste that occurs in the process of reconditioning the drum, analyzing the factors of the waste, and provide alternative solutions that need to be done in order to reduce waste. Research is only done on recondition drum process at MAP, for improvement focused only on 2 types of waste that is dominant based on WAM method, and product defect data used in research is data in 2017.

2. Methods

2.1. Lean Six Sigma

Lean and Six Sigma can be defined as business philosophy, systemic and systematic approaches to identify and eliminate waste or non-added activities through continuous radical (continuous improvement) to achieve sigma-level work, by streaming products (materials, work in process, output) and information using pull system (pull system) from internal and external customers for excellence and perfection produce only 3.4 defects for every single operation 3.4 DPMO (defect per million of opportunities).

2.2. Value Stream Mapping

Value stream mapping is a visual process mapping in which there is a flow of information and material flow in order to prepare better methods and performance [2]. The purpose of value stream mapping is to identify and eliminate waste sources by applying proposed-state value streams that can become a reality in the near term. The goal is to build a production chain in accordance with the concept of lean that each process is directly connected to the demand of the customer either with continuous flow or with pull system and every process cultivated as optimal as possible to produce in accordance with what the customer requested with the time and amount.

2.3. Process Activity Mapping

Process Activity Mapping is used to identify lead time and productivity of both physical product flow and information flow, not only within the scope of the company or also in other areas of the supply chain. The basic concept of this tool is to map each stage of activity that occurs from operations, transportation, inspection, delay and storage, then grouping into types of activities that begin with value adding activities, non value added activities and non value adding activities . The purpose of this mapping is to help understand the flow of processes, to identify waste, to identify whether a process can be rearranged more efficiently, to identify improvements in value added flows.

2.4. Waste Assessment Model

The Waste Assessment Model is a model developed to simplify the search for waste issues and identify to eliminate waste. This model illustrates the relationship between seven waste (O: overproduction, P: Processing, I: Inventory, T: Transportation, D: Defect, W: Waiting, and M: Motion).

2.5. Waste Relationship Matrix (WRM)

Waste Relationship Matrix (WRM) is a matrix used to analyze measurement criteria. The line on the matrix shows the effect of a particular waste on the other 6 wastes, while the column in the matrix shows the waste that is affected by other waste. The diagonal of the matrix is placed with the highest relationship value and by default, each type of waste will have a principal relationship with the waste itself.

2.6. Waste Assessment Questionaire (WAQ)

Waste Assessment Questionaire (WAQ) is used to identify and allocate waste that occurs on the production line. The assessment questionnaire consisted of 68 different questions, where the questionnaire was known for the purpose of determining waste

2.7. Research Methodology

The following stages of methodology in conducting research at PT. Mulya Adhi Paramita can be seen in Figure 1. This research was conducted with many methodes, from data collecting until data processing. Using questionaire for FMEA and direct measure for timing process in Current Value Stream Mapping and Future Value Stream Maping, in order to make the whole proces's resulth with high eficiency and efective.

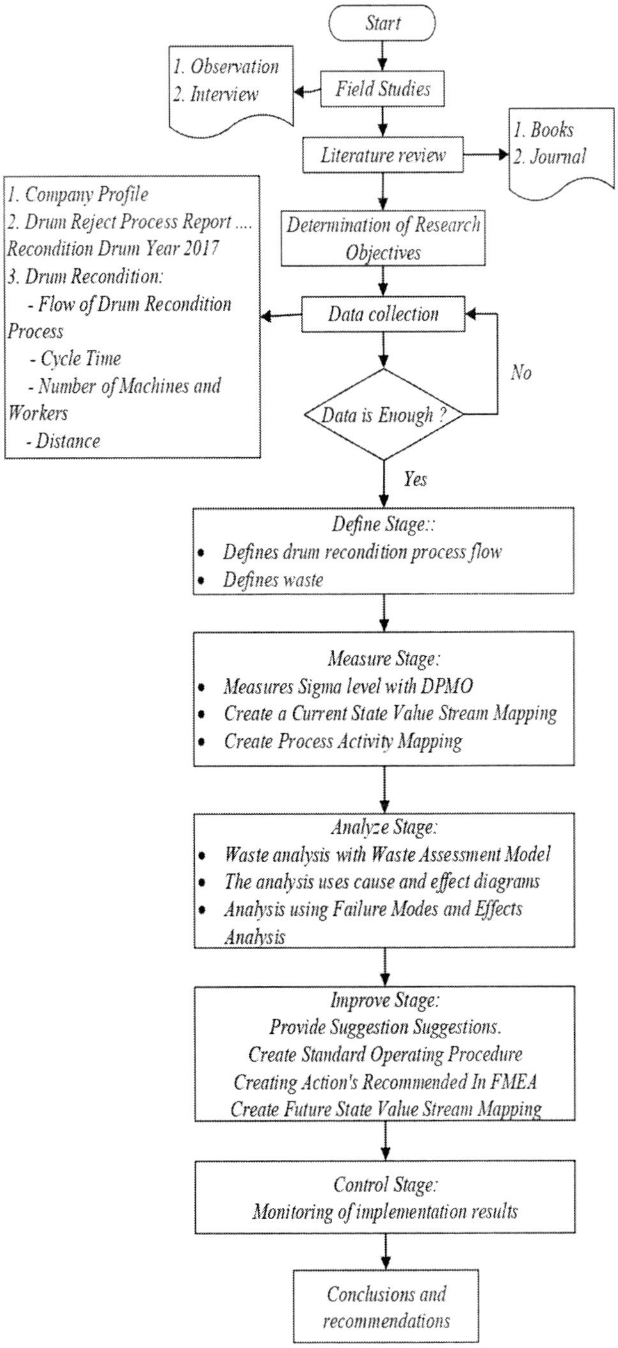

Figure 1. Flowchart of Research Methodology

3. Results and Discussion

3.1. Define

Define stage is the stage of stipulating the target of six sigma project which confirmed that drum recondition process with drain recondition process with stages of drain recondition process starting from drum used at its storage then goes to recondition process until product is done at its place of storage. The scheme process can be seen in Figure 2.

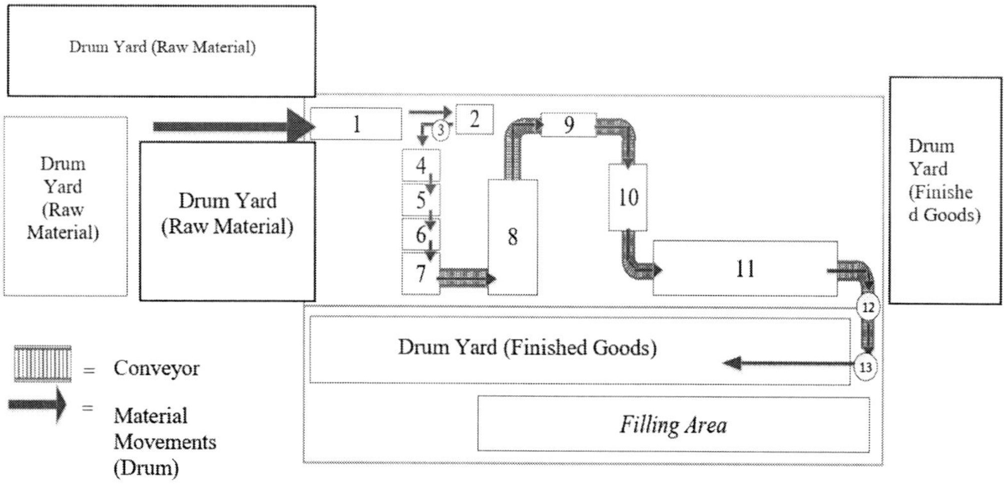

Figure 2. Flow of Drum Recondition Processes

Further defining the waste along with some of which are contained in the process of reconditioning iron drums in MAP, namely:

Waste Motion

Waste motion is type of waste that occurs due to the movement or machine that does not add value to goods and services for customers, but only increase the cost and time. Waste motion that occurs in the process of reconditioning drums, namely:

a. In storing the used drum on the drum yard, the worker must create a drum row and drum up the stratum.
b. The worker does not move the drum defect on the storage area so as to move the recurrent drum defect in the drain recondition area for blocking the movement and transfer of the WIP drum.
c. Workers put the drums that defect on the drum yard in vain so that it can mix with the former drum. This resulted in drum defect should be returned to the suppliers to be transferred again in the process of reconditioning because workers can Space considerations not distinguish the former drum raw material and drum defect.
d. The process of moving the drum into a storage place that is far enough distance and the number of manually.

3.2. Measure

The measure stage is the measurement stage to make it easier to know the current level of performance by using Defect Per Million Opportunity, Current Value Stream Mapping, and Process Activity Mapping. Here is a calculation of DPMO performed to determine the level of Sigma using drum defect drum process data in 2017 and Sigma value of 3.77 Sigma which means that defect is enough to be resolved but still needed further analysis. The results of DPMO calculations can be seen in Table 1.

Table 1. *Defect per million opportunities*

Information	Results
Defect	44.874 drum
Unit	637.993 drum
Opportunities	6
Defect per unit	0,07033
Defect per Opportunity	0,011722699
DPMO	11.722,69
Sigma level	3,77 Sigma

Furthermore, the mapping of Current Value Stream Mapping to know the flow of information relating to the process of reconditioning drums, stages of iron drain recondition process from used drums to finished products for consumers, and Value Added Time of 18.92 minutes and Non Value Added Time of 97.06 minute, Lead Time of 115.98 minutes, and the percentage of Process Cycle Efficiency (PCE) of 16.31% so that PCE <30% then the drain recondition process does not include lean.

Then create Process Activity Mapping is used to determine the proportion of activities or activities including Value Added (VA) and Non Value Added (NVA) and identify waste in the value stream and optimize the process so that more efficient and effective. Based on the activity mapping tool process it can know the values and activities that can be seen in Table 2 and Table 3. Total activities of 46 activities that have a total time of 91.847 minutes.

Table 2. Summary Process Activity Mapping By Value

Information	Number of Activities	Total Time (minutes)
Value Added	14	60,224
Non Value Added	4	9,741
Non Value Added Necessary	28	21,8826
TOTAL	46	91,8476

Table 3. Summary of Activity Mapping by Activity

Activities	Number of Activities	Total Time (minutes)	Percentage (%)
Operation	23	63,549	69,19
Transportation	16	18,230	19,85
Inspection	3	0,328	0,36
Storage	0	0	0
Delay	4	9,741	10,61
TOTAL	46	91,847	100

Based on the above table it can be seen that in the process of reconditioning drum for the proportion of operation time has the largest time that is for 63 minutes with a percentage of 69% of the total time. It can be concluded that the waste that occurs in the drain recondition process is only found in 4 activities that are non value added which is classified as Delay with 9.7 minutes of total time total of 91.8 minutes, this means the process of reconditioning drum in the current condition is quite efficient.

3.3. Analyze

The analyze stage is the stage to analyze the most dominant problem of waste analysis and the causal factors of waste in the drain recondition process. The process of waste analysis using waste assessment model method by conducting discussion with recondition drum workers to give an understanding about WRM and WAQ method, then the WRM and WAQ questionnaires were given to 5 correct reconditioning drum workers, 1 supervisor, 2 leader and 2 staff so they are able to answer the questionnaire correctly. The following is the result of the waste analysis process so that the final result is the rank of waste that can be seen and Figure 3.

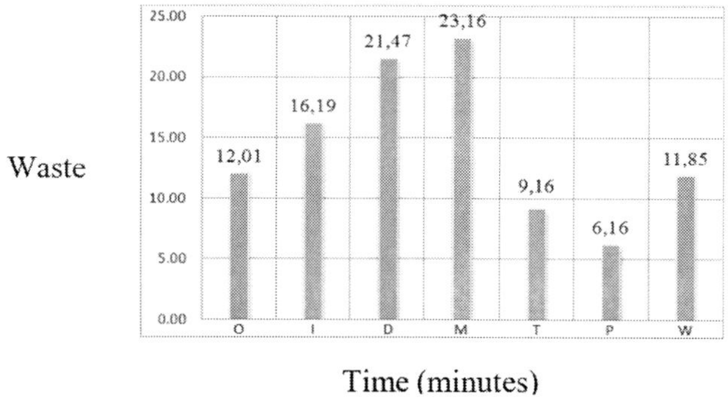

Figure 3. Rank Result of Waste Assessment Calculation

Based on the graph can be known waste with the greatest value is the motion of 23.16; defect of 21.47; inventory of 16.19; over production of 12.01; transportation of 9.16; waiting for 11.85, and process of 6.16. So that can be known 2 biggest waste which will become priority of repair is waste motion and defect.

3.4. Improve

The improvement stage is an improvement stage by determining the repair solution related to the causal factors that have been found in the analyze phase so that it can reduce the waste on the drain recondition process. The following suggestions for improvement are Future State Value Stream Mapping

Future State Value Stream Mapping is a condition of value stream mapping after the implementation of the improvement so that it can show the percentage comparison of process cycle efficiency before improvement that is 16.31% to 21.5% after implementation.

3.5. Control

At this stage the authors will implement proposed suggestions for improvements, Standard Operating Procedures, and Actions Recommended that have been made in the process of reconditioning used iron drums in PT. Mulya Adhi Paramita to overcome waste motion and defect in order to minimize waste in accordance with the purpose of this study. The result of the implementation for 10 working days is the reduction of defect percentage of 0.72% which means the implementation of the implementation is quite successful which can be seen in Table 4.

Table 4. Percentage of Defect Comparison

Before Implementation	After Implementation
Percentage defect = 7,03%	Percentage defect = 6,31 %
The existence of a decrease in magnitude 0,72 %	

DPMO calculation is then performed again using data defect after implementation for 10 working days. DPMO calculation results can be seen on Table 5.

Table 5. DPMO After Implementation

Information	Results
Defect	1727 drum.
Unit	27.372 drum.
Oportunities	6
Defect per unit	0,0630
Defect per Opportunity	0,01051561
DPMO	10515,61
Sigma Level	3,81 Sigma

Based on the above table it can be seen there is an increase in sigma value of 0.04 from 3.77 sigma to 3.81 sigma. This means that the implementation has done a good enough impact in improving the performance results of the drain recondition process.

4. Conclusion

The following conclusions can be found after doing this research, namely:

1. Waste contained in recondition drum at PT. Mulya Adhi Paramita based on the identification of the Waste Assessment Model with the second largest ranking of waste motion and waste defect such as drum failed paint, drum leak, drum in the water or wet, the drum of the inner oil and drum in the other material.
2. The factors causing waste motion are basically due to poor storage and other inefficient activities. While waste defect caused by various factors starting from human, machine, environment, and storage.
3. A repair solution used to reduce waste occurring in the reconditioning process of drums, especially waste motion and defect is to provide suggestions of improvements, rack design for drum storage, standard operating procedure, and action recommended on FMEA.

5. References

[1] Gasperz V 2007 *Lean Six Sigma for Manufacturing and Services Industries.* (Jakarta: PT. Gramedia Pustaka Utama)

[2] Hines P and Rich N 1997 The Seven Value Stream Mapping Tools. *International Journal of Operations & Production Management* **17**

[3] Rother M and Shook J 1999 *Learning to See: Value Stream Mapping to Add Value and Eliminate Muda,* 2nd Ed. (Massachussets: Lean Enterprise Institute, Brookline)

[4] Rawabdeh I 2005 *A Model for The Assessment of Waste in Job Shop Environments.* (New York: Engineering Internasional Journal)

[5] Besterfield D H 1994 *Quality Control.* 4th Ed. (New Jersey: Prentice-Hall International, Inc., Englewood Cliffs)

[6] Casadei G, Serra K and Tani 2007 Implementation of a direct control algorithm for induction motors based on discrete space vector modulation. *IEEE Transactions on Power Electronics* **15**(4) pp 769-77

[7] Kevin A, Lange S, Legget C and Beth B 2001 *Potential Failure and Effect Analysis.* (New Jersey: Sons and Inc)

[8] Sugiyono 2006 *Statistika untuk Penelitian.* (Bandung: CV. Alfabeta)

[9] Dimyati M 2006 *Belajar dan Pembelajaran.* (Jakarta: Rineka Cipta)

[10] Randers J M 1972 *Limits to Growth.* (New York: Universitas Books)

Measurement and proposal of improving Marketing Process to improve the Quality of Aftersales Services with Fuzzy Quality Function Deployment and Data Mining Methods in OV Agency

R Fitriana, W Kurniawan and M R Anwar

Quality Engineering Laboratory, Department of Industrial Engineering, Faculty of Industrial Technology, Universitas Trisakti, Indonesia
rinaf@trisakti.ac.id, wawan_bdb2@yahoo.com, anwaresta@gmail.com

Abstract. OV Agency is a marketing, sales and service office under PL Assurance. The issues contained in the OV Agency include a lack of understanding of the client against the types and benefits of unit-linked insurance products taken, causing the client to stop paying the premium (expired) or closing policy (surrender), the problems caused by the agency lacking a detailed explanation lead to misunderstanding by the customer. The purpose of this study is to analyze customer satisfaction and service quality conditions, using Fuzzy-Quality Function Deployment to improve service quality, and use data mining to obtain useful indications based on data obtained from questionnaires and interviews. Questionnaires were made with the results of discussions with OV Agency which were then distributed to 100 customer samples. The results of the questionnaire answers are converted into fuzzy numbers which are then processed by the QFD method to determine the technical response to be performed first. The company uses an optimistic index with a value of 0.5 in carrying out technical responses. The three main priorities of technical responses to be undertaken under the prescribed Fuzzy QFD method are to increase the frequency and quality of training for agents, improve quality standards within the company, and update the methods and techniques used in the training. Data mining is calculating to determine the overall level of customer satisfaction. The satisfactory results of OV Agency customer satisfaction rate, it can be seen from satisfactory posterior probability has the greatest value of 0.0017.
Keywords: Fuzzy, Quality Function Deployment, Data Mining

1. Introduction

OV agency is a marketing, sales and service operations office under the auspices of PL assurance. Services provided by this company include financial literacy, life insurance & income protection, corporate insurance, estate planning, wealth management, will & trust services. The study was limited to OV agency offices. In almost 3 years, OV agency has been ranked as the top four among more than 200 agencies in Indonesia.

Problems in the agency include client dissatisfaction with the type and benefits of unit link insurance products taken, which causes the client to stop paying premiums (lapsed) or close the policy (surrender). Another thing, the client also lacks understanding of the procedures and requirements of the claim, which causes the claim to be late or not paid. The things mentioned above are caused by consumers' misunderstandings when getting an explanation by an agent or an agent lacking a detailed explanation that causes misunderstanding by the customer.

The purpose of this study is to 1) Analysis of customer satisfaction conditions and current service quality at OV Agency; 2) Proposed use of Fuzzy Quality Function Deployment in marketing process at OV Agency to improve service quality; 3) Using data mining to obtain a relationship or pattern that provides useful indications based on data obtained from questionnaires and interviews.

QFD is a method used to design a process of products or services tailored to the wishes of consumers or customers (Voice of Customer or VoC) [1]. Meanwhile, according to [2] in [3], QFD method in principle is an effort made to translate what the consumer wants into what is produced by the company.

When using fuzzy calculations in the QFD matrix, the result of the weight calculation for each functional item is not an integer but a fuzzy number. Liou and Wang in [4] describe a perfect algorithm for fuzzy ranking methods. Their method of using integral values as a fuzzy index has resulted in efficient and consistent calculations. An adjustable index for integrated fuzzy numbers (ω) is used as an optimistic level in an improvement process.

According to Day in [3], to find out company problems related to product planning or service processes involving consumers, the HoQ matrix is used. Data mining is a semi-automatic process that uses statistical techniques, mathematics, artificial intelligence, and machine learning to extract and identify useful and useful knowledge information stored in large databases [5].

Table 1. Research Position

NO	TOPIC	SOURCE
1.	Integration between Fuzzy, Servqual, and Quality Function Deployment	[6]
2.	Integration between Business Intelligence, Fuzzy FMEA and Data Mining	[7]
3.	Integration between Business Intelligence, Quality Function Deployment, and Data Mining	[8]
4.	Integration between Fuzzy and Quality Function Deployment	[9]
5.	Integration between ANP Weighted Crisp and Fuzzy QFD	[10]
6.	Integration between Fuzzy, Quality Function Deployment, and Data Mining	This Research Position

Table 1 is a research table on QFD and Data Mining around the world. The position of this research among other studies around the world, can be seen from various aspects above. Can be seen from the aspect of Quality Control research methodology is a combination of fuzzy and method of quality function deployment, where QFD is applied simultaneously with fuzzy logic to reduce the subject subjectivity of respondents. Data mining method is also used to know the classification of satisfaction level on OV Agency.

2. Methodology

The study began through an interview with the senior agency director of OV Agency to get initial information about the company, the products offered, and problems that occurred. Then conducted research based on sales data and persistence data. In addition, the initial phase research was carried out with literature studies through the collection of theories related to the topic of research through reference to textbooks, e-books, and journals. Based on the results of both identified that the problem to be lifted is the lack of understanding of consumers when given an explanation by the agent or the agency lacks detailed explanations that cause misunderstandings by customers that cause customers to close the policy. From these problems determined the purpose of this study.

The next stage is to collect data consisting of primary and secondary data. Primary data consists of the results of the distribution of questionnaires, while secondary data consists of general company data, HR, and products offered. The next process is to analyze the current conditions of customer satisfaction with insurance products and services. Data processing is done by data mining classification method to determine the level of customer satisfaction. Processing questionnaire data and interviews using fuzzy quality fuction deployment. From the results of the analysis of Fuzzy-Quality Function Deployment (FQFD) can be made improvement proposals.

3. Result and Discussion

3.1 Data Mining

The data used in data mining processing using the data of questionnaire results of satisfaction level on products and insurance services that have undergone processing in the form of code changes satisfaction level to the level of satisfaction.

The initial stage in manual data mining calculation using bayes method is by determining the posterior classification probability first. Prior probability is obtained by dividing the number of one classification by the number of correspondents or can be seen in formula 1.

$$Prior\ Probability = \frac{x_i}{\sum_{i=1}^{n} x} \quad (1)$$

On insurance products and services have a posterior probability that has great value in satisfactory classification with values above 0.5. The overall posterior probability results data can be seen in Table 2.

Table 2. Probability of Prior Products and Insurance Services

Classification	Count	Prior Probability
Very Satisfying	17	0.17
Satisfying	56	0.56
Good Enough	26	0.26
Not Satisfying	1	0.01
Very Unsatisfying	0	0.00

The next stage after getting prior probability is to get the probability of interaction between each attribute with classification. Having obtained the probability of interaction of each attribute by classification, the next step in the use of classification by bayes method is to calculate P (At | K) or known as the probability of attribute interaction with classification. P (At | K) is obtained by multiplying each probability of interaction with the classification of respondents.

$$P\ (At|K) = P\ (At|K)_1 \times P\ (At|K)_2 \times \ldots \times P\ (At|K)_n \quad (2)$$

After that done the calculation of P (K) P (At | K). The calculation is by multiplication of the prior probability of the classification with the probability of the interaction already obtained.
Next calculate the posterior probability, posterior probability can be obtained by the following way:

$$Posterior\ Probability = \frac{P(K).P(At|K)}{P(K)} \quad (3)$$

The results of the posterior probability calculation of OV Agency insurance products and services using the Bayes method can be seen in Table 3.

Table 3. Results of Calculation of Posterior Probabilities

Classification	Prior Probability	Posterior Probability
Very Satisfying	0.17	7,05266E-12
Satisfying	0.56	0,001719532
Good Enough	0.26	7,56975E-10
Not Satisfying	0.01	2,60812E-32
Very Unsatisfying	0.00	0

The prior probability is the probability derived from the preliminary study data, whereas the posterior probability is the probability that has been corrected after additional information or after conducting the research. The conclusion of the bayes method uses the result of posterior probability data by looking at the result of posterior probability. The conclusion is taken when the highest posterior probability value in the classification or can be formulated as follows:

$$h_{MAP} = arg\ max\ P(x|h)\ P(h) \quad (4)$$

In this study, OV Agency insurance products and services based on data table 8 can be concluded that the level of service satisfaction provided by the company to respondents is at a satisfaction level of satisfied.

3.2 Quality Function Deployment

Voice of customer is an OV Agency customer's voice obtained from customer interviews, customer complaints, from researches ever conducted on the quality of service on insurance, and consultation with the company, to know the needs and desires of the real customers.

Quantitative data collection at this stage is done by conducting surveys using tools or instruments in the form of questionnaires distributed to customers of OV Agency through google form. Criteria of the intended respondents are customers who have or have had an insurance policy in PT. PL Assurance by using the services of OV Agency. For the initial stage of the dissemination of approximately 30 questionnaires as a pilot test. After the questionnaire was valid and reliable, 100 questionnaires were distributed to obtain the required data. After getting 100 respondents, the choice of accepting response on google form is closed.

This section shows how important each attribute is the voice of customer in the eyes of the respondent. The calculation is done by determining the mode of the attribute importance level questionnaire, which results can be seen in Table 4.

Table 4. Importance to Customer

No.	Code	*Voice of Customer* (Importance)	Modus
1	A1	Completeness of the Agent's tools, documents or tools while serving you	4
2	A2	Appearance & carriage Agent when interacting with you	4
3	B1	The accuracy of solutions or products provided by the agent according to the needs & desires	5
4	B2	Agent's ability to resolve requests & problems	5
5	B3	Knowledge & skills Agent in insurance financial services	5
6	B4	Agent's ability to understand your needs & wants	5
7	C1	Speed & alertness Agent in providing responses	5
8	C2	Clarity of time estimation of the Agent on a service process	5
9	D1	Agent's ability to listen to your needs & wants	5
10	D2	Availability of agent time in serving you	5
11	E1	The language the Agent uses is clear & easy to understand when serving you	4
12	E2	The hospitality shown by the Agent while serving you	5
13	E3	Ease of Agent to be contacted either face to face or by phone	5
14	F1	Your trust in agents for long-term service	5
15	G1	Security & convenience in doing transaction through Agent	5

The value of improvement ratio is obtained from the comparison between the company's target and the current satisfaction performance, where this value shows the weight of difficulties to make improvements in meeting customer needs. This value also determines which Voice of Customer attributes will later be followed up. If the ratio is less than 1 ($\leq$1), then the Voice of Customer (VoC) attribute does not need to be given a technical response anymore. A smaller ratio equal to 1 means that the level of customer satisfaction with an attribute is above or equal to the level of importance of the attribute so that it no longer needs to be given a technical response to improve the level of customer satisfaction.

Development of House of Quality matrix is done to answer customer's desire on service attributes at Onevison Agency, with aim to improve service quality and OV Agency performance. Because the technical response provided was related to service aspects, the determination was made by conducting interviews and discussions with the senior agency director of OV Agency. Besides that, the writer also gave advice on what actions should be taken by the OV Agency based on the results of a questionnaire which was distributed to 100 OV Agency agents. From here OV Agency provides a series of technical actions in response to all the customer's request.

Correlation matrix needs to be created to map relationships and interdependencies between technical responses. If we can map, then we can see if the dependence affects the technical response to meet customer needs. This relationship is illustrated by symbols. Empty means there is no relationship between technical responses, circle with content means having a strong positive relationship between technical responses, an empty circle means having a very strong relationship between technical responses. From the results of interviews with the senior agency director of the OV Agency, the technical correlation can be prepared as shown in Figure 1.

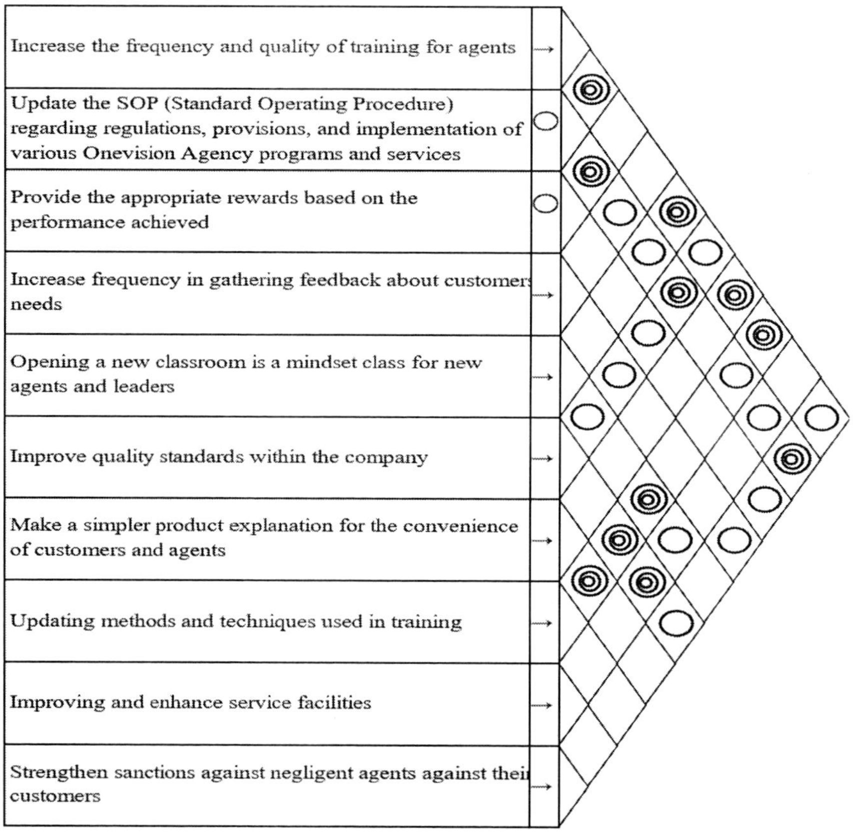

Figure 1. Technical Responses Correlation Matrix

Having obtained the level of importance and level of relationship between the technical response with the VoC attribute, then we can calculate the weighting of technical response by using fuzzy numbers. Each level of importance of the VoC attribute is broken down by its fuzzy numbers, Fuzzy numbers derived from the membership function of a trapezoidal fuzzy number, as can be seen in Figure 2. Trapezoidal fuzzy numbers are chosen because they are the easiest to understand and best suited for their use for QFD because the varying levels in determining the level of importance and level of relationship, these numbers can be seen in Table 5.

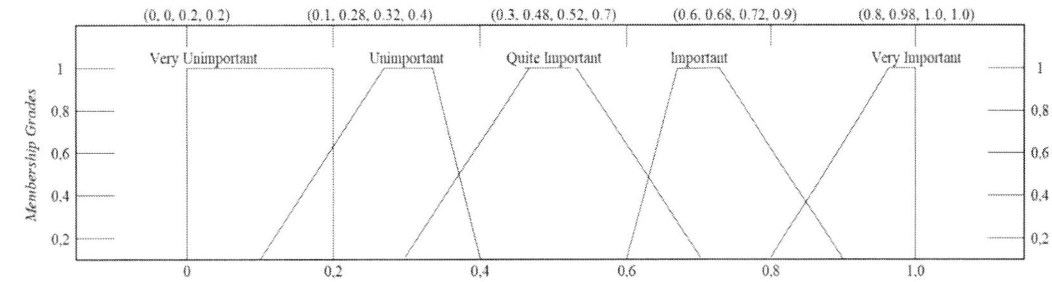

Figure 2. Trapezoidal fuzzy numbers for QFD

Table 5. Trapezoidal fuzzy numbers for QFD

Importance/Correlation	*Trapezoidal Fuzzy Numbers*			
	α	β	γ	δ
Very Unimportant/Very Weak	0	0	0.2	0.2
Unimportant/Weak	0.1	0.28	0.32	0.4
Quite Important/Strong Enough	0.3	0.48	0.52	0.7
Important/Strong	0.6	0.68	0.72	0.9
Very Important/Very Strong	0.8	0.98	1.0	1.0

Examples of solving steps to the constituent trapezoidal fuzzy numbers for the importance of the VoC "Speed & Alertness of Agents in providing responses" are:

- The value of interest rate mode of the VoC "The speed & alertness of the agent in responding/responding" is 5.
- The number 5 is the value of importance with the level of "Very Important".
- These attributes fall into very strong categories, so the trapezoidal fuzzy numbers of their constituents are 0.8, 0.98, 1, and 1 (α, β, γ, δ).

Here is the trapezoidal fuzzy number of the compilers of the VoC attributes for their importance:

Table 6. Trapezoidal Fuzzy Numbers Composer Importance Level of VoC

Num.	Code	Voice of Customer (Importance)	α	β	γ	δ
1	A1	Completeness of the Agent's tools, documents or tools while serving you	0.6	0.68	0.72	0.9
2	A2	Appearance & carriage Agent when interacting with you	0.6	0.68	0.72	0.9
3	B1	The accuracy of solutions or products provided by the agent according to the needs & desires	0.8	0.98	1.0	1.0
4	B2	Agent's ability to resolve requests & problems	0.8	0.98	1.0	1.0
5	B3	Knowledge & skills Agent in insurance financial services	0.8	0.98	1.0	1.0
6	B4	Agent's ability to understand your needs & wants	0.8	0.98	1.0	1.0
7	C1	Speed & alertness Agent in providing responses	0.8	0.98	1.0	1.0
8	C2	Clarity of time estimation of the Agent on a service process	0.8	0.98	1.0	1.0
9	D1	Agent's ability to listen to your needs & wants	0.8	0.98	1.0	1.0
10	D2	Availability of agent time in serving you	0.8	0.98	1.0	1.0
11	E1	The language the Agent uses is clear & easy to understand when serving you	0.6	0.68	0.72	0.9
12	E2	The hospitality shown by the Agent while serving you	0.8	0.98	1.0	1.0
13	E3	Ease of Agent to be contacted either face to face or by phone	0.8	0.98	1.0	1.0
14	F1	Your trust in agents for long-term service	0.8	0.98	1.0	1.0
15	G1	Security & convenience in doing transaction through Agent	0.8	0.98	1.0	1.0

After solving the level of importance and level of relationship into its trapezoidal fuzzy number respectively, then by adding (+) and fuzzy multiplication (x) will be obtained fuzzy trapezoidal number of the compiler of technical response with the formula 5.

$$W_j = (R_{1j} \otimes D_1) \oplus (R_{2j} \otimes D_2) \oplus \dots \oplus (R_{nj} \otimes D_n) \quad (5)$$

Where:
Wj : the result weight for the j-functional item
Rij : level of correlation between the j-functional items with the i-strategic plan
At : the importance of i-th strategic plan
$\oplus$: fuzzy sum
$\otimes$: fuzzy multiplication
The result of fuzzy numbers calculation of each technical response (α, β, γ, δ) can be seen in Table 7.

Table 7. Fuzzy Numbers Compiler for Each Technical Response

Num.	Technical Responses	*Fuzzy Numbers*			
		α	β	γ	δ
1	Increase the frequency and quality of training for agents	4.04	5.72	6.12	7.31
2	Update the SOP regarding regulations, provisions, and implementation of various OV Agency programs and services	1.44	2.27	2.48	3.20
3	Provide the appropriate rewards based on the performance achieved	2.24	3.68	4.04	5.20
4	Increase frequency in gathering feedback about customers' needs	2.40	3.55	3.72	3.90
5	Opening a new classroom is a mindset class for new agents and leaders	1.20	2.35	2.60	3.50
6	Improve quality standards within the company	3.36	5.15	5.48	6.20
7	Make a simpler product explanation for the convenience of customers and agents	1.20	1.80	1.96	2.50
8	Update methods and techniques used in training	3.16	4.66	5.00	5.91
9	Improve and enhance service facilities	1.44	2.00	2.16	2.70
10	Strengthen sanctions against negligent agents against their customers	2.32	3.80	4.08	4.70

After obtaining the four fuzzy trapezoidal numbers of the technical response, we can calculate the total integral value for W from the decrease (formula 3), that is the total integral value for fuzzy $\tilde{A}$ with an optimism index $\omega(I_T^{\omega}(\tilde{A}))$, can be described in the following formula:

$$I_T^{\omega}(\tilde{A}) = (1-\omega)\int_0^1 g_A^L(y)dy + \omega \int_0^1 g_A^R(y)dy \quad (6)$$

namely with the result of the formula as follows:

$$Wj = \frac{1}{2} \times \{(1-\omega) \times (\alpha_j + \beta_j) + \omega \times (\gamma_j + \delta_j)\} \quad (7)$$

The results of the calculation of the total integral values for each technical response for each level of optimism can be seen in Table 8.

Table 8. Total Integral Value of Technical Response

Num.	Technical Responses	W		
		ω = 0	ω = 0.5	ω = 1
1	Increase the frequency and quality of training for agents	5.27	6.16	7.05
2	Update the SOP (Standard Operating Procedure) regarding regulations, provisions, and implementation of various OV Agency programs and services	1.86	2.35	2.84
3	Provide the appropriate rewards based on the performance achieved	2.96	3.79	4.62
4	Increase frequency in gathering feedback about customers' needs	2.97	3.39	3.81
5	Opening a new classroom is a mindset class for new agents and leaders	1.78	2.41	3.05
6	Improve quality standards within the company	4.26	5.05	5.84
7	Make a simpler product explanation for the convenience of customers and agents	1.5	1.87	2.23
8	Update methods and techniques used in training	4.07	4.84	5.6
9	Improve and enhance service facilities	2.17	2.49	2.81
10	Strengthen sanctions against negligent agents against their customers	3.06	3.73	4.39

From the above calculation results, it can be sorted priority technical response that should be done to meet the Voice of Customer.

4. Conclusion

The results of the analysis of the initial conditions, the research questionnaire design, fuzzy-QFD calculation, and calculation results of data mining has been carried out, the conclusion as follows:

1. The calculation result using data mining method with Bayes classification customer of OV Agency generally located in satisfactory classification.
2. Based on customer satisfaction analysis and current service quality at OV Agency whose data obtained from questionnaire results, it can be decided that action is needed to improve the quality of service and product. It can be seen from the gap that exists between the level of importance and the level of satisfaction, where there is no value zero or minus the improvement ratio.
3. Based on the condition of the company and the results of interviews with senior agency directors from the OV Agency. It is recommended that companies use the optimistic index level (ω) with a value of 0.5 to make a decision for run a technical response.
4. Based on the results of data processing using Fuzzy-QFD method, obtained HOQ matrix used to see the level of customer satisfaction and technical response which should first be implemented on OV Agency.
5. Based on the HOQ matrix obtained from Fuzzy-QFD, it can be decided that the 3 main priorities for actions that should be implemented first are to increase the frequency and quality of training for agents, improve quality standards in the company, and update the methods and techniques used in training.

5. References

[1] Kiran D R 2016 *Total Quality Management: Key Concepts and Case Studies.* 1st Ed. (Butterworth-Heinemann Oxford)

[2] Cohen L 1995 *Quality Fuction Deployment. How to make QFD Work with You.* (Addison-Wesley)

[3] Jaiswal E S 2012 *Case Study on Quality Function Deployment (QFD) 2012 IOSR J. of Mechanical and Civil Eng.* **3** 6 (India: Mewar University India)

[4] Garcia M S and Lamata M T 2007 A modification of the index of liou and wang for ranking fuzzy number *Int. J. of Uncertainty, Fuzziness and Knowledge- Based Systems* **15** 04 pp.411-424

[5] Turban E 2005 *Decision Support System and Intelligent System.* (Yogyakarta: Penerbit Andi).

[6] Bayu A, Effendi M and Effendi U 2017 Customer Satisfaction Analysis with Fuzzy–Servqual Method and Quality Function Deployment (QFD) (Case Study at Café Right Time Malang) *J. Teknologi Pertanian* **18** 2 (Malang: Teknologi Pertanian Universitas Brawijaya)

[7] Fitriana R 2013 Rancang Bangun Sistem Intelejensi Bisnis Untuk Argoindsutri Susu Sekala Menengah di Indonesia. *Dissertation.* (Bogor: Institut Pertanian Bogor)

[8] Fitriana R, Saragih J and Firmansyah M A 2016 Business Intelligence System Model Proposals to Improve the Quality of Service at PT GIA. *Proc. of 9th Int. Seminar on Industrial Engineering and Management.*

[9] Nurhakim L 2008 *Analisis aspek pelayanan perusahaan perbankan nasional dengan metode fuzzy QFD* Skripsi Industrial Engineering University of Indonesia Depok Indonesia.

[10] Zaim S, Sevkli M, Camgöz-Akdağ H, Demirel O F, Yayla A Y and Delen D 2014 Use of ANP Weighted Crisp and Fuzzy QFD for Product Development. *Expert Systems with Applications* **41** 9 pp 4464-74

Design of Flat Shoes Quality Control System using PDCA (Case Study at PT DAT)

D K Sari[1], D Hetharia[2], D Saraswati[3] and R Marizka[4]

Industrial Engineering Department, Universitas Trisakti, Jakarta 11440, Indonesia
[1]debbie.kemala@trisakti.ac.id, [2]dorintaria@yahoo.com, [3]docki@trisakti.ac.id, [4]raniamarizka@gmail.com

Abstract. The D'Arcadia Treasure (DAT) is a Small and Medium Enterprises (SME) that made shoes from traditional Indonesian fabrics. The research objective is to design a structured and documented quality control system to make companies easier to control the production process according to the quality standards desired by consumers. The initial stage is to identify of DAT's Critical to Quality that will be the basis for compiling a Quality Plan, then using Statistical Process Control method for the control process and Cost of Quality identification. The results of the initial condition analysis are known there are 5 stages of the main process of making flat shoes as the object of research, namely spraying, pattern making, sewing, assembly, and finishing. Furthermore, the quality control system is designed to be divided into 3, namely raw material receipts, production processes, and finished goods equipped with quality control tools checksheet tables and instructions for use. The results of the implementation and analysis of Failure Mode Effect Analysis obtained the highest RPN value of 224 in the assembly process. The entire series of control processes are carried out with the Plan Do Check Action procedure which is a continuous cycle to be implemented in the company.
Keywords: Quality control system, Statistical Process Control, Failure Mode and Effect Analysis, Plan Do Check Action

1. Introduction

Quality has become an important component of every modern organization throughout the world and will remain an important part in efforts to improve performance [1]. Good quality according to the consumer's point of view if the product purchased is in accordance with the wishes, has benefits that are in accordance with the needs and equal to the sacrifices incurred by consumers. If the quality of the product cannot meet the desires and needs of consumers, then they will regard it as a poor quality product [2]. Quality control is engineering and management activities, by measuring the characteristics of product quality, comparing it with specifications or requirements, and taking appropriate corrective actions if there is a difference between the actual appearance and the standard [2]. Factors that influence quality control are process capability, specified specifications, acceptable levels of non-conformity and quality costs. Statistical Quality Control (SQC) is used to quality control process, while the quality cost calculation is intended to find out what costs must be incurred by the company in the process of controlling the quality of its products [2]. Quality control can be achieved with three approaches, namely the raw material approach, the quality of the production process, and the final product approach [3]. The application of Failure Mode and Effect Analysis (FMEA) aims to prevent problems occurring in processes and products. If used in design and manufacturing processes, FMEA can reduce failure or reduce costs by identifying and improving products and processes quickly during the development process [4]. Overall, quality control must be carried out through a continuous process. The continuous quality control process can be carried out by implementing the PDCA (plan-do-check-action) introduced by Dr. W. Edwards Deming [5]. This PDCA approach has been applied to analyze hospital services that recommend quality improvements in the X-ray service unit based on minimum standards of hospital services [6].

D'Arcadia Treasure (DAT) is an SME that produces handmade shoes for the DAT brand, most of which use traditional Indonesian fabrics such as batik, weaving, lurik, which are made by combining other materials such as canvas, denim, leather, sueade, jute, and so on. Other materials used are also synthetic or genuine leather for rubber insole and antislip for outsole. This business is also responsible for several other client brands made by DAT using their materials as needed. Some types of shoes produced are sandals, flat shoes, wedges, heels, clogs, sneakers, and boots, but the special output produced by DAT is flat shoes.

Figure 1. Flat shoes, the products observed

A touch of innovation that follows current trends or fashions, makes shoes unique and beautiful. Along with the development of its business, DAT products that have been marketed are often imitated by similar SMEs with the same selling price and even higher but lower quality. Design innovations continue to be developed so that almost every month DAT produces 24 Stock keeping Units (SKUs) or product numbers. The production target of 200 pairs of shoes per day by setting working hours is not limited and workers are paid per unit of product produced, as a result often over stock due to a marketing process that has not been maximized and the quality control and guarantee process cannot be carried out properly. The company has not implemented a quality control process with the right concept, so that when there are products that are not in accordance with the quality standards received by the customer, the company will replace the product with a new product but with a different model and not in accordance with the customer's wishes. Likewise with products that are not sold because defects can be returned and then sold at production prices.

This study aims to design a structured and documented quality control system, identify and monitor the process behavior to achieve the desired product quality using Statistical Quality Control (SQC) tools. The results are then implemented and analyzed by the Failure Mode Effect Analysis (FMEA) method. The implementation of the quality control process is then carried out with the Plan Do Check Action (PDCA) cycle.

2. Research methodology

The initial stage is to identify the DAT Critical to Quality (CTQ) for customers who will be the basis for compiling the quality plan (Quality Plan), then using the Statistical Process Control (SPC) method for the control process as well as cost of quality identification. The next stage is to design a quality control system that is in accordance with the conditions of the company and the production process carried out. Data retrieval is carried out by direct observation at the factory location and involving company owners and workers to fill in data forms in the form of production data, sales data, inventory data and defect and disability data. The results of data processing and analysis produce a system that can be used by D.A.T to control and guarantee quality by applying the use of check sheets as well as the instructional sheets made specifically for DAT with flat shoes products as objects in this study. The results of the design are then implemented and analyzed using the Failure Mode Effect Analysis (FMEA) method, in addition to the Cost of Quality calculated from the implementation. All stages of the research were then identified as a cycle of Plan Do Check Action (PDCA) which became a reference for companies to carry out the process of quality control continuously.

3. Results and Discussion

3.1. Production capacity

DAT makes products continuously with a total monthly production capacity of 2800 pairs with a maximum capacity of 8000. All production processes are done manually, which is why companies rely on human resources to reduce or increase volume. For standard shoe models 1 tailor can produce approximately 20 pairs of shoes per day, as well as 1 person assembler can produce approximately 20 pairs of shoes per day. DAT has 6 tailors and 6 fabricators who work for 6 working days, so that the minimum number of shoes that can be produced in a week is 720 pairs.

3.2. Type of Machine and Number of Machines

Making shoes that are done manually still requires several machines to facilitate the production process. Some of the machines owned by DAT are 6 sewing machines for horizontal sewing process, 1 leather set-up machine for leather which is useful for thinning the ends of the material, 1 cangklong machine for sewing upper curved parts, 1 compressor with spraygun for gluing the material, and 1 grinding machine to flatten sole shoes.

3.3. Analysis of Current Quality Control Conditions

A good quality control system is a quality control system that exists when receiving raw materials, production processes and finished products. In identifying the current quality control conditions will be discussed regarding the quality control of raw materials, production processes and finished goods that are applied in DAT at this time.

The quality control of raw materials in DAT has not been implemented. Based on the results of the interview with DAT Production, it was explained that the raw material received from the supplier had not been inspected by the company. Raw materials received directly are stored in the raw material warehouse. There are several raw materials that are inspected in the process. Inspection of the main raw materials, namely batik cloth, weaving and lurik is done during the pattern making process. The inspection is only done visually and there is no recording. Supporting raw materials are also inspected for outsole and insole materials. Outsole and Insole are inspected before the assembly process takes place. Both inspections are done visually and there is no recording. Batik fabrics, weaving and striated, require a quality control system with a 100% inspection process because the material is the main object and characteristic of DAT shoes. Outsole and Insole also require a quality control system with a 100% inspection process because these materials affect the comfort quality of DAT shoes.

The quality control of the production process applied by DAT is currently divided into two, namely the initial and final inspections. Initial inspection is carried out after the sewing process, while the final inspection is carried out during the finishing process. In the production process itself is divided into 5 main stages namely the process of spraying, pattern making, sewing, assembly and finishing. Based on the results of company information, during the spraying process, pattern making and assembly were not carried out.

The spraying process is a vulnerable process because it depends on weather conditions, if the weather does not support it will greatly affect the results of gluing and risk of causing material defects. The results of the spraying are not inspected so that good or bad quality material is continued to the next process. The pattern-making process is a fairly difficult process because it requires high accuracy in equating the pattern motifs right and left. The results of the pattern making continue to the sewing stage without prior inspection.

The initial inspection process is only done when entering the sewing process, where the parts examined are upper. On inspection the sewing results are carried out by checking the tidiness, the quality of the material and the similarity of the motif. The initial inspection phase is not recorded for products that are received or rejected, so the company cannot measure the stability of the process. The product received is continued to the assembly process and the rework process or scrap will not be continued. The next process is assembly, where the results are not inspected first but proceed to the finishing stage and inspected simultaneously. This stage is the final inspection which is a unit of quality control of the finished product.

Current quality control of finished goods is included in the final inspection carried out during the finishing process. Observations were made on material quality, gluing, stitching, assembly, and similarity in shoe size. At this stage only manual recording is done for the number of products that are feasible and not feasible to sell, but there are no known types of defects. Even records are not done routinely, companies have not implemented binding regulations to document inspection. This stage aims to find out the number of products that are successfully produced by the assembly operator and tailor, so that it can be known the work wages per day for the operator.

3.4. Designing a Quality Control System

Based on the results of the analysis of the quality control conditions applied by DAT, a more structured quality control system was designed. The overall design of the quality control system is divided into three sub-systems, namely the quality control system for raw materials, the production process and finished products. In the part of the quality control system, there are basic elements which consist of main and supporting raw materials. For the basic elements of the quality control system the production process is divided based on the stages of production, namely the results of spraying, pattern making, sewing and assembly processes. The last system is controlling the quality of finished products that have basic elements in the form of the results of the finishing process of production.
The general quality control system design in DAT can be seen in Figure 2.

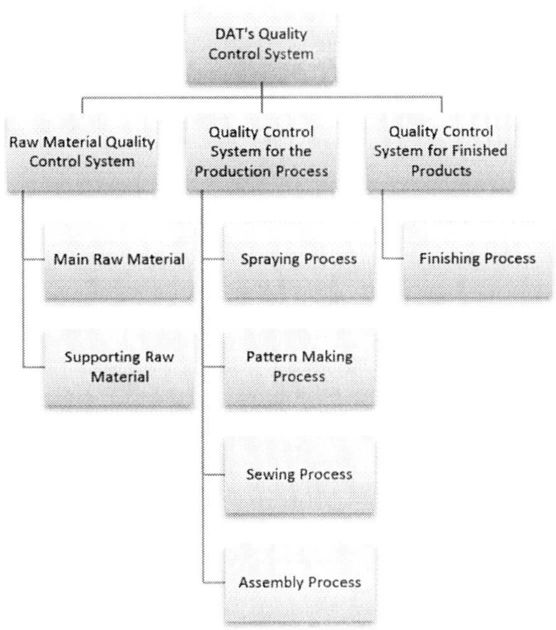

Figure 2. Designing a Quality Control System for DAT

3.5. Designing Check Sheet and Inspection Instructions

The quality control process is carried out using quality characteristics guidelines that have been established as Quality Plans and followed by the creation of inspection sheets and inspection instructions. Furthermore, each quality control operator is required to complete a check sheet in accordance with the inspection instructions. The quality characteristics that have been made are

IOP Conf. Series: Materials Science and Engineering **528** (2019) 012073 doi:10.1088/1757-899X/528/1/012073

followed by making a check sheet for each material. Example for the process for material quality control, the sheet contains the raw material code, the name of the raw material, the date of arrival and inspection, the name of the inspected material, the various characteristics of the quality, and the status of the recipient. This check sheet has two signs, namely "–" there is no defect, while the sign "√" means that there is a disability. Figure 3 is an example of a check sheet that is used to examine the main raw material, namely batik cloth.

d.a.t (Hand)made in Indonesia	D'ARCADIA TREASURE

Main Raw Materials Check Sheet

Raw Material Code	: 01
Name of Raw Material	: Kain Batik
Day / Date of Arrival	:
Day / Date of Examination	:

No.	Name of Material	Quality Characteristics			Status
		Dirty	**Candle stains**	**Fiber Thread Not Neat**	
1					
2					
3					
4					
5					
6					
7					
8					
9					
10					

Catatan :

Total Accepted :
Total Rejected :

Made by :	Checked by :	Approved by :
(..........................)	(..........................)	(..........................)

Figure 3. An example of a check sheet

The designed check sheet is followed by making inspection instructions. This inspection instruction aims to make it easier for the operator to inspect each fabric. Figure 4 is an example of inspection instructions for filling out a sheet checking the main raw material for batik cloth.

3.6. Failure Mode and Effect Analysis (FMEA)

Based on the results of system implementation, several types of failures in making flat shoes are obtained in the sewing and assembly process. Both processes obtained data from the implementation for 14 days from November 24th until December 7th, 2017. This process is a critical process in making flat shoes. In this study, FMEA is used to describe failures that occur in the type of failure in the sewing and assembly process and its causes. The results of FMEA on Table 1 are carried out to determine the priority of improvement on failures that occur by calculating the Risk Priority Number (RPN) value obtained from the multiplication of the three variables of severity, occurrence, and detectability to improve product quality. Data collection and ranking ranking of each FMEA criterion is obtained from the results of interviews with the DAT Production Head and obtained the highest RPN value of 234 in the assembly process because the assembly is less neat.

d . a . t (Hand)made in Indonesia	D'ARCADIA TREASURE
	INSPECTION INSTRUCTIONS

ABOUT: Process for Quality Control of Batik Fabrics

Aim : To control the quality of raw material for batik cloth and ensure that the raw material is in accordance with the desired requirements.
Executor :
Note : *Check sheet of the main raw material for batik cloth*
Method : 100% Inspection

Work Process Description

1. Batik fabric inspection coming from suppliers. Prepare check sheets for raw materials for

2. Cleaning inspection from batik cloth.
   2.1 If there are no stains or there are few stains on the material that can still be received, then mark " ━ " on the check sheet in the Dirty column.
   check sheet in the Dirty column. Separate the material and give a statement "Reject" in the columnStatus.

3. Wax stain inspection from batik cloth.
   3.1 If there is no wax stain or there is a slight wax stain on the material that is still
   3.2 If there is a wax stain on an unacceptable material, then mark "✔ ☐on the check sheet

4. Inspect the tidiness of fiber yarn from batik cloth.
   4.1 If the fiber yarn in the material is neat, then put a " ━ " mark on the check sheet in the
   4.2 If there is a fiber thread that is not neat on the material and cannot be accepted, then

5. Separate batik cloth that has been inspected
   5.1 Batik cloth is received, then stored in the warehouse of raw materials.
   5.2 Batik cloth that has Reject status, then returned to the supplier.

Made by :	Approved by :
(.........................)	(.........................)

Figure 4. An example of inspection instructions

Table 1. FMEA Implementation Results on Flat Shoes

No.	Process Function	Potential Failure Mode	Potential Effect(s) of Failure	S	Potential Cause(s)	O	Current Process Control	D	RPN
1	Stitching	Thread pulled	Wrinkled Upper	3	Thread pieces are too short	7	Visual inspection	8	168
		Stitches not neat	Stitches are not straight	3	Operator performance decreases	7	Periodically supervision	8	168
		Shrinking material	Material cannot be used	3	Bad material quality	6	Identify the ingredients that are easy to stretch	8	144
		Different left-hand motives	Different motive of a pair of shoes	4	The patern is less precise	5	Periodically supervision	8	160
		Material defects	Visual defects	4	Bad material quality	5	Identify types of ingredients that are easily fragile	8	160
2	Assembly	Uncluttered assemblies	Discomfort when used	4	Uncompressed sole mounting	7	Periodically supervision	8	**224**
		The stitches are open	Upper cannot be used	4	Careless Tailor operators	6	Periodically supervision	8	192
		Material defects	Visual defects	4	Bad material quality	5	Identify types of ingredients that are easily fragile	8	160
		Unsuitable upper back composition	Discomfort when used	5	The operator did not use the standard	5	Measurement of before and after assembly	6	150
		Gluing is not neat	Decreases aesthetic value	3	Gluing time is not long enough	5	Set the standard of drying time	8	120

3.7. Costs of Quality

The collection of data on the quality control system that has been implemented, followed by the calculation of the quality costs. The quality costs calculated in this case are the cost of valuation and the cost of internal failure. The assessment costs calculated are raw material inspection costs, spraying, sewing, assembly, finishing and inspection equipment production processes.

Inspection data for the production process carried out from November 24 to December 2, 2017, namely as many as 24 sheets of material on spraying, 445 pairs of flat shoes on sewing, 404 pairs of flat shoes on the assembly and 158 pairs of flat shoes in finishing, while for raw materials carried out since December 12 to 20, 2017 as many as 34 fabrics, 120 insoles, and 220 outsoles. Based on the data, the total assessment fee is Rp. 438,158.83. For internal failure costs, the calculation is done based on the condition of the defective product. Reject costs are issued for 3 pieces of the finishing process, while the rework costs are charged to the sewing process as many as 39 pieces and the assembly process is 36 pieces. There is no rejected material for raw materials. The total cost of internal failure from November 4 to November 8, 2017 is Rp. 849,220.63. Based on these calculations, obtained the total cost of quality from the results of implementation for 2 weeks for the production process and 1 week for raw material inspection that is equal to Rp. 1,287,379.46. The results of this quality cost calculation can be used as one of the quality control parameters for the company.

3.8. Quality Control Process with PDCA

Based on the results of the design and preparation of the check sheet, and the implementation phase that has been carried out, a quality control cycle is designed with the PDCA approach. This approach can be used as a guideline for companies to carry out the process of quality control continuously. Table 2 describes the PDCA approach to controlling the quality of flat shoes products.

Table 2. Flat Shoes Quality Control with PDCA

Control Process	Plan	Do	Check	Action
Acceptance of Raw Materials	Preparation of Quality Characteristics Guidelines (Quality Plan), making check sheets for the main raw materials and supporting raw materials, and making inspection instructions.	Completing check sheet according to inspection instructions	Check the filled check sheet and analyze the results	Perform separation of raw materials that are not in accordance with the quality characteristics to be further carried out the process of returning or repurchasing to suppliers.
Production Process	Preparation Quality Plan, making check sheets for spraying, pattern making, sewing and assembly processes as well as making inspection instructions	Completing check sheets according to inspection instructions.	Check the filled check sheet for semi-finished products and analyze the results	Separate the Semi-finished products that are not in accordance with quality characteristics into reject or rework products.
Finished Products	Preparation Quality Plan, making check sheets for finishing products in the production process and making inspection instructions	Completing the finished product check sheet according to inspection instructions	Check the filled check sheet for the finished product as a result of the finishing process.	The output of the finished product quality control system is the product received for packaging and storage as well as products that need improvement, or rejected products.

Source: primary data processed and verified by the owner of DAT, 2018

4. Conclusion

The results of data processing and preliminary analysis of the condition of the company can describe the process carried out by the company so far uncontrolled and not well documented so that the company is in desperate need of a plan for its production system and structured quality control planning. After planning and handling this quality control system can be implemented can facilitate the creation of a quality management system that can help companies document every activity in the production process and quality control. In addition, the company can also make improvements to production process errors and quality control that have not been well documented. At the stage of this research the following processes have been carried out:

- The results of the analysis of the initial conditions are known to have 5 stages of the main process of making flat shoes namely spraying, pattern making, sewing, assembly, and finishing. There are 2 stages of inspection, namely the results of sewing and finishing.
- The design of a divided quality control system 3, namely for the receipt of raw materials, production processes, and finished goods based on a quality plan that has been previously determined.
- The results of the quality control system design are then followed by the manufacture of quality control aids in the form of a checksheet table and instructions for use for each control process that has been planned so that it can be implemented.
- The implementation process is then continued with the analysis process using the Failure Mode Effect Analysis (FMEA) method which generate the highest RPN value of 224, namely in the assembly process where the assembly process is not neat.
- In the implementation process, the total cost of quality is obtained for 2 weeks for the production process and 1 week for raw material inspection, which is Rp. 1,287,379.46.
- The results of this control process will then be carried out by the Plan Do Check Action (PDCA) procedure.

5. References

[1] Evans J R and Lindsay W M 2016 *The Management and Control of Quality*. 10th Ed. (United State of America: Cengage Learning)

[2] Montgomery D C 2009 *Introduction to Statistical Quality Control*. 6th Ed. (New York: John Wiley & Sons Inc.)

[3] Assauri S 1998 *Manajemen Produksi*. (Jakarta: Universitas Terbuka)

[4] McDermott R E, Mikulak R J and Beauregard M R 2009 *The Basic of FMEA*. 2nd Ed. (New York: Productivity Press)

[5] Gryna F M 2001 *Quality Planning and Analysis From Product Development Through Use*. 4th Ed. (Singapore: Mc-Graw Hill Int. Edition)

[6] Ariani P D, Hari S N and Sari L 2013 *Analisis Pengendalian Kualitas Dengan Pendekatan P.D.C.A. (Plan-Do-Check-Act) Berdasarkan Standar Minimal Pelayanan Rumah Sakit Pada Rsud Dr. Adhyatma Semarang(Studi Kasus Pada Instalasi Radiologi)*. *Diponegoro Journal of Social and Politic* pp 1-12

Assistance Program for the Implementation of ISO 9001 2015: Case Study of Telkom Junior High School

Y Rohayati[1,2] and K Sari[1,3]

[1]Industrial Engineering, Telkom University
Jl. Telekomunikasi, Terusan Buah Batu, Bandung 40257, Indonesia
[2]yatirohayati@telkomuniversity.ac.id, [3]fajaromakartika@gmail.com

Abstract. *Yayasan Pendidikan Telkom* is one of the educational foundation that houses several of private education institutions ranging from elementary, secondary, to higher education. The foundation constantly improves the quality of education services as the most important component in carrying out institutional operations in order to always gain the trust of prospective students and the community. The efforts of the educational institutions to improve the quality of services, one of them by implementing a quality management system that refers to the latest edition of ISO 9001. Telkom Junior High School is one of the educational institutions that will now begin to implement ISO 9001: 2015. The teachers and school staff experience many obstacles in preparing to face the implementation of the quality management system. With the exploratory research approach, this study aims to produce an assistance model and describe the stages that will be passed by educational institutions in preparing their institutions to face ISO 9001: 2015. This research is the initial stage of studies that will be conducted and will be more focused on the preparation stage, while the discussion of the implementation phase will be carried out in future research. This assistance model is designed to help educational institutions to implement ISO 9001: 2015 according to plan, and especially so that the preparation process takes place quickly and directed.
Keywords: Quality Management System, ISO 9001: 2015, implementation, preparation phase, Telkom Junior High School

1. Introduction

For private education institutions, maintaining the quality of education services in a systematic manner is the most important component in carrying out institutional operations in order to gain the trust of prospective students and the community because private educational institutions are often chosen as secondary alternative, after government-owned or public schools. However, with the excellent quality of private education services, many have found that private education institutions are preferred over public education institutions. This is indicated by the increasing number of private education institutions. The efforts of educational institutions to improve the quality of services, one of them is by building a quality management system that refers to ISO 9001 and the latest standard is the 2015 edition. This is experienced by secondary schools organized by Telkom Education Foundation or *Yayasan Pendidikan Telkom* (YPT) which began to implement ISO 9001: 2015 for quality management systems. All educational institutions owned by YPT have the status of private schools, striving to have superior education services, one of which will be achieved through the implementation of ISO 9001: 2015.

In an effort to improve the quality of service of its educational institutions, YPT as the organizer of educational institutions, constantly enhances its services and management systems that are applied to all institutions it owns, one of which is for middle schools, which are branded as Telkom Junior High School.

The application of the ISO 9001 quality management system standard in the service industry is relatively new compared to the manufacturing industry, especially in educational institutions. The Quality Management System based on ISO 9001 has been well received and applied in many countries [1] [2], even studies that have examined the positive impact of its application have been widely carried out. One of them is about increasing customer satisfaction [3], standardizing work procedures [4]. In addition, some of them also discussed the obstacles faced in implementing ISO 9001 [5].

This study will examine about the important phases in the ISO 9001 implementation process, which is the preparation phase that is rarely discussed in the research. According to the experience of many institutions, preparation for ISO 9001 implementation often lasts long and protracted, so that there are institutions that then take shortcuts by submitting the preparation process entirely to other parties, such as consultants, to design the system. This results in the low involvement of people who are executors of the management system, and the sustainability of implementation is threatened.

This study aims to develop assistance model for ISO 9001 implementation that will improve the effectiveness of the preparation process, so that it can assist educational institutions in expediting and accelerating the preparation process. Before, there wasn't any research like this that contained an implementation model that discussed the phases of ISO 9001 Implementation in Educational Institutions.

Improving the quality of the process in order to create superior services is a necessity, it should be the aim of every organization that acts as a service provider. In services, service delivery is very much determined by the process that accompanies it, and the process that is organized with standard and is carried out consistently will provide opportunities to produce superior services.

The implementation of the Quality Management System in educational institutions is very much needed to guide the management of key and important processes to improve superior and consistent education services for users and other related parties.

[6] There are two types of motivation that underlie the organization in carrying out ISO 9001 certification: 1). Internal motivation, to improve the quality of the organization, and 2). External motivation, which is used as an amplifier in the effectiveness of promotion and marketing [7]. As a private education institution, according to Telkom Junior High School, ISO 9001 certification can be used as a tool to increase the effectiveness of the promotion that will be conducted. In addition, the implementation of ISO 9001 is a form of quality assurance that will increase trust for customers. In services, brand trust is a factor that will affect customer satisfaction and customer loyalty.

Initially, ISO 9001 was widely applied in the manufacturing industry, then, followed by the service industry. However, research on the application of ISO 9001 in educational institutions cannot be widely found. Several studies discussing the application of ISO 9001 in educational institutions were carried out in Portuguese [8], the United States and the United Kingdom [9].

2. Research Methods

This study aims to create a model that can be used as a reference or a standard for the preparation process until the implementation of Quality Management System ISO 9001:2015 at Telkom Junior High School. The research was conducted using the Exploratory Research approach which aims to gather a deep understanding of a problem. In this study, exploratory research was used to identify actions that need to be carried out in the process of preparation & implementation of the system based on real conditions in the object of research. The data generated are qualitative data. This study focused on the preparation phase for the implementation of ISO 9001: 2015 and it will correct the steps that have been taken which has resulted in the delay of the ISO 9001 implementation process, which is due to a less effective preparation process.

The object of this research is Telkom Junior High School, which is one of 54 educational institutions owned by YPT. This object is one of the institutions that have not succeeded in carrying out ISO 9001 implementation, despite having prepared beforehand.

Since there are more than one number of educational institutions and those institutions are in different cities, the first attempt is to choose a pilot project to make corrections to the ISO 9001 implementation process, and the selected institution is Telkom Junior High School in Bandung. It was chosen based on the fact that educational institutions with middle schools that have implemented ISO 9001 are still few and rare, so that they will provide added value and create excellence in providing educational services.

The preparation process that has been carried out in Telkom Junior High School contains training, which has been done several times. It has discussed the materials needed to implement, which are awareness, understanding ISO 9001 requirements, documentation system creation techniques for quality management systems, and risk preparation profiles. Even so, the level of readiness to start implementing ISO is still low, having proven not to have complete quality standards and structured documentation as required in ISO 9001.

The form of correction to the preparation process for ISO 9001 implementation at Telkom Junior High School was carried out in the form of Assistance, which was divided into 2 phases: mentoring for the preparation process and mentoring for the implementation process.

3. Discussion and Result

The Assistance Program is divided into two phases: Preparation phase and Implementation phase, each of which consists of several sub-phases. In this study there were two teams involved, which are the implementation team who came from Telkom Junior High School and the research team who acted as the Assistance Team. The Assistance Team will oversee the entire process, both in the Preparation Phase and the Implementation Phase.

The Preparation Phase consists of five sub-phases and the Implementation Phase consists of four sub-phases, which are explained into nine steps, as follows:

1. Measuring the implementation readiness, using a closed questionnaire, also to explore the obstacles faced in the preparation process that has been passed.
2. Preparing a schedule to deal with implementation which is directed at the internal quality audit implementation schedule, Management Review Meeting, and External Quality Audit. Scheduling is done with the Implementation Team that has been formed.
3. Identifying what requirements have been and have not been fulfilled which is done by filling out an open questionnaire and documentation review (desk evaluation)
4. Providing consultation to improve understanding needed especially in fulfilling the completeness in the documentation.
5. Checking the completeness of the fulfilment of requirements through documentation
6. Organizing an Internal Quality Audit which aims to examine the conformity between practices and requirements in ISO 9001. An important activity in AMI is to examine artefacts and other evidence that illustrates that a standard has been well travelled.
7. Following up on findings that must be corrected, which can be in the form of non-conformity, opportunity for improvement (OFI) or observation.
8. Organizing Review Management, in the form of meetings, to discuss several important matters.
9. Conducting an External Audit or Certification, in the form of surveillance, which will determine the eligibility for obtaining recognition has successfully implemented ISO 9001.

The nine steps are summarized in Table 1:

Table 1. Implementation Assistance Program ISO 9001:2015

Step	Distribution	Tools				
		Q	DR	I	C	T
A	**Preparation**					
1	Measurement of implementation readiness	√				
2	Arrangement of implementation timeline			√		
3	Identify fulfilment of requirements	√	√			
4	Improvement the understanding of ISO 9001				√	√
5	Fulfilment of ISO 9001 requirements		√	√	√	
B	**Implementation**					
1	Conducting AMI		Implementation			
2	Corrective action of non-conformity			Implementation		
3	Review management meeting				Implementation	
4	External audit (certification)					Implementation

Note: Q = Questionnaire; DR = Document Review; I = Interview; C = Consultancy; T = Training

In the preparation process, several tools are used to get information from objects to the researcher or provide information from the researcher to the object. This information is used to determine the steps that must be taken. Searching for information from objects is done by distributing questionnaires, conducting interviews, or conducting document reviews, while giving information from researchers to objects is done by providing consultation and training.

In this study, new preparation steps are included, while the implementation steps will be continued in further research.

In initiating the mentoring process, the readiness and identification of the obstacles faced are first measured, which lead to a slow and protracted preparation process. This measurement uses a questionnaire which consist of closed and open questions. It was distributed to 21 respondents (teacher and staff). The results obtained are described in Figure 1.

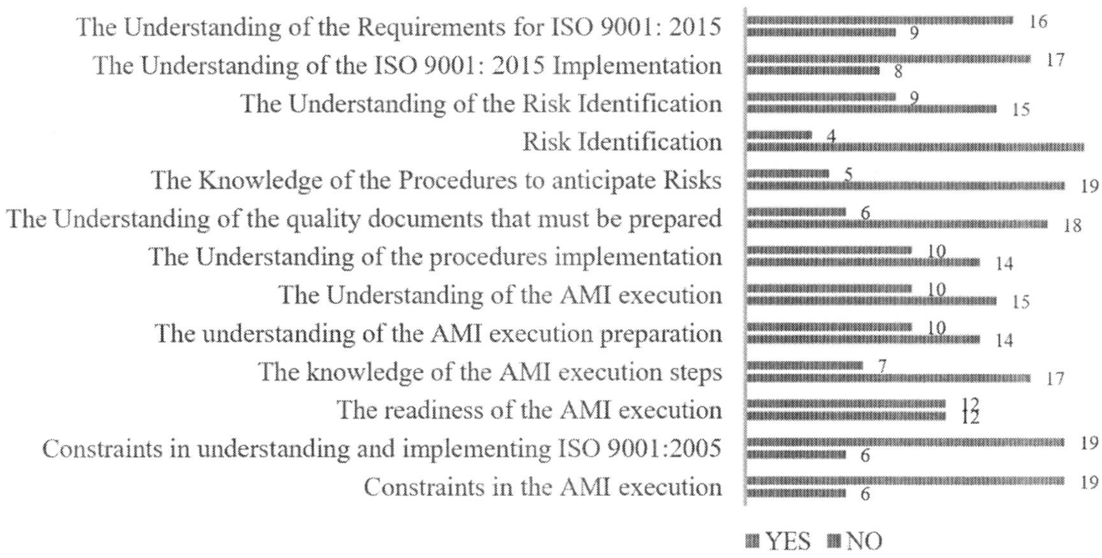

Figure 1. Implementation Readiness Measurement ISO 9001: 2015

The graph above shows the survey results to 21 respondents consisting of teachers and staff. In the survey there were 13 questions related to components or important things that must be understood in implementing ISO 9001: 2015, where there were only 2 answer choices (yes and no). Based on this picture it can be seen that in each question there are no answers, more than that in 8 of the 13 questions, YES answers are more than NO answers so it can be concluded that the understanding of teachers and staff at Telkom Junior High School for ISO: 9001 implementation is still very poor.

The readiness of Telkom Junior High School towards implementation ISO still has problems and this has caused the preparation time that had been carried out for a long time, exceeding the planned time. Figure 1 provides information about the low readiness of Telkom Junior High School to step in the implementation phase. The constraints that are still being faced must be formulated, and the solution is sought to be resolved immediately.

By using open questions, an exploration of the constraints faced is obtained, which results in a low level of vulnerability, as described in Figure 2.

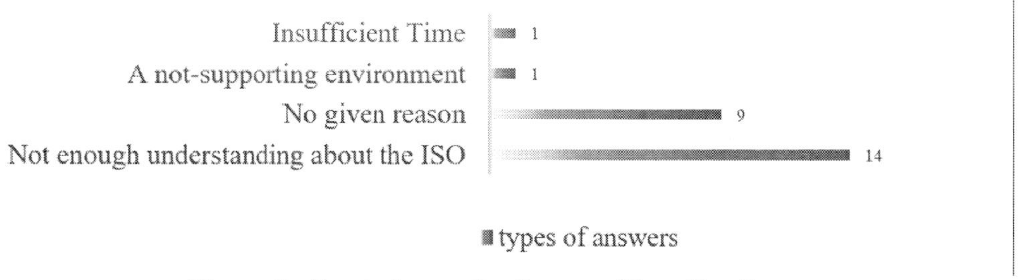

Figure 2. Constraints to the Causes of Low Readiness

The graph above shows the results of a survey of 21 respondents consisting of teachers and staff. In the survey there were 4 statements of constraints faced by respondents when facing ISO 9001: 2015. The constraints that most respondents experienced were lack of knowledge about how to implement ISO 9001: 2015

Reinforced from the interview results revealed that the preparation time that lasted longer than planned, due to their understanding of the requirements contained in ISO 9001 has not been well mastered, especially related to the preparation of risk profiles which is a realization of risk based thinking. These constraints require handling in the mentoring program, which is overcome by providing training and consultancy. The problem that is most not understood is related to the making of operating procedure standards, in the process performance measurement section, as well as the creation of a risk profile.

Carrying out the review document was first done with the help of a tool in the form of a check list, which must be asked and verified by the ISO 9001 Team. The check list contains 30 questions, listing the requirements in ISO 9001 that must be fulfilled in the form of providing operating procedures, policies and completeness standards document that must be owned. Based on the check list, it was found that Telkom Junior High School readiness in fulfilling the requirements and completeness of the new document reached 8/30, or 29.63, thus requiring follow-up. Based on the results of the check list, training and consultations are carried out, with the aim that the team can meet all requirements in full.

After being given the time to be able to meet all the shortcomings, the check list will be reused to ensure that the level of fulfilment of the requirements has reached 100%, so that it can go to the implementation phase.

When carrying out the Implementation phase to all standards and policies, it is possible to improve the standard that has been set, as a form of fine tuning. After the implementation lasts for three months, the Internal Quality Audit is conducted, to ensure that the standards that have been set have been implemented, and to explore opportunities for improving the quality of the process.

4. Conclusion

The implementation of ISO 9001 is one way to structure efforts to improve the quality and consistency of services and to systematically improve quality but the preparation process doesn't often run smoothly, or even tends to be protracted. Even though it has received training in knowledge received through training that discusses various materials related to implementation, there are various obstacles which have resulted in a longer process of preparation, which has the effect of decreasing motivation to implement. An assistance program was introduced to overcome the problem which was divided into two phases: The Preparation Phase and the Implementation Phase, but in this study, the discussion was limited only to the Preparation Phase.

Assistance Program in the Preparation phase applied in the study includes five sub-phases, and in its implementation, questionnaire tools, check lists and interviews are used to extract information from the Implementation Team, as well as consultation and training that serves to provide understanding and skills by the Team Companion to the Implementation Team, which is needed in the debriefing process during the preparation phase.

In order to be able to accommodate various conditions faced in the Preparation Phase, this Program will be applied to other educational institutions, so that a more complete and generalizable Assistance Program can be obtained. The assistance model that was generated in Telkom Junior High School can be a reference for other educational institutions to prepare the implementation of ISO 9001: 2015.

5. References

[1] Magd H. and Curry A., 2003 An empirical analysis of management attitudes towards ISO 9001:2000 in Egypt *The TQM Magazine* **Vol 15** No 6 pp 381-90

[2] Magd H.A.E., 2006 An investigation of ISO 9000 adoption in Saudi Arabia *Managerial Auditing Journal* **Vol 21** No 2 pp 132-47

[3] Cebeci U. and Beskese A., 2002 An approach to the evaluation of quality performance of the companies in Turkey *Journal of Industrial Engineering and Management* **Vol 17** Issue: 1/2 pp 92-100.

[4] Tari J.J., Azorin J.F.M, Heras I., 2012 Benefits of the ISO 9001 and ISO 14001 standards: A literature review *Journal of Industrial Engineering and Management* **Vol 5** No 2 pp 297-322

[5] Zeng S.X. and Tian P., 2007 Overcoming barriers to sustainable implementation of the ISO 9001 system *Managerial Auditing Journal* **Vol 22** No 3 pp 244-254

[6] Sampaio P., Saraiva P., Rodrigues A.G., 2009 ISO 9001 certification research: question, answer and approaches *International Journal of Quality & Reliability Management* **Vol 26,** No 1 pp 38-58

[7] Magd H., Kadasah, N. and Curry A., 2003 ISO 9000 implementation: a study of manufacturing companies in Saudi Arabia *Managerial Auditing Journal* **Vol 18** No 4 pp 313-22

[8] Thonhauser T. and Passmore D.L., 2006 ISO 9000 in Education: A Comparison between the United States and England *Research in Comparative & International Education* **Vol 1** No 2 pp 156-173

[9] Gamboa A.J. and Melao M.F., 2012 The impacts and success factors of ISO 9001 in education: Experiences from Portuguese vocational schools *International Journal of Quality & Reliability Management* **Vol 29** No 4 pp 384-401

Analysis of Increasing Quality of Surfactant Powder at Oversize Parameter Using PDCA and FMEA in PT BCCI

R Prasetyani[1], A T Huda[1] and D R Ningtyas[1]

[1] Jurusan Teknik Industri, Fakultas Teknik, Universitas Pancasila Jl. Lenteng Agung Srengseng Sawah, Jagakarsa, Jakarta Selatan, Indonesia
rini.prasetyani@univpancasila.ac.id, ataufiqul01@gmail.com, desinta@univpancasila.ac.id

Abstract. Quality of product is the important thing for PT BCCI in order to compete with other chemical industry of surfactant. In the production process of surfactant powder the particle size of the product is bigger than screener with mesh size 30 will produce oversize reaching 30% from every batch of products, indicates there is an issue that to be fixed immediately. This research was to find out the root of problem and action to improve the quality product and increase the output. PDCA and FMEA method were used to resolve these problems. The results are priority problem that needs to be improved based on the highest RPN value is the machine on the product evaporation system is the less working Vacuum maximum and the product temperature is less high. The product temperature need increase to 60°C which was previously 58°C and pressure on Vacuum <50 mbar. After improvement was applied the results obtained where the value of Cpk increased from 0.163 in the previous study to 3.609 and Oversize decreased from 30% to 20% in each batch process, this is in accordance with the target to be issued with the Oversize percentage below 30% in each process.
Keywords: PDCA, FMEA, surfactant

1. Introduction

PT BCCI is PT BASF Care Chemicals Indonesia located at Jalan Raya Jakarta Bogor KM 31.2 Cimanggis Depok and hereinafter referred to as PT BCCI in the writing of this Final Project.
Based on the previous research and analysis on the process of making powder surfactant in the Dryer Grinder system at PT BCCI, it was found Cpk 0.163, from the process capability calculation, it can be seen that the process capability has not met the requirements where the capability value can be said to be good if the Cpk value 1. there is a problem in the process of making surfactant powder caused by equipment that is not optimal, one of the production process equipment is vacuum and temperature, vacuum is very influential on the process of making powder surfactants (Grinder dryers), where vacuum or vacuum is needed to make or taking water content in raw material with initial water content of 4% to 0.9% in powder surfactant products. Product temperature is also much needed for evaporation in the manufacture of this powder surfactant, with the temperature expected to evaporate the water content of the raw material easily taken by vacuum system, with the process expected to reduce the product oversize. However, the temperature is also very sensitive due to the characteristics of surfactant products that are easy to.

From June 2017 there still a lot of oversize around 30% per batch. Powder surfactants with oversized parameters that do not meet the specifications (particle size) will be reworked again to achieve results that meet specifications, but require longer processing time and the company must pay for the rework process. Therefore, it is necessary to systematically investigate the causes of size mismatches in powder surfactant products so that improvements can be made to quality and production results by reducing oversize 20% of each batch.

2. Methods

Data Processing and Analysis are carried out with a phase of the PDCA cycle that relates to problems that occur during the testing process. In addition, the results of the summary of the estimated values before and after the improvement are also displayed. All stages of the PDCA cycle are the Plan Phase, Do Stage, Check Phase and Action Stage was applied in full, with the following explanation[1]:

1. Plan Phase:
   a. Create a fishbone diagram[2] to find out the root causes of oversized mismatches based on brainstorming data.
   b. Perform Failure Mode and Effect Analysis (FMEA)[3][4] to determine the factors that most influence the defect to parameterize, choose the priority of improvement based on the highest RPN (Risk Priority Number) value and provide a repair proposal.
2. Do Phase : Implementing corrective actions.
3. Check Phase:
   a. Conduct monitoring to ensure that the proposals made provide an effect of improvement leading to the expected targets and processes.
   b. Make additional improvement proposals for other problems that may arise, if needed. Comparing implementation actions after repairs and before repairs.
4. Action Stage: Standardize the process in the form of the SOP (Standard Operating Procedure) so that the proposed improvements can be carried out effectively and efficiently.

At the conclusion stage, a conclusion is drawn which contains a summary of the results of data processing and analysis based on the research objectives by providing suggestions that can build or improve the company's performance based on the analysis that has been made.

3. Result and Discussion

3.1. Plan Phase

3.1.1. Data Collection. The data that were collected is size of surfactant powder which oversize as a product out of specification. Table 1 shows the amount of oversized product in February 2018.

3.1.2. Cause of Identification Brainstorming Results. To find out the factors that influence the occurrence of Oversize Powder surfactant as a product out of specification, Cause and Effect Analysis are done by brainstorming through questionnaires to respondents that are directly related to the process.

3.2. Do Phase

3.2.1. Cause and Effect Analysis. From the results of the analysis with the previous brainstorming method, it can be identified that there are four categories that cause excessive oversize products in PT BCCI, as shown in Table 1.

Table 1. Oversize Data in february 2018

Day	Batch Number				
	1	2	3	4	5
1	350	354	352	354	352
2	355	354	350	352	352
3	354	354	352	354	352
4	355	350	352	350	352

Table 1. Oversize Data in february 2018 (cont.)

Day	Batch Number				
	1	2	3	4	5
5	350	352	350	354	352
6	354	350	352	350	354
7	352	354	350	354	352
8	350	354	350	354	352
9	350	354	352	354	352
10	355	354	350	352	352
11	354	354	352	354	352
12	355	350	352	350	352
13	350	352	350	354	352
14	354	350	352	350	354
15	352	354	350	354	352
16	350	354	350	354	352
17	352	350	354	352	354
18	352	355	354	350	352

Table 2. Excessive identification of oversize causes factors.

No	Category	Causes Factors
1	Human	Didn't check the water content of raw material which used for process
2	Machine	- Machine operated fully 24 hour - Not optimal in vacuum process - Temperature of product evaporation not maksimal
3	Material	Water content of raw material was out of specification
4	Method	- There was no schedule of cleaning of filter grinder - The method of evaporation product not optimal

The results in Table 2 are then drawn into fishbone diagrams as shown in Figure 1.

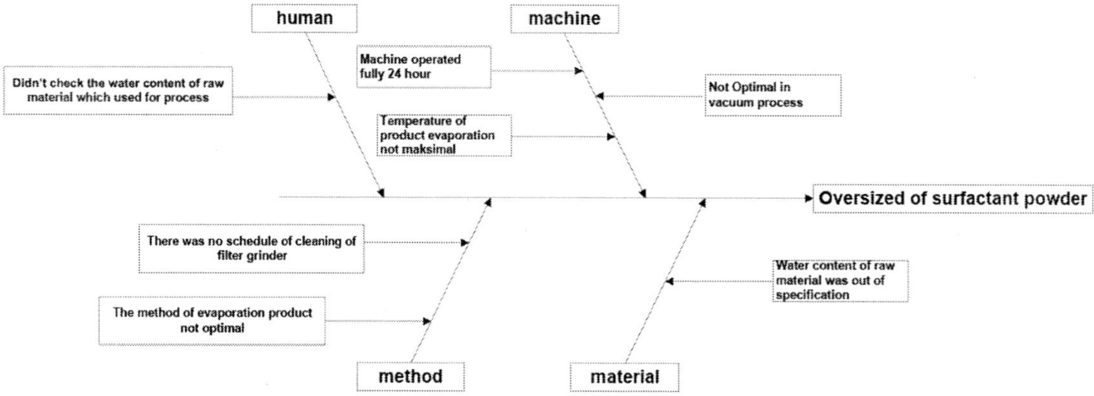

Figure 1. Fishbone diagram

*3.2.2. Failure Mode Effect and Analysis.*From the results of Cause and Effect Analysis FMEA is conducted to determine the priority of the proposed improvements as well as the application for the proposed improvements that will be made. To do FMEA, previously, brainstorming was conducted with several employees at PT BCCI (Production Leader Group and Quality Assurance Staff) to determine the value of Severity, Occurrence, and Detection. The values of each characteristic can be seen in tables 3, 4, and 5.

Table 3. Severity table (S).

Severity level	Criteria	Rating
Very high	Failure can made much of oversize product	10
High	Failure can made oversize product not uniform, in product size	7
Moderate	Failure can influence yield of product output	5
Low	Failure can made oversize product but still in specification	3
In Control	Failure not affected in product quality	1

Table 4. Occurence table (O).

Occurence level	Criteria	Possibility of failure	Rating
Very high	Failure is almost inevitable	2500 kg per 8000kg	10
High	Failure continues to recur	2000 kg per 8000 kg	7
Moderate	Failure occurs occasionally	1500 kg per 8000 kg	5
Low	Failure relatively few	1000 kg per 8000kg	3
In Control	Failure is not possible	500 kg per 8000 kg	1

Table 5. Detection Table (D).

Occurence level	Criteria	Rating
Very high	The possibility of problems is very high, prevention methods are not effective, the cause is still repeated	10
High	The possibility of problems is high, prevention methods not fully effective, the cause is still repeated	7
Moderate	The possibility of problems is moderate, prevention methods sometimes allow problem occur	5
Low	The possibility of problems is low	3
In Control	The possibility of problems is effective, not allow problem occur	1

Table 6. FMEA.

Category	Failure Mode	Failure Effects	Failure causes	Failure control	S	O	D	RPN
Human	Product was rough	Size of particle size out of specification	Didn't check the water content of raw material which used for process	Chose raw material with low water content	5	7	5	175

Table 6. FMEA (cont.)

Category	Failure Mode	Failure Effects	Failure causes	Failure control	S	O	D	RPN
Machine	The formation of powder is not perfect	Powder product was rough	Vacuum is not maximal, vacuum filter is dirty, product evaporation temperature is less high	Increase product temperature, making schedul cleaning filter grinder	10	7	5	350
Material	Much of oversize	Lumpy and rough product	Water content of Raw material was high	Large size of particles are Oversize much	7	5	5	175
Method	Particle size of product was rough	Product evaporation is less than optimal Vacuum> 50 mbar	Product temperature is not high High vacuum presure	Increase the product temperature and maximize the vacuum grinder	7	7	7	343

From the results of the RPN (Risk Priority Number) assessment on the FMEA, the causal factor with the highest RPN from the equipment machine category is less than the maximum temperature and vacuum. Therefore, improvements will be focused on these factors.

The implementation was carried out in February 2018. From the CPK results showed that the increase in output of surfactant powder products was very significant up to 100 kg per batch, the product oversized was 10%, ie before repair Average 480 kg/batch to 350 kg/batch, according to the expected target.

3.3. Phase (check) implementation of repair

3.3.1. Normality Test. Normality tests were carried out on the data presented in table 1 of the data from the analysis of oversized data on 06.07.08 in February 2018 with a total of 90 data. The stages of testing the normality of the data using the Kolmogorov-Smirnov[4] method manually are as follows:
Calculation :
Is Known:
$X_i = 352{,}16$
i = 1,64
n = 90
Calculation $\bar{\bar{x}}$
$\bar{\bar{x}} = \frac{\sum \bar{x}}{n} = 352{,}16$
Calculation $\bar{R}$
$\bar{R} = \frac{\sum R}{n} = 3{,}83$
Calculate the X control chart boundary line

A_2 = 0,577 (Constant Table)

UCL $= \bar{\bar{x}} + A_2.\bar{R}$

= 352,16 + (0,577 x 3,83)

= 354,37

CL $= \bar{\bar{x}}$

= 352,16

LCL $= \bar{\bar{x}} - A_2.\bar{R}$

= 352,16 - (0,577 x 3,83)

= 349,94

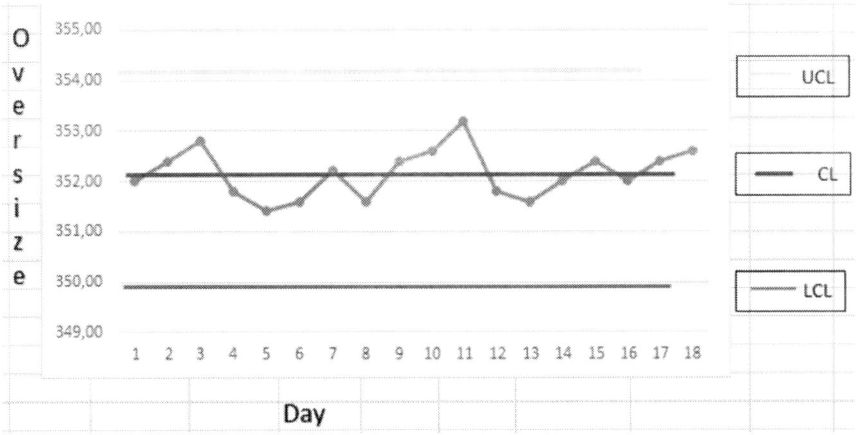

Figure 2. Control map X.

Calculates the R control map boundary

D_4 = 2,114 (Constant Table)

D_3 = 0

UCL $= D_4 x \bar{R}$

= 2,114 x 3,83

= 8,10

CL $= \bar{R}$

= 3,83

LCL $= D_3 x \bar{R}$

= 0 x 3,83

= 0,00

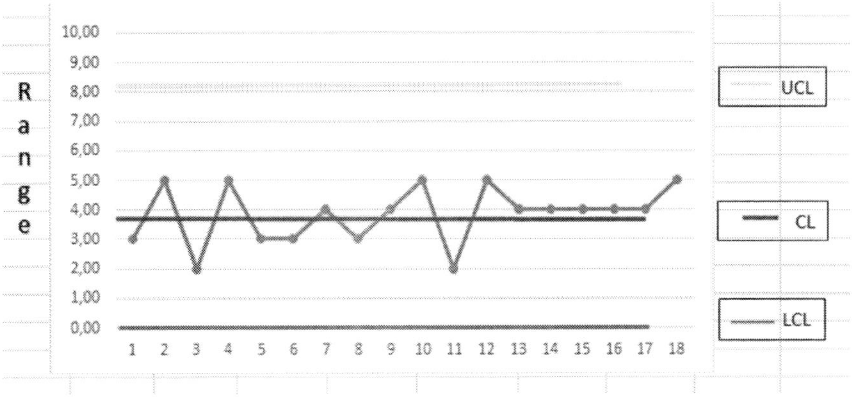

Figure 3. Control map R.

3.3.2. Process Capability. The specifications set by the company for the upper specification limits (USL) and lower specification limits (LSL) are 4.3 and 0, thus the process capability ratio (Cp) and process capability index (Cpk) can be calculated as follows:

$S = \frac{R}{d_2}$ (the value of d2 is taken from the table of constants on the control map where for the sample size 5 the value is 2,326)

$= \frac{3.83}{2.326}$

$= 1,648$

CP $= \frac{\text{USL}-\text{LSL}}{6\text{S}}$

$= \frac{370-0}{6\,(1,648)}$

$= 0$

CPU $= \frac{\text{USL}-\text{X}}{3\text{S}}$

$= \frac{370-352,16}{3(1,648)}$

$= 3,609$

CPL $= \frac{\text{X}-\text{LSL}}{3\text{S}}$

$= \frac{352,16-0}{3(1,648)}$

$= 71,227$

Cpk = Min (CPL; CPU) = 3,609

3.4. Stage (action) doing combating work procedures

At this stage standardization of work procedures are carried out from the causes that have been corrected, namely:

1. Creating a new Work Instruction by increasing the product temperature from 58°C to 60°C for maximum product evaporation.
2. Make a Schedule Cleaning filter grinder to get the maximum vacuum pressure.

4. Conclusion

Some conclusions can be drawn as follows:

The root cause of the problem of excessive oversize based on the highest RPN value in FMEA (Failure Mode Effect Analysis) is in the machine or tool category that is the maximum evaporation temperature and the vacuum is not working optimally.

The proposed improvements that have been implemented for oversize problems are raising the product temperature to 60°C in the process and making the cleaning filter grinder scheduling to get the maximum pressure <50 mbar.

From the calculation of process capability, after the improvement has fulfilled the requirements, the capability value said to be good or fulfilling the requirements is the Cpk ≥1. Therefore the CPK results after implementation are 3.609.

5. Reference

[1] Gasperz V 2002 *Metode analisis untuk peningkatan kualitas* (Jakarta: Pustaka Utama)
[2] Ishikawa K 1985 *What Is Total Quality Control? The Japanese Way,* translated by David J. Lu (New Jersey : Prentice-Hall)
[3] Carlson C S 2012 *Effective FMEAs* (New Jersey : A John Wiley & Sons, Inc)
[4] Stamatis D H 2003 *Failure Mode Effect Analysis* (America Society for Quality: Milwaukee)
[5] Irwan, Haryanto Didi 2015 *Pengendalian Kualitas Statistik (Pendekatan Teoritis dan Aplikatif)* (Bandung : Alfabeta)

Quality and Reliability Engineering in Service Industry: A Proposed Alternative Improvement Framework

M A Hadiyat, R D Wahyudi, Y Sari and E Herowati

Industrial Engineering, University of Surabaya
arbi@staff.ubaya.ac.id

Abstract. Quality and reliability engineering technique are widely used as improvement tools in manufacturing industries, involving the design experiments, quality function deployment (QFD), and survival analysis, focuses on improving the lifecycle of products. However, started at early last decade, some researches in improving the service lifecycle in some service industry consider to implements design of experiment in service industry, followed by other tools such as QFD, and the reliability engineering tools. This research proposed alternatives incorporated quality and reliability engineering methods in evaluating and improving the lifecycle of service. Starting with evaluating the weaknesses of service using SERVQUAL method, finding the improvement plan by implementing QFD, designing the robust service design using Taguchi designed experiment methods, and then implementing survival analysis for ensuring the reliability of designed services. This framework has successfully implemented in designing and evaluating existing airport service system with some recommendation generated in.
Keywords: Service life cycle, SERVQUAL, QFD, Taguchi robust design, survival analysis

1. Introduction

Continuous improvement in service industry has important role in raising up customer satisfaction and the way to keep the service deployed excellently. Some research shows that the service design proposed by various method become critical point, not because of its outstanding solution, but things that are more important are how to keep the service life cycle longer, i.e. keeping customer satisfaction survive over time. In addition, evaluation of deployed service design needs more comprehensive by considering innovative tools to strength the improvement path.

Starting by measuring gaps between expected and perceived service, the SERVQUAL method proposed by [1] and [2] perfectly identifies service attributes that should be improved. The negative gaps opens the opportunity for improvement at corresponding SERVQUAL attributes. Once the improvement plan found, the QFD (quality function deployment) helps the service provider in generating solution alternatives with some weighting calculation for selecting the best ones. (see [3] and [4]). The combination between SERVQUAL and QFD for service improvement has successfully implemented in various business field.

Otherwise, regardless of evaluating deployed service using SERVQUAL, some researcher proposed to design the service directly, especially for new type of services. Engineering tools for designing service includes kansei method [5], and robust taguchi design ([6] and [7]) commonly mentioned as robust quality engineering [8]. These methods focus on generating treatment in service refer to what the customer has desired, including the emotional aspect.

Once the service design has deployed and re-visiting service evaluation process, periodical SERVQUAL based survey still performed by service provider in order to get the feedback and re-evaluate the negative gaps. Of course, those gaps in every SERVQUAL survey taken would gives dynamic responses from customers since every survey samples different respondent. This tends to the continuous change in every time service provider proposed the improvement plan, in other words, the service design that recently been deployed but must be replaced with new ones in only short term period. Otherwise, service provides has not get enough information when the service design should been replaced with the new one, or how long the deployed service felt unsatisfied by customers in certain period. If a deployed service still gives satisfaction for customer, service provider should not rashly replace it. This problem can be solved by implementing reliability-engineering concept in evaluating how long a deployed service will gradually tends to un-satisfaction of customers.

The aim of this research is proposing alternative method for continuing the improvement and sustaining the service deployment in the term of service lifecycle. First, starting by evaluating existing service, then generate the improvement plan, followed by designing the robust service design by using quality engineering tools, finished by predicting how long the designed service will remain satisfy the customer by adopting reliability engineering principles, and then back to first step above.

2. Literature review

In this paper, quality and reliability methods are proposed to implement, capturing the dynamic of customer satisfaction with SERVQUAL, then improve it by using hard engineering tools.

2.1. SERVQUAL and QFD

Many research implement the integration of SERVQUAL and QFD in evaluating and improving service, as in [3] and [4]. This method simply measure five gaps for evaluating service performance, including gaps between perceived and expected attributes based on SERVQUAL dimensions [1]. Figure 1 shows the SERVQUAL model.

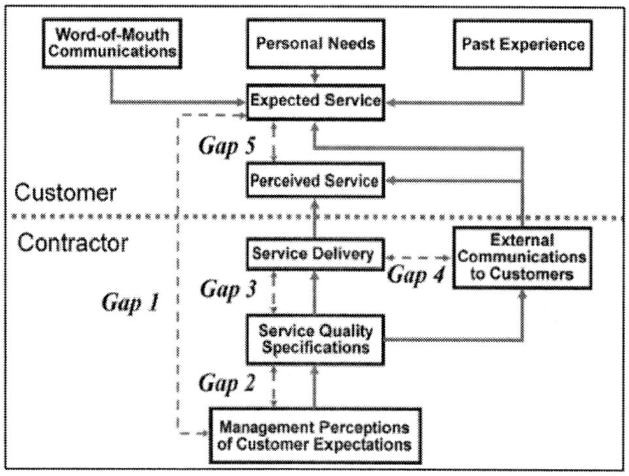

Figure 1. SERVQUAL gaps, calculating by difference between expected and perceived service (taken from [1])

Negative gaps represents unfulfilled customer expectation tends to unsatisfactory. The QFD then take place accommodating those negative gaps (as "what" part), and translate it to the house of quality then generating the alternative solutions (in "how" part). Weighting process between "what" and "how" part in QFD calculates the relationship between gaps and solution, and prioritize them.

2.2. Taguchi robust design

The Taguchi robust design has widely used in hard engineering, especially in selecting best combination of machining parameter to maximize or minimize the desired responses. Many researches successfully adopted Taguchi method, and the results have been confirmed for improvement methods options in the field of quality engineering (see [8] and [9]), including some Taguchi model modifications [10]. Based on design experiments method in service by [11], Taguchi simplifies the procedures and number of experiment run by adopting orthogonal array experiment design and transforming experiment responses to the signal-to-noise ratio (SNR) characteristic. There are three types of SNR for optimizing the responses [8]; larger the better (as used in this paper), nominal the best, and smaller the better

$$SNR = -10 \, log\left[\frac{1}{n}\sum_{i=1}^{n}\frac{1}{y^2}\right] \quad (1)$$

Ignoring the complex mathematical and statistical assumptions, Taguchi method accepted by practical users in many factories, including Service Company. [6] and [7] initiates to use Taguchi method in designing the service deployed to customers. Factors in Taguchi treated as service design attributes being improved, with the factors levels represents alternatives for solutions (see [12]). Treating factor level as solution alternatives, the experiment conducted by confirming the design to the customers. The customer's answer assumed as data responses in Taguchi, then optimizes it by using Taguchi's response table and graph. The selected factor level as optimization result then applied and adopted by service provider as new service design

2.3. Survival analysis
Survival analysis or commonly mentioned as reliability analysis, often used for predicting lifetime of an electrical or mechanical based component [13]. Implementation of this method started by conducting reliability experiment i.e. measuring the lifetime of component from firstly started until failure condition reached, then these data are fitted to certain statistical probability density function. Lifetime prediction stated as probability of survival in certain continuous time, explained below;

a. Measure the survival or reliability data t (age of components, based on reliability experiment)
b. Fit the data to probability density function (PDF) $f(t)$, the most appropriate PDF then selected for failure prediction
c. The fitted PDF then used for predicting time to reach the failure condition and mean time to failure (MTTF). This prediction is stated as reliability function $R(t)$, where

$$R(t) = P(t > T) = 1 - F(t) \quad (2)$$

$$F(t) = \int_0^t f(t)dt \quad (3)$$

$$MTTF = E(t) = \int_0^\infty R(t)\,dt = \int_0^\infty t\,f(t)\,dt \quad (4)$$

Treating age of component t before failure as customer satisfaction before unsatisfactory condition, [14] and [15] have performing this survival analysis applied to customer satisfaction. Then, completing those both paper, [16] even was fitting Bayesian mixture probability density function for combining satisfaction variables with duration of customer being experience the service. Those research shows that the survival analysis as part of reliability engineering has successfully applied to evaluate of service duration before "failure"; in this case, failure represents customer unsatisfactory.

3. Proposed framework

Incorporating the SERVQUAL, QFD, Taguchi robust design, and survival analysis becomes framework for improving service deployed to customers, started from identifying step, improving, designing, and evaluating. These steps form the cycle or continuous improvement in terms of service lifecycle. Figure 2 shows this cycle, and explained below;

1. Creating questionnaire, based on SERVQUAL attributes and dimensions to evaluate existing service deployed to customer. Questions includes the expected and perceived attributes answered by respondent, and the difference between them represents customer gap (fifth gap)
2. Calculate the customer's gap. All negative attributes gaps will lead for next improvement.
3. The Quality Function Deployment (QFD) take place in accommodating all negative gaps, store in in the "what" part, then generating solution as the "how" part works in QFD. This steps

produce the improvement plans, and QFD will weights and prioritizes it for selecting critical solutions related to most negative gaps

4. Selected improvement plan then forms the factors in Taguchi design. Each factor consists of two levels, i.e. options for each factor to be deployed in service design. Taguchi method require each factor levels (options) must be mutually exclusive without some overlap in it.
5. Designing the deployed service, by using optimized Taguchi robust design. All the factors and their levels assigned to Taguchi's orthogonal array, as if it an experimental design. Conducting experiment by creating questionnaire consist of all combination of factors levels, refer to orthogonal array provided before. Respondent will answer the questionnaire; this activity represent the perception of customer in order to involve them in designing the service.

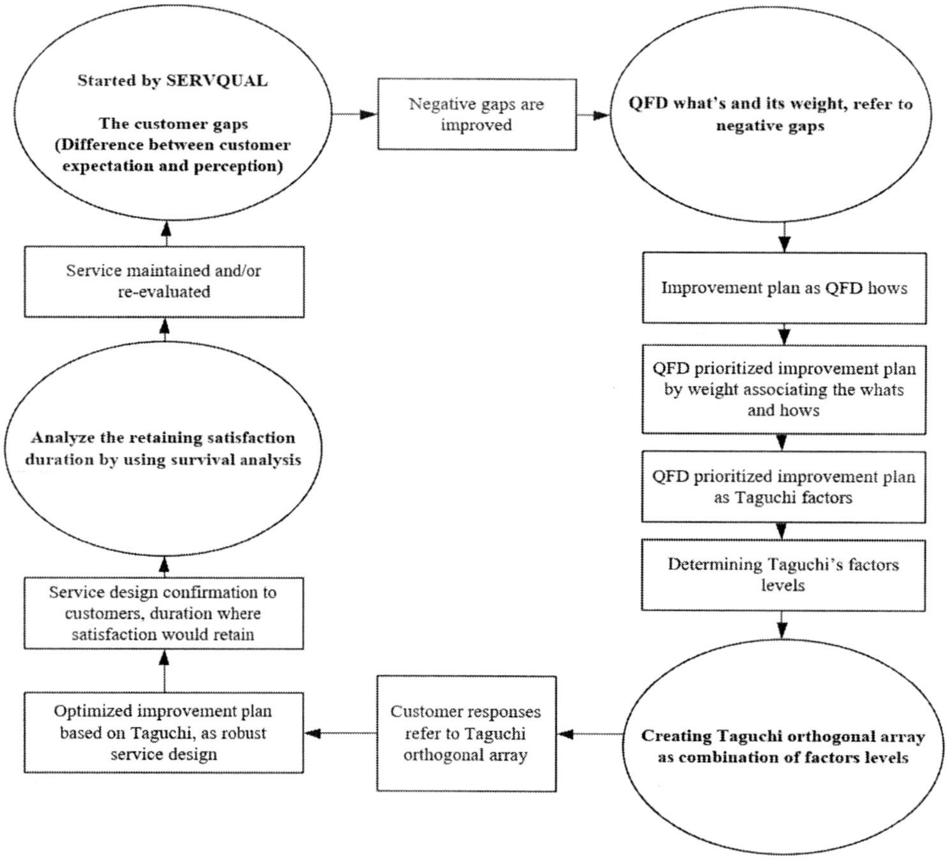

Figure 2. Framework for service continuous improvement (modified form [12] and [16])

6. Optimized improvement plan. Analyzing Taguchi experiment data, transforming it to Taguchi's S/N ratio, and finding the best combination of service design option, based on what customer has answered and confirmed. Optimization process uses Taguchi's response graph and table
7. Selected improvement plan as result of Taguchi then deployed to customers. Service provider execute this step as part of their tasks, this plan plays the role as new service design.
8. Along with deployed new improvement plan, service provides evaluate the robustness of it, by using simple two points questionnaire, i.e.; how satisfy the customer with this designed service, and how long customer has been experiencing this designed service. The interaction between both questions are then measured and produced the survival data.

9. Survival data then captured by modeling them with statistical probability distribution. The best fitted probability distribution then uses for predicting the duration of being "failure", i.e. the unsatisfactory condition of customer.
10. Once the failure condition reached, service provider should re-evaluate the deployed service design, by re-survey the customer based on SERVQUAL attributes and dimension. Then, these steps return to step 1 above.

4. Result and discussion

The framework in Figure 2, has been implemented in a case study where the object is airport service provider in Abdul Rahman Saleh airport, Malang, Indonesia (airport code MLG). Regularly, MLG airport services the passengers before and after they take the flight. Considering that MLG airport is small airport only for domestic flight, the service provider still treats the customer well and periodically improve the service. Follows steps above, result of improvement framework explained in this sub part.

Table 1. SERVQUAL attributes and their negative gaps

SERVQUAL ATRIBUTES	Negative Customer Gaps
Public facilities look and cleanness	-0.41
Waiting room layout and comfortable	-2.67
Availability of mini store or mart	-1.27
Communications facilities (Wi-Fi, free internet kiosk, etc.)	-0.54
Sprightly entrance door officer	-0.85
Willingness to prompted service from officer	-0.44
Complete disability and woman facilities	-1.69
Sufficient number of waiting seat	-1.14

4.1. SERVQUAL and QFD analysis

The questionnaire design accommodates the SERVQUAL dimensions and attributes similar to [1] with Likert scale answer. There are 20 attributes in 5 dimensions, measuring gap between expected and perceived service (customer gap). Table 1 shows attributes with negative customer gap, which should be improved by service provider. Assigning all negative gap attributes into house of quality in QFD ("what" part), service provider has generate solution alternatives as "how" part (see Figure 3). Calculating relationship weight between "what" and "how", improvement plans (i.e. selected "how") then chosen and assumes them as factors in Taguchi method. Only selected plans that should be confirmed to customers because of their critical perception considered to be assigned in Taguchi.

4.2. Taguchi robust design service

Factors in Taguchi method represent the improvement plan selected from QFD (see Table 2). Each factors divided into two levels improvement option, where the mutually exclusive between levels should be fulfilled; this will ensure there will be no overlaps in each solution option. Next, factors are assigned to orthogonal array design of experiment as shown in Table 3, consist of factor level combination that should be conducted. Additional survey to customers are conducted as well as carrying out Taguchi experiment, and customer answers for Taguchi questionnaire in Table 4 represent experiment responses. By calculating Taguchi S/N ratio as in table 5, optimized solution was reached. This result represents the new service design, i.e. grouped seat based on airline name, ergonomics non minimalist seat design equipped by electricity plug in it, and provide some minimart and vending machine at certain locations. This service design completes another improvement plan that are not selected in Taguchi. Next, service provider deploy the new service design obtained from Taguchi method. It should have more robustness to customer satisfaction because they are involved in the decision of new service design. Once the new service design deployed, service provider then evaluate the lifecycle of it, in other words, service provider conducting periodical survey where the questions lead to the data analyzed in survival analysis

Relative Weight	Weight / Importance	Quality Characteristics (a.k.a. "Functional Requirements" or "Hows") / Demanded Quality (a.k.a. "Customer Requirements" or "Whats")	maximizes waiting room layout	standardize the airport facilities	provide mini store for snack and beverages	speed up the check in queue	periodically evaluate officer's service	provide smartphone charger station	upgrade the facilities for disability	periodically maintain internet system	additional person for speeding up service	maximizes waiting room seat
26.4	24.4	Waiting room layout and comfortable	⊙		O							⊙
13.9	12.8	Public facilities look and cleanness		⊙	O				O		⊙	
13.5	12.5	Availability of mini store or mart			⊙							
12.7	11.8	Sprightly entrance door officer				O	⊙				O	
11.1	10.3	Complete disability and woman facilities	O	⊙				⊙	⊙			
7.7	7.1	Sufficient number of waiting seat	⊙	O					O			⊙
7.6	7.1	Communications facilities		O				O		⊙		
7.0	6.5	Willingness to prompted service from officer					⊙				O	
		Max Relationship Value in Column	9	9	9	3	9	9	9	9	9	9
		Weight / Importance	340.0	271.2	242.6	38.2	177.6	123.0	164.9	68.7	184.2	306.7

Figure 3. The House of Quality in QFD analysis

Table 2. Taguchi robust design factors and levels

Factor code	Taguchi Factors (Selected from QFD)	Mutually exclusive difference between factor levels	Level 1 (solution option 1)	Level 2 (solution option 2)
A	maximizes waiting room layout	seat layout	grouped seat based on airline name	additional seat by moving indoor garden outside waiting room
B	maximizes waiting room seat	Waiting room capacity and seat design	minimalist seat design, centralized electricity plug	ergonomics non minimalist seat design, equipped by electricity plug in it
C	provide mini store for snack and beverages	kiosk design	provide centralized minimart at 1st floor (check in room)	provide some minimart and vending machine at certain locations

Table 3. Orthogonal array as designed experiment

Trial No.	Factors			Customers Response Mean	S/N ratio (larger he better)	
	A	B	C			
	Column no					Larger the better S/N ratio =
	1	2	3			
1	1	1	1	2.57	4.9576	$-10\ log\left[\frac{1}{n}\sum_{i=1}^{n}\frac{1}{y^2}\right]$
2	1	2	2	4.10	11.5353	
3	2	1	2	2.67	5.3917	
4	2	2	1	3.83	10.7788	

4.3. Survival analysis

Last steps in this proposes framework, evaluating the service lifecycle by adapting survival analysis, by measuring time or duration of customer being experienced the service until unsatisfied condition (See Table 6). Unsatisfied has been reached when customer answer second question in scale 1 to 5; otherwise, scale 6 to 10 represent satisfied condition. Duration of being unsatisfied then measured in years, as the customers could distinguish their satisfaction level by years of service experience.

Table 4. Taguchi questionnaire design

According to your expected service, please rate these service option combinations from 1 (unwanted) until 5 (most wanted)				
Treatment number	A: Waiting room layout	B: Waiting room seats	C: Mini stores	Your response (scale 1 to 5)
Treatment 1	L1: grouped seat based on airline name	L1: minimalist seat design, centralized electricity plug	L1: provide centralized minimart at 1st floor (check in room)	......
Treatment 2	L1: grouped seat based on airline name	L2: ergonomics non minimalist seat design, equipped by electricity plug in it	L2: provide some minimart and vending machine at certain locations	......
Treatment 3	L2: additional seat by moving indoor garden outside waiting room	L1: minimalist seat design, centralized electricity plug	provide some minimart and vending machine at certain locations	......
Treatment 4	L2: additional seat by moving indoor garden outside waiting room	L2: ergonomics non minimalist seat design, equipped by electricity plug in it	L1: provide centralized minimart at 1st floor (check in room)	......

Table 5. Taguchi optimization result

Level	Factors		
	A	B	C
1	8.246**	5.175	7.868
2	8.085	11.157**	8.463**
difference	0.161	5.982	0.595
Rank	3	1	2
** selected factor level combination being deployed			

Table 6. Survival analysis questionnaire

According to your experience for this service, please answer these questions numerically	
How long you have been experiencing this service (in years)	.... years
How satisfy are you at current day	unsatisfied 1 2 3 4 5 6 7 8 9 10 satisfied

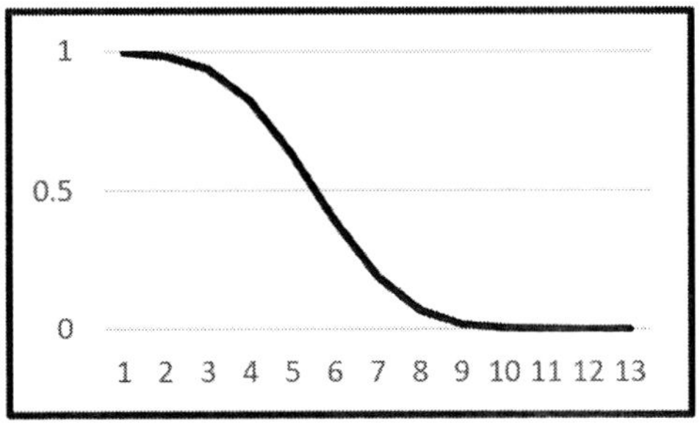

Figure 4. Survival function. $R(t)$ as Y axis, duration (years) as X axis. $R(t)$ represent probability of service being survive at certain period t

Fitting probability distribution for duration data, normal distribution selected as best fitted, so prediction of service lifecycle before unsatisfied condition are based on this, as in (5) and (6).

$$f(t) = \frac{1}{\sigma\sqrt{2\pi}} \exp\left[-\frac{(t-\mu)^2}{2\sigma^2}\right] \tag{5}$$

$$R(t) = \int_t^{\infty} f(t)dt \tag{6}$$

Mean time to failure in (4) are calculated, average duration of being fall down into unsatisfied is around 5.541 years. Continuing with graphing the survival function *R*(*t*) as in Figure 4, recommendation for next new improved service design should be taken by service provider at the around fifth years after current service has been deployed.

5. Conclusion

This proposed framework has successfully implemented with case study. Existing airport service has evaluated, and new service design has generated for ensuring customer satisfaction still in high condition until certain period where some re-evaluation and re-designing should be performed. Once the customer satisfaction fails into "failure" condition, then this framework could be re-implemented continuously.

6. References

[1] Parasuraman A, Zeithaml V A, Berry L L 1988 SERVQUAL: a multi-item scale for measuring consumer perceptions of the service quality *J. of Retailing* **64** p 12
[2] Lovelock C, Patterson P, Wirtz J 2015 *Services Marketing* (Melbourne – Pearson Australia)
[3] Sahney S, Banwet D K, Karunes S 2004 A SERVQUAL and QFD approach to total quality education *Int. J. of Productivity and Performance Management* **53** p 143
[4] Andronikidis A, Georgiou A C, Gotzamani K, Kamvysi K 2009 The application of quality function deployment in service quality management *The TQM Journal* **21** p 319
[5] Hartono M 2012 Incorporating service quality tools into Kansei Engineering in services: A case study of Indonesian tourists *Procedia economics and finance* **4 p** 201
[6] Kumar A, Motwani J, Otero L 1996 An application of Taguchi's robust experimental design technique to improve service performance *Int. J. of Quality & Reliability Management* **13** 85
[7] Raajpoot N, Javed R, Koh K 2008 Application of Taguchi design to retail service *Int. J. of Commerce and Management* **18** p 184
[8] Belavendram N 1995 *Quality by Design: Taguchi Techniques for Industrial Experimentation* (Hertfordshire UK: Prentice-Hall) p 68
[9] Roy R 2001 *Design of experiments using the Taguchi approach* (New York: Wiley-Interscience) p 78
[10] Hadiyat M A and Wahyudi R D 2013 Integrating Steepest Ascent for the Taguchi Experiment: A Simulation Study *Int. J. of Technology* **4** p 280
[11] Antony J, Coleman S, Montgomery D C, Anderson M J, Silvestrini RT 2010 Design of experiments for non-manufacturing processes: benefits, challenges and some examples *Proc. Institution of Mechanical Engineers Part B: Journal of Engineering Manufacture* **225** p 2078
[12] Hadiyat A 2018 Incorporating Servqual-QFD with Taguchi Design for optimizing service quality design *IOP Conf. Series: Materials Science and Engineering* **316** 012045
[13] Ebeling C 2010 *An Introduction to Reliability and Maintainability Engineering* (Waveland)
[14] Chen Y, Zhang Hand Zhu P 2009 Study of customer lifetime value model based on survival analysis *Proc. of WRI World Congress on Computer Science and Information Engineering* (Los Angeles: IEEE) pp 266–270
[15] Lu J 2003 Modeling customer lifetime value using survival analysis: An application in the telecommunications industry *Proc. of SAS User Group International Conference 28* (Washington: SAS Institute) pp 120–128
[16] Hadiyat A, Wahyudi R D, Sari Y 2017 Survival analysis for customer satisfaction: A case study *IOP Conf. Series: Materials Science and Engineering* **273** 012029

Measuring Acceptance Level of Online Service for Business Permit in Surabaya using Technology Acceptance Model

I M Ronyastra[1], Gunawan and E K Muhammad

Industrial Engineering Department, University of Surabaya, Surabaya, Indonesia
[1]ronyastra@staff.ubaya.ac.id, gunawan@staff.ubaya.ac.id, erlangyeah@yahoo.com

Abstract. Surabaya is known nationally as the pioneer of e-government in Indonesia, as time goes Surabaya's e-government started to provide online services for permits and licenses. One of the service that they provide is business license known as Surat Izin Usaha Perdagangan/Tanda Daftar Perusahan (SIUP/TDP). With this service, citizens are able to take care of such licenses online. This research discussed about citizen's level of acceptance of the online service and identifying factors that influence the acceptance of online SIUP/TDP in Surabaya using Technology Acceptance Model (TAM). The factors that influenced the acceptance level were measured through questionnaire survey for 100 respondents who already experienced the service. The data is then analysed through structural equation modelling (SEM). The result shown that 8 of 9 hypotheses are failed to be rejected. Three solutions can be proposed to improve the level of acceptance of the service namely: animated guidance, better socialization campaign, and scheduled site maintenance.
Keywords: E-government, Level of Acceptance, Technology Acceptance Model, Structural Equation Modelling

1. Introduction

The development of information technology has impacted many sectors including the management process of a country. The use of information technology in this case is often referred to as electronic government (e-government) which is a new phenomenon because previously, information technology was more widely used for internal government affairs. E-government is an external approach that connects government agents with external stakeholders such as citizens, business owners, and among government agents [1]. With the existence of e-government, citizens have 24-hour access to government services, other than that another perceived benefit is a closer relationship between government and citizens, increased transparency, reduced costs and increased quality and speed of service [2] .

These benefits make the Indonesian government encourage and support the implementation of e-government in every service area of the Indonesian government as stated in Presidential Instruction No. 3 of 2003 which instructed that e-Government was intended to guarantee the integration of electronic document management and processing systems and information in developing a transparent public service system. Furthermore, it is emphasized in Presidential Regulation No. 54 of 2015 by assigning the director of informatics application as the manager of the e-government process.

In the e-government implementation, Surabaya municipal government is recognized as the best example and become the model for the others. With their Surabaya Single Window (SSW) platform, the government can serve many kinds of services to the citizens which includes business permit i.e. SIUP/TDP. Despite the recognition, it seems to attract only a few researchers to academically analysing the implementation success. Dewi and Mujahidin [3] analysed SSW using a method called Government Adoption Model, while Bisma[4] interested in analysing the transactional service in e-government. This research measured the acceptance level from the citizens toward the SIUP/TDP online services and proposed some improvement actions for weak performing aspects: The acceptance level is measured according to the classic Technology Acceptance Model (TAM) with some modifications as proposed by Belanche et.al.[5].

2. Literature Review

Literature on e-government are grouped into two broad categories based on the point of view used: supply and demand [1]. Researches in the supply group mainly discussing obstacles experienced in e-government implementation to reach its objective, while the demand group focusing from users point of view such as measuring satisfaction level and financial impact after the implementation of e-government. E-Government is a government process that involves the use of communication and information technology, especially web-based applications to provide access to information and services that are faster, easier, and more efficient to the public[6]. More concisely, according to the OECD in [7] e-government can be defined as the use of communication and information technology for better governance. Our research took the demand side when trying to assess the acceptance level of the e-government in Surabaya.

For Indonesia specifically, only a few literatures take it as the object of the research. Prahono and Elidjen [8] measured e-government role and its relation with public administration reform where they took 32 samples from e-government websites and found out that through e-government the public administration reformation has started although at a slow pace. Surabaya municipal e-government has become the object for Bisma [4] and Dewi and Mujahidin [3].

As the objective of this research is to measure the performance of e-government implementation, so a metric is needed. Previous research included in the supply category offers several different measurement models. As summarized by Fath-Allah et.al [9] from about 25 different models to measure the maturity of e-government, the most important stage of maturity can be grouped into four stages: presence, interaction, transaction, and integration. E-government performance measurement conducted by Huang and Benyoucef [10] used 13 guidelines and 13 criteria to measure the usability and credibility levels of e-government sites. Choi et.al [11] proposed a model called the STOPE framework. This framework considers 5 dimensions in e-government to be measured including: Strategy, Technology, Organization, People, and Environment. Each dimension is then broken down into 4-5 sub dimensions so that the measurement process can be done easier. As a bridge to the gap between theory and practice, the STOPE Framework is combined with the Analytical Hierarchical Process (AHP) so that the results can be used as a basis for taking action by the government in improving the quality of e-government services.

Prahono and Elidjen [8] measures Index of Reform (IR) which is an index prepared by considering 4 parameters related to e-government, namely citizen document services, business licensing services, transparency of planning, and financial transparency. Each parameter is then given an assessment between 0-4 and then aggregated into an IR which also has a scale of 0-4. The results revealed that only 15.6% of the websites had score between 2.75 and 4.00, Surabaya City ranked highest among the 32 samples taken. Bisma [4] used SEM-PLS to analyse factors affecting e-government transactional service adoption and identified that social norms have positive relations with user intention to use the service. The respondents were Surabaya, Sidoarjo, and Malang citizens. The Surabaya Single Window (SSW) is the object taken by Dewi and Mudjahidin [3]. They analysed the adoption level using Government Adoption Model (GAM) which consist of 11 explanatory factors that influenced the level of adoption in both static and interaction maturity stage.

Belanche et.al [5] proposed a TAM based framework to assess adoption level of e-government services. They integrated trust factor into the TAM framework and testing 4 new hypotheses in addition to the traditional 5 TAM related hypotheses. They tested their framework for e-government services in Spain. TAM also used by Hamid et.al [12] when they measured the continuance intention to use e-government in Malaysia. They analysed the effect of perceived usefulness and perceived ease of use and found out that both factors positively related to the intention. Belanche et.al [5] mentioned that their framework has only tested in Spain with Spanish speaking respondents, it is necessary to test the framework in other country. We tried to adopt the framework in our research for Surabaya e-government services because we share the same believe that trust is a major factor in the acceptance level of such services.

3. Methodology

3.1. Data Collection

To collect the data, we adapt questionnaire from Belanche et.al [5] and translated them into Bahasa Indonesia. The indicators are presented in Table.1 where we asked the respondents on 5 dimensions and 18 indicators besides the demographic questions. The respondents are contacted in the public service mall UPTSA Surabaya and they fill the questionnaire online via a tablet computer provided by the surveyor. We surveyed 30 respondents as our pre-sample to check against the validity and reliability of the instrument which both are confirmed valid and reliable. Afterwards, we continue to survey another 70 respondents to make up the 100 minimum respondents.

Table 1. Questionnaire Indicators

Variable	Notation	Indicator
Perceived Usefulness	PU1	Useful for me
	PU2	Speed up my activities
	PU3	Save my money
	PU4	Save my time
Perceived Ease of Use	PEO1	The site is easy to find in the internet
	PEO2	The steps are easy to be understood
	PEO3	Easy to fill in the admission data
	PEO4	Easy to upload documents
Trust	T1	Official service from Surabaya government
	T2	Reliable functionality (never error)
	T3	Will not divulge my personal data
Attitude	AT1	Give a pleasant experience
	AT2	Give a modern experience
	AT3	Is a good idea
	AT4	Interesting
Intention to use	INT1	Using it now
	INT2	Recommend to other
	INT3	Willing to use in another time

3.2. Analysis Process

The data collected are then analysed using Structural Equation Modelling (SEM) which started with model development as shown in Figure 1. The process continued with confirmatory factor analysis to analyse the loading factor on the outer model before moving on the inner model analysis. The final step is to test 9 hypotheses to check the relationship between indicators and variables.

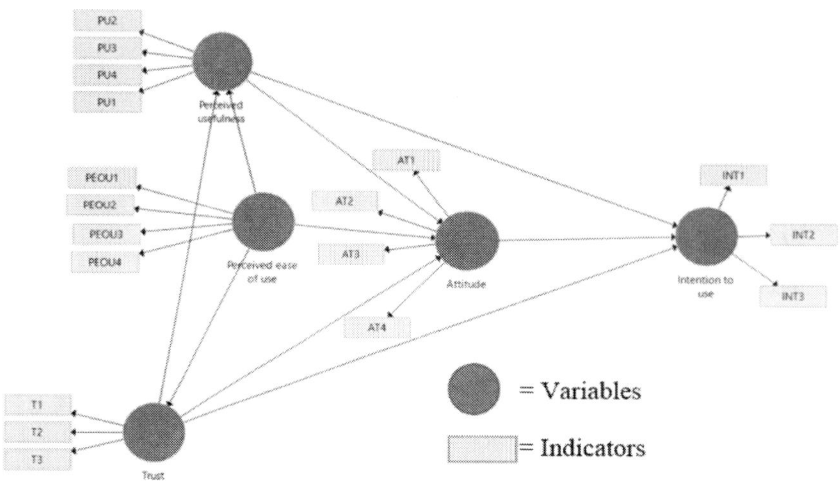

Figure 1. Structural Equation Model

4. Result and Discussion

4.1. Outer Model

Outer model is a model that focuses on the relationship between the latent variables and each indicator. The average variance extracted in Table 2 suggests some values is below 0.5 (before removal) which means the variables' validity is not sufficient. In order to make the variable valid, we must remove the indicators that have high loading factor to the variable. We removed the AT4 (Interesting) and PU2 (Speed up my activities) because we think the respondents have difficulties to express their subjective judgment on the questions that may be too general for them. After the removal, the variables become valid.

Table 2. Outer Model AVE before removal of AT4 and PU2

Variables	**Average Variance Extracted**	
	before removal	after removal
Attitude	0,488	0,526
Intention to use	0,564	0,564
Perceived ease of use	0,554	0,554
Perceived usefulness	0,484	0,556
Trust	0,557	0,554

We also observe from Table 3, each indicator has better correlation to its dedicated variable than to the other variables suggesting that there is no problem with the discriminant validity. In Table 4, we observed some Cronbach's Alpha values fall under 0.7 threshold which requires us to observe the composite reliability to check the reliability of the model. The composite reliability shown a good reliability because all the values are above the 0.7 threshold, so we can confirm that the outer model is reliable.

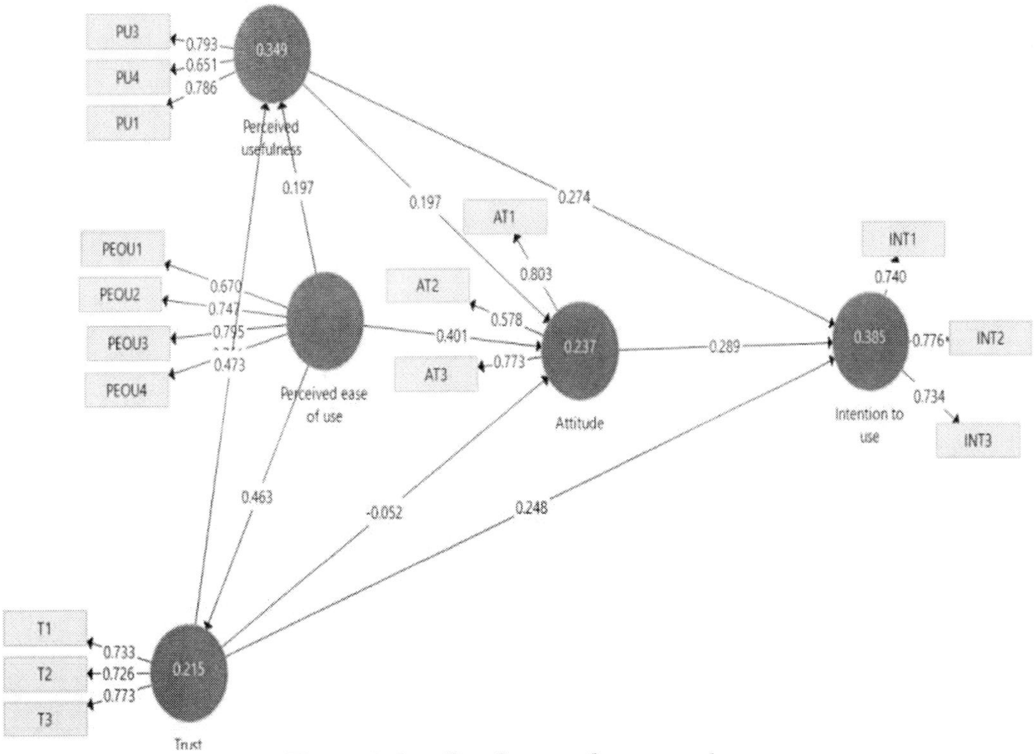

Figure 2. Loading factors after removal

4.2. Inner Model

The inner model is analysed by observing the regression values from independent variables toward the dependent variables. The R^2 values presented in Table 5 suggesting that only less than 40% of variations in the independent variables are explained by the dependent variables indicating there should be other variables to be identified as the better explanatory variable.

Table 3. Cross Loading

	Variable				
Indi.	AT	IU	PEOU	PU	T
AT1	0,803	0,402	0,373	0,267	0,198
AT2	0,578	0,217	0,259	0,176	0,206
AT3	0,773	0,309	0,351	0,272	0,139
INT1	0,274	0,762	0,39	0,386	0,308
INT2	0,329	0,758	0,371	0,457	0,442
INT3	0,393	0,733	0,337	0,291	0,293
PEOU1	0,306	0,377	0,674	0,239	0,44
PEOU2	0,296	0,403	0,755	0,317	0,349
PEOU3	0,395	0,373	0,788	0,4	0,349
PEOU4	0,363	0,284	0,756	0,263	0,221
PU1	0,32	0,448	0,436	0,784	0,446
PU3	0,24	0,374	0,235	0,793	0,464
PU4	0,162	0,295	0,225	0,652	0,342
T1	0,079	0,264	0,296	0,384	0,74
T2	0,241	0,34	0,322	0,327	0,721
T3	0,21	0,419	0,401	0,519	0,772

Table 4. Reliability Testing

	Cronbach's Alpha	Composite Reliability
Attitude	0,549	0,765
Intention to use	0,615	0,795
Perceived ease of use	0,731	0,832
Perceived usefulness	0,607	0,789
Trust	0,608	0,789

Table 5. R-Square

	R square
Attitude	0,237
Intention to use	0,386
Perceived Usefulness	0,348
Trust	0,216

4.3. Hypotheses Testing

We tested 9 hypotheses in this research as Belanche et.al[5] did. The result presented in Table 6 is mostly confirming their finding except for hypothesis 8 which tested the effect of trust variable towards the attitude variable. Our result suggest that trust is neither significant nor positively affect the attitude variable. So, we could now graph our research model as in Figure 3 where we do not put arrow sign from Trust towards Attitude. The difference findings may imply some differences between sample taken in Spain and sample taken in Surabaya that impacting the acceptance level of the e-government services.

Table 6. Hypotheses Testing

Hyp.	Relations	Original Sample (O)	P Values	Conclusion
H1	Attitude → Intention to use	0,289	0,017	Significant and Positively Affects
H2	Perceived Usefulness →Attitude	0,197	0,01	Significant and Positively Affects
H3	Perceived Usefulness → Intention to use	0,274	0,007	Significant and Positively Affects
H4	Perceived ease of use → Attitude	0,401	0,003	Significant and Positively Affects
H5	Perceived ease of use → Perceived usefulness	0,197	0,038	Significant and Positively Affects
H6	Perceived ease of use → Trust	0,463	0	Significant and Positively Affects
H7	Trust →Perceived usefulness	0,473	0	Significant and Positively Affects
H8	Trust --> Attitude	-0,052	0,309	Not Significant and Not Positively Affects
H9	Trust --> Intention to use	0,248	0,017	Significant and Positively Affects

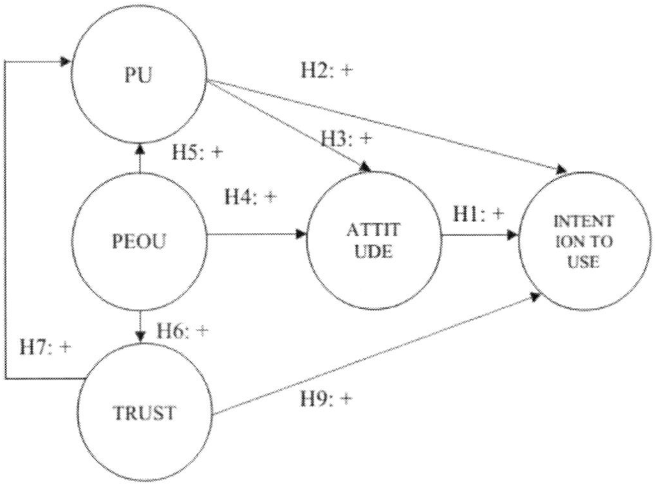

Figure 3. Research Model after SEM

4.4. Service improvement
Based on the analysis, some ideas can be suggested to the municipal government to improve the service offered especially in the perceived usefulness, perceived ease of use, and trust variables. The lowest loading factor in PU is from PU4 (save my time) which could be improved by preparing a detailed step-by-step tutorial video on the e-government website so the user does not need to come to the offline office. Improvement for PEOU1 can be done through a better socialization campaign so users can easily find the link to e-government websites. And lastly, to improve trust variable, the system need to have a regular maintenance schedule to ensure the reliability of the system (T2).

5. Conclusion
This research has confirmed 8 out of 9 hypothesis formulated by Belanche et.al [5] regarding the factors that have effect on the acceptance level of online business permit service in Surabaya Municipal. The research rejected the hypothesis that trust has significant and positive effect on attitude toward the service, however trust variable has significance and positive effect toward the perceived usefulness and intention to use. This implies that trust variable is an important aspect that need attention from the government. The survey also indicates that the service need to improve performance of indicator PU4, PEOU1, and T2.

This research only took samples from one type of service offered from around 24 different services available in the Surabaya Single Window which indicates that this research can be extended to learn the experience of other services' user. Another limitation is the point of contact is at the offline office which may well filled by those users who cannot use the service online which may lead to some bias in their judgment toward the system. Further research could help eliminating those limitations.

6. References
[1] Das A, Singh H and Joseph D 2017 A longitudinal study of e-government maturity. *Inf. Manag.* **54**(4) pp 415–426

[2] Kurfalı M, Arifoğlu A, Tokdemir G and Paçin Y 2017 Adoption of e-government services in Turkey. *Comput. Human Behav.* **66** pp 168–178

[3] Dewi L A S and Mudjahidin 2014 Analisis Penerapan Aplikasi Surabaya Single Windows Pemerintah Kota Surabaya Menggunakan Government Adoption Model (GAM). *J. Tek. POMITS* **3**(2) pp 210–215

[4] Bisma R 2017 Faktor Adopsi Layanan E-Government Jenis Layanan Transaksi. *J. Inf. Eng. E.* **01** pp 83–88

[5] Belanche D, Casaló L V and Flavián C 2012 Integrating trust and personal values into the

technology acceptance model: The case of e-government services adoption. *Cuad. Econ. y Dir. la Empres.* **15**(4) pp 192–204

[6] Sang S, Lee J and Lee J 2009 E-government adoption in ASEAN: the case of Cambodia," *Internet Research* **19**(5) pp 517–534

[7] Lee J 2010 10 year retrospect on stage models of e-Government: A qualitative meta-synthesis. *Gov. Inf. Q.* **27**(3) pp 220–230

[8] Prahono A and Elidjen 2015 Evaluating the Role e-Government on Public Administration Reform: Case of Official City Government Websites in Indonesia. *Procedia-Procedia Comput. Sci.* **59**(Iccsci) pp 27–33

[9] Fath-allah A, Cheikhi L, Al-qutaish R E and Idri A 2014 E-Government Maturity Models. *Int. J. Softw. Eng. Apl.* **5**(3) pp 71–91

[10] Huang Z and Benyoucef M 2014 Usability and credibility of e-government websites. *Gov. Inf. Q.* **31**(4) pp 584–595

[11] Choi H, Park M J, Rho J J and Zo H 2014 Rethinking the assessment of e-government implementation in developing countries from the perspective of the design–reality gap: Applications in the Indonesian e-procurement system. *Telecomm. Policy* **40**(7) pp 644–660

[12] Hamid A A, Razak F Z A, Bakar A A and Abdullah W S W 2015 The Effects Of Perceived Usefulness And Perceived Ease Of Use On Continuance Intention To Use E-Government. *Procedia Econ. Financ.* **35**(October) pp 644–649

Value Engineering Towards the Design of Bread Production Process Tools

Ayu Bidiawati [1,*] Inna Kholidasari [1] Bintang Manggala Elani [1]

[1]Department of Industrial Engineering, Faculty of Industrial Technology, Universitas Bung Hatta, Padang, Indonesia
*)ayubidiawati@bunghatta.ac.id

Abstract. In a bakery, an operator needs a working tool for bread production process. To avoid injury and inconvenience the tool should be suitable and appropriate for the operator's body posture and workload. Operators' assistive device produced is trolley rack. Value engineering approach is carried out on this tool to identify and develop its functions in achieving the balance among cost, reliability and appearance of the product. The work-plan is divided into several development stages: creative stage, analysis stage, development stage and recommendation stage. Product value is determined by comparing between product benefits and product costs. This study employs Function Analysis System Technique (FAST) method which consists of 5 stages, namely information, creative, analysis, development and presentation/ recommendation stages. The recommendation stage explains the best alternative design. The results show that alternative II has the highest performance value which is 97.16 and the highest rating of 0.45. It is rated as the best design and has good product durability.
Keywords: work tools, value engineering, FAST method, performance

1. Introduction

The research focused on home industries that produce various types of bread. In bread production process, workers are found inconvenient during the process of transferring bread from the oven (baking) to the cart, where the design has not considered ergonomic aspects, so if workers work for a long period of time, he/she will have musculoskeletal problems. This also affects work productivity of operator. Productivity is comparison between the results achieved with the role of workers in certain period of time. If there is no improvement in the development of work system using assistive devices suitable with the workers' wishes, there will be no progress in work productivity. Referring to these problems, a solution was developed in form of new alternatives with the aim of creating a safe, comfortable and easy to use tool. This bread production process employs 6 workers. During the production process, some operators work in standing position and perform the same hand movements repeatedly (repetitive). To minimize non-ergonomics problem from the bread production process, work risk identification was carried out with Rapid Upper Limb Assessment (RULA) method [1] and work risk identification with OCRA method [2]. Assessment of work posture using the OCRA and RULA methods showed that operator used the right hand more dominant while working so the right hand experiences fatigue faster. The identification results with RULA method showed that work stations that have the highest risk are in the stirring, milling, printing and burning work stations. This is due to bent body posture and unbalanced legs of operator while doing his work [1]. The identification results using OCRA method showed that work stations with the highest risk are found in 3 work stations namely milling, printing and baking work stations [2].

The operator aid tool is designed in form of trolley rack. The design of this trolley rack is made by considering value engineering calculations, to identify and develop the function of the rack. This is to get a balance among cost, reliability and appearance of the trolley rack. The important thing discussed in this research is functional value of the product components and how to increase the value or benefits that will be obtained from a product. This utilization can be achieved if there is production costs reduction, additional product value or both. For cost reduction, focus will be on designing product components using materials, product forms, manufacturing processes and assembly processes. This value engineering method can be developed to minimize existing products or design new products. For the industry to become more competitive we must act upon the challenges faced by the market more and more demanding and competitive in a globalized world [3].

The purpose of this value engineering technique is to improve efficiency and effectiveness in product design and development to minimize possible errors. Lack of understanding of consumer needs characteristics is because information quality received is not good enough so that it can lead to errors in designing the product. In addition, product development process will not run well if the aspects of habit, an environment that does not support creativity, lack of understanding and knowledge of costs are still the same. It will hamper the design process. Therefore, to anticipate errors in designing the bread production process aid, value engineering concept is applied to control costs by analyzing value towards its function without eliminating the desired quality and reliability. Value Engineering is a systematic system design technique to identify the functions needed, applying values and developing alternatives to achieve the best functional balance among cost, condition, and performance of system or product [4]. Value engineering concept aims to improve efficiency and effectiveness in carrying out the design by describing errors in design. According to Lawrence D. Miles [5], value engineering is a creative and systematic approach with the aim of reducing or eliminating unnecessary costs. Whereas according to the Society of American, value engineering is a technique that is applied systematically to determine the function of a product or service that determine monetary value of the function and fulfill it with minimum total cost.

However, according to Zimmerman and Hart [6], value engineering is a technique and management that uses a systematic approach to achieve the best functional balance between costs and performance of a product. Value Engineering is a system approach aimed to achieve the desired functions of a product, process, system or service at minimum overall cost, and having maximum performance with consistency without in any way affecting the quality, reliability, performance and safety of the product, process, system or service [7]. Value is defined as the ratio between product performance and the costs needed to get performance. Product's good performance will not necessarily produce high value if the cost required is very high. Therefore, product value can be increased in three ways, namely reducing costs while maintaining performance, improving performance while maintaining costs, improving performance and lowering costs [8]. By referring to several definitions, value engineering can be interpreted as creative and systematic management technique by identifying and developing the functions of object or service to achieve cost balance, reliability and appearance of a product or system. Value engineering is a method that emphasizes the analysis of various item functions or system through systematic and directed approach to achieve optimum use and financing for each rupiah spent [8] [5] [10][6][4]. So the results achieved in value engineering can be in form of alternative use of materials, new designs, etc. by considering alternative that have more value. Value engineering is a powerful problem solving tool that can reduce while maintaining or improving performance and quality requirements [11].

The main purpose of creating a product is basically to sell products quickly, with maximum benefits and provide satisfaction to consumers. Thus the product designers should not create excessive product functions or use of production materials that is not useful and raise the price. So the idea must be developed with a starting point from minimize cost, time, and material [12]. Information stage of value engineering is basic foundation for any value investigation. In this stage, all important information is collected to understand carefully the object being investigated. The information is then analyzed to find object functions, so that it can be clarified as a primary or secondary function [13]. In this study the function stage for problem solving techniques is by identifying the functions needed using Function Analysis System's Technique (FAST). Objective of value engineering job plan creative stage is to produce various alternatives that can fulfill or carry out the main functions by using creative techniques. This creative stage can only be started if the desired function has been determined and evaluated. At this stage there is no analysis to ideas issued by each team member so that all ideas will be accommodated to fulfill basic functions without seeing progress first [13]. Ideas can be developed more broadly by doing so in a group whose members are from different fields of work known as brainstorming. Analysis stage of value engineering job plan is to evaluate alternatives provided in creative stage. Evaluation results are used to determine useful alternatives for further study which will be given the greatest potential for cost savings. Thought based on strict rules that do not apply in previous stages, at this stage the ideas and thoughts that have emerged before will analyzed and criticized.

In this development stage, development program was made until it became a complete proposal. The development program is based on detailed plan from evaluated ideas that is useful to obtain all relevant information in order to develop the program into an acceptable proposal and to implement it. Every aspect that is relevant to capability, design, quality, manufacturing, packaging, and marketing must be understood as an effort to change the idea that has been evaluated into a proposal that can be submitted [9]. Recommendation stage is the final stage in value engineering work plan which aims to convince decision makers to be able to decide whether the selected design is capable and good to do [13]. Analytical Hierarchy Process (AHP) is a decision support model that will disentangle multi-factor or multi-criteria problems into a hierarchy. According to [14], hierarchy is defined as representation of complex problem in multi-level structure where the first level is the goal, followed by the level of factors, criteria, and sub criteria and so on down to the last level of the alternative. The procedure in AHP method consists of several stages according to Suryadi and Ramdhan in Taminanto [15], namely arranging hierarchy of the problems faced and determining priority of the elements.

2. Method

Research stage was carried out to create value engineering work plan through several stages, namely:

a. Information stage using FAST method

Information stage is regarding matters relating to the design of bread production process aids. The components of this tool are trolley frames, trolley pads and trolley handles. Then from information obtained, it is analyzed using FAST Method to find object functions, so that it can be clarified as primary or secondary function. This function analysis is viewed from the tools that operator is expected to provide comfort and practicality at work.

b. Creative stage with Brainstorming

At this stage, several designs are made that help to develop and repair existing components, regarding materials, dimension, and shape that is possible to be repaired. At this stage other alternative ideas are raised, to master the same function.

c. Analysis stage with AHP method

This process is related to the selection and decision making that will allow solutions that can be implemented. In this analysis stage, several mechanisms of Hurt (2003) are carried out as follows:

- Profit and loss Analysis.
- Determination of priority criteria.
- Weighting analysis on criteria.
- Evaluation matrix.
- Weighting criteria on modification alternatives
- Performance of selected alternative.

d. Development stage

The aim of development is to choose the best alternative from selected alternatives in previous stage by comparing the results of performance calculations in previous stage. The success of this stage depends on two aspects, namely product performance, and the costs spent. So that at this stage value calculation will be taken by comparing performance value to the cost.

e. Recommendation stage.

This stage is to explain the best alternative designs.

3. Result and Discussion

3.1 Information Stage

Information obtained from initial design analysis, where initial conditions of bread production process aids have weaknesses in components; material, dimensions or shape of the component, the size and shape dimensions are not yet in accordance with anthropometry. From the results of interviews conducted, information data obtained based on design requirements analysis that the system and operator's work position are not good enough and cause work complaints. The equipment that is used now does not last long because its dimensions are not ergonomic. Then the information is analyzed using Function Analysis System Technique (FAST) method with the following results in Table 1 and also shown in Figure 1.

Table 1. Function of the Bread Production Process Tools

Component	Function		Type
	Verb	Noun	
The Bread Production Process Tools	To Give	Comfort and Practicality	Secondary (S)
	Give	The Place for Bread Production Process Tools	Primary (P)

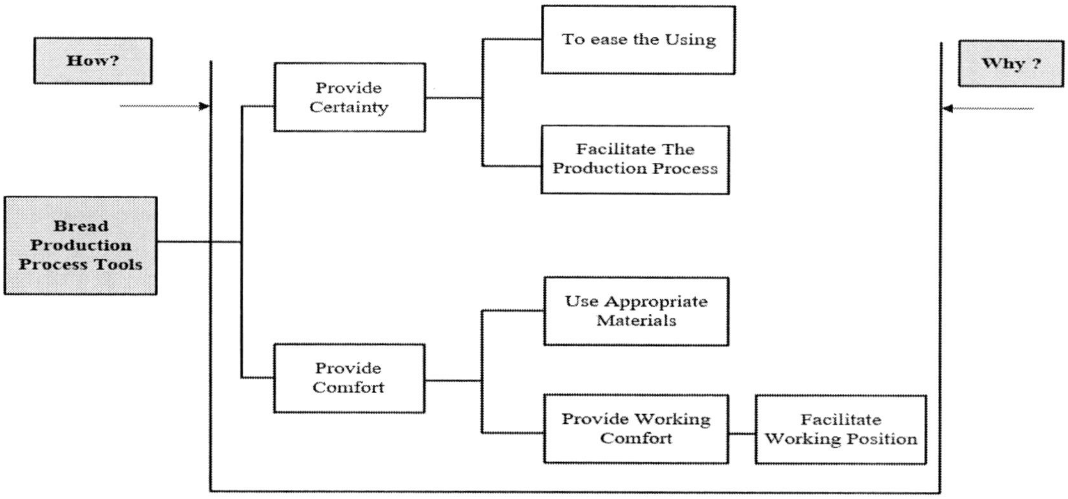

Figure 1. Chart of FAST for Production Process Tools

3.2 Creative Stage

Imperfection to generate new ideas is one of main causes in proposed alternative comparison. The proposed alternative is obtained from component reduction, simplification or modification while maintaining the object's main function. Based on the weaknesses and conditions of existing equipment, alternative design of assistive devices is developed in accordance with anthropometric data. This is shown in Figure 2.

IOP Conf. Series: Materials Science and Engineering **528** (2019) 012078 doi:10.1088/1757-899X/528/1/012078

Alternatives	Improvements
Alternative I 	1. The design tool is made from iron plate and iron ace 2. The design tool is made with a size of 90 cm x 80 cm x 103 cm 3. The design tool is painted in silver 4. The design tool has economic life > 4 years 5. Rectangular design tool 6. The design tool is made to move & place items 7. 14. The design tool is given a rubberized grip
Alternative II 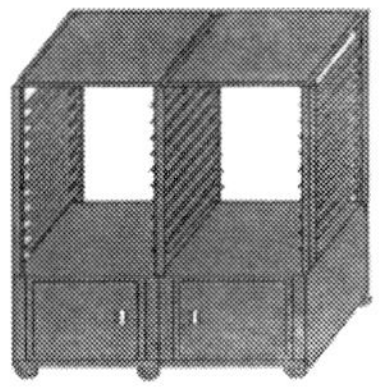	1. The design tool is made of steel plate 2. The design tool is made with a size of 80 cm x 80 cm x 262cm 3. The design tool is painted in gray 4. The design tool has economic life > 4 years 5. The design tool has a square shape 6. The design tool serves to move goods 7. The design tool is given a foamy push handle

Figure 2. Modification alternatives

3.3 Analysis Stage

The process at this analysis stage uses AHP method to give weighting to selection criteria for the design of this aid. A characteristic of operator necessity for this bread production process is based on customer's voice, namely the trolley rack. It has good design and size as needed, the material used is stronger and sturdy, and trolley frame is made permanent so it is not easily damaged during production process, and the trolley has wheels so that it is easier to push, providing a baking sheet. Assessment of criteria priority determination is carried out through questionnaires containing questions concerning the order of priority level. This is shown in Figure 3. The criteria considered for profit and loss analysis are form design, size design, material type, performance, material costs and ease of obtaining material with the following results:

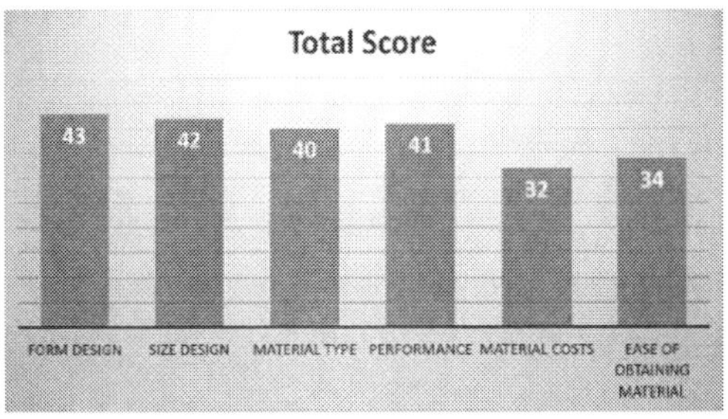

Figure 3. Criteria Priority Determination

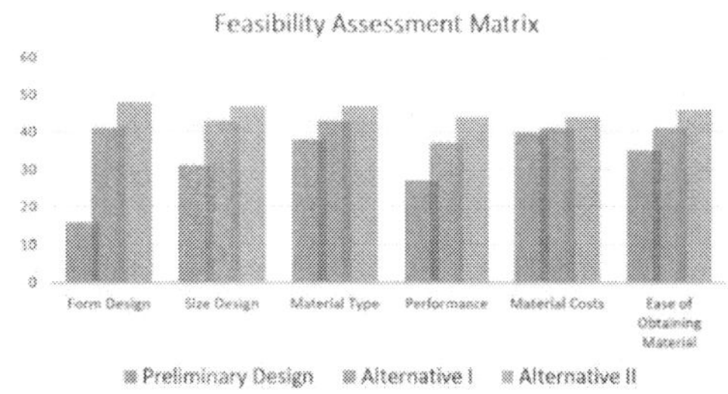

Figure 4. Feasibility Assessment Matrix

The evaluation matrix has two alternative modifications and initial design alternative, by converting from verbal language to nominal language on Likert scale. This is shown in Figure 4. The evaluation matrix results of the best ranking alternatives and other modification alternatives as well as initial design, based on questionnaires that have been distributed and filled directly by respondents are as follows in Figure 5.

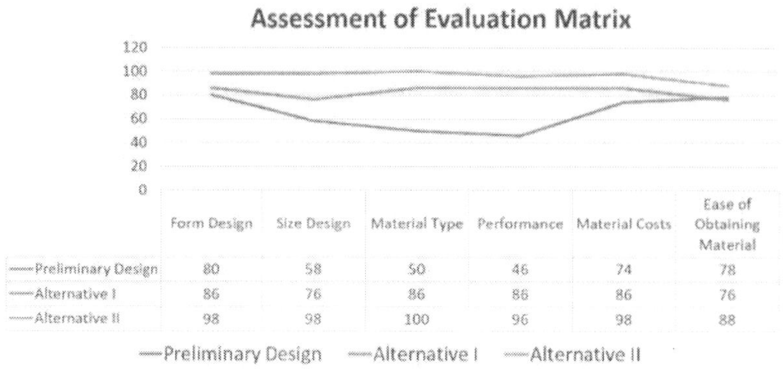

Figure 5. Evaluation Assessment Matrix

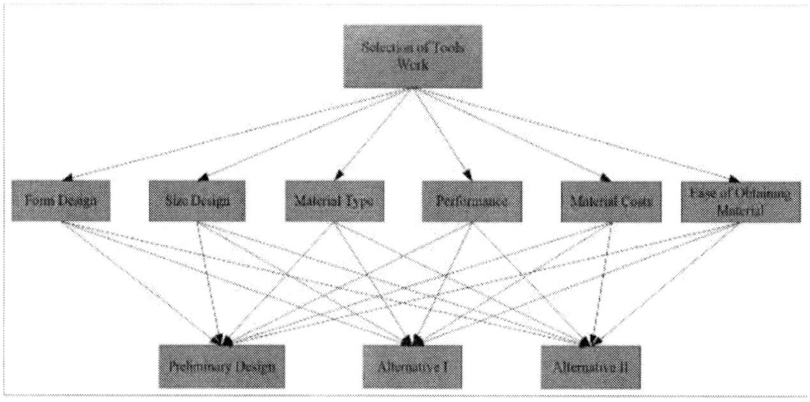

Figure 6. Alternative Selection Hierarchy

Calculation of weighting criteria in modified alternative is conducted to calculate the performance of each criterion based on pairing comparison method in accordance with importance level referring to qualitative scale in the decision support system [14]. Pairing matrix is used to normalize weighting by dividing each entry with the number of columns so that the weight of each criterion (Eigen Vector) is finally obtained. Furthermore, it is determined how well the consistency value of the data is. This is shown in Figure 6. Processing with AHP method uses manual formulation with data sources from questionnaire. Calculation of hierarchical weighting factors for all criteria. This is the results of a combined preference analysis of 10 respondents. After the paired geomean comparison values on the criteria are obtained, then a pairing matrix is made for all criteria as shown in table 2.

Table 2. The Paired Comparison Matrix

Criteria	**1**	**2**	**3**	**4**	**5**	**6**
1	1	1.1161	1.4051	2.2333	2.0939	1.7118
2	0.89605735	1	1.5683	3.8730	1.6438	2.8071
3	0.71169312	0.63763311	1	3.2746	1.6808	1.3904
4	0.44776788	0.25819778	0.30538081	1	3.1958	2.5561
5	0.47757773	0.60834651	0.59495478	0.129107	1	1.9663
6	0.58418039	0.35623954	0.71921749	0.391221	0.0856939	1
Total	4.11727646	3.97651694	5.59295308	11.0850317	10.1228694	11.4317

With the elements in each column divided by the number of columns concerned, the normalized relative weights will be obtained. The value of eigen vector is generated from average relative weights for each row shown in table 3.

Tabel 3. Priority Matrix for All Criteria

Criteria	1	2	3	4	5	6	Vector Eigen	% Weight
1	0.2428790	0.2806728	0.2512269	0.2014699	0.2068485	0.1497415	0.21504313	21.5
2	0.2176335	0.2514764	0.2804064	0.3493901	0.1623848	0.2455540	0.24405243	24.41
3	0.1728553	0.1603497	0.1787964	0.2954074	0.1660399	0.1216267	0.17730443	17.73
4	0.1087534	0.0649306	0.0546010	0.0902117	0.3157010	0.2235975	0.17426427	17.43
5	0.1159936	0.1529848	0.1063758	0.0282282	0.0987862	0.1720042	0.10524145	10.52
6	0.1418852	0.0895858	0.1285935	0.0352927	0.0502396	0.0874761	0.08409428	8.41
Total	0.1666667	0.1666667	0.1666667	0.1666667	0.1666667	0.1666667	1	100

The maximum eigen value (λ maximum) is shown in table 4.

Table 4. Maximum Eigen Value

Criteria	Sum of Entry	Matrix *Priority*	ɣ Max (A/B)
1	1.4900381	0.21504313	6.92901977
2	1.6980303	0.24405243	6.95764554
3	1.3078646	0.17730443	7.37637858
4	0.9389969	0.17426427	5.38835012
5	0.6817816	0.10524145	6.47826118
6	0.5458783	0.08409428	6.49126552
	Total		39.6209207
	ɣ Maximum		6.60348678

The consistent index values obtained are:
For n = 6, RI = 1.24 (Saaty table, 1993), then:

$$\mathbf{CI}\frac{(\gamma\ \mathbf{max}-\mathbf{n})}{(n-1)} = \frac{(6{,}6035-6)}{(6-1)} = \mathbf{0.1207} \qquad \mathbf{CR} = \frac{\mathbf{CI}}{\mathbf{RI}} = \frac{0{,}1207}{1{,}24} = \mathbf{0.0973}$$

Because CR < 0.100 means that the respondent is consistent.

Performance value calculation for each selected modification alternative and initial alternative through multiplication of the weight obtained for each criterion with the final results obtained from questionnaire is shown in Table 5.

Table 5. Performance Value Calculation

Modification Alternative	Criteria						Pn	Rank
	Form Design	Size Design	Material Type	Performance	Material Costs	Ease of Obtaining material		
Weight	0.2150	0.2441	0.1773	0.1743	0.1052	0.0841		
Preliminary Design	80	58	50	46	74	78	63.02	3
Alternative I	86	76	86	86	86	76	83.73	2
Alternative II	98	98	100	96	98	88	97.16	1

3.4 Development Stage

The development stage depends on product performance and the costs spent so value calculation takes performance value when compared to the cost. Calculation of initial design costs is based on costs calculation required in bread production process. Calculation of alternative design cost components is taken from the results of the best alternative design tools selection. Value calculation is based on comparison between performance (Pn) and the costs incurred in making alternative modifications (C) so performance value of modification design of bread production aid tool is obtained. The value of all alternative design tools for the bread production process is shown in table 6.

Table 6. Modification Alternative Value

Modification Alternative	Pn (Performance)	N	Cn (Cost)	Vn (*Value*)	Rank
Preliminary Design	63.02	8251	440,000	1	3
Alternative I	83.73	8251	5,000,000	0.38	2
Alternative II	97.16	8251	1,800,0000	0.45	1

3.5 Recommendation Stage

The final results of this stage get the best alternative design, namely alternative II. This is the best alternative tool to help bread production process, compared to other alternatives. Alternative II gets the highest performance value of 97.16 with a value of 0.45. Alternative II gets the best design and has good durability.

4. Conclusion

- Results of the value engineering approach through FAST analysis shows the recommended bread production process tools have the best values based on functional descriptions.
- The criteria that affect the modification tools based on the level of importance are form design (18.53%), size design (18.10%), material type (17.24%), performance (17.67%), material costs (13.79%), and ease of obtaining material (14.66%).
- For selected alternatives shows best weighting criteria: form design size (0.2150), size design (0.2441), material type (0.1773), performance (0.1743), material costs (0.1052), and ease of obtaining material (0.0841).
- The highest performance is obtained by alternative II with a Pn value of 97.16 and Vn value of 0.45.

5. References

[1] Bidiawati A and Setiawati L 2017 The Risk Assessment Work of Operator Based on Rapid Upper Limb Assessment Method (RULA) To Improvement of Work System. *Proceeding the 15th International Conference on QIR (Quality in Research)* ISSN: 1411-1484 (Bali: Faculty of Engineering Universitas Indonesia) pp 1049 – 1057

[2] Bidiawati A, Setiawati L and Kurnia Y 2018 Tools Design of Bread Production Process To Minimize Musculoskeletal Disorders Based On OCRA Mehod. *International Journal of Engineering & Technology* **7**(2.29), Website: www.sciencepubco.com/index.php/IJET pp 106 – 109

[3] de Santis D S, Dedini F G and Marcicano D S 2016 Lean Manufacturing and Value Engineering, Two Concepts for Sustainable Management. *Journal of Textile Science & Engineering* **6**(6) ISSN: 2165–8064 pp 1–6

[4] Park R J 1998 *Value Engineering: A Plan for Invention*. (Michigan: St. Lucie Press)

[5] Miles M B and Huberman A M 2006 *Qualitative Data Analysis: A Sourcebook of New Methods*. (Baverly Hills: SAGE)

[6] Zimmerman L W and Hart G D 1982 *Value Engineering, A Practical Approach for Owner, Designers and Contractors*. 14th Ed. (New York, USA: Van Nostrand Reinhold Company)

[7] Kumar S L, Nallusamy S and Modak R 2018 Execution of value Engineering Approach for Design Development and Cost Reduction of Monoblock Pump. *ARPN Journal of Engineering and Applied Sciences* **13**(5) ISSN : 1819-6608 pp 1734 – 1741

[8] Annappa C M and Panditrao K S 2014 Application of Value Engineering for Cost Reduction of Household Furniture Product - A Case Study. *International Journal of Innovative Research in Science, Engineering and Technology*, ISSN: 2319-8753 **3**(10)

[9] Crum L W 1971 *Value Engineering The Organized Search for Value*. (Longman Group Limited)

[10] Dell'Isola A 1982 *Value Engineering in the Construction Industry*. (New York: Van Nostrand Reinhold)

[11] Ramani B and Pitoda J 2017 A Critical Literature Review on Application of Value Engineering in Building Construction Project. *International Journal of Creative Research Thoughts (IJCRT)* **5**(4) ISSN: 2320–2882, www.ijcrt.org. pp 1461-1464.

[12] Wahono H 2015 Penerapan Value Engineering Pekerjaan Baja Profil Terhadap Baja Ringan pada Pembangunan Persekolahan Eben Haezer Manado. *Jurnal Sipil Stastik* **3**(3) (Manado: Universitas Sam Ratulangi)

[13] Chandra S C V S 2014 *Maximizing Construction Project And Investment Budget Efficiency With Value Engineering*. (Jakarta: PT. Elex Media Komputind)

[14] Saaty T L 1993 *Pengambilan Keputusan Bagi Para Pemimpin, Proses Hirarki Analitik untuk Pengambilan Keputusan dalam Situasi yang Kompleks*. (Jakarta: Publisher PT. Pustaka Binaman Pressindo)

[15] Taminanto 2012 *Sistem Pendukung Keputusan dengan Metode* Analitycal Heirarchy Process (AHP) *untuk Penentuan Prestasi Kinerja Dokter pada RSUD. Sukarji*, Publisher Infokes **2**(1) pp 1-15

Effects of cocoa clones and fermentation times on physical and chemical characteristics of cocoa beans *(Theobroma cacao* L.*)*

A Assa, Rosniati and M R Yunus

Center for Plantation Based Industry, Agency for Research and Development of Industry, Ministry of Industry, Makassar, South Sulawesi, Indonesia
emma_bbihp@yahoo.com

Abstract. The effects of cocoa clones and fermentation time on the physical and chemical characteristics of cocoa beans has been examined. This study used cocoa pods of superior clones, namely Sulawesi 1, Sulawesi 2, and MCC 02 clones cultivated in East *Luwu*, South Sulawesi, Indonesia. Prior to the fermentation, the mature cocoa pods were stored for six days. Fermentation took place for 24, 48, 72, 96, and 120 hours using fermentation boxes made of *styrofoam* with the capacity of box is 7 kg of wet beans. The drying of the cocoa beans used sun drying method. The results showed that the interaction between fermentation time and type of clones have significantly affected moisture content, proportion of cocoa nibs and shells, and fat content; but the interaction of the both had no significant effect on pH of cocoa nibs and FFA content. In general, the Sulawesi 2 clone had better physical and chemical characteristics than those of the Sulawesi 1 and MCC 02 clones.
Keywords: fermentation, physical, chemical, cocoa beans, clones

1. Introduction

Cocoa (*Theobroma cacao* L.) is cultivated mainly in West Africa, Central and South America and Asia. Based on the ICCO 2017 data, eight largest cocoa producing countries today are Ivory Coast, Ghana, Indonesia, Nigeria, Brazil, Equador, and Malaysia. The total world cocoa production in 2015/2016, is estimated 3.972 thousand tonnes, in which Indonesia contributed for about 8.0 % of the total world production [1].

Cocoa bean is the main raw material for chocolate manufacture, and therefore it is very important to maintain cocoa bean quality and apply strick requirement, so the chocolates produced safe for consumption. The government of Indonesia has established SNI 2323-2008 standards for cocoa beans, in which the bean quality is classified according to physical, chemical, and organoleptic performances of the cocoa beans [2].

One of the efforts to increase productivity and quality of the cocoa beans is by using excellent cocoa clones. Beside having larger bean sizes and higher fat contents, the cocoa also relatively resistant towards fungi, insect and other plant diseases. In South Sulawesi in primarily, some of such the cocoa clones have been developed and cultivated in this area since several years ago, namely Sulawesi 1 and Sulawesi 2 clones [3] and local clones namely MCC 01 and MCC 02 clones [5]. One of the cocoa production centers of this types of the cocoa clones is East *Luwu* Regency, South Sulawesi Province, Indonesia with production around 480 ton a year [6].

The low quality of the Indonesia cocoa beans, especially cocoa beans originating from South Sulawesi relates to the relatively high acidity along with weak in flavor [3], [7]. In addition, it relates to inconsistency in quality and mostly unfermented or partially fermented. As we know, the aims of fermentation is to form flavor precursor (combination of aroma and taste), to reduce bitterness and astringency, and to improve physical performance, such as color of the cocoa beans [8]. During fermentation process, the cocoa beans experience physical and chemical changes which will influence the cocoa bean quality.

Since fermentation time is a key factor in the cocoa bean fermentation, it is required to apply the appropriate fermentation times. One of the factors affects this fermentation time is the variety of the cocoa [9]. The other factors are the pulp content and the layer height of the cocoa beans in the fermentation boxes [3].

The aims of this study was to examine the effects of cocoa clones and fermentation times on the physical and chemical characteristics of the cocoa beans. The study used cocoa pods of the superior clones, namely Sulawesi 1, Sulawesi 2, and MCC 02 clones, cultivated in East *Luwu*, South Sulawesi, Indonesia.

2. Materials and Methods

2.1. *Preparation of the cocoa beans*

In this study the cocoa pods of the Sulawesi 1, Sulawesi 2, and MCC 02 clones were procured from Sahar Cocoa Village (SCV) farm, a plasma cocoa farm of Mars Symbioscience Indonesia Ltd., located in *Maliowowo* Village, District of Angkona, East *Luwu,* South Sulawesi, Indonesia.

The mature cocoa pods were first storaged for six days before the beans were removed from the pod husk. The cocoa beans were then separated from the placenta. Fermentation time of 24, 48, 72, 96, and 120 hours used fermentation boxes made of *styrofoam* with the capacity of 7 kg wet beans, each. After fermentation, the cocoa beans were sun-dried for 5 days, until reaching 6 to 7% moisture contents.

2.2. *Preparation of the samples and analysis*

Preparation and analysis of the cocoa bean samples, as dried fermented beans, were conducted at Laboratory of Chemical and Microbiologycal Testing, Center for Plantation Based Industry, in Makassar, South Sulawesi, Indonesia.

Assessment for the physical and chemical characteristics of the cocoa beans referred to *SNI Biji Kakao*: 2323-2008 for moisture, pH of nibs, proportion of cocoa nibs, proportion of shell, fat contents, and FFA contents.

3. Result and Discussion

The ANOVA summary of the effects of the clone types and fermentation times on the physical and chemical characteristics of the cocoa beans are presented in Table 1.

Table 1. ANOVA summary of the effect of clones and fermentation times on the physical and chemical characteristics of the cocoa beans (Sulawesi 1, Sulawesi 2, and MCC 02 clones)

Variable	Characteristic of Physical and Chemical Quality in Cocoa Beans					
	Moisture	pH of nibs	Proportion of cocoa nibs	Proportion of shells	Fat content	FFA content
Clone (C)	0.44^{*}	0.14^{*}	71.25^{*}	73.89^{*}	27.61^{*}	0.20^{*}
Fermentation (FT)	0.26^{*}	4.39^{*}	53.65^{*}	52.87^{*}	4.77^{*}	0.44^{*}
Interaction (C x FT)	1.13^{*}	0.20^{ns}	16.57^{*}	17.39^{*}	0.39^{*}	0.04^{ns}

* = significant at $p < 0.05$ ns = not significant

The ANOVA analysis showed that the clone types and fermentation times and their interaction significantly affected the physical and chemical characteristics of the cocoa beans at $p < 0.05$, except for the pH of nibs and free fatty acid levels.

Further analysis using Duncan's test to examine the effects on the physical and chemical characteristics of the cocoa beans is presented in Table 2.

Table 2. Duncan summary test for the effects of the clones and fermentation times on the physical and chemical characteristics of the cocoa beans (Sulawesi 1, Sulawesi 2, and MCC 02 clones)

Variables	Characteristic of Physical and Chemical Quality in Cocoa Beans					
	Moisture	**pH of nibs**	**Proportion of cocoa nibs**	**Proportion of shells**	**Fat content**	**FFA content**
Clones						
Sulawesi 1	6.40[b]	5.41[a]	81.08[d]	18.90[a]	49.88[d]	0.73[d]
Sulawesi 2	6.62[a]	5.36[a]	83.50[b]	16.48[b]	50.28[c]	0.78[c]
MCC 02	6.45[b]	5.25[b]	84.84[a]	15.08[c]	51.93[b]	0.85[b]
Fermentation time (hours)						
24	6.40[c]	6.05[a]	80.58[e]	1.44[a]	50.02[e]	0.63[e]
48	6.47[b]	5.19[c]	82.63[c]	17.28[c]	50.54[d]	0.68[d]
72	6.50[b]	5.10[d]	81.59[d]	18.38[b]	50.95[c]	0.82[c]
96	6.63[a]	4.95[e]	83.86[b]	16.14[d]	51.03[b]	0.91[b]
120	6.33[d]	5.39[b]	85.22[a]	14.75[e]	51.17[a]	0.96[a]

The values followed by different notation (a, b, c, d) are significantly different at $p < 0.05$

a. Moisture

The result of analysis showed that he highest moisture content in Sulawesi clones 2 (6.62) followed by MCC 02 (6.45) and Sulawesi 1 (6.40). The highest moisture obtained at 96 hours fermentation is 6.63 and is significantly different from other fermentation time treatments. The fermentation time of 48 hours (6.47) and 72 hours (6.50) shows the same moisture or not significantly different. The lowest moisture was obtained at 120 hours fermentation time of 6.33.

The moisture content of dried cocoa beans is full fermented which is less than 7.5%. Table 2 shows that the moisture content of dry cocoa beans obtained ranged from 6.33 to 6.62 for the three clones. This means that the moisture of the three clones has met the requirements of *SNI 2323 Biji Kakao: 2008* which is a maximum of 7.5% [2]. The moisture of more than 8% causes the cocoa beans to be easily attacked by fungi and insects, thus increasing the risk of cocoa beans damage, but if the moisture of less than 5% will cause the cocoa beans to break easily [10], [11]. In addition to the fermentation process, the moisture is also affected by the way of drying. The drying process can cause the moisture of cocoa beans to decrease from 60% to 7% [12], [13].

b. pH of cocoa nibs

The highest pH of cocoa nibs obtained from Sulawesi 1 (5.41) clones followed by Sulawesi 2 (5.36) and MCC 02 (5.25). The highest pH of cocoa nibs at 24 hours fermentation was 6.05 and the lowest was at 96 hours fermentation time of 4.95.

The pH of the cocoa nibs is an important parameter to determine the acidity of fermented cocoa beans. The acidity of the cocoa nibs greatly determines the flavor of chocolate because it is not only related to the taste of the acid, but also determines reaction of flavor compounds during roasting [14], [3]. Biehl *et al.* [15] reported that flavor formation is more likely to occur in seeds at pH of 5.0 - 5.5 than at pH of 4.0 - 4.5. The pH of cocoa nibs that are too high (near neutral pH) will produce a low flavor. This is because cocoa beans which have high pH are likely to have mouldy beans. Whereas if the pH is too low (very acid < 4.5) caused by excessive production of acetic acid it will also produce a low flavor. This occurs because of a strong acidification process so that the exopeptidase enzyme conducts proteolysis and produces a low flavor [17], [18].

Yusianto *et al.* [19] and Biehl *et al.* [16] states that fruit ripening affects the pH value or acidity of nibs. The process of fruit ripening causes some pulp to be used for the respiration process and some pulp content evaporates. This results in a reduction in alcohol formation during fermentation because the substrate is reduced and the fermentation conditions are not aerobic, resulting in the formation of acetic acid only slightly so that the acid content of the dried cocoa beans becomes low.

In this study it was found that the interaction between differences in clones and fermentation time had no significant effect on the pH of cocoa nibs. This is because the chemical changes that occur during fermentation will still take place during drying and it is suspected that the type of acid diffused into the cocoa nibs has a pH that is relatively the same at each fermentation time.

The cocoa nib of the Sulawesi 1 clone has higher pH than that of the Sulawesi 2 and the MCC 02 clones. This is because the bean size of the Sulawesi 1 clone is relatively smaller compared to the bean sizes of the two other clones causing better oxidation process of the acid compounds in fermentation. This was parallel to the Duncan *et al.* finding [16]. However, the three clones had quite low acidity levels of the beans, indicated by pH values above 5.0.

c. Proportion of cocoa nibs

Duncan's test results showed the effect of clones and fermentation time were significantly different on the proportion of cocoa nibs. The highest proportion of cocoa nibs was obtained at MCC 02 clone (84.84%) followed by Sulawesi 2 (83.50%) and Sulawesi 1 (81.08%). The highest proportion of cocoa nibs at 120 hours fermentation was 85.22% and the lowest at 24 hours fermentation was 80.58%.

According to Wood & Lass [12] and Afoakwa *et al.,* [20], important factors that influence bean size are temperature and rainfall distribution during the fruit development process. High temperatures and low rainfall can cause lower bean levels. The number of nib contained in cocoa beans is a major concern for the cocoa industry or processor because the high content of the nibs will produce more cocoa fat yield.

d. Proportion of shells

Duncan's test results showed the effect of clones and fermentation time were significantly different on the proportion of cocoa shells. The highest proportion of cocoa shells in Sulawesi 1 clone (18.90%) followed by Sulawesi 2 (16.48%) and MCC 02 (15.08%). The highest proportion of cocoa shells obtained at 24 hours fermentation is 19.44% and is significantly different from other fermentation time treatments. The lowest proportion of cocoa shells level at 120 hours fermentation is 14.75%.

The proportion of cocoa shells content is influenced by the type of clone, fermentation time and drying. The slime layer of unfermented cocoa beans will have a slippery mucus layer and the structure of nibs tends to be solid and hard [19]. High of proportion of nibs content cause the energy and time needed to evaporate water to get out through thicker layers of cocoa shells for relatively longer.

e. Fat content

Duncan's test results showed that the effect of clones and fermentation time were significantly different on the fat content of dry cocoa beans. The highest fat content in the MCC 02 clone (51.93%) was followed by Sulawesi 2 (50.28%) and MCC 02 (49.88%). The range of fat content in Indonesian cocoa beans is 49 - 52% [4].

The highest fat content obtained at 120 hours fermentation is 51.17% and is significantly different from other treatments of fermentation time. The lowest fat content at 24 hours fermentation is 50.02%. In general, the fat content of the three clone types increased with the increase of fermentation time. This is because the fermentation process can cause a decrease in the content of non-fat ingredients such as proteins, polyphenols and decomposed carbohydrates so that relatively fat levels will increase [21], [13].

The cocoa beans of the MCC 02 clone contained higher fat contents (5,39%) compared with the two other cocoa beans of the Sulawesi 1 clone (49.88%) and of the Sulawesi 2 clone (50.28%). The higher cocoa bean size of the MCC 02 clone most probably caused this high fat content. As stated by

[3] the cocoa bean size contributes to the fat content. The higher the bean size, the higher its fat content is, and vice versa.

Cocoa butter from Indonesia is famous for having a higher level of hardness than cocoa butter from Africa. If used in making chocolate, it will produce chocolate that is not easy to melt because of the higher level of hardness of fat [4].

f. FFA content

The presence of free fatty acids (FFA) in cocoa butter must be avoided because this is one indicator of quality damage. The results of FFA levels analysis of the three clones ranged from 0.63 to 0.96%. The highest FFA level was in MCC 02 clone (0.85%) followed by Sulawesi 2 (0.78%) and MCC 02 (0.73%). The highest FFA levels were obtained at 120 hours of fermentation, which was 0.96% and the lowest at 24 hours fermentation was 0.63%.

The interaction between clone differences and fermentation time had no significant effect on FFA levels. The same thing was reported by Ramlah [22] that fermentation time did not give a significant effect on FFA levels. Free fatty acids generally appear when dried cocoa beans are stored in a warehouse that is less clean and moist.

Free fatty acids are free fatty acids, not bound as triglycerides. FFA levels that can be received according to the Codex Allimentarius are <1.75% and the best is <1.3% ([23], [4]. Cocoa beans are considered to begin to suffer damage if FFA levels are greater than 1.3%.

Free fatty acids can be formed through oxidation reactions or hydrolized enzymes that come from broken seeds or poor storage so that they break down. Free fatty acids that are too high will cause undesirable taste and are not good for heart and blood vessel health because they cause high cholesterol triggered by the buildup of FFA in the blood. Cocoa beans with moldy interior will produce fat with high FFA levels [4].

4. Conclusion

The types of the cocoa clones and the fermentation times and their interaction affected on moisture content, proportion of cocoa nibs, proportion of shells, and fat content, but the interaction of the both did not significantly affected on the pH of cocoa nibs and the FFA content.

Acknowledgments

In this study, the cocoa pods and other facilities were provided by *Sahar Cocoa Village* (SCV) Farming and Mars Symbioscience Indonesia Ltd., particularly in the preparation of the cocoa beans.

5. References

[1] ICCO 2017 *ICCO Quarterly Bulletin of Cocoa Statistics* **XLIII**(2)

[2] Badan Standardisasi Nasional 2008 *SNI Biji Kakao: 2323-2008*. ICS.67.140.30.

[3] Wahyudi T, Panggabean T R and Pujiyanto 2013 *Panduan Lengkap Kakao Manajemen Agribisnis dari Hulu hingga Hilir* (Jakarta: Penebar Swadaya)

[4] Wahyudi, T, Pujiyanto and Misnawi 2015 *Kakao: Sejarah, Botani, Proses Produksi, Pengolahan, dan Perdagangan* (Yogyakarta: Gadjah Mada University Press)

[5] Dinas Perkebunan Provinsi Sulawesi Selatan 2014 *Pelepasan Klon Kakao Unggul Lokal MCC 01 dan MCC 02 Asal Kabupaten Luwu Utara, Sulawesi Selatan*. (Pemerintah Kabupaten Luwu Utara bekerja sama dengan Pusat Penelitian Kopi dan Kakao Indonesia)

[6] Sekretariat Jenderal-Kementerian Pertanian 2016 *Outlook Kakao: Komoditas Pertanian Subsektor Perkebunan*. (Jakarta: Pusat Data dan Sistem Informasi Pertanian)

[7] Duncan R J E, Gofrey G, Yap T N, Pettipher G L and Tharumarajah T 1989 Improvement of Malaysian cocoa bean flavor by modification of harvesting, fermentation and drying methods the Sime-Cadbury process. *The Planters* **65** pp 157-173

[8] Misnawi 2003 Influence of Cocoa Polyphenols and Enzyme Reactivation on the Flavour Development of Unfermented and Under-fermented Cocoa Beans. *Thesis* Doctor of Phylosophy (Selangor: Universiti Putra Malaysia)
[9] Amin S 2005 *Teknologi Pasca Panen Kakao Untuk Masyarakat Perkakaoan Indonesia* (Jakarta: Badan Pengkajian dan Penerapan Teknologi Press)
[10] Towaha J, Anggraini D A and Rubiyo 2012 Keragaan mutu biji kakao dan produk turunannya pada berbagai tingkat fermentasi: studi kasus di Tabanan Bali. *Pelita Perkebunan* **28**(3) pp 166-183
[11] Basri Z 2010 Mutu biji kakao hasil sambung samping. *Media Litbang Sulteng* III pp 112-118
[12] Wood G A R and Lass R A 1985 *Cocoa.* (London: Longman Group Ltd)
[13] Camu N, de Winter N, Addo S K, Takrana J S, Berdern H and De Vuyst L 2008 Fermentation of cocoa beans: influence of microbial activities and polyphenol concentration on the flavour of chocolate *J. Sci. Food Agric.* **88** pp 2288-2297
[14] Cortes T R, Jimenes G R and Lepe M R 2012 Isolation and characterization of acetic acid bacteria in cocoa ferementation *African J. Micro. Res.* **6** pp 339-347
[15] Biehl B, Brunner E, Passern D, Quesnel V C and Adomako D 1985 Acidification, proteolysis and flavor potential in fermenting cocoa beans *J. Sci. Food Agric.* **36** pp 583–598
[16] Biehl B, Meyer B, Crone G, Pullman L and Said M B 1989 Chemical and physical changes in the pulp during ripening and pod storage of cocoa pods. *J. Sci. Food Agric.* **48** pp 180-208
[17] Yuliatin N 2007 *Pengaruh Suhu dan Lama Refermentasi Terhadap Pembentukan Senyawa Calon Pembentuk Ciatarasa pada Biji Kakao (Theobroma cacao L.) Underfermented.* (Jember: Fak. Teknologi Pertanian Universitas Jember)
[18] Susijahadi and Jinap S 1998 Isolation of yeast having potential in alcohol fermentation from fermented cocoa beans *Proc. Malaysian Science and Technologies Congress* (Kuala Terenggano Darul Imal Malaysia)
[19] Yusianto dan Wahyudi T 1991 Pengolahan kakao mulia dengan metode Sime Cadbury: pengaruh lama penyimpanan buah, lama fermentasi dan sifat fisiko kimia. *Pelita Perkebunan* **7**(2) pp 48-56
[20] Afoakwa E O, Quao J, Takrama J, Budu A S and Saalia F K 2011 Chemical composition and physical quality characteristics of Ghanaian cocoa beans as affected by pulp pre-conditioning and fermentation. *J. Food Science Technol.* **47** pp 3-11
[21] de-Brito E S, Garcia P N H, Gallao M I, Angelo L C, Pedro S F and Marcia R B 2000 Structural and chemical change in cocoa *(Theobroma cacao L.)* during fermentation, drying, and roasting. *J. Sci. Food Agric.* **81** pp 281-288
[22] Ramlah S 2013 Hubungan antara lama fermentasi dengan mutu biji kakao. *Prosiding Seminar Nasional Teknologi Industri Kakao dan Hasil Perkebunan Lainnya* (Balai Besar Industri Hasil Perkebunan) pp 153-161
[23] Mulato S, Widyotomo S, Misnawi, Suharjono E 2010 *Pengolahan Produk Primer dan Sekunder.* (Jember: ID Puslitkoka)

Cooperation between power plant in East Kalimantan by integrating renewable energy power plant

Muslimin[1], Willy Tambunan[2] and Wahyuda[2,3]

[1]Electrical Engineering, Universitas Mulawarman, Samarinda, Indonesia
[2]Industrial Engineering, Universitas Mulawarman, Samarinda, Indonesia
[3]wahyuda@gmail.com

Abstract. Government power plants cannot supply all electricity demand in East Kalimantan Province especially if there is damage to the power plant with the largest production capacity, so cooperation with other suppliers needs to be done. There are four stakeholders that can supply electricity, namely State Electricity Company (SEC), Independent Power Producer (IPP), Rent, and Excess Capacity (EC). Each supplier has different power generation characteristics that lead to differences in fuel costs and emissions. This research proposes 4 mathematical models. The models are scenarios 4 forms of cooperation between electrical stakeholders in East Kalimantan taking into consideration the potential of renewable energy (RE) power plants. This mathematical model is solved with a heuristic algorithm solver excel. The result, the best scenario by integrating three power plants owned by SEC, IPP and RE. Whereas collaboration with the smallest emissions is SEC + Rent + RE.
Keywords: Electricity supply chain, Power Plant, Optimization, Renewable Energy, Supplier cooperation

1. Introduction

East Kalimantan is still experiencing a power outage. This is due to the unbalanced between supply and demand. There are 4 power suppliers in East Kalimantan: State Electricity Company (SEC), Independent Power Producer (IPP), Rent, and Excess Capacity (EC). While the center of demand spread over 12 different locations.

Each supplier has a different power plant. This difference leads to differences in costs and emissions. Costs and emissions are heavily dependent on the fuel used, where in this case there are still many who use oil and coal. Coal-fired power plants are usually cheaper but have a negative impact on the environment [1].

Costs and emissions can be reduced in various ways, one with the appropriate allocation of loads to each power plant using economic dispatch as in [2], [3]. Alternatively, using a renewable energy-fueled power plant as in [4]. Economic dispatch was originally used to minimize the cost of power plants in the interconnected system [5], [6]. Furthermore, economic dispatch is developed into economic dispatch, which also minimizes emissions [7] - [10].

Economic Dispatch has been widely used for power plant allocation with various constraints and solutions with various methods. However, ED has not been able to provide a detailed allocation for implementation in collaboration among suppliers. Therefore, this study proposes 4 types of a collaboration of power suppliers that have different power plant characteristics. Collaboration is modeled mathematically with the approach of the transport-economic dispatch model as in [3].

Computing is done with an excel solver. The combination between economic dispatch and transportation makes the computational outreach more detailed than the usual economic dispatch.

2. Methods

This research was conducted using 5 stages: Problem formulation, Literature review, modeling, Experiment, and Analysis (Figure 1).

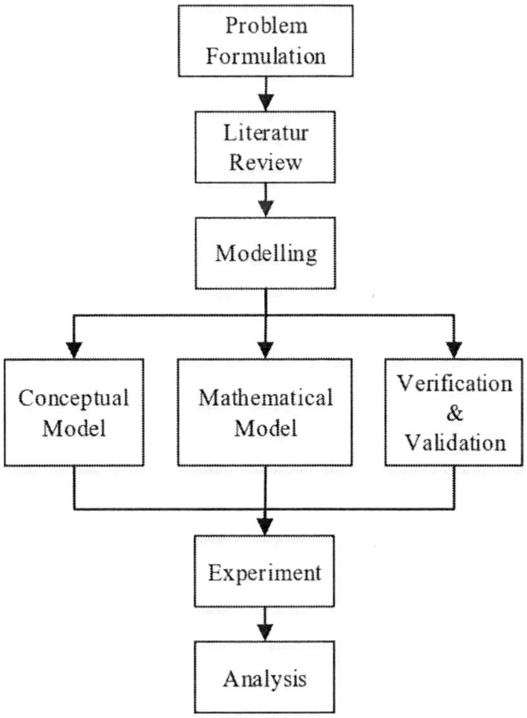

Figure 1. Research methodology

Formulation of the problem. How is the best collaboration of power suppliers in East Kalimantan to get the smallest total cost and emissions.
Literature study. Literature uses a journal published by Elsevier. Topics about economic dispatch, optimization, collaboration.
Modelling. There are 3 stages of modeling. First, the conceptual model to facilitate in designing the details of mathematical models. Second, a mathematical model to illustrate supplier collaboration. There are 4 mathematical models in this research: SEC + RE, SEC + IPP + RE, SEC + Rent + RE, and SEC + EC + RE.

Model 1: SEC+RE

This model is a collaboration of SEC and RE power plants. The mathematical model is as follows:

$$Minimize \sum_{i=1}^{N} F_i\,(P_i) = a_i\,P_i^2 + b_i P_i + c_i + C_{RE} \quad (1)$$

Constraints

$$\sum_{i=1}^{N} P_i + P_{RE} = PD \quad (2)$$

$$Pmin_i \leq P_i \leq Pmax_i \quad (3)$$

$$P_{REmin} \leq P_{RE} \leq P_{REmax} \quad (4)$$

Model 2: SEC+IPP+RE

This model is a collaboration of the SEC, IPP and RE power plants. The mathematical model is as follows:

$$Minimize\ \sum_{i=1}^{N} F_i\ (P_i) + \sum_{i=1}^{N} F_{iIP}\ (P_{iIP}) = a_i\ P_i^2 + b_i P_i + c_i + a_{iIP}\ P_{iIP}^2 + b_{iIP} P_{iIP} + c_{iIP} + C_{RE} \quad (5)$$

Constraints

$$\sum_{i=1}^{N} P_i + \sum_{i=1}^{N} p_{iIP} + P_{RE} = PD \quad (6)$$

$$P_{imin} \leq P_i \leq P_{imax} \quad (7)$$

$$P_{iIPmin} \leq P_{iIP} \leq P_{iIPmax} \quad (8)$$

$$P_{REmin} \leq P_{RE} \leq P_{REmax}$$

Model 3: SEC+Rent+RE

This model is a collaboration of the SEC, Rent and RE power plants. The mathematical model is as follows:

$$Minimize \sum_{i=1}^{N} F_i\ (P_i) + \sum_{i=1}^{N} F_{iR}\ (P_{iR}) + P_{RE} = a_i\ P_i^2 + b_i P_i + c_i + a_{iR}\ P_{iR}^2 + b_{iR} P_{iR} + c_{iR} + C_{RE} \quad (9)$$

Constraints

$$\sum_{i=1}^{N} P_i + \sum_{i=1}^{N} p_{iR} + P_{RE} = PD \quad (10)$$

$$P_{imin} \leq P_i \leq P_{imax} \quad (11)$$

$$P_{iRmin} \leq P_{iR} \leq P_{iRmax} \quad (12)$$

$$P_{REmin} \leq P_{RE} \leq P_{REmax}$$

Model 4: SEC+EC+RE

This model is a collaboration of SEC, EC and RE power plants. The mathematical model is as follows:

$$Minimize \sum_{i=1}^{N} F_i\ (P_i) + \sum_{i=1}^{N} F_{iE}\ (P_{iE}) + P_{RE} = a_i\ P_i^2 + b_i P_i + c_i + a_{iE}\ P_{iE}^2 + b_{iE} P_{iE} + c_{iE} + C_{RE} \quad (13)$$

Constraints

$$\sum_{i=1}^{N} P_i + \sum_{i=1}^{N} p_{iE} + P_{RE} = PD \quad (14)$$

$$P_{imin} \leq P_i \leq P_{imax} \quad (15)$$

$$P_{iEmin} \leq P_{iE} \leq P_{iEmax} \quad (16)$$

$$P_{REmin} \leq P_{RE} \leq P_{REmax} \quad (17)$$

Description of the formula:

$F_i\ (P_i)$: SEC fuel costs
$F_{iIP}\ (P_{iIP})$: IPP fuel costs
$F_{iR}\ (P_{iR})$: Rent fuel costs
$F_{iE}\ (P_{iE})$: Excess Capacity fuel costs
P_{RE} : Total production of renewable energy power plants
PD : Demand
P_i : total the i-power plant production
P_{imin} : The minimum amount that the i-power plant must produce
P_{imax} : Maximum capacity that can be produced by i-power plant
P_{RE} : Total production of renewable energy
P_{REmin} : The minimum amount that must be produced by renewable energy
P_{REmax} : Maximum production capacity of power plant renewable energy
P_{iIP} : Total IPP production
P_{iIPmin} : The minimum amount that must be produced by the IPP power plant
P_{iIPmax} : Maximum production capacity that IPP can produce
P_{iR} : Total Rent production
P_{iRmin} : The minimum amount that must be produced by the Rent power plant
P_{iRmax} : Maximum production capacity that Rent can produce

P_{iE} : Total Excess Capacity production
P_{iEmin} : The minimum amount that must be produced by the EC power plant
P_{iEmax} : Maximum production capacity that EC can produce

Analysis
The analysis was based on experiments on four scenarios (mathematical models). The result is a comparison of different types of collaboration between suppliers. The parameters used are cost and emission.

3. Result and Discussion

3.1. Results
This study yielded 2 main comparisons, ie supplier collaboration without integration with renewable energy, and collaboration between suppliers with renewable energy integration.

3.1.1. Collaboration of suppliers without renewable energy. When collaboration is done without involving renewable energy, the result is as follows:

Table 1. Collaboration of suppliers without RE

No	Scenario	Peak Load		Low Load	
		Cost (Rp)	Emission (kg)	Cost (Rp)	Emission (kg)
1	SEC	1,159,917.5	173.7	671,491.6	105.3
2	SEC+IPP	273,136.5	161.8	153,746.1	99.8
3	SEC +Rent	634,620.1	155.9	227,274.5	86.5
4	SEC +EC	753,529.9	166.7	332,655.4	98.5

Experiments were conducted on 4 scenarios. Each scenario consists of 2 conditions: Peak Load and Low Load. Scenario 1: Only use PLN power plant. Scenario 2: PLN + IPP, Scenario 3: PLN + Rent, and scenario 4: PLN + Excess capacity.

Scenario 1: In peak load condition, fuel cost Rp. 1,159,917.5 with emissions of 173.7 kg, whereas in low load conditions required fuel costs Rp. 671,491.6 with emission of 105.3 kg. Scenario 2: In peak load condition, fuel cost Rp. 273,136.5 with emission of 161.8 kg, whereas in low load condition required fuel cost Rp. 153,746.1 with emission of 99.8 kg. Scenario 3: In peak load condition, fuel cost Rp. 634,620.1 with emission of 155.9 kg, whereas in low load condition it takes fuel cost Rp. 227,274.5 with emission 86.5 kg. Scenario 4: In peak load condition, fuel cost Rp. 753.529.9 with emission 166.7 kg, whereas in low load condition required fuel cost Rp. 332,655.4 with an emission of 98.5 kg.

3.1.2. Supplier collaboration with renewable energy. When suppliers collaborate with renewable energy, the results are as follows:

Table 2. Supplier collaboration involves renewable energy

No	Scenario	Peak Load		Low Load	
		Cost (Rp)	Emissions (kg)	Cost (Rp)	Emissions (kg)
1a	SEC+RE	1,111,121.8	171.0	176,827.0	90.9
2a	SEC +IPP+RE	266,252.0	160.1	113,411.4	93.6
3a	SEC +Rent+RE	606,778.9	148.1	161,356.3	79.0
4a	SEC +EC +RE	718,084.0	163.7	117,237.6	89.8

Scenario 1a: In peak load condition, fuel cost Rp. 1,111,121.8 with emission 171 kg, whereas in low load condition needed fuel cost Rp. 266,252.0 with emission of 105.3 kg. Scenario 2a: In peak load condition, fuel cost Rp. 273,136.5 with emission 160.1 kg, whereas in low load condition required fuel cost Rp. 113.411.4 with emission of 93.6 kg. Scenario 3a: In peak load condition, fuel cost Rp. 606,778.9 with emissions of 148.1 kg, whereas in low load conditions required fuel costs Rp. 161,356.3 with emission of 79.0 kg. Scenario 4a: In peak load condition, fuel cost Rp. 718,084.0 with the emission of 163.7 kg, whereas in low load condition required fuel cost Rp. 117.237.6 with emissions of 89.8 kg. Graphically can be seen in Figure 2.

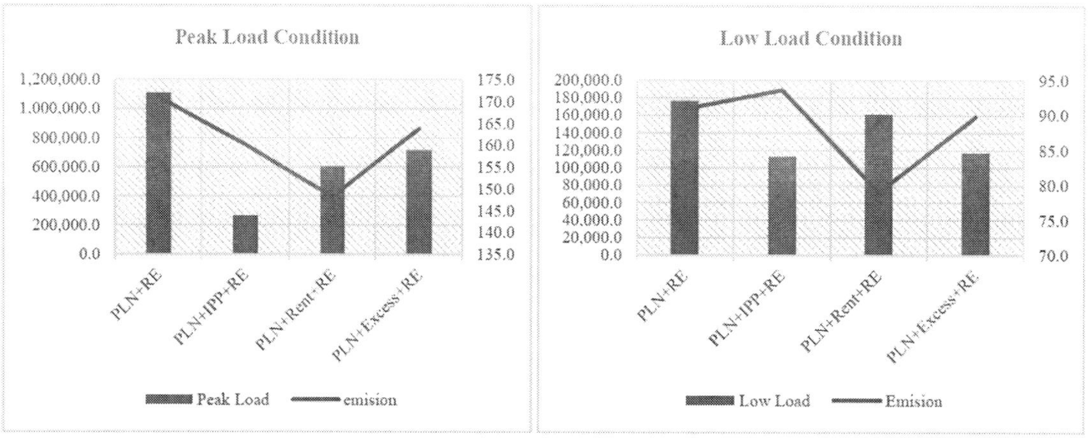

Figure 2. Cost and emission of 4 types of scenarios under peak and low load conditions.

3.2. Discussion

The entry of Renewable Energy into supplier collaboration has reduced total costs and emissions. Costs and emissions reductions can be seen in Table 3.

Table 3. Reduced costs and emissions due to Renewable Energy

Scenario	Peak Load		Low Load	
	Cost (Rp)	Emissions (kg)	Cost (Rp)	Emissions (kg)
SEC+RE	48,795.7	2.7	494,664.6	14.4
SEC +IPP+RE	6,884.5	1.7	40,334.7	6.2
SEC +Rent+RE	27,841.1	7.8	65,918.3	7.4
SEC +EC+RE	35,445.9	2.9	215,417.8	8.8

Collaboration SEC + IPP + RE resulted in decreased cost Rp. 6,884.5 (2.52%) during peak load, whereas when low load resulted decreased cost Rp. 40,334.7 (26.23%). Similarly, emissions, when peak load dropped 1.7 kg (1.02%), while when low load down 6.2 kg (6.2%) as in Table 4.

Table 4. Percentage reduction in costs and emissions due to Renewable Energy

Scenario	Peak Load		Low Load	
	Cost (Rp)	Emissions (%)	Cost (Rp)	Emissions (kg)
SEC+RE	4.21	1.54	73.67	13.71
SEC +IPP+RE	2.52	1.02	26.23	6.20
SEC +Rent+RE	4.39	5.01	29.00	8.61
SEC +EC+RE	4.70	1.77	64.76	8.88

The SEC + Rent + RE collaboration resulted in a decrease in the cost of Rp. 27,841.1 (4.9%) during peak load, while when low load resulted decreased cost Rp. 65,918.3 (29%). Similarly, emissions, when peak load down 7.8 kg (4.39%), while when low load down 7.4 kg (8.61%).

4. Conclusion

Collaboration among suppliers in East Kalimantan is absolutely necessary. The best collaboration is SEC + IPP + RE. This collaboration yields the smallest total fuel cost to meet all demands. In addition to providing the smallest total cost, this collaboration also generates greater energy reserves than any other collaboration, making it safer from the point of view of continuity of power supply in the event of damage to one of the power plants, whereas collaboration produces the smallest emissions when combining SEC, Rent and RE power plants. However, if there is damage to the power plant with the largest capacity, all existing scenarios were not able to anticipate it.

5. References

[1] Wahyuda and Santosa B 2015 Dynamic Pricing in Electricity: Research Potential in Indonesia. *Procedia Manuf.* **4** pp 300–306
[2] Wahyuda, Santosa B and Rusdiansyah A 2018 Cost analysis of an electricity supply chain using modification of price based dynamic economic dispatch in wheeling transaction scheme. *IOP Conf. Ser. Mater. Sci. Eng.*
[3] Wahyuda, Santosa B and Rusdiansyah A 2017 Load allocation of power plant using multi echelon economic dispatch. *AIP Conf. Proc.* 020007
[4] Mahdi F P, Vasant P, Kallimani V, Watada J, Fai P Y S and Abdullah-Al-Wadud M 2018 A holistic review on optimization strategies for combined economic emission dispatch problem. *Renew. Sustain. Energy Rev.* **81** (March) pp 3006–3020
[5] Prasad R and Nath K 2016 DPD: An intelligent parallel hybrid algorithm for economic load dispatch problems with various practical constraints. *Expert Syst. Appl.* **63** pp 295–309
[6] Secui D C 2015 A new modified artificial bee colony algorithm for the economic dispatch problem. *Energy Convers. Manag.* **89** pp 43–62
[7] Zhou J, Wang C, Li Y, Wang P, Li C, Lu P and Mo L 2017 A multi-objective multi-population ant colony optimization for economic emission dispatch considering power system security. **45** pp 684–704
[8] Ghasemi M, Aghaei J, Akbari E, Ghavidel S and Li L 2016 A differential evolution particle swarm optimizer for various types of multi-area economic dispatch problems. **107** pp 182–195
[9] Abdelaziz A Y, Ali E S and Elazim S M A 2016 Combined economic and emission dispatch solution using Flower Pollination Algorithm. *Int. J. Electr. Power Energy Syst.* **80** pp 264–274
[10] Zou D, Li S, Li Z and Kong X 2017 A new global particle swarm optimization for the economic emission dispatch with or without transmission losses. *Energy Convers. Manag.* **139** pp 45–70

Integrated Logistics and Transportation Routing in Rural Logistics System

T S Sinaga[1,2] **and S N Bahagia**[1]

[1]Faculty of Industrial Technology, Bandung Institute of Technology, Bandung, Indonesia
[2]Department of Industrial Engineering, Faculty of Engineering, University of Sumatera Utara, Medan, Indonesia
tutie_rani@yahoo.co.id, senatornurb@yahoo.co.id

Abstract. The agricultural distribution problem of Rural Logistics System is addressed in this paper to improve farmers' income by minimizing annual total cost. The system consists of three entities: (i) logistics terminal, (ii) distribution warehouses, and (iii) industries. The logistics terminal collects agricultural products from farmers' crops, which are stored in distribution warehouses, which are further distribution based on demand of industries. This study developed non-linear mix integer programming model to minimize the annual total costs by incorporating (i) the approach of coordinated planning system, (ii) concept of echelon stock, (iii) single time cycle policy and (iv) transportation route. Many constraints were developed using various criteria. Feasible solution to the mathematical model was obtained by using heuristics method by Clarke and Wright.
Keywords: integration, echelon concept, annual cost, nonlinear mix integer programming

1. Introduction

Contribution of the agricultural sector to the Indonesian economy continues to increase, both on gross domestic product and employment creation for people. However, improvement of this sector is not always accompanied by the increasing of farmers' income. The tendency is brokers collect the products directly from the gate of farmers to make low selling price of agricultural product in rural area. Therefore, Government focuses on the improvement strategy of logistics system in rural area, such as by reducing cost to bring benefits to farmers' income.

Rural Logistics System can solve the problems of distribution of agricultural products, the availability of products and improve the living standards of farmers [1]. The Rural Logistics System aims to increase farmers' income, reduce the risk of agricultural products, and encourage integrated rural-urban development [2], Rural Logistics Systems can also create new markets as the urban market becomes saturated [3].

Rural Logistics System is the same as the logistics system in general, handles all logistical activities that are related to the delivery of products from or to rural area to increase farmers' income. The Rural Logistics System is almost identical to the urban logistics system, it simply refers to the production of rural communities and other economic activities such as the provision of transportation, packaging, processing, storage and other related activities [4]. Rural Logistics System is supply chain networks involving upstream and downstream farmers, outlets, logistics terminal, wholesalers, industries or

IOP Conf. Series: Materials Science and Engineering **528** (2019) 012081 doi:10.1088/1757-899X/528/1/012081

retailers, and final consumers who need agricultural products or services. There are two types of product flow included in the rural logistics discussion which are agricultural logistics and rural business logistics [5].

Issue of rural logistics in Indonesia is how to bridge the interests of farmers in rural areas to get a reasonable and stable selling price of agricultural products while still protecting the interests of consumers in urban areas in terms of availability and quality. Therefore, implementation of rural logistics systems is very important to supports the government's efforts to realize an effective integrated logistics system, so it can enhance the welfare of the community. The Government has built Sub Logistics Terminal (STA) as one of many entities in Rural Logistics System, however in this research we focused only to 3 echelons of supply chain. Distribution network from logistics terminal to industries of a rural logistics system is shown in Figure 1 below:

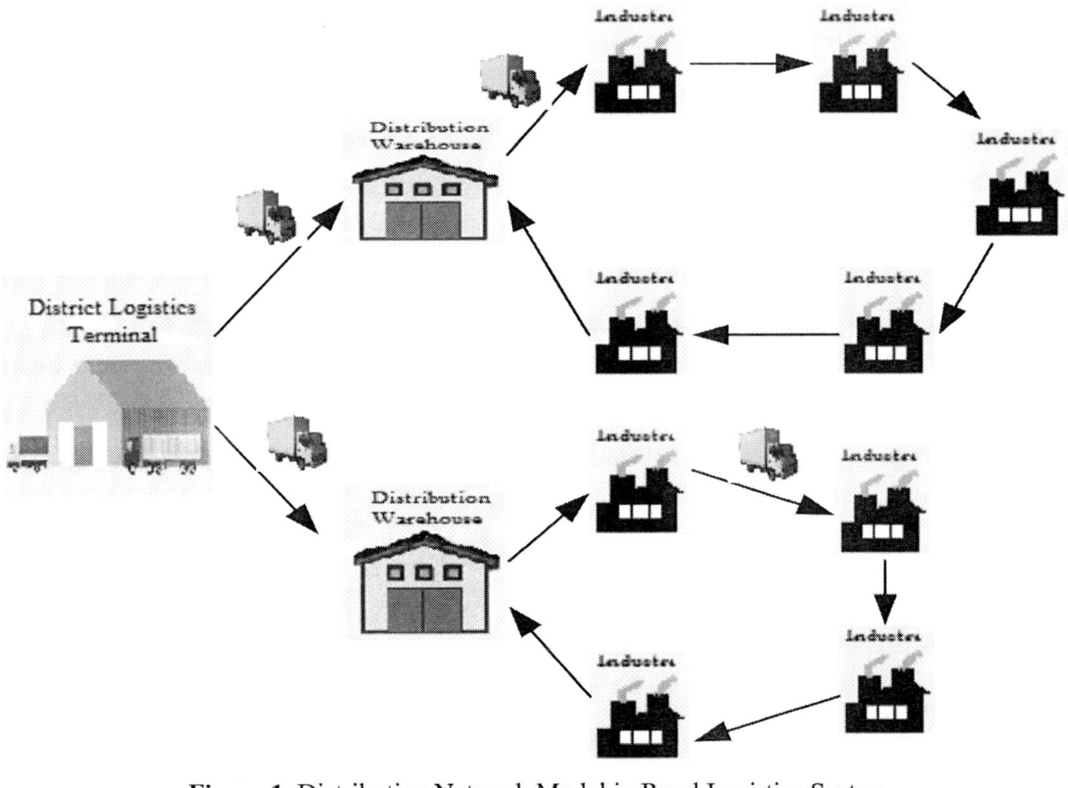

Figure.1. Distribution Network Model in Rural Logistics System

The figure shows that rural logistics system has a district logistics terminal and multi distribution warehouses to meet the demands from many industries that scattered away from rural locations. Agricultural products from smallholder farmers collected to logistics terminal for sorting and repackaging products, accordance with its standard. Furthermore, agricultural products were sent to Distribution Warehouse before delivered to industries that need them.

Relationship among entities in supply chain needs to be considered, because it is realized that industry is one of the targets to reduce cost of the system [6]. This discussion would be very helpful in determining the business strategy of the Rural Logistics System in order to provide better service to customer, by implementing competitive prices and increasing farmers' income. This model answers the most important question in logistics management: how many products that must be ordered from industries to distribution warehouse and from distribution warehouse to Logistics terminal? Since rural logistics system is the most famous and applicable in many countries, numerous conceptual models of rural logistics system have been developed by researchers and academicians. However, a model to answer this problem has never been developed before.

This paper is organized as follows. Section 2 describes methodology of research, including component of model are introduced. Section 3 presents mathematical model and an algorithm to search feasible solution. In Section 4, some numerical examples are presented to illustrate the effectiveness of the model. Finally, this paper gives the conclusion in Section 5.

2. Methodology

Mathematical model of rural logistics system was developed from an optimization model of integrated three echelon supply chain system [7] and accommodated optimal model of transportation with balanced workload [8]. Based on integrated concept [7], solving logistics problems requires an integrated systemic approach using the following steps i.e. : developed a coordinated planning, implementation of echelon stock concept and single time cycle policy. Liu [8] suggested to schedule a set of routes to deliver the commodities at minimum total distance travelled, considering vehicle capacity and balance of vehicle's workload, to minimize the number of vehicles used. These heuristic approach consists of three phases. First, tracing an initial feasible solution using a savings-based method. The second is improvement of route using five steps of one-point movement. Last phase is balancing workload and delivery time. As a constraint, in this paper route of vehicle did not consider balances of vehicle's workload.

2.1. Component of Model

In this section component of model was investigated. This paper focused to Annual Cost of Rural Logistics System, including inventory costs and transportation costs. Problem definition for this model is "how is the rural logistics system that can guarantee product's availability and increase farmers' income by minimizing annual cost?" Structural aspects from this model consists only three entities which are: a logistics terminal, distribution warehouses, and industries. Its functional aspect considered the flow of agricultural product, information, and money among the entities.

The performance criteria used in this model is minimization annual cost of the Rural Logistics System, including order cost, holding cost, shortages cost, opportunity cost, and the transportation cost at the logistics terminal and distribution warehouse. Expected performance criteria and the decision variables set in the developed model are:

1. At Logistics Terminal: annual cost, order lot sizing and order period
2. At Distribution Warehouse: quantity orde and , order frequency, frequency of delivery, and unit of product that can be served by a certain vehicle.

The assumptions underlying the models in this paper are 1) Demand has a normal distribution 2) Demand that cannot fulfilled will be lost (lost sales) and 3) If products are inconsistent with consumer needs, it will be returned and this may result in risk of damage. There are various constraints with rural network problem, such as limited vehicle, single item, fixed cost of vehicles, etc. In this paper, vehicle used both revised driving route from logistics terminal to distribution warehouse and actual driving route at distribution warehouse to customers.

2.2. Model Formulation

Mathematical model for integration system in rural logistics is described as follows

a. Index Parameters Set
   - I : number of industries
   - J : *number* of distribution warehouses (DW)
   - K : number of vehicles used
   - U : number of nodes in transport route $U = I \cup J$

b. Indices
   - d : length of distances between node in distribution warehouse area

g : number of products to serve by vehicle
h : number of demands from industry
i : index for industry
j : index for distribution warehouse
k : index of vehicle
m : origin number of routing node
n : destination number of routing node

c. Problem Parameters Set

C_o : *annual* cost (Rp/year)
C_d : *Annual* cost at distribution warehouse (Rp/years)
CP_j : *total* inventory cost per year at distribution warehouse (Rp/years)
CT_j : total *transportation* cost per period at distribution warehouse (Rp/years)
A_j : order cost at DW_j (Rp/Order)
D_j : demand at DW_j (unit/years)
D_{ij} : demand from $industry_i$ to DW_j (unit/years)
h_m : demand have to be delivered to industry from DW_j (unit/delivery)
μ_j : mean of demand DW_j from DW_j to logistics terminal, in certain leadtime
μ_{ij} : mean of demand to delivery from DW-j to industry-i. (unit/delivery)
σ : standard deviation of demand to delivery from DW-j to industry-i. (unit/delivery)
θ_m : mean of demand delivered from DW_j to node m (unit/delivery)
H_j : holding cost per unit per year at DW_j (Rp/unit/years)
z : z value on standard normal distribution for α level $P(x=\theta_m)$
L_j : lead time from DW_j to logistics terminal (years)
S_j : shortage cost per unit per period (Rp/unit/delivery)
P_j : opportunity cost per unit per period (Rp/unit/delivery)
l : leadtime of Distribution Warehouse
A_o : order cost at logistics terminal (order/year)
D_o : *demand* at logistics terminal (unit/year)
H_o : holding cost at logistics terminal (Rp/unit/tahun)
$f(g_{jk})$: demand function to be served by vehicle-k at DW_j
v_{jk} : transport volume of vehicle-k at DW_j (piece/vehicle)
os_k : set up cost vehicle-k used at DW_j (Rp/delivery)
ob_j : loading and unloading cost at DW_j for each delivery (Rp/unit/delivery)
ot_{mn} : transportation unit cost from node m to node n at DW_j (Rp/km/delivery)
op_j : opportunity cost per unit at DW_j (Rp/unit/delivery)
ok_j : reverse cost per unit at DW_j (Rp/unit/ delivery)
d_{mn} : length of distance from node m to node n at DW_j (km)
x_{mnk} : 1 if vehicle k travels directly from node n to m , 0 for otherwise.
y_{jk} : 1 if vehicle k travels directly from DW j to customer i, 0 for otherwise

d. Decision Variables

1. At Logistics Centre:
   Q_o : order lot sizing (unit)
   T_o : order period (years)
2. At Distribution Warehouse:
   Q_j : quantity order (units)
   N_j : order frequency to logistics terminal
   N_l : frequency of delivery from DW_j to industry i in leadtime (delivery/period)
   N_t : frequency of product delivery from DW_j to industry i per years (delivery/period)
   g_{jk} : unit of product to serve by vehicle-k at DW_j (unit/vehicle)

IOP Conf. Series: Materials Science and Engineering **528** (2019) 012081 doi:10.1088/1757-899X/528/1/012081

The objective function of the model is to minimize the annual total cost (C_{tot}) on a certain rural logistic system. Total annual cost can be calculated by summing annual costs at the logistics terminal (Co) and annual cost at DW (C_d).

$$C_{tot} = C_0 + C_d \tag{1}$$

Furthermore each component's cost is described and calculated in the following way:
The annual cost incurred at distribution warehouse was the sum of the annual inventory cost (CP_j) with the annual transportation cost on all DWj (CT_j).

$$C_d = \sum_{j\in J}(CP_j + CT_j) \tag{2}$$

$$CP_j = A_j x \frac{D_j}{Q_j} + H_j[\frac{Q_j}{2} + R_j - L_j.D_j] + \frac{D_j}{Q_j}.N_l.S_j.\sum_{k\in K_j}\int_{jk}^{\infty}(g_{jk} - v_{jk}).f(g_{jk})dg_{jk} + \frac{D_j}{Q_j}.N_l.P_j.\sum_{k\in K_j}\int_0^{V_{jk}}(v_{jk} - g_{jk}).f(g_{jk})dg_{jk} \tag{3}$$

$$CT_j = N_t.\sum_{k\in K_j}(os_k + ob_j.v_{jk} + \sum_{m\in U}\sum_{n\in U}Ot_{mn}.d_{mn}.x_{mnk} + op_j.\int_{jk}^{\infty}(g_{jk} - v_{jk}).f(g_{jk})dg_{jk} + ok_j\int_0^{V_{jk}}(v_{jk} - g_{jk}).f(g_{jk})dg_{jk} \tag{4}$$

The annual costs incurred at the logistics terminal were only order cost, inventory cost, and transportation cost, due to the level of demand. Demand calculation in logistics terminal was justified only based on Distribution Warehouse needs.

$$C_o = A_o.\frac{D_o}{Q_o} + H_o[\frac{Q_o}{2} + R_o - \sum_{j\in J}(L_o + L_j)D_j] + \sum_{j\in J}\frac{D_j}{Q_j}(os_o + ot_o.Q_j + d_{oj}.oc_{oj}) \tag{5}$$

Subject to:

1. $D_o = \sum_{j\in J} D_j$ $\quad \forall j \in J$
2. $\frac{Qo}{D_o} = N_j\frac{Q_j}{D_j}$ $\quad \forall j \in J$
3. $D_j = \sum_{i\in I} D_{ij}$ $\quad \forall j \in J$
4. $D_{ij} = N_t.\mu_{ij}$ $\quad \forall j \in J$
5. $\mu_{ij} + z\sigma_{ij} = \theta_m$ $\quad \forall m \in I_j$
6. $\sum_{k\in K_j}\sum_{m\in U} x_{mnk} = 1$ $\quad \forall j \in I_j$
7. $\sum_{m\in U} x_{mnk} - \sum_{m\in U} x_{nmk} = 0$ $\quad \forall n \in U;\ \forall k \in K_j$
8. $\sum_{m\in U} x_{mnk} = y_{jk}$ $\quad \forall n \in U;\ \forall k \in K_j$
9. $\sum_{m\in U} x_{nmk} = y_{jk}$ $\quad \forall n \in U;\ \forall k \in K_j$
10. $\sum_{m\in I_j}\theta_m\{\sum_{n\in U} x_{mnk}\} = v_{jk}$ $\quad \forall k \in K_j$
11. $\sum_{m\in I_j}\{\sum_{n\in U} x_{mnk}\} - u\{\sum_{m\in U} y_{jk}\} \le 0$ $\quad \forall k \in K_j$
12. $\sum_{m\in S}\sum_{n\in S}\sum_{k\in K} x_{mnk} \le |s| - 1$ $\quad 2 \le |s| \le I_j;\ \forall S \in I_j$
13. $x_{mnk}, x_{nmk} = 0\ or\ 1$ $\quad \forall m,n \in U;\ \forall k \in K_j$
14. $Q_o \ge 0$
15. $Q_j \ge 0$ $\quad \forall j \in J$
16. $N_j \ge 0$ $\quad \forall j \in J$, integer
17. $v_{jk} \ge 0$ $\quad \forall k \in K_j$
18. $y_{jk} = 1\ if\ \sum_{m\in U}\sum_{n\in U} x_{mnk} > 1\ or\ o\ for\ other$ $\quad \forall k \in K_j$
19. $\sum_{m\in I_j} h_m\{\sum_{n\in U} x_{mnk}\} = g_{jk}$ $\quad \forall k \in K_j$

Constraint (1) ensures that demand of warehouse is fulfilled, and (2) present single cycle time policy to the model. Constraint (3) states that demand of a distribution warehouse is a sum form industries' demand and (4) is be used to find demand of each industry. Constraint (5) show sum of demand delivered from DW_j to consumer *i*. Constraint (6) ensures that convenience industry is placed on exactly one route. Constraint (7) shows the flow constraints. Constraint (8) and (9) ensure vehicles are assigned to a route from distribution warehouse to the industry, so at least one vehicle link goes to the depot and one leaves the industry. Constraints (10) and (11) impose the maximal workload that can be by brought each vehicle. Constraint (12) ensures that the number of vehicle route must be smaller or equal to the number of nodes that enter into the solution. Constraints (14)-(17) are non negativity. Constraint (18) is to choice vehicle route. Constraints (19) is number of vehicle load.

2.3. Solution Procedure

The proposed annual cost problem or rural logistics system needs an effective solution method to obtain the optimal solution. The formulation of optimization model is nonlinear mix integer programming problems. Since integration inventory with vehicle routing are larger-scale application, it is very difficult to solve it with ordinary NLP method, hence developed heuristics procedure to solve the problem. In this research, Q_o* can be obtained if $\frac{\partial C_{tot}}{\partial Q_o} = 0$ for N_j. Nj* can determined using heuristic approach [7] by assuming that Nj is a continuous number. The determination of x_{mnk} value did not use the heuristic method proposed by Liu [8]. It will be developed by using heuristic method that was proposed by Clarke and Wright [9]. This approach can solve the mathematical model and has capability to produce feasible solutions.

3. Result and Discussion

In this section, a rural logistics system is proposed to test this model. The system has a logistics terminal and five distribution warehouses in each region. Each warehouse unit must fulfill demand of process industries. Agricultural products are supplied to the logistics terminal and forwarded to distribution warehouses before being distributed to industries in the region. The hypothetical data was used in this model test. Furthermore, the computation of heuristic procedure was constructed by Visual Basic to improve efficiency and quality of solution. The implementation results for the logistics system conditions considered are as follows:

1. The optimal solution of vehicles route for each distribution warehouse can be seen in Table 1 below:

Table 1. Feasible Solution for Route of Vehicle, Workload, and Distance of Route at Distribution Warehouse

Distribution Warehouse	Vehicle	Route	workload (unit)	Distance (km)
1	1	DW1-3-4-2-DW1	111,954	59
2	1	DW2-2-3-DW2	113,424	28
	2	DW2-1-5-4-DW2	120,488	88
3	1	DW3-1-3-4-DW3	106,27	65
	2	DW3-2-5-6-DW3	116,564	93
4	1	DW4-2-4-DW4	111,072	50
	2	DW4-1-3-DW4	81,564	49
5	1	DW5-1-3-DW5	96,66	36
	2	DW5-4-5-DW5	117,148	53
	3	DW5-2-6-DW5	81,762	67

2. Detail feasible solutions for quantity order, order frequency and annual cost on distribution warehouses are shown Table 2.

Table 2. feasible solution for quantity order, order frequency and annual cost

Level	N_j	Q_j* (unit)	T_j* (year)	TC_j	
DW1	2	1.655	0,1769	Rp	2.371.842,52
DW2	2	2.845	0,1852	Rp	2.865.931,41
DW3	2	1.627	0,1695	Rp	2.687.246,74
DW4	2	1.639	0,1625	Rp	2.468.167,40
DW5	2	2.299	0,1916	Rp	1.939.932,99
Total (C_d)				**Rp**	**12.333.121,05**

The total annual cost for a rural logistics system is Rp 61.134.868. 83, with order quantity at logistics terminal is 21.178 units, length of cycle 0.376 years, and annual cost Rp. 48.801.747,78.

In this paper, we analysed a mathematical model of the annual cost problem. The constraints introduced were capacity of vehicle and risk cost when delivered products did not fit customer requirements, for instance shortage cost and reverse cost. To determine Nj*, we used Y1 and Y2 as influence factors and they manifest the difference from previous model [7]. The inventory rule in this condition is applicable when producers coime to consumers with unclear information about their needs so enhance the risk of excess supply.

The inventory rule in this condition is when producers come to consumers with unclear information about the needs of consumers and producers so enhance the risk of excess supply.

4. Conclusion

A nonlinear mix integer programming model was proposed to solve annual cost problem in Rural Logistics System. The model incorporated coordinated planning, stock echelon concept, single cycle time policy, and transportation route. Heuristic procedures were developed to solve the formulated mathematical model. The solutions of numeric example were obtained by computation effort according to the proposed procedure. The research contribution is on the more effective and reliable algorithm, regarding the solution quality and run time reduction. Potential future research can be done by considering farmers as an entity of Rural Logistics System with a stochastic demand.

Acknowledgment

Authors would like to thank the Ministry of Finance, Indonesia for the BUDI-DN/LPDP Scholarship.

5. References

[1] Yun Z and Lingyun S 2011 Rural logistics demand forecasting based on the BP neural network method. *Proc. of Int. Conference on Management and Service Science* pp 1-5

[2] Xu C H and Huang J 2014 Research on manufacturing engineering in agriculture with construction of China's rural logistics system. *Applied Mechanics and Materials* **540** pp 532-536

[3] Viswanadham N D and Ramakrishna D 2007 *Achieving Rural & Global Supply Chain Excellence, The Indian Way, Center for Global Logistics and Manufacturing Strategies.* (Gachibowli: India School of Business) pp 15 – 34

[4] Zhou Y L, Shao W Q, Bai D M and Su L U 2016 Study on The Modes for The Development of Rural Logistics in The Process of New Urbanization. *Proc. of Conference Logistics Sci-Tech.* **4** (May) pp 23–32

[5] Zhang, Xi-Zhou, Chen L and Liu J 2010 The Mechanism of Rural Business Logistics Operation in Mainland China A Research Based on Home Appliances Going to Countryside Management and Service Science (MASS). *Proc. of Int. conference on Management and Service Science China* pp 1-5

[6] Stank, Theodore P, Traichai and Patrick A 1998 Logistics Strategy, Organizational Design, And Performance in A Cross-Border Environment. *Transportation Research Part E (Logistic and Transp. Rev)* **34**(1) pp 75-86
[7] Nur Bahagia S 1999 Model optimasi integral sistem rantai nilai tiga eselon. *Proceedings of Seminar Sistem Produksi* (Bandung: Departement of Industrial Engineering-ITB) **4** pp 1-9
[8] Liu C M, Chang T C and Huang L F 2006 Multi-Objective Heuristics for the Vehicle Routing Problem. *International Journal of Operations Research* **3**(3) pp 173-181
[9] Clarke G and Wright J 1964 Scheduling of vehicles from a central depot to a number of delivery points. *Operations Research* **12**(4) pp 568-581

Development of vendor management and e-Procurement systems using android platform

B Angrian[1] and T R Sahroni[1]

[1]Industrial Engineering Department, BINUS Graduate Program – Master of Industrial Engineering, Bina Nusantara University 11480, Jakarta, Indonesia
taufik@binus.edu

Abstract. Supply chain management has become a crucial aspect of the performance of industries and companies as it manages the flow of goods, services, and information in order to fulfil customers need while reducing the total costs. In order to reduce the production total cost, e-procurement is introduced. E-procurement means of reducing cost is by shortening cycle time and create a simpler procurement process. The aim of this research is to develop a vendor management and e-procurement system using the Android platform. The developed system is hoped to be able to even shorten the cycle time, ease vendor selection and provide a more flexible and convenience procurement process. Through data measurement and documentation review, the developed system was tested against these three parameters for their validity. In the end, this research summarizes the test result and conclude based on the test result that those parameters are achieved in the developed vendor management and e-procurement system.
Keywords: Technical, Economic, Feasibility, Solar, Cathodic

1. Introduction

Supply chain management has been the most interesting field for engineers and researcher in developing the interaction of vendor, industry, and customer. Supply chain management has been introduced in order to minimize the length of delivery time to customer and to reduce the cycle time in producing the services or products. The main purpose of the supply chain management is to reduce the total cost in supply chain activities. One of the supply chain activities is procurement. Procurement is the act of obtaining or buying goods and services process that include preparation and processing of a demand as well as the end receipt and approval of payment [1]. Procurement is vital to company especially in manufacturing company as procurement is the key for determining the overall cost and quality of the produced product. Due to increasing of competitiveness nowadays, company is demanded to be able to make a fast and accurate decision in choosing qualified and appropriate vendor in procurement process. Other than that, flexibility and time needed in responding the tender from the vendor is also a priority.

Nowadays, the application of information technology has been widely used in manufacturing industries. With the vast development of information technologies, traditional procurement process is slowly being replaced with the newer electronic procurement (E-procurement) process. E-procurement is part of the business-to-business (B2B) commerce being conducted on the internet, in which buyer can make a direct purchase from the vendor through their website by using software packages or through e-marketplace, e-hub and trading exchange. By using e-procurement, companies can lower transaction costs, overall goods costs, labor costs, faster ordering and delivery times [2]. However, the e-procurement process is still considered having some flaws such as inflexible and slow response by some vendors and having a hard time selecting a qualified and appropriate vendor in tender bidding process caused by the lack of vendor information. To counter the existing flaws, mobile based e-procurement system is suggested. Through the newly designed e-procurement system that consists of developing framework for e-procurement system which is directly integrated with the vendor management system with vendor scoring system, hoping to reduce the flaws of the previous procurement system.

The main essential of supply chain management is the optimization of material flows and associated information flows involved with an organization's operations. The three types of flow that need to be managed are product and service flow, information flow, and financial flow [3]. One of the supply chain activities that will affect the total cost and overall companies' performance is procurement. Procurement is the process of obtaining supply or process of purchasing goods and services from vendor for the production needs [2]. With the development of information technology, e-procurement is introduced and is slowly replacing the traditional procurement process. E-procurement is a collective term for a range of different technologies that can be used to automate the internal and external processes associated with the sourcing and ordering process of goods and services [4]. E-procurement is often considered similar to E-commerce, but still they have big difference such as e-Commerce often faces a large number of individual consumers, while e-procurement usually involves dealings with companies [5]. Study confirmed that through using automated procedure for purchasing through e-procurement enables a company to save up to 8%-12% of total purchases [6].

The process of procurement is a critical process for almost all company especially manufacturing company as procurement determines the quality, cost, delivery of raw materials and most importantly the ongoing production operation of manufacturing company. To ensure everything goes smoothly depend on a right and appropriate vendor selection process. The selection and evaluation of vendors have been a widely discussed topic in the literature. Based on previous study, vendor is ought to satisfy a widely known standard of the 7 Rights Rule, which is: right product, condition, quantity, time, customer, place and costs [7]. The most popular criterion considered by the decision makers for evaluating and selecting the most appropriate vendor is quality, followed by delivery, price/cost, manufacturing capability, service, management, technology, research and development, finance, flexibility, reputation, relationship, risk, and safety and environment [8]. There are four most influential criteria that should be noted and evaluated in order to acquire the best and appropriate vendor to take on the procurement, which are: items quality, items delivery, items cost and item productivity [9].

A detail review and comparative analysis of different Smartphone's operating systems has been made. The compared operating system consist of android, iOS, Symbian, Blackberry, Windows Phone, WebOS, Ubuntu and Firefox. Out of all the compared operating system, android operating system gets 80.7% and is the best Smartphone OS in the world today [10]. Android is an open-source mobile operating system developed by google that runs on the Linux Kernel. Android Mobile Application Development is based on Java language codes, as it allows developers to code in the Java language. With the vigorous development through Android, mobile applications have been widely used on the various mobile devices. Android mobile applications are evolving at a meteor pace to give a rich and fast user experience [11].

Currently, most of industries have been focused on how to manage the vendor database in order to select the best vendor with good qualifications to procure their services or products. However, there is lack of information due to the number of vendor that has been increasing significantly until now. In general, the problem statements of this paper to support the procurement response from the vendor during procurement process, lack of vendor information for selection in procurement process, and inflexibility in procurement process. From these reasons, the aim of this paper is to develop a vendor management and e-procurement systems using android platform in order to carry out the problem in current procurement process.

2. Materials and Methods

In this study, there are two main parts of system development. Figure 1 shows the vendor management and e-procurement frameworks that to be developed using android platform.

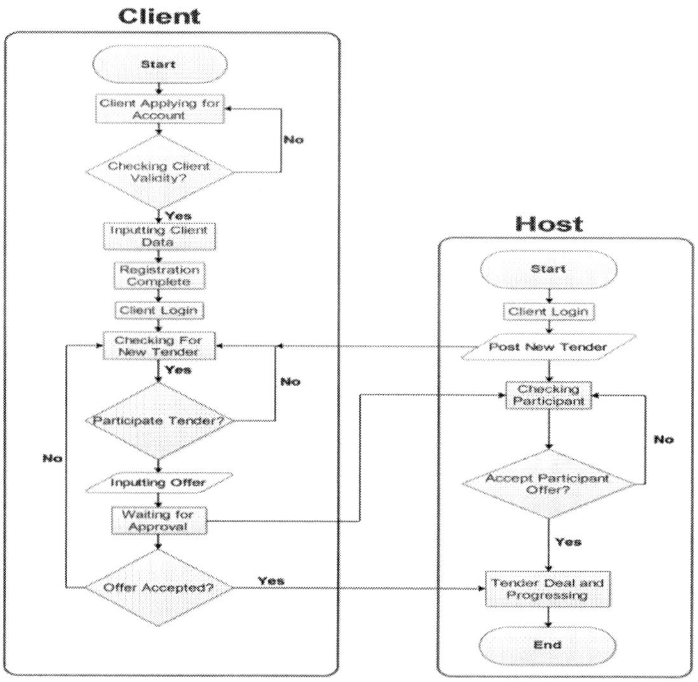

Figure 1. Vendor management and e-procurement framework

An account is required before client can access the server and take part in bidding process. The account consists of a username and password. Clients are unable to register on their own for a new account. So if clients want to take part in the tender bidding process, they will need to apply an account from the host. The host will then request all the information and documents needed to check for the validity of the client. Host will then input all the client information into the database and create an account for the client. After installing the software, the client can start checking for their performance, new tender and also participate in a tender bidding process. When a new tender is available, it will be displayed in the software. By accessing the new tender, client will be able to see the detail of the new tender such as due date and contents. A price offering option is also available in the tender screen. By inputting the offered price, the client will be able to take part in the tender bidding process. After taking part in the bidding process, the client needs to wait for the host to accept the offer. If the offer is accepted, it will be shown in the active tender list.

Host also need an account to access the server. Same as the client the account consists of a username and password. After login in, the host can check for every registered client's information, active tender, new tender that needs approval and also posting new tender. Host need to fill in the due date and the content of the tender before posting the tender. After the posting completed, the host will need to wait for the client to participate in the new tender. If new participant is available, it will be shown in the tender approval and host will be able to pick the best participant based on their score and offered price.

In order to overcome the problem of slow procurement response from vendor in procurement process, a parameter for measurement is needed. Procurement process time cycle is then assigned to be measured in this problem. In this parameter, an approximately time needed for vendor to be notice and respond to new tender is measured. The main purpose is for showing the needed time for vendor to respond to new tender. For the problem of lack of vendor information for selection in procurement process, a parameter for measurement is needed. A vendor scoring mechanism is then design to overcome this problem. By applying the vendor scoring mechanism in the system is hoped to overcome the lack of vendor information problem. The applied vendor scoring system will be discussed here. The discussion consists of how the vendor scoring system work and help the vendor selection process. The purpose of this discussion is to show how the vendor scoring system eases the vendor selection process.

In order to overcome the problem of inflexibility in procurement process, a parameter for measurement is needed. Inflexibility and convenience of the system is assigned as the parameters for this problem. These parameters are measured by conducting journal research and articles. The main purpose is for showing the use of mobile phone can increase the efficiency in procurement process. Flexibility in this parameter is addressed by easy and fast respond from the vendor in the bidding process by using the mobile phone.

3. Results and Discussions

The host software consists of three main screens which are all vendor lists screen, ongoing tender screen, and tender approval screen. The user can select each main screen by sliding to the left or right. User can also directly entering the new tender posting screen from every main screen by clicking on the new tender button that is available in every main screen. Figure 2 shows the host software main screen.

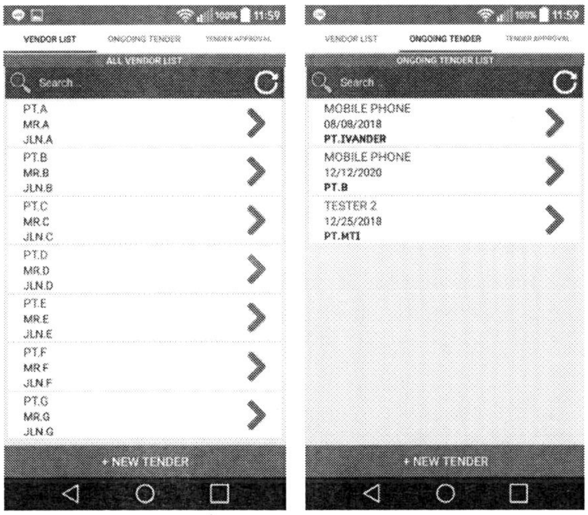

Figure 2. Host software main screen

In the first main screen, the all vendor list screen, host can see a list of all registered vendor in the screen and by selecting each one of the vendor, host can see its information such as vendor company name, owner, score of overall performance, and history. Apart from vendor information, there are two buttons at the bottom of the screen which are SMS and call button that can directly contact the vendor's person in charge. In the second main screen, the ongoing tender list screen, host can check on the entire list of active tender. All the tender that have been approved by the host will be mark as active and be shown here. By selecting on each tender, Host can see its detail such as tender title, tender due date, tender detail, vendor that is working on this tender and offered price. A complete tender button is located at the bottom of the screen for ending the tender and removes the tender from the active tender list.

In the third main screen, the tender approval screen, host can check on the entire list of posted and unapproved tender. When the host selects one of the posted tender, it can check on the tender title, tender due date, tender detail and a list of vendor that participate in the tender bidding. In the participant list, the host can approve one of the vendors to handle the tender based on the score and offered price. A tender that is approved will be removed from the approval list and moved to the active tender list. Host will be directed to new tender post screen if the new tender button is pressed. In the new tender post screen, host will see a few brackets that need to be filled in order to post a new tender. The brackets are tender title, tender due date, and tender detail. After host has filled all of them, host can submit and post the tender using a post button that is located at the bottom of the screen. When the post buttons is pushed and has successfully posted the tender, the new posted tender will be moved to the approval tender list to wait for new participant. The host will then be automatically directed back to the main screen. The software consists of three main screen which are welcome screen, ongoing tender screen, and tender participation screen. Clients can select each main screen by sliding to the left or right. Figure 3 shows the client software main screen.

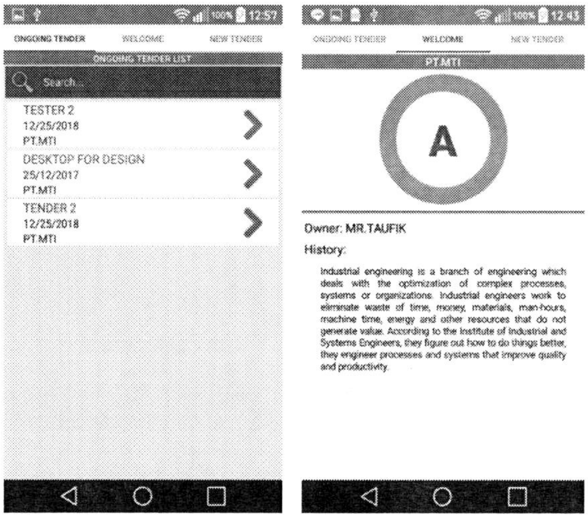

Figure 3. Client software main screen

In the first main screen, the ongoing tender list, clients can check for the list of currently on going tender with the host. All the tender that the client participate in and get approve by the host will be shown here. By selecting each of the tender, clients can check for its information such as tender title, tender due date, tender details and price offered for the tender. In the second main screen, the welcome screen, clients will be able to see their company's owner name, history and overall performances scores. In the third main screen, the tender participation screen, clients can check for the list of new tender that is posted by the host and has not been assigned to a vendor yet. When accessing the tender, the client will see the tender title, tender due date, tender detail. Apart from that, a price offering option is available which consist of a brackets for price offers and a participate button located at the lower part of the screen. By filling in the bracket with best offer and hit the participate button, vendor will automatically participating in the tender bidding with the offered price. After successfully participating in the bidding, clients need to wait the approval from the host.

In order to measure the time needed for vendor to notice and respond to the new tender, a simple test is done. In an ideal condition where the Smartphone is connected to the internet and having a good signal reception, a SMS and E-mail blast is sent from the host to every registered vendor in the system to notify them to check the new posted tender. Upon receiving the SMS or E-mail, vendors can immediately launch the vendor management software from their phone to check the detail of the new tender. If the vendor knows the exact market price, they can immediately participate in the tender bidding by offering their best price for the tender. Excluding any interruption that prevent vendor to access the Smartphone such as Smartphone's battery is depleted, left in the office or home, and broken, the approximate time needed for vendor to aware and check the new tender is less than three minutes. Another five minutes is needed for vendor to confirm the tender market price and participate in the tender bidding. The total approximately time is less than ten minutes. Considering the needed time, using e-tender and e-bidding on Smartphone is proven able to shorten the overall procurement cycle time.

A vendor scoring system is implemented in the vendor management software to assist in vendor selection. The most popular criterion considered by the decision makers for evaluating and selecting the most appropriate vendor is quality, followed by delivery, price/cost, manufacturing capability, service, management, technology, research and development, finance, flexibility, reputation, relationship, risk, and safety and environment [8]. The four most influential criteria in evaluating the vendor are: item qualities, item delivery, items costs and item productivity. Therefore, the vendor scoring system uses these four criteria to assess the performances of each vendor [9]. According to Parthiban, Zubar, & Garge, the weight for each of the criteria is as follow [9]:

Table 1. Performance criteria weight and maximum score

Performance Criteria	Weight	Maximum Score
Quality	4.0	35.71 points
Cost	3.2	28.57 points
Delivery	2.4	21.43points
Productivity	1.6	14.29 points
Total	11.2	100 points

With the maximum score of 100 points, each criterion will be assigned a maximum score based on the weight. The total score of each vendor will be graded in a four alphabetic form with different score range as shown in the Table 2.

Table 2. Score range and value

Grade	Value	Colour	Score range
A	Best	Green	80 – 100 points
B	Good	Yellow	60- 79 points
C	Warning	Red	40 – 59 points
D	Bad	Black	< 39 points

The calculated score will then be implemented into the vendor management software for vendor score check and assisting with vendor selection in procurement process. By showing the vendor grade directly in the vendor management software can speed up the selection process as the host need not extra effort in looking for the vendor performance in the other department or software. Figure 4 shows the vendor performance and selection screenshot.

IOP Conf. Series: Materials Science and Engineering **528** (2019) 012082 doi:10.1088/1757-899X/528/1/012082

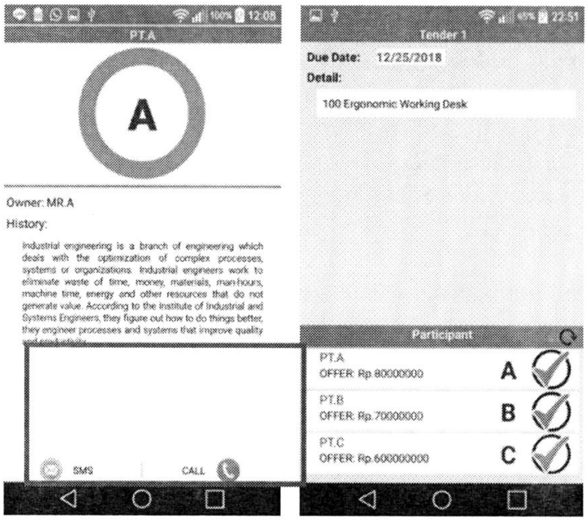

Figure 4. Vendor performance and selection

Due to the vast development over years, today Smartphone we are using is more powerful than most of the desktop 10 years ago. The Smartphone progression is causing the shipments of Smartphone increasing every year. In 2011, market results shown that 73 million more units of Smartphone are being sold that year hence resulting the first time Smartphone outsold personal computers [12]. According to Goldman, having a Smartphone is like having both a mobile phone and computer at the same time [13] because the Smartphone enable user to do things that they never thought possible without laptop or desktop [14]. Apart from to be able to do almost everything that a laptop or desktop can do, Smartphone are able to be used on any place and any time without the need to find a fixed docking place. Hence, with the flexibility nature, the users have become more dependent to the Smartphone than before [15].

Convenience is about adding comfort or saving time in doing any work such as useful, handy or helpful device, article or service. A Smartphone is powerful enough to offer various functionalities comparable to a computer or laptop. With the various available functionalities, there are some examples of user in using the functionalities in their work such as: tax practitioners who are able to provide adequate tax advice to their client even without in their office thus making their job more convenient and effective. Some of the information worker would even use their Smartphone as their primary working device until a personal computer is necessary. Judging at the ways Smartphone is being used in the example, flexibility, time saving and usefulness of usage on Smartphone can be added as the advantages of the Smartphone. Time saving and flexibility is having the ability in aiding the job when and where ever they are. As for the usefulness is having the ability to complete tasks quickly and performs various pc-related tasks. A UK telecommunication industry called Ofcom released a statistic of Smartphone usage in the United Kingdom [12]:

a) 37% of adults and 60% of teens admit they are highly addicted to their Smartphone.
b) 51% of adults and 65% of teens say they have used their Smartphone while socializing with others.
c) 23% of adults and 34% of teens have used their Smartphone during mealtimes.
d) 22% of adult and 47% of teens admitted using or answering their Smartphone while in the bathroom

The result shows 57% of users are using their Smartphone during their mealtimes and 69% of users admitted using their Smartphone while in the bathroom. Even though it is not a good circumstance for using Smartphone but still by looking from another point of view, these circumstances has proven that Smartphone usage is more flexible and convenience than desktop or laptop.

4. Conclusion

This research found that by using the developed vendor management and e-procurement system on android Smartphone, companies can shorten the overall procurement cycle time. The procurement cycle time is shortened by implemented a vendor scoring system in the system hence user can made a fast and accurate vendor selection directly from the system without needing extra effort in looking for the vendor performance in the other department or software. Apart from shortening cycle time, the developed system is proved to be more flexible and convenience in assisting the user in procurement process. Flexibility is addressed to the ability of able to be used on any place and any time without the need to find a fixed docking place. As for convenience is addressed to adding comfort or saving time in doing any work such as useful, handy or helpful device, article or service.

Acknowledgments

The authors thanks to the Bina Nusantara University, Jakarta, Indonesia

5. References

[1] Chaffey D 2009 *E-Business and E-Commerce Management: Strategy, Implementation and Practice.* 4th Ed. (London, England: Pearson Education Limited)

[2] Russel R S and Taylor B W 2014 *Operations and Supply Chain Management.* 8th Ed. (Wiley International Edition)

[3] Fitzsimmons J A and Fitzsimmons M J 2011 *Service Management: Operations, Strategy, Information Technology.* 7th Ed., S. Mattson, Ed. (New York: McGraw-Hill)

[4] Bof F and Previtali P 2010 National models of public (e)-procurement in Europe. *Journal of e-Government Studies and Best Practices* 14

[5] Johnson M E and Whang S 2002 E-Business and Supply Chain Management: An Overview and Framework. *Tuck School of Business Working Paper* **20**

[6] Piera C, Roberto C, Giuseppe C and Teresa M 2014 E-procurement and E-supply Chain: Features and Development of E-collaboration. *Future Software Engineering and Multimedia Engineering* **6** pp 8-14

[7] Zak J 2015 Comparative analysis of multiple criteria evaluations of suppliers in different industries. *Transportation Research* pp 809-819

[8] Ho W, Xu X and Dey P K 2010 Multi-criteria decision making approaches for supplier evaluation and selection: A literature review. *European Journal of Operational Research* **202**(1) pp 16-24

[9] Parthiban D P, Zubar H A and Garge C P 2012 A Multi Decision Criteria Making Approach for Supplier Selection. *Procedia Engineering* pp 2312-2328

[10] Bala K, Sharma S and Kaur G 2015 A Study on Smartphone based Operating System. *International Journal of Computer Applications* **121**(1) pp 17-22

[11] Holla S and Katti M M 2012 Android based mobile application development and security. *International Journal of Computer Trends and Technology* **3**(3) pp 486-490

[12] Sarwar M and Soomro T R 2013 Impact of Smartphone's on Society. *European Journal of Scientific Research* **98**(2) pp 216-226

[13] Goldman S M 2010 Transformers. *Journal of Consumer Marketing* **27**(5) pp 469-473

[14] Suki N M 2013 Students' dependence on smart phones: The influence of social needs, social influences and convenience. *Campus-Wide Information Systems* **30**(2) pp 124-134

[15] Arif I and Aslam W 2013 Students' dependence on smart phone and its effect on purchase behavior. *South Asian Journal of Global Business Research* **5**(2) pp 285-302

[16] Harun A, Soon L T, Kassim A W and Sulong R 2015 Smartphone Dependency and Its Impact on Purchase Behavior. *Asian Social Science* **11**(26) pp 196-211

Calculation of Raw Material Costs for the Palm Oil Supply Chain Value Added Using Modified Hayami Method

Syarif Hidayat

Department of Industrial Engineering, Universitas Al Azhar Indonesia, Jakarta, Indonesia
syarif_hidayat@uai.ac.id

Abstract. The stakeholders in the palm oil supply chain (POSC) are the smallholder farmers who produce fresh fruit bunch (FFB), the traders, CPO factory who produces CPO from FFB, refinery who transformed CPO into frying oil, the distributors/retailers, and the consumers. The palm oil tree regularly produces FFB from its third until 24th year of life.The output of the farmers is measured in its weight (tonnes), so are the outputs of the trader, the CPO factory, the refinery, and the distributors/retailers. However the input to the farmer must be measured differently because the main raw materials are bought in the form of seedling or young palm-oil tree, to be planted on the field and regularly bear FFB to be later sold in tons. The objective of this study is to build a calculation model to allocate the raw material prices of the seedling. Original Hayami method to calculate single company added value was modified to facilitate calculation of the added value of supply chain network of companies. The calculation is based on the crude palm oil (CPO) factory capacity of processing 30 tons of FFB per hour, calculated for a year. The life-cycle-cost approach was used to calculate the appropriate cost of the raw material. Using this approach the raw material cost is IDR 1,012/kg, and the the total POSC added-value is IDR 13,011 per kg of product. It is expected that the modified Hayami added value calculation method can be later applied to larger and more complex industries.
Keywords: value added, supply chain, palm oil, Hayami method, life-cycle-cost

1. Introduction

A supply chain is a chain of interdependent companies operating in sequence and cooperate in handling, improving and controlling the flows of goods, money and information beginning with the supplier in the upstream all the way downstream until the end consumers [4] [8]. The principal roles of the supply chain is to add value to the products by moving them from one to another location, or to perform the modification processes [3]. The value adding processes may be applied to the quality, costs, delivery activities, flexibilities in sending the products, and innovations [7]. One of the most important supply chains in Indonesia is the palm oil supply chain (POSC). Export of palm oil and its derivatives has always been increasing. In 2010 the export value was USD 15,6 billions, which has shown an increase of 34,6 % compared to 2009. The export tax was US$ 2,8 billion. This value came second only to oil export.

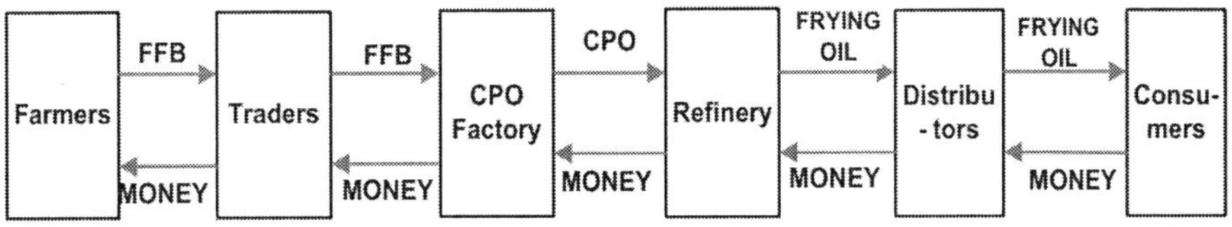

Figure 1. Palm oil supply chain actors

Figure 1 shows the 6 actors in the POSC. The smallholder farmers sell their fresh fruit bunch (FFB) to Crude Palm Oil (CPO) factory through traders. CPO factory converts the FFB into CPO. CPO is sold to the refinery, who converts CPO into frying oil and sends the product to the distributors. In this study only frying oil refinery is discussed, although CPO can be converted into many types of popular consumer products. The distributors subsequently sell them to the consumers through the retailers. This continuous process is required to ensure the consistent quantitative operating levels along the POSC from the upstream to the downstream. The output of the farmers are measured in its weight (tonnes or kgs), so are the outputs of the trader, the CPO factory, the refinery, and the distributors/retailers. However, the input to the farmer can not be measured in its weight because the main raw materials are in the form of seedling or young palm-oil trees bought in units, to be planted on the field and regularly bear FFB to be later sold in tons. Therefore there must be a specific procedure to calculate the raw material costs in rupiah. Original Hayami method to calculate single company added value was modified to facilitate calculation of the added value of supply chain network of companies [1].

The objective of this study is to build a model to allocate the raw material prices of the seed or young palm oil tree, while considering the successive investment and operational risk levels.

2. Method

Added value is a form of company's and supply chain performance measurement. It is defined as the margin in a certain stage and can be measured as Profit = total revenue – total raw materials and processing costs [5]. Total revenue is basically the multiplication of the unit selling price (UP) times the quantity sold (Q), while the processing costs are the sum of the total fixed cost and total variable cost. Hayami method [1] [2] is appropriate for use in a chain of companies linked in a supply chain. With this method we can measure the added value along the chain, output values, and some productivity measures, as well as the return to the investors. The method is easy to understand and use, and it provides sufficient useful information to the investors and the workers. The weakness of the original method is that it only measures the added value in one production cycle, one type of commodity and one business actor. Therefore the original method needs to be modified to suit the condition of a sequence of companies in a supply chain. Table 1 shows the three parts the original Hayami method. Formula listed under the column "Value" is used to compute the related variables listed under column "Variables".

In line with the objective, this study is organized into several parts as follows:

(1) Set the assumptions for the calculation. This is necessary in building a calculation model which later may be changed or adapted to suit the condition of the business.
(2) Modify the Hayami process to facilitate sequence of companies. This is necessary to enable calculation of a sequence of companies instead of a single company.
(3) Calculate the required raw materials in volumes and prices for the farmers. This is needed as the nature, the format or unit of the products in the chain may be different.
(4) Calculate the added values for the whole palm oil supply chain.

Figure 2 shows the procedure to calculate the raw material costs for use in calculating the added value along the POSC. The first step performed is the original method and n capacity of the CPO factory. Calculating the required FFB, area of the palm-oil estate, investment and operational costs are performed based on the assumptions of the productivities of the estate, and the CPO and frying-oil factories. The price of the palm-oil seeds is calculated using *Life Cycle Analysis* method with productive life of palm-oil of 25 years and its *Net Present Value (NPV)*. Then the process is continued with the CPO and frying oil produced, the added values for each actor and their comparison. Also calculated is the return for each of the *stakeholders* [5].

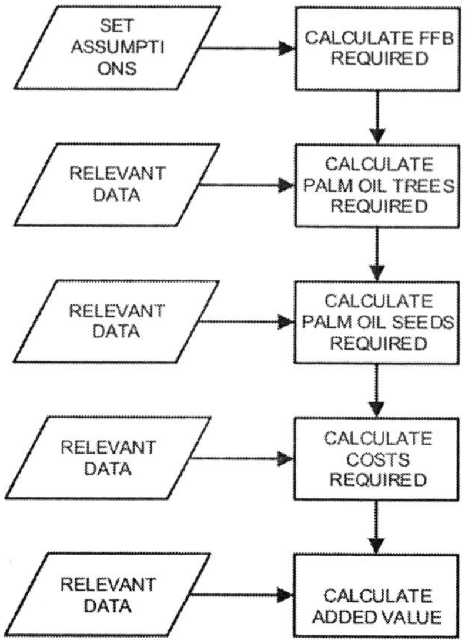

Figure 2. Flowchart for calculation

Table 1. Original Hayami Added-Value Worksheet

No.	Variables	Unit	Value
I. Output, Input, and Prices			
1	Output	kg	(1)
2	Main raw material	kg	(2)
3	Number of Main Direct Labor (Manday)	MD	(3)
4	Conversion Factor		(4) = (1) / (2)
5	Direct Labor Cost Coefficient (Manday/Kg)	MD/kg	(5) = (3) / (2)
6	Output price	IDR/kg	(6)
7	Direct Labor Costs (IDR/Manday)	IDR/MD	(7)
II. Income and Added Value			
8	Raw Material Costs	IDR/kg	(8)
9	Other Input Costs	IDR/kg	(9)
10	Output value (sales)	IDR	(10) = (4) x (6)
11	a. Added Value	IDR/kg	(11a) = (10) – (8) – (9)
	b. Added Value Ratio (%)	%	(11b) = (11ab) / (4b)
12	a. Direct labor cost income	IDR/kg	(12a) = (5) * (7)
	b. Direct labor cost ratio	%	(12b) = (12a) / (11a) x 100
13	a. Profit = added value	IDR/kg	(13a) = (11a) – (12a)
	b. Profit in %	%	(13b) = (13a) / (10) x 100
III. Reward for the Production Owner			
14	Margin	IDR/kg	(14) = (10) – (8)
	a. Direct labor income	%	(14a) = (12a) / (14) x 100
	b. Contribution from other input (%)	%	(14b) = (9) / (14) x 100
	c. Company Profit (%)	%	(14c) = (13a) / (14) x 100

Note: MD = mandays

All information and data needed for this study are obtained and identified from recent literatures and by interviewing relevant managers in the subsequential palm oil industries. The quantitative results are taken from the authors previous study in the POSC [2]. In this study we used the Hayami Method wich are adopted with the supply chain characteristics of palm oil agroindustry, and try to reduce or delete the weaknesses of the method. With the modification, the basis for the calculation is the money (in rupiah) and not the weight of the products. The time basis for the process is a year to enable a total investment and annual expenditures. The original Hayami method was only applied on one single company and one single product and one single production cycle [1].

The field data used for this study was obtained from some different areas of palm oil plantation in Sumatra, because no single company run the whole cycle of palm oil products. Data for the downstream companies were obtained from different companies. The primary palm oil field data were obtained from palm oil estates and processing companies in Tungkal Ulu Jambi, Dumai, Aceh Singkil, Lampung and Medan, while secondary data on frying oil refinery were obtained from distributors/retailers or from Biro Pusat Statistics (BPS) reports. For each agent in the supply chain we made business assumptions to ensure data consistency, i.e. the base for the calculation is the capacity of processing 30 ton FFB/hour of the CPO factory. The material balance along the complete network of supply chain was adapted to the amount of CPO produced, raw material and investment costs, processing costs, and the energy/utility and transportation costs.

3. Result and Discussion

3.1. Set the assumptions for the calculation.

Table 2 shows the assumptions for plam oil estate figures taken from Hidayat [2]. To maintain the consistency of calculation along the supply chain, the unit used is the most frequent and dominant units. In this case the dominant unit used is the weight of the product i.e. tons or kgs. The basis for overall calculation is the amount of FFB needed by the CPO factory which is 30 tons FFB/hour. This is then projected for the need for a year, and the required area of palm oil estate to supply this amount. Consequently the number of farmers and traders are calculated. The quantity of frying-oil yields from the Frying-Oil factory is calculated. All related costs are included and considered. Note: All data are assumptions only.

Table 2. Assumptions for the farmers estate

	Estate life time	25	year
	Working day/ year	300	day
	manhours/day	20	hours
	FFB required / year	90,000	ton
	FFB Production /year	90,000,000	kg
	FFB Production one life cycle	2,250,000,000	kg
	FFB Production /ha	29,680	kg
Estate area:	FFB Estate	3,032	ha
	Morbidity rate	1	
	Projected area of PO Estate	4,094	
	Number of seeds /ha	135	trees
Seedling cost:	seedling required	552,645	trees
	Price of seedling / unit	25,000	IDR
	Seedling costs for 25 years	13,816,121,968	IDR
	Seedling costs/year	552,644,879	IDR
	NPV	8	
	Seedling costs at year-4	4,328,093,632	IDR
	Seedling costs per kg FFB	48	IDR
Life cycle costs:	Operational costs	514,361,000	for 25 yrs
	Investment costs/ha	20,059,000	
	Life cycle costst/ha	534,420,000	
	Total life cycle costs	1,620,545,822,102	3032 ha
	Operational costs/year	64,821,832,884	
Investment costs/ha:	Quoted from SK DirjenBun 2011	20,059,000	IDR
	Land price/ha	122,472,000	IDR
	Total investment costs/ha	142,531,000	IDR
	Cost Breakdown		
Production costs:	Operational costs (IDR)	62,388,800,539	IDR
	Non Production costs		
	Investment costs (IDR)	17,288,126,685	IDR

The table shows the result of the calculation of related variables in the palm oil business.

3.2. Modify the Hayami process to facilitate sequence of companies

The general modification process to the Hayami method is shown by comparing Table 1 into Table 3. The first step performed is the original method and capacity of the CPO factory. Calculating the required FFB, area of the palm-oil estate, investment and operational costs are performed based on the assumptions of the productivities of the estate, the CPO and frying-oil factories. The price of the palm-oil seeds is calculated using Life Cycle Analysis method with productive life of palm-oil of 25 years and its Net Present Value (NPV). Then the process is continued with the CPO and frying oil produced, the added values for each actor and their comparison. Also calculated is the return for each of the stakeholders.

The modified Hayami Method is shown in Table 3. The top portion shows the summary of value added along the integrated supply chain. The first part of the method collects all facts and calculate all the revenues, products, inputs, outputs and all prices and costs. The second part the method calculates the level of revenues and useful business ratios based on the inputs from the first part. On the third part the method provide returns to the investors/owners in unit of percentages (%) for the investors and the workers. Note: All data are assumptions only.

The detail explanation of the modified worksheet follows:

a. For the top portion of the worksheet: Palm Oil Supply Chain Interaction: the first 3 rows show the prices of the input raw material, product selling prices, and the total added value per kg of the product. The figures show the values in sequence along the supply chain actors.
b. For Part I: Output, Input and Prices, the information on the output was made in two rows i.e. in weight of tons and in rupiah. This is because the form of the material or product can not be processed in its original from, i.e. the palm-oil tree can not be processed, but it must be planted in order to yield the fruits (FFB).
c. For Part II: Income and Added Value. The raw material costs is already entered in the top part of the worksheet. Other Input costs (line 10) refer to all required activities and products needed other than the direct production or operations. Values are entered using rupiah as the unit. Added value is the difference between the sold FFB process minus the raw material and other input costs. The ratios (in percentages) show the percentage of the added value from the product costs.

Table 3. Modified Hayami Worksheet

No.	Variables	Unit	Value
Palm Oil Supply Chain Interaction			
1	Raw Material Costs (IDR/kg)	IDR/kg	(1)
2	Product selling price	IDR/kg	(2)
3	Total Added Value per kg	IDR/kg	(3) = (2 last actor) - (1)
I. Output, Input, and Prices			
4	a. Output, Sales Volumes (kg)	kg	(4a)
	b. Output, Sales Values (IDR)	IDR	(4b)
5	Total Cost of Main Raw Materials (IDR)	IDR	(5)
6	Number of Main Direct Labor (Manday)	MD	(6)
7	Conversion Factor		(7) = (4b) / (5)
8	Direct Labor Cost Coefficient (Manday/Kg)	MD/kg	(8) = (4b) / (6)
9	Direct Labor Costs (Rp/Manday)	IDR/MD	(9)
II. Income and Added Value			
10	a. Other Input Costs (Production)	IDR	(10a)
11	a. Added Value (IDR)	IDR	(11a) = (4b) - (5+10)
	b. Added Value Ratio (%)	%	(11b) = (11ab) / (4b)
III. Reward for the Production Owner			
12	Margin (IDR/Kg)	IDR	(12) = (4b) - (5)
	a. Contribution from other input (%)	%	(12a) = (10) / (12)*100%
	b. Company Profit (%)	%	(12b) = (11a) / (12)*100%
	Note: MD = mandays		

d. For Part III: Reward for the Production Owner.
Return for the Production Owner (= margin) is calculated as the difference between the output values minus the contribution of the production owner aside from the raw materials/seeds which are used in the production proses.

3.3. Calculate the required raw materials in volumes and prices for the farmers,
Calculation formula added value using modified Hayami Method was done as follows:
1. Six actors are involved in POSC: farmers, trader, CPO factory, frying-oil refinery, distributors/retailers and the consumers. In the original Hayami method there was only one actor.
2. The processing capacity of the CPO factory is 30 ton FFB/hour. Assuming 300 working-days and effective 8 hours /day, the factory needs 180.000.000 kg FFB/year. Assuming each hectare of palm oil estate produces 29,680 kg of FFB/hectare per year, that amount of FFB is to be supplied by palm-oil estate of 6.065 ha. With each farmer owning 2 hectares of estate then we need 3.032 farmers.
3. While we assume that the most efficient production of palm oil was produced in its period of 18 years of a total of 25 years of life, we must take into account and calculate all the investment, expenditures and operational and harvesting costs. These are all indicated on Table 2.
4. Products processed in the original Hayami method was only one type, while in the modified method are 6 types namely palm-oil tree, FFB, Palm kernel oil (PKO), CPO, Frying-oil, stearin and PFAD. Each of this product excepting the palm-oil tree have the same unit of business transaction namely tons or kgs. The palm-oil tree as the sole source of raw material must be bought in the form of tree and is not measured in kg but in unit of tree. Therefore, only the appropriated cost of the productive tree in terms of rupiah should be calculated and used in the supply chain added calue calculation.
5. The original Hayami method life-cycle is less than 1 year. The modified method spans this into the most productive years of the tree, which is 16 years. However, the practical life time of the palm-oil tree is 25 years. The CPO factory and frying-oil refinery life time is around the same period.
6. The cost of the seedling or the young palm-oil tree is calculated as shown on Table 4 as follows:

Table 4. Calculation/allocation for seedling cost per year

Variable	Quantity	Unit
Estate lifetime	25	yrs
Seedling costs:		
Needed seedling for 25 yrs	552,645	unit
Price of seedling/unit	15,000	IDR
Seedling costs for 25 yrs	8,289,673,181	IDR
Seedling costs/yr	331,586,927	IDR
Estate area	3,032	ha
Life cycle costs		
Operational costs/yr/ha	411,488,800	IDR
Operational costs/ha	10,287,220,000	IDR
Investment costs/ha	20,059,000	IDR
Life cycle costs/ha	10,307,279,000	IDR
Total life cycle costs	4,544,231,764,781	IDR
Operational costs/yr	181,769,270,591	IDR
Total seedling+life cycle costs	4,552,521,437,962	IDR
Total Production /yr	180,000,000	kg/yr
Seedling costs /kg product	25,292	IDR
Seedling costs /kg product/yr	1,012	IDR

3.4. Calculate the added values for the whole palm oil supply chain
Table 5 shows the complete Hayami Method to calculate the added-value along the POSC. It is essentially an extension of Table 3 by adding the next actors in the chain. The output of the first actor (farmer) becomes the input to the second actor (the trader). The selling price of teh farmer becomes the

Table 5. Complete Hayami Method for POSC

No.	Variables	Unit	Farmer	Trader	CPO Factory	Refinery	Distributor	Consumer
Palm Oil Supply Chain Interaction								
1	Raw Material Costs (IDR/kg)	IDR/kg	1,012	1,220	1,270	7,200	13,000	14,023
2	Product selling price	IDR/kg	1,220	1,270	7,200	13,000	14,023	
3	Total Added Value per kg	IDR/kg	208	50	5,930	5,800	1,023	
I. Output, Input, and Prices								
4	a. Output, Sales Volumes (kg)	kg	180,000,000	180,000,000	43,200,000	32,832,000	32,832,000	
	b. Output, Sales Values (IDR)	IDR	219,600,000,000	241,905,833,333	311,040,000,000	426,816,000,000	460,417,500,000	
5	Total Cost of Main Raw Materials (IDR)	IDR	458,490,566	219,600,000,000	241,905,833,333	311,040,000,000	426,816,000,000	
6	Number of Main Direct Labor (Manday)	MD	6	6	140	140	10	
7	Conversion Factor		479	1.10	1.29	1.37	1.08	
8	Direct Labor Cost Coefficient (Manday/Kg)	MD/kg	0.0000000131	0.000000000027	0.000000000579	0.000000000450	0.000000000023	
9	Direct Labor Costs (IDR/Manday)	Rp/MD	231,000,000	99,000,000	3,905,320,000	1,974,000,000	141,000,000	
II. Income and Added Value								
10	Other Input Costs (IDR/kg)	IDR/kg	154,520,733,154	276,833,333	13,005,501,000	75,722,802,704	366,733,333	
11	a. Added Value (Rp)	Rp	64,620,776,280	22,029,000,000	56,128,665,667	40,053,197,296	33,234,766,667	216,066,405,910
	b. Added Value Ratio (%)	%	29%	9%	18%	9%	7%	
III. Reward for the Production Owner								
12	Margin (IDR/Kg)	IDR/kg	219,141,509,434	22,305,833,333	69,134,166,667	115,776,000,000	33,601,500,000	
	b. Contribution from other input (%)	%	71%	1%	19%	65%	1%	
	c. Company Profit (%)	%	29%	99%	81%	35%	99%	

input buying cost for the trader. This is iterated all the way along the chain until the most downstream (the consumer). Note: All data are assumptions only.

The six actors in the POSC faces different levels of investment, technology and risks. The obtained added-value therefore should be different, but properly graded. This modified method was used to manage six types of products namely FFB, PKO, CPO, frying-oil, stearin and PFAD with diferent levels of selling prices. For simplicity of this report, Table 4 only shows the first product i.e. the frying-oil. The calculated appropriate cost of the seedling or young palm oil tree is IDR 1,012 per kg. This becomes the input cost into the POSC. While the end selling price to the consumer is IDR 14,023 per kg, then the total POSC added-value is IDR 14,023 – IDR 1,012 = IDR 13,011 per kg of product, which is the frying oil. Note: All data are assumptions only.

4. Conclusion

The original Hayami method has been successfully modified to facilitate more correlated products, multi-year horizon, and more than one actor involved in the supply-chain. The six actors in the POSC faces different levels of investment, technology and risks. The obtained added-value therefore should be different, but properly graded. This modified method was used to manage six types of products namely FFB, PK, CPO, frying-oil, stearin and PFAD wth diferent levels of selling prices.

With the assumed industrial processing capacity of CPO factory of 30 ton FFB/hour and FFB selling price of IDR 1.220/kg we obtain the comparative added values of Farmers: Trader: CPO Factory: Frying-Oil Factory: Distributor = 29%: 9%: 18%: 9%: 7%. With the assumption that each farmer owns 2 hectares of palm-oil estate, the added-value for each farmer is IDR 208 per kg.

5. References

[1] Hayami Y, Kawagoe T, Morooka Y and Siregar M 1987 *Agricultural Marketing and Processing in Upland Java. A Perspective from a Sunda Village.* (Bogor: The CPGRT Centre)

[2] Hidayat S, Marimin, Suryani A, Sukardi and Yani M 2012b Modifikasi Metode Hayami untuk Perhitungan Nilai Tambah pada Rantai Pasok Agroindustri Kelapa Sawit. *Jurnal Teknologi Industri Pertanian* (Bogor: IPB) **22**(1) pp 22-31

[3] Janvier J A M 2012 A New introduction to supply chains and supply chain management: Definitions and theories perspective. *J Int Business Research* **5**(1) pp 194-207

[4] Preckel P V, Gray A, Boehlje M and Kim S 2004 Risk and value chains: Participant sharing of risk and rewards. *J Chain and Network Services* **4**(1) pp 25-32

[5] Salvatore D 2004 *Managerial Economics in a Global Economy with Economic Applications Card.* 5nd ed. (New York: Thompson Learning)

[6] Silva C A, Baker D, Shepherd A W, Jenane C and Cruz S M 2009 *Agro-Industries for Development.* (Rome: FAO)

[7] Trienekens J H 2011 Agricultural value chains in developing countries; A framework for analysis. *J Int Food and Agribusiness Management Review* **14**(2) pp 51-82

[8] Van der Vorst J G A J, Silva C A and Trienekens J H 2007 *Agro-Industrial Supply Chain Management: Concepts and Applications.* (Wageningen: Department of Social Sciences, Wageningen University)

IOP Conf. Series: Materials Science and Engineering **528** (2019) 012084 doi:10.1088/1757-899X/528/1/012084

Multi Echelon Distribution Model for Electric Market Deregulation Collaboration Strategy in East Kalimantan

Irwan Gani[1], Wahyuda[2], Budi Santosa[3] and Muliati[1]

[1]Economics Faculty, Universitas Mulawarman, Samarinda, Indonesia
[2]Industrial Engineering, Universitas Mulawarman, Samarinda, Indonesia
[3]Industrial Engineering, Institut Teknologi Sepuluh Nopember, Surabaya, Indonesia
irwan.gani@feb.unmul.ac.id, wahyuda@gmail.com, muliati@feb.unmul.ac.id

Abstract. Electric market deregulation aims to provide flexibility for customers to have many suppliers and low prices. However, market deregulation also provides opportunity for certain parties to manipulate supply, so there is a scarcity of products that result in price increases. Therefore, this research proposes a collaborative strategy for electricity market deregulation using the Multi Echelon distribution model applied in East Kalimantan with dummy data. Collaboration strategy made based on optimization of mathematical models in two stages and three scenarios. The simulation was carried out using Excel Solver covering three regions, dynamic time and estimated price fluctuations over three periods. As a result, Gencos gets the biggest profit when serving basic load. Whereas wheeling occurs, the D4 region is a strategic area that generates the largest profit compared to the D1 and D6 regions.
Keywords: Supply Chain, Optimization, Electricity, Transportation Model, Deregulation Market, Multi Echelon Distribution

1. Introduction

Electricity market deregulation has changed the vertical integration of electric power production systems into three separate systems namely Gencos, Transcos and Discos. Gencos are companies that have power plant to produce electricity, while Transcos are those that own transmission network to send electricity from producers to consumers. The last part of this system is Discos, which consist of companies that are responsible in distributing the electricity to end customer. The separation of these three systems are aimed to create competition among Genco companies so that electricity can be produced at low cost, and ensure continuity of electric supply.

Based on literature electricity rates are influenced by three factors, i.e., fuel costs, losses and transmission costs [1], while the continuity of supply depends on the amount of production reserves at the power plant [2]. Fuel costs depend the characteristics of the power plant [3]. The difference in power plant characteristics among Gencos companies causes the allocation of loads needs to be integrated. This integration can be done by using Economic Dispatch process as in [1]–[3].

Electricity produced by the power plant is sent to customers through Transcos. Gencos should achieve cost minimization in order to avoid additional cost in the form of losses. System integration between Gencos and Discos can be done with Price Based Dynamic Economic Dispatch (PBDED) [7], [8]. PBDED has succeeded in creating integration between the two with low cost, even though the results are still lacking in detail [9]. More detailed PBDED obtained by combining PBDED models with multi echelon [1]. Merging these two models successfully integrates Gencos and Transcos to deliver electricity to customers (DISCOS). The result of application of this model is very dependent on the characteristics of supply and demand in respective area.

This paper applied a combined PBDED model with multi echelon in the East Kalimantan electricity system. The electricity system in respective area consists of interconnected power plants through 150 kv and 20 kv transmission networks. Each power plants are situated in sparsely area as well as serving wide range location of end customer. The result of this research is providing infomation of optimal allocation of power plants with parameters of costs and emissions using Excel Solver.

2. Methods

This research was conducted using 5 stages: Problem formulation, Literature review, modeling, Simulation, and Analysis (Figure 1).

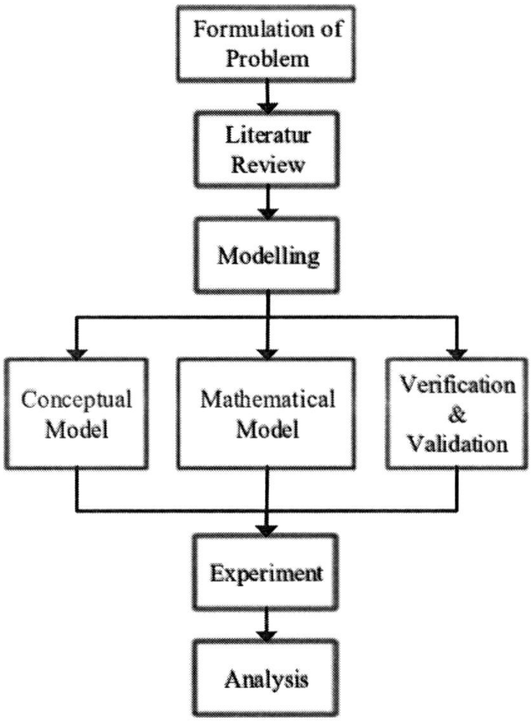

Figure 1. Research methodology

Formulation of the problem. What is the best collaboration of power suppliers in East Kalimantan to get the smallest total cost and emission?

Literature study. All of journals used as literatur in this research published by Elsevier on Economic Dispatch, optimization, and collaboration topics.

Modeling. There are three stages of modeling. First, build the conceptual model to facilitate in designing details of mathematical models; Second, provide mathematical model to illustrate supplier collaboration. Third, Verification and validation. So, there are four mathematical models in this research: SEC + RE, SEC + IPP + RE, SEC + Rent + RE, and SEC + EC + RE.

PBDED Model and Multi Echelon Distribution

The PBDED model as in[4] is aimed to maximize profit with the following objective functions:

$$maximize\ PF = RV - TC \quad (1)$$

In this case,

$$TC = \sum_{t=1}^{T}\sum_{i=1}^{N}\left[C_i\,(P_{(i,t)} + ST_t\right] \quad (2)$$

$$RV = \sum_{t=1}^{T}\sum_{i=1}^{N}\left[\sigma_g(t)\,(P_{(i,t)}I_{(i,t)}\right] \quad (3)$$

C_i: production cost of unit i, $P_{(i,t)}$: Output of generator i at time t. $I_{(i,t)}$ commit or not commit at time t. ST_t start-up cost at time t. t: time of dispatch. i: index generator. N is the number of generating units. $\sigma_g(t)$ load forecasting at time t. $C_i\ (P_{(i,t)})$ generation cost of unit i

1. Demand Constraint

$$\sum P_{(i,t)}\, I_{(i,t)} \le D_t \quad t = 1, \ldots, Tm \quad (4)$$

2. Generator Constraint

$$P_{i,min} \le P_{(i,t)}I_{(i,t)} \le P_{i,max} \quad (5)$$

$$-DR_i \le P_i - P_i^0 \le UR_i \quad (6)$$

$$\max\{P_i^0 - DR_i, P_{i,min}\} \le P_i \le min\{P_i^0 + UR_i, P_{i,max}\} \; (7)$$

The purpose of the PBDED model is to maximize profit not just minimizing fuel costs as in (Columbus & Simon, 2013)

$$maximize\ PF = RV - TC \quad (7)$$

In this case,

$$TC = \sum_{t=1}^{T}\sum_{i=1}^{N}\left[C_i\,(P_{(i,t)} + ST_t\right] \quad (8)$$

$$RV = \sum_{t=1}^{T}\sum_{i=1}^{N}\left[\sigma_g(t)\,(P_{(i,t)}I_{(i,t)}\right] \quad (9)$$

C_i: production cost of unit i, $P_{(i,t)}$: Output of generator i at time t. $I_{(i,t)}$ commit or not commit at time t. ST_t start-up cost at time t. t: time of dispatch. i: index generator. N is the number of generating units. $\sigma_g(t)$ load forecasting at time t. $C_i\ (P_{(i,t)})$ generation cost of unit i

1. Demand Constraint

$$\sum P_{(i,t)}\, I_{(i,t)} \le D_t \quad t = 1, \ldots, Tm \quad (10)$$

2. Generator Constraint

$$P_{i,min} \le P_{(i,t)}I_{(i,t)} \le P_{i,max} \quad (11)$$

$$-DR_i \le P_i - P_i^0 \le UR_i \quad (12)$$

$$\max\{P_i^0 - DR_i, P_{i,min}\} \le P_i \le min\{P_i^0 + UR_i, P_{i,max}\} \; (7)$$

While additional constraints are as in [5]:

$$F = \sum_{t=1}^{T}\sum_{i=1}^{I} F_{it}(P_{it}) \quad (13)$$

$$F_w = \sum_{t=1}^{T}\sum_{i=1}^{I} F_{it}(P_{it})_w \quad (14)$$

$$C_w = F_w - F + C_{ijt}^y + C_{jkt}^y \quad (15)$$

$$F_{it}(P_{it}) = a_i P_{it}^2 + b_i P_{it} + c_i \quad (16)$$

$$C_{ijt}^y = \frac{P_{ijt}^y}{P_{ijt}} z_{ij} \quad (17)$$

$$C_{jkt}^y = \frac{P_{jkt}^y}{P_{jkt}} z_{jk} \quad (18)$$

$$E_{it}(P_{it}) = \alpha_i P_{it}^2 + \beta_i P_{it} + \delta_i \quad (19)$$

Equation (7) is used to model the cost of fuel with a regular load, whilst equation (8) applied to accomodate the cost of fuel when wheeling occurs. The 9th equation is aimed to calculate the transmission cost. In order to model the fuel cost we used a quadratic function in equation (10). Whereas the last two equation i.e. equation (11) and (12) were MW-mile method to calculate the transmission of 150 KV and 20 KV

Simulation. Simulation in this research is divided into two stages. In the first stage we did two calculation, which is allocation and distribution of electricity using basic loads, and calculation of allocation and distribution of electricity with additional demand loads. The second stages consist of collaborative scenarios. Scenario 1, we put additional demand of power in D1 region; Scenario 2, additional demand in the D4 region; and Scenario 3, additional demand in D6 region.

Analysis. This section contains Results and Discussion. In Results section we provide data and computational process whilst in Discussion various finding will be presenter.

3. Results and Discussion

3.1. Results

In the calculation we used dummy data (as shown in Table 1). Simulation carried out dynamically using three periods by dividing demand into two scenarios namely regular and wheeling. The estimated selling price of electricity varies over three periods.

Table 1. Dummy data for 3 periods

	Demand			Forecasted	
Period	Regular	Wheeling	R+W	price	Revenue
1	9,000	200	9,200	1,000	200,000
2	9,400	150	9,550	1,300	195,000
3	9,900	100	10,000	1,250	125,000
Total	28,300	450	28,750	3,550	520,000

Table 2. Results of electricity market collaboration simulation

		WHEELING								
		D1			D4			D6		
	Regular Load	Min. Cost	Min. Emission	Min. Losses	Min. Cost	Min. Emission	Min. Losses	Min. Cost	Min. Emission	Min. Losses
Fuel Cost	16,986,879	28,440,769	49,419,721	38,544,976	17,748,035	41,149,713	50,309,190	17,748,035	41,528,235	43,354,120
Fuel Cost Wheeling		11,453,890	32,432,842	21,558,098	761,156	24,162,834	33,322,311	761,156	24,541,356	26,367,241
Transmission cost	1,406,095	186,168	262,960	178,323	4,688	240,035	290,727	4,688	245,837	251,747
Distribution cost	544,628	106,249	136,674	85,889	15,420	159,612	204,929	31,941	171,227	171,662
Total cost wheeling		11,746,307	32,832,476	21,822,310	781,264	24,562,482	33,817,968	797,785	24,958,421	26,790,650
Revenue	33,595,000	34,115,000	34,115,000	34,115,000	34,115,000	34,115,000	34,115,000	34,115,000	34,115,000	34,115,000
Profit	14,657,398	5,381,815	-15,704,355	-4,694,189	16,346,857	-7,434,360	-16,689,846	16,330,337	-7,830,299	-9,662,529
Emission		6,431,270	3,405,086	3,810,006	6,381,379	2,627,183	2,725,668	6,381,379	2,567,821	2,651,158
Losses		10,425	8,443	6,921	6,954	6,412	6,358	6,954	6,415	6,358

Regular load defined as routine load which is usually served by the electrical system in East Kalimantan. With this particular load the cost of fuel Rp. 16,986,879, the cost of transmission were Rp. 1,406,095, and the cost of distribution Rp. 544,628. Revenue generated from this scheme were 33,595,000 with profit of 14,657,398 (as shown in Table 2).

When an additional contract occurs in the D1 area, there are three objectives considered:

1. Cost Minimization. Power plant allocation were aimed to minimize costs required cost of fuel as much as Rp. 28,440,769, transmission costs Rp. 186,168 and distribution costs Rp. 106,249. The profit gained from this scheme was Rp. 5,381,815.
2. Emission Minimization. Within this scheme we found that power plant requires a fuel cost of Rp. 49,419,721, transmission costs of Rp. 186,168, and distribution costs Rp. 106,249. Profit generated from this simulation was Rp. -15,704,355.
3. Loss Minimization. The allocation of a power plant with this purpose generates total cost as much as Rp. 38,544,976, transmission costs Rp. 178,323, and distribution costs Rp. 85,889. The profit generated from this scheme was Rp. -4,694,189

Similar to our treatment to D1 area we set three objectives in D4 area:
1. Cost Minimization. In achieving this goal we should allocated costs of fuel as much as Rp. 17,748,035, set transmission costs in Rp. 4,688 and distribution costs Rp. 15,420. Profit expected from this allocation was Rp. 16,346,857.
2. Emission Minimization. Power plant required fuel cost of Rp. 41,149,713, transmission costs of Rp. 240,035, and distribution costs Rp. 159,612 for this purpose. With this scheme profit gained was Rp. -7,434,360.
3. Loss Minimization. profit generation expected from this scheme was Rp. -16,689,846 with total cost Rp. 50,309,190, transmission costs Rp. 290,727, and distribution costs Rp. 204,929.

The last treatment for D6 area were the same as the previous area. The result were:
1. Cost Minimization. Power plant allocation with the aim of minimizing costs requires a fuel cost of Rp. 17,748,035, transmission costs Rp. 4,688 and distribution costs Rp. 31,941. Profit gained in this scheme was Rp. 16,330,337.
2. Minimize emissions. Power plant allocation with this purpose requires a fuel cost of Rp. 41,528,235, transmission costs of Rp. 245,837, and distribution costs Rp. 171,227. Profit expected was Rp. -7,830,299.
3. Loss Minimization. The allocation of a power plant with this purpose generates a total cost of Rp. 43,354,120, transmission costs Rp. 251,747, and distribution costs Rp. 171,662. Profit generated was Rp. -9,662,529

3.2. Discussion

Collaboration between power plants to service basic loads requires a total fuel cost of Rp. 16,986,879. This cost required to produce 35,152 MW of electricity to meet the demand of 28,300 MW for three periods. It can be seen that that the amount of production is greater than demand due to losses occured in transmission and distribution networks (Equations 4 and 10). Under this skenario revenue gained was Rp. 33,595,000 and profit generated as much as Rp. 14,657,398. This basic load scheme is used as basis reference when we set additional demand in D1, D4, and D6 regions.

Additional demand in D1 of 450 MW affected revenue to increase by Rp. 520,000. However, along with increasing in production, the cost of fuel was also rising in Rp. 11,453,890, from Rp. 16,986,879 to Rp. 28,440,769. The imbalance between increasing in fuel costs and revenue caused a decreasing in profit as much as Rp. 9,275,583. Revenue increases that are not proportional to the increase in profit are due to the equation of fuel costs (Equation 13) and the improper determination of selling prices of products. As a result tariffs for additional products must be distinguished from product rates for basic loads.

Additional demand for D4 of 450 MW caused an increasing in revenue of Rp. 520,000. The consequence was increasing in cost of production which is caused by increasing in fuel from Rp. 16,986,879 to Rp. 17,748,035 (Rp. 761,156). Compare to the first scheme, revenue gained from this area was smaller than the increase in fuel costs. However, there was an increase in profit of Rp. 1,689,459 due to additional demand in D4 region has changed the allocation structure of the power plant production. The nearer distance of power plant within this scheme production has caused the decreasing in cost of transmission and distribution.

Additional demand for D6 of 450 MW caused revenue to increase by Rp. 520,000, but increased production also increased fuel costs from Rp. 16,986,879 to Rp. 17,748,035 (Rp. 761,156 difference). In this scheme the increasing of revenue generated was smaller than the increasing of fuel costs. However, we found significant increase in profit as much as of Rp. 1,672,939,- duet o additional demand in the D4 region has changed the allocation structure of the power plant production. The closer location of power plant in the production has caused significant decreasing in cost of transmission and distribution.

3.2.1. *Wheeling in the D1 region*

Table 3. Scenario for cost minimization, emission minimization, and losses minimization in region D1

	Minimize of cost			Minimize of Emission			Minimize of losses		
Per.	Cost (Rp)	Demand (MW)	Production (MW)	Cost (Rp)	Demand (MW)	Production (MW)	Cost (Rp)	Demand (MW)	Production (MW)
1	6,936,256	9,200	12,540	12,871,474	9,200	11,607	10,999,378	9,200	11,607
2	9,306,439	9,550	13,673	17,130,549	9,550	13,507	14,895,975	9,550	13,507
3	12,198,075	10,000	14,750	19,417,698	10,000	13,788	12,649,623	10,000	12,266
Tot.	28,440,769	28,750	40,963	49,419,721	28,750	38,901	38,544,976	28,750	37,379

The first scenario, minimizing costs. This scenario requires a total cost of Rp. 28,440,769 to produce electricity as much as 40,963 MW. Scenario 2, emission minimization. This scenario requires a fuel cost of Rp. 49,419,721 to produce electricity 38,901 MW. Scenario 3, Minimizing losses. This scenario requires a fuel cost of Rp. 38,544,976 to produce electricity as much as 37,379 MW, according to Table 3.

Additional demand causes an increase in the amount of production, and transmission / distribution costs. The first scenario requires an additional fuel cost of Rp. 11,453,890. The cost of wheeling 500 kv is Rp. 186,168, while the wheeling cost of 150 kv is Rp. 106,249. So that the total additional costs if there is wheeling in the D1 area of Rp. 11,746,307, detailed results in Table 4.

Table 4. Additional wheeling costs for region D1

	Minimize of cost	Minimize of Emission	Minimize of losses
Fw - F (Rp)	11,453,890	32,432,842	21,558,098
wheeling 500 KV (Rp)	186,168	262,960	178,323
wheeling 150 KV (Rp)	106,249	136,674	85,889
Additional cost (Rp)	11,746,307	32,832,476	21,822,310

3.2.2. *Wheeling in the D4 region*

Table 5. Scenario for cost minimization, emission minimization, and losses minimization in region D4

Per	Minimize of cost			Minimize of Emission			Minimize of losses		
	Cost (Rp)	Demand (MW)	Production (MW)	Cost (Rp)	Demand (MW)	Production (MW)	Cost (Rp)	Demand (MW)	Production (MW)
1	5,136,670	9,200	11,351	12,446,306	9,200	11,152	15,061,762	9,200	11,097
2	5,892,102	9,550	11,905	13,293,906	9,550	11,737	16,089,169	9,550	11,737
3	6,719,263	10,000	12,447	15,409,501	10,000	12,273	19,158,259	10,000	12,273
Tot	17,748,035	28,750	35,704	41,149,713	28,750	35,162	50,309,190	28,750	35,108

Table 5 present simulation in three different scenarios. The first scenario, minimizing costs requires total cost of Rp. 17,748,035 to produce electricity as much as 35,704 MW whist the scenario 2, i.e. emission minimization need fuel cost of Rp. 41,149,713 to produce electricity 35,162 MW. In the last scenario, minimizing losses, the cost required to produce as much as 35,108 MW electricity was Rp. 50,309,190.

It is known that additional demand has caused an increase in amount of production, and transmission / distribution costs. The first scenario requires an additional fuel cost of Rp. 761,156. The cost of wheeling 500 kv is Rp. 4,688, while in 150 kv the cost increased to Rp. 15,420. So that the total additional costs if there is wheeling in the D4 area was Rp. 781,264 (detail calculation provided in Table 6).

Table 6. Additional wheeling costs for region D4

	Minimize of cost	Minimize of Emission	Minimize of losses
Fw - F (Rp)	761,156	24,162,834	33,322,311
wheeling 500 KV (Rp)	4,688	240,035	290,727
wheeling 150 KV (Rp)	15,420	159,612	204,929
Additional cost (Rp)	781,264	24,562,482	33,817,968

3.2.3. *Wheeling in the D6 region*

Table 7. Scenario for cost minimization, emission minimization, and losses minimization in region D4

Per	Minimize of cost			Minimize of Emission			Minimize of losses		
	Cost (Rp)	Demand (MW)	Production (MW)	Cost (Rp)	Demand (MW)	Production (MW)	Cost (Rp)	Demand (MW)	Production (MW)
1	5,136,670	9,200	11,351	12,763,635	9,200	11,155	13,634,872	9,200	11,097
2	5,892,102	9,550	11,905	13,774,139	9,550	11,737	13,527,739	9,550	11,737
3	6,719,263	10,000	12,447	14,990,461	10,000	12,273	16,191,509	10,000	12,273
Tot	17,748,035	28,750	35,704	41,528,235	28,750	35,165	43,354,120	28,750	35,108

The first scenario is cost minimization. This scenario required total cost of Rp. 17,748,035 to produce electricity as much as 35,704 MW. In scenario 2, emission minimization, the fuel cost expected was Rp. 41,528,235 in order to produce 35,165 MW electricity. In scenario 3, minimizing losses, we found that the system requires the fuel cost of Rp. 43,354,120 to produce electricity as much as 35,108 MW (detail calculation in Table 6).

Additional demand causes an increase in amount of production, and transmission / distribution costs. The first scenario requires an additional fuel cost of Rp. 761,156 with cost of wheeling 500 kv as much as Rp. 31,941. So that the total additional costs if there is wheeling in the D6 area of Rp. Rp. 797,785 (see Table 8)

Table 8. Additional wheeling costs for region D6

	Minimize of cost	Minimize of Emission	Minimize of losses
Fw - F (Rp)	761,156	24,541,356	26,367,241
wheeling 500 KV (Rp)	4,688	245,837	251,747
wheeling 150 KV (Rp)	31,941	171,227	171,662
Additional cost (Rp)	797,785	24,958,421	26,790,650

4. Conclusion

Gencos gets the highest profit when serving basic load. Additional demand resulted additional revenue, however it caused decreasing in profit. This happens because additional revenue is not proportional to the increased additional production costs. The current price level should be evaluated considering the fluctuations in demand in the electricity market deregulation. Region 4 (D4) can be used as a distribution center for electricity customers outside the basic load. This region can generate the biggest profit with the smallest transmission and distribution costs.

Environmentally friendly electricity production is more expensive because low-emission plants require fuel-powered plants, which is more expensive than conventional power plant. Nevertheless, this high cost production can also be reduced by reducing losses, even though the smaller losses do not automatically lead to the cheaper total cost cheaper since it depend on the characteristics of the generator involved.

5. References

[1] Wahyuda, Santosa B and Rusdiansyah A 2018 Cost analysis of an electricity supply chain using modification of price based dynamic economic dispatch in wheeling transaction scheme. *IOP Conf. Ser. Mater. Sci. Eng.*

[2] Wahyuda and Santosa B 2015 Dynamic Pricing in Electricity: Research Potential in Indonesia. *Procedia Manuf.* **4** pp 300–306

[3] Bhattacharjee V and Khan I 2018 A non-linear convex cost model for economic dispatch in microgrids. *Appl. Energy* **222** (January) pp 637–648

[4] Zhou J, Wang C, Li Y, Wang P, Li C, Lu P and Mo L 2017 A multi-objective multi-population ant colony optimization for economic emission dispatch considering power system security. **45** pp 684–704

[5] Mahdi F P, Vasant P, Kallimani V, Watada J, Fai P Y S and Abdullah-Al-Wadud M 2018 A holistic review on optimization strategies for combined economic emission dispatch problem. *Renew. Sustain. Energy Rev.* **81**(March) pp 3006–3020

[6] Zhou J, Wang C, Li Y, Wang P, Li C, Lu P and Mo L 2017 A multi-objective multi-population ant colony optimization for economic emission dispatch considering power system security. *Appl. Math. Model.* **45** pp 684–704

[7] Liao G 2012 Integrated Isolation Niche and Immune Genetic Algorithm for solving Bid-Based Dynamic Economic Dispatch. *Electr. Power Energy* **42** pp 264–275

[8] Columbus C C and Simon S P 2012 Profit based unit commitment : A parallel ABC approach using a workstation cluster q. *Comput. Electr. Eng.* **38**(3) pp 724–745

[9] Wahyuda, Santosa B and Rusdiansyah A 2017 Load allocation of power plant using multi echelon economic dispatch. *AIP Conf. Proc.* 020007

[10] Columbus C C and Simon S P 2013 Profit based unit commitment for GENCOs using parallel NACO in a distributed cluster. *Swarm Evol. Comput.* **10** pp 41–58

Analysis of Supply Chain Network of FBS Production in Small and Medium Textile and Textile Product (TTP) Industry

N Nurhasanah[1], M Devana[1], B Aribowo[1], R Safitri[2], B Samiono[3], CF Lutfia[1], P Kalifa[1], SW Fauzia[1] and A Supriyanto[2]

[1]Industrial Engineering Department, Faculty of Science and Technology, Universitas Al Azhar Indonesia, Jakarta 12110, Indonesia
[2]Informatic Engineering Department, Faculty of Science and Technology, Universitas Al Azhar Indonesia, Jakarta 12110, Indonesia
[3]Management Department, Faculty of Economy, Universitas Al Azhar Indonesia, Jakarta 12110, Indonesia
nunungnurhasanah74@gmail.com

Abstract. Textile and textile product industries are one of prioritized industries to be developed for its strategic roles in national economy, e.g. contributing to foreign exchange, absorbing a large number of labours, and serving to meet national clothing need. TTP industry has a relatively complex supply chain. Problems in supply chain in companies implementing supply-chain management (SCM) that may occur, among others, are those related to supply-chain mapping. This SCM is used to carry out monitoring, controlling, and identifying position of an organization relative to its competitor and to objectives that they want to achieve. One of companies engaging in TTP industry is FBS Small and Medium Industry (SMI) owned by Mr. Rico since 2004. The company is a make-to-stock manufacturer which produce boys clothing ranging from jeans, t-shirts, to shirts. We conducted interview with experts in the company. Data was collected from the interview with the experts (CEO of the company and his reliable assistant). Based on discussion of FBS SMI supply-chain, it was concluded that basically the flow from suppliers to consumers was same in each production area. Raw materials are transferred from Kebayoran Headquarter to each production areas spread across Java Island, e.g. Jakarta, Pandeglang, Central Java, and Sukabumi, with different CMTs. It is expected that supply-chain mapping analysis of this company is able to clarify the flow of production from raw materials to finished goods reached by consumers.
Keywords: supply-chain management, TTP industry, CMTs

1. Introduction

Textile and textile product (TTP) industries are one of prioritized industries to be developed for its strategic roles in national economy, e.g. contributing to foreign exchange, absorbing a large number of labours, and serving to meet national clothing need. TTP industry has a relatively complex supply chain. Problems in supply chain in companies implementing SCM that may occur, among others, are those related to supply-chain mapping. This supply-chain management system is used to carry out monitoring, controlling, and identifying position of an organization relative to its competitor as well as to objectives that they want to achieve. One of companies engaging in TTP industry is FBS Small and Medium Industry (SMI) owned by Mr. Rico since 2004.

2. Methods
The first step carried out in this research was observation and literature study. Observation is conducted at FBS Headquarters at Kebayoran, South Jakarta. At there, we saw varieties of products, the processes from start (raw material) to end (finish goods/FG), the places for production, the beginning business story starts and introduced directly by the owner, And study author do research on all aspects required in research. Followed by the second step, e.g. identification of problems arising in supply-chain management of FBS Small and Medium Industry. Subsequently, problem statements were composed. Data was collected from raw material suppliers, production areas, CMIT, and retails of the FBS SMI. The data is from the owner, Mr. Rico and his right-hand man, Mr. Mukhlas. Data processing was performed to the collected data by mapping out supply-chain of the FBS SMI. Data analysis was done to the results obtained from data processing which is mapping out supply-chain of the FBS SMI production areas such as Jakarta, Sukabumi, Pandeglang and Central Java. In the final step, conclusion was drawn from the analysis and discussion of the results.

3. Results and Discussion

3.1. Supply-Chain Management (SCM) Network of FBS Small and Medium Industry
In order to map out supply-chain flow of FBS SMI, we started by conducting interview with experts from the company. Results of this interview showed list of suppliers, production areas, the CMT, and stores or retails where the products are being sold, e.g. Tanah Abang Market Block A.

3.2. List of Suppliers of FBS SMI
- Mandiri Sukses
- HN
- Putra Nagatama
- PD Naga Sakti
- Kwalitas Super Tekstil Pty Ltd

3.3. List of Suppliers and CMT

Table 1. Production areas and CMT

Production Areas	CMT	Production Areas	CMT
Jakarta	Joglo	Sukabumi	Sukamaju
	Cipadu		Padabeunghar
	Petukangan		Parakanlima
	Grogol		Narogong
Pandeglang	Petir		Pasir Batok
	Baros		Bojong Waru
	Cibokor Nanggor		Ciherang
	Cibuah		Ciguyang
Central Java	Pemalang		
	Purbalingga		

3.4. Outline of Supply-Chain Network from Supplier to Consumer

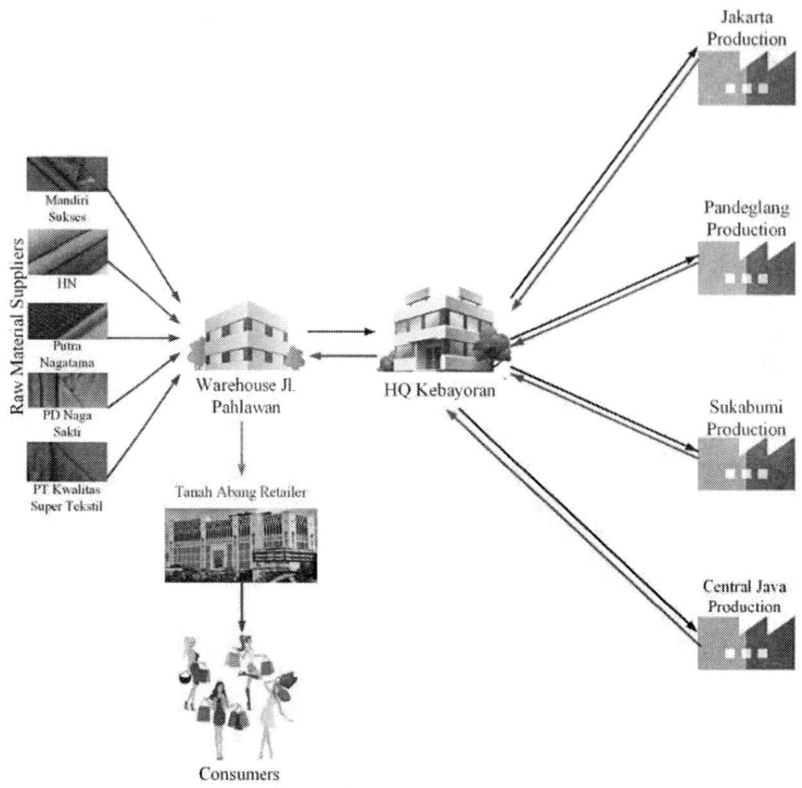

Figure 1. Outline of Supply-Chain Network of FBS SMI

The leftmost part of Figure 1 above shows the list of cloth suppliers of FBS SMI. Clothes from these suppliers are sent to a warehouse at Jl. Pahlawan where raw materials of FBS factory are stored. After being received in the warehouse at Jl. Pahlawan, the raw materials are sent to Headquarter at Kebayoran for cutting. The raw materials are then divided into particular amounts before being picked up by their production (dressmaking) places of children clothes which are located in different areas, e.g. Jakarta, Pandeglang, Sukabumi, and Central Java. The finished goods are sent back to Kebayoran for quality control (QC). After passing QC process, the finished goods are returned to the warehouse at Jl. Pahlawan for laundry and ironing (finishing), followed by packing. Once the packing is done, the FGs are distributed immediately to stores/retails of FBS at Tanah Abang Market Block A where consumers purchase them.

3.5. Mapping of Supply-Chain Network at Jakarta Production Area

The leftmost part of Figure 2 above shows the list of cloth suppliers of FBS SMI. Cloths from these suppliers are sent to a warehouse at Jl. Pahlawan where raw materials of FBS factory are stored. After being received in the warehouse at Jl. Pahlawan, the raw materials are sent to Headquarter at Kebayoran for cutting. The raw materials are then divided into particular amounts before being picked up by their production (dressmaking) places of children clothes which are located in four different places of Cut, Make and Trim (CMT) in Jakarta, e.g. Cipadu CMT, Grogol CMT, Petukangan CMT, and Joglo CMT. The FGs are sent back to Kebayoran for quality control (QC). After passing QC process, the FGs are returned to the warehouse at Jl. Pahlawan for laundry and ironing (finishing), followed by packing. Once the packing is done, the FGs are distributed immediately to stores/retails of FBS at Tanah Abang Market Block A where consumers purchase them.

IOP Conf. Series: Materials Science and Engineering **528** (2019) 012085 doi:10.1088/1757-899X/528/1/012085

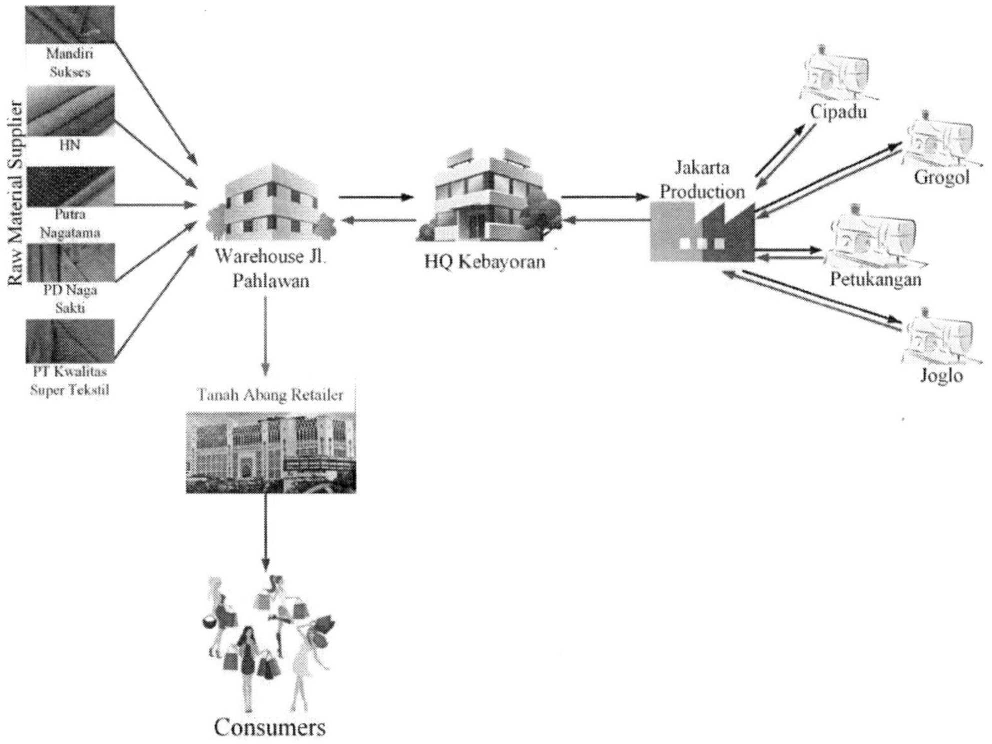

Figure 2. Supply-Chain Network at Jakarta Production Area

3.6. Mapping of Supply-Chain Network in Pandeglang Production Area

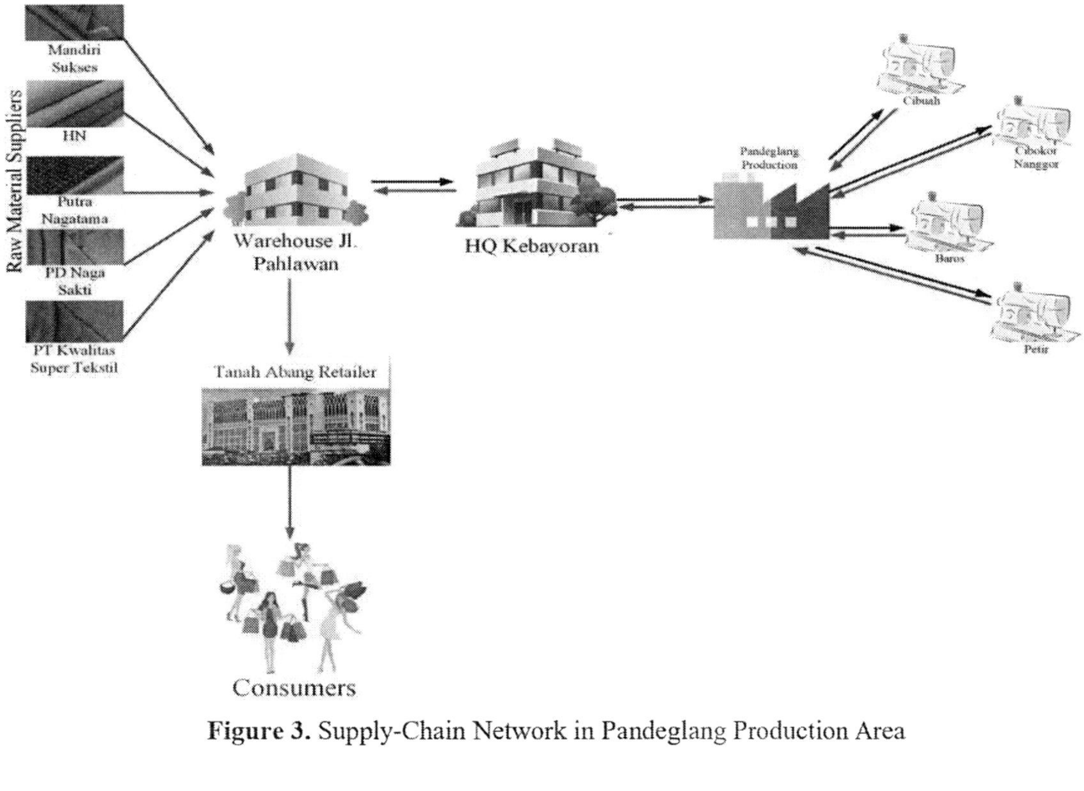

Figure 3. Supply-Chain Network in Pandeglang Production Area

The leftmost part of Figure 3 above shows the list of cloth suppliers of FBS SMI. Cloths from these suppliers are sent to a warehouse at Jl. Pahlawan where raw materials of FBS factory are stored. After being received in the warehouse at Jl. Pahlawan, the raw materials are sent to Headquarter at Kebayoran for cutting. The raw materials are then divided into particular amounts before being picked up by their production (dressmaking) places of children clothes which are located in four different places of CMT in Pandeglang, e.g. Petir CMT, Baros CMT, Cibuah CMT, and Cibokor Naggor CMT. The FGs are sent back to Kebayoran for quality control (QC). After passing QC process, the FGs are returned to the warehouse at Jl. Pahlawan for laundry and ironing (finishing), followed by packing. Once the packing is done, the FGs are distributed immediately to stores/retails of FBS at Tanah Abang Market Block A where consumers purchase them.

3.7. *Mapping of Supply-Chain Network in Sukabumi Production Area*

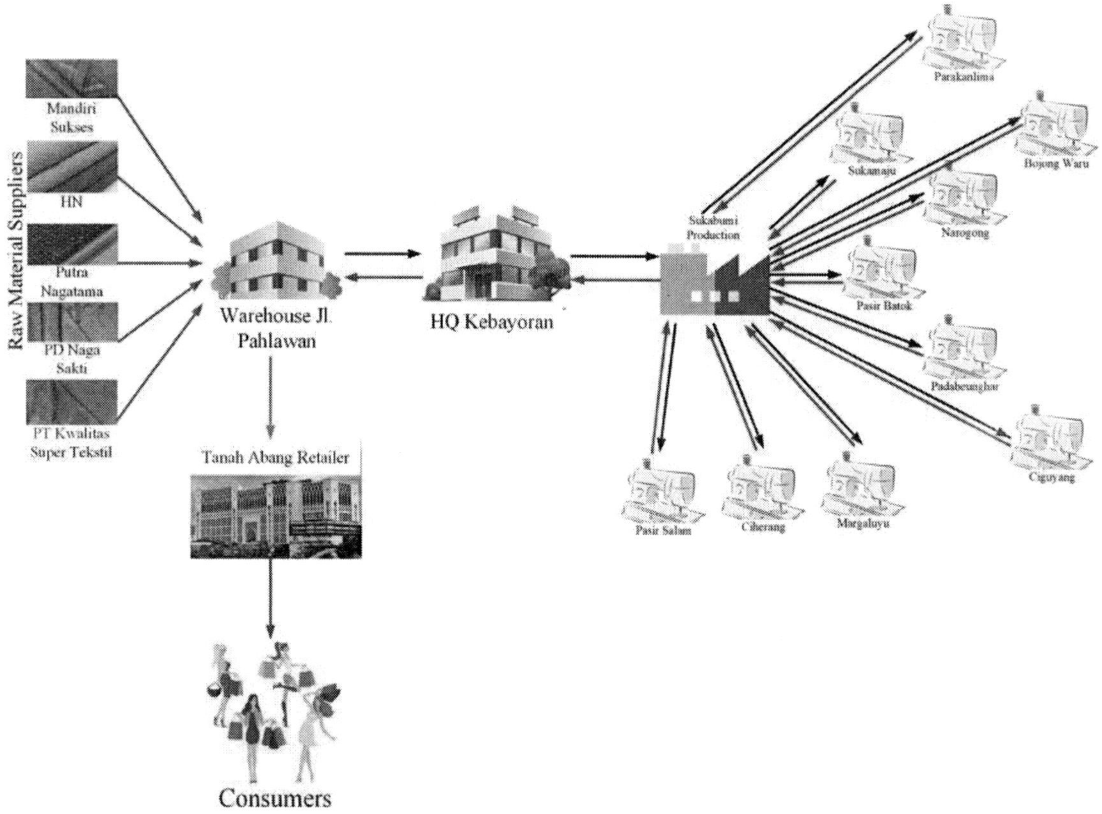

Figure 4. Supply-Chain Network in Sukabumi Production Area

The leftmost part of Figure 4 above shows the list of cloth suppliers of FBS SMI. Cloths from these suppliers are sent to a warehouse at Jl. Pahlawan where raw materials of FBS factory are stored. After being received in the warehouse at Jl. Pahlawan, the raw materials are sent to Headquarter at Kebayoran for cutting. The raw materials are then divided into particular amounts before being picked up by their production (dressmaking) places of children clothes which are located in ten different places of CMT in Sukabumi, e.g. Parakanlima CMT, Sukamaju CMT, Bojong Waru CMT, Narogong CMT, Pasir Batok CMT, Padabeunghar CMT, Ciguyang CMT, Margaluyu CMT, Ciherang CMT, and Pasir Salam CMT. The FGs are sent back to Kebayoran for quality control (QC). After passing QC process, the FGs are returned to the warehouse at Jl. Pahlawan for laundry and ironing (finishing), followed by packing. Once the packing is done, the FGs are distributed immediately to stores/retails of FBS at Tanah Abang Market Block A where consumers purchase them.

4. Conclusion

Based on discussion of supply-chain management of FBS SMI above, it is concluded that basically the flow from suppliers to consumers was same in each production area. Raw materials are transferred from Kebayoran Headquarter to each production areas spread across Java Island, e.g. Jakarta, Pandeglang, Central Java, and Sukabumi, with different CMTs. It is expected that supply-chain mapping analysis of this company is able to clarify the flow of production from raw materials to finished goods reached by consumers.

Acknowledgement

Thanks to Directorate of Research and Community Service of the Ministry of Research and Technology of the Republic of Indonesia which funded this research activity in 2018 with an Applied Product Research scheme.

5. References

[1] Anatan L and Ellitan L 2008 *Supply Chain Management. Theory and Application.* (Bandung: Alfabeta)
[2] Chopra and Meindl P S 2001 *Supply chain management: Strategy, planning, and operations*. (New Jersey: Prentice Hall)
[3] Ganeshan, Harison and Hugos 2003 An Introduction to Supply Chain. *Supply Chain Journal* (New York)
[4] Indrajit R E and Djokopranoto R 2002 *Konsep Management Supply Chain: Cara Baru Memandang Mata Rantai Penyediaan Barang*. (Jakarta: PT Gramedia Wiiasarana Indonesia)
[5] Nurhayati dkk 2012 Pengertian Industri Kecil dan Menengah. *Jurnal Administrative Reform.* **2**(3) pp 1341-1354
[6] Pujawan I N 2005 *Supply Chain Management.* (Surabaya: Guna Widya)
[7] Supply-Chain Council Team 2006 *Supply Chain Operation Reference Model Version 8.0*. (Supply Chain Council, inc.)

IOP Conf. Series: Materials Science and Engineering **528** (2019) 012086 doi:10.1088/1757-899X/528/1/012086

Supply Chain Risk Management on Wooden Toys Industries by using House of Risk (HOR) and Analytical Network Process (ANP) Method

W N Tanjung[1], R S Khodijah[2], S Hidayat[3], E Ripmiatin[4], S A Atikah[5] and S S Asti[6]

[1,2,3,5,6]Industrial engineering department, Faculty of science and technology, Universitas Al Azhar Indonesia, Jakarta 12110, Indonesia
[4]Informatic engineering department, Faculty of science and technology, Universitas Al Azhar Indonesia, Jakarta 12110, Indonesia
[1]widya@uai.ac.id, [2]ratustk@gmail.com

Abstract. Good risk management will reduce the impact and risk of an event. In this study discussed the risks that may arise in the wooden toy industry supply chain and how to manage these risks. The calculation divided into three phases; there are risk mapping, risk assessment, and risk management. House of Risk (HOR) is used in calculating potential risks and Analytical Network Process (ANP) is used to choose the best strategy for supply chain risk management. Followed by HOR calculation, risk that has the greatest value and selected as potential risk is the risk of price/cost fluctuations with an ARP value of 432. The strategy of this risk is grouped based on 2 causes, namely raw material price fluctuations and poor management. With ANP calculations, the strategy for raw material price fluctuations are understanding and determining the quantity of material used (28.14%) and combining woods with various commodities (23.23%). Then the alternative for poor management are doing strategic financial accounting (54.48%) and doing financial planning (33.07%).
Keywords: risks, strategy, supply chain risk management, house of risk, analytical network process

1. Introduction

The more intense competition requires companies to be able to compete in their fields. This has triggered the company to reduce the mistakes and prevent risks in any sector. One of the most important sectors of the company in the continuity of its business processes is their supply chain sector. The supply chain is a network of companies that work together to create and deliver a product to the end user [14]. Supply Chain Management (SCM) is a systematic and strategic coordination of traditional business functions within and across companies in a supply chain to develop the company's long-term performance and overall supply chain [13]. In another way, SCM was defined as an activity of managing activities in order to obtain the raw material into goods in process or semi-finished goods and finished goods and then send the product to consumers through a distribution system [9]. In the supply chain process, there are various risks that can affect the supply chain flow and it's caused the processes cannot run properly. The risk to the supply chain can be defined as an event caused by an imbalance between demand and supply [7]. Therefore, risks in the supply chain also defined as disruption of information and resource flows in supply chain networks due to uncertain terminations and variations [2]. Handling of risks that disrupt the supply chain sector in the supply chain sector must be overcome with good risk management. Risk management model, which measures the effects that will be caused by risk, is part of the risk management process, which covers the entire process of risk analysis and risk evaluation [12]. In general, the supply chain risk management process consists of risk identification, risk analysis, risk evaluation and risk mitigation. Risk identification is suggested as a fundamental step in the risk management process [6].

SCRM is a concept that is used to control and coordinate the risks that can be generated and mitigation actions on the company's supply chain. So with regard to the risks in supply chain management, SCRM plays an important role in keeping the supply chain system from being disturbed by the risks that might occur. Several studies have been conducted on the topic of supply chain risk identification and evaluation such as, hazard operability (HAZOP) analysis, FMEA and fuzzy theory, and Fuzzy FMEA [14]. Supply Chain Risk Management is a blend of the concept of Supply Chain Management with Risk Management. The risks to the supply chain can be defined as a place of events caused by an imbalance between demand and supply. Supply chain disruptions can lead to problems such as lead time, stock out, inability to meet customer demand, and increased costs [4][3]. The magnitude of risk can be measured by considering two fundamental parameters of risk, namely the possibility of risk and risk severity [3]. However, it must be realized also that the extent of certain risks is also highly dependent on many factors involved, such as human factors, workplace factors, material and equipment factors, etc. that are difficult to measure and handle in the traditional way [17].

In previous study, risks that may occur in the wooden toys industries are delays in material delivery schedules, suppliers unable to meet material needs, errors in inspection procedures when material arrives, errors in calculating material requirements, material stock out during the production process, and other risks [8]. However, to solve these problems some methods need to be developed. House of Risk (HOR) is a method can be used in this case. HOR Model is used to identify risk agent and focused in preventive actions [2][14]. HOR is a modification of FMEA (Failure Modes and Effect of Analysis) and a quality house model (HOQ) to prioritize the risk sources which are the first to be chosen for the most effective actions in order to reduce the potential risks from risk sources [16]. After that, strategy for risk management can be done using the Analytical Network Process (ANP) method. The ANP is a generalization of the Analytic Hierarchy Process (AHP), by considering the dependence between the elements of the hierarchy. Many call issues can not be structured hierarchically as a result of they involve the interaction and dependence of higher-level components in an exceedingly hierarchy on lower-level components. Therefore, ANP is represented by a network, rather than a hierarchy [15].

2. Methods

The first stage in this study is identifying risks that often occur in the company's supply chain as be seen in Figure 1. The data used is qualitative and the data is obtained through interviews with selected experts who are considered to understand the company's supply chain. Then the identification of risks that have the potential to appear in the company's supply chain regarding the risks that occur, the source of the causes of risk, where the risk is and how the risk arises.

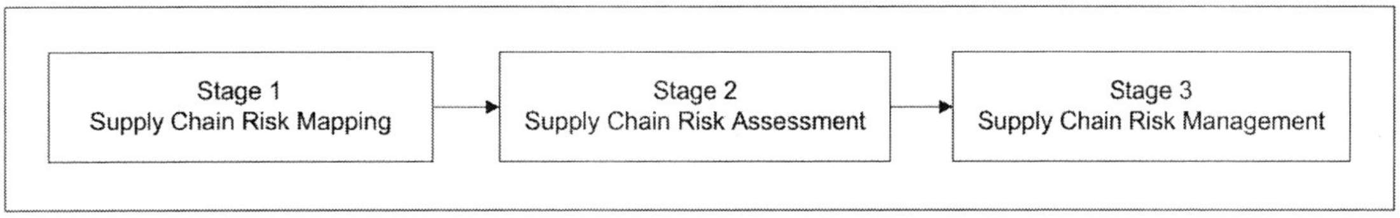

Figure 1. Supply Chain Risk Management Stages

The second stage is an assessment of the risks identified. Assessment is done with a 1-10 rating by experts. The assessment is in the form of severity of the risk event, the occurrence of the risk agent, and the correlation of the risk agent and risk event. Then it is calculated the value of Aggregate Risk Potential

(ARP) as a value to determine the risk of a priority agent that needs to be carried out risk management. Then the ARP calculation's for identifying the highest ARP value. After obtaining a risk based agent on the highest ARP value can be seen in the Pareto diagram, the risk agent is managed by identifying what strategies must be carried out to mitigate the risk. Here's the formula from ARP.

$$\mathrm{ARPj} = \mathrm{Oj} \times \sum(\mathrm{Si} \times \mathrm{Rij}) \tag{1}$$

Information :
ARPj : Aggregate Risk Potential
Oj : Occurance
Si : Severity
Rij : Corellation between Risk Event dan Risk Agent

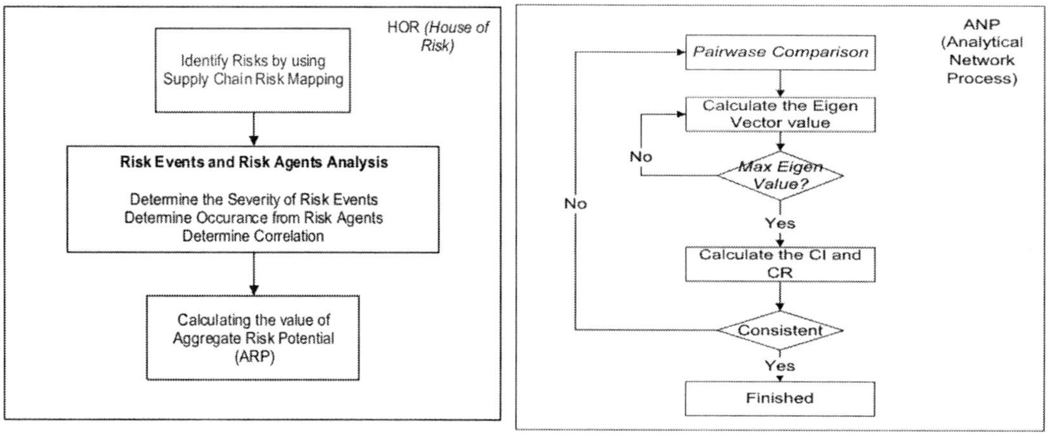

Figure 2. HOR and ANP Methodology

The third stage, then it was identified several alternative strategies related to risk mitigation actions based on journal references and then adjusted to the situation of the company's objects. After several alternative strategies have been identified, the best alternative strategies are chosen by using Analytical Network Process (ANP) method and the steps for all methods can be seen in figure 2.

The data needed is qualitative data obtained from the same expert to provide an assessment of each of the strategies that have been identified in the form of pairwase comparison. Then an expert assessment is done using pairwase comparison by bringing together the assessment of the three experts using Geometric Mean (GM). Here's the formula from GM.

$$\mathrm{GM} = (\mathrm{R1 \, x \, R2 \, x \ldots x \, Rn})^{1/n} \tag{2}$$

Information :
GM : Geometric mean
R : Value of respondents questionnaire n
n : Number of respondent

Then calculate the eigen value of the vector, choose the largest eigen vector, and calculate the consistency index and consistency ratio. If eigen vector is consistent, continued with supermatrix identification and ranking the results of the calculation of the strategy. Here's the formula from CI and CR.

$$CI = (\lambda\mathrm{max} - n)/(n-1) \tag{3}$$

Information :
CI : Consistency Index
λ : Eigen vector value
n : Number of alternative/criteria

$$CR = \mathrm{CI}/\mathrm{RI} \tag{4}$$

Information :

IOP Conf. Series: Materials Science and Engineering **528** (2019) 012086 doi:10.1088/1757-899X/528/1/012086

CR : Consistency Ratio
RI : Random index (n=3; RI=0,52)

3. Result and Discussion

3.1 Risk Identification and Assessment

Before the selection of priority risk, risk identification is carried out based on all activities in the supply chain conducted by wooden toys industries. Then the risks are classified based on sub-risks. The following is the classification and assessment of these risks.

Table 1. Risk Events, Risk Agents, and Corellation Assessment

Symbol	Risk Event	Severity	Symbol	Risk Agent	Occurance	Relation RA & RE
E1	Demand Risk	6	A1	Competitor Moves	6	2
			A2	Delays in Delivery to Customers	6	9
			A3	Forecast Error	8	3
			A4	Market Saturation	4	7
E2	Environment Risk	6	A5	Macroeconomic Uncertainty	8	5
			A6	Natural Disasters	4	7
			A7	Policy Uncertainty	5	2
			A8	Social, Culture & Politic Uncertainty	5	3
E3	Financial Risk	7	A9	Cost/Price Risks	8	9
			A10	Exchange Rate Risk	3	3
E4	Information Risk	6	A11	Breakdown of IT Infrastructure	3	5
			A12	Distorted Information	4	3
			A13	Inadequate Information Security	4	3
			A14	Information Delay	5	3
			A15	Wrong Choice of Communication	4	2
E5	Operational Risk	7	A16	Capacity Inflexibility	6	3
			A17	Design Changes	7	7
			A18	Disruption in Production	5	9
			A19	Inventory Risks	6	2
			A20	Variability in Production Process	5	2
E6	Supply Risk	7	A21	Dependency on Single Supplier	4	3
			A22	Inflexibility of Supplier	6	3
			A23	Poor Delivery Performance	3	7
			A24	Supplier Poor Quality	5	9
			A25	Supplier Bankruptcy	4	9

Risk identification is obtained from one of the journal references [10] namely the supply chain risk category in general, then adjustments to the situation in the wooden toy industry supply chain. After obtaining an assessment from each expert, the average of the three values is calculated to get one value rounded down, it is done to make it easier to do the next calculation. As for risk assessment, a 1-10 scale is used to facilitate experts in assessing severity, occurrence and correlations. Then the Aggregate Risk Potential is calculated on the HOR method. Following are the calculations.

3.2 ARP Calculation and Pareto Diagram

Table 2. ARP calculation on the HOR Method

Risk Event	*Risk Agent* A1	A2	A3	A4	A5	A6	A7	A8	A9	...	A17	A18	A19	A20	A21	A22	A23	A24	A25	*Seve-rity*
E1	2	9	3	7																5
E2					5	7	2	3			*									5
E3									9											6
E4																				5
E5											7	9	2	2						6
E6															3	3	7	9	9	6
Occurance	6	6	8	4	8	4	5	5	8	...	7	5	6	5	4	6	3	5	4	
ARPj	60	270	120	140	200	140	50	75	432	...	294	270	72	60	72	108	126	270	216	

ARP calculation results can be seen in Table 2, where the calculation involves the risk events symbolized by E and the risk agent symbolized by A. Then the severity, occurance and relations that have been obtained from the opinions of the three experts are used in this calculation. To get the ARP value, the following equation is used. Here are some examples of ARP calculations using the equation (1).

$ARP_1 = 6 \times \sum [\, 5\,(2)] = 60$
$ARP_2 = 6 \times \sum [\, 5\,(9)] = 270$
$ARP_3 = 8 \times \sum [\, 5\,(3)] = 120$

After obtaining ARP values from each risk agent, then the ARP values are ranked according to the largest ARP value. There is pareto diagrams to make it easier to see ARP calculation results visually. Here is the pareto diagram.

Figure 3. ARP Pareto Chart

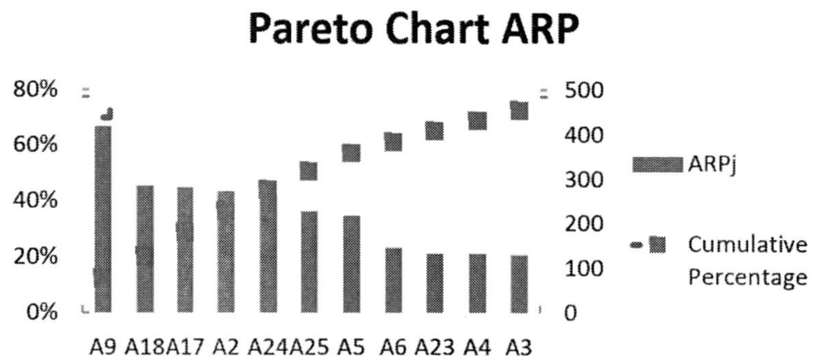

Pareto diagram is obtained from the results of the plot of ARP calculation data and cumulative percentage, so that the results obtained that the largest ARP value is A9 or risk agent price or cost with 432. In other words, the price or cost risk agent is a risk that is considered urgent so that further management by wooden toys industries.

3.3 Alternative Strategies Identification and Assessment

The result of using HOR is price or cost risk, the risk is used as a reference to identify several alternative strategies for managing the risks that have been obtained. After identifying alternative strategies from journal references, adjusting the alternative strategies to the situation in the wooden toy industry supply chain. Then the best alternative strategy is chosen by using the ANP method. Based on several sources, some of the causes of risk of price fluctuations include fluctuations in raw material prices and poor management. Here are some alternative strategies that have been identified from each of the causes of the risk of price fluctuations.

ANP calculations are must be done. Then an assessment by the same expert was conducted regarding the comparison of each alternative strategy by the same three experts. The assessment is arranged into pairwise comparison matrix. To get one value from the three expert judgments, a geometric calculation is calculated to make an assessment. Following are the results of the geometric mean calculation.

Table 3. Pairwase Comparison Matrix of Price Fluctuations Strategies

	T1 : Track price changes	**T2 : Establish clear rules with suppliers**	**T3 : Understand and determine the quantity of material used**	**T4 : Set a different savings goal**	**T5 : Combining woods with various commodities**
T1 : Track price changes	1	0.8434	0.3057	0.3625	0.7211
T2 : Establish clear rules with suppliers	1.1856	1	1.1696	0.3625	0.7539
T3 : Understand and determine the quantity of material used	3.2711	0.8550	1	2.7589	0.7211
T4 : Set a different savings goal	2.7589	2.7589	0.3625	1	0.7539
T5 : Combining woods with various commodities	1.3867	1.3264	1.3867	1.3264	1

Table 4. Pairwase Comparison Matrix of Poor Management Strategies

	M1 : Do strategic financial accounting	**M2 : Recruit external accountants**	**M3 : Do financial planning**
M1 : Do strategic financial accounting	1	4.2172	1.71
M2 : Recruit external accountants	0.2371	1	0.3625
M3 : Do financial planning	0.584804	2.7589	1

3.4 Geometric Mean Calculation

A study sometimes uses a geometric mean or a mean to find out the results of individual evaluations from several experts based on the questions posed in the form of comparisons (pairwase comparison) [1]. Geometric mean can be calculated by the following equation. Here are some examples of GM calculations using the equation (1).

$GM_{T1T2} = (3 \times 0{,}2 \times 1)^{1/3} = 0{,}8434$

$GM_{T1T3} = (0{,}2 \times 1 \times 0{,}1429)^{1/3} = 0{,}3057$

$GM_{T1T4} = (0{,}3333 \times 1 \times 0{,}1429)^{1/3} = 0{,}3625$

3.5 Consistency Index Calculation

A consistency ratio is a ratio that states whether the judgment provided by the expertise is consistent or not [5]. It is recommended that it should be less than or equal to 0.10. Inconsistency may be an adjustment needed to improve the consistency of the comparisons. But the adjustment should not be as

large as the judgment itself, nor small that it would have no consequence [10]. Previously the calculation of consistency index was done using the following equation (2). Here are the CI calculations.
$CI_{poor\ management} = (3{,}0013 - 3)/(3 - 1) = 0{,}000699$

3.6 Consistency Ratio Calculation
The calculation of consistency ratio by using the following equation (3). Here are the CR calculations.
$CR_{poor\ management} = 0{,}000699/0{,}52 = 0{,}00062$

3.7 Results of Alternative Strategy Selection
After calculating the CI and CR from both of these strategy assessments, the results of the two strategy experts from the three experts have a consistency level below 10%, which means that expert judgment shows consistent answers that can be used in this study. Then the determination of alternative strategies is ideal for mitigating the risk of price or cost fluctuations.

In Figure 3, an alternative strategy is obtained from the two causes of price or cost risk. In the alternative raw material price fluctuations strategy denoted by T, the largest eigenvalue or normals value is T3 (Understand and determine the quantity of material used) with 28.14% and T5 (Combining woods with various commodities) with 23.23%. As for the alternative poor management strategies denoted by M, the highest eigenvalue or normals value is M1 (Doing strategic financial accounting) with 54.48% and M3 (Doing financial planning) with 33.07%.

4. Conclusion
Based on the results of the study on the determination of priority risks from several risks that exist in the Wooden toys industries, obtained several risks that are classified based on 6 risk events with 25 risk agents. The results of data processing using the HOR (House of Risk) method are obtained by the risk agent with the largest ARP, namely A9 or price or cost risk agent with a value of 432. It can be concluded that the risk agent must be prioritized and considered urgent in the supply chain Wooden toys industries is a price or cost risk agent.

The risk management actions are selected using ANP. The strategy of price or cost risk is grouped based on 2 causes, namely raw material price fluctuations and poor management. The strategy of raw material price fluctuations consists of 5 alternative strategies notated by T, while the poor management strategy consists of 3 alternative strategies notated by M. For raw material price fluctuations, T3 is chosen (Understanding and determining the quantity of material used) with 28.14% and T5 (Combining woods with various commodities) with 23.23%. The alternative poor management strategies are M1 (Doing strategic financial accounting) with 54.48% and M3 (Doing financial planning) with 33.07%. Therefore, the alternative strategy chosen can be used as an action to manage price or cost risk in the hope of reducing or preventing the occurrence of price or cost risks in the wooden toys industries.

Acknowledgements
This work was supported in part by Simlitabmas Ristekdikti followed SK No.3/E/KPT/2017 and contract with LP2M Al Azhar Indonesia University under agreement No.064/SPK/A-01/UAI/IV/2018.

5. References
[1] Ascarya 2005 Analytic Network Process (ANP) : Pendekatan Baru Studi Kualitatif. Makalah Dipresentasikan pada Seminar Intern Program Magister Akuntansi Fakultas Ekonomi di Universitas Trisakti, Jakarta
[2] Badariah N, Surjasa D, and Trinugraha Y 2013 Analisa Supply Chain Risk Management Berdasarkan Metode Failure Mode and Effects Analysis (FMEA). *Jurnal Teknik Industri*. 110-118
[3] Chopra S and Shodi M S 2007 Managing Risk to Avoid Supply-Chain Breakdown. *Sloan Management Review* pp 53-61

[4] Chopra S and Meindl P 2013 *Supply Chain Management: Strategy, Planning, and Operations.* (New Jersey: Prentice Hall Inc)
[5] Ekawati R, Trenggonowati D L and Aditya V D 2018 Penilaian Performa Supplier Menggunakan Pendekatan Analytic Network Process (ANP). *Journal Industrial Services* pp 151-158
[6] Hallikas and Virolainen V M 2004 Risk management processes in supplier networks. *International Journal of Production Economics* pp 47–58
[7] Handayani D I 2016 Potensi Risiko Pada Supply Chain Risk Management. *Spektrum Industri* pp 1-108
[8] Himawan S 2017 *Pengelolaan Risiko Rantai Pasok Menggunakan Metode Fuzzy House of Risk (FHOR).* (Jakarta: Universitas Al Azhar Indonesia)
[9] J Heizer and B Rander 2004 *Manajemen Operasi.* (Jakarta: Salemba)
[10] Jafarnejad M, Ebrahimi M, Abbaszadeh M A, dan Abtahi S M 2014 Risk Management in Supply Chain using Consistent Fuzzy Preference Relations. *International Journal of Academic Research in Business and Social Sciences* pp 77-89
[11] Jüttner U 2005 Supply Chain Risk Management: Understanding the Business Requirements from a Practitioner Perspective. *Int. J. of Logistics Management* pp 120-141
[12] Nasution S, Arkeman Y, Soewardi K, and Djanata T 2014 Identifikasi dan Evaluasi Risiko Menggunakan Fuzzy FMEA pada Rantai Pasok Argoindustri Udang. *Jurnal Riset Industri* pp 135-146
[13] Pujawan I N and Geraldin L H 2009 House of Risk: a Model for Proactive Supply Chain Risk Management. *Jurnal Bussiness Process Management* **15**(6) pp 953-966.
[14] Pujawan I N 2010 *Supply Chain Management.* 2nd Ed. (Surabaya: Guna Widya)
[15] Saaty T and Vargas L 2006 Decision Making with *the Analytic Network Process.* Economic, Political, Social and Technological Applications with Benefits, Opportunities, Costs and Risks. (Pittsburgh: RWS Publication)
[16] Ulfah M, Maarif M S, Sukardi and Raharja S 2015 Analisis dan Perbaikan Manajemen Risiko Rantai Pasok Gula Rafinasi dengan House of Risk Analysis. *Jurnal Teknologi Industri Pertanian* (Bogor: Fakultas Teknologi Pertanian IPB)
[17] Zeng J, An M and John S N 2007 Application of A Fuzzy Based Decision Making Methodology to Contruction Project Risk Assessment. *Int. J. of Project Management* pp 589-900

Risk Mitigation for Agricultural Products Distribution in Agro-business Terminal Mantung, Kabupaten Malang

T Oktiarso[1] and A H K Nadira[1]

[1] Industrial Engineering Department, Universitas Ma Chung, Jl. Villa Puncak Tidar N-01, Malang, Indonesia
oktiarso@yahoo.com, oktiarso@gmail.com

Abstract. Distribution process of the horticultural products in Kabupaten Malang is affected by many factors. The distribution factors need to be identified to find the potential risks arise from every problem, in order to anticipate the barriers and losses. This research is conducted using the Supply Chain Operation Reference (SCOR) method in identifying risks and their risk agents, and using the House of Risks (HOR) method in analysing every risk to determine the mitigation strategies. The results for STA Mantung discovered 15 risks with 23 risk agents. From these 23 risk agents, 12 major risk agents were selected and analysed to determine risk mitigation strategies. There are 14 horticultural mitigation strategies. With these risk management and mitigation strategy the STA Mantung shall be better prepared to face them.
Keywords: SCOR, HOR, Agricultural

1. Introduction

Agricultural products distributions always affected by many factors. [1] found twelve variables affected on distribution of food grain in Punjab, India. [1] Their research found that for six month distributions of food grain, storage anf time of distributions dominates the factors in the distributions. Factors affected in agricultural product distribution have to measure to know the performance of the distribution activity. Thera are five categories for performance measurement which devided into 17 criteria for these five categories [2]. The categories and criteria is aimed to measure the performance of distriibution channels in distributing commodities from farmers to end customers. Because agricultural products are perishable, they have risk in distributing from farmers to end customers. Yeboah, et al, found that there are eight risk factors affected in agricultural distributions [3]. Market effects, logistic, weather, biological related and government policy is the most affected in distributions of agricultural products. A solution to overcome risks is contract between distributions channels in agricultural products marketing. [4] Contracts will hel each distribution channels to minimize risk of deterioting the products. HOR and SCOR is the effective method to find out the risk agents and mitigations. [5]

Kabupaten Malang is one of the districts that are producers of agricultural products to various regions. Agricultural products from Malang Regency are rice, vegetables, tubers, and fruits. Some sub-districts which are the main producers in Kabupaten Malang are Dau, Poncokusumo, Ngantang and Pagelaran. As one of the producers, farmers, traders and suppliers must carry out a good distribution process evenly. The agricultural products of Kabupaten Malang are distributed to various regions in Indonesia, both in Kabupaten Malang itself, to outside the province of East Java. However, in the process of distributing agricultural products, there are still some problems. One of them is that some regions in Kabupaten Malang have difficulties in making maximum sales for their agricultural products. Therefore, the process of distributing agricultural products is considered to be less than optimal considering that this is still the case in the distribution process. As one of the producers, farmers, traders and suppliers must carry out a good distribution process evenly. The agricultural products of Kabupaten Malang are distributed to various regions in Indonesia, both in Kabupaten Malang itself, to outside the province of East Java. However, in the process of distributing agricultural products, there are still some problems. One of them is that some regions in Kabupaten Malang have difficulties in making maximum sales for their agricultural products. Therefore, the process of distributing agricultural products is considered to be less than optimal considering that this is still the case in the distribution process.

Because problems are still often found in the supply chain of agricultural products in Malang Regency, it is necessary to identify and analyze risk to find the risks arising from any existing problems in order to anticipate the losses and obstacles that can be caused. Risk management and supply chain risk analysis is defined as supply chain management. This supply chain management is able to find the cause of the problems caused through the process of identification and analysis of the distribution process. After an analysis, an appropriate strategy will be found to maximize the distribution supply chain, in this case the distribution of agricultural products from vegetable products in Malang Regency.

The research aimed to examine the risks involved in every decision taken in the supply chain of agricultural product distribution in Kabupaten Malang in order to minimize problems that would arise to the cause of each problem. This research was conducted at the Technical Unit (UPT) of the Sub-Department of Agribusiness (STA) Mantung, Pujon, Kabupaten Malang. This is because the process of entering and selling, until distribution is carried out by the perpetrators of the horticulture type vegetable supply chain at the UPT STA Mantung using SCOR (Supply Chain Operations Reference) and HOR (House of Risk).

2. Methods

Preliminary observations were made by visiting markets in Kabupaten Malang. Discussions were held with a trader, supplier and retailer from problems that often arise during the process of distributing vegetables. The results of the discussion found several problems that occurred, such as the distribution of products to the market as a final place of sale for agricultural products still not maximally accommodating all agricultural products, causing a lot of agricultural products that could not be sold. Not only the two problems that arise but there are still many other problems that occur in the process of distributing vegetables in Kabupaten Malang.

The next step is to observe UPT STA Mantung, Pujon, Kabupaten Malang, to confirm the problems that occur in the distribution chain of vegetable products in Kabupaten Malang.

The Supply Chain Operation Reference (SCOR) method is used to measure the activities of horticulture product distribution. The SCOR method used is level 1, which defines the scope and content of the supply chain. At this level the performance of the supply chain will be seen. Supply chain management using the SCOR method is divided into several parts of the supply chain, namely planning (plan), procurement (source), making (make), delivery (deliver), and return [6]. The following are the functions of each supply chain division:

1. Planning (Plan)
   This section describes activities related to developing plans to operate supply chains.
2. Procurement (Source)
   This section describes ordering, scheduling, shipping and receiving of goods until the service is received.
3. Making (Make)
   This section describes activities related to material conversion or the creation of product.
4. Delivery (Deliver)

This section describes the process of meeting the demand for goods of the supply chain.

5. Returns

This section describes activities related to the return of goods or products for various reasons.

Data obtained from interview and field study, the activities of the supply chain of the vegetable distribution process at UPT STA Mantung Kabupaten Malang. This supply chain mapping uses the SCOR method which has 5 elements of supply chain division namely, plan, source, make, deliver, and return. The mapping or distribution of supply chain activities is used to make it easier to see the relationship between each chain of functions and see potential risks.

The House of Risk is used to measure the risk in distribution of horticulture product in STA Mantung. The results of the development and modification of the HOQ model that serves to determine risk agents and determine priorities in prevention [5]. Table 1 shows HOR phase 1:

Table 1. HOR phase 1

Business Process	Risk (Ei)	Risk Agent			*Severity*
		A1	A2	A3	
Plan	E1	R11	R12	R13	S1
Source	E2	R21			S2
Make	E3				S3
Deliver	E4				S4
Return	E5				S5
Occurrence (O_i)		O1	O2	O3	
ARP		ARP1	ARP2	ARP3	

After performing the first phase of the HOR model for seven steps, the next step is to create a second phase HOR model. Table 2 shows HOR phase 2:

Table 2. HOR phase 2

Risk Agent	ARP	Mitigation Strategy		
		SM1	SM2	SM3
A1	ARP1	F11	F12	
A2	ARP2			
A3	ARP3			
A4	ARP4			
Total Effectively Strategy		TE1	TE2	TE3
Strategies difficulty level		D1	D2	D2
Total Effectiveness of Difficulty Ratios		ETD1	ETD2	ETD3
Priority Rating of Mitigation Strategies				

By choosing the risk agent with the highest rating or rating using the Pareto diagram the potential risk priority is defined. A priority rating from the risk agent is doing to identify the appropriate mitigation strategy using the Ishikawa diagram for each of the risk agents which have been previously selected. The next step is to give an assessment of the correlation between the selected risk agents and mitigation strategies.

3. Result and Discussion

The research is focused on the distribution supply chain for traders or village collectors in STA Mantung Kabupaten Malang and on some vegetables like mustard greens, cabbage, carrots, leeks, and celery. Because the five vegetables are the most common vegetables found in STA Mantung Kabupaten Malang, the identification process of SCOR method is needed to classify supply chain elements. These elements are plan, source, make, deliver, and return. The following are the criteria or activities in the STA Mantung distribution supply chain in Kabupaten Malang which are included in the SCOR elements:

Table 3 SCOR elements for ativities in STA Mantung

Plan	Plan of delivery schedules Plan of vegetables Plan of the amount of vegetables Plan of fees or payments Vegetable quality requirements
Source	Checking of the supply of vegetables Checking of the supply of vegetables Check of the quality of vegetables Schedule for checking Check of costs or payments
Make	The process of weighing vegetables Packaging process Buying and selling process
Deliver	Process of sending vegetables Vehicle selection Selection of travel routes
Return	Return qualification Return process

Interviews were conducted with authorized three office staff in the STA **Man**tung and two collectors. The following are identified risks:

Table 4 List of risk on horticulture products distributions in STA Mantung

Plan	E1	Increase in the price of vegetables from the main producers
	E2	Collector / trader profits are reduced
	E3	Incompatibility of the types of vegetables ordered
Source	E4	Incompatibility of the quality of vegetables ordered
	E5	Too many vegetables
Make	E6	There are broken vegetables
	E7	Depreciation of the quantity and quality of vegetables
	E8	Decreasing the quality of vegetables due to packaging
	E9	Order cancellation
	E10	The number of collectors or traders
	E11	The amount of vegetables wasted
Deliver	E12	Decrease in quality due to the shipping process
	E13	Errors in the type or amount of vegetables sent
	E14	Damaged vehicles
	E15	Delays in delivery

After getting a correlation assessment, the next step is the calculation of aggregate risk potential (ARP). The ARP calculation could be shown as table below

Table 5 Risk Agent Sequence Based on ARP Calculation

Risk Agent	ARP Value	Ranks	Risk Agent	ARP Value	Ranks
ARP15	2156	1	ARP11	910	11
ARP2	1908	2	ARP7	692	12
ARP6	1820	3	ARP19	594	13
ARP9	1536	4	ARP17	568	14
ARP1	1421	5	ARP3	468	15
ARP10	1404	6	ARP16	378	16
ARP23	1280	7	ARP21	356	17
ARP20	1172	8	ARP5	351	18
ARP8	1056	9	ARP13	344	19
ARP22	1024	10			

Analysis using fishbone diagrams or causal diagrams, a number of mitigation strategies or actions are obtained for each risk agent. The following are mitigation strategies for each risk agent that have been analyzed:

Table 6 Mitigation Strategy

Codes	Types	Strategy	Explanation
SM1	*Reactive*	Sort the main producers	Sort the main producers (farmers) by looking at the best quality vegetables.
SM2	*Proactive*	Make basic specifications of vegetable quality	Make patented basic specifications for the quality of vegetables that will be accepted or rejected, so that every trader has good vegetable quality.
SM3	*Proactive*	Making a *cooling storage*	Make cooling storage to maintain the quality of vegetables to avoid shrinkage. Cooling storage usually can maintain the quality of vegetables for ± 2 weeks.
SM4	*Avoid*	Fast sales	A sure and fast sale could be done by recording all requests or sales that have been previously done in order to estimate sales that must be done in the future. This is done so that not much vegetables should be wasted.
SM5	*Avoid*	Avoiding congested routes	Avoiding solid delivery routes by studying other alternative routes.
SM6	*Reactive*	Choose the right time to send	Choose the right time to make deliveries by avoiding heavy hours of vehicles.
SM7	*Proactive*	Training on the treatment of horticultural products	Conduct training for HR at STA Mantung Malang Regency in the treatment or handling of horticultural products, especially vegetables.
SM8	*Proactive*	Make inventory plans and seasonal calendars	Make inventory planning from previous requests or sales so as not to harm traders or collectors at STA Mantung. In addition, create a seasonal calendar for some vegetables that often experience scarcity or up and down commodity prices.

Table 6 Mitigation Strategy (cont.)

Codes	Types	Strategy	Explanation
SM9	*Proactive*	Conduct training on human error	Conduct training to avoid and reduce human error of human resources in STA Mantung.
SM10	*Avoid Eliminate*	Choose and make design facilities that are environmentally friendly	Choosing facilities that are environmentally friendly, easy and safe to be used by HR at STA Mantung. In addition, it adds a place to sell for auction traders.
SM11	*Avoid*	Selection of the right type of vehicle	Choosing the right vehicle so as not to damage the vegetables in the course of shipping. It is better to use a vehicle that does not require too many vegetables to be stacked.
SM12	*Proactive*	Complete recording of consumer information	Carrying out complete and clear information and administration regarding consumer order information so that errors do not occur in meeting consumer demand. The main information needed is the name, address, number and type of vegetables, and delivery schedule.
SM13	*Avoid*	Confirm the main producer	Confirm in advance to the main producer (farmer) regarding the availability of types and quantities of vegetables to the quality in accordance with the request.
SM14	*Proactive*	Regular inspection of vehicle feasibility	Conduct routine inspections for the feasibility of vehicles in shipping vegetables so as not to harm various parties. This inspection can be done for 6 months or once a year.
SM8	*Proactive*	Make inventory plans and seasonal calendars	Make inventory planning from previous requests or sales so as not to harm traders or collectors at STA Mantung. In addition, create a seasonal calendar for some vegetables that often experience scarcity or up and down commodity prices.

The final step in processing the second phase of HOR data is to provide an action rating or risk mitigation strategy. This rating is obtained from the total ratio of the effectiveness of the difficulty. The highest total difficulty effectiveness ratio becomes the highest ranking of risk mitigation actions. Following is the ranking of risk mitigation actions that have been identified:

Table 7 Risk Mitigation Action Rating

Mitigation Strategy		Ranks
SM11	Selection of the right type of vehicle	1
SM1	Sort the main producers	2
SM4	Sure and fast sales	3
SM3	Make cooling storage	4
SM7	Training on the treatment of horticultural products	5
SM5	Avoiding congested routes	6
SM6	Choose the right time to send	7
SM12	Complete recording of consumer information	8
SM2	Make basic specifications of vegetable quality	9
SM13	Confirm the main producer	10
SM9	Conduct training on human error	11
SM10	Choose design facilities that are environmentally friendly	12
SM8	Make inventory plans and seasonal calendars	13
SM14	Regular inspection of vehicle feasibility	14

4. Conclusion

Based on the risk identification using the SCOR method,15 risks were found in the distribution of vegetables in STA Mantung, with details, three risks to planning, two risks to procurement, six risks to manufacture, and four risks to shipping. Identification of risk agents was found as many as 23 risk agents causing the risk that had been identified previously. Furthermore, after processing the data using HOR, the selected risk agent ranks to carry out risk mitigation strategies to reduce, avoid, move, and eliminate risks. Based on the selection of risk agents, twelve risk agents were found that needed mitigation strategies.

Acknowledgments

The researcher expressed his gratitude to the Ministry of Technology and Higher Education for the opportunity and support in carrying out this research using national institutional strategic research grants.

5. References

[1] Grover A and Chopra S 2017 *International Journal on Food System Dynamics* **8**(4) pp 298-316
[2] Castro J A O and Jaimes W A 2017 *J. of Industrial Eng. and Management* **10**(4) pp 687-710
[3] Yeboh N Y, Feng Y, Daniel O S and Joseph N B 2014 *Journal of management and Strategy* **5**(2) pp 31-48
[4] Yi H and Li Y 2013 *E3 Jornal of Business Management and Economics* **4**(7) pp 166-172
[5] Pujawan I N and Geraldine L H 2009 *Business Process Management* **15**(6) pp 953-967
[6] Supply Chain Council 2010 *Supply Chain Operations Reference Model.* (United States of America: Supply Chain Council)

IOP Conf. Series: Materials Science and Engineering **528** (2019) 012088 doi:10.1088/1757-899X/528/1/012088

The relation of Indonesia's strategic industry principles and Supply Chain Operations Reference (SCOR) Performance Attribute

JokoSulistio and Afrizal Bayu Alfatih

Universitas Islam Indonesia
jokosulistio@uii.ac.d, 14522266@students.uii.ac.id

Abstract. Article 33 of the1945 Constitution clearly states that all vital factors of production, called strategic industries, are controlled by the state and to be use for the prosperity of Indonesian people. This article implies that the strategic industries supply chain should be managed accordingly. The fourth amendment of the 1945 Constitution offers a macro framework as principles to conduct the supply chain. They are Togetherness, Efficiency of justice, Continuity, Environmental perspective and Self-Sufficiency. Furthermore, the framework can be used to build performance measurement indicators for the supply chain. Currently the method that is widely used to measure supply chain performance is the Supply Chain Operation Reference (SCOR). This study studied the relations of the strategic industries principle and performance indicators within the SCOR. Twelve professionals and researchers are interviewed as respondents to investigate those relations. More than 90% respondents agreed that; (1) the principle of togetherness is strongly related to Reliability, Responsiveness and Agility; (2) Efficiency of justice is strongly related to Responsiveness; (3) Continuity is strongly related to Agility; (4) Environmental Perspective is strongly related to Green Supply Chain, and; (5) Self-Sufficiency is strongly related to Responsiveness. Therefore, SCOR performance measurement attribute is related with the strategic industries principles.
Keywords: SCOR, Supply Chain, Performance Measurement, The 1945 Constitution of the Republic of Indonesia

1.Introduction

Indonesia is a unitary state, in form of Republic and based on law. One aspect that is given legal supervision is Indonesia's economic system. The Indonesian economic system is economic democracy which stated in Article 33 of the 1945 Constitution. The article clearly states that all factors of production that have an important role in community life in the territory of the Republic of Indonesia controlled by the state and used for the prosperity of the Indonesian people. The effort to interpret Article 33 of the 1945 Constitution was carried out to study the Constitutional Court's ruling in relation to the Judicial Review Act No. 7/2004, Law No. 22/2001, and Law No. 20/2002. Furthermore, the factors of production as referred to in Article 33 of the 1945 Constitution cannot be left on the market mechanism. the government must keep the market efficient and provide benefits for the Indonesian people. One of the effort is through pricing policies [1].

Earth, water, and all the wealth in it is controlled by the State and used to the greatest extent for the prosperity of the people, and the national economy is organized based on economic democracy. Article

33 of the 1945 Constitution also has principles that must be considered in carrying out economic democracy. These principles are of togetherness, efficiency, justice, sustainability, environmental insight, independence. The following is an explanation of these principles: 1) The principle of togetherness is the process of activities or performance of business people and consumers who together in the welfare of both, where people who run business sell products and consumers buy them to meet their needs, 2) Fair Licensing is the consumer has the right to get products those sold by businesses without any social and economic conditions and differences. 3) Continuity is the process of continuing relations between business people and consumers to meet the needs and welfare of the lives of both parties. 4) Environmental insight is the production process carried out by business people in utilizing natural resources and economic resources without causing damage to the environment, 5) Self-reliance is that business people themselves carry out their own strategies in product development to obtain maximum profit.

The business entities owned by Indonesian government is called Badan Usaha Milik Negara (BUMN). The companies ought to create value for the welfare of people's lives. Nonetheless, BUMNs are poorly performed and considered lacking in service. For example, company such as PT. PLN (Persero) is obligated to spread electricity for the people, but in fact they experienced network trouble which led reduction to the value perceived by the people. To solve the problem, the supply chain management should be placed under the microscope. The supply chain of production factors that control the livelihood of many people from upstream to downstream needs to be managed for the greatest prosperity of the people. Supply Chain Management (SCM) is all activities involved in selecting suppliers, procuring, and all other logistics activities. This includes coordination and collaboration with partner networks such as suppliers, agents, third parties and consumers [2]. To improve its performance, management should assign parameters and indicators to measure it. In general, supply chain measurement can be categorized into two activities, namely (1) supply chain and metric measurement, and (2) supply chain performance improvement. organizations need to develop a supply chain performance measurement system that is strong but still flexible to anticipate changes [3]. Performance measurement is classified into four, namely (1) financial costs, (2) quality, (3) time, and (4) flexibility. It is not recommended to use all the measurements it offers but to choose the most important measurement for the organization [4].

In this case, strategic industries supply chain performance should consider the principles of economic democracy. Such efforts are highly challenging due its level of abstraction. The development of supply chain performance measurement, in the context of article 33 of the 1945 Constitution, has never been done before. Supply chain performance measurement methods development studies has been carried out by many researchers. A framework or new steps in measuring the performance of a company are obtained. By utilizing the capabilities of the Multi-Regional Input-Output framework to deal with the complex and global nature of supply chains, current work presents a robust environmental sustainability performance measurement model supported by several different methods [4]. A case study analysis uses a Supply Chain Operational Reference (SCOR) combined with a Balance Score Card (BSC). The aim is to propose a SCM performance measurement framework for small and medium enterprises using qualitative and quantitative approaches. The study produced an integrated supply chain measurement framework to evaluate and plan activities within Small and Medium Enterprises (SMEs). The resulting framework includes measurements that are tangible and intangible. Included in tangible measurements are: cost, time, capacity, productivity and utility. While intangible measurements include effectiveness, reliability, availability and flexibility[5].

However, none of the supply chain performance development was deducted from the principles of economic democracy as the constitution mandate. Moreover, does the principles even relate to supply chain performance measurement? This research objective is to provide relationship between the principles of economic democracy and supply chain measurement. Supply Chain Operational Reference (SCOR) performance attributes are used to investigate that relations.

2. Methods

This is a qualitative research but use quantitative analysis for several data in the form of observations or data Research locations are strategic companies in Indonesia. This research begins with interviews with professors from the Faculty of Law of a distinguish university in Indonesia. The interview intended was to have deep understanding regarding the principles of economic democracy stated in Article 33 of the 1945 Constitution. Hypothesis or opinion that is still a temporary answer, then the truth will be proven empirically / real. For this reason, researchers conduct data collection. Data collection is carried out on certain populations that have been determined by researchers. In this study, researchers used sample data obtained from the population.

The sampling technique used by researchers is nonprobability sampling. This technique is a sampling technique that does not provide equal opportunities / opportunities for each element or member of the population to be selected as a sample. The sample technique used by researchers to select samples specifically and are part of nonprobability sampling is quota sampling and sampling purpose. Quota sampling is a technique for determining a sample of a population that has certain characteristics to the amount (quota) that the researcher wants. So in this study, researchers determined 12 members of the sample to be used. The sample in this study is the principle used in the search for a relationship between Article 33 of the 1945 Constitution and the principles of Supply Chain Management Performance Measurement. The number of sample members that are most appropriate to be used in the study is depending on the level of accuracy or error desired. The smaller the error rate, the greater the number of sample members used as data sources. A decent sample size in a simple study, namely the number of members as many as 10 samples[6]. So with the number of members of 10 samples, this research can be said to be feasible.

Sampling purpose is a sample determination technique with certain considerations. Researchers in this study have established a classification that is suitable to be used as a sample, namely the principles contained in Article 33 of the 1945 Constitution concerning economic democracy and developing principles in Supply Chain Management Performance Measurement. The data collection conducted in this study was observation, interviews, and questionnaires. The data which is a collection of 30 questions data given to 12 experts / samples, with a period of work for 3-4 years or more and understand the Supply Chain Management Performance Measurement.

In the early stages of implementation, the researcher seek additional information through experts in the field of business law before following a questionnaire regarding the relationship between the principles of Article 33 of the 1945 Constitution in the context of Supply Chain Management Performance Measurement. The content of the questionnaire is about the relationship between principles that contains 30 questions that must be filled. Observation is observing the results or responses of the respondents in determining the relationship between the principles of economic democracy in the 1945 Constitution in the context of Supply Chain Management Performance Measurement.

3. Research Results and Discussion

Observations were made by distributing questionnaires containing 30 questions of principle relationships that would be given to 12 expert respondents in the field of Supply Chain Management. This study uses SCOR level 2 principle which contains Reliability, Responsiveness, Agility, Cost, Assets, and Green Supply Chain. This principle is carried out on the grounds that the level of performance measurement of a company can be seen from this principle. The following is a description of the principles in Supply Chain Management: (1) Reliability is the level of reliability and consistency of a company to carry out its production process. As well as the ability to provide services in accordance with the promises offered, (2) Responsiveness is the response or alertness of the company in helping customers and providing fast and responsive services, which include: alertness of employees in serving customers, speed of employees in handling transactions, and complaint handling customer or consumer., (3) Agility is how fast the supply chain responds to changes in the environment, customer preferences, competitive forces. This is a measure of how a company adjusts

IOP Conf. Series: Materials Science and Engineering **528** (2019) 012088 doi:10.1088/1757-899X/528/1/012088

their supply chain to this change and then how quickly it is able to achieve it, (4) Cost is the amount of costs associated with the process for Planning, Source, Delivery and Return, and is calculated as Sales - Profit - Costs for Serving (eg, marketing, sales, administration), (5) Assets are assets (resources) owned by a business entity that can be measured clearly using a unit of money and its sorting system based on how quickly the changes are converted into cash units, (6) Green Supply Chain is to consider the environmental effects of all products and processes, including environmental influences derived from goods / products and processes ranging from raw materials to finished products, and the final disposal of these products. Here are the respondents with the responses they gave:

Table 1. The Inter-Principle Relations of Economic Democracy in the Context of Supply Chain.

Principles of Economic Democracy of the 1945 Constitution Article 33	SCOR Principles in Supply Chain Management	(%)	Information
Principles of Togetherness	Reliability	92,8 %	Strongly Related
	Responsiveness	92,9 %	Strongly Related
	Agility	92,9 %	Strongly Related
	Cost	57,2 %	Related
	Assets	50 %	Related
	Green SC	100 %	Strongly Related
Efficiency of justice	Reliability	85,7 %	Related
	Responsiveness	92,6 %	Strongly Related
	Agility	71,5 %	Related
	Cost	85,7 %	Related
	Assets	57,2 %	Related
	Green SC	85,7 %	Related
Continuity	Reliability	85,7 %	Related
	Responsiveness	85,7 %	Related
	Agility	92,9 %	Strongly Related
	Cost	42,9 %	Related
	Assets	57,1 %	Related
	Green SC	100 %	Related
Environmental Perspective	Reliability	85,8 %	Related
	Responsiveness	84,3 %	Related
	Agility	78,6 %	Related
	Cost	50 %	Related
	Assets	0 %	Not Related
	Green SC	100 %	Strongly Related
Self-Sufficiency	Reliability	50 %	Not Related
	Responsiveness	92,8 %	Strongly Related
	Agility	57,1 %	Related
	Cost	0 %	Not Related
	Assets	57,1 %	Related
	Green SC	64,4 %	Not Related

4. Discussion

Determining the relationship between Article 33 of the 1945 Constitution regarding economic democracy towards the Supply Chain Management in the context of Performance Measurement has been determined by 11 respondents who are experts in Supply Chain Management. The following is a relationship that has been determined based on the answers of the respondents: (1) The principle of Mutuality has a relationship with Reliability (92,8%), Responsiveness (92,9%), Agility (92,9%), Cost

(57,2%), Green SC (100%). Explanation of relationship: In this case, the principle of togetherness has a strong relationship with the principles of Supply Chain in the context of Performance Measurement. This is corroborated because the principle of togetherness is a plan or purpose intended for the service of many people or consumers so as to create a good lead relationship from the company to consumers. Supply chain principles will also be achieved if the principle of togetherness is carried out well. Another meaning in question is that every company goal carried out must be based on togetherness in providing services so that consumers will remain confident in the company's performance. In this connection, the principle of Assets is not included in the principle relationship because according to some respondents the asset is an inventory owned by the company so that it is more internal to the company. Respondent argument: Many respondents argue that this principle of togetherness is important in supply chain for performance measurement. each stakeholder in the supply chain must work together to achieve a high level of performance measurement of the company. Researcher's argument: It is true that the principle of togetherness related to all aspects. All and everything if not done together it will be difficult to get results. The cooperation in question is the cooperation of fellow workers of the company, even companies with consumers who work together on product transactions. (2) Fair Efficiency has a relationship with Reliability (85,7%), Responsiveness (92,6%), Agility (71,5%), Cost (85,7%), Assets (57,2%), Green SC (85,7%). Explanation of relationship: Efficiency in fairness is one of the principles of economic democracy which emphasizes equalization or does not differentiate on one side. The party in question is the consumer or other party concerned in the operation of the company. Whereas for the relationship of fairness efficiency with the supply chain principle of respondents who have filled out the questionnaire, they agree with the relationship between these principles. Relationships are related to all aspects of the supply chain so that it must be done fairly so that disputes do not occur because there are differences in service. Respondent argument: products that have good performance will provide services to all parties in quality that is always consistent at all levels of consumers. Researcher's argument: Justice is a unifying one nation, even beyond that to create peace. Justice in this connection is the service of the company to consumers that is carried out fairly. Services provided do not see the social and economic level of consumers. Companies in Indonesia should be like this. (3) Sustainable Principles have a relationship with Reliability (85,7%), Responsiveness (85,7%), Agility (92,9%), Assets (57,1%), Green SC (100%). Explanation of the relationship: The principle of sustainability is one of the principles of economic democracy which means that everything that is considered good or beneficial must be continued and products that are not reliable will be wasteful in utilizing input, which results in disruption of sustainability. However, the Cost principle in the Supply Chain is not included in this relationship. This is because the cost or price can change at any time depending on the product inventory, therefore it cannot be linked to the principle of sustainability. Respondent argument: For sustainability, it is very necessary to deal with environmental changes. Research argument: In this connection actually interrelated because the nature continues or continues with an unknown time. All kinds of activities carried out by the company where the activity does not harm the company must be continued for the service of the affected community. (4) Environmental insight has a relationship with Reliability (85,8%), Responsiveness (84,3%), Agility (78,6%), Green SC (100%). Explanation of the relationship: Environmental insight is the principle of economic democracy that enters into the environment in which the company is founded. All activities carried out by the company must pay attention to the surrounding ecosystem. There are many advantages if the environment is sustainable and unpolluted, such as the comfort of workers at work and consumers who increasingly believe in the performance of the company that produces healthy and good products. All services to the community are also not free from environmental insight. This is because the concept of reliability must be oriented towards sustainable output so that it must consider aspects of the environment that are beautiful and comfortable. In this connection, the principle of cost does not enter into the relationship because the environment does not have a relationship with price and vice versa. So that the respondent does not agree if the cost has a relationship with the principles of environmentally sound. Respondent argument: Green supply chain is basically the principle of paying attention to environmental sustainability. Researcher's argument: In this connection, the principle of insight can be a key principle because the company is established. A good environment and a beautiful ecosystem will not cause harm to the company and society. A good environment will bring benefits both in terms of product sales and consumer confidence in the services provided by the company. Therefore, the environment must be maintained and preserved. (5) The Principle of Independence has a relationship with Responsiveness (92,8%), Agility (57,1%), Assets (57,1%). Explanation of relationship: The principle of independence is a way or strategy of the company in orienting the company's performance in order to get the level of trust from consumers. In this connection there are only two supply chain principles related to the principle of independence, namely agility and assets. The purpose of the relationship with agility is that the company has a special strategy and special strength in providing services. Even this method varies in each company. While for the purpose of the relationship with assets, the company has inventory or inventory in the production system, so that the company in its own use of these assets in providing services. This is all done by the company without any external factors interfering in it. Respondent's argument: The greater the value of assets, the easier it will be to establish the principle of independence. Researcher's argument: This relationship can be said to be a relationship that has the least connection. This is because the principle of independence has the intention that all kinds of company activities are carried out by the company internally. The strategic influence used by the company causes this principle to be less related to the supply chain principle.

5. Conclusions

More than 90% respondents agreed that; (1) the principle of togetherness is strongly related to Reliability, Responsiveness and Agility; (2) Efficiency of justice is strongly related to Responsiveness; (3) Continuity is strongly related to Agility; (4) Environmental Perspective is strongly related to Green Supply Chain, and; (5) Self-Sufficiency is strongly related to Responsiveness. Therefore, SCOR performance measurement attribute is related with the strategic industries principles in accordance to Article 33 of the 1945 Constitution. This method can be used by researchers to further development of supply chain performance measurement. This method can be used as a reference for all companies running in Indonesia so that implementing the appropriate service rules stated in Article 33 of the 1945 Constitution on economic democracy.

6. References

[1] Magnar K, Junaenah I and Taufik G A Tafsir MK Atas Pasal 33 UUD 1945: Studi Atas Putusan MK Mengenai Judicial Review Terhadap UU No. 7/2004, UU No. 22/2001, Dan UU No. 20/2002 *J. Konstitusi* **7** pp 111–80

[2] Hsiau H J and Lin C W R 2009 A fuzzy pert approach to evaluate plant construction project scheduling risk under uncertain resources capacity *J. Ind. Eng. Manag.* **2** pp 31–47

[3] Gopal P R C and Thakkar J 2012 A review on supply chain performance measures and metrics: 2000-2011 *Int. J. Product. Perform. Manag.* **61** pp 518–47

[4] Acquaye A, Ibn-Mohammed T, Genovese A, Afrifa G A, Yamoah F A and Oppon E 2018 A quantitative model for environmentally sustainable supply chain performance measurement *Eur. J. Oper. Res.* **269** pp 188–205

[5] Thakkar J, Kanda A and Deshmukh S G 2009 Supply chain performance measurement framework for small and medium scale enterprises *Benchmarking An Int. J.* **16** pp 702–23

[6] Roscoe 1982 *Research Methods For Business* (New York: McGraw Hill)

IOP Conf. Series: Materials Science and Engineering **528** (2019) 012089 doi:10.1088/1757-899X/528/1/012089

A framework for the impact of lean six sigma on supply chain performance in manufacturing companies

Gihon Davilia Pardamean Gultom[1] **and Eric Wibisono**[2]

Department of Industrial Engineering, University of Surabaya
[1] gihon.gultom@gmail.com, [2] ewibisono@staff.ubaya.ac.id

Abstract. The use of lean six sigma in manufacturing industry has been more intense in the past decade. This method comprises two concepts: *lean* and *six sigma*, each with its own objectives. Lean is a concept that aims to optimize the value stream process by eliminating all kind of waste. Meanwhile, the goal of *six sigma* is to increase process capability in the value stream by aiming for zero defects and reduced process variation. The supply chain performance of manufacturing companies is determined by the ongoing internal activities reflected in the value stream, therefore both lean and six sigma applications are hypothesized to have an impact on the performance. Based on the reviewed literature, this study aims to produce a framework that can be used to explain the impact of lean and six sigma on supply chain performance. In the proposed framework, lean indicators are reflected in just-in-time, quality management and employee involvement, whereas six sigma indicators are characterized by the six sigma variables. At the other end of the equation, supply chain performance is defined by four constructs which are supplier selection, production, delivery and logistic performance, and warehouse.
Keywords: lean, six sigma, supply chain performance

1. Introduction

The rapid economic growth has resulted in an increasingly tight competition both in manufacturing and service industries. The increasing market demand, approximately 671% from 1970 [1], has put companies in fierce competition to stay ahead among other competitors in order to meet the market demand. Along with this global development, manufacturing companies should not only think about production but also start focusing on supply chain efficiency and effectiveness because they cover all activities from the procurement stage of raw materials to the delivery of finished goods [2]. Moreover, supply chain can be defined as an integrated chain in a network from upstream to downstream [3]. Supply chain management is a concept that integrates all parties into the overall existing value chain in the system that needs to be managed as company's asset.

Many companies have seen transition from proven practices such as re-engineering, six sigma, total quality management (TQM), data mining, lean six sigma and lean start-ups [4]. Lean six sigma is no longer a foreign method for manufacturing companies as many have applied it to improve product quality [5]. Lean six sigma is an integration of lean and six sigma, defined as a business philosophy, a systemic and systematic approach by identifying and eliminating waste or non-value-added activities through radical continuous improvement [6]. In addition, according to [2], six sigma is a business strategy and systematic methodology, which leads to breakthroughs in profitability through quantum acquisition in the quality of product or service, customer satisfaction, productivity, effectiveness and efficiency.

A number of research related to lean and six sigma in supply chain management have been carried out, such as design for six sigma (DFSS) to design information technology solutions so that inter-departmental information communication with supply chain operations can be carried out effectively [7]. Furthermore, information sharing between functions is encouraged to improve supply chain performance [8]. Another research argues that an important issue in a supply chain is the incapacity of information and the flow of processes between each interconnected function both from external and internal parties. Another study implemented lean concepts to optimize the use of warehouse layout because of its important role in the supply chain [10]. There are more studies related to the application of lean and six sigma in supply chain, but there are still few studies discussing the impact of lean and six sigma implementation on supply chain performance.

Recent research related to the relationship of lean to supply chain performance is carried out in [11] where the authors tested the effect of lean on sustainable supply chain by considering the green aspect. In this research, sustainable supply chain was measured using three categories, namely sustainable supplier selection, sustainable production and sustainable delivery, and logistics performance. Meanwhile, the research for six sigma was conducted by [12] who discuss the relationship between the implementation of six sigma and the performance of organizations. Both studies, however, were conducted separately, i.e. without integrating lean and six sigma concepts. Given the above gap, this study aims to fill the gap by establishing a framework to explain the relationship between lean six sigma and supply chain performance.

2. Method

Literature review is used as a method in this paper. We collected the key concepts and synthesized the ideas from the literature, and used them as components to develop our proposed framework.

2.1. The benefit of lean and six sigma implementation

The application of lean six sigma is caused by many reasons such as the variations that occur and produce a range of values for certain activities, processes, products or services [13]. Lean sigma means doing things in a simple and efficient way, but still providing superior quality and rapid service. In addition, its application can increase the profitability of the business, increase the effectiveness and efficiency of all operations to meet the customer's and even beyond customers' expectations [14].

2.2 Lean

Lean is adopted from the principles of Kaizen, kanban and just-in-time [15]. This method is identical to minimization of seven wastes, namely (i) over-production, (ii) defect/reject, (iii) unnecessary inventory, (iv) inappropriate processing, (v) excessive transportation, (vi) waiting/idle, and (vii) unnecessary motion [12]. While many researchers have classified waste in manufacturing environment, there is potential waste found in logistics such as inventory, transportation, space and facilities, time, packaging, administration and knowledge [16].

2.3 The relationship between lean and supply chain performance

Lean has been studied independently in relation to various aspects of the supply chain such as sourcing, manufacturing and logistics [17]. Supply chain performance is believed to be prone of disruptions by a number of variations such as demand fluctuations, number of shipments and product quality [18]. In this case, some variations can be seen in factors such as lead times, fill rates, and on-time deliveries [19]. Based on these theories, hypotheses related to lean relationships with supply chain performance were developed [11]. The hypotheses structure is shown in Figure 1.

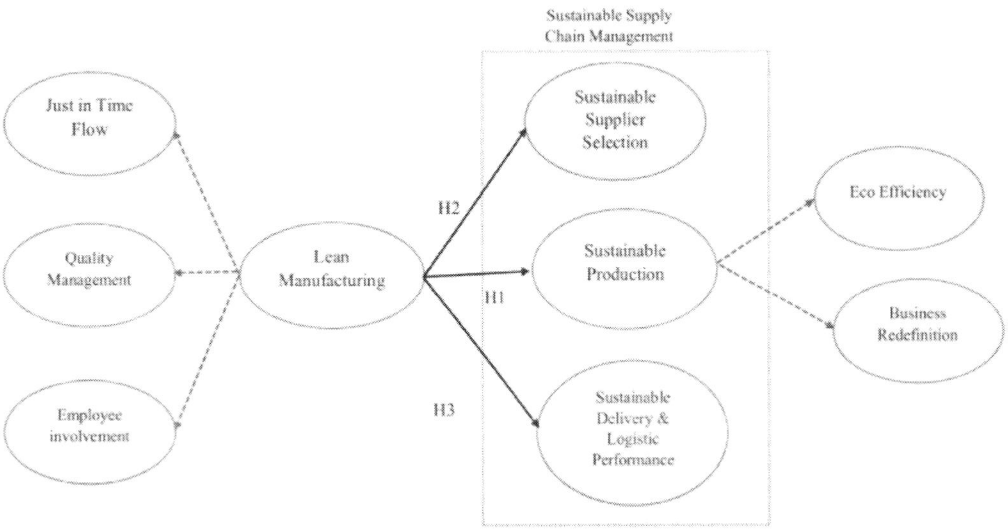

Figure 1. The hypotheses of lean relationship to supply chain performance [11]

2.3.1 The hypothesis on supplier selection. An efficient and flexible supply chain allows companies to be able to choose the right supplier [20]. Supplier selection is an important consideration because purchasing decisions have a major impact on the entire supply chain system [21]. Moreover, "supplier involvement" is an important component of lean production [22]. Based on those theories, it can be concluded that lean has a relationship with the supplier selection.

2.3.2 The hypothesis on sustainable production. Lean has been widely accepted by manufacturing industry to improve performance both in the production process and SCM and to minimize the existing waste [23]. Moreover, lean has a very strong relationship with production. In practice, the considered aspects in determining the right strategy in the supply chain include acquiring materials, converting raw materials into end products and shipping products.

2.3.3 The hypothesis on sustainable delivery and logistic performance. Authors in [11] hypothesized that there is a significant negative relationship between lean and sustainable delivery and logistics performance. Lean distribution causes rapid movement in logistics so that it increases emissions, viewed from the green aspect. It would, however, save logistics costs if balanced with good scheduling.

2.4 Six sigma

Six sigma was developed by Motorola in the 1980s. Six sigma is a process that aims to deliver more value to the customers and stakeholders by focusing on improving product quality and company productivity. DMAIC (*define, measure, analyze, improve* and *control*) is a standard approach suggested in six sigma implementation. The *define* phase is the first stage that defines the problem to be solved or the process improvement goal. A good problem statement must identify the customer and the CTQ (critical to quality). The *measure* phase focuses on how to measure the internal processes that affect the CTQ. The *analyze* phase focuses on the question of why excessive defects, errors, or variations occur. The *improve phase* gathers ideas and solutions to solve the problems. The proposed ideas are then evaluated and selected to get the most promising and affecting idea to the selected CTQ. The *control* phase focuses on how to maintain continuous improvement, including placing devices in their place and to ensure that the main variables remain within the maximum area that can be received in the modified process.

2.4.1 The relationship between six sigma and supply chain performance. A research was conducted on the implementation of six sigma to improve supply chain management operations on Samsung companies with the aim of improving the process, quality and synchronization of the company's value chain which includes logistics, sales and customer service [24, 27]. The success achieved by Samsung was due to all planned projects that could be executed properly. In addition, a framework on the critical factors that influence the success in six sigma projects was developed in [5]. Another empirical research carried out by [8] discusses the effects of implementing six sigma on operational performance, customer satisfaction and cultural change. One important study that determined the relationship between six sigma and organizational performance was conducted in [12]. In this research, the authors use knowledge indicators with socialization, externalization, combination and internalization variables to measure the success of six sigma projects towards company performance. The relationship model can be seen in Figure 2.

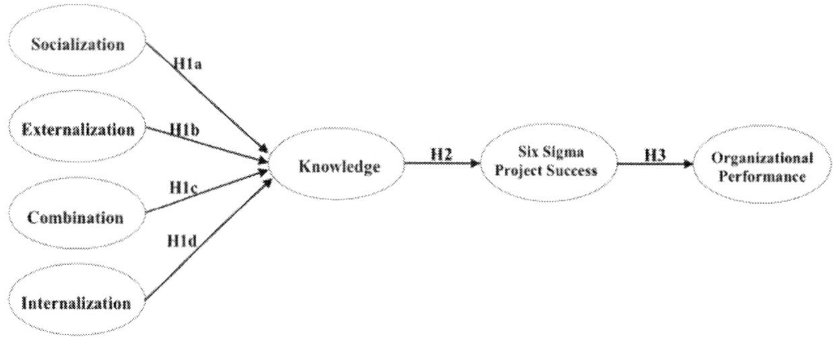

Figure 2. A model of the relationship between six sigma and organizational performance [25]

2.4.2 The hypothesis of six sigma model on company performance. Six hypotheses are proposed in [12]. These are:

- H1a: the socialization has a positive effect on knowledge in six sigma DMAIC project;
- H1b: the externalization has a positive effect on knowledge in six sigma DMAIC project;
- H1c: the combination has a positive effect on knowledge in six sigma DMAIC project;
- H1d: the internalization has a positive effect on knowledge in six sigma DMAIC project;
- H2: the knowledge creations in Six Sigma DMAIC phases have a positive effect on the success of six sigma project;
- H3: the success of six sigma project has a positive effect on organizational performance.

2.5 Warehouse

The warehouse operational coordination is very critical in the supply chain of all manufacturing industries. Realizing the role of a warehouse that could impact supply chain performance, a new strategy to optimize the warehouse with lean principles is suggested in [2] and three variables are proposed, namely providing a most economical storage in relation to retrieval time, use of space, providing maximum flexibility in order to meet changing storage and handling requirements, and providing economical path to collect orders of customers. In 1988, a study was carried out in the United Kingdom concluding that 55% of all warehouse operations costs could be associated with order taking [25].

3. Results and discussion

This study aims to produce a framework to explain the impact of lean and six sigma applications to supply chain performance. Two models from [5] and [11] are integrated into one framework for the said purpose. In addition, warehouse attributes are added to the supply chain performance. This section explains the process in model development related to lean, six sigma and supply chain performance based on the hypotheses discussed in the literature review.

3.1. Lean model

Lean is the extension of *just-in-time* (JIT) [15]. Moreover, the link of JIT and quality management to green supply chain and organization performance is considered in [26]. All indicators used in this study were equated with the indicators used by [11] as shown in Table 1.

Table 1. Lean indicators

Construct	Code	Definition
Just-In Time	JT1	In order to achieve continuous product flow, processes and layout are redesigned
	JT2	Bottlenecks and buffers are removed, Kanban is implemented
Quality Management	QM1	Worker driven continuous improvement (lean six sigma, quality circle, etc.)
	QM2	Quality is built in design and methods such as mistake proofing (poka yoke)
Employee Involvement	EI1	Suggestions per employee as compared to others in industry
	EI2	Training hours of new production workers
	EI3	Importance to open and honest communication
	EI4	Workers are empowered to stop assembly line if they observe any mistake

3.2. Six sigma model

The impact of six sigma on organizational performance is studied in [25]. However, this study uses six sigma measurement indicators proposed by [13]. The indicators can be seen in Table 2.

Table 2. Six sigma indicators

Construct	Code	Definition
Six sigma method (SSM)	SM1	The project was developed under the six sigma method
	SM2	The method used for conducting the project was DMAIC or DMADV (for DFSS projects)
	SM3	All DMAIC steps were completely fulfilled

3.3. Supply chain performance model

Supply chain performance (SCP) measured using indicators from [11] is divided into sustainable supplier selection, sustainable production, and sustainable delivery and logistics performance. Acknowledging its important role in the supply chain and its linkage to the SCP [10], warehouse factors are included in the SCP model. The indicators can be seen in Table 3.

Table 3. Supply chain performance indicators

Construct	Code	Definition
Supplier Selection	SS1	Jointly develop environmental management solutions
	SS2	Regular audits for environmental compliance and and practices of supplier's operations
	SS3	Combined teams to reduce material use and reuse recycled materials

IOP Conf. Series: Materials Science and Engineering **528** (2019) 012089 doi:10.1088/1757-899X/528/1/012089

Table 3. Supply chain performance indicators (cont.)

Construct	Code	Definition
Production (business redefinition)	BR1	Designing product for easy disassemble or reuse
	BR2	Introduction innovative products with sustainability consideration
	BR3	Strive to develop sustainable technology
Delivery & logistic performance	DL1	Environmental consideration while selecting mode of transport
	DL2	Preference to new & more fuel-efficie vehicles
	DL3	Consolidation of freight flow to reduce environmental impact
Warehouse	W1	Economical storage in relation to retrieving time, use of space.
	W2	Flexibility in order to meet changing storage and handling requirements
	W3	Economical path to collect orders of customers

3.4. Framework development

Based on all the constructs and indicators above, a framework is proposed that can be used to explain the relationship between lean and six sigma to supply chain performance. The framework is shown Figure 3. Eight proposed hypotheses are formulated based on the reviewed theories. It is important to note here that the proposed hypotheses above are yet to be tested with empirical data. The hypotheses are:

- H1 lean has a positive relationship to supplier selection;
- H2: lean has a positive relationship with production;
- H3: lean has a negative relationship with delivery and logistics;
- H4: lean has a positive relationship with warehouse;
- H5: six sigma has a positive relationship with supplier selection;
- H6: six sigma has a positive relationship with production;
- H7: six sigma has a positive relationship with delivery & logistics;
- H8: six sigma has a positive relationship with warehouse.

4. Conclusion and remarks for future research

This paper develops a framework to explain the relationship between lean and six sigma applications to supply chain performance. Eight lean indicators are derived from three constructs and three six sigma indicators came from one construct. On the other side of the equation, twelve indicators from four constructs are proposed to represent the supply chain performance. For further research, it is suggested to perform an empirical study by taking samples from manufacturing companies to validate this framework.

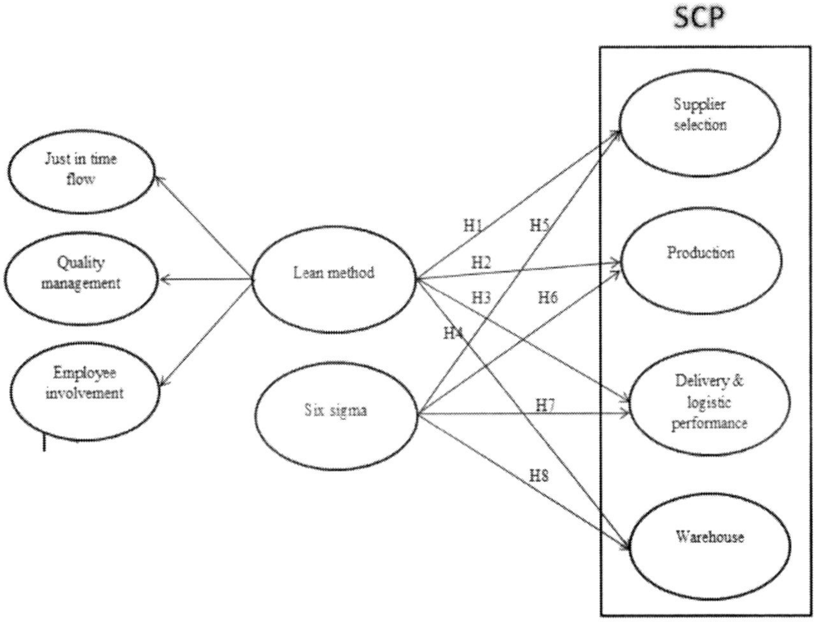

Figure 3. Proposed framework for the relationship between lean six sigma and supply chain performance

5. References

[1] Drake D F and Spinler S 2013 OM forum—Sustainable operations management: An enduring stream or a passing fancy? *Manuf. Serv. Oper. Manag.* **15**(4) 689–700

[2] Dharmapriya U and Kulatunga A 2011 New strategy for warehouse optimization – Lean warehousing *Int. Conf. Ind. Eng. Oper. Manag.* 513–519

[3] Christopher M and Rutherford C 2004 Creating supply chain resillience through agile six sigma *Critical Eye* June/August 24–28

[4] Benner M J and Tushman M L 2015 Reflections on the 2013 decade award—"Exploitation, exploration, and process management: The productivity dilemma revisited" Ten years later *Acad. Manag. Rev.* **40**(4) 497–514

[5] Madlberger M 2009 What drives firms to engage in interorganizational information sharing in supply chain management? *Int. J. e-Collab.* **5**(2) 18–42

[6] Kwak Y and Anbari F 2006 Benefits, obstacles, and future of six sigma *Technovation* **26**(5–6) 708–715

[7] Meyer D A, Loch C H and Pich M T 2002 Managing project uncertainty: From variation to chaos *MIT Sloan Manage. Rev.* **43**(2) 60–67

[8] Lee K L 2002 *Critical Success Factors of Six Sigma Implementation and the Impact on Operations Performance* (Ph.D. Dissertation) Cleveland State University

[9] Li S S Roa, Nathan R T S and Rao S S 2006 The impact of supply chain management practices on competitive advantages and organizational performance *Omega-Int. J. Manage. Sci.* **34**(2) 107–124

[10] Antony J and Banuelas R 2002 Key ingredients for the effective implementation of Six Sigma program *Meas. Bus. Excell.* **6**(4) 20–27

[11] Jakhar S K, Rathore H and Mangla S K 2018 Is lean synergistic with sustainable supply chain? An empirical investigation from emerging economy *Resour. Conserv. Recy.* **139** 262–269

[12] Simpson D F and Power D J 2005 Use the supply relationship to develop lean and green suppliers *Supply Chain Manag.* **10**(1) 60–68

[13] Marzagão D S L and Carvalho 2016 Critical success factors for six sigma projects *Int. J. Proj. Manag.* **34**(8) 1505–18
[14] Antony J and Banuelas R 2001 A strategy for survival *Manuf. Eng.* **80**(3) 119–121
[15] Carvalho H, Duarte S and Cruz Machado V 2011 Lean, agile, resilient and green: Divergencies and synergies *Int. J. Lean Six Sigma* **2**(2) 151–179
[16] Goldsby T and Martichenko R 2005 *Lean Six Sigma Logistics: Strategic Development to Operational Success* J. Ross Publishing, Inc.
[17] Pullin J 2002 Success comes in threes *Professional Engineering* **15**(8) p28
[18] Mitchell E M and Kovach J V 2016 Improving supply chain information sharing using Design for Six Sigma *Eur. Res. Manag. Bus. Econ.* **22**(3) 147–154
[19] Lambert D M and Pohlen T L 2001 Supply chain metrics *Int. J. Logist. Manag.* **12**(1) 1–19
[20] Ugarte G M, Golden J S and Dooley K J 2016 Lean versus green: The impact of lean logistics on greenhouse gas emissions in consumer goods supply chains *J. Purch. Supply Manag.* **22**(2) 98–109
[21] Guo C and Li X 2014 A multi-echelon inventory system with supplier selection and order allocation under stochastic demand *Int. J. Prod. Econ.* **151** 37–47
[22] Lewis M A 2000 Lean production and sustainable competitive advantage *Int. J. Oper. Prod. Man.* **20**(8) 959–978
[23] Indrawati S and Ridwansyah M 2015 Manufacturing continuous improvement using lean six sigma: An iron ores industry case application *Procedia Manufacturing* 528–534
[24] Yang H M, Choi B S, Park B J, Suh H S, Chae M and Bongsug C 2007 Supply chain management six sigma: A management innovation methodology at the Samsung Group *Supply Chain Manag.* **12**(2) 88–95
[25] Sin A B, Zailani S, Iranmanesh M and Ramayah T 2015 Structural equation modelling on knowledge creation in Six Sigma DMAIC project and its impact on organizational performance *Int. J. Prod. Econ.* **168** 105–117
[26] Zhu Q and Sarkis J 2004 Relationships between operational practices and performance among early adopters of green supply chain management practices in Chinese manufacturing enterprises *J. Oper. Manag.* **22**(3) 265–289
[27] Xia W and Wu Z 2005 Supplier selection with multiple criteria in volume discount environments *Omega* **35**(5) 494–504

IOP Conf. Series: Materials Science and Engineering **528** (2019) 012090 doi:10.1088/1757-899X/528/1/012090

Supply Chain Management (SCM) – Is it Value Addition towards Academia?

Prof. Dr. Md. Mamun Habib[1] and Ikram Hasan[2]

[1,2]BRAC School of Business (BBS), BRAC University, Bangladesh
[1]University of Texas – Arlington (UTA), USA
[1]mamunhabib@gmail.com, [1]mamunhabib@bracu.ac.bd, [1]mohammad.habib@uta.edu
[2]ikramhasan.evan@bracu.ac.bd, [2]ikramhasan.shavan@gmail.com

Abstract. This keynote paper illustrates the theory and evolution of Supply Chain Management (SCM) as well as highlights chronological prospective of SCM in terms of time frame in different areas of manufacturing and service industries. The objective of SCM is to incorporate activities across and within organizations for providing the customer/stakeholders value. Supply Chain Management (SCM) has been widely researched in numerous application domains during the last few decades. Despite the popularity of SCM research and applications, considerable confusion remains as to its meaning, there are several attempts made by researchers and practitioners to appropriately define SCM. This paper demonstrates Educational Supply Chain Management, Integrated Tertiary Educational Supply Chain Management (ITESCM) model, as the application of SCM in the service industry, which would add the values towards academia. ITESCM model were identified and confirmed by 493 respondents, representing experts and administrators, faculty, staffs of the university, employers, graduates, etc. The resulting model was subsequently evaluated for accuracy and validity by multiple linear regressions (MLR) analysis and the structural equation modeling (SEM) technique. The research model provides a novel approach for decision makers of each supply chain components to review and appraise their performance toward fulfillment of ultimate goals i.e. producing high-caliber graduates and high-impact research outcomes, which represent two main contributions, for the betterment of the end customer, i.e., the society.
Keywords: Supply chain management, academia, education, SCM, research, ITESCM.

1. Introduction

Researchers usually focused on Supply chain management (SCM) issues in profit organizations during last decade. Research objectives may include adding value, reducing cost, or slashing response time in various parties involved in the manufacturing supply chain. However, very few studies were attempted in non-profit organizations, i.e. service industries. An extremely scarce number of research papers focused on SCM in the academia [17], [25].

Ref. [17] states that a profit organization attempts to maximize profits, whereas a non-profit organization considers monetary returns of less importance. Their major objectives may include improved literacy rate, better quality of life, equal opportunities for all genders or races, etc. The revenues gained by a non-profit organization would be used primarily to balance the expenditure of the organization. Due to conflicting objectives, managing a successful profit organization may be drastically different from a non-profit organization [25]. Recently, an increasingly large number of research studies highlight the criticalness of SCM as a means to assuring organizational success.

SCM assists the business organization to compete in the dynamic international market. The objective of SCM is to incorporate activities across and within organizations for providing the customer value. This should also be applicable to the academia, which represents a type of non-profit organizations. The goal is to provide the society value by producing high quality graduates and research outcomes. An integrated educational supply chain involves coordination and information sharing up and down the process among all stakeholders. With technology facilitating information flow, a coordinated supply chain can be designed to meet the strategic, planning, and operating objectives of the educational institutions. It also means establishing effective and feasible relationships both inside and outside the organization [8].

SCM is needed for various reasons: improving operations, better outsourcing, increasing profits, enhancing customer satisfaction, generating quality outcomes, tackling competitive pressures, increasing globalization, increasing importance of E-commerce, and growing complexity of supply chains. Supply chains are relatively easy to define for manufacturing industries, where each participant in the chain receives inputs from a set of suppliers, processes those inputs, and delivers them to a different set of customers. With educational institutions, one of the primary suppliers of process inputs is customers themselves, who provide their bodies, minds, belongings, or knowledge as inputs to the service processes [8], [17], [24].

This exploratory study reveals the following objectives: a) Analysis the overview of SCM through different citations and the evolution of SCM, b) Presentation of ITESCM as SCM practices on tertiary educational institutions

2. Evolution of SCM

The supply chain literature review was conducted to study the past researches. Before the 1950s, logistics was thought of in military terms It had to do with procurement, maintenance, and transportation of military facilities, materials, and personnel. The study and practice of physical distribution and logistics emerged in the 1960s and 1970s [8], [17], [25].

The logistics era prior to 1950 has been characterized as the "dormant years," when logistics was not considered a strategic function. Around 1950s changes occurred that could be classified as a first "Transformation." The importance of logistics increased considerably, when physical distribution management in manufacturing firms was recognized as a separate organizational function. The SCM concept was coined in the early 1980s by consultants in logistics [8]. The authors emphasized that the supply chain must have been viewed as a single entity and that strategic decision-making at the top level was needed to manage the chain in their original formulation. This perspective is shared with logisticians as well as channel theorists in marketing [17].

SCM has become one of the most popular concepts within management in general since its introduction in the early 1980s [17]. A number of journals in manufacturing, distribution, marketing, customer management, transportation, integration, etc. published articles on SCM or SCM-related topics. The evolution of SCM continued into the 1990s due to the intense global competition [25].

Ref. [8] went as far as claiming there was a paradigm shift within the management literature: "One of the most significant changes in paradigm of modern business management is that individual businesses no longer compete as solely autonomous entities, but rather as supply chains. Business management has entered the era of inter-network competition and the ultimate success of a single business will depend on management's ability to integrate the company's intricate network of business relationships."

Ref. [17] adopted SCM in the National Health Service. In fact, it was the first paper of SCM in the service industry. Ref. [34] illustrated the customer supplier duality in the service organizations as it pertained to SCM in the service industry. Ref. [33] explored supply chain application to the service industry. Ref. [32] proposed an educational supply chain as a tool for strategic planning in tertiary education. The study was based on a survey among employers and students.

Survey findings revealed that integration and coordination among students and employers should have been promoted. Ref. [31] explored a framework for SCM based on several service industries including automobile, grocery, computers, book publishing etc. According to the case study conducted at the City University of Hong Kong, ref. [19] defined educational supply chain as the 'Student' and the 'Research' supply chain.

Ref. [17] represents the first large scale empirical study that systematically investigate input of the university, output of the university through educational SCM. This exploratory research addresses the education supply chain, the research supply chain, and educational management as major constituents in an Integrated Tertiary Educational Supply Chain Management (ITESCM) model. Its applicability was successfully verified and validated through survey data from leading tertiary educational institutions around the world. Redesigned ITESCM Model was developed in 2012 which is more users friendly [21], [24]. The emergence and evolution of SCM may be depicted as a timeline shown in Figure 1.

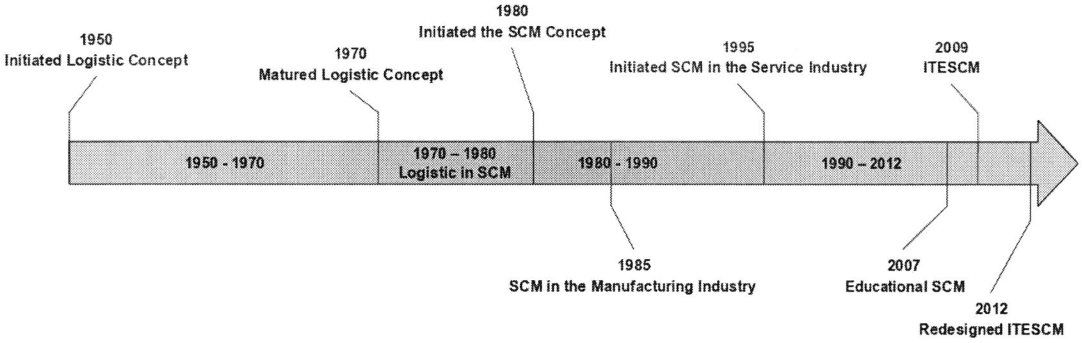

Figure 1. Evolutionary Timeline of Supply Chain Management

3. Research Methodology

The analysis of this paper is based on both primary and secondary data. First part of this paper, evolution of SCM evolved Secondary data sources, particularly online databases, books, journals, conference papers, etc. On the other hand, 2nd part of this paper furnished based on the analysis of literature, past theoretical frameworks, interviews with stakeholders. ITESCM model constructs were identified and confirmed by 493 respondents, representing university administrators, faculty and staffs, employers, and graduates. Its applicability was successfully verified and validated through survey data from leading tertiary educational institutions around the world.

4. SCM Practices

Supply chain management practice on the service industry has been depicted in this paper. ITESCM model denotes supply chain managements for the universities which represents one of the service industries. One of the main goals of an educational supply chain is to improve the well-being of the end customer or the society. To achieve this goal, educational institutions need to have a certain degree of knowledge about the partners in their supply chains including suppliers, customers, and the society. The performance of the supply chain management depends on the seamless coordination of all supply chain stakeholders to ensure attainment of desirable outcomes.

The ITESCM, which stands for Integrated Tertiary Educational Supply Chain Management, model represents supply chain management for the academia [10], [15], [17], [20]. This model depicts the integrated form of educational supply chain and educational management for the universities. Educational supply chain also consists of education supply chain and research supply chain. This paper is the revised version of ITESCM model, which represents academic supply chain management for the universities.

In academic supply chain management, raw materials are students as well as internal and external projects. Finished products are graduates and research outcomes [8], [11], [9]. Suppliers, supplied inputs, the service provider, customers, supplied outputs, and the consumer have been identified in the integrated supply chain for the universities. Fig. 4 illustrates an education supply chain and a research supply chain, which together form the integrated supply chain for the universities [11],[12].

Initially, researcher develops a conceptual framework of Integrated Tertiary Educational Supply Chain Management (ITESCM) model. Also using research it is possible to designs an integrated supply chain for the universities. Co-ordination and information sharing involve in integrated supply chain up and down the process. Researcher depicts holistic view of educational supply chain in Figure 2 for providing the clear conception of the conceptual framework. Single-level, bi-directional simplified form of supply chain management has been formulated in this framework for the universities, as shown in Figure 3. Since a single party is unable to do anything in the higher educational institutions, the researcher involves different parties to achieve final outcomes [28], [29], [30].

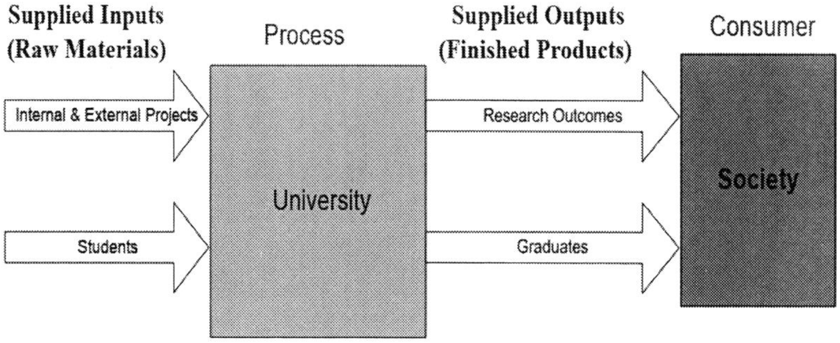

Figure 2. Holistic View of Educational Supply Chain

However, determine the supplier and customer of the intangible product in the service industry is very difficult. Suppliers, the service provider, customers, and the consumer have been identified in this research. This exploratory study also identifies supplied inputs, customer-consuming output (O/P), customer-supplying input (I/O) and finally supplied outputs [28].

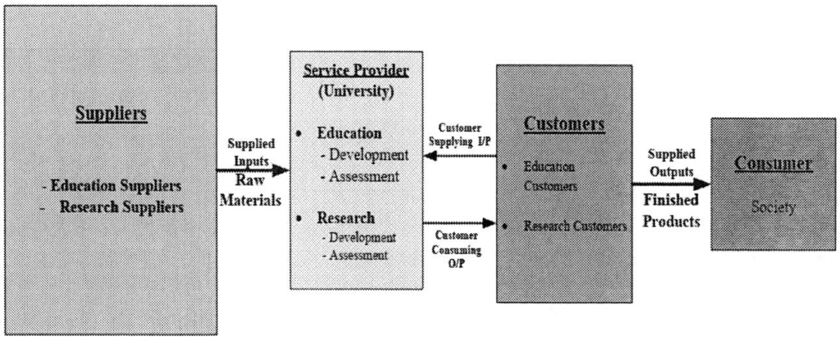

Figure 3. Simplified Form of Supply Chain Management for the Universities

In this paper, authors intend to redesign ITESCM model that is the revised form of original ITESCM. That model would be easily explicable and research equations are friendly for users who intend to apply in practical field of tertiary educational institutions. Fig. 5 illustrates redesigned ITESCM model [28], [29], [30].

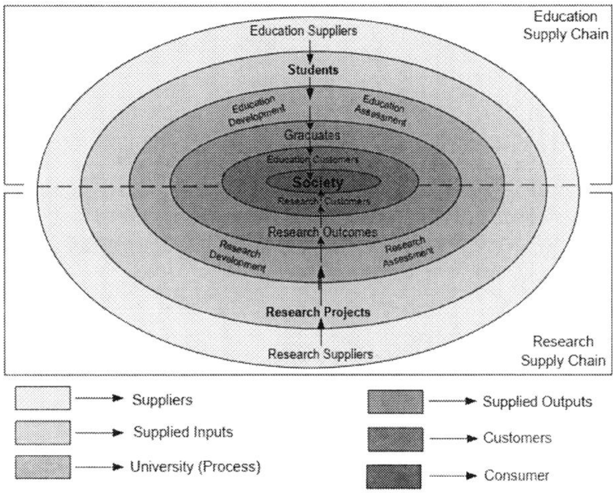

Figure 4. An Integrated Supply Chain for the Universities

Different Factors in the Universities

According to the concept of three decision levels, including strategic, planning and operating, in SCM, this concept would be adopted for the higher educational institutions [24]. To accomplish proper teaching and research works in the universities; different factors have to need analyzed. Four factors, namely faculty capabilities, facilities, programs establishment, university culture [8], [9], [10], [13], [14], [29], [30] will be illustrated in this section.

Programs Establishment (PE): Programs establishment would be occurred for the education and research in terms of development and assessment in the universities. Universities design different programs, to enhance the diversification in education development and establish various programs to assess the development. Universities also intend different programs to increase the diversification in research development and research assessment. Universities have to attempt product differentiation, i.e. programs establishment. Hands-on experience, industrial placements, social demand, provision of IT facilities, and innovative academic methods all demonstrate attempts to differentiate programs establishment [8], [9].

Faculty Capabilities (FC): Faculty members establish good communication, provide rich environment for classroom observation, model best practices, create opportunities for reflection, and support students' participation in curriculum planning, teaching and research. Traditionally, university faculty members are evaluated according to the three major criteria: teaching, research, and services [12], [8], [9].

University Culture (UC): The concept of organizational culture would be applicable for the universities by the name of University Culture. However, the type of the university culture will fully depends on the university management or administrator. In fact, university culture is the personality of the university [9], [15].

Facilities (FA): Universities offer a wide range of modern facilities to their students. These include state of the art lecture halls, libraries, laboratories and IT services to ensure that students are provided with an environment in which they can learn, both successfully and comfortably. Lecture rooms are principally conducted using state-of-the-art distance learning technology, online education, e-learning via Internet. Online databases, e-journal, digital library, etc. represents modern research facilities in the universities [8], [11].

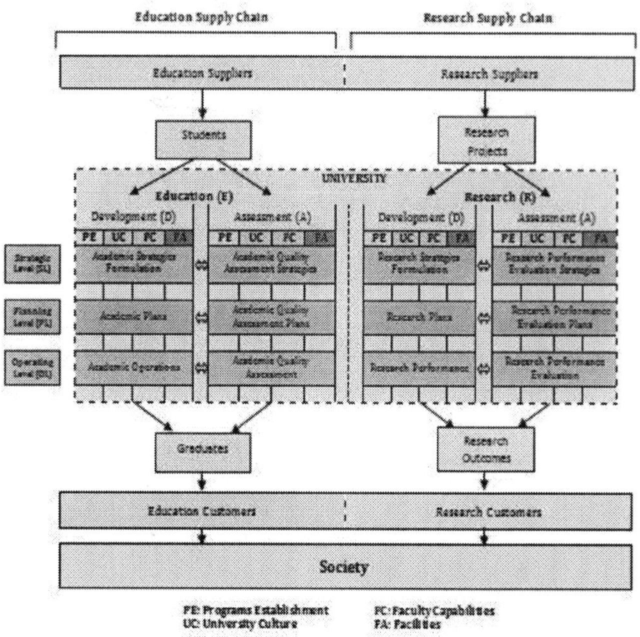

Figure 5. Redesigned ITESCM Model

5. Redesigned ITESCM Model evaluation

The researchers pointed out six hypotheses and three models. Hypothesis H_1 stands for graduates and hypothesis H_2 for research outcomes. Hypotheses H_3, H_4, H_5 and H_6 for supplied outputs. The following hypotheses are established.

H_1 There is a relationship between education development and education assessment with graduates

H_2 There is a relationship between research development and research assessment with research outcomes

H_3 There is a relationship between graduates and education customers

H_4 There is a relationship between research outcomes and research customers

H_5 There is a relationship between education customers and the society

H_6 There is a relationship between research customers and the society

Model A – Graduates

Model A contains Education Development (E_d) and Education Assessment (E_a). There are four subgroups, including programs establishment (E_{dPE}), university culture (E_{dUC}), faculty capabilities (E_{dFC}), and facilities (E_{dFA}), respectively in Education Development. Similarly four subgroups are available in Education Assessment.

MLR Equations

$$E_d = 0.63E_{dPE} + 0.70E_{dUC} + 0.65E_{dFC} + 0.65E_{dFA} \quad (1)$$

$$E_a = 0.68E_{aPE} + 0.74E_{aUC} + 0.69E_{aFC} + 0.66E_{aFA} \quad (2)$$

$$Graduates = 0.97E_d + 0.92E_a \quad (3)$$

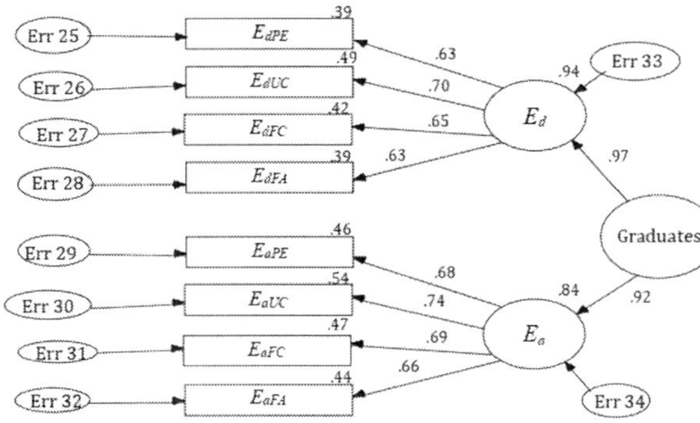

Figure 6. AMOS Graphics Output of Model A (Standardized Estimates)

Equation (3) depicts that education development is highly contributed to produce quality graduates in the universities.

Graduates = $0.97E_d + 0.92E_a$
$= 0.97\,[0.63E_{dPE} + 0.70E_{dUC} + 0.65E_{dFC} + 0.65E_{dFA}] + 0.92\,[0.68E_{aPE} + 0.74E_{aUC} + 0.69E_{aFC} + 0.66E_{aFA}]$
$= 0.61E_{dPE} + 0.68E_{dUC} + 0.63E_{dFC} + 0.61E_{dFA} + 0.63E_{aPE} + 0.68E_{aUC} + 0.63E_{aFC} + 0.61E_{aFA}$

The above equation shows university culture at education development and education assessment is highly contributed to produce the graduates in the universities.

Model Fit Index: Chi-square = 8.936 (Ratio of relative chi-square close to 5 indicates reasonable fit) [2], Degrees of freedom = 19, Probability level = .000, RMSEA = 0.127 NFI = 0.880, CFI = 0.891 (NFI and CFI values close to 1 indicate a very good fit) [8], [9].

The above equations, graphics output, Model fit index, and above all statistical discussion on AMOS magnifies that hypothesis H_1 fails to reject and states that there are significant relationship between education development and graduates as well as education assessment and graduates.

Model B - Research Outcomes

Model B contains Research Development (R_d) and Research Assessment (R_a). There are four subgroups, including

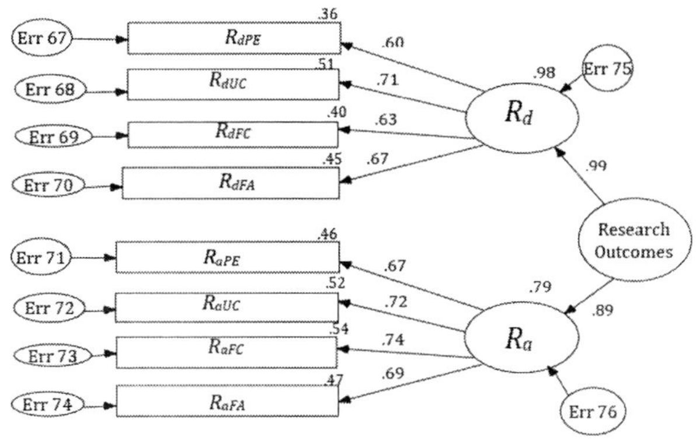

Figure 7. AMOS Graphics Output of Model B (Standardized Estimates)

programs establishment (R_{dPE}), university culture (R_{dUC}), faculty capabilities (R_{dFC}), and facilities (R_{dFA}), respectively in Research Development. Similarly four subgroups are available in Research Assessment.

MLR Equations

$$R_d = 0.60R_{dPE} + 0.71R_{dUC} + 0.63R_{dFC} + 0.67R_{dFA} \quad (4)$$

$$R_a = 0.67R_{aPE} + 0.72R_{aUC} + 0.64R_{aFC} + 0.69R_{aFA} \quad (5)$$

$$Research\ Outcomes = 0.99R_d + 0.89R_a \quad (6)$$

From the research findings, equation (6) depicts that research development is highly contributed to produce research outcomes in the universities. From Equation (6),

$$Research\ Outcomes = 0.99\ [0.60R_{dPE} + 0.71R_{dUC} + 0.63R_{dFC} + 0.67R_{dFA}] + 0.89\ [0.67R_{aPE} + 0.72R_{aUC} + 0.64R_{aFC} + 0.69R_{aFA}]$$

$$= 0.59R_{dPE} + 0.70R_{dUC} + 0.62R_{dFC} + 0.66R_{dFA} + 0.60R_{aPE} + 0.64R_{aUC} + 0.66R_{aFC} + 0.61R_{aFA} \quad (7)$$

From the research results of equation (7), university culture and facilities in research development as well as faculty capabilities in research assessment are highly contributed to produce the research outcomes in the universities.

Model Fit Index: Chi-square = 9.991, Degrees of freedom = 19, Probability level = .000, RMSEA = 0.135, NFI = 0.872, CFI = 0.883

The above equations, graphics output, Model fit index and above all statistical discussion on AMOS rectifies that hypothesis H_2 fails to reject and states that there are significant relationship between research development and research outcomes as well as research assessment and research outcomes.

Model C - Supplied Outputs

Model C is representing the inter relationships among different variables to justify the hypotheses H_3, H_4, H_5 and H_6 by SEM through AMOS.

MLR Equations

$$Society = 0.61EducationCustomers + 0.61\ ResearchCustomers$$

$$= 0.61\ [0.34Graduates] + 0.61\ [0.15ResearchOutcomes]$$

$$= 0.21Graduates + 0.09ResearchOutcomes \quad (8)$$

From the research findings, the society consists of graduates and research outcomes. The author defined the society as the function of graduates and research outcomes.

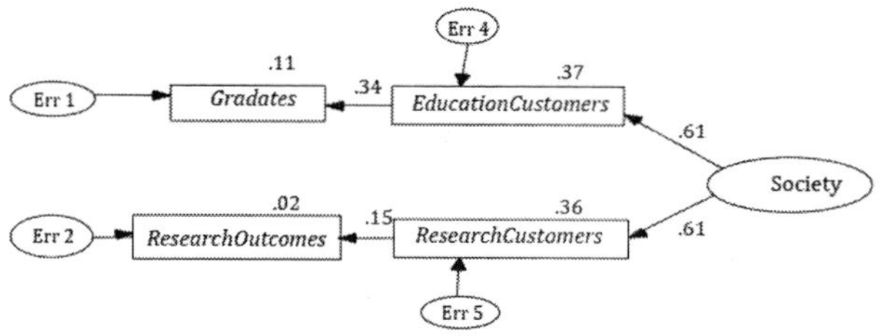

Figure 8. AMOS Graphics Output of Model C (Standardized Estimates)
$Society = f\,(Graduates,\ ResearchOutcomes)$

The equation (8) represents that graduates are highly contributed to the society. This equation also depicts that education customers and research customers have equal contribution to the society.

$$Society = 0.21\ [0.97E_d + 0.92E_a] + 0.09\ [0.99R_d + 0.89R_a]$$

$$= 0.20E_d + 0.19E_a + 0.09\ R_d + 0.08R_a \quad (9)$$

The above equation represents the relationship between the society and education development, education assessment, research development, research assessment. Education development and then education assessment are highly contributed to the society.

$Society = 0.20\ [0.63E_{dPE} + 0.70E_{dUC} + 0.65E_{dFC} + 0.65E_{dFA}] + 0.19\ [0.68E_{aPE} + 0.74E_{aUC} + 0.69E_{aFC} + 0.66E_{aFA}] + 0.09\ [0.60R_{dPE} + 0.71R_{dUC} + 0.63R_{dFC} + 0.67R_{dFA}] + 0.08[0.67R_{aPE} + 0.72R_{aUC} + 0.64R_{aFC} + 0.69R_{aFA}]$

$$Society = 0.126E_{dPE} + 0.14E_{dUC} + 0.13E_{dFC} + 0.126E_{dFA} + 0.129E_{aPE} + 0.141E_{aUC} + 0.131E_{aFC} + 0.125E_{aFA} + 0.054R_{dPE} + 0.064R_{dUC} + 0.057R_{dFC} + 0.06R_{dFA} + 0.054R_{aPE} + 0.058R_{aUC} + 0.059R_{aFC} + 0.055R_{aFA} \quad (10)$$

Model Fit Index: Chi-square = 5.494, Degrees of freedom = 3, Probability level = 0.001, RMSEA = 0.096, NFI = 0.896, CFI = 0.911

Equations, graphics output, Model fit index, and above all statistical discussion on AMOS states that there are significant relationships between graduates and education customers, research outcomes and research customers. There are also significant relationships among education customers, research customers and the society. Therefore, hypotheses H_3, H_4, H_5 and H_6 fail to reject.

6. Implications of Redesigned ITESCM

Integrated Tertiary Educational Supply Chain Management (ITESCM) model was developed by Habib in 2009 [8], [9]. Due to receiving feedback from academicians and practitioners, the researchers attempt to revise ITESCM model to comply it in real-life application for different universities in the world. Redesigned model is user friendly and easy to understand for current university administrators and prospective investigators.

If the researcher chooses 5 (strongly agree of 5 point Likert Scale) for each function in equation (10), in that case, the maximum value of equation (10) will equal to **7.545**. On the other hand, if the investigator selects 1 (strongly disagree) for each function in the equation, in that case, the minimum value of equation (10) will equal to **1.509**. Then, the researcher suggests cut off the value for the function of the society at fifty percent is **4.527** to indicate the value can be accepted.

$$UniversityOutcomes = \frac{Society - 1.509}{7.545 - 1.509} * (100 - 0) \quad (11)$$

The resulting suitability index, ***UniversityOutcomes*** in equation (11) ranges from 0% to 100% with 0% being the least favorable and 100% being the most suitable. The index of at least 50% may serve as a rough acceptance criterion for the well-being society [9], [26], [27].

7. Conclusion

This paper encompasses the evolution of SCM in terms of time frame for the manufacturing and service industries. In addition, this research represents the first large scale empirical study that systematically investigate input, output and process of the tertiary academic institutions through redesigned ITESCM model. This empirical study based on 493 respondents from all stakeholders, including experts and administrators, faculty members and staffs of the university, employers, graduates, etc. The hypotheses testing and SEM technique through AMOS were also applied.

This paper proposes the model of academic supply chain management for the tertiary educational institutions. This model links educational management with general business management. From a managerial point of view, this research provides a novel approach to developing and assessing supply chain management application in the academia, which represents a service industry.

8. References

[1] Arbuckle J L 2005 *Amos™ 6.0 User's Guide* (USA: Amos Development Corporation)

[2] Bentler P M 1990 Comparative fit indexes in structural models. *Psychological Bulletin* **107** pp 238–246

[3] Cutler B D, Moberg C R, Gross A and Speh T W 1998 Identifying antecedents of Information exchange within supply chains. *Int. J. of Physical Distribution & Logistics Management* (UK) **32**(9) pp 755-770
[4] Ebel R L 1551 Estimation of the reliability of ratings. *Psychometrika* **16** pp 407-424
[5] Habib M and Jungthirapanich 2008 An integrated framework for research and education supply chain for the Universities. *IEEE Int. Conference on Management of Innovation and Tech.* (Thailand) pp 1027-1032, ISBN 978-1-4244-2329-3
[6] Habib M and Jungthirapanich C 2009 Integrated educational management for the Universities. *The Journal of China-USA Business Review* (USA: David Publishing Company) **8**(8) pp 25-38 ISSN 1537-1514
[7] Habib M and Jungthirapanich C 2009 Research Framework of Educational Supply Chain Management for the Universities. *IEEE Int. Conference on Eng. Management and Service Sciences EMS* (China) ISBN 978-1-4244-4638-4
[8] Habib M 2009 *An Integrated Educational Supply Chain Management (ITESCM)*. Ph.D. Dissertation (Thailand: Graduate School of Information Technology, Assumption University of Thailand)
[9] Habib M 2010 *Supply Chain Management for Academia.* (Germany: LAP Lambert Academic Publishing) ISBN 978-3-8433-8026-3
[10] Habib M 2010 An Empirical Research of ITESCM (Integrated Tertiary Educational Supply Chain Management) Model. *Management and Services*. Sciyo.com ISSN 978-953-307-118-3.
[11] Habib M and Jungthirapanich C 2010 An Empirical Research of Educational Supply Chain for the Universities. *The 5th IEEE Int. Conf. on Management of Innovation and Tech.* (Singapore) ISBN: 978-1-4577-0738-4
[12] Habib M and Jungthirapanich C 2010 An Empirical Study of Educational Supply Chain Management for the Universities. *INFORMS Int. Conf. on Industrial Eng. and Operations Management* (Bangladesh) ISBN 978-984-33-0989-1
[13] Habib M and Jungthirapanich 2010 An Empirical Research of Integrated Educational Management for the Universities. *The 2nd IEEE Int. Conf. on Information Management and Eng.* (China) ISBN 978-1-4244-5263-7
[14] Habib M 2011 An Exploratory Study of Supply Chain Management for Tertiary Educational Institutions. *IEEE Int. Tech. Management Conference (ITMC) 2011a*, (San Jose, California USA) ISBN 978-1-61284-950-8.
[15] Habib M 2011 An Exploratory Research on Educational Supply Chain Management. *The IEEE Int. Conf. on Industrial Eng. and Eng. Management (IEEM)* (Singapore) ISBN 978-1-4577-0738-4
[16] Habib M 2011 An Exploratory Analysis of Educational Management for the Universities. *Int. J. of Eng. Business Management (IJEBM)* (Croatia) **3**(3) ISSN 1847-9790
[17] Habib M 2011 *Supply Chain Management (SCM): Theory and Evolution.* Dr. Md. Mamun Habib (Editor). *Supply Chain Management–Applications and Simulations.* [InTech Open Access] (Croatia) ISBN 978-953-307-250-0
[18] Harris R 2009 *Decision Making Techniques*. July 3, 1998. Available: www.virtualsalt.com
[19] Lau A K W 2007 *Educational supply chain management: a case study.* (Emerald Group Publishing Limited) ISSN 1074-8121 **15**(1) pp 15-27
[20] Pathik B B, Habib Md M and Chowdhury M T 2012 *Analysis of Educational Supply Chain Management Model: A Case Study Approach.* (Submitted) *Proc. of the 2012 Int. Conf. on Industrial Eng. and Operations Management* (Turkey)
[21] Pathik B B, Habib Md M and Chowdhury M T 2012 A Descriptive Study on Supply Chain Management Model for the Academia. (Submitted) *The 6th IEEE Conf. of Management of Innovation and Tech.* (Indonesia)

[22] Wheaton, Muthén B, Alwin D F and Summers G F 1997 Assessing reliability and stability in panel models. *Sociological methodology*. D R Heise, ed. (San Francisco: Jossey-Bass) pp 84–136
[23] Habib, Md. Mamun and Goncharuk A G 2012 *Performance Measurement through Supply Chain Management: A Case Study on Academia* (Germany: LAP Lambert Academic Publishing) ISBN 978-3-8473-7021-5
[24] Pathik B B and Habib Md M 2012 Redesigned ITESCM Model: An Academic SCM for the Universities. *Int. J. of Supply Chain Management* (UK: ExcelingTech Publisher) **1**(1)
[25] Habib Md M 2010 Supply Chain Management: Theory and its Future Perspectives. *Int. J. of Business, Management and Social Sciences (IJBMSS)* **1**(1) ISSN 2249-7463
[26] Habib Md M and Pathik B B 2012 Academic Supply Chain Management for Tertiary Educational Institutions. *The IEEE Int. Conf. on Industrial Eng. and Eng. Management (IEEM)* (Hong Kong)
[27] Habib M 2014 Supply Chain Management (SCM): Its Future Implications. *Open Journal of Social Sciences* **2**(9) pp 238-246 ISSN: 2327-5952 (Print), 2327-5960 (Online)
[28] Govindaraju B, Jeyasingam J, Habib Md M, Letchmana U and Radhakrishnan R 2017 The Impact of Supply Chain Management Practices on the Performance of Private Universities in Malaysia. *Int. J. of Supply Chain Management (IJSCM)* **6**(3) pp 22-35 ISSN: 2050-7399 (Online), 2051-3771 (Print)
[29] Paguio D P and Habib Md M 2017 A Proposed Supply Chain Management Model for Teacher Education Institutions: A Structural Equation Modeling. *Int. J. of Supply Chain Management (IJSCM)* **6**(2) pp 15-26 ISSN: 2050-7399 (Online), 2051-3771 (Print)
[30] Govindaraju B, Jeyasingam J, Habib Md M 2016 Education Supply Chain Management Model to Achieve Sustainability in Private Universities in Malaysia: A Review. *Int. J. of Supply Chain Management (IJSCM)* **5**(4) ISSN: 2050-7399 (Online), 2051-3771 (Print)
[31] Cigolini R, Cozzi M and Perona M 2004 A new framework for supply chain management. *Int. J. of Operations & Production Management* **24**(1) pp 7-41
[32] O'Brien E M and Kenneth R 1996 Educational supply chain: a tool for strategic planning in tertiary education? *Marketing Intelligence & Planning* **14**(2) pp 33-40
[33] Kathawala Y and Abdou K 2003 Supply chain evaluation in the service industry: a framework development compared to manufacturing. *Managerial Auditing J.* **18**(2) pp 140-149
[34] Sampson S E 2000 Customer-supplier duality and bidirectional supply chains in service organization. *Int. J. of Service Industry Management* **11**(4) pp 348-364